LIFE

The Science of Biology

TENTH EDITION

LIFE

The Science of Biology
TENTH EDITION

DAVID
SADAVA
The Claremont Colleges

DAVID M.
HILLIS
University of Texas

H. CRAIG
HELLER
Stanford University

MAY R.
BERENBAUM
University of Illinois

SINAUER

MACMILLAN

THE COVER
The sea slug *Elysia crispata*. This animal is able to carry out photosynthesis using chloroplasts incorporated from the algae it feeds on (see back cover). Photograph © Alex Mustard/Naturepl.com.

THE FRONTISPIECE
Red-crowned cranes, *Grus japonensis*, gather on a river in Hokkaido, Japan. ©Steve Bloom Images/Alamy.

LIFE: The Science of Biology, Tenth Edition
Copyright © 2014 by Sinauer Associates, Inc. All rights reserved.
This book may not be reproduced in whole or in part without permission.

ADDRESS EDITORIAL CORRESPONDENCE TO:
Sinauer Associates, Inc., 23 Plumtree Road, Sunderland, MA 01375 U.S.A.

www.sinauer.com
publish@sinauer.com

ADDRESS ORDERS TO:
MPS / W. H. Freeman & Co., Order Dept., 16365 James Madison Highway, U.S. Route 15, Gordonsville, VA 22942 U.S.A.

EXAMINATION COPY INFORMATION: 1-800-446-8923

Courier Corporation, the manufacturer
of this book, owns the *Green Edition* Trademark

Library of Congress Cataloging-in-Publication Data

Life : the science of biology / David Sadava ... [et al.]. -- 10th ed.
    p. cm.
Includes bibliographical references and index.
ISBN 978-1-4292-9864-3 (casebound) — 978-1-4641-4122-5 (pbk. : v. 1) —
ISBN 978-1-4641-4123-2 (pbk. : v. 2) — ISBN 978-1-4641-4124-9 (pbk. : v. 3)
1. Biology--Textbooks. I. Sadava, David E.
QH308.2.L565 2013
570--dc23   2012039164

Printed in U.S.A.
First Printing December 2012
The Courier Companies, Inc.

To all the educators who have worked tirelessly
for quality biology education

The Authors

DAVID HILLIS   MAY BERENBAUM   CRAIG HELLER   DAVID SADAVA

DAVID SADAVA is the Pritzker Family Foundation Professor of Biology, Emeritus at the Keck Science Center of Claremont McKenna, Pitzer, and Scripps, three of The Claremont Colleges. In addition, he is Adjunct Professor of Cancer Cell Biology at the City of Hope Medical Center in Duarte, California. Twice winner of the Huntoon Award for superior teaching, Dr. Sadava has taught courses on introductory biology, biotechnology, biochemistry, cell biology, molecular biology, plant biology, and cancer biology. In addition to *Life: The Science of Biology and Principles of Life*, he is the author or coauthor of books on cell biology and on plants, genes, and crop biotechnology. His research has resulted in many papers coauthored with his students, on topics ranging from plant biochemistry to pharmacology of narcotic analgesics to human genetic diseases. For the past 15 years, he has investigated multidrug resistance in human small-cell lung carcinoma cells with a view to understanding and overcoming this clinical challenge. At the City of Hope, his current work focuses on new anti-cancer agents from plants. He is the featured lecturer in "Understanding Genetics: DNA, Genes and their Real-World Applications," a video course for The Great Courses series.

DAVID M. HILLIS is the Alfred W. Roark Centennial Professor in Integrative Biology and the Director of the Dean's Scholars Program at the University of Texas at Austin, where he also has directed the School of Biological Sciences and the Center for Computational Biology and Bioinformatics. Dr. Hillis has taught courses in introductory biology, genetics, evolution, systematics, and biodiversity. He has been elected to the National Academy of Sciences and the American Academy of Arts and Sciences, awarded a John D. and Catherine T. MacArthur fellowship, and has served as President of the Society for the Study of Evolution and of the Society of Systematic Biologists. He served on the National Research Council committee that wrote the report *BIO 2010: Transforming Undergraduate Biology Education for Research Biologists*. His research interests span much of evolutionary biology, including experimental studies of viral evolution, empirical studies of natural molecular evolution, applications of phylogenetics, analyses of biodiversity, and evolutionary modeling. He is particularly interested in teaching and research about the practical applications of evolutionary biology.

H. CRAIG HELLER is the Lorry I. Lokey/Business Wire Professor in Biological Sciences and Human Biology at Stanford University. He has taught in the core biology courses at Stanford since 1972 and served as Director of the Program in Human Biology, Chairman of the Biological Sciences Department, and Associate Dean of Research. Dr. Heller is a fellow of the American Association for the Advancement of Science and a recipient of the Walter J. Gores Award for excellence in teaching and the Kenneth Cuthberson Award for Exceptional Service to Stanford University. His research is on the neurobiology of sleep and circadian rhythms, mammalian hibernation, the regulation of body temperature, the physiology of human performance, and the neurobiology of learning. He has done research on a huge variety of animals and physiological problems, including from sleeping kangaroo rats, diving seals, hibernating bears, photoperiodic hamsters, and exercising athletes. Dr. Heller has extended his enthusiasm for promoting active learning via the development of a two-year curriculum in human biology for the middle grades, through the production of Virtual Labs—interactive computer-based modules to teach physiology.

MAY BERENBAUM is the Swanlund Professor and Head of the Department of Entomology at the University of Illinois at Urbana-Champaign. She has taught courses in introductory animal biology, entomology, insect ecology, and chemical ecology and has received teaching awards at the regional and national levels from the Entomological Society of America. A fellow of the National Academy of Sciences, the American Academy of Arts and Sciences, and the American Philosophical Society, she served as President of the American Institute for Biological Sciences in 2009 and currently serves on the Board of Directors of AAAS. Her research addresses insect–plant coevolution and ranges from molecular mechanisms of detoxification to impacts of herbivory on community structure. Concerned with the practical application of ecological and evolutionary principles, she has examined impacts of genetic engineering, global climate change, and invasive species on natural and agricultural ecosystems. In recognition of her work, she received the 2011 Tyler Prize for Environmental Achievement. Devoted to fostering science literacy, she has published numerous articles and five books on insects for the general public.

Contents in Brief

Preface

Biology is a constantly changing scientific field. New discoveries about the living world are being made every day, and more than 1 million new research articles in biology are published each year. Beyond the constant need to update the concepts and facts presented in any science textbook, in recent years ideas about how best to educate the upcoming generation of biologists have undergone dynamic and exciting change.

Although we and many of our colleagues had thought about the nature of biological education as individuals, it is only recently that biologists have come together to discuss these issues. Reports from the National Academy of Sciences, Howard Hughes Medical Institute, and College Board AP Biology Program not only express concern about how best to instruct undergraduates in biology, but offer concrete suggestions about how to design the introductory biology course—and by extension, our book. We have followed these discussions closely and have been especially impressed with the report "Vision and Change in Undergraduate Biology Education" (visionandchange.org). As participants in the educational enterprise, we have answered the report's call to action with this textbook and its associated ancillary materials.

The "Vision and Change" report proposes five core concepts for biological literacy:

1. Evolution
2. Structure and function
3. Information flow, exchange, and storage
4. Pathways and transformations of energy and matter
5. Systems

These five concepts have always been recurring themes in Life, but in this Tenth Edition we have brought them even more "front and center."

"Vision and Change" also advocates that students learn and demonstrate core competencies, including the ability to apply the process of science using quantitative reasoning. Life has always emphasized the experimental nature of biology. This edition responds further to these core competency issues with a new working with data feature and the addition of a statistics primer (Appendix B). The authors' multiple educational perspectives and areas of expertise, as well as input from many colleagues and students who used previous editions, have informed the approach to this new edition.

Enduring Features

We remain committed to blending the presentation of core ideas with an emphasis on introducing students to the process of scientific inquiry. Having pioneered the idea of depicting important experiments in unique figures designed to help students understand and appreciate the way scientific investigations work, we continue to develop this approach in the book's 70 **Investigating Life** figures. Each of these figures sets the experiment in perspective and relates it to the accompanying text. As in previous editions, these figures employ the structure Hypothesis, Method, Results, and Conclusion. We have added new information focusing on the individuals who performed these experiments so students can appreciate more fully that science is a human and very personal activity. Each Investigating Life figure has a reference to BioPortal (*yourBioPortal.com*), where discussion and references to follow-up research can be found. A related feature is the **Research Tools** figures, which depict laboratory and field methods used in biology. These, too, have been expanded to provide more useful context for their importance.

Some 15 years ago, *Life's* authors and publishers pioneered the use of **balloon captions** in our figures. We recognized then that many students are visual learners, and this fact is even truer today. *Life's* balloon captions bring the crucial explanations of intricate, complex processes directly into the illustration, allowing students to integrate information without repeatedly going back and forth between the figure, its legend, and the text.

We continue to refine our chapter organization. Our **opening stories** have always provide historical, medical, or social context to intrigue students and show how the subject of each chapter relates to the world around them. In the Tenth Edition, the opening stories all end with a question that is revisited throughout the chapter. At the end of each chapter the answer is presented in the light of material the student encountered in the body of the chapter.

A **chapter outline** asks questions to emphasize scientific inquiry, each of which is answered in a major section of the chapter. A **Recap** summarizes each section's key concepts and poses questions that help the student review and test their mastery of these concepts. The recap questions are similar in form to the learning objectives used in many introductory biology courses. The **Chapter Summaries** highlight each chapter's key figures and defined terms, while restating the major concepts presented in the chapter in a concise and student-friendly manner, with references to specific figures and to the activities and animated tutorials available in BioPortal.

At the end of the book, students will find a much-expanded glossary that continues *Life's* practice of providing Latin or Greek derivations for many of the defined terms. As students become gradually (and painlessly) more familiar with such root words, the mastery of vocabulary as they continue in their biological or medical studies will be easier. In addition, the popular **Tree of Life appendix** (Appendix A) presents the phylogenetic tree of life as a reference tool that allows students to place any group of organisms mentioned in the text into the context of the rest of life. The web-based version of Appendix A provides links to photos, keys, species lists, distribution maps, and other information (via the online database at DiscoverLife.org) to help students explore biodiversity in greater detail.

New Features

The Tenth Edition of *Life* has a different look and feel from its predecessors. The new color palette and more open design will, we hope, be more accessible to students. And, in keeping with our heightened emphasis on scientific inquiry and quantitative analysis, we have added **Working with Data** exercises to almost all chapters. In these innovative exercises, we describe the context and approach of a research paper that provides the basis of the analysis. We then ask questions that require students to analyze data, make calculations, and draw conclusions. Answers (or suggested possible answers) to these questions are included in BioPortal and can be made available to students at the instructor's discretion.

Because many of the questions in the Working with Data exercises require the use of basic statistical methods, we have included a **Statistics Primer** as the book's Appendix B, describing the concepts and some methods of statistical analysis. We hope that the Working with Data exercises and statistics primer will reinforce students' skills and their ability to apply quantitative analysis to biology.

We have added links to **Media Clips** in the body of the text, with at least one per chapter. These brief clips are intended to enlighten and entertain. Recognizing the widespread use of "smart phones" by students, the textbook includes **instant access (QR) codes** that bring the Media Clips, Animated Tutorials, and Interactive Summaries directly to the screen in your hand. If you do not have a smart phone, never fear, we also provide direct web addresses to these features.

As educators, we follow current discussions of pedagogy in biological education. The chapter-ending **Chapter Reviews** now contain multiple levels of questions based on Bloom's taxonomy: Remembering, Understanding and Applying, and Analyzing and Evaluating. Answers to these questions appear at the end of the book.

For a detailed description of the media and supplements available for the Tenth Edition, please turn to "*Life's* Media and Supplements" on page xvii.

The Ten Parts

PART ONE, THE SCIENCE OF LIFE AND ITS CHEMICAL BASIS   Chapter 1 introduces the core concepts set forth in the "Vision and Change" report and continues the much-praised approach of focusing on a specific series of experiments that introduces students to biology as an experimentally based and constantly expanding science. Chapter 1 emphasizes the principles of biology that are the foundation for the rest of the book, including the unity of life at the cellular level and how evolution unites the living world. Chapters 2–4 cover the chemical principles and building blocks that underlie life. Chapter 4 also includes a discussion of how life could have evolved from inanimate chemicals.

PART TWO, CELLS   The nature of cells and their role as the structural and functional basis of life is foundational to biology. These revised chapters include expanded explanations of how experimental manipulations of living systems have been used to discover cause and effect in biology. Students who are intrigued by the question "Where did the first cells come from?" will appreciate the updated discussion of ideas on the origin of cells and organelles, as well as expanded discussion of the evolution of multicellularity and cell interactions. In response to reviewer comments, the discussion of membrane potential has been moved to Chapter 45, where students may find it to be more relevant.

PART THREE, CELLS AND ENERGY   The biochemistry of life and energy transformations are among the most challenging topics for many students. We have worked to clarify such concepts as enzyme inhibition, allosteric enzymes, and the integration of biochemical systems. Revised presentations of glycolysis and the citric acid cycle now focus, in both text and figures, on key concepts and attempt to limit excessive detail. There are also revised discussions of the ecological roles of alternate pathways of photosynthetic carbon fixation, as well as the roles of accessory pigments and reaction center in photosynthesis.

PART FOUR, GENES AND HEREDITY   This crucial section of the book is revised to improve clarity, link related concepts, and provide updates from recent research results. Rather than being segregated into separate chapters, material on prokaryotic genetics and molecular medicine are now interwoven into relevant chapters. Chapter 11 on the cell cycle includes a new discussion of how the mechanisms of cell division are altered in cancer cells. Chapter 12 on transmission genetics now includes coverage of this phenomenon in prokaryotes. Chapters 13 and 14 cover gene expression and gene regulation, including new discoveries about the roles of RNA and an expanded discussion of epigenetics. Chapter 15 covers the subject of gene mutations and describes updated applications of medical genetics.

PART FIVE, GENOMES   This extensive and up-to-date coverage of genomes expands and reinforces the concepts covered in Part Four. The first chapter of Part Five describes how genomes are analyzed and what they tell us about the biology of prokaryotes and eukaryotes, including humans. Methods of DNA sequencing and genome analysis, familiar to many students in a general way, are rapidly improving, and we discuss these advances as well as how bioinformatics is used. This leads to a chapter describing how our knowledge of molecular biology and genetics underpins biotechnology—the application of this knowledge to practical problems and issues such as stem cell research. Part Five closes with a unique sequence of two chapters that explore the interface of developmental processes with molecular biology (Chapter 19) and with evolution (Chapter 20), providing students with a link between these two crucial topics and a bridge to Part Six.

PART SIX, THE PATTERNS AND PROCESSES OF EVOLUTION Many students come to the introductory biology course with ideas about evolution already firmly in place. One common view, that evolution is only about Darwin, is firmly put to rest at the start of Chapter 21, which not only illustrates the practical value of fully understanding modern evolutionary biology, but briefly and succinctly traces the history of "Darwin's dangerous idea" through the twentieth century and up to the present syntheses of molecular evolutionary genetics and evolutionary developmental biology—fields of study that uphold and support the principles of evolutionary biology as the basis for comparing and comprehending all other aspects of biology. The remaining sections of Chapter 21 describe the mechanisms of evolution in clear, matter-of-fact terms. Chapter 22 describes phylogenetic trees as a tool not only of classification but also of evolutionary inquiry. The remaining chapters cover speciation and molecular evolution, concluding with an overview of the evolutionary history of life on Earth.

PART SEVEN, THE EVOLUTION OF DIVERSITY Continuing the theme of how evolution has shaped our world, Part Seven introduces the latest views on biodiversity and the evolutionary relationships among organisms. The chapters have been revised with the aim of making it easier for students to appreciate the major evolutionary changes that have taken place within the different groups of organisms. These chapters emphasize understanding the big picture of organismal diversity—the tree of life—as opposed to memorizing a taxonomic hierarchy and names. Throughout the book, the tree of life is emphasized as a way of understanding and organizing biological information.

PART EIGHT, FLOWERING PLANTS: FORM AND FUNCTION The emphasis of this modern approach to plant form and function is not only on the basic findings that led to the elucidation of mechanisms for plant growth and reproduction, but also on the use of genetics of model organisms. In response to users of earlier editions, material covering recent discoveries in plant molecular biology and signaling has been reorganized and streamlined to make it more accessible to students. There are also expanded and clearer explanations of such topics as water relations, the plant body plan, and gamete formation and double fertilization.

PART NINE, ANIMALS: FORM AND FUNCTION This overview of animal physiology begins with a sequence of chapters covering the systems of information—endocrine, immune, and neural. Learning about these information systems provides important groundwork and explains the processes of control and regulation that affect and integrate the individual physiological systems covered in the remaining chapters of the Part. Chapter 45, "Neurons and Nervous Systems," has been rearranged and contains descriptions of exciting new discoveries about glial cells and their role in the vertebrate nervous system. The organization of several other chapters has been revised to reflect recent findings and to allow the student to more readily identify the most important concepts to be mastered.

PART TEN, ECOLOGY Part Ten continues *Life*'s commitment to presenting the experimental and quantitative aspects of biology, with increased emphasis on how ecologists design and conduct experiments. New exercises provide opportunities for students to see how ecological data are acquired in the laboratory and in the field, how these data are analyzed, and how the results are applied to answer questions. There is also an expanded discussion of aquatic biomes and a more synthetic explanation of how aquatic, terrestrial, and atmospheric components integrate to influence the distribution and abundance of life on Earth. In addition there is an expanded emphasis on examples of successful strategies proposed by ecologists to mitigate human impacts on the environment; rather than an inventory of ways human activity adversely affects natural systems, this revised Tenth Edition provides more examples of ways that ecological principles can be applied to increase the sustainability of these systems.

Exceptional Value Formats

We again provide *Life* both as the full book and as a set of paperback volumes. Thus, instructors who want to use less than the whole book can choose from these split volumes, each of which contains the book's front matter, appendices, glossary, and index.

- Volume I, *The Cell and Heredity*, includes: Part One, The Science of Life and Its Chemical Basis (Chapters 1–4); Part Two, Cells (Chapters 5–7); Part Three, Cells and Energy (Chapters 8–10); Part Four, Genes and Heredity (Chapters 11–16); and Part Five, Genomes (Chapters 17–20).

- Volume II, *Evolution, Diversity, and Ecology*, includes: Chapter 1, Studying Life; Part Six, The Patterns and Processes of Evolution (Chapters 21–25); Part Seven, The Evolution of Diversity (Chapters 26–33); and Part Ten, Ecology (Chapters 54–59).

- Volume III, *Plants and Animals*, includes: Chapter 1, Studying Life; Part Eight, Flowering Plants: Form and Function (Chapters 34–39); and Part Nine, Animals: Form and Function (Chapters 40–53).

Responding to student concerns, there also are two ways to obtain the entire book at a significantly reduced cost. The loose-leaf edition of *Life* is a shrink-wrapped, unbound, three-hole-punched version that fits into a three-ring binder. Students take only what they need to class and can easily integrate instructor handouts and other resources.

Life was the first comprehensive biology text to offer the entire book as a truly robust eBook, and we offer the Tenth Edition in this flexible, interactive format that gives students a different way to read the text and learn the material. The eBook integrates student media resources (animations, activities, interactive summaries, and quizzes) and offers instructors a powerful way to customize the textbook with their own text, images, web links, and, in BioPortal, quizzes, and other materials.

We are proud that our print edition is a greener *Life* that minimizes environmental impact. *Life* was the first introductory biology text to be printed on paper earning the Forest Stewardship Council label, the "gold standard in green paper," and it continues to be manufactured from wood harvested from sustainable forests.

Many People to Thank

One of the wisest pieces of advice ever given to a textbook author is to "be passionate about your subject, but don't put your ego on the page." Considering all the people who looked over our shoulders throughout the process of creating this book, this advice could not be more apt. We are indebted to the many people who help to make this book what it is. First and foremost among these are our colleagues, biologists from over 100 institutions. Before we set pen to paper, we solicited the advice of users of *Life*'s Ninth Edition, as well as users of other books. These reviewers gave detailed suggestions for improvements. Other colleagues acted as reviewers when the book was almost completed, pointing out inaccuracies or lack of clarity. All of these biologists are listed in the reviewer credits, along with the dozens who reviewed all of the revised assessment resources.

Once we began writing, we had the superb advice of a team of experienced, knowledgeable, and patient biologists working as development and line editors. Laura Green of Sinauer Associates headed the team and coordinated her own fine work with that of Jane Murfett, Norma Roche, and Liz Pierson

to produce a polished and professional text. We are especially indebted to Laura for her work on the important Investigating Life and new Working with Data elements. For the tenth time in ten editions, Carol Wigg oversaw the editorial process. Her positive influence pervades the entire book. Artist Elizabeth Morales again translated our crude sketches into beautiful new illustrations. We hope you agree that our art program remains superbly clear and elegant. Johannah Walkowicz effectively coordinated the hundreds of reviews described above. David McIntyre, photo editor extraordinaire, researched and provided us with new photographs, including many of his own, to enrich the book's content and visual statement. Joanne Delphia is responsible for the crisp new design and layout that make this edition of *Life* not just clear and readable but beautiful as well. Christopher Small headed Sinauer's production team and contributed in innumerable ways to bringing *Life* to its final form. Jason Dirks coordinated the creation of our array of media and instructor resources, with Mary Tyler, Mitch Walkowicz, and Carolyn Wetzel serving as editors for our expanded assessment supplements.

W. H. Freeman continues to bring *Life* to a wider audience. Associate Director of Marketing Debbie Clare, the regional specialists, regional managers, and experienced sales force are effective ambassadors and skillful transmitters of the features and unique strengths of our book. We depend on their expertise and energy to keep us in touch with how *Life* is perceived by its users. Thanks also to the Freeman media group for eBook and BioPortal production.

Finally, we thank our friend Andy Sinauer. Like ours, his name is on the cover of the book, and he truly cares deeply about what goes into it.

DAVID SADAVA

DAVID HILLIS

CRAIG HELLER

MAY BERENBAUM

Reviewers for the Tenth Edition

Between Edition Reviewers

Shivanthi Anandan, Drexel University

Brian Bagatto, The University of Akron

Mary Bisson, University at Buffalo, The State University of New York

Meredith Blackwell, Louisiana State University

Randy Brooks, Florida Atlantic University

Heather Caldwell, Kent State University

Jeffrey Carrier, Albion College

David Champlin, University of Southern Maine

Wesley Colgan, Pikes Peak Community College

Emma Creaser, Unity College

Karen Curto, University of Pittsburgh

John Dennehy, Queens College, The City University of New York

Rajinder Dhindsa, McGill University

James A. Doyle, University of California, Davis

Scott Edwards, Harvard University

David Eldridge, Baylor University

Joanne Ellzey, The University of Texas at El Paso

Douglas Gayou, University of Missouri

Stephen Gehnrich, Salisbury University

Arundhati Ghosh, University of Pittsburgh

Nathalia Glickman Holtzman, Queens College, The City University of New York

Elizabeth Good, University of Illinois at Urbana-Champaign

Harry Greene, Cornell University

Alice Heicklen, Columbia University

Albert Herrera, University of Southern California

David Hibbett, Clark University

Mark Holbrook, University of Iowa

Craig Jordan, The University of Texas at San Antonio

Walter Judd, University of Florida

John M. Labavitch, University of California, Davis

Nathan H. Lents, John Jay College of Criminal Justice, The City University of New York

Barry Logan, Bowdoin College

Barbara Lom, Davidson College

David Low, University of California, Davis

Janet Loxterman, Idaho State University

Sharon Lynn, The College of Wooster

Julin Maloof, University of California, Davis

Richard McCarty, Johns Hopkins University

Sheila McCormick, University of California, Berkeley

Marcie Moehnke, Baylor University

Roberta Moldow, Seton Hall University

Tsafrir Mor, Arizona State University

Alexander Motten, Duke University

Barbara Musolf, Clayton State University

Stuart Newfeld, Arizona State University

Bruce Ostrow, Grand Valley State University

Laura K. Palmer, The Pennsylvania State University, Altoona

Robert Pennock, Michigan State University

Kamini Persaud, University of Toronto, Scarborough

Roger Persell, Hunter College, The City University of New York

Matthew Rand, Carleton College

Susan Richardson, Florida Atlantic University

Brian C. Ring, Valdosta State University

Jay Rosenheim, University of California, Davis

Ben Rowley, University of Central Arkansas

Ann Rushing, Baylor University

Mikal Saltveit, University of California, Davis

Joel Schildbach, Johns Hopkins University

Christopher J. Schneider, Boston University

Paul Schulte, University of Nevada, Las Vegas

Leah Sheridan, University of Northern Colorado

Gary Shin, University of California, Los Angeles

Mitchell Singer, University of California, Davis

William Taylor, The University of Toledo

Sharon Thoma, University of Wisconsin, Madison

James F. A. Traniello, Boston University

Terry Trier, Grand Valley State University

Sara Via, University of Maryland

Curt Walker, Dixie State College

Fred Wasserman, Boston University

Alexander J. Werth, Hampden-Sydney College

Elizabeth Willott, University of Arizona

Accuracy Reviewers

Rebecca Rashid Achterman, Western Washington University

Maria Ambrosetti, Emory University

Miriam Ashley-Ross, Wake Forest University

Felicitas Avendaño, Grand View University

David Bailey, St. Norbert College

Chhandak Basu, California State University, Northridge

Jim Bednarz, Arkansas State University

Charlie Garnett Benson, Georgia State University

Katherine Boss-Williams, Emory University

Ben Brammell, Asbury University

Christopher I. Brandon, Jr., Georgia Gwinnett College

Carolyn J. W. Bunde, Idaho State University

Darlene Campbell, Cornell University

Jeffrey Carmichael, University of North Dakota

David J. Carroll, Florida Institute of Technology

Ethan Carver, The University of Tennessee at Chattanooga

Peter Chabora, Queens College, The City University of New York

Heather Cook, Wagner College

Hsini Lin Cox, The University of Texas at El Paso

Douglas Darnowski, Indiana University Southeast

Stephen Devoto, Wesleyan University

Rajinder Dhindsa, McGill University

Jesse Dillon, California State University, Long Beach

James A. Doyle, University of California, Davis

Devin Drown, Indiana University

Richard E. Duhrkopf, Baylor University

Weston Dulaney, Nashville State Community College

David Eldridge, Baylor University

Kenneth Filchak, University of Notre Dame

Kerry Finlay, University of Regina

Kevin Folta, University of Florida

Douglas Gayou, University of Missouri

David T. Glover, Food and Drug Administration

Russ Goddard, Valdosta State University

Elizabeth Godrick, Boston University

Leslie Goertzen, Auburn University

Elizabeth Good, University of Illinois at Urbana-Champaign

Ethan Graf, Amherst College

Eileen Gregory, Rollins College

Julie C. Hagelin, University of Alaska, Fairbanks

Nathalia Glickman Holtzman, Queens College, The City University of New York

Dianne Jennings, Virginia Commonwealth University

Jamie Jensen, Bringham Young University

Glennis E. Julian

Erin Keen-Rhinehart, Susquehanna University

Henrik Kibak, California State University, Monterey Bay

Brandi Brandon Knight, Emory University

Daniel Kueh, Emory University

John G. Latto, University of California, Santa Barbara

Kristen Lennon, Frostburg State University

David Low, University of California, Santa Barbara

Jose-Luis Machado, Swarthmore College

Jay Mager, Ohio Northern University

Stevan Marcus, University of Alabama

Nilo Marin, Broward College

Marlee Marsh, Columbia College South Carolina

Erin Martin, University of South Florida, Sarasota-Manatee

Brad Mehrtens, University of Illinois at Urbana-Champaign

Michael Meighan, University of California, Berkeley

Tsafrir Mor, Arizona State University

Roderick Morgan, Grand Valley State University

Jacalyn Newman, University of Pittsburgh

Alexey Nikitin, Grand Valley State University

Zia Nisani, Antelope Valley College

Laura K. Palmer, The Pennsylvania State University, Altoona

Nancy Pencoe, State University of West Georgia

David P. Puthoff, Frostburg State University

Brett Riddle, University of Nevada, Las Vegas

Leslie Riley, Ohio Northern University

Brian C. Ring, Valdosta State University

Heather Roffey, McGill University

Lori Rose, Hill College

Naomi Rowland, Western Kentucky University

Beth Rueschhoff, Indiana University Southeast

Ann Rushing, Baylor University

Illya Ruvinsky, University of Chicago

Paul Schulte, University of Nevada, Las Vegas

Susan Sharbaugh, University of Alaska, Fairbanks

Jonathan Shenker, Florida Institute of Technology

Gary Shin, California State University, Long Beach

Ken Spitze, University of West Georgia

Bruce Stallsmith, The University of Alabama in Huntsville

Robert M. Steven, The University of Toledo

Zuzana Swigonova, University of Pittsburgh

Rebecca Symula, The University of Mississippi

Mark Taylor, Baylor University

Mark Thogerson, Grand Valley State University

Elethia Tillman, Spelman College

Terry Trier, Grand Valley State University

Michael Troyan, The Pennsylvania State University, University Park

Sebastian Velez, Worcester State University

Sheela Vemu, Northern Illinois University

Andrea Ward, Adelphi University

Katherine Warpeha, University of Illinois at Chicago

Fred Wasserman, Boston University

Michelle Wien, Bryn Mawr College

Robert Wisotzkey, California State University, East Bay

Greg Wray, Duke University

Joanna Wysocka-Diller, Auburn University

Catherine Young, Ohio Northern University

Heping Zhou, Seton Hall University

Assessment Reviewers

Maria Ambrosetti, Georgia State University

Cecile Andraos-Selim, Hampton University

Felicitas Avendaño, Grand View University

David Bailey, St. Norbert College

Jim Bednarz, Arkansas State University

Charlie Garnett Benson, Georgia State University

Katherine Boss-Williams, Emory University

Ben Brammell, Asbury University

Christopher I. Brandon, Jr., Georgia Gwinnett College

Brandi Brandon Knight, Emory University

Ethan Carver, The University of Tennessee, Chattanooga

Heather Cook, Wagner College

Hsini Lin Cox, The University of Texas at El Paso

Douglas Darnowski, Indiana University Southeast

Jesse Dillon, California State University, Long Beach

Devin Drown, Indiana University

Richard E. Duhrkopf, Baylor University

Weston Dulaney, Nashville State Community College

Kenneth Filchak, University of Notre Dame

Elizabeth Godrick, Boston University

Elizabeth Good, University of Illinois at Urbana-Champaign

Susan Hengeveld, Indiana University Bloomington

Nathalia Glickman Holtzman, Queens College, The City College of New York

Glennis E. Julian

Erin Keen-Rhinehart, Susquehanna University

Stephen Kilpatrick, University of Pittsburgh

Daniel Kueh, Emory University

Stevan Marcus, University of Alabama

Nilo Marin, Broward College

Marlee Marsh, Columbia College

Erin Martin, University of South Florida, Sarasota-Manatee

Brad Mehrtens, University of Illinois at Urbana-Champaign

Darlene Mitrano, Christopher Newport University

Anthony Moss, Auburn University

Jacalyn Newman, University of Pittsburgh

Alexey Nikitin, Grand Valley State University

Zia Nisani, Antelope Valley College

Sabiha Rahman, University of Ottawa

Nancy Rice, Western Kentucky University

Brian C. Ring, Valdosta State University

Naomi Rowland, Western Kentucky University

Jonathan Shenker, Florida Institute of Technology

Gary Shin, California State University, Long Beach

Jacob Shreckengost, Emory University

Michael Smith, Western Kentucky University

Ken Spitze, University of West Georgia

Bruce Stallsmith, The University of Alabama in Huntsville

Zuzana Swigonova, University of Pittsburgh

William Taylor, The University of Toledo

Mark Thogerson, Grand Valley State University

Elethia Tillman, Spelman College

Michael Troyan, The Pennsylvania State University

Ximena Valderrama, Ramapo College of New Jersey

Sheela Vemu, Northern Illinois University

Suzanne Wakim, Butte College

Katherine Warpeha, University of Illinois at Chicago

Fred Wasserman, Boston University

Michelle Wien, Bryn Mawr College

Robert Wisotzkey, California State University, East Bay

Heping Zhou, Seton Hall University

LIFE's Media and Supplements

yourBioPortal.com

BioPortal is the online gateway to all of *Life*'s digital resources, including the fully interactive eBook, a wide range of student and instructor media resources, and powerful assessment tools. BioPortal includes the following features and resources:

Life, Tenth Edition eBook
(eBook also available stand-alone)

- Complete online version of the textbook
- Integration of all Media Clips, Activities, Animated Tutorials, and other media resources
- In-text links to all glossary entries, with audio pronunciations
- A flexible notes feature and easy text highlighting
- Searchable glossary and index
- Full-text search

Additional eBook features for instructors:

- *Content Customization*: Instructors can easily hide chapters or sections that they don't cover in their course, re-arrange the order of chapters and sections, and add their own content directly into the eBook.
- *Instructor Notes*: Instructors can annotate the eBook with their own notes and content on any page. Instructor notes can include text, Web links, images, links to BioPortal resources, uploaded documents, and more.

LearningCurve

New for the Tenth Edition, LearningCurve is a powerful adaptive quizzing system with a game-like format that engages students. Rather than simply answering a fixed set of questions, students answer dynamically-selected questions to progress toward a target level of understanding. At any point, students can view a report of how well they are performing in each topic area (with links to eBook sections and media resources), to help them focus on problem areas.

Student BioPortal Resources

DIAGNOSTIC QUIZZING. The pre-built diagnostic quizzes assesses student understanding of each section of each chapter, and generates a Personalized Study Plan to effectively focus student study time. The plan includes links to specific textbook sections, animated tutorials, and activities.

INTERACTIVE SUMMARIES. For each chapter, these dynamic summaries combine a review of important concepts with links to all of the key figures, Activities, and Animated Tutorials.

ANIMATED TUTORIALS. In-depth tutorials that present complex topics in a clear, easy-to-follow format that combines a detailed animation or simulation with an introduction, conclusion, and brief quiz.

MEDIA CLIPS. New for the Tenth Edition, these short, engaging video clips depict fascinating examples of some of the many organisms, processes, and phenomena discussed in the textbook.

ACTIVITIES. A range of interactive activities that help students learn and review key facts and concepts through labeling diagrams, identifying steps in processes, and matching concepts.

LECTURE NOTEBOOK. New for the Tenth Edition, the Lecture Notebook is included online in BioPortal. The Notebook includes all of the textbook's figures and tables, with space for note-taking, and is available as downloadable PDF files.

BIONEWS FROM SCIENTIFIC AMERICAN. BioNews makes it easy for instructors to bring the dynamic nature of the biological sciences and up-to-the minute currency into their course, via an automatically updated news feed.

BIONAVIGATOR. A unique visual way to explore all of the Animated Tutorials and Activities across the various levels of biological inquiry—from the global scale down to the molecular scale.

WORKING WITH DATA. Online versions of the Working with Data exercises that are included in the textbook.

FLASHCARDS AND KEY TERMS. The Flashcards and Key Terms provide an ideal way for students to learn and review the extensive terminology of introductory biology, featuring a review mode and a quiz mode.

INVESTIGATING LIFE LINKS. For each Investigating Life figure in the textbook, BioPortal includes an overview of the experiment featured in the figure with links to the original paper(s), related research or applications that followed, and additional information related to the experiment.

GLOSSARY. The full glossary, with audio pronunciations for all terms.

TREE OF LIFE. An interactive version of the Tree of Life from Appendix A. The online Tree links to a wealth of information on each group listed.

MATH FOR LIFE. A collection of mathematical shortcuts and references to help students with the quantitative skills they need in the biology laboratory.

SURVIVAL SKILLS. A guide to more effective study habits, including time management, note-taking, effective highlighting, and exam preparation.

Instructor BioPortal Resources

Assessment

- LearningCurve and Diagnostic Quizzing reports provide instructors with a wealth of information on student comprehension, by textbook section, along with targeted lecture resources for those areas requiring the most attention.

- Comprehensive question banks include questions from the Test Bank, LearningCurve, Diagnostic Quizzes, Study Guide, and textbook Chapter Review.

- Question filtering allows instructors to select questions based on Bloom's category and/or textbook section, in order to easily select the desired mix of question types.

- Easy-to-use assessment tools allow instructors to create quizzes and many other types of assignments using any combination of publisher-provided questions and those created by the instructor.

Media Resources
(see Instructor's Media Library below for details)

- Videos
- PowerPoint Presentations (Figures & Tables, Lecture, Editable Labels, Layered Art)
- Supplemental Photos
- Active Learning Exercises
- Instructor's Manual
- Lecture Notes
- Answers to Working with Data Exercises
- Course management features
- Complete course customization capabilities
- Custom resources/document posting
- Robust gradebook
- Communication Tools: Announcements, Calendar, Course Email, Discussion Boards

Student Supplements

Life, Tenth Edition Study Guide
(Paper, ISBN 978-1-4641-2365-8)

The *Life* Study Guide offers a variety of study and review resources to accompany each chapter of the textbook. The opening Big Picture section gives students a concise overview of the main concepts covered in the chapter. The Study Strategies section points out common problem areas that students may find more challenging, and suggests strategies for learning the material most effectively. The Key Concept Review section combines a detailed review of each section with questions that help students synthesize and apply what they have learned, including diagram questions, short-answer questions, and more open-ended questions. Each chapter concludes with a Test Yourself section that allows students to test their comprehension. All questions include answers, explanations, and references to textbook sections.

Life Flashcards App

Available for iPhone/iPad and Android, the *Life* Flashcards App is a great way for students to learn and review all the key terminology from the textbook, whenever and wherever they want to study, in an intuitive flashcard interface. Available in the iTunes App Store and Google Play.

CatchUp Math & Stats
Michael Harris, Gordon Taylor, and Jacquelyn Taylor
(ISBN 978-1-4292-0557-3)

Presented in brief, accessible units, this primer will help students quickly brush up on the quantitative skills they need to succeed in biology.

Student Handbook for Writing in Biology, Third Edition
Karen Knisely (ISBN 978-1-4292-3491-7)

This book provides practical advice to students who are learning to write according to the conventions in biology, using the standards of journal publication as a model.

Bioethics and the New Embryology: Springboards for Debate
Scott F. Gilbert, Anna Tyler, and Emily Zackin
(ISBN 978-0-7167-7345-0)

Our ability to alter the course of human development ranks among the most significant changes in modern science and has brought embryology into the public domain. The question that must be asked is: Even if we can do such things, should we?

BioStats Basics: A Student Handbook
James L. Gould and Grant F. Gould (ISBN 978-0-7167-3416-1)

Engaging and informal, *BioStats Basics* provides introductory-level biology students with a practical, accessible introduction to statistical research.

Inquiry Biology: A Laboratory Manual, Volumes 1 and 2

Mary Tyler, Ryan W. Cowan, and Jennifer L. Lockhart (Volume 1 ISBN 978-1-4292-9288-7; Volume 2 ISBN 978-1-4292-9289-4)

This introductory biology laboratory manual is inquiry-based—instructing in the process of science by allowing students to ask their own questions, gather background information, formulate hypotheses, design and carry out experiments, collect and analyze data, and formulate conclusions.

Hayden-McNeil Life Sciences Lab Notebook

(ISBN 978-1-4292-3055-1)

This carbonless laboratory notebook is of the highest quality and durability, allowing students to hand in originals or copies, not entire composition books. Contains Hayden-McNeil's unique white paper carbonless copies and biology-specific reference materials.

Instructor Media & Supplements

Instructor's Media Library

(Available both online via BioPortal and on disc; disc version ISBN 978-1-4641-2364-1)

The *Life*, Tenth Edition Instructor's Media Library includes a wide range of electronic resources to help instructors plan their course, present engaging lectures, and effectively assess their students. The Media Library includes the following resources:

TEXTBOOK FIGURES AND TABLES. Every figure and table from the textbook (including all photos and all un-numbered figures) is provided in both JPEG (high- and low-resolution) and PDF formats, in multiple versions.

UNLABELED FIGURES. Every figure is provided in an unlabeled format, useful for student quizzing and custom presentations.

SUPPLEMENTAL PHOTOS. The supplemental photograph collection contains over 1,500 photographs, giving instructors a wealth of additional imagery to draw upon.

ANIMATIONS. An extensive collection of detailed animations, all built specifically for Life, and viewable in either narrated or step-through mode.

VIDEOS. Featuring many new segments for the Tenth Edition, the wide-ranging collection of video segments help demonstrate the complexity and beauty of life.

POWERPOINT RESOURCES. For each chapter of the textbook, many different PowerPoint presentations are available, providing instructors the flexibility to build presentations in the manner that best suits their needs, including the following:

- Textbook Figures and Tables
- Lecture Presentation
- Figures with Editable Labels
- Layered Art Figures
- Supplemental Photos
- Videos
- Animations
- Active Learning Exercises

INSTRUCTOR'S MANUAL, LECTURE NOTES, and **TEST BANK** are available in Microsoft Word format for easy use in lecture and exam preparation.

MEDIA GUIDE. A PDF version of the Media Guide from the Instructor's Resource Kit, convenient for searching.

ACTIVE LEARNING EXERCISES. Set up for easy integration into lectures, each exercise poses a question or problem for the class to discuss or solve during lecture. Each also includes a multiple-choice element, for easy use with clicker systems.

ANSWERS TO WORKING WITH DATA EXERCISES. Complete answers to all of the Working with Data exercises.

Instructor's Resource Kit

(Binder, ISBN 978-1-4641-4131-7)

The *Life*, Tenth Edition Instructor's Resource Kit includes a wealth of information to help instructors in the planning and teaching of their course. The Kit includes:

INSTRUCTOR'S MANUAL

- *Chapter Overview*: A brief, high-level synopsis of the chapter.
- *What's New*: A guide to the revisions, updates, and new content added to the Tenth Edition.
- *Key Concepts & Learning Objectives*: New for the Tenth Edition, this section includes the major learning goals for the chapter, a detailed set of key concepts, and specific learning objectives for each key concept.
- *Chapter Outline*: All of the chapter's section headings and sub-headings.
- *Key Terms*: All of the important terms introduced in the chapter.

LECTURE NOTES. Detailed lecture outlines for each chapter, including references to relevant figures and media resources.

MEDIA GUIDE. A visual guide to the extensive media resources available with Life, including all animations, activities, videos, and supplemental photos.

Overhead Transparencies

(ISBN 978-1-4641-4127-0)

The set of overheads includes over 1,000 transparencies—including all of the four-color line art and all of the tables from the text—in two convenient binders. All figures have been formatted and color-enhanced for clear projection in a wide range of conditions. Labels and images have been resized for improved readability.

Test File

(Paper, ISBN 978-1-4292-5579-0)

The *Life*, Tenth Edition Test File includes over 5,000 questions and has been revised and reviewed for both accuracy and effectiveness. All questions are referenced to specific textbook headings and categorized according to Bloom's taxonomy. This allows instructors to easily build quizzes and exams with the desired mix of content, coverage, and question types (factual, conceptual, analyzing/applying, etc.). Each chapter includes a wide range of multiple choice and fill-in-the-blank questions, in addition to diagram questions that involve the student in working with illustrations of structures, graphs, steps in processes, and more.

Computerized Test Bank

(CD, ISBN 978-1-4641-4128-7)

The entire Test File, plus the Diagnostic Quizzes, Learning-Curve questions, Study Guide questions, and Textbook End-of-Chapter Review questions are all included in Wimba's easy-to-use Diploma program (software included). Designed for both novice and advanced users, Diploma allows instructors to quickly and easily create or edit questions, create quizzes or exams with a "drag-and-drop" feature (using any combination of publisher-provided and instructor-added questions), publish to online courses, and print paper-based assessments.

Figure Correlation Tool

An invaluable resource for instructors switching to *Life*, Tenth Edition from another textbook or from *Life*, Ninth Edition, this online tool provides correlations between all of the figures in *Life*, Tenth Edition and figures in other majors biology textbooks and *Life*, Ninth Edition.

Course Management System Support

As a service for *Life* adopters using Blackboard, WebCT, AN-GEL, or other course management systems, full electronic course packs are available.

Faculty Lounge for Majors Biology is the first publisher-provided website for the majors biology community that lets instructors freely communicate and share peer-reviewed lecture and teaching resources. The Faculty Lounge offers convenient access to peer-recommended and vetted resources, including the following categories: Images, News, Videos, Labs, Lecture Resources, and Educational Research. **majorsbio.facultylounge.whfreeman.com**

iclicker

Developed for educators by educators, iclicker is a hassle-free radio-frequency classroom response system that makes it easy for instructors to ask questions, record responses, take attendance, and direct students through lectures as active participants. For more information, visit **www.iclicker.com**.

LabPartner is a site designed to facilitate the creation of customized lab manuals. Its database contains a wide selection of experiments published by W. H. Freeman and Hayden-McNeil Publishing. Instructors can preview, choose, and re-order labs, interleave their own original experiments, add carbonless graph paper and a pocket folder, customize the cover both inside and out, and select a binding type. Manuals are printed on-demand. **www.whfreeman.com/labpartner**

The Scientific Teaching Book Series is a collection of practical guides, intended for all science, technology, engineering and mathematics (STEM) faculty who teach undergraduate and graduate students in these disciplines. The purpose of these books is to help faculty become more successful in all aspects of teaching and learning science, including classroom instruction, mentoring students, and professional development. Authored by well-known science educators, the Series provides concise descriptions of best practices and how to implement them in the classroom, the laboratory, or the department. For readers interested in the research results on which these best practices are based, the books also provide a gateway to the key educational literature.

Scientific Teaching

Jo Handelsman, Sarah Miller, and Christine Pfund (ISBN 978-1-4292-0188-9)

Transformations:
Approaches to College Science Teaching

Deborah Allen and Kimberly Tanner (ISBN 978-1-4292-5335-2)

Entering Research: A Facilitator's Manual

Workshops for Students Beginning Research in Science

Janet L. Branchaw, Christine Pfund, and Raelyn Rediske (ISBN 978-1-429-25857-9)

Discipline-Based Science Education Research:
A Scientist's Guide

Stephanie Slater, Tim Slater, and Janelle M. Bailey (ISBN 978-1-4292-6586-7)

Assessment in the College Classroom

Clarissa Dirks, Mary Pat Wenderoth, Michelle Withers (ISBN 978-1-4292-8197-3)

Contents

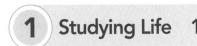

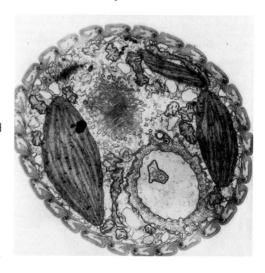

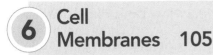
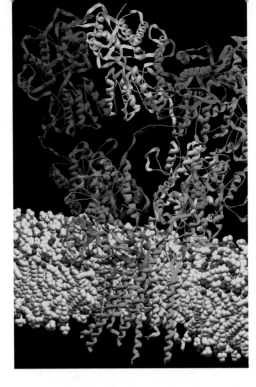
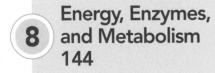

PART FOUR
Genes and Heredity

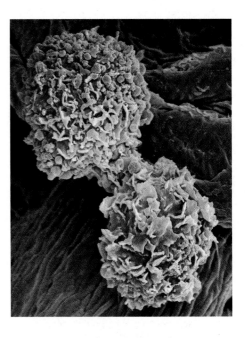

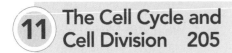

PART FIVE
Genomes

17 Genomes   352

16 Regulation of Gene Expression   328

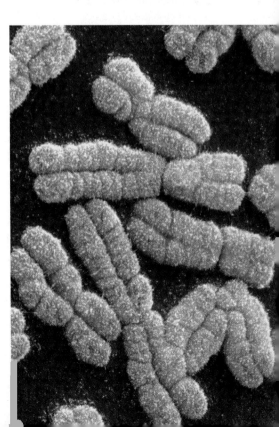

PART SIX
The Patterns and Processes of Evolution

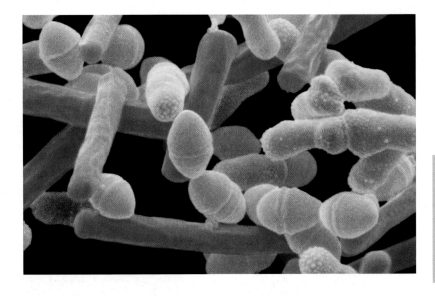

PART SEVEN
The Evolution of Diversity

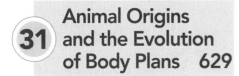

PART EIGHT
Flowering Plants: Form and Function

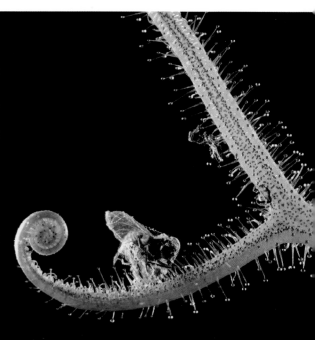

PART NINE
Animals: Form and Function

Immunology: Animal Defense Systems   856

Animal Reproduction   880

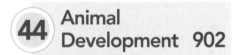

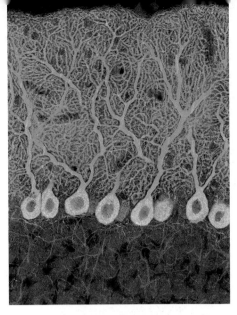

46 Sensory Systems   946

47 The Mammalian Nervous System: Structure and Higher Functions   967

48 Musculoskeletal Systems   986

49 Gas Exchange 1005

50 Circulatory Systems 1025

53 Animal Behavior 1093

PART TEN
Ecology

54 Ecology and the Distribution of Life 1121

55 Population Ecology 1149

56 Species Interactions and Coevolution 1169

PART ONE
The Science of Life and Its Chemical Basis

Studying Life

What's Happening to the Frogs? Tyrone Hayes grew up near the great Congaree Swamp in South Carolina collecting turtles, snakes, frogs, and toads. He is now a professor of biology at the University of California at Berkeley. In the laboratory and in the field, he is studying how and why populations of frogs are endangered by agricultural pesticides.

AMPHIBIANS—frogs, salamanders, and wormlike caecilians— have been around so long they watched the dinosaurs come and go. But for the last three decades, amphibian populations around the world have been declining dramatically. Today more than a third of the world's amphibian species are threatened with extinction. Why are these animals disappearing?

Tyrone Hayes, a biologist at the University of California at Berkeley, probed the effects of certain chemicals that are applied to croplands in large quantities and that accumulate in the runoff water from the fields. Hayes focused on the effects on amphibians of atrazine, a weed killer (herbicide) widely used in the United States and some other countries, where it is a common contaminant in fresh water (its use has been banned in the European Union). In the U.S., atrazine is usually applied in the spring, when many amphibians are breeding and thousands of tadpoles swim in the ditches, ponds, and streams that receive runoff from farms.

In his laboratory, Hayes and his associates raised frog tadpoles in water containing no atrazine and also in water with concentrations ranging from 0.01 parts per billion (ppb) up to 25 ppb. Concentrations as low as 0.1 ppb had a dramatic effect on tadpole development: it feminized the males. When these males became adults, their vocal structures—which are used in mating calls and thus are crucial for successful reproduction—were smaller than normal; in some, eggs were growing in the testes; some developed female sex organs. In other studies, normal adult male frogs exposed to 25 ppb had a tenfold reduction in testosterone levels and did not produce sperm. You can imagine the disastrous effects of such developmental and hormonal changes on the capacity of frogs to breed and reproduce.

But these experiments were performed in the laboratory, with a species of frog bred for laboratory use. Would the results be the same in nature? To find out, Hayes and his students traveled from Utah to Iowa, sampling water and collecting frogs. They analyzed the water for atrazine and examined the frogs. The only site where the frogs were normal was one where atrazine was undetectable. At all other sites, male frogs had abnormalities of the sex organs.

Like other biologists, Hayes made observations. He then made predictions based on those observations, and designed and carried out experiments to test his predictions.

Could atrazine in the environment affect species other than amphibians?

See answer on p. 18.

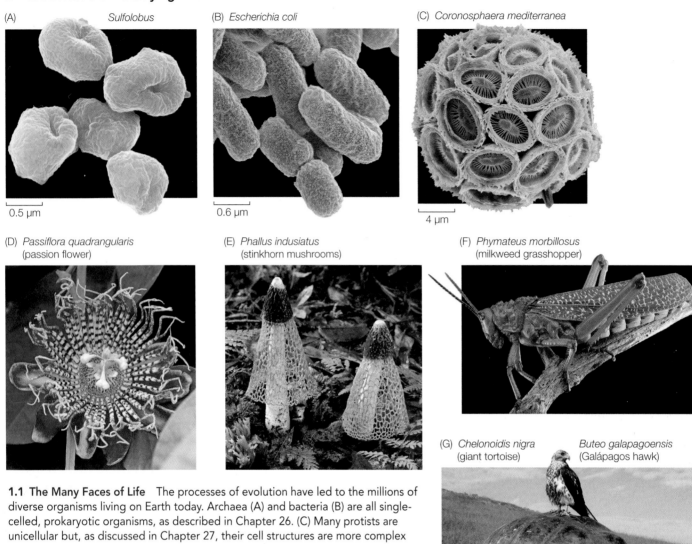

(A) *Sulfolobus*

(B) *Escherichia coli*

(C) *Coronosphaera mediterranea*

0.5 µm

0.6 µm

4 µm

(D) *Passiflora quadrangularis*
(passion flower)

(E) *Phallus indusiatus*
(stinkhorn mushrooms)

(F) *Phymateus morbillosus*
(milkweed grasshopper)

(G) *Chelonoidis nigra*
(giant tortoise)

Buteo galapagoensis
(Galápagos hawk)

1.1 The Many Faces of Life   The processes of evolution have led to the millions of diverse organisms living on Earth today. Archaea (A) and bacteria (B) are all single-celled, prokaryotic organisms, as described in Chapter 26. (C) Many protists are unicellular but, as discussed in Chapter 27, their cell structures are more complex than those of the prokaryotes. This protist has manufactured "plates" of calcium carbonate that surround and protect its single cell. (D–G) Most of the visible life on Earth is multicellular. Chapters 28 and 29 cover the green plants (D). The other broad groups of multicellular organisms are the fungi (E), discussed in Chapter 30, and the animals (F, G), covered in Chapters 31–33.

1.1 What Is Biology?

Biology is the scientific study of living things, which we call organisms (**Figure 1.1**). The living organisms we know about are all descended from a common origin of life on Earth that occurred almost 4 billion years ago. Living organisms share many characteristics that allow us to distinguish them from the nonliving world:

• Organisms are made up of a common set of chemical components, including particular carbohydrates, fatty acids, nucleic acids, and amino acids, among others.

• The building blocks of most organisms are **cells**—individual structures enclosed by plasma membranes.

• The cells of living organisms convert molecules obtained from their environment into new biological molecules.

• Cells extract energy from the environment and use it to do biological work.

• Organisms contain genetic information that uses a nearly universal code to specify the assembly of proteins.

• Organisms share similarities among a fundamental set of genes and replicate this genetic information when reproducing themselves.

• Organisms exist in populations that evolve through changes in the frequencies of genetic variants within the populations over time.

• Living organisms self-regulate their internal environments, thus maintaining the conditions that allow them to survive.

Taken together, these characteristics logically lead to the conclusion that all life has a common ancestry, and that the diverse organisms alive today all originated from one life form. If life had multiple origins, we would not expect to see the striking similarities across gene sequences, the nearly universal genetic code, or the common set of amino acids that characterizes every known living organism. Organisms from a separate origin of life—say, on another planet—might be similar in some ways to life on Earth. For example, such life forms would probably possess heritable genetic information that they could pass on to offspring. But we would not expect the details of their genetic code or the fundamental sequences of their genomes to be the same as or even similar to ours.

The list is necessarily simplified, and some forms of life may not display all of the listed characteristics all of the time. For example, the seed of a desert plant may go for many years without extracting energy from the environment, converting molecules, regulating its internal environment, or reproducing; yet the seed is alive. And there are viruses, which are not composed of cells and cannot carry out physiological functions on their own (they parasitize host cells to function for them). Yet viruses contain genetic information, and they mutate and evolve. So even though viruses are not independent cellular organisms, their existence depends on cells. In addition, it is highly probable that viruses evolved from cellular life forms. Thus most biologists consider viruses to be a part of life.

This book will explore the details of the common characteristics of life, how these characteristics arose, and how they work together to enable organisms to survive and reproduce. Not all organisms survive and reproduce with equal success, and it is through differential survival and reproduction that living systems evolve and become adapted to Earth's many environments. The processes of evolution have generated the enormous diversity of life on Earth, and evolution is a central theme of biology.

Life arose from non-life via chemical evolution

Geologists estimate that Earth formed between 4.6 and 4.5 billion years ago. At first the planet was not a very hospitable place. It was some 600 million years or more before the earliest life evolved. If we picture the 4.6-billion-year history of Earth as a 30-day month, life first appeared some time around the end of the first week (**Figure 1.2**).

When we consider how life might have arisen from nonliving matter, we must take into account the properties of the young Earth's atmosphere, oceans, and climate, all of which were very different than they are today. Biologists postulate that complex biological molecules first arose through the random physical association of chemicals in that environment. Experiments simulating the conditions on early Earth have confirmed that the generation of complex molecules under such conditions is possible, even probable. The critical step for the evolution of life, however, was the appearance of **nucleic acids**—molecules that could reproduce themselves and also serve as templates for the

1.2 Life's Timeline   Depicting the 4.6 billion years of Earth's history on the scale of a 30-day month provides a sense of the immensity of evolutionary time.

synthesis of **proteins**, large molecules with complex but stable shapes. The variation in the shapes of these proteins enabled them to participate in increasing numbers and kinds of chemical reactions with other molecules. These subjects are covered in Part One of this book.

Cellular structure evolved in the common ancestor of life

Another important step in the history of life was the enclosure of complex proteins and other biological molecules by membranes that contained them in a compact internal environment separate from the surrounding (external) environment. Molecules called fatty acids played a critical role because these molecules do not dissolve in water; rather they form membranous

(A)

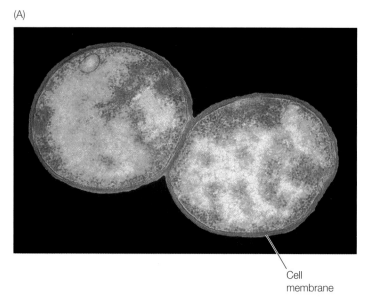

Cell
membrane

(B)

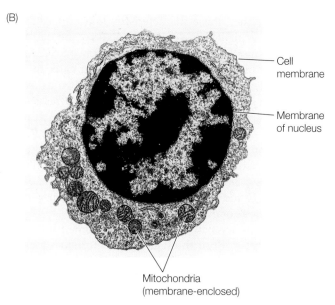

Cell
membrane

Membrane
of nucleus

Mitochondria
(membrane-enclosed)

1.3 Cells Are Building Blocks for Life   These photographs of cells were taken with a transmission electron microscope (see Figure 5.3) and enhanced with added color to highlight details. (A) Two pro-karyotic cells of an *Enterococcus* bacterium that lives in the human digestive system. Prokaryotes are unicellular organisms with genetic films and biochemical material enclosed inside a single membrane. (B) A human white blood cell (lymphocyte) represents one of the many specialized cell types that make up a multicellular eukaryote. Multi-ple membranes within the cell-enclosing outer membrane segregate the different biochemical processes of eukaryotic cells.

films that, when agitated, can form spherical structures. These membranous structures could have enveloped assemblages of biological molecules. The creation of an internal environment that concentrated the reactants and products of chemical reac-tions opened up the possibility that those reactions could be integrated and controlled within a tiny cell (**Figure 1.3**). Scien-tists postulate that this natural process of membrane formation resulted in the first cells with the ability to reproduce—that is, the evolution of the first cellular organisms.

For the first few billion years of cellular life, all the organ-isms that existed were unicellular and were enclosed by a single outer membrane. Such organisms, like the bacteria that are still abundant on Earth today, are called **prokaryotes**. Two main groups of prokaryotes emerged early in life's history: the **bacteria** and **archaea**. Some representatives of each of these groups began to live in a close, interdependent relationship with one another, and eventually merged to form a third ma-jor lineage of life, the **eukaryotes**. In addition to their outer membranes, the cells of eukaryotes have internal membranes that enclose specialized organelles within their cells. Eukaryote organelles include the nucleus that contains the genetic mate-rial and the mitochondria that power the cell. The structure of prokaryote and eukaryote cells and their membranes are the subjects of Part Two.

At some point, the cells of some eukaryotes failed to sepa-rate after cell division, remaining attached to each other. Such permanent colonial aggregations of cells made it possible for some of the associated cells to specialize in certain functions, such as reproduction, while other cells specialized in other functions, such as absorbing nutrients. This **cellular specializa-tion** enabled multicellular eukaryotes to increase in size and become more efficient at gathering resources and adapting to specific environments.

Photosynthesis allows some organisms to capture energy from the sun

Living cells require energy in order to function, and the bio-chemistry of the fundamental processes of energy conversion that drive life is covered in Part Three.

To fuel their cellular **metabolism** (energy transformations), the earliest prokaryotes took in small molecules directly from their environment and broke them down to their component atoms, thus releasing and using the energy contained in the chemical bonds. Many modern prokaryotes still function this way, and they function very successfully. But about 2.5 bil-lion years ago, the emergence of **photosynthesis** changed the nature of life on Earth.

The chemical reactions of photosynthesis transform the energy of sunlight into a form of biological energy that powers the synthesis of large molecules. These large mol-ecules can then be broken down to provide metabolic en-ergy. Photosynthesis is the basis of much of life on Earth today because its energy-capturing processes provide food for other organisms. Early photosynthetic cells were prob-ably similar to present-day prokaryotes called cyanobacteria (**Figure 1.4**). Over time, photosynthetic prokaryotes became so abundant that vast quantities of oxygen gas (O_2), which is a by-product of photosynthesis, began to accumulate in the atmosphere.

During the early eons of life, there was no O_2 in Earth's at-mosphere. In fact, O_2 was poisonous to many of the prokary-otes living at that time. As O_2 levels increased, however, those

(A)

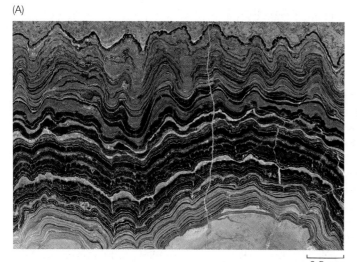

0.5 cm

(B)

Stromatolites form as small grains of sediment are cemented together by communities of microorganisms, especially cyanobacteria.

10 cm

1.4 Photosynthetic Organisms Changed Earth's Atmosphere (A) Colonies of photosynthetic cyanobacteria and other microorganisms produced structures called stromatolites that were preserved in the ancient fossil record. This section of fossilized stromatolite reveals layers representing centuries of growth. (B) Living stromatolites can still be found in appropriate environments.

organisms that *did* tolerate O_2 were able to proliferate. The abundance of O_2 opened up vast new avenues of evolution because **aerobic metabolism**—a biochemical process that uses O_2 to extract energy from nutrient molecules—is far more efficient than **anaerobic metabolism** (which does not use O_2). Aerobic metabolism allows organisms to grow larger and is used by the majority of organisms today.

Oxygen in the atmosphere also made it possible for life to move onto land. For most of life's history, UV radiation falling on Earth's surface was so intense that it destroyed any organism that was not well shielded by water. But the accumulation of photosynthetically generated O_2 in the atmosphere for more than 2 billion years gradually produced a thick layer of ozone (O_3) in the upper atmosphere. By about 500 million years ago, the ozone layer was sufficiently dense and absorbed enough of the sun's UV radiation to make it possible for organisms to leave the protection of the water and live on land.

Biological information is contained in a genetic language common to all organisms

The information that specifies what an organism will look like and how it will function—its "blueprint" for existence—is contained in the organism's **genome**: the sum total of all the DNA molecules contained in each of its cells. **DNA** (deoxyribonucleic acid) molecules are long sequences of four different subunits called **nucleotides**. The sequence of these four nucleotides contains genetic information. **Genes** are specific segments of DNA that encode the information the cell uses to create amino acids and form them into proteins (**Figure 1.5**). Protein molecules govern the chemical reactions within cells and form much of an organism's structure.

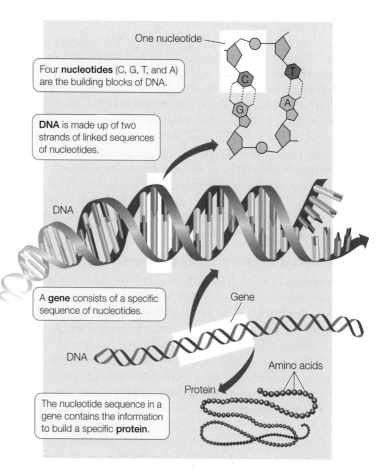

One nucleotide

Four **nucleotides** (C, G, T, and A) are the building blocks of DNA.

DNA is made up of two strands of linked sequences of nucleotides.

DNA

A **gene** consists of a specific sequence of nucleotides.

Gene

DNA

Amino acids

Protein

The nucleotide sequence in a gene contains the information to build a specific **protein**.

1.5 DNA Is Life's Blueprint The instructions for life are contained in the sequences of nucleotides in DNA molecules. Specific DNA nucleotide sequences comprise genes. The average length of a single human gene is 16,000 nucleotides. The information in each gene provides the cell with the information it needs to manufacture molecules of a specific protein.

By analogy with a book, the nucleotides of DNA are like the letters of an alphabet, and protein molecules are sentences. Combinations of proteins that form structures and control biochemical processes are the paragraphs. The structures and processes that are organized into different systems with specific tasks (such as digestion or transport) are the chapters of the book, and the complete book is the organism. If you were to write out your own genome using four letters to represent the four nucleotides, you would write more than 3 billion letters. Using the size type you are reading now, your genome would fill about 1,000 books the size of this one. The mechanisms of evolution are the authors and editors of all the books in the library of life.

All the cells of a multicellular organism contain essentially the same genome, yet different cells have different functions and form different structures—contractile proteins form in muscle cells, hemoglobin in red blood cells, digestive enzymes in gut cells, and so on. Therefore different types of cells in an organism must express different parts of the genome. How cells control gene expression in ways that enable a complex organism to develop and function is a major focus of current biological research.

The genome of an organism consists of thousands of genes. This entire genome must be replicated as new cells are produced. However, the replication process is not perfect, and a few errors, known as mutations, are likely to occur each time the genome is replicated. Mutations occur spontaneously; they can also be induced by outside factors, including chemicals and radiation. Most mutations are either harmful or have no effect, but occasionally a mutation improves the functioning of the organism under the environmental conditions it encounters.

The discovery of DNA in the latter half of the twentieth century and the subsequent elucidation of the remarkable mechanisms by which this material encodes and transmits information transformed biological science. These crucial discoveries are detailed in Parts Four and Five.

Populations of all living organisms evolve

A **population** is a group of individuals of the same type of organism—that is, of the same **species**—that interact with one another. **Evolution** acts on populations; it is the change in the genetic makeup of biological populations through time. Evolution is the major unifying principle of biology. Charles Darwin compiled factual evidence for evolution in his 1859 book *On the Origin of Species*. Darwin argued that differential survival and reproduction among individuals in a population, which he termed **natural selection**, could account for much of the evolution of life.

Although Darwin proposed that all organisms are descended from a common ancestor and therefore are related to one another, he did not have the advantage of understanding the mechanisms of genetic inheritance and mutation. Even so, he observed that offspring resembled their parents; therefore, he surmised, such mechanisms had to exist. Part Six will describe how Darwin's theory of natural selection is both supported and explained by the massive body of molecular genetic data elucidated during the twentieth century, and how these elements coincide and mesh in the modern field of evolutionary biology.

If all the organisms on Earth today are the descendants of a single kind of unicellular organism that lived almost 4 billion years ago, how have they become so different? As mentioned earlier, organisms reproduce by replicating their genomes, and mutations are introduced almost every time a genome is replicated. Some of these mutations give rise to structural and functional changes in organisms. As individuals mate with one another, the genetic variants stemming from mutation can change in frequency within a population, and the population is said to evolve.

Any population of a plant or animal species displays variation, and if you select breeding pairs on the basis of some particular trait, that trait is more likely to be present in their offspring than in the general population. Darwin himself bred pigeons, and was well aware of how pigeon fanciers selected breeding pairs to produce offspring with unusual feather patterns, beak shapes, or body sizes (see Figure 21.5). He realized that if humans could select for specific traits in domesticated animals, the same process could operate in nature; hence the term "natural selection" as opposed to artificial (human-imposed) selection.

How does natural selection function? Darwin postulated that different probabilities of survival and reproductive success would do the job. He reasoned that the reproductive capacity of plants and animals, if unchecked, would result in unlimited growth of populations, but we do not observe such growth in nature; in most species, only a small percentage of an individual's offspring will survive to reproduce. Thus any trait that confers even a small increase in the probability that its possessor will survive and reproduce would spread in the population.

Because organisms with certain traits survive and reproduce best under specific sets of conditions, natural selection leads to **adaptations**: structural, physiological, or behavioral traits that enhance an organism's chances of survival and reproduction in its environment (**Figure 1.6**). In addition to natural selection, evolutionary processes such as sexual selection (for example, selection due to mate choice) and genetic drift (the random fluctuation of gene frequencies in a population due to chance events) contribute to the rise of biodiversity. These processes operating over evolutionary history have led to the remarkable diversity of life on Earth.

Biologists can trace the evolutionary tree of life

As populations become geographically isolated from one another, they evolve differences. As populations diverge from one another, individuals in each population become less likely to reproduce with individuals of the other population. Eventually these differences between populations become so great that the two populations are considered different species. Thus species that share a fairly recent evolutionary history are generally more similar to each other than species

(A) *Dyscophus guineti*

(B) *Xenopus laevis*

(C) *Agalychnis callidryas*

(D) *Rhacophorus nigropalmatus*

1.6 Adaptations to the Environment The limbs of frogs show adaptations to the different environments of each species. (A) This terrestrial frog walks across the ground using its short legs and peglike digits (toes). (B) Webbed rear feet are evident in this highly aquatic species of frog. (C) This arboreal species has toe pads, which are adaptations for climbing. (D) A different arboreal species has extended webbing between the toes, which increases surface area and allows the frog to glide from tree to tree.

that share an ancestor in the more distant past. By identifying, analyzing, and quantifying similarities and differences between species, biologists can construct **phylogenetic trees** that portray the evolutionary histories of the different groups of organisms.

Tens of millions of species exist on Earth today; many times that number lived in the past but are now extinct. Biologists give each of these species a distinctive scientific name formed from two Latinized names—a **binomial**. The first name identifies the species' **genus** (plural *genera*)—a group of species that share a recent common ancestor. The second is the name of the species. For example, the scientific name for the human species is *Homo sapiens*: *Homo* is our genus, *sapiens* our species. *Homo* is Latin for "man," and *sapiens* is from the Latin word for "wise" or "rational." Our closest relatives in the genus *Homo* are the Neanderthals, *Homo neanderthalensis*. Neanderthals are now extinct and are known only from their fossil remains.

Much of biology is based on comparisons among species, and these comparisons are useful precisely because we can place species in an evolutionary context relative to one another. Our ability to do this has been greatly enhanced in recent decades by our ability to sequence and compare the genomes of different species. Genome sequencing and other molecular techniques have allowed biologists to augment evolutionary knowledge based on the fossil record with a vast array of molecular evidence. The result is the ongoing compilation of phylogenetic trees that document and diagram evolutionary relationships as part of an overarching tree of life, the broadest categories of which are shown in **Figure 1.7** and will be surveyed in more detail in Part Seven. (The tree is expanded in Appendix A, and you can also explore the tree interactively.)

Although many details remain to be clarified, the broad outlines of the tree of life have been determined. Its branching patterns are based on a rich array of evidence from fossils, structures, metabolic processes, behavior, and molecular

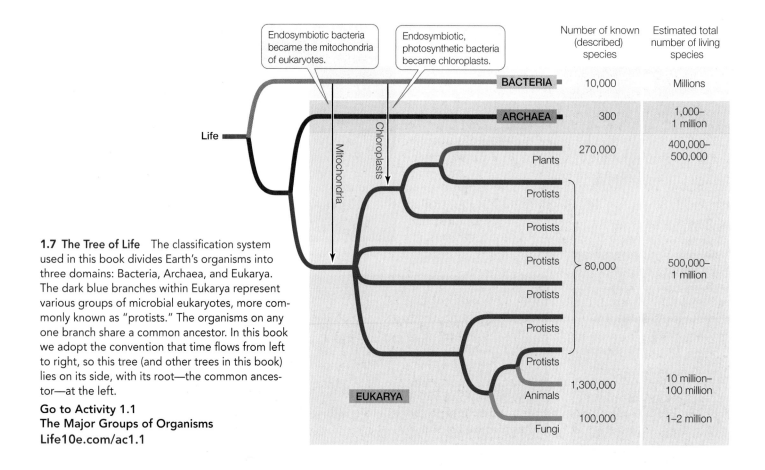

	Number of known (described) species	Estimated total number of living species
BACTERIA	10,000	Millions
ARCHAEA	300	1,000–1 million
Plants	270,000	400,000–500,000
Protists		
Protists		
Protists	80,000	500,000–1 million
Protists		
Protists		
Protists		
Animals	1,300,000	10 million–100 million
Fungi	100,000	1–2 million

Endosymbiotic bacteria became the mitochondria of eukaryotes.

Endosymbiotic, photosynthetic bacteria became chloroplasts.

1.7 The Tree of Life The classification system used in this book divides Earth's organisms into three domains: Bacteria, Archaea, and Eukarya. The dark blue branches within Eukarya represent various groups of microbial eukaryotes, more commonly known as "protists." The organisms on any one branch share a common ancestor. In this book we adopt the convention that time flows from left to right, so this tree (and other trees in this book) lies on its side, with its root—the common ancestor—at the left.

Go to Activity 1.1
The Major Groups of Organisms
Life10e.com/ac1.1

analyses of genomes. Two of the three main domains of life—Archaea and Bacteria—are single-celled prokaryotes, as mentioned earlier in this chapter. However, members of these two groups differ so fundamentally in their metabolic processes that they are believed to have separated into distinct evolutionary lineages very early. Species belonging to the third domain—Eukarya—have eukaryotic cells whose mitochondria and chloroplasts originated from endosymbioses of bacteria.

Plants, fungi, and animals are examples of familiar multicellular eukaryotes that evolved independently, from different groups of the unicellular eukaryotes informally known as protists. We know that plants, fungi, and animals had independent origins of multicellularity because each of these three groups is most closely

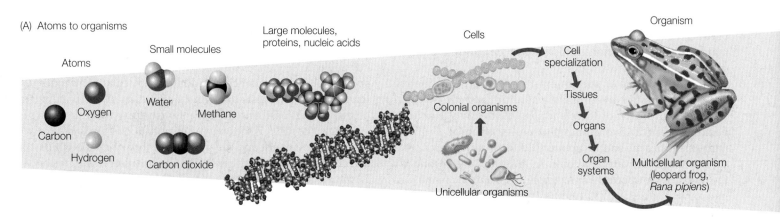

1.8 Biology Is Studied at Many Levels of Organization
(A) Life's properties emerge when DNA and other molecules are organized in cells, which form building blocks for organisms. (B) Organisms exist in populations and interact with other populations to form communities, which interact with the physical environment to make up the many ecosystems of the biosphere.

Go to Activity 1.2 The Hierarchy of Life
Life10e.com/ac1.2

related to different groups of unicellular protists, as can be seen from the branching pattern of Figure 1.7.

Cellular specialization and differentiation underlie multicellular life

Looking back at Figure 1.2, you can see that for more than half of Earth's history, all life was unicellular. Unicellular species remain ubiquitous and highly successful in the present, even though the diverse multicellular organisms, owing to their much larger size, may seem to us to dominate the planet.

With the evolution of cells specialized for different functions within the same organism, these differentiated cells lost many of the functions carried out by single-celled organisms, and a **biological hierarchy** emerged (**Figure 1.8A**). To accomplish their specialized tasks, assemblages of differentiated cells are organized into **tissues**. For example, a single muscle cell cannot generate much force, but when many cells combine to form the tissue of a working muscle, considerable force and movement can be generated. Different tissue types are organized to form **organs** that accomplish specific functions. The heart, brain, and stomach are each constructed of several types of tissues, as are the roots, stems, and leaves of plants. Organs whose functions are interrelated can be grouped into **organ systems**; the esophagus, stomach, and intestines, for example, are all part of the digestive system. The physiology of two major groups of multicellular organisms (land plants and animals) is discussed in detail in Parts Eight and Nine, respectively.

Living organisms interact with one another

Organisms do not live in isolation, and the internal hierarchy of the individual organism is matched by the external hierarchy of the biological world (**Figure 1.8B**). As mentioned earlier in this section, a group of individuals of the same species that interact with one another is a population. The populations of all the species that live and interact in a defined area (areas are defined in different ways and can be small or large) are called a **community**. Communities together with their abiotic (nonliving) environment constitute an **ecosystem**.

Individuals in a population interact in many different ways. Animals eat plants and other animals (usually members of another species) and compete with other species for food and other resources. Some animals prevent other individuals of their own species from exploiting a resource, be it food, nesting sites, or mates. Animals may also cooperate with members of their own species, forming social units such as a termite colony or a flock of birds. Such interactions have resulted in the evolution of social behaviors such as communication and courtship displays.

Plants also interact with their external environment, which includes other plants, fungi, animals, and microorganisms. All terrestrial plants depend on partnerships with fungi, bacteria, and animals. Some of these partnerships are necessary to obtain nutrients, some to produce fertile seeds, and still others to disperse seeds. Plants compete with each other for light and water and have ongoing evolutionary interactions with the animals that eat them. Through time, many adaptations have evolved in plants that protect them from predation (such as thorns) or that help then attract the animals that assist in their reproduction (such as sweet nectar or colorful flowers). The interactions of populations of plant and animal species in a community are major evolutionary forces that produce specialized adaptations.

Communities interacting over a broad geographic area with distinguishing physical features form ecosystems; examples

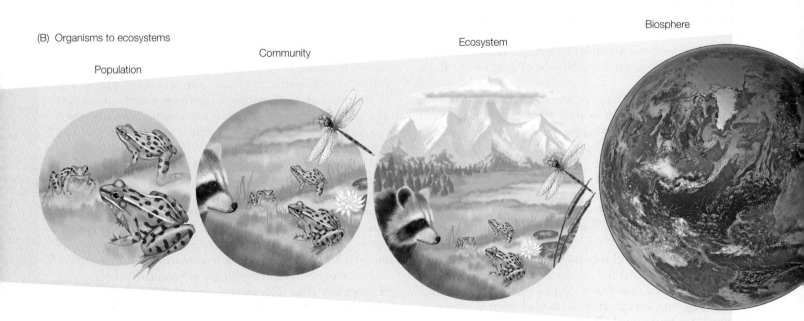

(B) Organisms to ecosystems

Population   Community   Ecosystem   Biosphere

(B) *Spermophilus parryii*

(A) *Propithecus verreauxi*

 Go to Media Clip 1.1
Leaping Lemurs
Life10e.com/mc1.1

1.9 Energy Can Be Used Immediately or Stored (A) Animal cells break down food molecules and use the energy contained in the chemical bonds of those molecules to do mechanical work, such as running and jumping. This composite image of a sifaka (a type of lemur from Madagascar) shows the same individual at five stages of a single jump. (B) The cells of this Arctic ground squirrel have broken down the complex carbohydrates in the plants it consumed and converted those molecules into fats. The fats are stored in the animal's body to provide an energy supply for the cold months.

include Arctic tundra, coral reef, and tropical rainforest. The ways in which species interact with one another and with their environment in populations, communities, and ecosystems is the subject of ecology, covered in Part Ten of this book.

Nutrients supply energy and are the basis of biosynthesis

Living organisms acquire nutrients from the environment. Nutrients supply the organism with energy and raw materials for carrying out biochemical reactions. Life depends on thousands of biochemical reactions that occur inside cells. Some of these reactions break down nutrient molecules into smaller chemical units, and in the process some of the energy contained in the chemical bonds of the nutrients is captured by high-energy molecules that can be used to do different kinds of cellular work.

One obvious kind of work cells do is mechanical—moving molecules from one cellular location to another, moving whole cells or tissues, or even moving the organism itself, as muscles do (**Figure 1.9A**). The most basic cellular work is the building, or synthesis, of new complex molecules and structures from smaller chemical units. For example, we are all familiar with the fact that carbohydrates eaten today may be deposited in the body as fat tomorrow (**Figure 1.9B**). Still another kind of work is the electrical work that is the essence of information processing in nervous systems.

The myriad biochemical reactions that take place in cells are integrally linked in that the products of one reaction are the raw materials of the next. These complex networks of reactions must be integrated and precisely controlled; when they are not, the result is malfunction and disease.

Living organisms must regulate their internal environment

The specialized cells, tissues, and organ systems of multicellular organisms exist in and depend on an **internal environment** that is made up of extracellular fluids. Because this environment serves the needs of the cells, its physical and chemical composition must be maintained within a narrow range of physiological conditions that support survival and function. The maintenance of this narrow range of conditions is known as **homeostasis**. A relatively stable internal (but extracellular) environment means that cells can function efficiently even when conditions outside the organism's body become unfavorable for cellular processes.

The organism's regulatory systems obtain information from sensory cells that provide information about both the internal and external conditions the organism is subject to at a given time. The cells of regulatory systems process and integrate this information and send signals to components of physiological systems, which can change in response to these signals so that the organism's internal environment remains reasonably constant.

The concept of homeostasis extends beyond the internal environment of multicellular organisms, however. In both unicellular and multicellular organisms, individual cells must regulate physiological parameters (such as acidity and salinity), maintaining them within a range that allows those cells to survive and function. Individual cells regulate these properties through actions of the plasma membrane that encloses them and are the cell's interface with its environment (either internal or external). Thus self-regulation to maintain a more or less constant internal environment is a general attribute of all living organisms.

The preceding section briefly outlined the major features of life—features that will be covered in depth in subsequent chapters of this book. Before going into the details of what we know about life, however, it is important to understand how scientists obtain information and how they use that information in broadening our understanding of Earth's diverse living organisms and putting this understanding to practical use.

1.2 How Do Biologists Investigate Life?

Scientific investigations are based on observation, data, experimentation, and logic. Scientists use many different tools and methods in making observations, collecting data, designing experiments, and applying logic, but they are always guided by established principles that allow us to discover new aspects about the structure, function, evolution, and interactions of organisms.

Observing and quantifying are important skills

Biologists have always observed the world around them, but today our ability to observe is greatly enhanced by technologies such as electron microscopes, rapid genome sequencing, magnetic resonance imaging, and global positioning satellites. These technologies allow us to observe everything from the distribution of molecules in the body to the movement of animals across continents and oceans.

Observation is a basic tool of biology, but as scientists we must also be able to quantify the information, or **data**, we collect as we observe. Whether we are testing a new drug or mapping the migrations of the great whales, applying mathematical and statistical calculations to the data we collect is essential. For example, biologists once classified organisms based entirely on qualitative descriptions of the physical differences among them. There was no way of objectively determining evolutionary relationships of organisms, and biologists had to depend on the fossil record for insight. Today our ability to quantify the molecular and physical differences among species, combined with explicit mathematical models

of the evolutionary process, enables quantitative analyses of evolutionary history. These mathematical calculations, in turn, facilitate comparative investigations of all other aspects of an organism's biology.

Scientific methods combine observation, experimentation, and logic

Textbooks often describe "*the* scientific method," as if there is a single, simple flow chart that all scientists follow. This is an oversimplification. Although flow charts such as the one shown in **Figure 1.10** incorporate much of what scientists do, you should not conclude that scientists necessarily progress through the steps of the process in one prescribed, linear order.

Observations lead to questions, and scientists make additional observations and often do experiments to answer those

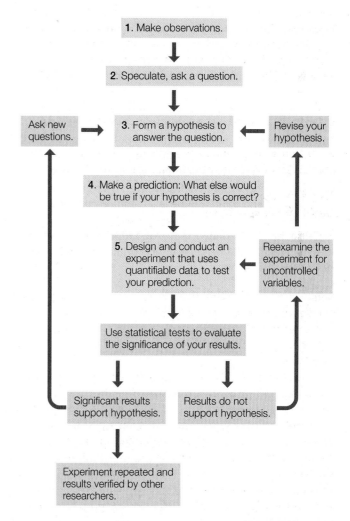

1.10 Scientific Methodology The process of observation, speculation, hypothesis, prediction, and experimentation is a cornerstone of modern science, although scientists may initiate their research at several different points. Answers gleaned through experimentation lead to new questions, more hypotheses, further experiments, and expanding knowledge.

questions. This hypothesis–prediction approach traditionally has five steps: (1) making observations; (2) asking questions; (3) forming hypotheses, which are tentative answers to the questions; (4) making predictions based on the hypotheses; and (5) testing the predictions by making additional observations or conducting experiments.

After posing a question, a scientist often uses **inductive logic** to propose a tentative answer. Inductive logic involves taking observations or facts and creating a new proposition that is compatible with those observations or facts. Such a tentative proposition is a **hypothesis** (plural *hypotheses*). In formulating a hypothesis, scientists put together the facts and data at their disposal to formulate one or more possible answers to the question. For example, at the opening of this chapter you learned that scientists have observed the rapid decline of amphibian populations worldwide and are asking why. Some scientists have hypothesized that a fungal disease is a cause; other scientists have hypothesized that increased exposure to ultraviolet radiation is a cause. Tyrone Hayes hypothesized that exposure to agricultural chemicals, specifically the widely used herbicide atrazine, could be a cause.

The next step in the scientific method is to apply a different form of logic—**deductive logic**—that starts with a statement believed to be true (the hypothesis) and then goes on to predict what facts would also have to be true to be compatible with that statement. Hayes knew that atrazine is commonly applied in the spring, when amphibians are breeding, and that atrazine is a common contaminant in the waters in which amphibians live as they develop into adults. Thus he predicted that frog tadpoles exposed to atrazine would show adverse effects of the chemical once they reached adulthood.

 Go to Animated Tutorial 1.1
Using Scientific Methodology
Life10e.com/at1.1

Good experiments have the potential to falsify hypotheses

Once predictions are made from a hypothesis, experiments can be designed to test those predictions. The most informative experiments are those that have the ability to show that the prediction is wrong. If the prediction is wrong, the hypothesis must be questioned, modified, or rejected.

There are two general types of experiments, both of which compare data from different groups or samples. A **controlled experiment** manipulates one or more of the factors being tested; **comparative experiments** compare unmanipulated data gathered from different sources. As described at the opening of this chapter, Tyrone Hayes and his colleagues conducted both types of experiments to test the prediction that the herbicide atrazine, a contaminant in freshwater ponds and streams throughout the world, affects the development of frogs.

INVESTIGATING**LIFE** ▦▦▦▦

1.11 Controlled Experiments Manipulate a Variable The Hayes laboratory created controlled environments that differed only in the concentrations of atrazine in the water. Eggs from leopard frogs (*Rana pipiens*) raised specifically for laboratory use were allowed to hatch and the tadpoles were separated into experimental tanks containing water with different concentrations of atrazine.[a]

HYPOTHESIS Exposure to atrazine during larval development causes abnormalities in the reproductive tissues of male frogs.

Method
1. Establish 9 tanks in which all attributes are held constant except the water's atrazine concentration. Establish 3 atrazine conditions (3 replicate tanks per condition): 0 ppb (control condition), 0.1 ppb, and 25 ppb.
2. Place *Rana pipiens* tadpoles from laboratory-reared eggs in the 9 tanks (30 tadpoles per replicate).
3. When tadpoles have transitioned into adults, sacrifice the animals and evaluate their reproductive tissues.
4. Test for correlation of degree of atrazine exposure with the presence of abnormalities in the gonads (testes) of male frogs.

Results

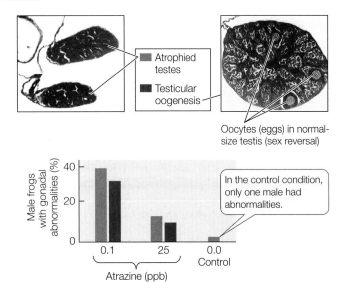

■ Atrophied testes
■ Testicular oogenesis

Oocytes (eggs) in normal-size testis (sex reversal)

In the control condition, only one male had abnormalities.

Male frogs with gonadal abnormalities (%)

Atrazine (ppb)

CONCLUSION Exposure to atrazine at concentrations as low as 0.1 ppb induces abnormalities in the gonads of male frogs. The effect is not proportional to the level of exposure.

Go to **BioPortal** for discussion and relevant links for all INVESTIGATING**LIFE** figures.

[a]Hayes, T. et al. 2003. *Environmental Health Perspectives* III: 568–575.

In a controlled experiment, we start with groups or samples that are as similar as possible. We predict on the basis of our hypothesis that some critical factor, or variable, has an effect on the phenomenon we are investigating. We devise some method to manipulate *only that variable* in an "experimental" group and compare the resulting data with data from an unmanipulated "control" group. If the predicted difference occurs, we then apply statistical tests to ascertain the probability that the manipulation created the difference (as opposed to the difference being the result of random

INVESTIGATING**LIFE**

1.12 Comparative Experiments Look for Differences among Groups To see whether the presence of atrazine correlates with testicular abnormalities in male frogs, the Hayes lab collected frogs and water samples from different locations around the U.S. The analysis that followed was "blind," meaning that the frogs and water samples were coded so that experimenters working with each specimen did not know which site the specimen came from.[a]

HYPOTHESIS Presence of the herbicide atrazine in environmental water correlates with gonadal abnormalities in frog populations.

Method
1. Based on commercial sales of atrazine, select 4 sites (sites 1–4) less likely and 4 sites (sites 5–8) more likely to be contaminated with atrazine.
2. Visit all sites in the spring (i.e., when frogs have transitioned from tadpoles into adults); collect frogs and water samples.
3. In the laboratory, sacrifice frogs and examine their reproductive tissues, documenting abnormalities.
4. Analyze the water samples for atrazine concentration (the sample for site 7 was not tested).
5. Quantify and correlate the incidence of reproductive abnormalities with environmental atrazine concentrations.

Results

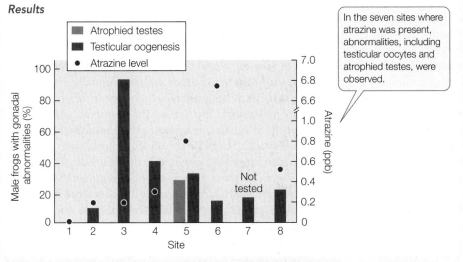

In the seven sites where atrazine was present, abnormalities, including testicular oocytes and atrophied testes, were observed.

CONCLUSION Reproductive abnormalities exist in frogs from environments in which aqueous atrazine concentration is 0.2 ppb or above. The incidence of abnormalities does not appear to be proportional to atrazine concentration at the time of transition to adulthood.

Go to **BioPortal** for discussion and relevant links for all INVESTIGATING**LIFE** figures.

[a]Hayes, T. et al. 2003. *Nature* 419: 895–896.

When his controlled experiments indicated that atrazine indeed affects reproductive development in frogs, Hayes and his colleagues performed a comparative experiment. They collected frogs and water samples from eight widely separated sites across the United States and compared the incidence of abnormal frogs from environments with very different levels of atrazine (**Figure 1.12**). Of course, the sample sites differed in many ways besides the level of atrazine present.

The results of experiments frequently reveal that the situation is more complex than the hypothesis anticipated, thus raising new questions. In the Hayes experiments, for example, there was no clear direct relationship between the *amount* of atrazine present and the percentage of abnormal frogs: there were fewer abnormal frogs at the highest concentrations of atrazine than at lower concentrations. There are no "final answers" in science. Investigations consistently reveal more complexity than we expect, so scientists must design systematic approaches to identify, assess, and understand that complexity.

Statistical methods are essential scientific tools

Whether we do comparative or controlled experiments, at the end we have to decide whether there is a difference between the samples, individuals, groups, or populations in the study. How do we decide whether a measured difference is enough to support or falsify a hypothesis? In other words, how do we decide in an unbiased, objective way that the measured difference is significant?

chance). **Figure 1.11** describes one of the many controlled experiments performed by the Hayes laboratory to quantify the effects of atrazine on male frogs.

The basis of controlled experiments is that one variable is manipulated while all others are held constant. The variable that is manipulated is called the **independent variable**, and the response that is measured is the **dependent variable**. A good controlled experiment is not easy to design because biological variables are so interrelated that it is difficult to alter just one.

A comparative experiment starts with the prediction that there will be a difference between samples or groups based on the hypothesis. In comparative experiments, however, we cannot control the variables; often we cannot even identify all the variables that are present. We are simply gathering and comparing data from different sample groups.

Significance can be measured with statistical methods. Scientists use statistics because they recognize that variation is always present in any set of measurements. Statistical tests calculate the probability that the differences observed in an experiment could be due to random variation. The results of statistical tests are therefore probabilities. A statistical test starts with a **null hypothesis**—the premise that any observed differences are simply the result of random differences that arise from drawing two finite samples from the same population. When quantified observations, or data, are collected, statistical methods are applied to those data to calculate the likelihood that the null hypothesis is correct.

More specifically, statistical methods tell us the probability of obtaining the same results by chance even if the null hypothesis were true. *We need to eliminate, insofar as possible, the chance that any differences showing up in the data are merely the result of random variation in the samples tested.* Scientists generally conclude that the differences they measure are significant if statistical tests show that the probability of error (that is, the probability that a difference as large as the one observed could be obtained by mere chance) is 5 percent or lower, although more stringent levels of significance may be set for some problems. Appendix B of this book is a short primer on statistical methods that you can refer to as you analyze data that will be presented throughout the text.

Discoveries in biology can be generalized

Because all life is related by descent from a common ancestor, shares a genetic code, and consists of similar biochemical building blocks, knowledge gained from investigations of one type of organism can, with thought and care, be generalized to other organisms. Biologists use **model systems** for research, knowing that they can extend their findings from such systems to other organisms. For example, our basic understanding of the chemical reactions in cells came from research on bacteria but is applicable to all cells, including those of humans. Similarly, the biochemistry of photosynthesis—the process by which all green plants use sunlight to produce biological molecules—was largely worked out from experiments on *Chlorella*, a unicellular green alga. Much of what we know about the genes that control plant development is the result of work on *Arabidopsis thaliana*, a relative of the mustard plant. Knowledge about how animals, including humans, develop has come from work on sea urchins, frogs, chickens, roundworms, mice, and fruit flies. Being able to generalize from model systems is a powerful tool in biology.

Not all forms of inquiry are scientific

Science is a unique human endeavor that has certain standards of practice. Other areas of scholarship share with science the practice of making observations and asking questions, but scientists are distinguished by *what they do with their observations* and *how they frame the answers*. Quantifiable data, subjected to appropriate statistical analysis, are critical in evaluating hypotheses (the Working with Data exercises you will find throughout this book are intended to reinforce this way of thinking). In short, scientific observation and evaluation is the most powerful approach humans have devised for learning about the world and how it works.

Scientific explanations for natural processes are objective and reliable because *a hypothesis must be testable* and *a hypothesis must have the potential of being rejected* by direct observations and experiments. Scientists must clearly describe the methods they use to test hypotheses so that other scientists can repeat their results. Not all experiments are repeated, but surprising or controversial results are always subjected to independent verification. Scientists worldwide share this process of testing and rejecting hypotheses, contributing to a common body of scientific knowledge.

If you understand the methods of science, you can distinguish science from non-science. Art, music, and literature all contribute to the quality of human life, but they are not science.

They do not use scientific methods to establish what is fact. Religion is not science, although religions have historically attempted to explain natural events ranging from unusual weather patterns to crop failures to human diseases. Most such phenomena that at one time were mysterious can now be explained in terms of scientific principles. Fundamental tenets of religious faith, such as the existence of a supreme deity or deities, cannot be confirmed or refuted by experimentation and are thus outside the realm of science.

The power of science derives from strict objectivity and absolute dependence on evidence based on *reproducible and quantifiable observations.* A religious or spiritual explanation of a natural phenomenon may be coherent and satisfying for the person holding that view, but it is not testable and therefore it is not science. To invoke a supernatural explanation (such as a "creator" or "intelligent designer" with no known bounds) is to depart from the world of science. Science does not necessarily say that religious beliefs are wrong; they are simply not part of the world of science, and many religious beliefs are untestable using scientific methods.

Science describes how the world works. It is silent on the question of how the world "ought to be." Many scientific advances that contribute to human welfare also raise major ethical issues. Recent developments in genetics and developmental biology may enable us to select the sex of our children, to use stem cells to repair our bodies, and to modify the human genome. Although scientific knowledge allows us to do these things, science cannot tell us whether or not we *should* do so or, if we choose to do them, how we should regulate them. Such issues are as crucial to human society as the science itself, and a responsible scientist does not lose sight of these questions or neglect the contributions of the humanities or social sciences in attempting to come to grips with them.

▓ RECAP **1.2**

Scientific methods of inquiry start with the formulation of hypotheses based on observations and data. Comparative and controlled experiments are carried out to test hypotheses.

- Explain the relationship between a hypothesis and an experiment. **See pp. 11–12 and Figure 1.10**
- What is controlled in a controlled experiment? **See pp. 11–12 and Figure 1.11**
- What features characterize questions that can be answered only by using a comparative approach? **See p. 13 and Figure 1.12**
- Explain why arguments must be supported by quantifiable and reproducible data in order to be considered scientific. **See pp. 13–14**
- Why can the results of biological research on one species often be generalized to very different species? **See p. 14**

The vast body of scientific knowledge accumulated over centuries of human civilization allows us to understand and manipulate aspects of the natural world in ways that no other species can. These abilities present us with challenges, opportunities, and above all, responsibilities.

1.3 Why Does Biology Matter?

Human beings exist in and depend on a world of living organisms. The oxygen in the air we breathe is produced by photosynthesis conducted by countless billions of individual organisms. The food that fuels our bodies comes from the tissues of other living organism. The fuels that drive our cars and power our electric plants are, for the most part, various forms of carbon molecules produced by living organisms—mostly millions of years ago. Inside and out, our bodies are covered in complex communities of living unicellular organisms, most of which help us maintain our health. There are also harmful species that invade our bodies and can cause mild to serious diseases, or even death. These interactions with other species are not limited to humans. Ecosystem function depends on thousands of complex interactions among the millions of species that inhabit Earth. In other words, understanding biological principles is essential to our lives and for maintaining the functioning of Earth as we know it and depend on it.

Modern agriculture depends on biology

Agriculture represents some of the earliest human applications of biological principles. Even in prehistoric times, farmers selected the most productive or otherwise favorable plants and animals to use as seed stock for propagation, and over generations farmers continued and refined these practices. His knowledge of this kind of artificial selection helped Charles Darwin understand the importance of natural selection in evolution across all of life.

In modern times, increasing knowledge of plant biology has transformed agriculture in many ways and has resulted in huge boosts in food production (**Figure 1.13**), which in turn has allowed the planet to support a far larger human population than it once could have. Over the past few decades, detailed knowledge of the genomes of many domestic species and the development of technology for directly recombining genes have allowed biologists to develop new breeds and strains of animals, plants, and fungi of agricultural interest. For example, new strains of crop plants are being developed that are resistant to pests or can tolerate drought. Moreover, understanding evolutionary theory allows biologists to devise strategies for the application of pesticides that minimize the evolution of pest resistance. And better understanding of plant–fungus relationships results in better plant health and higher productivity. These are just a few of the many ways that biology continues to inform and improve agricultural practice.

Biology is the basis of medical practice

People have speculated about the causes of diseases and searched for methods to combat them since ancient times. Long before the microbial causes of many diseases were known, people recognized that infections could be passed from one person to another, and the isolation of infected persons has been practiced as long as written records have been available.

Modern biological research informs us about how living organisms work, and about why they develop the problems and

1.13 A Green Revolution The agricultural advancements of the last 100 years have vastly increased yields and nutritional value of crops such as grains that sustain the expanding human population. In the last 30 years, these advancements have included genetic recombination techniques. Here a researcher with the U.S. Department of Agriculture works with a strain of "supernutritious" rice that provides high levels of the amino acid lysine.

infections that we call disease. In addition to diseases caused by infection of other organisms, we now know that many diseases are genetic—meaning that variants of genes in our genomes cause particular problems in the way we function. Developing appropriate treatments or cures for diseases depends on understanding the origin, basis, and effects of these diseases, as well understanding the consequences of any changes that we make. For example, the recent resurgence of tuberculosis is the result of the evolution of bacteria that are resistant to antibiotics. Dealing with future tuberculosis epidemics requires understanding aspects of molecular biology, physiology, microbial ecology, and evolution—in other words, many of the general principles of modern biology.

Many of the microbial organisms that are periodically epidemic in human populations have short generation times and high mutation rates. For example, we need yearly vaccines for flu because of the high rate of evolution of influenza viruses, the causative agent of flu. Evolutionary principles help us understand how influenza viruses are changing, and can even help us predict which strains of influenza virus are likely to lead to future flu epidemics. This medical understanding—which combines an application of molecular biology, evolutionary

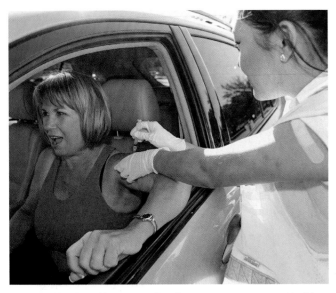

1.14 Medical Applications of Biology Improve Human Health
Vaccination to prevent disease is a biologically based medical practice that began in the eighteenth century. Today evolutionary biology and genomics provide the basis for constant updates to vaccines that protect humans from virus-borne diseases such as flu. In the developed world, vaccinations have become so commonplace that some are offered on a "drive-through" basis.

theory, and basic principles of ecology—allows medical researchers to develop effective vaccines and other strategies for the control of major epidemics (**Figure 1.14**).

Biology can inform public policy

Thanks to the deciphering of genomes and our newfound ability to manipulate them, vast new possibilities now exist for controlling human diseases and increasing agricultural productivity—but these capabilities raise ethical and policy issues. How much and in what ways should we tinker with the genes of humans and other species? Does it matter whether the genomes of our crop plants and domesticated animals are changed by traditional methods of controlled breeding and crossbreeding or by the biotechnology of gene transfer? What rules should govern the release of genetically modified organisms into the environment? Science alone cannot provide all the answers, but wise policy decisions must be based on accurate scientific information.

Biologists are increasingly called on to advise government agencies concerning the laws, rules, and regulations by which society deals with the increasing number of challenges that have a biological basis. As an example of the value of scientific knowledge for the assessment and formulation of public policy, consider a management problem. Scientists and fishermen have long known that Atlantic bluefin tuna (*Thunnus thynnus*) have a western breeding ground in the Gulf of Mexico and an eastern breeding ground in the Mediterranean Sea (**Figure 1.15**). Overfishing led to declining numbers of bluefin tuna,

(A)

1.15 Bluefin Tuna Do Not Recognize Boundaries   (A) Marine biologist Barbara Block attaches computerized data-recording tracking tags to a live bluefin tuna before returning it to the Atlantic Ocean, where its travels will be monitored. (B) At one time we assumed that bluefins from western- and eastern-breeding populations also fed on their respective sides of the Atlantic, so separate fishing quotas for each side (dashed line) in an attempt to speed recovery of the endangered western population. Now, however, tracking data have shown that the two populations *do not* remain separate after spawning, so in fact the arbitrary boundary and quotas do not protect the endangered population.

(B)

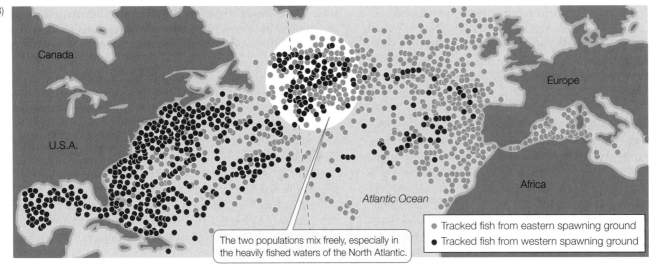

Canada

Europe

U.S.A.

Africa

Atlantic Ocean

The two populations mix freely, especially in the heavily fished waters of the North Atlantic.

● Tracked fish from eastern spawning ground
● Tracked fish from western spawning ground

(A) 1941

Riggs
Glacier

Muir
Glacier

(B) 2004

Riggs
Glacier

1.16 A Warmer World Earth's climate has been steadily warming for the last 150 years. The rate of this warming trend has also steadily increased, resulting in the rapid melting of polar ice caps, glaciers, and alpine (mountaintop) snow and ice. This photograph shows the effects of 63 years of climate change on two ancient, longstanding glaciers in Alaska. Over that time, Muir Glacier retreated some 7 kilometers and can no longer be seen from the original vantage point. Understanding how biological populations respond to such change requires integration of biological principles from molecular biology to ecosystem ecology.

especially in the western-breeding populations, to the point of these populations being endangered.

Initially it was assumed by scientists, fishermen, and policy makers alike that the eastern and western populations had geographically separate feeding grounds as well as separate breeding grounds. Acting on this assumption, an international commission drew a line down the middle of the Atlantic Ocean and established strict fishing quotas on the western side of the line, with the intent of allowing the western population to recover. Modern tracking data, however, revealed that in fact the eastern and western bluefin populations mix freely on their feeding grounds across the entire North Atlantic—a swath of ocean that includes the most heavily fished waters in the world. Tuna caught on the eastern side of the line could just as likely be from the western breeding population as the eastern; thus the established policy could not achieve its intended goal.

Policy makers take more things into consideration than scientific knowledge and recommendations. For example, studies on the effects of atrazine on amphibians have led one U.S. group, the Natural Resources Defense Council, to take legal action to have atrazine banned on the basis of the Endangered Species Act. The U.S. Environmental Protection Agency, however, must also consider the potential loss to agriculture that such a ban would create and thus has continued to approve atrazine's use as long as environmental levels do not exceed 30 to 40 ppb—which is 300 to 400 times the levels shown to induce abnormalities in the Hayes studies. Scientific conclusions do not always prevail in the political world. Some scientific conclusions may have more influence than others, however, especially when they indicate a strong possibility of negative effects on humans.

Biology is crucial for understanding ecosystems

The world has been changing since its formation and continues to change with every passing day. Human activity, however, is resulting in an unprecedented *rate* of change in the world's ecosystems. For example, the mining and consumption of fossil fuels is releasing massive quantities of carbon dioxide into Earth's atmosphere. This anthropogenic (human-generated) increase in atmospheric carbon dioxide is largely responsible for the rapid rate of climate warming recorded over the last 50 years (**Figure 1.16**).

Our use of natural resources is putting stress on the ability of Earth's ecosystems to continue to produce the goods and services on which our society depends. Human activities are changing global climates at an unprecedented rate and are leading to the extinctions of large numbers of species (such as the amphibians featured in this chapter). The modern, warmer world is also experiencing the spread of new diseases and the resurgence of old ones. Biological knowledge is vital for determining the causes of these changes and for devising policies to deal with them.

Biology helps us understand and appreciate biodiversity

Beyond issues of policy and pragmatism lies the human "need to know." Humans are fascinated by the richness and diversity of life, and most people want to know more about organisms and how they interact. Human curiosity might even be seen as an adaptive trait—it is possible that such a trait could have been selected for if individuals who were motivated to learn about their surroundings were likely to have survived and reproduced better, on average, than their less curious relatives.

1.17 Discovering Life on Earth These biologists are collecting insects in the top boughs of a spruce tree in the Carmanah Valley of Vancouver, Canada. Biologists estimate that the number of species discovered to date is only a small percentage of the number of species that inhabit Earth. To fill this gap in our knowledge, biologists around the world are applying thorough sampling techniques and new genetic tools to document and understand the Earth's biodiversity.

Far from ending the process, new discoveries and greater knowledge typically engender questions no one thought to ask before. There are vast numbers of questions for which we do not yet have answers, and the most important motivator of most scientists is curiosity.

Observing the living world motivates many biologists to learn more and to constantly collect new information (**Figure 1.17**). An intimate understanding of the **natural history** of a group of organisms—that is, how those organisms get their food, reproduce, behave, regulate their internal environments, and interact with other organisms—facilitates observations and provides a stronger basis for framing hypotheses about about those observations. The more information biologists have and the more the observer knows about general principles, the more he or she is likely to gain new insights from observing nature.

Most humans engage in activities that depend on biodiversity. You may be an avid birdwatcher, or enjoy gardening, or seek out particular species if you hunt or fish. Some people like to observe or collect butterflies, or mushrooms, or other groups of plants, animals, and fungi. Displays of spring wildflowers bring out throngs of human viewers in many areas of the world. Hiking and camping in natural areas full of diverse species are activities enjoyed by millions. All of these interests support the growing industry of eco-tourism, which depends on the observation of rare or unusual species. Learning about biology greatly increases our enjoyment of these activities.

RECAP 1.3

Biology informs us about the structure, processes, and interactions of the living organisms that make up our world. Informed decisions about food and energy production, health, and our environment depend on biological knowledge. Biology also addresses the human need to understand the world around us, and helps us appreciate the diverse planet we call home.

- Describe an example of how modern biology is applied to agriculture. **See p. 15**
- Why are some antibiotics not as effective for treating bacterial diseases as they were when the drugs were originally introduced? **See p. 15**
- What is an example of a biological problem that is directly related to global climate change? **See p. 17**

This chapter has provided a brief roadmap of the rest of the book. Thinking about the principles outlined here may help you to clarify and make sense of the pages of detailed description to come. At the end of the course you may wish to revisit Chapter 1 and see if you have a different perspective on the world of biology.

Could atrazine in the environment affect species other than amphibians?

ANSWER

An important aspect of the scientific process is the replication of experimental results. In some cases the exact same experiment is repeated in another laboratory by other investigators and the results are compared. In other cases the experiment is repeated on other species to test the generality of the findings.

Following the publications by Hayes and his students, other investigators tested the effects of atrazine on other species of amphibians as well as on vertebrates other than amphibians. Feminizing effects of atrazine have now been demonstrated in fish, reptiles, and mammals. These results are not surprising, because as you will learn in Chapters 41 and 43, the hormonal controls of sex development and function are the same, and therefore the effects of atrazine should generalize to other vertebrate species.

Biologists have now studied the molecular mechanisms of the effects of atrazine on the hormonal control of sex and found that very similar responses to atrazine are seen in fish and in cultures of human cells. So atrazine in the environment is increasingly a concern for the health of many other species—and that includes humans.

CHAPTER**SUMMARY** 1

 1.1   What Is Biology?

- **Biology** is the scientific study of living organisms, including their characteristics, functions, and interactions.

- All living organisms are related to one another through common descent. Shared features of all living organisms, such as specific chemical building blocks, a nearly universal genetic code, and sequence similarities across fundamental genes, support the common ancestry of life.

- Cells evolved early in the history of life. **Cellular specialization** allowed multicellular organisms to increase in size and diversity. **Review Figure 1.2**

- The instructions for a cell are contained in its **genome**, which consists of **DNA** molecules made up of sequences of **nucleotides**. Specific segments of DNA called **genes** contain the information the cell uses to make **proteins**. **Review Figure 1.5**

- **Photosynthesis** provided a means of capturing energy directly from sunlight and over time changed Earth's atmosphere.

- **Evolution**—change in the genetic makeup of biological **populations** through time—is a fundamental principle of life. Populations evolve through several different processes, including **natural selection**, which is responsible for the diversity of **adaptations** found in living organisms.

- Biologists use fossils, anatomical similarities and differences, and molecular comparisons of genomes to reconstruct the history of life. Three domains—**Bacteria**, **Archea**, and **Eukarya**—represent the major divisions, which were established very early in life's history. **Review Figure 1.7, ACTIVITY 1.1**

- Life can be studied at different levels of organization within a **biological hierarchy**. The specialized cells of multicellular organisms are organized into **tissues**, **organs**, and **organ systems**. Individual organisms form populations and interact with other organisms of their own and other species. The populations that live and interact in a defined area form a **community**, and communities together with their abiotic (nonliving) environment constitute an **ecosystem**. **Review Figure 1.9, ACTIVITY 1.2**

- Living organisms, whether unicellular or multicellular, must regulate their internal environment to maintain **homeostasis**, the range of physical conditions necessary for their survival and function.

 1.2   How Do Biologists Investigate Life?

- Scientific methods combine observation, gathering information (**data**), experimentation, and logic to study the natural world. Many scientific investigations involve five steps: making observations, asking questions, forming hypotheses, making predictions, and testing those predictions. **Review Figure 1.10**

- **Hypotheses** are tentative answers to questions. Predictions made on the basis of a hypothesis are tested with additional observations and two kinds of experiments, **comparative** and **controlled experiments**. Review Figures 1.11, 1.12, **ANIMATED TUTORIAL 1.1**

- Quantifiable data are critical in evaluating hypotheses. Statistical methods are applied to quantitative data to establish whether or not the differences observed could be the result of chance. These methods start with the **null hypothesis** that there are no differences. **See Appendix B**

- Biological knowledge obtained from a **model system** may be generalized to other species.

 1.3   Why Does Biology Matter?

- Application of biological knowledge is responsible for vastly increased agricultural production.

- Understanding and treatment of human disease requires an integration of a wide range of biological principles, from molecular biology through cell biology, physiology, evolution, and ecology.

- Biologists are often called on to advise government agencies on the solution of important problems that have a biological component.

- Biology is increasing important for understanding how organisms interact in a rapidly changing world.

- Biology helps us understand and appreciate the diverse living world.

Go to the Interactive Summary to review key figures, Animated Tutorials, and Activities
Life10e.com/is1

CHAPTER**REVIEW**

REMEMBERING

1. Which of the following is *not* an attribute common to all living organisms?
   a. They are made up of a common set of chemical components, including particular nucleic and amino acids.
   b. They contain genetic information that uses a nearly universal code to specify the assembly of proteins.
   c. They share sequence similarities among their genes.
   d. They exist in populations that evolve over time.
   e. They extract energy from the sun in a process called photosynthesis.

2. In describing the hierarchy of life, which of the following descriptions of relationships is *not* accurate?
   a. An organ is a structure consisting of different types of cells and tissues.
   b. A population consists of all of the different animals in a particular type of environment.
   c. An ecosystem includes different communities.
   d. A tissue consists of a particular type of cells.
   e. A community consists of populations of different species.

3. Which of the following is a property of a good hypothesis?
   a. It is a statement of facts.
   b. It is general enough to explain a variety of possible experimental outcomes.
   c. It is independent of any observations.
   d. It explains things that are not addressable by experimentation.
   e. It can be falsified by experiments.

4. Which of the following events was most directly responsible for increasing oxygen in Earth's atmosphere?
   a. The cooling of the planet
   b. The origin of eukaryotes
   c. The origin of multicellularity
   d. The origin of photosynthesis
   e. The origin of prokaryotes

5. Which of the following is a reason to use statistics to evaluate data?
   a. It enables you to prove that your hypothesis is correct.
   b. It enables you to exclude data that do not fit your hypothesis.
   c. It makes it possible to exclude the null hypothesis.
   d. It enables you to predict experimental results.
   e. It accounts for variation in scientific measurements.

▨ **UNDERSTANDING & APPLYING**

6. Why is it important in science to design and perform experiments that are capable of falsifying a hypothesis?

7. What is the significance of the fact that mitochondria and chloroplasts contain the DNA that instructs their form and function?

8. The results in Dr. Hayes's comparative experiments were more variable than the results from his controlled experiments. How would you explain this?

▨ **ANALYZING & EVALUATING**

9. Biologists can now isolate genes from organisms and decode their DNA. When the nucleotide sequences from the same gene in different species are compared, differences are discovered. How could you use those data to deduce the evolutionary relationships among the organisms in your comparison?

10. Mitochondria are cell organelles that have their own DNA and replicate independently of the cell itself. In most organisms, mitochondria are inherited only from the mother. Based on this observation, when might it be advantageous or disadvantageous to use mitochondrial DNA rather than nuclear DNA for studying evolutionary relationships among populations?

Go to BioPortal at **yourBioPortal.com** for Animated Tutorials, Activities, LearningCurve Quizzes, Flashcards, and many other study and review resources.

21 Mechanisms of Evolution

CHAPTER**OUTLINE**

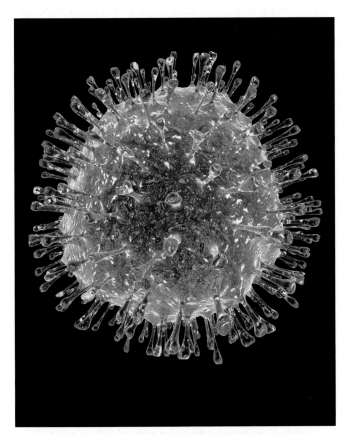

A Deadly Pathogen An artist's image of the H1N1 influenza virus that was the target of a recent flu vaccine. The spikes represent viral surface proteins, whose rapid evolution allows flu viruses to escape the host's immune system.

ON NOVEMBER 11, 1918, an armistice agreement signed in France signaled the end of World War I. But the death toll from four years of war was soon surpassed by the casualties of a massive influenza epidemic that began in the spring of 1918 among soldiers in a U.S. Army barracks. Over the next 18 months, this particular strain of flu virus spread across the globe, killing more than 50 million people worldwide—more than twice the number of combat–related deaths.

The 1918–1919 pandemic was noteworthy because the death rate among young adults—who are usually less likely to die from influenza than are the elderly or the very young—was 20 times higher than in flu epidemics before or since. Why was that particular virus so deadly, especially to typically hardy individuals? The 1918 flu strain triggered an especially intense reaction in the human immune system. This overreaction meant that people with strong immune systems were likely to be more severely affected.

In most cases, however, our immune system helps us fight viruses; this response is the basis of vaccination. Since 1945, programs to administer flu vaccines have helped keep the number and severity of influenza outbreaks in check. Last year's vaccine, however, will probably not be effective against this year's virus. New strains of flu virus are evolving continuously, ensuring genetic variation in the virus population. If these viruses did not evolve, we would become resistant to them, and annual vaccination would become unnecessary. But because the viruses do evolve, biologists must develop a new and different flu vaccine each year.

Vertebrate immune systems recognize proteins on the viral surface, so changes in these proteins mean that the virus can escape detection. Those virus strains with the greatest number of changes to their surface proteins are most likely to avoid detection and infect their hosts, and thus have an advantage over other strains. Biologists can observe evolution in action by following changes in influenza virus proteins from year to year.

We learn a great deal about the processes of evolution by examining rapidly evolving organisms such as viruses, and these studies contribute to the development of evolutionary theory. Evolutionary theory, in turn, is put to many practical uses, such as the development of better strategies for combating deadly diseases.

How do biologists use evolutionary theory to develop better flu vaccines?

See answer on p. 446.

21.1 What Is the Relationship between Fact and Theory in Evolution?

Biological populations change in their genetic makeup over time. This change in the genetic composition of populations over time is called **evolution**. We can, and do, observe evolutionary change on a regular basis, both in laboratory experiments and in natural populations. We measure the rates at which new mutations arise, observe the spread of new genetic variants through a population, and see the effects of genetic change on the form and function of organisms. In the fossil record, we observe the long-term morphological changes that are the result of underlying genetic changes. These underlying changes in the genetic makeup of populations drive the origin and extinction of species and fuel the diversification of life.

In addition to observing and recording physical changes over evolutionary time, biologists have accumulated a large body of evidence that has shown us *how* these changes occur and *what* evolutionary changes have occurred in the past. The resulting understanding of the mechanisms of evolutionary change is known as **evolutionary theory**.

We constantly apply evolutionary theory to the prevention and treatment of diseases, to the development of better agricultural crops and practices, and to the development of industrial processes that produce new molecules with useful properties. At a more basic level, evolutionary theory allows biologists to understand how life diversifies and how species interact. It also helps us make predictions about the biological world.

In everyday speech, people tend to use the word "theory" to mean an untested hypothesis, or even a guess. But "evolutionary theory" does not refer to any single hypothesis, and it certainly is not guesswork. A few scientists grasped the concept of evolutionary change even before Charles Darwin so clearly described his observations, presented his conclusions, and articulated an explanation for evolution in *On the Origin of Species*. The rediscovery of Gregor Mendel's experiments and the subsequent establishment of the principles of genetic inheritance early in the 1900s set the stage for vast amounts of research. By the end of the twentieth century, findings from many fields of biology firmly upheld Darwin's basic premises about the common ancestry of life and the role of natural selection as an important mechanism of evolution. Today a vast array of geological, morphological, and molecular data all support the factual basis of evolution.

When we refer to evolutionary theory, we are referring to our understanding of the mechanisms that result in genetic changes in populations over time. In many cases we can observe evolution directly, as with the influenza viruses described at the opening of this chapter. Over much longer periods, we can observe evolutionary changes in the fossil record. It is evolutionary theory, however, that allows us to apply our understanding of evolution to problems in medicine, agriculture, industry, and throughout biology.

 Go to Media Clip 21.1
Watching Evolution in Real Time
Life10e.com/mc21.1

Darwin and Wallace introduced the idea of evolution by natural selection

In the early 1800s, it was not yet evident to many people that life evolves. But several biologists had suggested that the species living on Earth had changed over time—that is, that evolution had taken place. Jean-Baptiste Lamarck, for one, presented strong evidence for the fact of evolution in 1809, but his ideas about *how* evolution occurred were not convincing. At that time, no one had yet envisioned a viable mechanism for evolution.

In the 1820s, a young Charles Darwin became passionately interested in the subjects of geology (with its new sense of Earth's great age) and natural history (the scientific study of how different organisms function and carry out their lives in nature). Despite these interests, he planned, at his father's behest, to become a doctor. But surgery conducted without anesthesia nauseated Darwin, and he gave up medicine to study at Cambridge University for a career as a clergyman in the Church of England. Always more interested in science than in theology, he gravitated toward scientists on the faculty, especially the botanist John Henslow. In 1831, Henslow recommended Darwin for a position on HMS *Beagle*, a Royal Navy vessel that was preparing for a survey voyage around the world (**Figure 21.1**).

Whenever possible during the five-year voyage, Darwin went ashore to study rocks and to observe and collect plants and animals. He noticed striking differences between the species he saw in South America and those of Europe. He observed that the species of the temperate regions of South America (Argentina and Chile) were more similar to those of tropical South America (Brazil) than they were to temperate European species. When he explored the islands of the Galápagos archipelago, west of Ecuador, he noted that most of the animals were endemic (found nowhere else) to the islands, although they were similar to animals found on the mainland of South America. Darwin also observed that the fauna of the Galápagos differed from island to island. He postulated that in earlier times some animals from mainland South America had arrived on the archipelago and had subsequently undergone different and distinctive changes on each of the islands. He wondered what might account for these changes.

When he returned to England in 1836, Darwin continued to ponder his observations. His ruminations were strongly influenced by the geologist Charles Lyell, who a few years earlier had popularized the idea that Earth had been shaped by slow-acting forces that were still at work. Darwin reasoned that similar thinking could be applied to the living world. Over the next decade, he developed the framework of an explanatory theory for evolutionary change based on three major propositions:

- Species are not immutable; they change over time.

- Divergent species share a common ancestor and have diverged from one another gradually through time (a concept Darwin termed **descent with modification**).

- Changes in species over time can be explained by **natural selection**: the differential survival and reproduction of individuals based on variation in their traits.

Charles Robert Darwin

21.1 Darwin and the Voyage of the *Beagle* The mission of HMS *Beagle* was to chart the oceans and collect oceanographic and biological information from around the world. The world map indicates the ship's path; the inset map shows the Galápagos Islands, whose organisms were an important source of Darwin's ideas on natural selection. The portrait is of Charles Darwin at age 27, shortly after the *Beagle* returned to England.

Go to Activity 21.1 Darwin's Voyage
Life10e.com/ac21.1

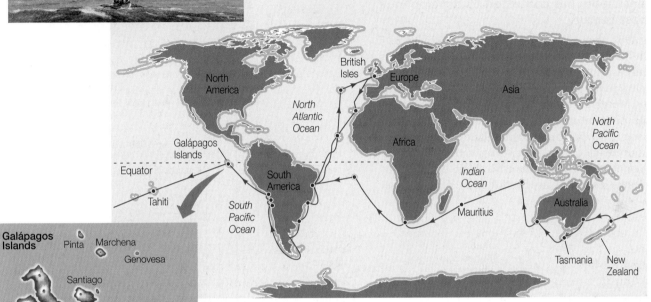

The first of these propositions was not unique to Darwin; several earlier authors had argued for the fact of evolution. A more revolutionary idea was his second proposition, that *divergent species are related to one another through common descent*. But Darwin is probably best known for his third proposition, that of natural selection.

Darwin realized that many more individuals of most species are born than survive to reproduce. He also knew that, although offspring usually resemble their parents, offspring are not identical to one another or to either parent. Finally, he was well aware of the fact that human breeders of plants and animals often selected their breeding stock based on the occurrence of particular traits. Over time, this selection resulted in dramatic changes in the appearance of the descendants of those plants or animals. In natural populations, wouldn't the individuals with the best chances of survival and reproduction be similarly "selected," and thus pass their traits on to the next generation? Darwin's simple but powerful idea was that nature did the selecting in natural populations on the basis of traits that resulted in greater survival and, eventually, greater likelihood of reproduction.

In 1844, Darwin wrote a long essay describing the role of natural selection as a mechanism of evolution, but he was reluctant to publish it, preferring to assemble more evidence. Darwin's hand was forced in 1858 when he received a letter and manuscript from another traveling English naturalist, Alfred Russel Wallace, who was studying the biota of the Malay Archipelago. Wallace asked Darwin to evaluate his manuscript, which included an explanation of natural selection almost identical to Darwin's. Darwin was at first dismayed, believing Wallace to have preempted his idea. Parts of Darwin's 1844 essay, together with Wallace's manuscript, were presented to the Linnean Society of London on July 1, 1858, thereby crediting both men for the idea of natural selection. Darwin then worked quickly to finish *On the Origin of Species*, which was published the following year.

Although Darwin and Wallace independently articulated the concept of natural selection, Darwin developed his ideas first. Furthermore, *On the Origin of Species* proved to be a stunning work of scholarship that provided exhaustive evidence from many fields supporting both the premise of evolution itself and the role of natural selection as a mechanism of evolution. Thus both concepts are more closely associated with Darwin than with Wallace.

 Go to Animated Tutorial 21.1
Natural Selection
Life10e.com/at21.1

The publication of *On the Origin of Species* in 1859 stirred considerable interest (and controversy) among scientists and the public alike. Scientists spent much of the rest of the nineteenth century amassing biological and paleontological data to test evolutionary ideas and to document the history of life on Earth. By 1900, the fact of biological evolution (defined at that time as change in the physical characteristics of populations over time) was established beyond any reasonable doubt. But the *genetic* basis of evolutionary change was not yet understood.

Evolutionary theory has continued to develop over the past century

Shortly after 1900, several individuals rediscovered Mendel's work (which had been published in 1866 but rarely read or cited) and the basic mechanisms of genetic inheritance began to be unraveled. In the first decades of the twentieth century, Thomas Hunt Morgan's studies on fruit flies led to his discovery of the role of chromosomes in inheritance. In the 1920s and early 1930s, the major principles of population genetics were established, the genetic basis of new variation (i.e., mutations) began to be understood, and mechanisms of evolution such as genetic drift were described (see Section 21.2). This work set

the stage for a "modern synthesis" of genetics and evolution that took place over the period 1936–1947. Some of the major contributors to this synthesis and a few of their books are listed in **Figure 21.2**.

Although chromosomes were soon understood to be the basis of genetic transmission in eukaryotes, their molecular structure remained a mystery until soon after the modern synthesis. Then, in 1953, James Watson and Francis Crick published their paper on the structure of DNA, opening the door to

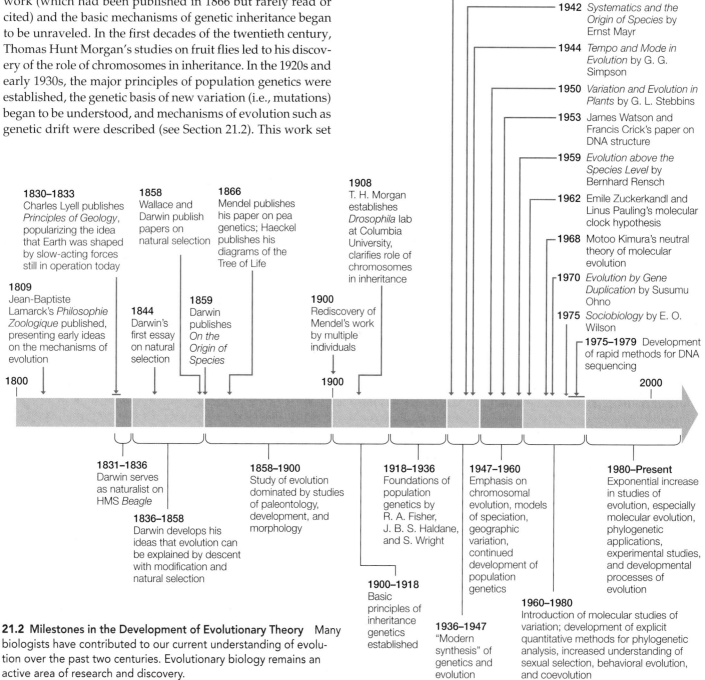

21.2 Milestones in the Development of Evolutionary Theory Many biologists have contributed to our current understanding of evolution over the past two centuries. Evolutionary biology remains an active area of research and discovery.

our current detailed understanding of molecular evolutionary mechanisms. By the 1960s, biologists could study and document changes in allele frequencies in populations over time (see Section 21.3). Most of this early work necessarily focused on variants of proteins that differed within and between populations and species because, even though the molecular structure of DNA was known, it was not yet practical to sequence long stretches of DNA. Nonetheless, many important advances occurred in evolutionary theory during this time. These advances were not focused solely on a genetic understanding of evolution. E. O. Wilson's 1975 book *Sociobiology*, for example, invigorated studies of the evolution of behavior (a subject that had fascinated Darwin).

In the late 1970s, several techniques were developed that allowed the rapid sequencing of long stretches of DNA, which in turn allowed researchers to ascertain the amino acid sequences of proteins. This capability opened a new door for evolutionary biologists, who could now explore the structures of genes and proteins and document evolutionary changes within and between species in ways never before possible. In the past three decades, well over a quarter of a million scientific papers on evolutionary observations, experiments, and theory have been published.

Genetic variation contributes to phenotypic variation

For a population to evolve, its members must possess heritable genetic variation, which is the raw material on which mechanisms of evolution act. In everyday life, we do not directly observe the genetic composition of organisms. What we see are **phenotypes**, the physical expressions of organisms' genes (including interactions among those genes). The features of a phenotype are its **characters**—eye color, for example. The specific form of a character, such as brown eyes, is a **trait**. A **heritable trait** is a trait that is at least partly determined by the organism's genes. The genetic constitution that governs a character is called its **genotype**. *A population evolves when individuals with different genotypes survive or reproduce at different rates.*

Different forms of a gene, known as **alleles**, may exist at a **locus** (a particular site on a chromosome). At any given locus, a single diploid individual carries no more than two of all the alleles found in the population. The sum of all copies of all alleles at all loci found in a population constitutes that population's **gene pool**. We can also refer to the "gene pool" for a particular locus (**Figure 21.3**). The gene pool contains the genetic variation that produces the phenotypic traits on which natural selection acts. Evolution can be defined as changes in the proportions of alleles in the gene pool over time. Thus, to understand evolution and the roles of various evolutionary mechanisms, we need to know how much genetic variation populations have, what the sources of that genetic variation are, and how genetic variation changes in populations over space and time.

The study of the genetic basis of evolution is made more difficult by the fact that genotypes alone do not determine all phenotypes. When one allele is dominant to another, for example, a particular phenotype can be produced by more than

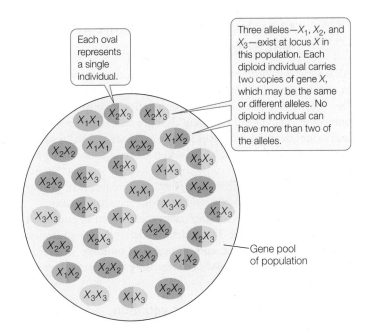

21.3 A Gene Pool A gene pool is the sum of all the alleles found in a population or at a particular locus in that population. This figure shows the gene pool for one locus, X, in a population of diploid organisms. The allele frequencies in this case are 0.20 for X_1, 0.50 for X_2, and 0.30 for X_3 (see Figure 21.10).

one genotype (e.g., *AA* and *Aa* individuals may be phenotypically identical). In addition, as we described in Section 20.4, a given genotype can produce different phenotypes depending on the environmental conditions encountered during development. For example, the cells of all the leaves on a tree or shrub are usually genetically identical, yet leaves of the same plant often differ in shape and size depending, for example, on the amount of ambient light they receive.

RECAP 21.1

Evolutionary change is directly observable in biological populations. Natural selection is one of the major mechanisms that results in evolution. It acts on genetic variation, which is required for evolutionary change to occur.

- How would you respond to someone who said that evolution was "just a theory"? **See pp. 428–430**

- Articulate the principle of natural selection and explain how natural selection leads to evolutionary change. **See pp. 428–429**

- What do you think were the most important of the discoveries and technological breakthroughs that have provided biologists with a more thorough understanding of evolution since Darwin's time? **See pp. 430–431 and Figure 21.2**

Although the importance of natural selection to evolution has been confirmed in many thousands of scientific studies, it is not the only process that drives evolution. In the next section we'll consider a more complete view of evolutionary processes and how they operate.

21.2 What Are the Mechanisms of Evolutionary Change?

Although the word "evolution" is often used in a general sense to mean simply "change," evolution in a biological context refers specifically to changes in the genetic makeup of populations over time. Developmental changes that occur in a single individual over the course of the life cycle are not the result of evolutionary change. Evolution is genetic change occurring in a **population**—a group of individuals of a single species that live and interbreed in a particular geographic area. It is important to remember that *individuals do not evolve; populations do.*

Natural selection is an important mechanism of evolution, but it does not act alone. Four additional processes—mutation, gene flow, genetic drift, and nonrandom mating—affect the genetic makeup of populations over time and thus can result in evolution.

Mutation generates genetic variation

The source of genetic variation is mutation. As described in Section 15.1, a **mutation** is any change in the nucleotide sequence of an organism's DNA. The process of DNA replication is not perfect, and some changes appear almost every time a genome is replicated. Mutations occur randomly with respect to their costs or benefits to the organism; it is natural selection acting on this random variation that results in adaptation. Most mutations are either harmful to their bearers (deleterious mutations) or have no effect (neutral mutations). But a few mutations are beneficial, and even previously deleterious or neutral alleles may become advantageous if environmental conditions change. In addition, mutation can restore genetic variation that other evolutionary mechanisms have removed. Thus mutation both creates and helps maintain genetic variation in populations.

Mutation rates can be very high, particularly in viruses, some of which have mutation rates as high as 10^{-3} changes per nucleotide per generation. The rapid evolution of influenza viruses challenges effective vaccine production, as we saw at the opening of this chapter. In some genes of eukaryotes, the mutation rate is much lower (on the order of 10^{-8} to 10^{-9} changes per base pair per generation). Even low overall mutation rates, however, create considerable genetic variation because each of a large number of genes may change and because many populations contain large numbers of individuals. For example, if the probability of a point mutation (an addition, deletion, or substitution of a single base) were 10^{-9} per base pair per generation, then each human gamete—the DNA of which contains 3×10^9 base pairs—would average three new point mutations ($3 \times 10^9 \times 10^{-9} = 3$), and each diploid zygote would carry an average of six new mutations. The current human population of about 7 billion people would thus be expected to carry about 42 billion new mutations (i.e., changes in the nucleotide sequences of their DNA that were not present one generation earlier). So even though the mutation rate in humans is low, human populations still contain enormous genetic variation on which other evolutionary mechanisms can act.

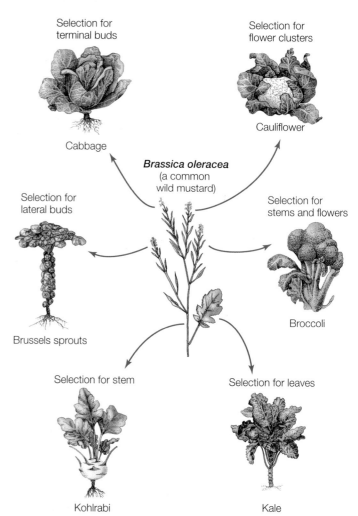

21.4 Many Vegetables from One Species All the crop plants shown here derive from a single wild mustard species. European agriculturalists produced these crop species by selecting and breeding plants with unusually large buds, stems, leaves, or flowers. The results substantiate the vast amount of variation present in the gene pool of the ancestral species.

Recall from Section 21.1 that the gene pool is the sum of the genetic variation in a population. Mutation adds new alleles to the gene pool. Biologists use two simple measures to characterize the variation in a given gene pool. The proportion of each allele in the gene pool is its **allele frequency**. Similarly, the proportion of each genotype among the individuals in the population is its **genotype frequency**. The calculations of allele and genotype frequencies in a population allow biologists to measure evolutionary change, as will be described in Section 21.3.

Selection acting on genetic variation leads to new phenotypes

As a result of mutation, the gene pools of nearly all populations contain variation for many characters. Selection on this variation can produce rapid evolutionary change, and evolution can proceed in many different directions. For example, **artificial selection**—the purposeful selection of specific phenotypes by humans—on different characters in a single European species of wild mustard has resulted in many different crop plants (**Figure 21.4**). Agriculturalists were able to achieve these results because the gene pool of the original mustard population

21.5 Artificial Selection Charles Darwin, who raised pigeons as a hobby, noted similar forces at work in artificial and natural selection. The "fancy" pigeons shown here represent 3 of the more than 300 varieties derived from the wild rock pigeon (*Columba livia*; left) by artificial selection on characters such as color and feather distribution.

contained variation in the alleles for the characters of interest (such as stem thickness or number of leaves).

As we noted in Section 21.1, Darwin compared this artificial selection by animal and plant breeders with natural selection. Many of Darwin's observations on the nature of variation and selection came from domesticated plants and animals. Darwin bred pigeons and thus knew firsthand the astonishing diversity in color, size, form, and behavior that breeders could achieve in these birds (**Figure 21.5**). He recognized close parallels between selection by breeders and selection in nature. Natural selection resulted in traits that helped organisms survive and reproduce more effectively; artificial selection resulted in traits that were preferred by the human breeders, for whatever reason.

Laboratory experiments also demonstrate the potential for selection to result in evolutionary change, often resulting in descendant populations containing phenotypes that were not present in the ancestral populations. In one such experiment, investigators, starting with a population of the fruit fly *Drosophila melanogaster* with intermediate numbers of abdominal bristles, bred two new populations. One population was selected for high and the other for low bristle numbers. After 35 generations, all flies in both the high- and low-bristle lineages had bristle numbers that fell well outside the range seen in the original population (**Figure 21.6**). In this experiment, artificial selection resulted in new combinations of the many different genes that were present in the original population, so that the phenotypic variation in subsequent generations fell outside that of the original population.

Natural selection works in much the same way. Slight trait differences among individuals increase the chance that a given individual will survive and reproduce, which increases the frequency of the favored trait in the next generation. A favored trait that evolves through natural selection is known as an **adaptation**; this word is used to describe both the trait itself and the process that produces the trait. Biologists regard an organism as being adapted to a particular environment when they can demonstrate that a slightly different organism reproduces and survives less well in that environment. To understand adaptation, biologists compare the performances of individuals that differ in their traits.

One consequence of natural seection is the purging of deleterious mutations from populations. Individuals with deleterious mutations are less likely to survive and reproduce, so they are less likely to pass their alleles on to the next generation. Biologists often distinguish between two broad categories of selection: **positive selection** (selection for beneficial changes) and **purifying selection** (selection against deleterious changes).

Gene flow may change allele frequencies

Few populations are completely isolated from other populations of the same species. Migration of individuals and movements of gametes (in pollen, for example) between populations—a phenomenon called **gene flow**—can change allele frequencies in a population. If the arriving individuals survive

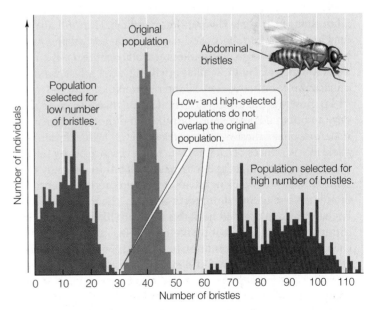

21.6 Artificial Selection Reveals Genetic Variation When investigators subjected *Drosophila melanogaster* to artificial selection for abdominal bristle number, that character evolved rapidly. The graph shows the number of flies with different numbers of bristles in the original population and after 35 generations of artificial selection for low and for high bristle numbers.

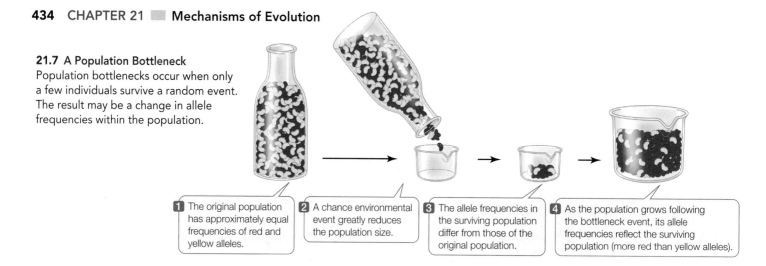

21.7 A Population Bottleneck
Population bottlenecks occur when only a few individuals survive a random event. The result may be a change in allele frequencies within the population.

1 The original population has approximately equal frequencies of red and yellow alleles.

2 A chance environmental event greatly reduces the population size.

3 The allele frequencies in the surviving population differ from those of the original population.

4 As the population grows following the bottleneck event, its allele frequencies reflect the surviving population (more red than yellow alleles).

and reproduce in their new location, they may add new alleles to the population's gene pool, or they may change the frequencies of alleles present in the original population.

About 35,000 years ago, modern humans expanded their range into the range of Neanderthals in Europe and western Asia. Although modern humans largely replaced the Neanderthals over the next 7,000 years, some interbreeding between these populations occurred, as evidenced by the presence of a small percentage of Neanderthal genes in modern non-African human populations. This incorporation of Neanderthal genes is an example of gene flow. Traits such as red hair (which was common in Neanderthal populations) may have entered modern human populations in this manner.

Genetic drift may cause large changes in small populations

In small populations, **genetic drift**—random changes in allele frequencies from one generation to the next—may produce large changes in allele frequencies over time. Harmful alleles may increase in frequency, and rare advantageous alleles may be lost. Even in large populations, genetic drift can influence the frequencies of neutral alleles (which do not affect the survival and reproductive rates of their bearers).

To illustrate the effects of genetic drift, suppose there are only two females in a small population of normally brown mice, and one of these females carries a newly arisen dominant allele that produces black fur. Even in the absence of any selection, it is unlikely that the two females will produce exactly the same number of offspring. Even if they do produce identical litter sizes and identical numbers of litters, chance events that have nothing to do with heritable traits are likely to result in differential mortality among their offspring. If, for example, each female produces one litter, but a flood envelops the black female's nest and kills her and her offspring, the novel allele could be lost from the population in just one generation. In contrast, if the brown female's litter is lost, then the frequency of the newly arisen allele (and phenotype) for black fur will rise dramatically in just one generation

Genetic drift is particularly potent when a population is reduced dramatically in size. Even populations that are normally large may occasionally pass through environmental conditions that only a small number of individuals survive, a situation known as a **population bottleneck**. The effect of genetic drift in

such a situation is illustrated in **Figure 21.7**, in which red and yellow beans represent two alleles of a gene. Most of the beans in the small sample of the "population" that "survives" the bottleneck event are, just by chance, red, so the new population has a much higher frequency of red beans than the previous generation had. In a real population, the red and yellow allele frequencies would be described as having "drifted."

A population forced through a bottleneck is likely to lose much of its genetic variation. For example, when Europeans first arrived in North America, millions of greater prairie-chickens (*Tympanuchus cupido*) inhabited the midwestern prairies. As a result of hunting and habitat destruction by the new settlers, the Illinois population of this species plummeted from about 100 million birds in 1900 to fewer than 50 individuals in the 1990s. A comparison of DNA from birds collected during the mid-twentieth century with DNA from the population surviving in the 1990s showed that Illinois prairie-chickens had lost most of their genetic diversity. Loss of genetic variation in small populations is one of the problems facing biologists who attempt to protect endangered species.

Genetic drift can have similar effects when a few pioneering individuals colonize a new region. Because of its small size, the colonizing population is unlikely to possess all the alleles found in the gene pool of its source population. The resulting reduction in genetic variation, called a **founder effect**, is equivalent to that in a large population reduced by a bottleneck. When a few humans migrated across the Bering Strait to colonize the Americas, for example, they brought with them a small sample of the genetic diversity that was present in Asian populations.

 Go to Animated Tutorial 21.2
Genetic Drift
Life10e.com/at21.2

Nonrandom mating can change genotype or allele frequencies

Mating patterns often alter genotype frequencies because the individuals in a population do not choose mates at random. For example, self-fertilization is common in many groups of organisms, especially plants. Any time individuals mate preferentially with other individuals of the same genotype (including themselves), homozygous genotypes will increase, and heterozygous genotypes will decrease, in frequency over time. The

Euplectes progne

21.8 What Is the Advantage? The extensive tail of the male African long-tailed widowbird inhibits its ability to fly. Darwin attributed the evolution of this seemingly nonadaptive trait to sexual selection.

opposite effect (more heterozygotes, fewer homozygotes) is expected when individuals mate primarily or exclusively with individuals of different genotypes.

Nonrandom mating systems that do not affect the relative reproductive success of individuals produce changes in genotype frequencies but not in allele frequencies, and thus do not, by themselves, result in evolutionary change in a population. However, nonrandom mating systems that result in differential reproductive success among individuals do produce allele frequency changes from one generation to the next. **Sexual selection** occurs when individuals of one sex mate preferentially with particular individuals of the opposite sex rather than at random.

Sexual selection was first suggested by Charles Darwin, who developed the idea to explain the evolution of conspicuous traits that would appear to inhibit survival, such as bright colors, long tails, and elaborate courtship displays in males of many species. He hypothesized that these features either improved the ability of their bearers to compete for access to mates (intrasexual selection) or made their bearers more attractive to members of the opposite sex (intersexual selection). The concept of sexual selection was either ignored or questioned for many decades, but recent investigations have demonstrated its importance.

Darwin argued that while natural selection typically favors traits that enhance the survival of their bearers or their descendants, sexual selection is primarily about successful reproduction. An animal that survives but fails to reproduce makes no contribution to the next generation. Thus sexual selection may favor traits that enhance an individual's chances of reproduction even if they reduce its chances of survival. For example, females may be more likely to see or hear males with a conspicuous trait (and thus be more likely to mate with those males), even though the conspicuous trait may increase the chances that the male will be seen or heard by a predator. The sexual signal may also indicate a successful genotype in the male. In many species of frogs, for example, females prefer males

INVESTIGATING**LIFE**

21.9 Sexual Selection in Action Behavioral ecologist Malte Andersson tested Darwin's hypothesis that excessively long tails evolved in male widowbirds because female preference for longer-tailed males increased their mating and reproductive success.[a]

HYPOTHESIS Female widowbirds prefer to mate with the male that displays the longest tail; longer-tailed males thus are favored by sexual selection because they will father more offspring.

Method   1. Capture males and artificially lengthen or shorten tails by cutting or gluing on feathers. In a control group, cut and replace tails to their normal length (to control for the effects of tail-cutting).
2. Release the males to establish their territories and mate.
3. Count the nests with eggs or young on each male's territory.

Results   Male widowbirds with artificially shortened tails established and defended display sites sucessfully but fathered fewer offspring than did control or unmanipulated males. Males with artificially lengthened tails fathered the most offspring.

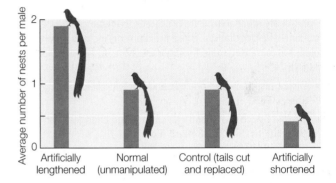

CONCLUSION Sexual selection in *Euplectes progne* has favored the evolution of long tails in the male.

Go to **BioPortal** for discussion and relevant links for all INVESTIGATING**LIFE** figures.

[a]Andersson, M. 1982. *Nature* 299: 818–820

with low-frequency calls. Male calls vary with body size, and a low-frequency call is indicative of a large-bodied frog. Frogs exhibit indeterminate growth—that is, they continue to grow indefinitely—so a large frog is a long-lived frog. In this case, the male's call represents what is known as an honest signal of the male's ability to survive in the local environment.

One trait that Darwin attributed to sexual selection is the remarkable tail of the male African long-tailed widowbird, which is longer than the bird's head and body combined (**Figure 21.8**) and impairs its ability to fly. Male widowbirds normally select and defend a territory where they perform courtship displays to attract females. Malte Andersson investigated whether sexual selection drove the evolution of widowbird tails (**Figure 21.9**). He captured male widowbirds and clipped the tails of some individuals while artificially lengthening the tails of others (by gluing on additional feathers). He then cut and re-glued the tail feathers of still other males, which served as controls. When released, both short- and long-tailed males successfully defended their display territories, indicating that a long tail does not confer an advantage in male–male competition.

However, males with artificially elongated tails attracted about four times more females than did males with shortened tails.

Why do female widowbirds prefer males with long tails? Biologists have developed several hypotheses. One possibility is that the ability to grow and maintain a costly feature such as a long tail may indicate that the male bearing it is vigorous and healthy. If so, then females that are attracted to long tails are indirectly attracted to vigorous, healthy males, which probably carry beneficial genes that should lead to high survivorship among their offspring. Another possibility is that females judge a male's body size by its overall length and they prefer larger males. It is also possible that a long tail stimulates the female visual system for reasons that are unrelated to any aspect of male health or vigor; the reasons why females prefer sexually selected traits are not always clear. What *is* clear is that female preferences often lead to the evolution of dramatic male adornments and mating displays.

─────────── **RECAP 21.2**

Evolutionary mechanisms are processes that change the genetic structure of a population. These mechanisms include natural selection, mutation, gene flow, genetic drift, and nonrandom mating.

- Distinguish between positive and purifying selection. **See p. 433**

- Explain how genetic drift can cause large changes in small populations. **See p. 434 and Figure 21.7**

- Describe some other examples (beyond those described in this book) of traits that may have evolved through sexual selection.

─────────────────────

The evolutionary mechanisms discussed so far act by changing the frequencies of alleles and genotypes in populations. How are these changes measured by biologists? In other words, how do we know that evolution is occurring?

21.3 How Do Biologists Measure Evolutionary Change?

Much of evolution occurs through gradual changes in the relative frequencies of different alleles in a population from one generation to the next. As we'll see in Chapter 24, major genetic changes can also be sudden, as happens, for example, when two formerly separated populations merge and hybridize. But in most cases, we can measure evolution by looking at gradual changes in allele and genotype frequencies (or of their respective phenotypes) in populations.

Evolutionary change can be measured by allele and genotype frequencies

To measure allele frequencies in a population precisely, we would need to count every allele at every locus in every individual in the population. Fortunately, we do not need to make such complete measurements because we can reliably estimate allele frequencies for a given locus by counting alleles

21.10 Calculating Allele and Genotype Frequencies Allele and genotype frequencies for a gene locus with two alleles in the population can be calculated using the equations in panel 1. When the equations are applied to two populations (panel 2), we find that the *frequencies of alleles A and a in the two populations* are the same, but the alleles are distributed differently between heterozygous and homozygous genotypes.

1 In any population, where N is the total number of individuals in the population:

$$\text{Frequency of allele } A = p = \frac{2N_{AA} + N_{Aa}}{2N} \quad \text{Frequency of allele } a = q = \frac{2N_{aa} + N_{Aa}}{2N}$$

Frequency of genotype $AA = N_{AA}/N$
Frequency of genotype $Aa = N_{Aa}/N$
Frequency of genotype $aa = N_{aa}/N$

2 Compute the allele and genotype frequencies for two separate populations of $N = 200$:

Population 1 (mostly homozygotes)	Population 2 (mostly heterozygotes)
$N_{AA} = 90$, $N_{Aa} = 40$, and $N_{aa} = 70$	$N_{AA} = 45$, $N_{Aa} = 130$, and $N_{aa} = 25$
$p = \dfrac{180 + 40}{400} = 0.55$	$p = \dfrac{90 + 130}{400} = 0.55$
$q = \dfrac{140 + 40}{400} = 0.45$	$q = \dfrac{50 + 130}{400} = 0.45$
Freq. AA = 90/200 = 0.45	Freq. AA = 45/200 = 0.225
Freq. Aa = 40/200 = 0.20	Freq. Aa = 130/200 = 0.65
Freq. aa = 70/200 = 0.35	Freq. aa = 25/200 = 0.125

in a sample of individuals from the population. The sum of all allele frequencies at a locus is equal to 1, so measures of allele frequency range from 0 to 1.

An allele's frequency is calculated using the following formula:

$$p = \frac{\text{number of copies of the allele in the population}}{\text{total number of copies of all alleles in the population}}$$

If only two alleles (we'll call them A and a) are found among the members of a diploid population, those alleles can combine to form three different genotypes: AA, Aa, and aa. A population with more than one allele at a locus is said to be polymorphic ("many forms") at that locus. Applying the formula above, as shown in **Figure 21.10**, we can calculate the relative frequencies of alleles A and a in a population of N individuals:

- Let N_{AA} be the number of individuals that are homozygous for the A allele (AA).

- Let N_{Aa} be the number that are heterozygous (Aa).

- Let N_{aa} be the number that are homozygous for the a allele (aa).

Note that $N_{AA} + N_{Aa} + N_{aa} = N$, the total number of individuals in the population, and that the total number of copies of both alleles present in the population is $2N$, because each individual is diploid. Each AA individual has two copies of the A allele, and each Aa individual has one copy of the A allele. Therefore, the total number of A alleles in the population is $2N_{AA} + N_{Aa}$. Similarly, the total number of a alleles in the population is $2N_{aa} + N_{Aa}$. If p represents the frequency of A and q represents the frequency of a, then

$$q = \frac{2N_{aa} + N_{Aa}}{2N}$$

and

$$p = \frac{2N_{AA} + N_{Aa}}{2N}$$

The calculations in Figure 21.10 demonstrate two important points. First, notice that for each population, $p + q = 1$, which means that $q = 1 - p$. So when there are only two alleles at a given locus in a population, we can calculate the frequency of one allele and obtain the second allele's frequency by subtraction. If there is only one allele at a given locus in a population, its frequency is 1: The population is then monomorphic ("one form") at that locus, and the allele is said to be **fixed**.

The second thing to notice is that population 1 (consisting mostly of homozygotes) and population 2 (consisting mostly of heterozygotes) have the same allele frequencies for A and a; thus they have the same gene pool for this locus. Because the alleles in the gene pool are distributed differently among individuals, however, the genotype frequencies of the two populations differ.

The frequencies of the different alleles at each locus and the frequencies of the different genotypes in a population describe that population's **genetic structure**. Allele frequencies measure the amount of genetic variation in a population; genotype frequencies show how a population's genetic variation is distributed among its members. Other measures, such as the proportion of loci that are polymorphic, are also used to measure variation in populations. With these measurements, it becomes possible to consider how the genetic structure of a population changes or remains the same over generations—that is, to measure evolutionary change.

Evolution will occur unless certain restrictive conditions exist

In 1908 the British mathematician Godfrey Hardy and the German physician Wilhelm Weinberg independently deduced the conditions that must prevail if the genetic structure of a population is to remain the same over time. If the conditions they identified do not exist, then evolution will occur. The resulting principle is known as **Hardy–Weinberg equilibrium**. Hardy–Weinberg equilibrium constitutes a model in which allele frequencies do not change across generations, and genotype frequencies can be predicted from allele frequencies. The expectations of Hardy–Weinberg equilibrium apply only to sexually reproducing organisms. Several conditions must be met for a population to be at Hardy–Weinberg equilibrium (which, you should notice, correspond inversely to the five principal mechanisms of evolution discussed in Section 21.2):

- *There is no mutation.* The alleles present in the population do not change and no new alleles are added to the gene pool.

- *There is no selection among genotypes.* Individuals with different genotypes have equal probabilities of survival and equal rates of reproduction.

- *There is no gene flow.* There is no movement of individuals or gametes into or out of the population or reproductive contact with other populations.

- *Population size is infinite.* The larger a population, the smaller will be the effect of genetic drift.

- *Mating is random.* Individuals do not preferentially choose mates with certain genotypes.

If these idealized conditions hold, two major consequences follow. First, the frequencies of alleles at a locus remain constant from generation to generation—that is, no evolutionary change occurs in the population. Second, following one generation of random mating, the genotypes occur at the following frequencies:

Genotype	AA	Aa	aa
Frequency	p^2	$2pq$	q^2

To understand why these consequences are important, start by considering a population that is *not* in Hardy–Weinberg equilibrium, such as generation I in **Figure 21.11**. This could occur, for example, if the initial population is founded by migrants from several other populations, thus violating the Hardy–Weinberg assumption of no gene flow. In this example, generation I has more homozygous individuals and fewer heterozygous individuals than would be expected under Hardy–Weinberg equilibrium (a condition known as heterozygote deficiency).

Even with a starting population that is not in Hardy–Weinberg equilibrium, we can predict that after a single generation of random mating, if the other Hardy–Weinberg assumptions are not violated, the allele frequencies will remain unchanged, but the genotype frequencies will return to Hardy–Weinberg expectations. Let's explore why this is true.

In generation I of Figure 21.11, the frequency of the A allele (p) is 0.55. Because we assume that individuals select mates at random, without regard to their genotype, gametes carrying A or a combine at random—that is, as predicted by the allele frequencies p and q. In this example, the probability that a particular sperm or egg will bear an A allele is 0.55. In other words, 55 out of 100 randomly sampled sperm or eggs will bear an A allele. Because $q = 1 - p$, the probability that a sperm or egg will bear an a allele is $1 - 0.55 = 0.45$.

To obtain the probability of two A-bearing gametes coming together at fertilization, we multiply the two independent probabilities of their occurrence:

$$p \times p = p^2 = (0.55)^2 = 0.3025$$

Therefore 0.3025, or 30.25 percent, of the offspring in generation II will have homozygous genotype AA. Similarly, the probability of two a-bearing gametes coming together is

$$q \times q = q^2 = (0.45)^2 = 0.2025$$

which means that 20.25 percent of generation II will have the aa genotype.

Generation I (Founder population)

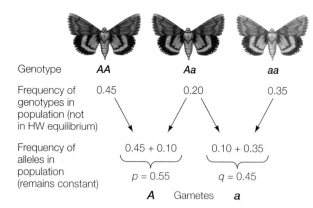

Under the assumptions of Hardy–Weinberg equilibrium, allele frequencies p and q remain constant from generation to generation. If Hardy–Weinberg assumptions are violated and the genotype frequencies in the parental generation are altered (say, by the loss of a large number of AA individuals from the population), then the allele frequencies in the next generation will be altered. However, based on the new allele frequencies, another generation of random mating will be sufficient to restore the genotype frequencies to Hardy–Weinberg equilibrium.

 Go to Animated Tutorial 21.3
Hardy–Weinberg Equilibrium
Life10e.com/at21.3

Generation II (Hardy–Weinberg equilibrium restored)

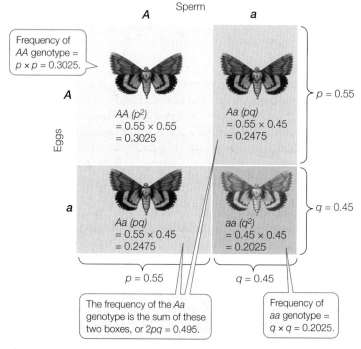

Frequency of AA genotype = $p \times p = 0.3025$.

AA (p^2)
= 0.55×0.55
= 0.3025

Aa (pq)
= 0.55×0.45
= 0.2475

$p = 0.55$

Aa (pq)
= 0.55×0.45
= 0.2475

aa (q^2)
= 0.45×0.45
= 0.2025

$q = 0.45$

$p = 0.55$ $q = 0.45$

The frequency of the Aa genotype is the sum of these two boxes, or $2pq = 0.495$.

Frequency of aa genotype = $q \times q = 0.2025$.

21.11 One Generation of Random Mating Restores Hardy–Weinberg Equilibrium Generation I of this population is made up of migrants from several source populations and so is not initially in Hardy–Weinberg equilibrium. After one generation of random mating, the allele frequencies are unchanged, and the genotype frequencies return to Hardy–Weinberg expectations. The lengths of the sides of each rectangle are proportional to the allele frequencies in the population; the areas of the rectangles are proportional to the genotype frequencies.

There are two ways of producing a heterozygote. An A sperm may combine with an a egg, the probability of which is $p \times q$; or an a sperm may combine with an A egg, the probability of which is $q \times p$. Consequently, the overall probability of obtaining a heterozygote is $2pq$, or 0.495 in this example. The frequencies of the AA, Aa, and aa genotypes in generation II of Figure 21.11 now meet Hardy–Weinberg expectations, and the frequencies of the two alleles (p and q) have not changed from generation I.

Deviations from Hardy–Weinberg equilibrium show that evolution is occurring

You probably have realized that populations in nature never meet the stringent conditions necessary to be at Hardy–Weinberg equilibrium—which explains why all biological populations evolve. Why, then, is this model considered so important for the study of evolution? There are two reasons. First, the expectations of Hardy–Weinberg equilibrium are useful for predicting the approximate genotype frequencies of a population from its allele frequencies. Second—and crucially—the model allows biologists to evaluate which mechanisms of evolution are acting on a particular population. Specific patterns of deviation from Hardy–Weinberg equilibrium can help us identify the various mechanisms of evolutionary change.

Natural selection acts directly on phenotypes

Although evolution is defined as changes in the genetic makeup of a population from one generation to the next, natural selection acts directly on the phenotype—that is, on the physical features expressed by an organism with a given genotype—and therefore acts only indirectly on the genotype. The reproductive contribution of a phenotype to subsequent generations relative to the contributions of other phenotypes is called its **fitness**.

Changes in reproductive rate do not necessarily change the genetic structure of a population. For example, if all individuals in a population experience the same increase in reproductive rate (during an environmentally favorable year, for instance), the genetic structure of the population will not change. Changes in numbers of offspring are responsible for increases and decreases in the *size* of a population, but only changes in the *relative* success of different phenotypes in a population will lead to changes in allele frequencies from one generation to the next. The fitness of individuals of a particular phenotype is a function of the probability of those individuals surviving multiplied by the average number of offspring they produce over their lifetimes. In other words, the *fitness of a phenotype is determined by the relative rates of survival and reproduction of individuals with that phenotype.*

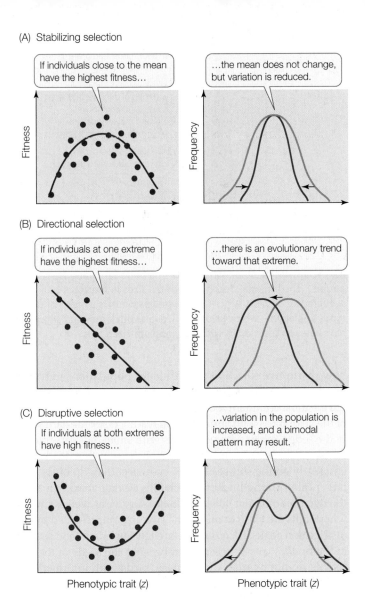

(A) Stabilizing selection

If individuals close to the mean have the highest fitness...

...the mean does not change, but variation is reduced.

(B) Directional selection

If individuals at one extreme have the highest fitness...

...there is an evolutionary trend toward that extreme.

(C) Disruptive selection

If individuals at both extremes have high fitness...

...variation in the population is increased, and a bimodal pattern may result.

Phenotypic trait (*z*)

Phenotypic trait (*z*)

21.12 Natural Selection Can Operate in Several Ways The graphs in the left-hand column show the fitness of individuals with different phenotypes for the same character. The right-hand graphs show the distribution of the phenotypes in the population before (light green) and after (dark green) the influence of selection.

Natural selection can change or stabilize populations

Until now, our discussion has focused on changes in alleles at a single genetic locus. Phenotypic traits controlled by a single locus are often distinguished by discrete qualities (black versus white, or smooth versus wrinkled), and are called qualitative traits. Many traits, however, are influenced by alleles at more than one locus. Such traits are likely to show continuous, quantitative variation rather than discrete qualitative variation, and so are known as quantitative traits. Body size, for example, is influenced by genes at many loci as well as by the environment (nutrition, for example). Therefore the distribution of body sizes of individuals in a population is likely to resemble a continuous bell-shaped curve.

Natural selection can act on characters with quantitative variation in any one of several different ways, producing quite different results (**Figure 21.12**):

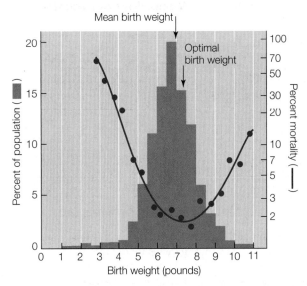

Mean birth weight

Optimal birth weight

21.13 Human Birth Weight Is Influenced by Stabilizing Selection Babies that weigh more or less than average are more likely to die soon after birth than are babies with weights closer to the population mean.

- **Stabilizing selection** preserves the average characteristics of a population by favoring average individuals.
- **Directional selection** changes the characteristics of a population by favoring individuals that vary in one direction from the mean of the population.
- **Disruptive selection** changes the characteristics of a population by favoring individuals that vary in both directions from the mean of the population.

STABILIZING SELECTION If the smallest and largest individuals in a population contribute fewer offspring to the next generation than do individuals closer to the average body size, then stabilizing selection is operating on body size (see Figure 21.12A). Stabilizing selection reduces variation in populations, but it does not change the mean. Natural selection frequently acts in this way, countering increases in variation brought about by sexual recombination, mutation, or gene flow. Rates of phenotypic change in many species are slow because natural selection is often stabilizing. Stabilizing selection operates, for example, on human birth weight. Babies who are lighter or heavier at birth than the population mean die at higher rates than babies whose weights are close to the mean (**Figure 21.13**).

DIRECTIONAL SELECTION Directional selection is operating when individuals at one extreme of a character distribution contribute more offspring to the next generation than other individuals do, shifting the average value of that character in the population toward that extreme. We noted in Section 21.2 that, in the case of a single locus, selection for a particular genetic variant is referred to as positive selection. Positive selection for many genetic variants at many loci, or for new combinations of those variants, results in the overall directional selection of a quantitative trait. By favoring one phenotype over another, directional selection results in an increase of the frequencies of alleles that produce the favored phenotype (as with the surface proteins of influenza discussed at the opening of this chapter).

21.14 A Result of Directional Selection In the American Southwest, long horns were advantageous for defending calves from attacks by predators, so cows with longer horns were more likely to raise calves successfully. As a result, horn length in feral herds of cattle increased between the early 1500s and the 1860s, leading to the Texas Longhorn breed. This evolutionary trend has been maintained by modern ranchers practicing artificial selection.

If directional selection operates over many generations, an evolutionary trend is seen in the population (see Figure 21.12B). Evolutionary trends can be reversed if the environment changes so that different phenotypes are favored. Or they can be halted when an optimal phenotype is reached or when trade-offs oppose further change (see Section 21.5). The character then undergoes stabilizing selection.

The long horns of Texas Longhorn cattle (**Figure 21.14**) are an example of a trait that has evolved through directional selection. Texas Longhorns are descendants of cattle brought to the New World by Christopher Columbus, who picked up a few cattle in the Canary Islands and brought them to the island of Hispaniola in 1493. The cattle multiplied, and their descendants were taken to the mainland of Mexico. Spaniards exploring what would become Texas and the southwestern United States brought these cattle with them, some of which escaped and formed feral herds. Populations of feral cattle increased greatly over the next few hundred years, but they faced heavy predation from bears, mountain lions, and wolves, especially on the young calves. Cows with longer horns were more successful in protecting their calves against attacks. Over a few hundred years, the average horn length in the feral herds increased considerably. In addition, these cattle evolved resistance to the endemic diseases of the Southwest, as well as higher fecundity and longevity. Texas Longhorns often live and produce calves well into their twenties—about twice as long as many breeds of cattle that have been artificially selected by humans for traits such as high fat content or high milk production (which are examples of artificial directional selection).

DISRUPTIVE SELECTION Disruptive selection is operating when individuals at opposite extremes of a character distribution contribute more offspring to the next generation than do individuals close to the mean. Directional selection increases variation in the population (see Figure 21.12C).

The strikingly bimodal (two-peaked) distribution of bill sizes in the black-bellied seedcracker, a West African finch, illustrates

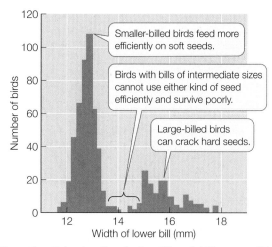

21.15 Disruptive Selection Results in a Bimodal Character Distribution The bimodal distribution of bill sizes in the black-bellied seedcracker of West Africa (*Pyrenestes ostrinus*) is a result of disruptive selection, which favors individuals with larger and smaller bill sizes over individuals with intermediate-sized bills.

how disruptive selection can influence populations in nature (**Figure 21.15**). The seeds of two types of sedges (marsh plants) are the most abundant food source for these finches during part of the year. Birds with large bills can readily crack the hard seeds of the sedge *Scleria verrucosa*. Birds with small bills can crack *S. verrucosa* seeds only with difficulty; however, they feed more efficiently on the soft seeds of *S. goossensii* than do birds with larger bills. Young finches whose bills deviate markedly from the two predominant bill sizes do not survive as well as finches whose bills are close to one of the two sizes represented by the distribution peaks. Because there are few abundant food sources in the finches' environment, and because the seeds of the two sedges do not overlap in hardness, birds with intermediate-sized bills are less efficient in using either one of the species' principal food sources. Disruptive selection therefore maintains a bimodal bill-size distribution.

RECAP 21.3

Hardy–Weinberg equilibrium describes the theoretical conditions for a non-evolving population. Deviations from Hardy–Weinberg expectations provide information about how evolution is occurring in a given population. Natural selection can both change and stabilize phenotypes within populations.

- Why is the concept of Hardy–Weinberg equilibrium important even though the assumptions on which it is based are never completely met in nature? **See pp. 437–438**
- Explain why natural selection that acts on a phenotype results in changes in genotype frequencies. **See p. 438**
- Describe the differences between stabilizing, directional, and disruptive selection, giving examples of each. **See pp. 439–440 and Figure 21.12**

Genetic drift, stabilizing selection, and directional selection all tend to reduce genetic variation within populations. Nevertheless, as we have seen, most populations harbor considerable genetic variation. What processes produce and maintain genetic variation within populations?

21.4 How Is Genetic Variation Distributed and Maintained within Populations?

Genetic variation is the raw material on which mechanisms of evolution act. In this section we will discuss several factors—neutral mutations, sexual recombination, frequency-dependent selection, and heterozygote advantage—that affect how genetic variation is established, how it is distributed among individuals, and how it is maintained within populations.

Neutral mutations accumulate in populations

An allele that does not affect the fitness of an organism—that is, an allele that is no better or worse than alternative alleles at the same locus—is called a **neutral allele**. Neutral alleles are added to a population over time through mutation, providing the population with considerable genetic variation. The frequencies of neutral alleles are not affected directly by natural selection. Even in large populations, neutral alleles may be lost, or may increase in frequency, purely by genetic drift.

Much of the phenotypic variation we are able to observe is not neutral. However, modern techniques enable us to measure neutral variation at the molecular level and provide the means to distinguish it from adaptive variation. Section 24.2 will describe how variation in neutral molecular traits can be used to study divergence among genes, populations, and species.

Sexual recombination amplifies the number of possible genotypes

In asexually reproducing organisms, each new individual is genetically identical to its parent unless there has been a mutation. When organisms reproduce sexually, however, offspring differ from their parents not only because they result from the combination of genetic material from two different gametes, but also through crossing over and independent assortment of chromosomes during meiosis, as described in Section 11.5. Sexual recombination generates an endless variety of genotypic combinations that increase the evolutionary potential of populations—a long-term advantage of sex. Although many species reproduce asexually most of the time, few are strictly asexual over long periods of evolutionary time. Almost all have some means of achieving genetic recombination.

The evolution of the mechanisms of meiosis and sexual recombination were crucial events in the history of life. Exactly how these attributes arose is puzzling, however, because sex has at least three striking disadvantages in the short term:

- Recombination breaks up adaptive combinations of genes.
- Sex reduces the rate at which females pass genes on to their offspring.
- Dividing offspring into separate sexes greatly reduces the overall reproductive rate.

To see why this last disadvantage exists, consider an asexual female that produces the same number of offspring as a sexual female. Let's assume that both females produce two offspring, but that 50 percent of the sexual female's offspring will be males

(and thus only contribute sperm). In this next (F_1) generation, both asexual females will produce two more offspring each, but there is only one sexual F_1 female to produce offspring. Thus the effective reproductive rate of the asexual lineage is twice that of the sexual lineage.

The evolutionary problem is to identify the advantages of sex that can overcome such short-term disadvantages. Several hypotheses have been proposed to explain the existence of sex, none of which are mutually exclusive. One is that sexual recombination facilitates repair of damaged DNA, because breaks and other errors in DNA on one chromosome can be repaired by copying the intact sequence from the homologous chromosome.

Another advantage is that sexual reproduction permits the elimination of deleterious mutations. As Section 13.4 described, DNA replication is not perfect. Errors are introduced in every generation, and most of these errors result in lower fitness. Asexual organisms have no mechanism to eliminate deleterious mutations. Hermann J. Muller noted that the accumulation of deleterious mutations in a nonrecombining genome is like a genetic ratchet. The mutations accumulate—"ratchet up"—at each replication. A mutation occurs and is passed on when the genome replicates, then two new mutations occur in the next replication, so three mutations are passed on, and so on. Over time, the least-mutated class of individuals is lost from the population as new mutations occur. Deleterious mutations cannot be eliminated except by the death of the lineage or a rare back mutation. This accumulation of deleterious mutations in lineages that lack genetic recombination is known as **Muller's ratchet**. In sexual species, on the other hand, genetic recombination produces some individuals with more of these deleterious mutations and some with fewer. The individuals with fewer deleterious mutations are more likely to survive. Thus sexual reproduction allows natural selection to eliminate particular deleterious mutations from the population over time.

Another advantage of sex is the great variety of genetic combinations it creates in each generation. Sexual recombination does not directly influence the frequencies of alleles; rather, *it generates new combinations of alleles on which natural selection can act*. It expands variation in a character influenced by alleles at many loci by creating new genotypes. For example, genetic variation can be a defense against pathogens and parasites. Most pathogens and parasites have much shorter life cycles than their hosts and can rapidly evolve counteradaptations to host defenses. Sexual recombination can give the host's defenses a chance to keep up.

Frequency-dependent selection maintains genetic variation within populations

Natural selection often preserves variation as a polymorphism (the presence of two or more variants of a character in the same population). When the fitness of a given phenotype depends on its frequency in a population, a polymorphism may be maintained by a process known as **frequency-dependent selection**. *Perissodus microlepis*, a small fish that lives in Lake Tanganyika in East Africa, provides an example of frequency-dependent selection.

P. microlepis feeds on the scales of other fish, approaching its prey from behind and dashing in to bite off several scales

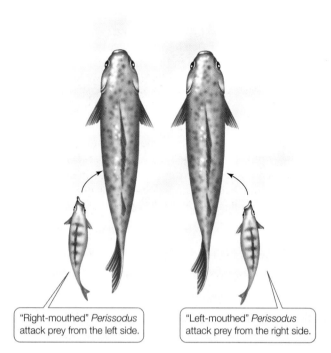

"Right-mouthed" *Perissodus* attack prey from the left side.

"Left-mouthed" *Perissodus* attack prey from the right side.

21.16 A Stable Polymorphism Frequency-dependent selection maintains equal proportions of left- and right-mouthed individuals of the scale-eating fish *Perissodus microlepis*.

from the prey's flank. Because of an asymmetrical jaw joint, the mouth of this scale-eating species opens either to the right or to the left; the direction is genetically determined (**Figure 21.16**). "Right-mouthed" individuals always attack from the victim's left, and "left-mouthed" individuals always attack from the victim's right. The distorted mouth enlarges the area of teeth in contact with the prey's flank, but only if the scale-eater attacks from the appropriate side.

Prey fish are alert to approaching scale-eaters, so attacks are more likely to be successful if the prey must watch both flanks. Vigilance by prey thus favors equal numbers of right- and left-mouthed scale-eaters in a population, because if attacks from one side were more common than the other, prey fish would pay more attention to potential attacks from that side. Over an 11-year study of *P. microlepis* in Lake Tanganyika, the genetic polymorphism was found to be stable, and the two phenotypes of the scale-eaters remained at about equal frequencies.

Heterozygote advantage maintains polymorphic loci

In many cases, different alleles of a particular gene are advantageous under different environmental conditions. Most organisms experience a wide variety of environmental conditions over time. A night is dramatically different from the preceding day. A cold, cloudy day differs from a clear, hot one. Day length and temperature change seasonally. For many genes, a single allele is unlikely to perform well under all these conditions. In such cases, heterozygous individuals (with two different alleles) are likely to outperform individuals that are homozygous (with only one of those two alleles).

21.17 A Heterozygote Mating Advantage Among butterflies of the genus *Colias*, males that are heterozygous for two alleles of the PGI enzyme can fly farther under a broader range of temperatures than males that are homozygous for either allele. Does this ability give heterozygous males a mating advantage?[a]

HYPOTHESIS Heterozygous male *Colias* will have proportionally greater mating success than homozygous males.

Method 1. For each of two *Colias* species, capture mated female butterflies in the field. In the laboratory, allow them to lay eggs.

2. Determine the genotypes of the females and their offspring, and thus the genotypes of the fathers.

3. Compare the frequency of heterozygotes among successfully mating males with the frequency of heterozygotes among all viable males (i.e., males captured flying with females).

Results For both species, the proportion of heterozygotes among the males that mated successfully was higher than the proportion of heterozygotes among all viable males.

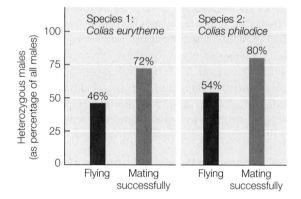

CONCLUSION Heterozygous *Colias* males have a mating advantage over homozygous males.

Go to **BioPortal** for discussion and relevant links for all INVESTIGATING**LIFE** figures.

[a]Watt, W. B. et al. 1985. *Genetics* 109:157–175.

Colias butterflies of the Rocky Mountains live in environments where dawn temperatures are often too cold, and afternoon temperatures too hot, for the butterflies to fly. Populations of these butterflies are polymorphic for a gene that encodes the enzyme phosphoglucose isomerase (PGI), which influences how well a butterfly flies at different temperatures. Butterflies with certain PGI genotypes can fly better during the cold hours of early morning; others perform better during midday heat. The optimal body temperature for flight is 35°C–39°C, but some butterflies can fly with body temperatures as low as 29°C or as high as 40°C. During spells of unusually hot weather, heat-tolerant genotypes are favored; during spells of unusually cool weather, cold-tolerant genotypes are favored.

Heterozygous *Colias* butterflies can fly over a greater range of temperatures than homozygous individuals, which should give them an advantage in foraging and finding mates. A test of this prediction found that heterozygous males did indeed have a mating advantage, and further, that this advantage maintains the polymorphism in the population (**Figure 21.17**).

WORKING WITH**DATA:**

Do Heterozygous Males Have a Mating Advantage?

Original Paper

Watt, W. B., P. A. Carter and S. M. Blower. 1985. Adaptation at specific loci. IV. Differential mating success among glycolytic allozyme genotypes of *Colias* butterflies. *Genetics* 109: 157–175.

Analyze the Data

Ward Watt and his colleagues tested the hypothesis that males with two different alleles for the PGI enzyme (heterozygotes) were more likely to mate successfully with females than were homozygous males. They reasoned that the heterozygous males could fly farther under a broader range of temperatures than could homozygous males, and that this ability would give heterozygous males greater access to receptive females. To test this hypothesis, they needed to know the frequency of heterozygotes among successfully mating males, and they needed to compare that frequency with the frequency of heterozygotes among males in the general population (i.e. all the potential mates available to females). To estimate the frequency of heterozygotes among mating males, Watt et al. collected mated female butterflies in the field and allowed them to lay eggs in the laboratory. They hatched the eggs and determined the genotypes of the offspring, as well as the genotypes of the females. Using this information, they could determine the genotypes of the males that fathered the larvae. They then compared the estimated frequency of heterozygotes among the successful fathers with the frequency of heterozygotes among all viable males in the population. Samples of their data are given in the table.

QUESTION 1

If we assume that the proportions of each genotype among mating males should be the same as the proportions seen among all viable males, what is the number of *mating males* expected to be heterozygous in each sample?

QUESTION 2

Use a chi-square test (see Appendix B) to evaluate the significance of the difference in the observed and expected numbers of heterozygous and homozygous individuals among the mating males. The critical value ($P = 0.05$) of the chi-square distribution with one degree of freedom is 3.841. Are the observed numbers of genotypes among mating males significantly different ($P < 0.05$) from the expected numbers in these samples?

QUESTION 3

The investigators determined the genotypes of enough larvae from each batch of eggs to judge the genotype of the father with 99% certainty. How many larvae did they need to measure to achieve that level of certainty?

Hint: If the female is homozygous—say, genotype *ii*—the number needed is small. However, if a female is heterozygous—say, genotype *ij*—and only *ii* and *ij* progeny are found among her offspring, more larvae need to be genotyped. In this particular case, the father can be only *ii* or *ij*. If he were *ij*, the probability that any one offspring is *not jj* = 0.75, so the chance of getting only *ii* and *ij* among *n* offspring is 0.75^n. What value of *n* is required to reduce the probability of error in determining the father's genotype to 0.01?

Species	All viable males[a]		Mating males	
	Heterozygous/ total	Percent heterozygous	Heterozygous/ total	Percent heterozygous
C. philodice	32/74	43.2	31/50	62.0
C. eurytheme	44/92	47.8	45/59	76.3

[a]"Viable males" are males captured flying with females (hence with the potential to mate)

Go to BioPortal for all WORKING WITH**DATA** exercises

Of course, the heterozygous genotype can never become fixed in the population, because the offspring of two heterozygotes will always include both classes of homozygotes in addition to heterozygotes.

Genetic variation within species is maintained in geographically distinct populations

Much of the genetic variation within species is preserved as differences among members living in different places (populations). Populations often vary genetically because they are subjected to different selective pressures in different environments. Environmental conditions may vary significantly even over short distances. For example, in the Northern Hemisphere, temperature and soil moisture differ dramatically between north- and south-facing mountain slopes. In the Rocky Mountains of Colorado, the proportion of ponderosa pines (*Pinus ponderosa*) that are heterozygous for a particular peroxidase enzyme is particularly high on south-facing slopes, where temperatures fluctuate dramatically, often on a daily

basis. This heterozygous genotype performs well over a broad range of temperatures. On north-facing slopes and at higher elevations, where temperatures are cooler and fluctuate less strikingly, a peroxidase homozygote, which has a lower optimal temperature, is much more frequent.

Plant species may also vary geographically in the chemicals they synthesize to defend themselves against herbivores. Some individuals of the white clover (*Trifolium repens*) produce the poisonous chemical cyanide. Poisonous individuals are less appealing to herbivores—particularly mice and slugs—than are nonpoisonous individuals. However, clover plants that produce cyanide are more likely to be killed by frost, because freezing damages cell membranes and releases cyanide into the plant's own tissues.

In European populations of *Trifolium repens*, the frequency of cyanide-producing individuals increases gradually from north to south and from east to west (**Figure 21.18**). Such a pattern of gradual change in phenotype across a geographic gradient is known as **clinal variation**. In the white clover cline,

The proportion of cyanide-producing individuals increases gradually along a gradient from colder to milder winters.

−13.3°C

White lines (isotherms) connect points with equal January mean temperatures.

−8.9°C

4.4°C

0°C

2.0°C   −4.4°C

8.0°C

Plants produce cyanide   Plants do not produce cyanide

21.18 Geographic Variation in a Defensive Chemical   The proportion of cyanide-producing individuals in European populations of white clover (*Trifolium repens*) depends on winter temperatures.

poisonous plants make up a large proportion of populations only in areas where winters are mild. Cyanide-producing individuals are rare where winters are cold, even though herbivores graze clovers heavily in those areas.

▇▇▇▇▇▇▇▇▇▇▇▇▇▇▇▇▇▇▇▇▇▇ **RECAP** 21.4

Neutral mutations, sexual recombination, frequency-dependent selection, and heterozygote advantage all act to maintain considerable genetic variation in most populations. Variation within species is also maintained among geographically distinct populations.

- Why is sexual reproduction is so prevalent in nature, despite its having at least three short-term evolutionary disadvantages? **See p. 441**
- How does frequency-dependent selection act to maintain genetic variation in a population? **See pp. 441–442**

The mechanisms of evolution have produced a remarkable variety of organisms. There are organisms that have adapted to nearly every environment on Earth. This natural variation, along with the success of breeders attempting to produce desired traits in domesticated plants and animals, suggests that evolution can produce a wide variety of adaptive traits. But are there limits to the adaptations evolution can produce?

21.5 What Are the Constraints on Evolution?

We would be mistaken to assume that evolutionary mechanisms can produce any trait we might imagine. Evolution is constrained in many ways. Lack of appropriate genetic variation, for example, prevents the development of many potentially favorable traits. If the allele for a given trait does not exist in a population, that trait cannot evolve, even if it would be highly favored by natural selection. Most possible combinations of genes and genotypes have never existed in any population and so have never been tested under natural selection.

In addition, constraints are imposed on organisms by the dictates of physics and chemistry. The size of cells, for example, is constrained by the stringencies of surface area-to-volume ratios (see Figure 5.2). The ways in which proteins can fold are limited by the bonding capacities of their constituent molecules (see Section 2.2). And the energy transfers that fuel life must operate within the laws of thermodynamics (see Section 8.1). Keep in mind that evolution works within the boundaries of these universal constraints as well as the constraints described in this section.

Developmental processes constrain evolution

As Section 20.5 explained, developmental constraints on evolution are paramount because *all evolutionary innovations are modifications of previously existing structures*. Human engineers seeking to power an airplane can start "from scratch" to design a completely new type of engine (powered by jet propulsion) to replace the previous type (powered by propellers). Evolutionary changes, however, cannot happen in this way. Current phenotypes of organisms are constrained by historical conditions and past selective pressures.

A striking example of such developmental constraints is provided by the evolution of fishes that spend most of their time on the sea bottom, where a ventrally flattened body is advantageous. One such lineage, the bottom-dwelling skates and rays, shares a common ancestor with sharks, whose bodies are already somewhat ventrally flattened and whose skeletal frame is made of flexible cartilage. Skates and rays evolved a body type that further flattened their bellies, allowing them to swim along the ocean floor (**Figure 21.19A**).

By contrast, plaice, sole, and flounder are bottom-dwelling descendants of deep-bellied, laterally flattened ancestors with bony skeletons. The only way these fishes can lie flat is to flop over on their sides. Their ability to swim is thus curtailed, but their bodies can lie still and are well camouflaged. During development, one eye of these flatfishes moves so that both eyes are positioned on the same side of the body (**Figure 21.19B**). Such shifts in eye position have evolved several times, and shifts have happened in both directions (that is, both left- and right-eyed flatfishes have evolved independently). Small shifts in the position of one eye probably helped ancestral flatfishes see better, resulting in the body forms found today. This path to producing a flattened body may not be optimal, but the fishes' developmental capabilities constrain the pathways that evolution can take.

(A) *Taeniura lymma*

(B) *Bothus lunatus*

21.19 Two Solutions to a Single Problem　(A) This stingray, whose ancestors were dorsoventrally flattened, lies on its belly. Stingrays' bodies are symmetrical around the dorsal backbone. (B) The flounder, whose ancestors were laterally flattened, lies on its side. (The backbone of this individual is at the right.) Flounders' eyes migrate during development so that both are on the same side of the body.

Trade-offs constrain evolution

Adaptations frequently impose both costs and benefits. For an adaptation to be favored, the fitness benefits it confers must exceed the fitness costs it imposes—in other words, the **trade-off** must be worthwhile. For example, there are metabolic costs associated with developing and maintaining certain conspicuous features (such as antlers or horns) that males use to compete with other males for access to females. The fact that these features are common in many species suggests that the benefits derived from possessing them must outweigh the costs.

As a result of trade-offs, many traits that are adaptive in one context may be maladaptive in another. Consider the rough-skinned newt and one of its predators, the common garter snake (**Figure 21.20A**). The newt sequesters a potent neurotoxin called tetrodotoxin (TTX) in its skin. TTX paralyzes nerves and muscles by blocking sodium channels (see Section 6.3). Most vertebrates—including many garter snakes—will die if they eat a rough-skinned newt. But some garter snakes can eat rough-skinned newts and survive: TTX-resistant sodium channels have evolved in the nerves and muscles of such individuals. However, for several hours after eating a newt, TTX-resistant snakes can

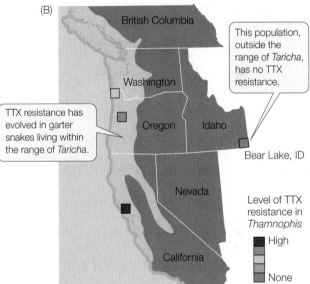

21.20 Resistance to a Toxin Comes at a Cost　(A) Garter snakes (*Thamnophis sirtalis*) prey on rough-skinned newts (*Taricha granulosa*). Rough-skinned newts defend themselves by sequestering a neurotoxin, TTX, in their skin. In turn, TTX-resistant sodium channels have evolved in some snake populations, allowing the snakes to eat toxic newts but resulting in slower movement by the snakes. (B) High TTX resistance in garter snakes is found only in regions where snake and newt populations overlap (tan area).

Go to Animated Tutorial 21.4
Assessing the Costs of Adaptation
Life10e.com/at21.4

move only slowly, and they never move as fast as nonresistant snakes. Resistant snakes are thus more vulnerable to their own predators than are TTX-sensitive snakes. This vulnerability leads to selection *against* TTX-resistant sodium channels in garter snake populations that occur outside the range of rough-skinned newts, even though there is selection *for* TTX resistance in many areas where newts are present (**Figure 21.20B**).

Short-term and long-term evolutionary outcomes sometimes differ

The short-term changes in allele frequencies within populations that we have emphasized in this chapter, often termed **microevolutionary** changes, are an important focus of study for evolutionary biologists. These changes can be observed directly, they can be manipulated experimentally, and they demonstrate the actual processes by which evolution occurs. By themselves, however, they may not be sufficient to predict long-term, or **macroevolutionary**, changes.

Long-term patterns of evolutionary change can be strongly influenced by events that occur so infrequently (such as a meteorite impact) or so slowly (such as continental drift) that they are unlikely to be observed during short-term studies. The evolutionary mechanisms at work may change over time with changing environmental conditions. Even among the descendants of a single ancestral species, different lineages may evolve in different directions. Additional types of evidence—evidence demonstrating the effects of rare and unusual events on trends in the fossil record—must be gathered if we wish to understand the course of evolution over billions of years. We will consider these long-term aspects of evolution in the remaining chapters of this section.

 RECAP 21.5

Developmental processes constrain evolution because all evolutionary innovations are modifications of previously existing structures. An adaptation can evolve only if the fitness benefits it confers exceed the fitness costs it imposes.

- Describe an example of an evolutionary trade-off in which the advantages of an adaptation outweigh its costs in the long run. **See p. 445 and Figure 21.19**

- How could the presence of a great deal of genetic variation within a population increase the chances that some members of the population would survive an unprecedented environmental change? Why is there no guarantee that this would be the case?

How do biologists use evolutionary theory to develop better flu vaccines?

ANSWER

Many different strains of influenza virus circulate among human populations and other vertebrate hosts each year, but only a few of those strains survive and produce descendants. One of the ways in which influenza strains differ is in the configuration of proteins on their surface. These surface proteins are the targets of recognition by the host immune system. When changes occur in the surface proteins of an influenza virus, the host immune system may no longer recognize the invading virus, so that virus is more likely to replicate successfully. Those virus strains with the greatest number of changes to their surface proteins are most likely to escape detection by the host immune system, and are therefore most likely to spread among the host population and result in future flu epidemics. In other words, there is positive selection for change in the surface proteins of influenza viruses.

By comparing the survival and proliferation rates of influenza virus strains that have different gene sequences coding for surface proteins, biologists can study the adaptation of the viruses over time. If biologists can predict which of the currently circulating flu virus strains are most likely to escape host detection, then they can identify the strains that are most likely to be involved in upcoming influenza epidemics and can target those strains for vaccine production.

How can biologists make such predictions? By determining the ratio of synonymous to nonsynonymous substitutions in genes that encode viral surface proteins, biologists can detect which codon changes (i.e., mutations) are under positive selection (using methods we will discuss in Section 24.2). They can then assess which of the currently circulating flu strains show the greatest number of changes in these positively selected codons. It is these flu strains that are most likely to survive, proliferate, and lead to the flu epidemics of the future, so they are the logical targets for new vaccines. This practical application of evolutionary theory leads to more effective flu vaccines—and thus fewer illnesses and influenza-related deaths each year.

 CHAPTER**SUMMARY 21**

21.1 What Is the Relationship between Fact and Theory in Evolution?

- **Evolution** is genetic change in populations over time. Evolution can be observed directly in living populations as well as in the fossil record of life.

- **Evolutionary theory** refers to our understanding of the mechanisms of evolutionary change.

- Charles Darwin in best known for his ideas on the common ancestry of divergent species and on **natural selection** (the differential survival and reproduction of individuals based on variation in their traits) as a mechanism of evolution. **See ANIMATED TUTORIAL 21.1, ACTIVITY 21.1**

- Since Darwin's time, many biologists have contributed to the development of evolutionary theory, and rapid progress in our understanding continues today. **Review Figure 21.2**

- For a population to evolve, its members must possess heritable genetic variation.

21.2 What Are the Mechanisms of Evolutionary Change?

- **Mutation** is the source of the genetic variation on which mechanisms of evolution act.

- The term **adaptation** refers both to a trait that evolves through natural selection and to the process that produces such traits.

continued

- Within populations, natural selection acts to increase the frequency of beneficial alleles (**positive selection**) and to decrease the frequency of deleterious alleles (**purifying selection**).

- Movement of individuals or gametes between populations results in **gene flow**.

- In small populations, **genetic drift**—the random loss of individuals and the alleles they possess from one generation to the next—may produce large changes in allele frequencies over time and greatly reduce genetic variation. **See ANIMATED TUTORIAL 21.2**

- **Population bottlenecks** occur when only a few individuals survive a random event, resulting in a drastic shift in allele frequencies within the population and the loss of genetic variation. Similarly, a population established by a small number of individuals colonizing a new region may lose genetic variation via a **founder effect. Review Figure 21.7**

- Nonrandom mating may result in changes in genotype and allele frequencies in a population.

- **Sexual selection** results from differential reproductive success based on individuals' phenotypes. **Review Figure 21.9**

 How Do Biologists Measure Evolutionary Change?

- Allele frequencies measure the amount of genetic variation in a population. Genotype frequencies show how a population's genetic variation is distributed among its members. Together, allele and genotype frequencies describe a population's **genetic structure. Review Figure 21.10**

- **Hardy–Weinberg equilibrium** predicts genotype frequencies from allele frequencies in the absence of evolution. Deviation from these frequencies indicates that evolutionary mechanisms are at work. **Review Figure 21.11, ANIMATED TUTORIAL 21.3**

- Natural selection can act on characters with quantitative variation in three different ways. **Review Figure 21.12**

- **Stabilizing selection** acts to reduce variation without changing the mean value of a trait.

- **Directional selection** acts to shift the mean value of a trait toward one extreme.

- **Disruptive selection** favors both extremes of trait values, resulting in a bimodal character distribution.

 How Is Genetic Variation Distributed and Maintained within Populations?

- Neutral mutations, sexual recombination, frequency-dependent selection, and heterozygote advantage can all maintain genetic variation within populations.

- **Neutral alleles** do not affect the fitness of an organism, are not affected by natural selection, and may accumulate or be lost by genetic drift.

- Despite its short-term disadvantages, sexual reproduction generates countless genotypic combinations that increase the evolutionary potential and survivorship of populations.

- A polymorphism may be maintained by **frequency-dependent selection** when the fitness of a genotype depends on its frequency in a population.

- Genetic variation within species may be maintained by the existence of genetically distinct populations over geographic space. A gradual change in phenotype across a geographic gradient is known as **clinal variation. Review Figure 21.18**

 What Are the Constraints on Evolution?

- Developmental processes constrain evolution because all evolutionary innovations are modifications of previously existing structures.

- Most adaptations impose costs as well as benefits. An adaptation can evolve only if the benefits it confers exceed the costs it imposes. **Review Figure 21.20, ANIMATED TUTORIAL 21.4**

 Go to the Interactive Summary to review key figures, Animated Tutorials, and Activities
Life10e.com/is21

CHAPTER**REVIEW**

▧ REMEMBERING

1. Long-horned cattle have greater difficulty moving through heavily forested areas compared with cattle that have short or no horns, but long-horned cattle are better able to defend their young against predators. This contrast is an example of
   a. an adaptation.
   b. genetic drift.
   c. natural selection.
   d. a trade-off.
   e. none of the above

2. Which statement about allele frequencies is *not* true?
   a. The sum of all allele frequencies at a locus is always 1.
   b. If there are two alleles at a locus and we know the frequency of one of them, we can obtain the frequency of the other by subtraction.
   c. If an allele is missing from a population, its frequency in that population is 0.
   d. If two populations have the same allele frequencies at a locus, they must have the same proportion of homozygotes at that locus.
   e. If there is only one allele at a locus, its frequency is 1.

3. Which of the following is *not* an assumption of Hardy–Weinberg equilibrium?
   a. There is no migration between populations.
   b. Natural selection is not acting on the alleles in the population.
   c. Mating is random.
   d. Multiple alleles must be present at every locus.
   e. All of the above

4. Laboratory selection experiments with fruit flies have demonstrated that
   a. bristle number is not genetically controlled.
   b. bristle number is not genetically controlled, but changes in bristle number are caused by the environment in which the fly is raised.
   c. bristle number is genetically controlled, but there is little variation on which natural selection can act.
   d. bristle number is genetically controlled, but selection cannot result in flies having more bristles than any individual in the original population had.
   e. bristle number is genetically controlled, and selection can result in flies having more, or fewer, bristles than any individual in the original population had.

5. Disruptive selection maintains a bimodal distribution of bill size in the black-bellied seedcracker because
   a. bills of intermediate sizes are difficult to form.
   b. the species' two major food sources differ markedly in size and hardness.
   c. males use their large bills in displays.
   d. migrants introduce different bill sizes into the population each year.
   e. older birds need larger bills than younger birds.

UNDERSTANDING & APPLYING

6. In what ways does artificial selection by humans differ from natural selection? Can you give some examples of a trait that might be favored by artificial selection in agriculture, but selected against by natural selection in a wild population?

7. As far as we know, natural selection cannot adapt organisms to future events. Yet many organisms appear to respond to natural events before they happen. For example, many mammals go into hibernation while it is still quite warm. Similarly, many birds leave the temperate zone for their southern wintering grounds long before winter has arrived. How do you think such "anticipatory" behaviors evolve?

8. As more humans live longer, many people face degenerative conditions such as Alzheimer's disease that (in most cases) are linked to advancing age. Assuming that some individuals may be genetically predisposed to successfully combat these conditions, is it likely that natural selection alone would act to favor such a predisposition in human populations? Why or why not?

ANALYZING & EVALUATING

9. The following sample lists the genotype at locus *A* for 10 individuals in a diploid population. Based on this sample, answer the questions that follow.
   a. Sample: *AA, AA, Aa, Aa, Aa, Aa, aa, aa, aa, aa*
   b. What is the observed frequency of allele *a*? The observed frequency of allele *A*?
   c. What are the observed frequencies of genotypes *aa, Aa,* and *AA*?
   d. After one generation of random mating, what would be the Hardy–Weinberg expectations for the frequencies of genotypes *aa, Aa,* and *AA*?
   e. What are some of the reasons you might expect the observed genotype frequencies to differ from the Hardy–Weinberg expectations?

10. Imagine you are studying a color polymorphism in a species of mice; in this species, some individuals have black coats and some have white coats. You want to know if the mice are mating randomly or if there is mate selection based on coat color. You decide to examine genotype frequencies at *a locus that is unrelated to coat color*. You collect the following data from mice sampled in a single location that potentially represents a single breeding population.
    a. In a sample of 25 white-coated mice: 1 mouse has genotype *aa*, 4 have genotype *Aa*, and 20 have genotype *AA*.
    b. In a sample of 25 black-coated mice: 24 mice have genotype *aa* and 1 has genotype *Aa*.

    Do your data support random mating between black mice and white mice? Can you think of other explanations for your data that do not involve nonrandom mate selection? How might you test and distinguish these various hypotheses?

Go to BioPortal at **yourBioPortal.com** for Animated Tutorials, Activities, LearningCurve Quizzes, Flashcards, and many other study and review resources.

22 Reconstructing and Using Phylogenies

Multiple Fluorescences The reef-building coral *Acropora millepora* shows both cyan and red fluorescences. This photograph was taken under a microscope that affects the colors we see; the colors are perceived differently by marine animals in their natural environment.

GREEN FLUORESCENT PROTEIN (GFP) was discovered in 1962 when Osamu Shimomura, an organic chemist and marine biologist, led a team that was able to extract the protein from the tissues of the bioluminescent jellyfish *Aequorea victoria* and purify it. Some 30 years after its initial discovery, Martin Chalfie had the idea (and the technology) to link the gene for GFP to other protein-coding genes so that the expression of specific genes of interest could be visualized in glowing green within cells and tissues of living organisms (see Figure 18.4). This work was extended by Roger Tsien, who changed some of the amino acids within GFP to create fluorescent proteins of several distinct colors. Different-colored fluorescent proteins meant that the expression of a number of different proteins could be visualized and studied in the same organism at the same time. These three scientists were awarded the 2008 Nobel Prize in Chemistry for the isolation of GFP and its development for visualizing gene expression.

Although Tsien was able to produce different-colored proteins, he could not produce a *red* protein. This was frustrating; a red fluorescent protein would have been particularly useful to biologists because red light penetrates tissues more easily than do other colors. Tsien's work inspired Mikhail Matz to look for new fluorescent proteins in corals (which are relatives of the jellyfishes). Among the different coral species he studied, Matz found proteins that fluoresced in various shades of green, cyan (blue-green)—and red.

How had fluorescent red pigments evolved among the corals, given that the necessary molecular changes had eluded Tsien? To answer this question, Matz sequenced the genes of the fluorescent proteins and used these sequences to reconstruct the evolutionary history of the amino acid changes that produced different colors in different species of corals.

Matz's work showed that the ancestral fluorescent protein in corals was green, and that red fluorescent proteins evolved in a series of gradual steps. His analysis of evolutionary relationships allowed him to retrace these steps. Such an evolutionary history, often depicted as a branching diagram of relationships among lineages, is called a phylogeny.

The evolution of many aspects of an organism's biology can be studied using phylogenetic methods. This information is used in all fields of biology to understand the structure, function, and behavior of organisms.

How are phylogenetic methods used to resurrect protein sequences from extinct organisms?

See answer on p. 464.

22.1 What Is Phylogeny?

Phylogeny is the history of evolutionary relationships among organisms or their genes. A **phylogenetic tree** is a diagram that portrays a reconstruction of that history. Phylogenetic trees are commonly used to depict the evolutionary histories of species, populations, and genes. Each branching point (or **node**) in a phylogenetic tree represents a point at which lineages diverged in the past. In the case of species, these splits represent past speciation events, when one lineage divided into two. Thus a phylogenetic tree can trace evolutionary relationships from the ancient common ancestor of a group of species through the various speciation events (when lineages split) up to the present populations of the organisms (**Figure 22.1**). Over the past several decades, phylogenetic trees have become important tools for studying and describing evolutionary patterns and for applying evolutionary theory throughout biology. You will need to understand phylogenetic trees to comprehend many articles and books about biology, including this one.

A phylogenetic tree may be used to portray the evolutionary history of all life forms; of a major evolutionary group (such as the insects); of a small group of closely related species; or in some cases, even of individuals, populations, or genes within a species. The common ancestor of all the organisms in the tree forms the **root** of the tree.

The phylogenetic trees in this book depict time flowing from left (earliest) to right (most recent) (**Figure 22.2A**). It is also common practice to draw trees with the earliest times at the bottom. The timing of a splitting event in a lineage is shown by the position of a node on the time axis, (sometimes called the divergence axis). These splits represent events in which one lineage diverged into two, such as a speciation event (for a tree of species), a gene duplication event (for a tree of genes), or a transmission event (for a tree of viral lineages transmitted through a host population). The divergence axis may have an explicit scale or simply show the relative timing of splitting events.

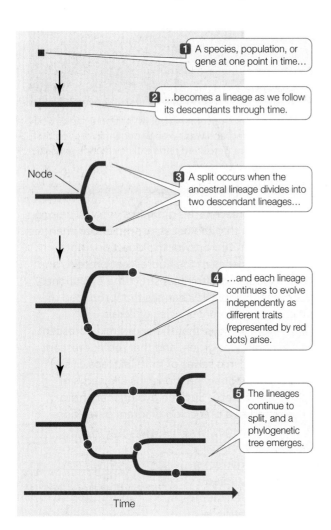

22.1 The Components of a Phylogenetic Tree Evolutionary relationships among organisms can be represented in a treelike diagram.

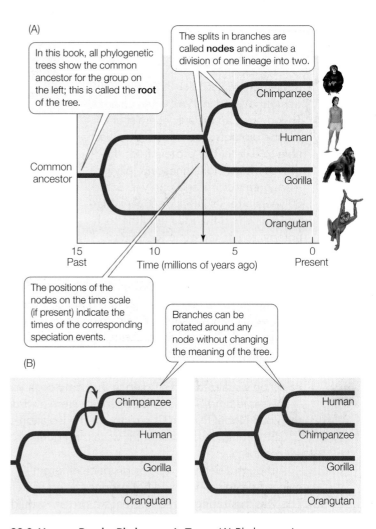

22.2 How to Read a Phylogenetic Tree (A) Phylogenetic trees can be produced with time scales, as shown here, or with no indication of time. If no time scale is shown, then the trees are only meant to depict the relative order of divergence events. (B) Lineages can be rotated around a given node, so the vertical order of taxa is largely arbitrary.

In this book, the order of nodes along the horizontal (time) axis has meaning, but the vertical distance between the branches does not. Vertical distances are adjusted for legibility and clarity of presentation; they do not correlate with the degree of similarity or difference between groups. Note too that lineages can be rotated around nodes in the tree, so the vertical order of lineages is also largely arbitrary (**Figure 22.2B**). The important information in the tree is the branching order along the time axis, as this indicates when the various lineages last shared a common ancestor.

Any species or group of species that we designate or name is called a **taxon** (plural taxa). Some examples of familiar taxa include humans, primates, mammals, and vertebrates (note that in this series, each taxon in the list is also a member of the next, more inclusive taxon). Any taxon that consists of an ancestor and all of its evolutionary descendants is called a **clade**. Clades can be identified by picking any point on a phylogenetic tree and then tracing all the descendant lineages to the tips of the terminal branches (**Figure 22.3**). Two species that are each other's closest relatives are called **sister species**; and any two clades that are each other's closest relatives are called **sister clades**.

Before the 1980s, phylogenetic trees tended to be seen only in the literature on evolutionary biology, especially in the area of **systematics**: the study and classification of biodiversity. But

almost every journal in the life sciences published during the last few years contains phylogenetic trees. Trees are widely used in molecular biology, biomedicine, physiology, behavior, ecology, and virtually all other fields of biology. Why have phylogenetic studies become so important?

All of life is connected through evolutionary history

In biology, we study life at all levels of organization—from genes, cells, organisms, populations, and species to the major divisions of life. In most cases, however, no individual gene or organism (or other unit of study) is exactly like any other gene or organism that we investigate.

Consider the individuals in your biology class. We recognize each person as an individual, but we know that no two are exactly alike. If we knew everyone's family tree in detail, the genetic similarity of any pair of students would be more predictable. We would find that more closely related students have more traits in common (from the color of their hair to their susceptibility or resistance to diseases). Similarly, biologists use phylogenies to make comparisons and predictions about shared traits across genes, populations, and species.

One of the great unifying concepts in biology is that all life is connected through its evolutionary history. The complete evolutionary history of life is known as the **tree of life**. Biologists estimate that there are tens of millions of species on Earth. Only about 1.8 million have been formally described and named. New species are being discovered and named all the time, and phylogenetic trees are continually being reviewed and revised. Thus our knowledge of the tree of life is far from complete, even for known species. Yet knowledge of evolutionary relationships is essential for making comparisons in biology, so biologists construct phylogenetic trees for groups of interest as the need arises. The evolutionary relationships among species, as shown in the tree of life, also form the basis for biological classification. This evolutionary framework allows biologists to make many predictions about the behavior, ecology, physiology, genetics, and morphology of species that have not yet been studied in detail.

Comparisons among species require an evolutionary perspective

When biologists make comparisons among species, they observe traits that differ within the group of interest and try to ascertain when those traits evolved. In many cases, investigators are interested in how the evolution of a trait depends on environmental conditions or selection pressures. For instance, scientists have used phylogenetic analyses to discover changes in the genome of HIV that confer resistance to particular drug treatments. The association of a particular genetic change in HIV with a particular drug treatment provides a hypothesis about the evolution of resistance that can be tested experimentally.

Any features shared by two or more species that have been inherited from a common ancestor are said to be **homologous**. Homologous features may be any heritable traits, including DNA sequences, protein structures, anatomical structures, and

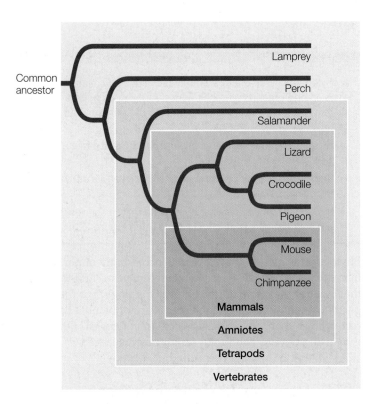

22.3 Clades Represent an Ancestor and All of Its Evolutionary Descendants All clades are subsets of larger clades, with all of life as the most inclusive taxon. In this example, the groups called mammals, amniotes, tetrapods, and vertebrates represent successively larger clades. Only a few species within each clade are represented on the tree.

even some behavior patterns. Traits that are shared across a group of interest are likely to have been inherited from a common ancestor. For example, all living vertebrates have a vertebral column, and all known fossil vertebrates had a vertebral column; thus the vertebral column is judged to be homologous in all vertebrates.

In tracing the evolution of a trait, biologists distinguish between ancestral and derived traits. A trait that was present in the ancestor of a group is known as an **ancestral trait** for that group. A trait found in a descendant that differs from its ancestral form is a **derived trait**. Derived traits that are shared among a group of organisms and are viewed as evidence of the common ancestry of that group are called **synapomorphies** (*syn*, "shared"; *apo*, "derived"; *morph*, "form," referring to the "form" of a trait). Thus the vertebral column is considered a synapomorphy—a shared, derived trait—of the vertebrates.

A particular trait may be ancestral or derived, depending on our phylogenetic point of reference. For example, all birds have feathers, which are highly modified scales. We infer from this fact that feathers were present in the common ancestor of modern birds, and therefore we consider the presence of feathers to be an *ancestral* trait for any group of modern birds (such as the songbirds). Feathers are not present in any other living animals, although there is fossil evidence for the presence of feathers in many extinct species of theropod dinosaurs. If we were reconstructing the phylogeny of all vertebrates, the presence of feathers would be a *derived* trait that informs us about the close evolutionary relationships between birds and their extinct theropod relatives.

Not all similar traits are evidence of relatedness, however. Similar traits can develop in distantly related groups of organisms for either of the following reasons:

- Independently evolved traits subjected to similar selection pressures may become superficially similar, a phenomenon called **convergent evolution**. For example, although the wing bones of bats and birds are homologous, having been inherited from a common ancestor, the wings of bats and the wings of birds—both adaptations for flight—are not homologous because they evolved independently from the forelimbs of different nonflying ancestors (**Figure 22.4**).

- A character may revert from a derived state back to an ancestral state in an event called an **evolutionary reversal**. For example, most frogs lack teeth in the lower jaw, but the ancestor of frogs did have such teeth. Teeth have been regained in the lower jaw of one South American species and thus represent an evolutionary reversal in that species.

Similar traits in distantly related taxa generated by convergent evolution or by evolutionary reversals are called homoplastic traits, or **homoplasies**.

Go to Media Clip 22.1
Morphing Arachnids
Life10e.com/mc22.1

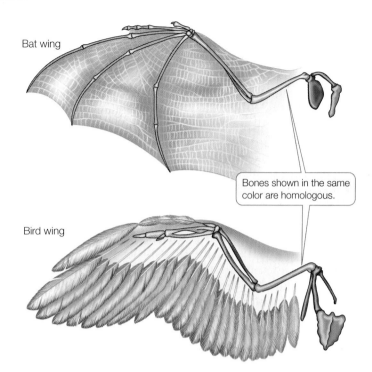

Bat wing

Bird wing

Bones shown in the same color are homologous.

22.4 The Bones Are Homologous, the Wings Are Not
The supporting bone structures of both bat wings and bird wings are derived from a common four-limbed ancestor and are thus homologous. However, the wings themselves—an adaptation for flight—evolved independently in the two groups.

■ RECAP 22.1

A phylogenetic tree is a description of evolutionary relationships among organisms or their genes. All living organisms share a common ancestor and are related through the phylogenetic tree of life.

- Describe the different elements of a phylogenetic tree. **See p. 450 and Figure 22.2**

- Explain the difference between an ancestral and a derived trait. **See p. 450**

- How might similar traits arise independently in species that are only distantly related? **See p. 452 and Figure 22.4**

Phylogenetic analyses have become increasingly important to many types of biological research in recent years, and they are the basis for the comparative nature of biology. For the most part, however, evolutionary history cannot be observed directly. How, then, do biologists reconstruct the past? One way is by using phylogenetic analyses to construct a tree.

22.2 How Are Phylogenetic Trees Constructed?

To illustrate how a phylogenetic tree is constructed, consider the eight vertebrate animals listed in **Table 22.1**: lamprey, perch, salamander, lizard, crocodile, pigeon, mouse, and chimpanzee.

TABLE 22.1

Eight Vertebrates and the Presence or Absence of Some Shared Derived Traits

Taxon	Derived Trait							
	Jaws	Lungs	Claws or nails	Gizzard	Feathers	Fur	Mammary glands	Keratinous scales
Lamprey (outgroup)	–	–	–	–	–	–	–	–
Perch	+	–	–	–	–	–	–	–
Salamander	+	+	–	–	–	–	–	–
Lizard	+	+	+	–	–	–	–	+
Crocodile	+	+	+	+	–	–	–	+
Pigeon	+	+	+	+	+	–	–	+
Mouse	+	+	+	–	–	+	+	–
Chimpanzee	+	+	+	–	–	+	+	–

We will assume initially that any given derived trait evolved only once during the evolution of these animals (that is, there has been no convergent evolution) and that no derived traits were lost from any of the descendant groups (there has been no evolutionary reversal). For simplicity, we have selected traits that are either present (+) or absent (–).

In a phylogenetic analysis, the group of organisms of primary interest is called the **ingroup**. As a point of reference, an ingroup is compared with an **outgroup**, a species or group that is closely related to the ingroup but is known to be phylogenetically outside it; the root of the tree is located between the ingroup and the outgroup. Any trait that is present in both the ingroup and the outgroup must have evolved before the origin of the ingroup and thus must be ancestral for the ingroup. In contrast, traits that are present only in some members of the ingroup must be derived traits within that ingroup. As we will see in Chapter 33, a group of jawless fishes called the lampreys is thought to have separated from the lineage leading to the other vertebrates before the jaw arose. Therefore we have included the lamprey as the outgroup for our analysis. Because derived traits are traits acquired by other members of the vertebrate lineage *after* they diverged from the outgroup, any trait that is present in both the lamprey and the other vertebrates is judged to be ancestral.

We begin by noting that the chimpanzee and mouse share two derived traits—mammary glands and fur—that are absent in both the outgroup and the other species of the ingroup. We then infer that mammary glands and fur are derived traits that evolved in a common ancestor of chimpanzees and mice after that lineage separated from the lineages leading to the other vertebrates. In other words, we provisionally assume that mammary glands and fur evolved only once among the animals in our ingroup. These traits are synapomorphies that unite chimpanzees and mice (as well as all other mammals, although we have not included other mammalian species in this example). By the same reasoning, we can infer that the other shared derived traits are

synapomorphies for the various groups in which they are expressed. For instance, keratinous scales are a synapomorphy of the lizard, crocodile, and pigeon.

Table 22.1 also tells us that, among the animals in our ingroup, the pigeon has a unique trait: the presence of feathers. As we discussed in Section 22.1, feathers are a synapomorphy of birds and some of their extinct dinosaur relatives. But since we only have one bird (and no extinct species) in this example, the presence of feathers provides no clues concerning relationships among the eight species of vertebrates we have sampled. Gizzards, however, are found in both birds and crocodiles, so this trait is evidence of a close relationship between birds and crocodilians.

By combining information about the various synapomorphies, we can construct a phylogenetic tree. We infer, for example, that mice and chimpanzees, the only two animals that share fur and mammary glands in our example, share a more recent common ancestor with each other than they do with pigeons and crocodiles. Otherwise we would need to assume that the ancestors of pigeons and crocodiles also had fur and mammary glands but subsequently lost them—unnecessary additional assumptions.

Figure 22.5 shows a phylogenetic tree for the vertebrates listed in Table 22.1, based on the shared derived traits we examined and the assumption that each derived trait evolved only once. This particular tree was easy to construct because the animals and characters we chose met the assumptions that derived traits appeared only once and were never lost after they appeared. Had we included a snake in the group, our second assumption would have been violated, because we know that the lizard ancestors of snakes had limbs that were subsequently lost. We would need to examine additional characters to determine that the lineage leading to snakes separated from the one leading to lizards long after the lineage leading to lizards separated from the others. In fact, the analysis of several characters shows that snakes evolved from burrowing lizards that became adapted to a subterranean existence.

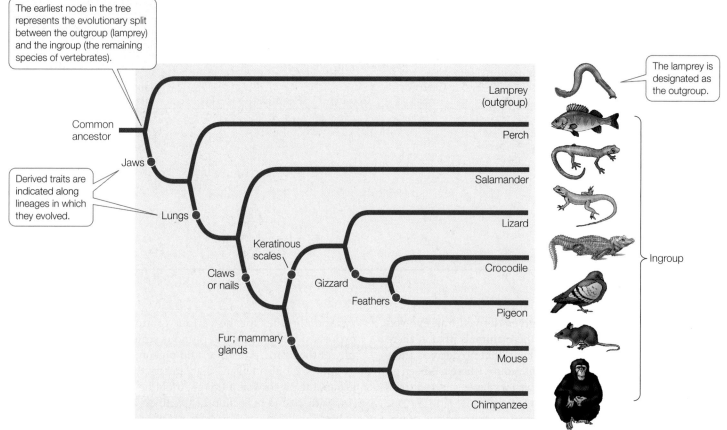

The earliest node in the tree represents the evolutionary split between the outgroup (lamprey) and the ingroup (the remaining species of vertebrates).

The lamprey is designated as the outgroup.

Derived traits are indicated along lineages in which they evolved.

22.5 Constructing a Phylogenetic Tree This phylogenetic tree was constructed from the information given in Table 22.1 using the parsimony principle. Each clade in the tree is supported by at least one shared derived trait, or synapomorphy.

Go to Activity 22.1 Constructing a Phylogenetic Tree
Life10e.com/ac22.1

Parsimony provides the simplest explanation for phylogenetic data

The phylogenetic tree shown in Figure 22.5 is based on only a very small sample of traits. Typically, biologists construct phylogenetic trees using hundreds or thousands of traits. With larger data sets, we would expect to observe some traits that have changed more than once, and thus we would expect to see some convergence and evolutionary reversal. How do we determine which traits are synapomorphies and which are homoplasies? One way is to invoke the principle of parsimony.

In its most general form, the principle of **parsimony** states that the preferred explanation of our observations is the simplest explanation. Applying the parsimony principle to the construction of phylogenetic trees entails minimizing the number of evolutionary changes that need to be assumed over all characters in all taxa in the tree. In other words, the best hypothesis under the parsimony principle is the one that requires the fewest homoplasies. This application of parsimony is a specific case of a general principle of logic called Occam's razor: the best explanation is the one that fits the data best while making the fewest assumptions.

We apply the parsimony principle in constructing phylogenetic trees not because all evolutionary change occurs parsimoniously, but because it is logical to adopt the simplest explanation that can account for the observed data. More complicated explanations are accepted only when the evidence requires them. Phylogenetic trees represent our best estimates about evolutionary relationships. They are continually modified as additional evidence becomes available.

Phylogenies are reconstructed from many sources of data

Naturalists have constructed various forms of phylogenetic trees for more than 150 years. In fact, the only figure in the first edition of *On the Origin of Species* was a phylogenetic tree. Tree construction has been revolutionized, however, by the advent of computer software for trait analysis and tree construction, allowing us to consider far more data than could ever before be processed. By combining these methods with the massive comparative data sets being generated through studies of genomes, biologists are learning details about the tree of life at a remarkable pace (see Appendix A: The Tree of Life).

Any trait that is genetically determined, and therefore heritable, can be used in a phylogenetic analysis. Evolutionary relationships can be revealed through studies of morphology, development, the fossil record, behavioral traits, and molecular traits such as DNA and protein sequences. Let's take a closer look at the types of data used in modern phylogenetic analyses.

Go to Animated Tutorial 22.1 Using Phylogenetic Analysis to Reconstruct Evolutionary History
Life10e.com/at22.1

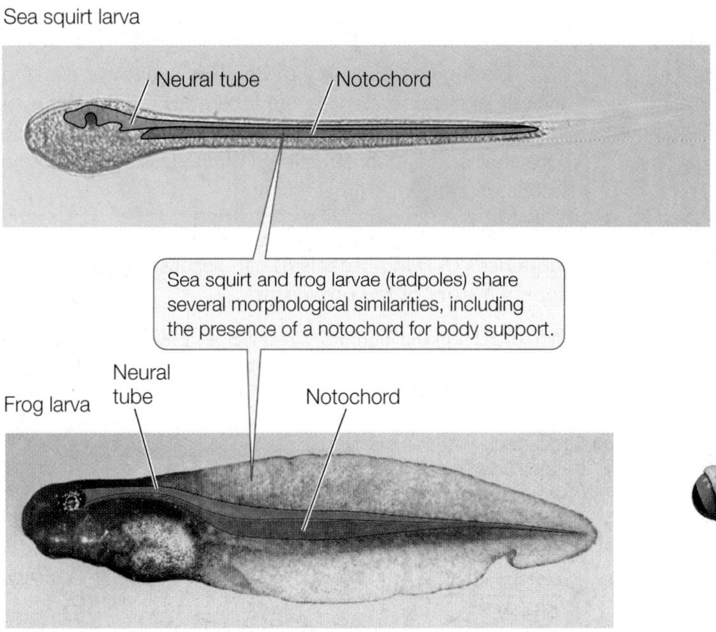

Sea squirt larva

Adult

Sea squirt and frog larvae (tadpoles) share several morphological similarities, including the presence of a notochord for body support.

Frog larva

Adult

Despite the similarity of their larvae, the morphology of adult frogs and sea squirts provides little evidence of the common ancestry of these two groups.

22.6 Development Reveals the Evolutionary Relationship between Sea Squirts and Vertebrates All chordates—a taxonomic group that includes sea squirts and frogs—have notochords at some stage of their development. The larvae share similarities that are not apparent in the adults. Such similarities in development can provide useful evidence of evolutionary relationships. The notochord is lost in adult sea squirts. In adult frogs, as in all vertebrates, the vertebral column replaces the notochord as the support structure.

MORPHOLOGY An important source of phylogenetic information is **morphology**: the presence, size, shape, and other attributes of body parts. Since living organisms have been observed, depicted, collected, and studied for millennia, we have a wealth of recorded morphological data as well as extensive museum and herbarium collections of organisms whose traits can be measured. New technological tools, such as the electron microscope and computed tomography (CT) scans, enable systematists to examine and analyze the structures of organisms at much finer scales than was formerly possible.

Most species are described and known primarily by their morphology, and morphology provides the most comprehensive data set available for many taxa. The features of morphology that are important for phylogenetic analysis are often specific to a particular group of organisms. For example, the presence, development, shape, and size of various features of the skeletal system are important for the study of vertebrate phylogeny, whereas floral structures are important for studying the relationships among flowering plants.

Morphological approaches to phylogenetic analysis have some limitations, however. Some taxa exhibit little morphological diversity despite great species diversity. For example, the phylogeny of the leopard frogs of North and Central America would be difficult to infer from morphological differences alone because the many species look very similar, despite important differences in their behavior and physiology. At the other extreme, few morphological traits can be compared across distantly related species (earthworms and mammals, for instance). Furthermore, some morphological variation has

an environmental (rather than a genetic) basis and so must be excluded from phylogenetic analyses. An accurate phylogenetic analysis often requires information beyond that supplied by morphology.

DEVELOPMENT Similarities in developmental patterns may reveal evolutionary relationships. Some organisms exhibit similarities in early developmental stages only. The larvae of marine creatures called sea squirts, for example, have a flexible gelatinous rod in the back—the notochord—that disappears as the larvae develop into adults. All vertebrate animals also have a notochord at some time during their development (**Figure 22.6**). This shared structure is one of the reasons for inferring that sea squirts are more closely related to vertebrates than would be suspected if only adult sea squirts were examined.

PALEONTOLOGY The fossil record is another important source of information on evolutionary history. Fossils show us where and when organisms lived in the past and give us an idea of what they looked like. Fossils provide important evidence that helps us distinguish ancestral from derived traits. The fossil record can also reveal when lineages diverged and began their independent evolutionary histories. Furthermore, in groups with few species that have survived to the present, information on extinct species is often critical to an understanding of the large divergences among the surviving species. The fossil record does have limitations, however. Few or no fossils have been found for some groups, and the fossil record for many groups is fragmentary.

BEHAVIOR Some behavioral traits are culturally transmitted and some are inherited. If a particular behavior is culturally transmitted, it may not accurately reflect evolutionary relationships (but may nonetheless reflect cultural connections). Bird songs, for instance, are often learned and may be inappropriate traits for phylogenetic analysis. Frog calls, however, are genetically determined and appear to be acceptable sources of information for reconstructing phylogenies.

MOLECULAR DATA All heritable variation is encoded in DNA, so the complete genome of an organism contains an enormous set of traits (the individual nucleotide bases of DNA) that can be used in phylogenetic analyses. In recent years, DNA sequences have become among the most widely used sources of data for constructing phylogenetic trees. Comparisons of nucleotide sequences are not limited to the DNA in the cell nucleus. Eukaryotes have genes in their mitochondria as well as in their nuclei; plant cells also have genes in their chloroplasts. The chloroplast genome (cpDNA), which is used extensively in phylogenetic studies of plants, has changed slowly over evolutionary time, so it is often used to study relatively ancient phylogenetic relationships. Most animal mitochondrial DNA (mtDNA) has changed more rapidly, so mitochondrial genes have been used extensively to study evolutionary relationships among closely related animal species (the mitochondrial genes of plants evolve more slowly). Many nuclear gene sequences are also commonly analyzed, and now that many entire genomes have been sequenced, they too are used to construct phylogenetic trees. Information on gene products (such as the amino acid sequences of proteins) is also widely used for phylogenetic analyses, as we will see in Chapter 24.

Mathematical models expand the power of phylogenetic reconstruction

As biologists began to use DNA sequences to construct phylogenetic trees in the 1970s and 1980s, they developed explicit mathematical models describing how DNA sequences change over time. These models account for multiple changes at a given position in a DNA sequence. They also take into account different rates of change at different positions in a gene, at different positions in a codon, and among different nucleotides (see Section 24.1). For example, transitions (changes between two purines or between two pyrimidines) are usually more likely than are transversions (changes between a purine and pyrimidine).

Mathematical models can be used to compute how a tree might evolve given the observed data. **Maximum likelihood** methods identify the tree that is most likely to have produced the observed data, given the assumptions of the model. Maximum likelihood methods can be used for any kind of characters, but they are most often used with molecular data, for

INVESTIGATING**LIFE**

22.7 Testing the Accuracy of Phylogenetic Analysis To test whether analysis of gene sequences can accurately reconstruct evolutionary phylogeny, we must have an unambiguously known phylogeny to compare against the reconstruction. Will the observed phylogeny match the reconstruction?[a]

HYPOTHESIS A phylogeny reconstructed by analyzing the DNA sequences of living organisms can accurately match the known evolutionary history of the organisms.

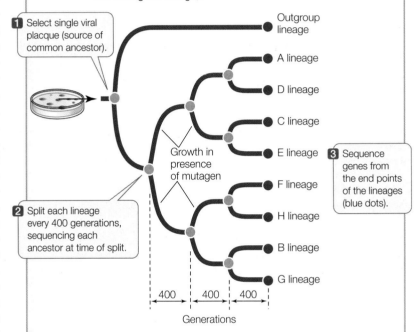

Method In the laboratory, researchers produced an unambiguous phylogeny of nine viral lineages, enhancing the mutation rate to increase variation among the lineages.

1 Select single viral placque (source of common ancestor).

2 Split each lineage every 400 generations, sequencing each ancestor at time of split.

Growth in presence of mutagen

3 Sequence genes from the end points of the lineages (blue dots).

Outgroup lineage
A lineage
D lineage
C lineage
E lineage
F lineage
H lineage
B lineage
G lineage

400 | 400 | 400

Generations

Viral sequences from the end points of each lineage (blue dots) were subjected to phylogenetic analysis by investigators who were unaware of the history of the lineages or the gene sequences of the ancestral viruses. These investigators reconstructed the phylogeny and ancestral DNA sequences based solely on their analyses of the descendants' genomes.

Results The true phylogeny and ancestral DNA sequences were accurately reconstructed solely from the DNA sequences of the viruses at the tips of the tree branches.

CONCLUSION Phylogenetic analysis of DNA sequences can accurately reconstruct evolutionary history.

Go to **BioPortal** for discussion and relevant links for all INVESTIGATING**LIFE** figures.

[a]Hillis, D. M. et al. 1992. *Science* 255: 589–592.

which explicit mathematical models of evolutionary change are easier to develop. The principal advantages of maximum likelihood analyses are that they incorporate more information about evolutionary change than do parsimony methods, and they are easier to treat in a statistical framework. The principal disadvantages are that they are computationally intensive and require explicit mathematical models of evolutionary change (which are difficult to develop for some kinds of characters).

WORKING WITH**DATA:**

Does Phylogenetic Analysis Correctly Reconstruct Evolutionary History?

Original Papers

Hillis, D. M., J. J. Bull, M. E. White, M. R. Badgett, and I. J. Molineux. 1992. Experimental phylogenetics: Generation of a known phylogeny. *Science* 255: 589–592.

Bull, J. J., C. W. Cunningham, I. J. Molineux, M. R. Badgett, and D. M. Hillis. 1993. Experimental molecular evolution of bacteriophage T7. *Evolution* 47: 993–1007.

Analyze the Data

Refer to the description of Hillis and colleagues' experiment with T7 virus below and in Figure 22.7. The full DNA sequences for the viral lineages produced in this experiment are thousands of nucleotides long. However, 23 of the nucleotide positions are shown in the table below, and you can use these data to repeat the researchers' analysis. Each nucleotide position represents a separate character.

QUESTION 1

Construct a phylogenetic tree from the nucleotide positions using the parsimony principle (see Section 22.2 and the examples in Table 22.1 and Figure 22.5). Use the outgroup to root your tree. Assume that all changes among nucleotides are equally likely.

QUESTION 2

Using your tree from Question 1, can you reconstruct the DNA sequences of the ancestral lineages?

QUESTION 3

Why did the investigators use a blind study design, in which the true identities of the viral lineages were not revealed until the analyses were complete? What potential for bias were they avoiding?

QUESTION 4

Transitions are mutations that change one purine to the other (G ↔ A) or one pyrimidine to the other (C ↔ T), whereas transversions exchange a purine for a pyrimidine or vice versa (e.g., A → C or T; C → A or G). Which kind of mutation predominates in this phylogeny? Why might this be the case?

Lineage	\multicolumn{23}{c}{Character at position}
	1	2	3	4	5	6	7	8	9	10	11	12	13	14	15	16	17	18	19	20	21	22	23
A	T	C	G	G	G	C	C	C	C	C	C	C	A	A	C	C	G	A	T	A	C	A	A
B	C	C	G	G	G	T	C	C	C	T	C	C	G	A	T	T	A	G	C	G	T	G	G
C	C	C	G	G	G	C	C	C	T	C	C	T	A	A	C	C	G	G	T	A	C	A	A
D	T	C	A	G	G	C	C	C	C	C	C	C	A	A	C	C	G	A	T	A	C	A	A
E	C	T	G	G	G	C	C	C	C	C	C	T	A	A	C	C	G	G	T	A	C	A	A
F	C	T	G	A	A	C	C	C	C	C	C	C	G	A	C	T	G	G	C	G	C	G	G
G	C	C	G	G	G	T	T	C	C	T	C	C	G	A	T	T	A	G	C	G	C	G	G
H	C	C	G	G	A	C	C	C	C	C	C	C	G	C	C	T	G	G	C	G	C	G	G
Outgroup	C	C	G	G	G	C	C	T	C	C	T	C	G	A	C	C	G	G	C	A	C	G	G

Go to **BioPortal** for all WORKING WITH**DATA** exercises

The accuracy of phylogenetic methods can be tested

If phylogenetic trees represent reconstructions of past events, and if many of these events occurred before any humans were around to witness them, how can we test the accuracy of phylogenetic methods? Biologists have conducted experiments both in living organisms and with computer simulations that have demonstrated the effectiveness and accuracy of phylogenetic reconstruction methods.

In one such experiment, David Hillis, James Bull, and their colleagues at the University of Texas used a single viral culture of bacteriophage T7 as a starting point and allowed lineages to evolve from this ancestral virus in the laboratory (**Figure 22.7**). The initial culture was split into two separate lineages, one of which became the ingroup for analysis; the other lineage became the outgroup used for rooting the tree. Mutagens were added to the viral cultures to increase the mutation rate so that the amount of change and the degree of homoplasy would be typical of the organisms analyzed in average phylogenetic analyses. The lineages in the ingroup were split in two after every 400 generations and samples of the virus were saved for analysis at each of these branching points. The lineages were allowed to evolve until there were eight lineages in the ingroup. The investigators then sequenced samples from the end points of the eight lineages as well as from the ancestors at the branching points. They then gave the sequences from the end points to other investigators to analyze, without revealing the known history of the lineages or the sequences of the ancestral viruses.

After the phylogenetic analysis was completed, the investigators asked two questions: Did phylogenetic methods reconstruct the known history correctly, and were the sequences of the ancestral viruses reconstructed accurately? The answer in both cases was yes: the branching order of the lineages was reconstructed exactly as it had occurred, more than 98 percent of the nucleotide positions of the ancestral viruses were reconstructed correctly, and 100 percent of the amino acid changes in the viral proteins were reconstructed correctly.

The experiment shown in Figure 22.7 demonstrated that phylogenetic analysis was accurate under the conditions tested, but it did not examine all possible conditions. Other experimental studies have taken other factors into account, such as the sensitivity of phylogenetic analysis to convergence under similar environments or to highly variable rates of evolutionary change.

Computer simulations based on mathematical models of evolutionary change have also been used extensively to study the effectiveness of phylogenetic analysis. These studies too have confirmed the accuracy of phylogenetic methods and have been used to refine those methods and extend them to new applications.

RECAP 22.2

Phylogenetic trees can be constructed by using the parsimony principle to find the simplest explanation for phylogenetic data. Maximum likelihood methods incorporate more explicit mathematical models of evolutionary change to reconstruct evolutionary history.

- Describe the process of reconstructing a phylogenetic tree. **See p. 453 and Figure 22.5**

- What are two methods biologists have used to test whether phylogenetic trees provide accurate reconstructions of evolutionary history? **See pp. 457–458 and Figure 22.7**

Biologists in many fields now routinely reconstruct the phylogenetic relationships of organisms. Let's examine some of the many uses of these phylogenetic trees.

22.3 How Do Biologists Use Phylogenetic Trees?

Information about the evolutionary relationships among organisms is useful to scientists investigating a wide variety of biological questions. In this section we will illustrate how phylogenetic trees can be used to ask questions about the past and to compare aspects of the biology of organisms in the present.

Phylogenetic trees can be used to reconstruct past events

Reconstruction of past events is important for understanding many biological processes. In the case of zoonotic diseases (diseases caused by infectious organisms transmitted to humans from another animal host), for example, it is important to understand when, where, and how the disease first entered a human population. Human immunodeficiency virus (HIV) is the cause of such a zoonotic disease: acquired immunodeficiency syndrome, or AIDS. Phylogenetic analyses have become important for studying the transmission of viruses such as HIV. They are also important for understanding the present global diversity

of such viruses and for determining their origins in human populations.

A broad phylogenetic analysis of immunodeficiency viruses shows that humans acquired these viruses from two different hosts: HIV-1 from chimpanzees, and HIV-2 from sooty mangabeys (**Figure 22.8**). HIV-1 is the common form of the virus in human populations in central Africa, where chimpanzees are hunted for food, and HIV-2 is the common form in human populations in western Africa, where sooty mangabeys are hunted for food. Thus it seems likely that these viruses entered human populations through hunters who cut themselves while skinning chimpanzees or sooty mangabeys. The relatively recent global pandemic of AIDS occurred when these infections in local African populations rapidly spread through human populations around the world.

In recent years, phylogenetic analysis has become important in forensic investigations that involve viral transmission events. For example, phylogenetic analysis was critical for a criminal investigation of a physician who was accused of purposefully injecting blood from one of his HIV-positive patients into his former girlfriend in an attempt to kill her. The phylogenetic analysis revealed that the HIV strains present in the girlfriend were a subset of those present in the physician's patient (**Figure 22.9**). Other evidence was needed, of course, to connect the physician to this purposeful transmission event, but the phylogenetic analysis was important to support the contention that the virus had been transmitted from the patient to the victim.

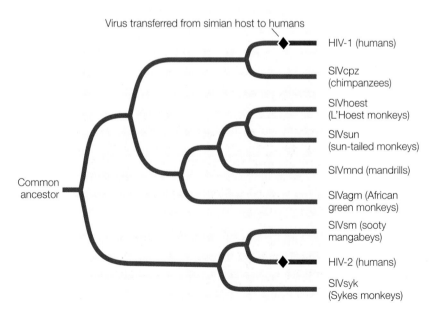

22.8 Phylogenetic Tree of Immunodeficiency Viruses   The evolutionary relationships of immunodeficiency viruses show that these viruses have been transmitted to humans from two different simian hosts: HIV-1 from chimpanzees and HIV-2 from sooty mangabeys. (SIV stands for simian immunodeficiency virus.)

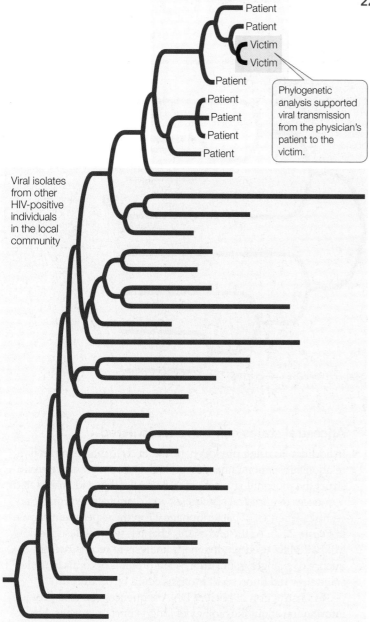

Patient
Patient
Victim
Victim

Phylogenetic analysis supported viral transmission from the physician's patient to the victim.

Patient
Patient
Patient
Patient
Patient

Viral isolates from other HIV-positive individuals in the local community

22.9 A Forensic Application of Phylogenetic Analysis This phylogenetic analysis demonstrated that strains of HIV virus present in a victim (shown in red) were a phylogenetic subset of viruses isolated from a physician's patient (shown in blue). This analysis was part of the evidence used to show that the physician drew blood from his HIV-positive patient and injected it into the victim in an attempt to kill her. The physician was found guilty of attempted murder by the jury.

Phylogenies allow us to compare and contrast living organisms

Male swordtails—a group of fishes in the genus *Xiphophorus*—have a long, colorful tail extension, and their reproductive success is closely associated with this appendage. Males with a long sword are more likely to mate successfully than are males with a short sword (an example of sexual selection; see Sections 21.2 and 23.5). Several explanations have been advanced for the evolution of this structure, including the hypothesis that the sword simply exploits a preexisting bias

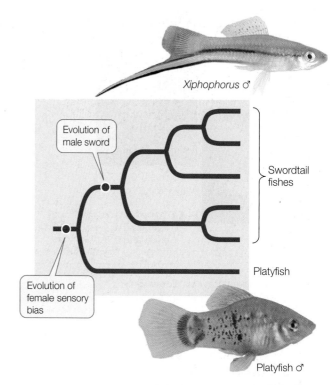

Xiphophorus ♂

Evolution of male sword

Swordtail fishes

Platyfish

Evolution of female sensory bias

Platyfish ♂

22.10 The Origin of a Sexually Selected Trait The tail extension of male swordtails (genus *Xiphophorus*) apparently evolved through sexual selection, as females mated preferentially with males that had long "swords." Phylogenetic analysis reveals that the platyfishes split from the swordtails before the evolution of the sword. The independent finding that female platyfish prefer male platyfish with an artificial sword further supports the idea that this appendage evolved as a result of a preexisting preference in females.

in the female sensory system—i.e., that female swordtails had a preference for males with long tails even before the tails evolved (perhaps because females assess the size of males by their total body length—including the tail—and prefer larger males).

To test this sensory exploitation hypothesis, phylogenetic reconstruction was used to identify the relatives of swordtails that had split most recently from their lineage before the evolution of swords. These closest relatives turned out to be the platyfishes, another group of *Xiphophorus*. Even though male platyfish do not normally have swords, when researchers attached artificial swordlike structures to the tails of some male platyfish, female platyfish preferred those males, thus providing support for the sensory exploitation hypothesis (**Figure 22.10**).

Phylogenies can reveal convergent evolution

Like most animals, many flowering plants (angiosperms) reproduce by mating with another individual of the same species. But in many angiosperm species, the same individual produces both male and female gametes (contained within pollen and ovules, respectively). Self-incompatible plant species have mechanisms to prevent fertilization of the ovule by the

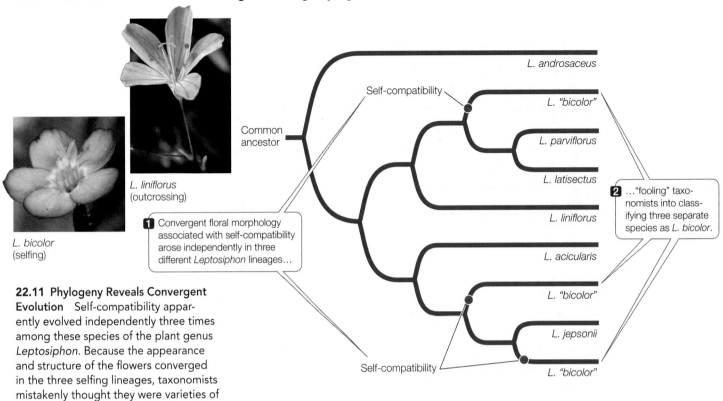

22.11 Phylogeny Reveals Convergent Evolution Self-compatibility apparently evolved independently three times among these species of the plant genus *Leptosiphon*. Because the appearance and structure of the flowers converged in the three selfing lineages, taxonomists mistakenly thought they were varieties of the same species.

individual's own pollen, and so must reproduce by outcrossing with another individual. Individuals of some species, however, regularly fertilize their ovules using their own pollen; they are referred to as self-fertilizing, or selfing, species, and their gametes as self-compatible.

The evolution of angiosperm fertilization mechanisms was examined in *Leptosiphon*, a genus in the phlox family that exhibits a diversity of mating systems and pollination mechanisms. The self-incompatible (outcrossing) species of *Leptosiphon* have long petals and are pollinated by long-tongued flies. In contrast, the self-pollinating species have short petals and do not require insect pollinators to reproduce successfully. Using nuclear ribosomal DNA sequences, investigators reconstructed the phylogeny of a subgroup of this genus (**Figure 22.11**). They then determined whether the gametes of each species were self-compatible by artificially pollinating flowers with the plant's own pollen or with pollen from other individuals and observing whether viable seeds formed.

The reconstructed phylogeny suggests that self-incompatibility is the ancestral state and that self-compatibility evolved three times within this group of *Leptosiphon*. The change to self-compatibility eliminated the plants' dependence on pollinators and was accompanied by the evolution of reduced petal size. Indeed, the striking morphological similarity of the flowers in the self-compatible taxa led to their being classified as members of a single species (*L. bicolor*). Phylogenetic analysis, however, showed them to be members of three distinct lineages. From this information, we can infer that self-compatibility and its associated floral structure are the result of convergent evolution in the three independent lineages that had been called *L. bicolor*.

Ancestral states can be reconstructed

In addition to using phylogenetic methods to infer evolutionary relationships among lineages, biologists can use them to reconstruct the morphology, behavior, or nucleotide and amino acid sequences of ancestral species (as was demonstrated for the ancestral sequences of bacteriophage T7 in the experiment shown in Figure 22.7). At the end of this chapter, we will describe how Mikhail Matz used phylogenetic analysis to reconstruct the sequence of changes in the fluorescent proteins of corals to understand how red fluorescent proteins could be produced.

Reconstruction of ancient DNA sequences can also provide information about the biology of long-extinct organisms. For example, phylogenetic analysis was used to reconstruct an opsin protein found in the ancestral archosaur (the most recent common ancestor of birds, dinosaurs, and crocodiles). Opsins are pigment proteins involved in vision; different opsins (with different amino acid sequences) are excited by different wavelengths of light. A team of investigators used a phylogenetic analysis of opsins from living vertebrates to estimate the amino acid sequence of the visual pigment that existed in the ancestral archosaur. A protein with that sequence was then constructed in the laboratory. The investigators tested the reconstructed opsin and found a significant shift toward the red end of the spectrum in the light sensitivity of this protein compared with that of most modern opsins. Modern species that exhibit similar opsin sensitivity are adapted for nocturnal vision, so the investigators inferred that the ancestral archosaur might have been active at night. Thus, reminiscent of the movie *Jurassic Park*, phylogenetic analyses are being used to reconstruct extinct species, one protein at a time.

 Go to Animated Tutorial 22.2
Phylogeny and Molecular Evolution
Life10e.com/at22.2

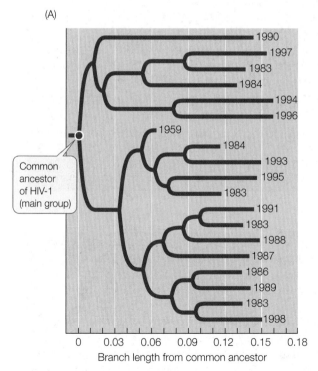

22.12 A Molecular Clock for the Protein Hemoglobin Amino acid replacements in hemoglobin have occurred at a relatively constant rate over nearly 500 million years of evolution. The graph shows the relationship between the time of divergence and the proportion of amino acids that have changed for 13 pairs of vertebrate hemoglobin proteins. The average rate of change represents the molecular clock for hemoglobin in vertebrates.

Molecular clocks help date evolutionary events

For many applications, biologists want to know not only the order in which evolutionary lineages diverged, but also the timing of those splits. In 1965, Emile Zuckerkandl and Linus Pauling hypothesized that rates of molecular change were constant enough that they could be used to predict evolutionary divergence times—an idea that has become known as the molecular clock hypothesis.

Of course, different genes evolve at different rates, and there are also differences in evolutionary rates among species related to generation times, environments, efficiencies of DNA repair systems, and other biological factors. Nonetheless, among closely related species, a given gene usually evolves at a reasonably constant rate, so the protein encoded by that gene does as well (**Figure 22.12**). A **molecular clock** is the average rate at which a given gene or protein accumulates changes, and this rate of change can be used to gauge the time of a particular split in the phylogeny. Molecular clocks must be calibrated using independent data, including the fossil record, known times of divergence, or biogeographic dates (such as the dates for separations of continents). Using such calibrations, times of divergence have been estimated for many groups of species that have diverged over millions of years.

Molecular clocks are used not only to date ancient events but also to study the timing of comparatively recent events. For example, most samples of HIV-1 have been collected from humans only since the early 1980s, although a few isolates from medical biopsies are available from as early as the 1950s. But biologists were able to use the observed changes in HIV-1 over the past several decades to extrapolate back to the common ancestor of a group of HIV-1 samples and estimate when HIV-1 first entered human populations from chimpanzees (**Figure 22.13**). Their molecular clock was calibrated using samples from the 1980s and

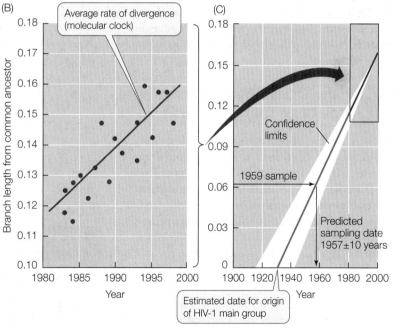

22.13 Dating the Origin of HIV-1 in Human Populations (A) A phylogenetic tree for samples of the main group of HIV-1 virus. The dates indicate the years in which the samples were taken. (For clarity, only a small fraction of the samples that were examined in the original study are shown.) (B) A plot of sample year versus genetic divergence from the common ancestor provided an average rate of divergence, or a molecular clock. (C) The molecular clock was used to date a sample taken in 1959 (as a test of the clock) and to estimate the date of origin of the HIV-1 main group (about 1930).

1990s, then tested using samples from the 1950s. As shown in Figure 22.13C, a sample from 1959 was dated by the molecular clock at 1957 ± 10 years. Extrapolation back to the common ancestor of the samples suggested a date of origin for this group of viruses of about 1930. Although AIDS was unknown to Western medicine until the 1980s, this analysis shows that HIV-1 was present (probably at a very low frequency) in human populations in Africa for at least a half-century before its emergence as a global pandemic. Biologists have used similar analyses to conclude that immunodeficiency viruses have been transmitted repeatedly into human populations from multiple primates for more than a century (see also Figure 22.8).

RECAP 22.3

Phylogenetic trees are used to reconstruct the past history of lineages, to determine when and where traits arose, and to make biological comparisons among genes, populations, and species. They can also be used to reconstruct ancestral traits and to estimate the timing of evolutionary events.

- Explain how phylogenetic trees can help determine the number of times a particular trait evolved. **See pp. 459–460 and Figure 22.11**

- How did the reconstruction of ancestral traits help biologists explain the evolution of visual pigment proteins? **See p. 460**

- How do molecular clocks add a time dimension to phylogenetic trees? **See p. 461 and Figure 22.12**

All of life is connected through evolutionary history, and the relationships among organisms provide a natural basis for making biological comparisons. For these reasons, biologists use phylogenetic relationships as the basis for organizing life into a coherent classification system, described in the next section.

22.4 How Does Phylogeny Relate to Classification?

The biological classification system in widespread use today is derived from a system developed by the Swedish biologist Carolus Linnaeus in the mid-1700s. Linnaeus developed a naming system called **binomial nomenclature** that has allowed scientists throughout the world to refer unambiguously to the same organisms by the same names (**Figure 22.14**).

22.14 Many Different Plants Are Called Bluebells
All three of these distantly related plant species are called "bluebells." Binomial nomenclature allows us to avoid the ambiguity of such common names and communicate exactly what is being described. (A) *Campanula rotundifolia*, found on the North American Great Plains, belongs to a larger group of bellflowers. (B) *Endymion non-scriptus*, English bluebell, is related to hyacinths. (C) *Mertensia virginica*, Virginia bluebell, belongs in a very different group of plants known as borages.

Linnaeus gave each species two names, one identifying the species itself and the other the genus to which it belongs. A **genus** (plural **genera**) is a group of closely related species. Optionally, the name of the taxonomist who first proposed the species name may be added at the end. Thus *Homo sapiens* Linnaeus is the name of the modern human species. *Homo* is the genus to which the species belongs, and *sapiens* identifies the particular species in the genus *Homo*; Linnaeus proposed the species name *Homo sapiens*. The name of the genus is always capitalized, and the name identifying the species is always lowercased. Both names are italicized, whereas common names (humans in this case) of organisms are not. Rather than repeating the name of a genus that is used several times in the same discussion, biologists often spell it out only once and abbreviate it to the initial letter thereafter (*D. melanogaster* rather than *Drosophila melanogaster*, for example).

As noted earlier, any group of organisms that is treated as a unit in a biological classification system, such as the genus *Drosophila*, or all insects, is called a taxon. In the Linnaean system, species and genera are further grouped into a hierarchical system of ranked taxonomic categories. The taxon above the genus in the Linnaean system is the **family**. The names of animal families end in the suffix "–idae." Thus Formicidae is the family that contains

(A) *Campanula rotundifolia*

(B) *Endymion non-scriptus*

(C) *Mertensia virginica*

all ant species, and the family Hominidae contains humans and our recent fossil relatives as well as our closest living relatives, the chimpanzees and gorillas. Family names are based on the name of a member genus but are not italicized; Formicidae is based on the genus *Formica*, and Hominidae is based on *Homo*. The same rules are used in classifying plants, except that the suffix "-aceae" is used for plant family names instead of "-idae." Thus Rosaceae is the family that includes the genus roses (*Rosa*) and other genera closely related to *Rosa*.

In the Linnaean system, families are grouped into **orders**, orders into **classes**, classes into **phyla** (singular phylum), and phyla into **kingdoms**. However, Linnaean classification is subjective—there are no explicit criteria for deciding if a particular taxon should be treated as an order or a family, for example—and today Linnaean terms are used largely for convenience. Although families are always grouped within orders, orders within classes, and so forth, there is nothing that makes a "family" in one group equivalent (in number of genera or in evolutionary age, for instance) to a "family" in another group.

Linnaeus recognized the overarching hierarchy of life, but he developed his system before evolutionary thought had become widespread. Biologists today recognize the tree of life as the basis for biological classification, and they often name taxa without placing them in any Linnaean rank. But regardless of whether they rank organisms in the Linnaean hierarchy or use unranked taxon names, modern biologists use evolutionary relationships as the basis for distinguishing and naming taxa.

Evolutionary history is the basis for modern biological classification

Biological classification systems are used to express relationships among organisms. The kind of relationship we wish to express influences which features we use to classify organisms. If, for instance, we were interested in a system that would help us decide what plants and animals were desirable as food, we might devise a classification system based on tastiness, ease of capture, and the number of edible parts each organism possessed. Early Hindu classifications of organisms were designed according to these criteria. Such systems served the needs of the people who developed them, but they are not adequate for formal scientific classification.

Biologists today use systems of classification to express the evolutionary relationships of organisms. Taxa are expected to be **monophyletic**, meaning that the taxon contains an ancestor and all descendants of that ancestor, and no other organisms (**Figure 22.15**). In other words, every taxon should be a complete branch on the tree of life (a clade; see Figure 22.3). Although biologists seek to describe and name only monophyletic taxa, the detailed phylogenetic information needed to do so is not always available. A group that does not include its

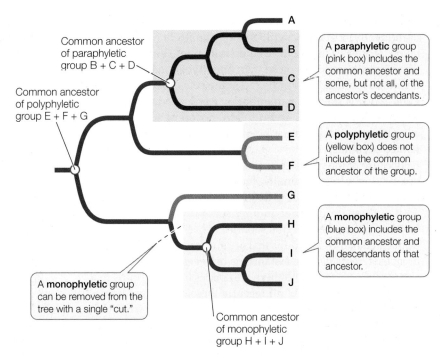

22.15 Monophyletic, Polyphyletic, and Paraphyletic Groups Monophyletic groups are the basis of taxa in modern biological classifications. Polyphyletic and paraphyletic groups are not appropriate for use in classifications because they do not accurately reflect evolutionary history.

Go to Activity 22.2 Types of Taxa Life10e.com/ac22.2

common ancestor is called a **polyphyletic** group. A group that does not include all the descendants of a common ancestor is called a **paraphyletic** group.

A true monophyletic group (i.e., a clade) can be removed from a phylogenetic tree by a single "cut" in the tree, as shown in Figure 22.15. Note that there are many monophyletic groups on any phylogenetic tree, and that these groups are successively smaller subsets of larger monophyletic groups. This hierarchy of taxa, with all of life as the most inclusive taxon and many smaller taxa within larger taxa down to individual species, is the modern basis for biological classification.

Virtually all taxonomists now agree that polyphyletic and paraphyletic groups are inappropriate as taxonomic units because they do not correctly reflect evolutionary history. The classifications used today still contain such groups, however, because some organisms have not been evaluated phylogenetically. As mistakes in prior classifications are detected, taxonomic names are revised and polyphyletic and paraphyletic groups are eliminated from the classifications.

Several codes of biological nomenclature govern the use of scientific names

Several sets of explicit rules govern the use of scientific names for organisms. Biologists around the world follow these rules voluntarily to facilitate communication and dialogue. Although there may be dozens of common names for an organism in many different languages, the rules of biological nomenclature are designed so that there is only one correct scientific name for any single recognized taxon and so that (ideally) a given scientific name applies to only a single taxon (that is, each scientific name is unique). Sometimes the same species is named more than once (when more than one taxonomist has taken up the task); the rules specify that the valid name is the first name that was proposed. If the same name is inadvertently given to two different species, then a replacement name must be given to the species that was named second.

Because of the historical separation of the fields of zoology, botany (including, originally, the study of fungi), and microbiology, different sets of taxonomic rules were developed for each of these groups. Yet another set of rules for classifying viruses emerged later. This separation has resulted in many duplicated names in groups that are governed by different sets of rules: *Drosophila*, for instance, is both a genus of fruit flies and a genus of fungi, and there are species in both groups that have identical names. Until recently these duplicated names caused little confusion, since traditionally biologists who studied fruit flies were unlikely to read the literature on fungi (and vice versa). Today, however, given the use of large, universal biological databases (such as GenBank, which includes DNA sequences from across the tree of life), it is increasingly important that each taxon have a unique name. Taxonomists are now working to develop common sets of rules that can be applied across all living organisms. For example, a universal system, known as the PhyloCode, that emphasizes the hierarchical and unified nature of the tree of life has been proposed for classifying all species of life.

 RECAP 22.4

Biologists organize and classify life by identifying and naming monophyletic groups. Several sets of rules govern the use of scientific names so that each species and higher taxon can be identified and named unambiguously.

- Explain the difference between monophyletic, paraphyletic, and polyphyletic groups. **See p. 463 and Figure 22.15**
- Why do biologists prefer monophyletic groups in formal classifications? **See p. 463**

Now that we have seen how evolution occurs and how phylogenies can be used to study evolutionary relationships, we are ready to consider the process of speciation. Speciation is what leads to the branching events on the tree of life, and it is the process that results in the millions of species that constitute biodiversity.

22.16 Evolution of Fluorescent Proteins of Corals Mikhail Matz and his colleagues used phylogenetic analysis to reconstruct the sequences of fluorescent proteins that were present in the extinct ancestors of modern corals. They then expressed these proteins in bacteria and plated the bacteria in the form of a phylogenetic tree to show how the colors evolved over time.

How are phylogenetic methods used to resurrect protein sequences from extinct organisms?

ANSWER

Most genes and proteins of organisms that lived millions of years ago have decomposed in the fossil remains of these species. Nonetheless, the sequences of many ancient genes and proteins can be reconstructed by the methods described in this chapter. As we saw in Section 22.3, just as we can reconstruct the morphological features of a clade's ancestors, we can reconstruct their DNA and protein sequences—if we have enough information about the genomes of their descendants. Biologists have reconstructed gene sequences from species that have been extinct for millions of years. Using this information, a laboratory can reconstruct real proteins that correspond to those sequences. This is how Mikhail Matz and his colleagues were able to resurrect fluorescent proteins from the extinct ancestors of modern corals, then visualize the colors produced by these proteins in the laboratory (**Figure 22.16**).

Biologists have even used phylogenetic analysis to reconstruct some protein sequences that were present in the common ancestor of life mentioned in Section 22.1. These hypothetical protein sequences were then made into actual proteins in the laboratory. When biologists measured the temperature optima for these resurrected proteins, they found that the proteins functioned best in the range of 55°C–65°C. This result is consistent with hypotheses that life evolved in a high-temperature environment.

To reconstruct protein sequences from species that have been extinct for millions or even billions of years, biologists use detailed mathematical models that take into account much of what we have learned about molecular evolution, as described in Section 22.2. These models incorporate information on rates of replacement among different amino acids in proteins, information on different substitution rates among nucleotides, and changes in the rate of molecular evolution among the major lineages of life.

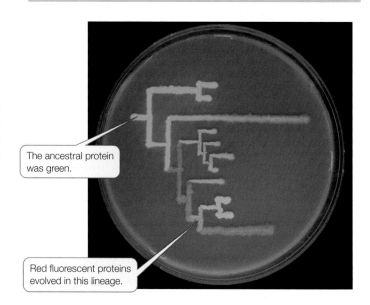

The ancestral protein was green.

Red fluorescent proteins evolved in this lineage.

CHAPTERSUMMARY 22

 22.1   What Is Phylogeny?

- **Phylogeny** is the history of evolutionary relationships among organisms or their genes. Groups of evolutionarily related species are represented as branches in a **phylogenetic tree**. Review Figures 22.1, 22.2

- Named species and groups of species are called **taxa**. A taxon that consists of an ancestor and all of its evolutionary descendants is called a **clade**. Review Figure 22.3

- **Homologies** are similar traits that have been inherited from a common ancestor. **Review Figure 22.4**

- A derived trait that is shared by two or more taxa and is inherited from their common ancestor is called a **synapomorphy**.

- Distantly related species may show similar traits that do not result from common ancestry. **Convergent evolution** and **evolutionary reversals** can give rise to such traits, which are called **homoplasies**.

 22.2   How Are Phylogenetic Trees Constructed?

- Phylogenetic trees can be constructed from synapomorphies using the logic of **parsimony**. Review Figure 22.5, ACTIVITY 22.1, ANIMATED TUTORIAL 22.1

- Sources of phylogenetic information include **morphology**, patterns of development, the fossil record, behavioral traits, and molecular traits such as DNA and protein sequences.

- Phylogenetic trees can also be constructed with **maximum likelihood** methods, which find the tree most likely to have generated the observed data.

- Phylogenetic methods have been tested in both experimental and simulation studies, and have been shown to be accurate under a wide variety of conditions.

 22.3   How Do Biologists Use Phylogenetic Trees?

- Phylogenetic trees are used to make comparisons among living organisms. **Review Figure 22.10**

- Phylogenetic trees are used to reconstruct the past and to understand the origin of traits. **Review Figure 22.11**

- Biologists can use phylogenetic trees to reconstruct ancestral states. **See ANIMATED TUTORIAL 22.2**

- Phylogenetic trees may include estimates of divergence times of lineages determined by **molecular clock** analysis. **Review Figure 22.13**

 22.4   How Does Phylogeny Relate to Classification?

- Biologists use phylogenetic relationships to organize life into a coherent classification system.

- Taxa in modern classifications are expected to be **monophyletic** groups. **Paraphyletic** and **polyphyletic** groups are not considered appropriate taxonomic units. **Review Figure 22.15, ACTIVITY 22.2**

- Several sets of rules govern the use of scientific names, with the goal of providing unique and universal names for taxa.

Go to the Interactive Summary to review key figures, Animated Tutorials, and Activities
Life10e.com/is22

CHAPTER**REVIEW**

REMEMBERING

1. Phylogenetic trees may be constructed for
   a. genes.
   b. species.
   c. major evolutionary groups.
   d. viruses.
   e. all of the above.

2. A shared derived trait, used as the basis for identifying a monophyletic group, is called
   a. a synapomorphy.
   b. a homoplasy.
   c. a parallel trait.
   d. a convergent trait.
   e. a phylogeny.

3. Convergent evolution and evolutionary reversal are two sources of
   a. homology.
   b. parsimony.
   c. synapomorphy.
   d. monophyly.
   e. homoplasy.

4. Taxonomists strive to describe and name only taxa that are
   a. monophyletic.
   b. paraphyletic.
   c. polyphyletic.
   d. homoplastic.
   e. monomorphic.

5. Which of the following groups have separate sets of rules for nomenclature?
   a. Animals
   b. Plants and fungi
   c. Bacteria
   d. Viruses
   e. All of the above

6. If two scientific names are proposed for the same species, how do taxonomists decide which name should be used?
   a. The name that provides the most accurate description of the organism is used.
   b. The name that was proposed most recently is used.
   c. The name that was used in the most recent taxonomic revision is used.
   d. The first name to be proposed is used.
   e. Taxonomists use whichever name they prefer.

▨ **UNDERSTANDING & APPLYING**

7. What is the problem with the classification shown below? How could the limits of Genus A and/or Genus B be modified so that both genera are monophyletic?

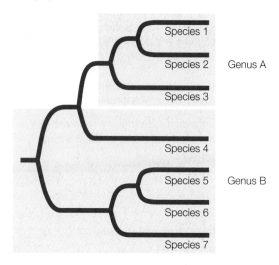

8. How are fossils helpful in identifying ancestral and derived traits of organisms? Describe an example.

9. In the figure shown below, what is the estimated average rate of change (expressed as the proportion of amino acid change per million years)?

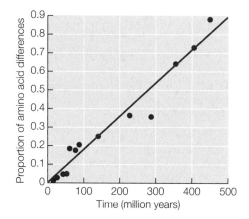

▨ **ANALYZING & EVALUATING**

10. West Nile virus kills birds of many species and can cause fatal encephalitis (inflammation of the brain) in humans and horses. The virus was first isolated in Africa (where it is thought to be endemic) in the 1930s, and by the 1990s it had been found throughout much of Eurasia. West Nile virus was not found in North America until 1999 (when it was first detected in New York), but since that time it has spread rapidly across most of the United States. Use the phylogenetic tree of West Nile virus isolates shown below to construct a hypothesis about the origin of the virus lineage that was introduced into the United States. The isolates are identified by their place and date of isolation.

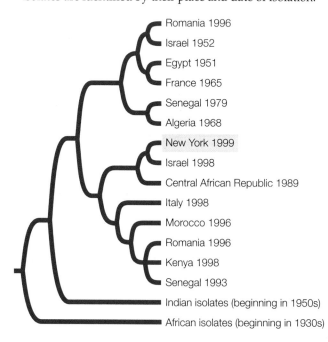

Go to BioPortal at **yourBioPortal.com** for Animated Tutorials, Activities, LearningCurve Quizzes, Flashcards, and many other study and review resources.

Speciation

CHAPTER**OUTLINE**

Many Species from One   This composite photograph shows a few of the nearly 1,000 species of haplochromine cichlids that are endemic to Lake Malawi, all of which arose from a single founder species.

NOT QUITE 2 MILLION YEARS AGO, a tectonic split in the Great Rift Valley of East Africa led to the formation of Lake Malawi, which lies between the modern countries of Malawi, Tanzania, and Mozambique. A few fish species entered the newly formed lake, including individuals of one species known as a haplochromine cichlid. Today the descendants of these "founding fish" include nearly a thousand distinct species of haplochromine cichlids. All of them are endemic to this single large and deep freshwater lake—they are not found anywhere else in the world. This vast array of cichlid species makes Lake Malawi the world's most diverse lake in terms of its fish fauna. How did so many different species arise from a single ancestral species in less than 2 million years?

By studying the history and timing of speciation events in Lake Malawi, biologists have pieced together some of the processes that led to so many cichlid species. The earliest haplochromine cichlids to enter the new lake encountered diverse habitats, as some of its shores were rocky while others were sandy. Different populations of the original cichlid species quickly adapted; fish in rocky habitats adapted to breeding and living in rocky conditions, and those in sandy habitats evolved specializations for life over sand. These changes resulted in an early speciation event.

Within each of these two major habitat types, there were numerous opportunities for diet specialization. Some populations of cichlids became rock scrapers, others became bottom feeders, fish predators,

scale-eaters, pelagic (open water) zooplankton eaters, or plant-feeding specialists. Each of these feeding specializations required different mouth morphology. The offspring of fish that bred with fish of similar morphology were more likely to survive than were fish with two very different parental morphologies. These differences in fitness led to the formation of many more new species, each adapted to a different feeding mode.

The Lake Malawi cichlids continued to diverge and form new species. Male cichlids competed for the attention of females through their bright body colors. Diversification of the body colors of males, and of the preferences of females for different body colors, led to more and more new species, each isolated from the other by their sexual preferences. Today biologists are studying the genomes of Lake Malawi cichlids to understand the details of the genetic changes that have given rise to so many species over so little time.

Can biologists study the processes of speciation in the laboratory?

See answer on p. 482.

23.1 What Are Species?

Biological diversity does not vary in a smooth, incremental way. People have long recognized groups of similar organisms that mate with one another, and they have noticed that there are usually distinct morphological differences between these groups. A group of organisms that can mate with one another and produce fertile offspring is commonly called a **species** (note that this is both the plural and singular form of the word). Species are the result of the process of **speciation**: the divergence of biological lineages and the emergence of reproductive isolation between those lineages.

Although "species" is a useful and common term, its usage varies among biologists who are interested in different aspects of speciation. Different biologists think about species differently because they ask different questions. How can we recognize and identify a species? How do new species arise? How do species remain distinct, especially from their recent close relatives? Why do rates of speciation differ among groups of organisms? In answering these questions, biologists focus on different attributes of species, leading to several ways of thinking about what species are and how they form. Most of the various **species concepts** proposed by biologists are simply different ways of approaching the question "What are species?"

We can recognize many species by their appearance

Someone who is knowledgeable about a group of organisms, such as birds or flowering plants, can usually distinguish the different species found in a particular area simply by looking at them. Standard field guides to birds, mammals, insects, and wildflowers are possible only because many species change little in appearance over large geographic distances (**Figure 23.1A**).

More than 250 years ago, Carolus Linnaeus developed the system of binomial nomenclature by which species are named today (see Section 22.4). Linnaeus described and named

thousands of species, but because he knew nothing about the genetics or the mating behavior of the organisms he was naming, he classified them on the basis of their appearance alone. In other words, Linnaeus used a **morphological species concept**, a construct that assumes that a species comprises individuals that "look alike" and that individuals that do not look alike belong to different species. Although Linnaeus could not have known it, the members of most of the groups he classified as species look alike because they share many alleles of the genes that code for their morphological features.

Using morphology to define species has limitations. Members of the same species do not always look alike. For example, males, females, and young individuals do not always resemble one another closely (**Figure 23.1B**). Furthermore, morphology is of little use in the case of cryptic species—instances in which two or more species are morphologically indistinguishable but do not interbreed (**Figure 23.2**). Biologists therefore cannot rely on appearance alone in determining whether individual organisms are members of the same or different species. Today, biologists use several additional types of information—especially behavioral and genetic data—to differentiate species.

Reproductive isolation is key

The most important factor in the divergence of sexually reproducing lineages from one another is the evolution of **reproductive isolation**, a state in which two groups of organisms can no longer exchange genes. If individuals of group A mate and reproduce only with one another, then group A constitutes a distinct species within which genes recombine. In other words, group A is an independent evolutionary lineage—a separate branch on the tree of life.

It was his recognition of the importance of reproductive isolation that brought evolutionary biologist Ernst Mayr to propose the **biological species concept**: *"Species are groups of actually or potentially interbreeding natural populations which are reproductively isolated from other such groups."* The phrase "actually or potentially" is an important element of this definition.

(A)

Aix sponsa
Male, Florida

Aix sponsa
Male, California

(B)

Aix sponsa
Female

23.1 Not All Members of the Same Species Look Alike (A) It is easy to identify these two male wood ducks as members of the same species, even though they are found on opposite coasts 2,000 miles apart. Despite their geographic separation, the two individuals are morphologically very similar. (B) Wood ducks are sexually dimorphic, which means the female's appearance is quite different from that of the male.

(A)

Hyla versicolor

(B)

Hyla chrysoscelis

23.2 Cryptic Species Look Alike but Do Not Interbreed These two species of gray tree-frogs cannot be distinguished by their external morphology, but they do not interbreed even when they occupy the same geographic range. *Hyla versicolor* is a tetraploid species (four sets of chromosomes), whereas *H. chrysoscelis* is diploid (two sets of chromosomes). And, although they look alike, the males have distinctive mating calls; female frogs recognize and mate with males of their own species based on these calls.

"Actually" says that the individuals live in the same area and interbreed with one another. "Potentially" says that even though the individuals do not live in the same area, and therefore do not interbreed, other information suggests that they *would* do so if they were able to get together. This widely used species concept does not apply to organisms that reproduce asexually, and it is limited to a single point in evolutionary time.

The lineage approach takes a long-term view

Evolutionary biologists often think of species as branches on the tree of life. This idea can be termed a **lineage species concept**. In this framework for thinking about species, one species splits into two descendant species, which thereafter evolve as distinct lineages. A lineage species concept allows biologists to consider species over evolutionary time.

A **lineage** is an ancestor–descendant series of populations followed over time. Each species has a history that starts with a speciation event by which one lineage is split into two and ends either with extinction or with another speciation event, at which time the species produces two descendant species. The process of lineage splitting is usually gradual, taking thousands of generations to complete. At the other extreme, an ancestral lineage may be split in two within a few generations (as happens with polyploidy, which we'll discuss in Section 23.3). The gradual nature of most splitting events means that at a single point in time, the final outcome of the process may not be clear. In these cases, it may be difficult to predict whether the incipient species will continue to diverge and become fully isolated from one another, or whether the two lineages will merge again in the future.

The different species concepts are not mutually exclusive

Many named variants of these three major classes of species concepts exist. These various concepts are not incompatible; they simply emphasize different aspects of species or speciation. The morphological species concept emphasizes the practical aspects of recognizing species, although it sometimes results in underestimation or overestimation of actual number of species. Mayr's biological species concept emphasizes that reproductive isolation is what allows sexual species to evolve independently of one another. The lineage species concept embraces the idea that sexual species are maintained by reproductive isolation, but extends the concept of a species as a lineage over evolutionary time. The lineage species concept is also able to accommodate species that reproduce asexually.

Virtually all species exhibit some degree of genetic recombination among individuals, even if recombination events are relatively rare. Significant reproductive isolation between species is therefore necessary for lineages to remain distinct over evolutionary time. Furthermore, reproductive isolation is responsible for the morphological distinctness of most species because mutations that result in morphological changes cannot spread between reproductively isolated species. Therefore, no matter which species concept we emphasize, the evolution of reproductive isolation is important for understanding the origin of species.

RECAP 23.1

Species are distinct lineages on the tree of life. Speciation is usually a gradual process as one lineage divides into two. Over time, lineages of sexual species remain distinct from one another because they have become reproductively isolated.

- Explain how the various species concepts emphasize different attributes of species. **See pp. 468–469**
- Why is the biological species concept not applicable to asexually reproducing organisms? **See pp. 468–469**
- Explain the role of reproductive isolation in each of the species concepts discussed in this section? **See pp. 468–469**

Although Charles Darwin titled his groundbreaking book *On the Origin of Species*, it included very little about speciation as we understand it today. Darwin devoted most of his attention to demonstrating that individual species are altered over time by natural selection. The remaining sections of this chapter discuss the many aspects of speciation that biologists have learned about since Darwin's time.

23.2 What Is the Genetic Basis of Speciation?

Not all evolutionary changes result in new species. A single lineage may change over time without giving rise to a new species. Speciation requires the interruption of gene flow within a species whose members formerly exchanged genes. But if a genetic change prevents reproduction between individuals of a species, how can such a change spread through a species in the first place?

Incompatibilities between genes can produce reproductive isolation

If a new allele that causes reproductive incompatibility arises in a population, it cannot spread through the population because no other individuals will be reproductively compatible with the individual that carries the new allele. So how can one reproductively cohesive lineage ever split into two reproductively isolated species? Several early geneticists, including Theodosius Dobzhansky and Hermann Joseph Muller, developed a genetic model to explain this apparent conundrum (**Figure 23.3**).

The Dobzhansky–Muller model is quite simple. First, assume that a single ancestral population is subdivided into two separate populations by some barrier to gene flow (by the formation of a new mountain range, for instance), and that these two populations then evolve as independent lineages. In one of the two populations, a new allele (*A*) arises and becomes fixed. In the other population, another new allele (*B*) becomes fixed *at a different gene locus*. Neither new allele at either locus results in any loss of reproductive compatibility. However, the two new forms of these two different genes have never occurred together in the same individual or population. Recall that the products of many genes must work together in an organism. It is possible that the new protein forms encoded by the two new alleles will not be compatible with each other. If individuals from the two populations come back together after these genetic changes, they may still be able to interbreed. However, the hybrid offspring may have a new combination of genes that is functionally inferior to that of either parent, or even lethal. This will not happen with all new combinations of genes, but over time, isolated populations will accumulate many allele differences at many gene loci. Some combinations of these differentiated genes will not function well together in hybrids. Thus genetic

incompatibility between the two isolated populations will develop over time.

Many empirical examples support the Dobzhansky–Muller model. This model works not only for pairs of individual genes, but also for some kinds of chromosomal rearrangements. Bats of the genus *Rhogeessa*, for example, exhibit considerable variation in centric fusions of their chromosomes. The chromosomes of the various species contain the same basic chromosomal arms, but in some species two acrocentric (one-armed) chromosomes have fused at the centromere to form larger, metacentric (two-armed) chromosomes. A polymorphism in centric fusion causes few, if any, problems in meiosis because the respective chromosomes can still align and assort normally. Therefore, a given centric fusion can become fixed in a lineage. However, if a *different* centric fusion becomes fixed in a second lineage, then hybrids between individuals of each lineage will not be able to produce normal gametes in meiosis (**Figure 23.4**). Most of the closely related species of *Rhogeessa* display different combinations of these centric fusions and are thereby reproductively isolated from one another.

Reproductive isolation develops with increasing genetic divergence

As pairs of species diverge genetically, they become increasingly reproductively isolated (**Figure 23.5**). Both the rate at which reproductive isolation develops and the mechanisms that produce it vary from group to group, as we'll see in the next two sections of this chapter. Reproductive incompatibility has been shown to develop gradually in many groups of plants, animals, and fungi, reflecting the slow pace at which incompatible genes accumulate in each lineage. In some cases, complete reproductive isolation may take millions of years. In other cases (as with the chromosomal fusions of *Rhogeessa* described above), reproductive isolation can develop over just a few generations.

Partial reproductive isolation has evolved in strains of *Phlox drummondii* artificially isolated by humans. In 1835, Thomas Drummond, after whom this species of garden plant is named, collected seeds in Texas and distributed them to nurseries in

23.3 The Dobzhansky–Muller Model In this simple two-locus version of the model, two lineages from the same ancestral population become physically separated from each other and evolve independently. A new allele becomes fixed in each descendant population, but at a different locus. Neither of the new alleles is incompatible with the ancestral alleles, but the two new alleles in the two different genes are incompatible with each other. Thus the two descendant lineages are reproductively incompatible.

Go to Animated Tutorial 23.1
Speciation Simulation
Life10e.com/at23.1

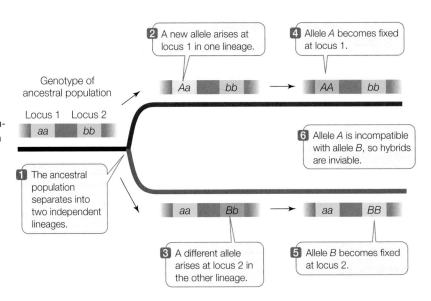

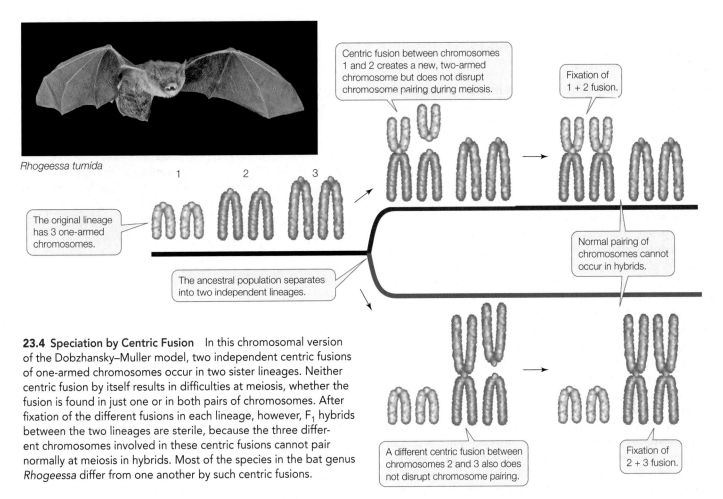

Rhogeessa tumida

23.4 Speciation by Centric Fusion In this chromosomal version of the Dobzhansky–Muller model, two independent centric fusions of one-armed chromosomes occur in two sister lineages. Neither centric fusion by itself results in difficulties at meiosis, whether the fusion is found in just one or in both pairs of chromosomes. After fixation of the different fusions in each lineage, however, F₁ hybrids between the two lineages are sterile, because the three different chromosomes involved in these centric fusions cannot pair normally at meiosis in hybrids. Most of the species in the bat genus *Rhogeessa* differ from one another by such centric fusions.

Europe. The European nurseries established more than 200 true-breeding strains of *P. drummondii* that differed in flower size, flower color, and plant growth form. The breeders did not select directly for reproductive incompatibility between

strains, but in subsequent experiments in which strains were crossed, biologists found that reproductive compatibility between strains (as measured by seed production) had been reduced by 14 to 50 percent, depending on the cross—even though the strains had been isolated from one another for less than two centuries.

23.5 Reproductive Isolation Increases with Genetic Divergence Among pairs of *Drosophila* species, the more the species differ genetically, the greater their reproductive isolation from each other. Each dot represents a comparison of one species pair. Such positive relationships between genetic distance and reproductive isolation have been observed in many groups of plants, animals, and fungi.

RECAP 23.2

When two parts of a population become isolated from each other by some barrier to gene flow, they begin to diverge genetically. The Dobzhansky-Muller model describes how new alleles or chromosomal arrangements that arise in the two descendent lineages can lead to genetically incompatibility, and hence reproductive isolation, of the two lineages.

- How can centric fusions occur in one *Rhogeessa* lineage without causing any reproductive difficulties, when different centric fusions in two different lineages of *Rhogeessa* lead to disruption of meiosis in hybrid individuals? **See p. 470 and Figure 23.4**

- What empirical evidence can you cite in support of the idea that genetic divergence of populations leads to loss of reproductive compatibility? **See pp. 470–471 and Figure 23.5**

We have now seen how the splitting of an ancestral population leads to genetic divergence and reproductive incompatibility in the two descendant lineages. How do populations become separated in the first place?

23.3 What Barriers to Gene Flow Result in Speciation?

Many biologists who study speciation have concentrated on geographic processes that can result in the division of an ancestral species. Splitting of the geographic range of a species is one obvious way of achieving such a division, but it is not the only way.

Physical barriers give rise to allopatric speciation

Speciation that results when a population is divided by a physical barrier is known as **allopatric speciation** (Greek *allos*, "other"; *patria*, "homeland"). Allopatric speciation is thought to be the dominant mode of speciation in most groups of organisms. The physical barrier that divides the range of a species may be a body of water or a mountain range for terrestrial organisms or dry land for aquatic organisms—in other words, any type of habitat that is inhospitable to the species. Such barriers can form when continents drift, sea levels rise and fall, glaciers advance and retreat, or climates change. The populations separated by such barriers are often, but not always, initially large. The lineages that descend from these founding populations evolve differences for a variety of reasons, including mutation, genetic drift, and adaptation to different environments in the two areas. As a result, many pairs of closely related **sister species**—species that are each other's closest relatives—may exist on either side of the geographic barrier. An example of a physical geographical barrier that produced many pairs of sister species was the Pleistocene glaciation that isolated freshwater streams in the eastern highlands of the Appalachian Mountains from streams in the Ozark and Ouachita Mountains (**Figure 23.6**). This splitting event resulted in many parallel speciation events among isolated lineages of stream-dwelling organisms.

Allopatric speciation may also result when some members of a population cross an existing barrier and establish a new, isolated population. Many of the more than 800 species of *Drosophila* found in the Hawaiian Islands are restricted to a single island. We know that these species are the descendants of new populations founded by individuals dispersing among the islands when we find that the closest relative of a species on one island is a species on a neighboring island rather than a species on the same island. Biologists who have studied the chromosomes of these fruit flies estimate that speciation in this group of *Drosophila* has resulted from at least 45 such founder events (**Figure 23.7**).

23.6 Allopatric Speciation Allopatric speciation may result when an ancestral population is divided into two separate populations by a physical barrier and those populations then diverge. (A) Many species of freshwater stream fishes were distributed throughout the central highlands of North America in the Pliocene epoch (about 3–5 million years ago). (B) During the Pleistocene, glaciers advanced and isolated fish populations in the Ozark and Ouachita Mountains to the west from fish populations in the highlands of the Appalachian Mountains to the east. Numerous species diverged as a result of this separation, including the ancestors of the four pairs of sister species shown here.

(A) Pliocene

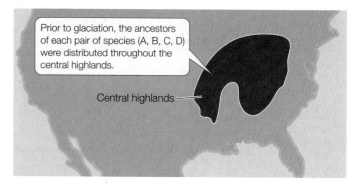

Prior to glaciation, the ancestors of each pair of species (A, B, C, D) were distributed throughout the central highlands.

Central highlands

(B) Pleistocene

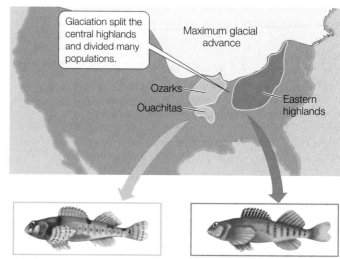

Glaciation split the central highlands and divided many populations.

Maximum glacial advance

Ozarks
Ouachitas
Eastern highlands

A₁ Missouri saddled darter
Etheostoma tetrazonum

A₂ Variegated darter
E. variatum

B₁ Bleeding shiner
Luxilus zonatus

B₂ Warpaint shiner
L. coccogenis

C₁ Ozark minnow
Notropis nubilus

C₂ Tennessee shiner
N. leuciodus

D₁ Ozark madtom
Noturus albater

D₂ Elegant madtom
N. elegans

 Go to Animated Tutorial 23.2
Speciation Mechanisms
Life10e.com/at23.2

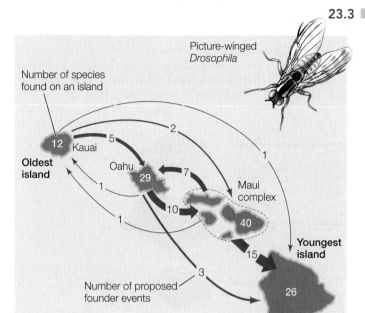

23.7 Founder Events Lead to Allopatric Speciation The large number of species of picture-winged *Drosophila* in the Hawaiian Islands is the result of founder events: the founding of new populations by individuals dispersing among the islands. The islands, which were formed in sequence as Earth's crust moved over a volcanic "hot spot," vary in age.

 Go to Animated Tutorial 23.3
Founder Events and Allopatric Speciation
Life10e.com/at23.3

The 13 species of finches found in the islands of the Galápagos archipelago, some 1,000 km off the coast of Ecuador, are one of the most famous examples of allopatric speciation. Darwin's finches (as they are usually called, because Darwin was the first scientist to study them) arose in the Galápagos from a single South American finch species that colonized the islands. Today the Galápagos species differ strikingly not only from their closest mainland relative, but also from one another (**Figure 23.8**). The islands are sufficiently far apart that the birds move among them only infrequently. In addition, environmental conditions differ widely from island to island. Some islands are relatively flat and arid; others have forested mountain slopes. Over millions of years, finch lineages on the different islands have differentiated to the point that when occasional immigrants do arrive from other islands, they either do not breed with the residents or, if they do, the resulting offspring do not survive as well as the offspring of established residents. The genetic distinctness of each finch species from the others and the genetic cohesiveness of the individual species are thus maintained.

Sympatric speciation occurs without physical barriers

Although geographic isolation is usually required for speciation, speciation can also occur in the absence of a physical barrier. Speciation without physical isolation is called **sympatric speciation** (Greek *sym*, "together with"). But how can such speciation happen? Given that speciation is usually a gradual

process, how can reproductive isolation develop when individuals have frequent opportunities to mate with one another?

DISRUPTIVE SELECTION Sympatric speciation may occur with some forms of disruptive selection (see Section 21.3) if individuals with different genotypes have a preference for distinct microhabitats where mating takes place. For example, sympatric speciation via disruptive selection appears to be taking place in the apple maggot fly (*Rhagoletis pomonella*) of eastern North America. Until the mid-1800s, *Rhagoletis* flies courted, mated, and deposited their eggs only on hawthorn fruits. About 150 years ago, European immigrants introduced apple trees into the region, and some flies began to lay their eggs on apples. Apple trees are closely related to hawthorns, but the smell of the fruits differs, and the apple fruits appear earlier in the season than those of hawthorns. Some early-emerging female *Rhagoletis* laid their eggs on apples, and over time, a genetic preference for the smell of apples evolved among early-emerging insects. When the offspring of these flies sought out apple trees for mating and egg deposition, they mated with other flies reared on apples, which shared the same genetic preferences.

Today the two groups of *Rhagoletis pomonella* in eastern North America appear to be on the way to becoming distinct species. One group mates and lays eggs primarily on hawthorn fruits, the other on apples. The incipient species are partially reproductively isolated because they mate primarily with individuals raised on the same fruit and which emerge at the same of year. In addition, the larvae of the apple-feeding flies now grow more rapidly on apples than they originally did. Sympatric speciation that arises from such host-plant specificity may be widespread among insects, many of which feed on only one plant species.

POLYPLOIDY The most common means of sympatric speciation is **polyploidy**, which results from the duplication of sets of chromosomes within individuals. Polyploidy can arise either from chromosome duplication in a single species (**autopolyploidy**) or from the combining of the chromosomes of two different species (**allopolyploidy**).

An autopolyploid individual originates when (for example) two accidentally unreduced diploid gametes (each with two sets of chromosomes) combine to form a tetraploid individual (with four sets of chromosomes). Tetraploid and diploid individuals of the same species are reproductively isolated because their hybrid offspring are triploid and thus are usually sterile; those offspring rarely produce viable gametes because their chromosomes do not segregate evenly during meiosis (**Figure 23.9**). So a tetraploid individual usually cannot produce viable offspring by mating with a diploid individual—but it *can* do so if it self-fertilizes or mates with another tetraploid. Thus polyploidy can result in complete reproductive isolation in two generations—an important exception to the general rule that speciation is a gradual process.

Allopolyploids may be produced when individuals of two different (but closely related) species interbreed. Such hybridization often disrupts normal meiosis, which can result in chromosomal doubling. Allopolyploids are often fertile because

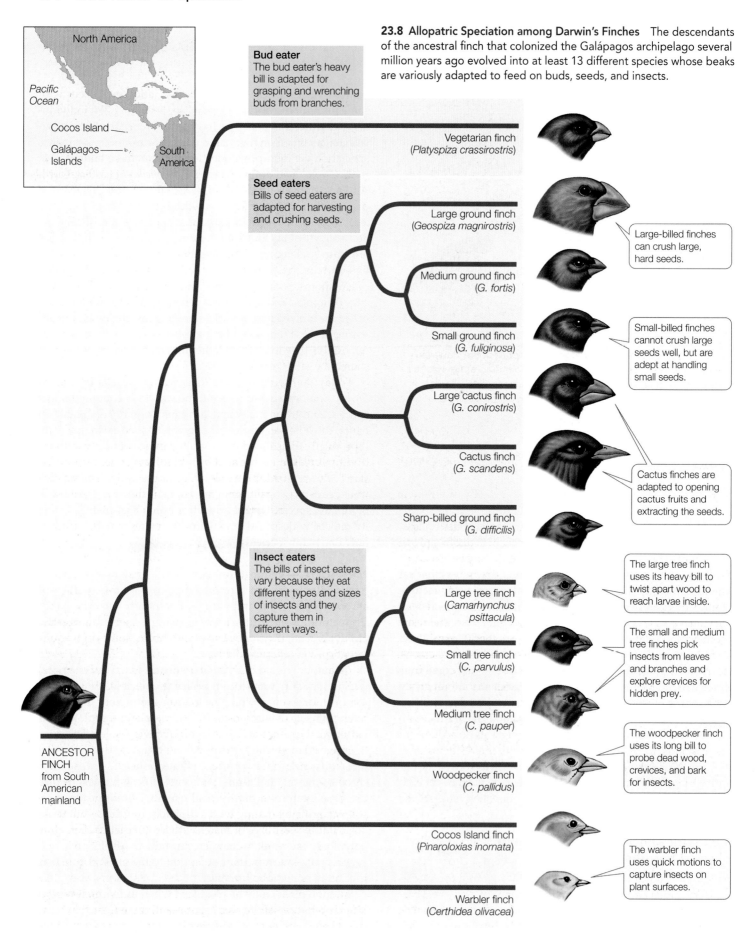

23.8 Allopatric Speciation among Darwin's Finches The descendants of the ancestral finch that colonized the Galápagos archipelago several million years ago evolved into at least 13 different species whose beaks are variously adapted to feed on buds, seeds, and insects.

North America

Pacific Ocean

Cocos Island

Galápagos Islands

South America

Bud eater
The bud eater's heavy bill is adapted for grasping and wrenching buds from branches.

Seed eaters
Bills of seed eaters are adapted for harvesting and crushing seeds.

Vegetarian finch
(*Platyspiza crassirostris*)

Large ground finch
(*Geospiza magnirostris*)

Large-billed finches can crush large, hard seeds.

Medium ground finch
(*G. fortis*)

Small ground finch
(*G. fuliginosa*)

Small-billed finches cannot crush large seeds well, but are adept at handling small seeds.

Large cactus finch
(*G. conirostris*)

Cactus finch
(*G. scandens*)

Cactus finches are adapted to opening cactus fruits and extracting the seeds.

Sharp-billed ground finch
(*G. difficilis*)

Insect eaters
The bills of insect eaters vary because they eat different types and sizes of insects and they capture them in different ways.

Large tree finch
(*Camarhynchus psittacula*)

The large tree finch uses its heavy bill to twist apart wood to reach larvae inside.

Small tree finch
(*C. parvulus*)

The small and medium tree finches pick insects from leaves and branches and explore crevices for hidden prey.

Medium tree finch
(*C. pauper*)

Woodpecker finch
(*C. pallidus*)

The woodpecker finch uses its long bill to probe dead wood, crevices, and bark for insects.

ANCESTOR FINCH from South American mainland

Cocos Island finch
(*Pinaroloxias inornata*)

The warbler finch uses quick motions to capture insects on plant surfaces.

Warbler finch
(*Certhidea olivacea*)

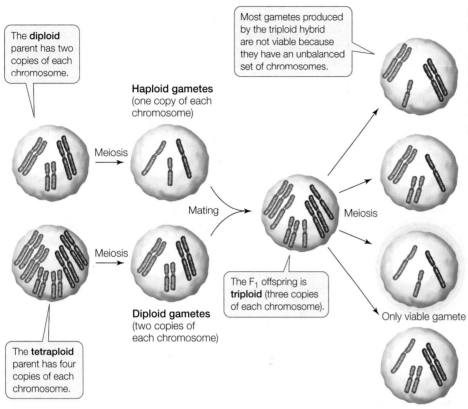

The **diploid** parent has two copies of each chromosome.

Haploid gametes (one copy of each chromosome)

Meiosis

Most gametes produced by the triploid hybrid are not viable because they have an unbalanced set of chromosomes.

Mating

Meiosis

Meiosis

Diploid gametes (two copies of each chromosome)

The **tetraploid** parent has four copies of each chromosome.

The F$_1$ offspring is **triploid** (three copies of each chromosome).

Only viable gamete

23.9 Tetraploids Are Reproductively Isolated from Their Diploid Ancestors Even if the triploid offspring of a diploid and a tetraploid parent survives and reaches sexual maturity, most of the gametes it produces have aneuploid (unbalanced) numbers of chromosomes. Such triploid individuals are effectively sterile. (For simplicity, the diagram shows only three homologous chromosomes. Most species have many more chromosomes, so viable gametes are extremely rare.)

each of the chromosomes has a nearly identical partner with which to pair during meiosis.

Speciation by polyploidy has been particularly important in the evolution of plants, although it has contributed to speciation in animals as well (such as the treefrogs in Figure 23.2). New species arise by polyploidy more easily among plants than among animals because many species of plants can reproduce by self-fertilization. In addition, if polyploidy arises in several offspring of a single parent, the siblings can fertilize one another. Botanists estimate that about 70 percent of flowering plant species and 95 percent of fern species are the result of recent polyploidization. Some of these species arose from hybridization between two species followed by chromosomal duplication and self-fertilization. Other species diverged from polyploid ancestors, so that the new species share their ancestors' duplicated sets of chromosomes.

RECAP 23.3

Allopatric speciation results from the separation of populations by geographic barriers; it is the dominant mode of speciation among most groups of organisms. Sympatric speciation may result from disruptive selection that results in ecological isolation, but polyploidy is the most common cause of sympatric speciation among plants.

- Explain why an effective barrier to gene flow for one species may not effectively isolate another species. **See p. 472**

- What are some obstacles to sympatric speciation? **See p. 473**

- What is the difference between allopolyploidy and autopolyploidy? **See pp. 473–474**

Most populations separated by a physical barrier become reproductively isolated only slowly and gradually. If two incipient species once again come into contact with each other, what keeps them from merging back into a single species?

23.4 What Happens When Newly Formed Species Come into Contact?

As we saw in Section 23.2, once a barrier to gene flow is established, reproductive isolation will begin to develop through genetic divergence. Over many generations, differences accumulate in the isolated lineages, reducing the probability that individuals from each lineage will mate successfully with individuals in the other when they come back into contact. In this way, reproductive isolation can evolve as a by-product of the genetic changes in the two diverging lineages.

Reproductive isolation may be incomplete when incipient species come back into contact, however, in which case some hybridization will occur. If hybrid individuals are less fit than non-hybrids, selection will favor parents that do not produce hybrid offspring. Under these conditions, selection will result in the strengthening, or **reinforcement**, of mechanisms that prevent hybridization.

Mechanisms that prevent hybridization from occurring are called **prezygotic isolating mechanisms**. Mechanisms that reduce the fitness of hybrid offspring are called **postzygotic isolating mechanisms**. Postzygotic isolating mechanisms result in selection against hybridization, which in turn leads to the reinforcement of prezygotic isolating mechanisms.

Prezygotic isolating mechanisms prevent hybridization

Prezygotic isolating mechanisms, which come into play before fertilization, can prevent hybridization in several ways.

MECHANICAL ISOLATION Differences in the sizes and shapes of reproductive organs may prevent the union of gametes from different species. With animals, there may be a match between the shapes of the reproductive organs of males and females of the same species, so that reproduction between individuals with mismatched reproductive structures is not physically possible. In plants, mechanical isolation may involve a pollinator. For example, some orchid species produce flowers that look and smell like the females of particular species of bee or wasp (**Figure 23.10**). When a male insect visits and attempts to mate with a flower (thinking it is a female of his species), his mating behavior results in the transfer of pollen to and from his body by appropriately configured anthers and stigmas on the orchid. Other insects, which may visit the flower but do not attempt to mate with it, do not trigger this transfer of pollen.

TEMPORAL ISOLATION Many organisms have distinct mating seasons. If two closely related species breed at different times of the year (or different times of day), they may never have an opportunity to hybridize. For example, in sympatric populations of three closely related leopard frog species, each species breeds at a different time of year (**Figure 23.11**). Although there is some overlap in the breeding seasons, the opportunities for hybridization are minimized.

23.10 Mechanical Isolation through Mimicry Many orchid species maintain reproductive isolation by means of flowers that look and smell like females of of one—and only one—bee or wasp species. A male insect of the correct species must land on the flower and attempt to mate with it; only males of this particular species are physically configured to collect and transfer the orchid's pollen. The constraints of this method of pollen transfer reproductively isolate the plant from related orchid species that attract different insect pollinators. The species shown here are the two players in one such interspecific relationship; see Figure 56.11 for another example.

BEHAVIORAL ISOLATION Individuals may reject, or fail to recognize, individuals of other species as potential mating partners. For example, the mating calls of male frogs of related species diverge quickly (**Figure 23.12**). Female frogs respond to mating calls from males of their own species but ignore the calls of other species, even closely related ones. The evolution of female preferences for certain male coloration patterns among the cichlids of Lake Malawi, described at the opening of this chapter, is another example of behavioral isolation.

Sometimes the mate choice of one species is mediated by the behavior of individuals of other species. For example, whether or not two plant species hybridize may depend on the food preferences of their pollinators. The floral traits of plants, including their color and shape, can enhance reproductive isolation either by influencing which pollinators are attracted to their flowers or by influencing where pollen is deposited on the bodies of their pollinators. A plant whose flowers are pendant (hanging downward; **Figure 23.13A**) will be pollinated by an animal with different physical characteristics than will a plant whose flowers grow upright (**Figure 23.13B**). Because each pollinator prefers (and is adapted to) a different type of flower, pollinators will rarely transfer pollen from one plant species to the other.

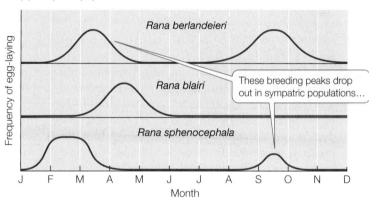

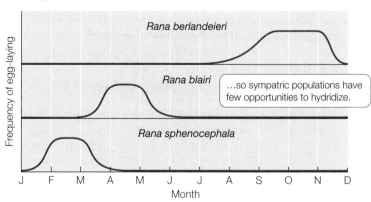

23.11 Temporal Isolation of Breeding Seasons (A) The peak breeding seasons of three species of leopard frogs (*Rana*) overlap when the species are physically separated (allopatry). (B) Where two or more species of *Rana* live together (sympatry), overlap between their peak breeding seasons is greatly reduced or eliminated. Selection against hybridization in areas of sympatry helps reinforce this prezygotic isolating mechanism.

Gastrophryne olivacea

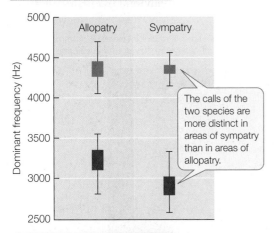

The calls of the two species are more distinct in areas of sympatry than in areas of allopatry.

Gastrophryne carolinensis ■

23.12 Behavioral Isolation in Mating Calls The males of most frog species produce species-specific calls. The calls of the two closely related frog species shown here differ in their dominant frequency (a high-frequency sound wave results in a high-pitched sound; a low frequency results in a low-pitched sound). Female frogs are attracted to the calls of males of their own species.

 Go to Media Clip 23.1
Narrowmouth Toads Calling for Mates
Life10e.com/mc23.1

Such isolation by pollinator behavior is seen in the mountains of California in two sympatric species of columbines (*Aquilegia*) that have diverged in flower color, structure, and orientation. *Aquilegia formosa* (**Figure 23.13C**) has pendant flowers with short spurs (spikelike, nectar-containing structures) and is pollinated by hummingbirds. *A. pubescens* (**Figure 23.13D**) has upright, lighter-colored flowers with long spurs and is pollinated by hawkmoths. The difference in pollinators means that these two species are effectively reproductively isolated even though they populate the same geographic range.

HABITAT ISOLATION When two closely related species evolve preferences for living or mating in different habitats, they may never come into contact during their respective mating periods.

(A)

(B)

(C) *Aquilegia formosa*

(D) *A. pubescens*

23.13 Reproductive Isolating Mechanisms May Be Mediated by Species Interactions (A) This hummingbird's morphology and behavior are adapted for feeding on nectar from pendant flowers. (B) The nectar-extracting proboscis of this hawkmoth is adapted to flowers that grow upright. (C) *Aquilegia formosa* flowers are normally pendant and are pollinated by hummingbirds. (D) Flowers of *A. pubescens* are normally upright and are pollinated by hawkmoths. In addition, their long floral spurs appear to restrict access by some other potential pollinators.

The *Rhagoletis* flies discussed in Section 23.3 experienced such habitat isolation, as did the cichlid fishes that first adapted to rocky or sandy habitats upon entering Lake Malawi, as described at the opening of this chapter.

GAMETIC ISOLATION The sperm of one species may not attach to the eggs of another species because the eggs do not release the appropriate attractive chemicals, or the sperm may be unable to penetrate the egg because the two gametes are chemically incompatible. Thus, even though the gametes of two species may come into contact, the gametes never fuse into a zygote.

Gametic isolation is extremely important for many aquatic species that spawn (release their gametes directly into the environment). It has been extensively studied in sea urchins. A protein known as bindin is found in sea urchin sperm and functions in attaching ("binding") the sperm to eggs. All sea urchin species studied produce this egg-recognition protein, but the bindin gene sequence diverges so rapidly that it has become species-specific—that is, sperm can attach only to eggs of the same species, so no interspecific hybridization occurs.

Postzygotic isolating mechanisms result in selection against hybridization

Genetic differences that accumulate between two diverging lineages may reduce the survival and reproductive rates of hybrid offspring in any of several ways:

- *Low hybrid zygote viability.* Hybrid zygotes may fail to mature normally, either dying during development or developing phenotypic abnormalities that prevent them from becoming reproductively capable adults.

- *Low hybrid adult viability.* Hybrid offspring may have lower survivorship than non-hybrid offspring.

- *Hybrid infertility.* Hybrids may mature into infertile adults. For example, the offspring of matings between horses and donkeys—mules—are sterile. Although otherwise healthy, mules produce no descendants.

Natural selection does not directly favor the evolution of postzygotic isolating mechanisms. But if hybrids have low fitness, then individuals that breed only within their own species will leave more surviving offspring than will individuals that interbreed with another species. Therefore, individuals that avoid interbreeding with members of other species will have a selective advantage, and any trait that contributes to such avoidance will be favored.

Donald Levin of the University of Texas has studied reinforcement of prezygotic isolating mechanisms in flowers of the genus *Phlox*. Levin noticed that most individuals of *P. drummondii* in most of the range of the species in Texas have pink flowers. However, where *P. drummondii* is sympatric with its close relative, the pink-flowered *P. cuspidata*, most *P. drummondii* have red flowers. No other *Phlox* species has red flowers. Levin performed an experiment whose results showed that reinforcement may explain why red flowers are favored where the two species are sympatric (**Figure 23.14**).

Likely cases of reinforcement are often detected by comparing sympatric and allopatric populations of potentially hybridizing species, as in the case of *Phlox*. If reinforcement is occurring, then sympatric populations of closely related species are expected to evolve more effective prezygotic reproductive barriers than do allopatric populations of the same species. As Figure 23.11 shows, the breeding seasons of sympatric populations of different leopard frog species overlap much less than do those of allopatric populations. Similarly, the frequencies of the frog mating calls illustrated in Figure 23.12 are more divergent in sympatric populations than in allopatric populations. In both cases, there appears to have been natural selection against hybridization in areas of sympatry.

23.14 Flower Color Reinforces a Reproductive Barrier in *Phlox* Most *Phlox drummondii* flowers are pink, but in regions where they are sympatric with *P. cuspidata*—which is always pink—most *P. drummondii* individuals are red. Most pollinators preferentially visit flowers of one color or the other. In this experiment, Donald Levin explored whether flower color reinforces a prezygotic reproductive barrier, lessening the chances of interspecific hybridization.[a]

HYPOTHESIS Red-flowered *P. drummondii* are less likely to hybridize with *P. cuspidata* than are pink-flowered *P. drummondii*.

Method 1. Introduce equal numbers of red- and pink-flowered *P. drummondii* individuals into an area with many pink-flowered *P. cuspidata*.

2. After the flowering season ends, measure hybridization by assessing the genetic composition of the seeds produced by *P. drummondii* plants of both colors.

Results Of the seeds produced by pink-flowered *P. drummondii*, 38% were hybrids with *P. cuspidata*. Only 13% of the seeds produced by red-flowered individuals were genetic hybrids.

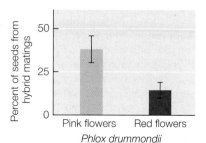

CONCLUSION *P. drummondii* and *P. cuspidata* are less likely to hybridize if the flowers of the two species differ in color.

Go to **BioPortal** for discussion and relevant links for all INVESTIGATING**LIFE** figures.

[a]Levin, D. A. 1985. *Evolution* 39: 1275–1281.

Hybrid zones may form if reproductive isolation is incomplete

Unless reproductive isolation is complete, closely related species may hybridize in areas where their ranges overlap, resulting in the formation of a hybrid zone. When a hybrid zone first forms, most hybrids are offspring of crosses

WORKING WITH**DATA:**

Does Flower Color Act as a Prezygotic Isolating Mechanism?

Original Paper

Levin, D. A. 1985. Reproductive character displacement in *Phlox. Evolution* 39: 1275–1281.

Analyze the Data

Donald Levin proposed that *Phlox drummondii* has red flowers only in locations where it is sympatric with pink-flowered *Phlox cuspidata* because having red flowers decreases interspecific hybridization. To test this hypothesis, Levin introduced equal numbers of red- and pink-flowered *P. drummondii* individuals into an area with many pink-flowered *P. cuspidata*. At the end of the flowering season, he assessed the genetic composition of the seeds produced by *P. drummondii*. The results are shown in the table (below, right).

QUESTION 1

Check the 95% confidence intervals for the proportion of hybrid seeds in red- and pink-flowered *P. drummondii* (shown graphically in Figure 23.14). There are many websites available for calculating confidence intervals; a good one is the Vassar College statistical computation site, VassarStats.net. You can go to this site and select "Proportions" from the left-hand menu, then select "The Confidence Interval of a Proportion." What are the numerical values of the 95% confidence intervals?

QUESTION 2

You can see that the proportions of hybrids among the seeds of red- versus pink-flowered samples are significantly different because the 95% confidence intervals do not overlap. To quantify the significance of this difference, use the website suggested in Question 1, but select "Significance of the Difference between Two Independent Proportions" from the "Proportions" menu. What null hypothesis are you testing in this case? (See Appendix B if you need help.) What is the *P*-value of getting results at least as different as these two samples if your null hypothesis is true?

QUESTION 3

How would you extend or improve the experimental design of this study? What kinds of additional test sites or conditions would you want to examine? How might replicate or control sites make the study more convincing?

Morph (flower color)	Number of seeds (progeny)		
	P. drummondii	Hybrid	Total
Red	181 (87%)	27 (13%)	208
Pink	86 (62%)	53 (38%)	139

Go to BioPortal for all WORKING WITH**DATA** exercises

between purebred individuals of the two hybridizing species. However, subsequent generations include a variety of individuals with varying proportions of their genes derived from the original two species, so hybrid zones often contain recombinant individuals resulting from many generations of hybridization.

Detailed genetic studies can tell us much about why narrow hybrid zones may persist for long periods between the ranges of two species. In Europe, the hybrid zone between two toad species of the genus *Bombina* has been studied intensively. The fire-bellied toad (*B. bombina*) lives in eastern Europe; the closely related yellow-bellied toad (*B. variegata*) lives in western and southern Europe. The ranges of the two species overlap in a long but very narrow zone stretching 4,800 kilometers from eastern Germany to the Black Sea (**Figure 23.15**). Hybrids between the two species suffer from a range of defects, many of which are lethal. Those hybrids that survive often have skeletal abnormalities, such as misshapen mouths, ribs that are fused to vertebrae, and a reduced number of vertebrae. By following the fates of thousands of toads from the hybrid zone, investigators found that a hybrid toad, on average, is only half as fit as a purebred individual of either species. The hybrid zone remains narrow because there is strong selection against hybrids and because adult toads do not move over long distances. The zone has persisted for hundreds of years, however, because individuals of both species continue to move short distances into it, continually replenishing the hybrid population.

23.15 A Hybrid Zone The narrow zone (shown in orange) in which fire-bellied toads meet and hybridize with yellow-bellied toads has been stable for hundreds of years.

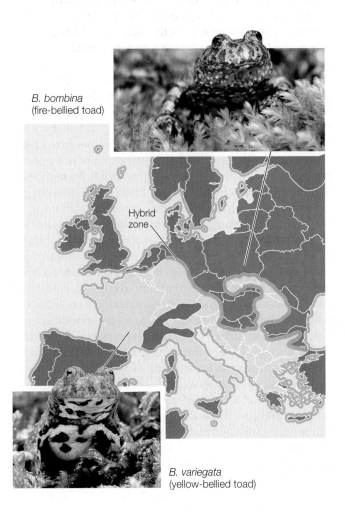

B. bombina
(fire-bellied toad)

Hybrid
zone

B. variegata
(yellow-bellied toad)

▌RECAP 23.4

Reproductive isolation may result from prezygotic or postzygotic isolating mechanisms. Lower fitness of hybrids can lead to the reinforcement of prezygotic isolating mechanisms.

- Distinguish among five types of prezygotic isolating mechanisms. **See pp. 476–477**
- Why are postzygotic isolating mechanisms said to *reinforce* prezygotic isolating mechanisms? **See pp. 478–479**
- Why don't most narrow hybrid zones, such as the one between *Bombina bombina* and *B. variegata*, get wider over time? **See p. 479 and Figure 23.15**

Some groups of organisms have many species, others only a few. Hundreds of species of *Drosophila* evolved in the small area of the Hawaiian Islands over about 20 million years. In contrast, there are only a few species of horseshoe crabs in the world, and only one species of ginkgo tree, even though these latter groups have persisted for hundreds of millions of years. Why do different groups of organisms have such different rates of speciation?

23.5 Why Do Rates of Speciation Vary?

Many factors influence the likelihood that a lineage will split to form two or more species. Therefore, rates of speciation (the proportion of existing species that split to form new species over a given period) vary greatly among groups of organisms. What are some of the factors that influence the probability of a given lineage splitting into two?

Several ecological and behavioral factors influence speciation rates

DIET SPECIALIZATION   Populations of species that have specialized diets may be more likely to diverge than are those with more generalized diets. To investigate the effects of diet specialization on rates of speciation, Charles Mitter and colleagues compared species richness in some closely related groups of true bugs (hemipterans). The common ancestor of these groups was a predator that fed on other insects, but a dietary shift to herbivory (eating plants) evolved at least twice in the groups under study. Herbivorous bugs typically specialize on one or a few closely related species of plants, whereas predatory bugs tend to feed on many different species of insects. High diversity of host plant species can thus lead to a correspondingly high species diversity among herbivorous specialists. The Mitter et al. study showed that among these insects, the herbivorous groups do indeed have many more species than the related predatory groups (**Figure 23.16**).

POLLINATION   Speciation rates are faster in animal-pollinated than in wind-pollinated plants. Animal-pollinated groups have, on average, 2.4 times as many species as related groups pollinated by wind. Among animal-pollinated plants, speciation rates are correlated with pollinator specialization. In columbines (*Aquilegia*), the rate of evolution of new species has been about three times faster in lineages that have long nectar spurs than in

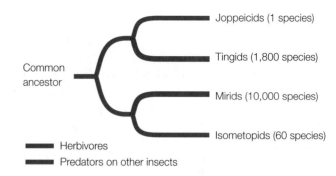

23.16 Dietary Shifts Can Promote Speciation   Herbivorous groups of hemipteran insects have speciated several times faster than closely related predatory groups.

lineages that lack spurs. Why do nectar spurs increase the speciation rate? Apparently it is because spurs restrict the number of pollinator species that visit the flowers, thus increasing opportunities for reproductive isolation (see Figure 23.13).

SEXUAL SELECTION   It appears that the mechanisms of sexual selection (see Section 21.2) result in high rates of speciation, as we saw in the case of the cichlids of Lake Malawi. Some of the most striking examples of sexual selection are found in birds with polygynous mating systems. Bird-watchers travel thousands of miles to Papua New Guinea to witness the mating displays of male birds of paradise, which have long, brightly colored tail feathers (**Figure 23.17A**) and look distinctly different

(A) *Paradisaea minor*

(B) *Manucodia comrii*

23.17 Sexual Selection Can Lead to Higher Speciation Rates (A) Birds of paradise and (B) manucodes are closely related bird groups of the South Pacific. Speciation rates are much higher among the sexually dimorphic, polygynous birds of paradise (33 species) than among the sexually monomorphic, monogamous manucodes (5 species).

from females of their species—a phenomenon called sexual dimorphism. Males assemble at display grounds called leks, and females come there to choose a mate. After mating, the females leave the display grounds, build their nests, lay their eggs, and feed their offspring with no help from the males. The males remain at the lek to court more females.

The closest relatives of the birds of paradise are the manucodes (**Figure 23.17B**). Male and female manucodes differ only slightly in size and plumage (they are sexually monomorphic). They form monogamous pair bonds, and both sexes contribute to raising the young. There are only 5 species of manucodes, compared with 33 species of birds of paradise. By itself, this one comparison would not be convincing evidence that sexually dimorphic clades have higher rates of speciation than do monomorphic clades. However, when biologists examined all the examples of birds in which one clade is sexually dimorphic and the most closely related clade is sexually monomorphic, the sexually dimorphic clades were significantly more likely to contain more species. But why would sexual dimorphism be associated with a higher rate of speciation?

Animals with complex sexually selected behaviors are likely to form new species at a high rate because they make sophisticated discriminations among potential mating partners. They distinguish members of their own species from members of other species, and they make subtle discriminations among members of their own species on the basis of size, shape, appearance, and behavior. Such discriminations can greatly influence which individuals are most successful in mating and producing offspring, so they may lead to rapid evolution of behavioral isolating mechanisms among populations.

DISPERSAL ABILITY   Speciation rates are usually higher in groups with poor dispersal abilities than in groups with good dispersal abilities because even narrow barriers can be effective in dividing a species whose members are highly sedentary. Until recently, the Hawaiian Islands had about 1,000 species of land snails, many of which were restricted to a single valley. Because snails move only short distances, the high ridges that separate the valleys were effective barriers to their dispersal. Unfortunately, introductions of other species and changes in habitat have resulted in the recent extinction of most of these unique Hawaiian land snails.

Rapid speciation can lead to adaptive radiation

The rapid proliferation of a large number of descendant species from a single ancestor species is called an **evolutionary radiation**. Evolutionary radiations often occur when a species colonizes a new area, such as an island archipelago that contains no other closely related species, because of the large number of open ecological niches. If such a rapid proliferation of species results in an array of species that live in a variety of environments and differ in the characteristics they use to exploit those environments, it is referred to as an **adaptive radiation**.

Several remarkable adaptive radiations have occurred in the Hawaiian Islands. In addition to its 1,000 species of land snails, the native Hawaiian biota includes 1,000 species of flowering plants, 10,000 species of insects, and more than 100 bird species. However, there were no amphibians, no terrestrial reptiles, and only one native terrestrial mammal (a bat) on the islands until humans introduced additional species. The 10,000 known native species of insects on Hawaii are believed to have evolved from about 400 immigrant species; only 7 immigrant species are believed to account for all the native Hawaiian land birds. Similarly, as we saw earlier in this chapter, an adaptive radiation in the Galápagos archipelago resulted in the 13 species of Darwin's finches, which differ strikingly in the size and shape of their bills and, accordingly, in the food resources they use (see Figure 23.8).

The 28 species of Hawaiian sunflowers called silverswords are an impressive example of an adaptive radiation in plants (**Figure 23.18**). DNA sequences show that these species share a relatively recent common ancestor with a species of tarweed from the Pacific coast of North America. Whereas all mainland

Dubautia menziesii

Madia sativa (tarweed)

Argyroxiphium sandwicense

Wilkesia gymnoxiphium

23.18 Rapid Evolution among Hawaiian Silverswords   The Hawaiian silverswords, three closely related genera of the sunflower family, are believed to have descended from a single common ancestor (a plant similar to the tarweed *Madia sativa*) that colonized Hawaii from the Pacific coast of North America. The four plants shown here are more closely related than they appear to be based on their morphology.

tarweeds are small, upright herbs (non-woody plants such as *Madia sativa*; see Figure 23.18), the silverswords include shrubs, trees, and vines as well as both upright and ground-hugging herbs. Silversword species occupy nearly all the habitats of the Hawaiian Islands, from sea level to above the timberline in the mountains. Despite their extraordinary morphological diversification, all silverswords are genetically very similar.

The Hawaiian silverswords are more diverse in size and shape than the mainland tarweeds because their tarweed ancestors first arrived on islands that harbored very few plant species. In particular, there were few trees and shrubs because such large-seeded plants rarely disperse to oceanic islands. Trees and shrubs have evolved from non-woody ancestors on many oceanic islands. On the mainland, however, tarweeds live in ecological communities that contain many tree and shrub species in lineages with long evolutionary histories. In those environments, opportunities to exploit the "tree" way of life have already been preempted.

RECAP **23.5**

Dietary specialization, pollinator specialization, sexual selection, and poor dispersal abilities are correlated with high rates of speciation. Open ecological niches present opportunities for adaptive radiations.

- How can pollinator specialization in plants and sexual selection in animals increase rates of speciation? **See pp. 480–481**

- Why do adaptive radiations often occur when a founder species invades an isolated geographic area? **See p. 481**

23.19 Evolution in the Laboratory For their experiments on the evolution of prezygotic isolating mechanisms in *Drosophila melanogaster*, Rice and Salt built an elaborate system of varying habitats contained within vials inside a large fly enclosure. Some groups of flies developed preferences for widely divergent habitats and became reproductively isolated within 35 generations.

The processes described in this chapter, operating over billions of years, have produced a world in which life is organized into millions of species, each adapted to live in a particular environment and to use environmental resources in a particular way. In the next chapter we consider how species evolve at the level of their genes and genomes.

Can biologists study the process of speciation in the laboratory?

ANSWER

Although speciation usually takes thousands or millions of years, and although it is typically studied in natural settings such as Lake Malawi, some aspects of speciation can be studied and observed in controlled laboratory experiments. Most such experiments use organisms with short generation times, in which evolution is expected to be relatively rapid.

William Rice and George Salt conducted an experiment in which fruit flies (*Drosophila melanogaster*) were allowed to choose food sources in different habitats. The habitats—where mating also took place—were vials in different parts of an experimental cage (**Figure 23.19**). The vials differed in three environmental factors: (1) light; (2) the direction (up or down) in which the fruit flies had to move to reach food; and (3) the concentrations of two aromatic chemicals, ethanol and acetaldehyde. In just 35 generations, the two groups of flies that chose the most divergent habitats had become reproductively isolated from each other, having evolved distinct preferences for the different habitats.

The experiment by Rice and Salt (see *American Naturalist* 131: 911–917, 1988) demonstrated an example of habitat isolation as a prezygotic isolating mechanism. Even though the different habitats were in the same cage, and individual fruit flies were capable of flying from one habitat to the other, habitat preferences were inherited by offspring from their parents, and populations from the two divergent habitats did not interbreed. Similar habitat isolation is thought to have resulted in the early split between cichlids that preferred the rocky versus the sandy shores of Lake Malawi. In controlled experiments like this one, biologists can observe many aspects of the process of speciation directly.

23.1 **What Are Species?**

- **Speciation** is the divergence of biological lineages and the emergence of reproductive isolation between those lineages.

- The **morphological species concept** distinguishes species on the basis of physical similarities and differences.

- The **biological species concept** distinguishes species on the basis of **reproductive isolation.**

- The **lineage species concept** recognizes evolutionarily independent lineages as species, allowing biologists to consider species over evolutionary time.

23.2 **What Is the Genetic Basis of Speciation?**

- Speciation usually results from the interruption of gene flow within a population.

- The Dobzhansky–Muller model describes how reproductive isolation between two physically isolated populations can develop through the accumulation of incompatible genes or chromosomal arrangements. **Review Figures 23.3, 23.4, ANIMATED TUTORIAL 23.1**

- Reproductive isolation increases with increasing genetic divergence between populations. **Review Figure 23.5**

23.3 **What Barriers to Gene Flow Result in Speciation?**
See ANIMATED TUTORIAL 23.2

- **Allopatric speciation**, which results when populations are separated by a physical barrier, is the dominant mode of speciation in most groups of organisms. This type of speciation may follow founder events, in which some members of a population cross a barrier and found a new, isolated population. **Review Figures 23.6–23.8, ANIMATED TUTORIALS 23.2 and 23.3**

- **Sympatric speciation** results when the genomes of two groups diverge in the absence of physical isolation. Such divergence can result from disruptive selection if individuals with different genotypes prefer distinct microhabitats.

- Sympatric speciation can occur within two generations via **polyploidy**. Polyploidy may arise from chromosome duplications within a species (**autopolyploidy**) or from hybridization that results in combining the chromosomes of two species (**allopolyploidy**). Review Figure 23.9

23.4 **What Happens When Newly Formed Species Come into Contact?**

- **Prezygotic isolating mechanisms** prevent hybridization; **postzygotic isolating mechanisms** reduce the fitness of hybrids.

- Postzygotic isolating mechanisms lead to **reinforcement** of prezygotic isolating mechanisms by natural selection. **Review Figures 23.11, 23.12, 23.14**

- Hybrid zones may form and persist if reproductive isolation between species is incomplete. **Review Figure 23.15**

23.5 **Why Do Rates of Speciation Vary?**

- Dietary specialization, pollinator specialization, sexual selection, and dispersal ability all influence speciation rates. **Review Figure 23.16**

- **Evolutionary radiation** refers to the rapid proliferation of descendant species from a single ancestor species. This often occurs following colonization, when new species may rapidly move into unoccupied ecological niches in a process known as **adaptive radiation**.

See ACTIVITY 23.1 for a concept review of this chapter

Go to the Interactive Summary to review key figures, Animated Tutorials, and Activities
Life10e.com/is23

CHAPTER**REVIEW**

REMEMBERING

1. Which of the following is *not* a condition expected to favor allopatric speciation?
   a. Continents drift apart and separate previously connected lineages.
   b. A mountain range separates formerly connected populations.
   c. Different environments on two sides of a barrier cause populations to diverge.
   d. The range of a species is separated by loss of intermediate habitat.
   e. Tetraploid individuals arise in one part of the range of a species.

2. Which of the following is *not* a potential prezygotic reproductive barrier?
   a. Temporal segregation of breeding seasons
   b. Differences in chemicals that attract mates

   c. Hybrid infertility
   d. Spatial segregation of mating sites
   e. Sperm that cannot penetrate an egg

3. A common means of sympatric speciation is
   a. polyploidy.
   b. hybrid infertility.
   c. temporal segregation of breeding seasons.
   d. spatial segregation of mating sites.
   e. imposition of a geographic barrier.

4. Which of the following is often associated with higher rates of speciation?
   a. Sexual dimorphism in birds
   b. Diet specialization in insects
   c. Poor dispersal ability
   d. Animal pollination in plants
   e. All of the above

UNDERSTANDING & APPLYING

5. The Dobzhansky–Muller model of speciation suggests that divergence among alleles at *different* gene loci leads to genetic incompatibility between species. Why is genetic incompatibility between two alleles at the *same* locus considered less likely?

6. Why do some combinations of chromosomal centric fusions cause problems in meiosis? Can you diagram what would happen at meiosis in a hybrid of the divergent lineages shown in Figure 23.4?

7. Assume that the reproductive isolation seen in the *Phlox* strains discussed in Section 23.2 results from lethal combinations of incompatible alleles at several loci among the various strains. Given this assumption, why might the reproductive isolation seen among these strains be partial rather than complete?

8. If allopatric speciation is the most prevalent mode of speciation, what do you predict about the geographic distributions of many closely related species?

ANALYZING & EVALUATING

9. In each of the species of columbine shown in Figure 23.13, the orientation of the flowers and the length of flower spurs are associated with a particular type of pollinator (hummingbirds or hawkmoths). Columbine flowers vary in other ways as well; for example, they differ in color, and probably in odor. What experiments could you design to determine the traits that various pollinators use to distinguish among the flowers of different columbine species?

10. The different finch species in the phylogeny shown in Figure 23.8 have all evolved on islands of the Galápagos archipelago within the past 3 million years. Molecular clock analysis (see Chapter 24) has been used to determine the dates of the various speciation events in that phylogeny. Geological techniques for dating rock samples (see Chapter 25) have been used to determine the ages of the various Galápagos islands. The table below shows the number of species of Darwin's finches and the number of islands that have existed in the archipelago at several times during the past 4 million years.

Time (mya)	Number of islands	Number of finch species
0.25	18	14
0.50	18	9
0.75	9	7
1.00	6	5
2.00	4	3
3.00	4	1
4.00	3	0

a. Plot the number of species of Darwin's finches and the number of islands in the Galápagos archipelago (dependent variables) against time (independent variable).

b. Are the data consistent with the hypothesis that isolation of populations on newly formed islands is related to speciation in this group of birds? Why or why not?

c. If no more islands form in the Galápagos archipelago, do you think that speciation by geographic isolation will continue to occur among Darwin's finches? Why or why not? What additional data could you collect to test your hypothesis (without waiting to see if speciation occurs)?

Go to BioPortal at **yourBioPortal.com** for Animated Tutorials, Activities, LearningCurve Quizzes, Flashcards, and many other study and review resources.

24 Evolution of Genes and Genomes

Adapting to a Murky Environment The elephant-nose fish (*Gnathonemus petersi*), a river-dwelling species from West Africa, is one of many fishes in which weakly discharging electric organs have evolved via duplications and modifications in sodium channel proteins.

SOME FISHES GENERATE HIGH-VOLTAGE electric discharges (up to 650 volts) that they use to stun their prey; perhaps the best known examples are the electric eels of Central and South America. Several other fish species produce weaker electric discharges. Most of these weakly electric fishes live in murky water, where visual cues are limited. They use electric signals to locate (but not to stun) their prey. Electric signals also allow them to communicate with other individuals of their own species.

Electric organs have evolved independently in several fish lineages. How did these organs evolve? Consider first the physical basis of the electric discharges. Voltage-gated sodium channels are large protein complexes that underlie the generation and propagation of rapid electric signals in nerve, muscle, and heart tissues. Electric signals are transmitted along nerves to muscles as the sodium channels embedded in cell membranes are stimulated to open. These channels control the concentration of positively charged sodium ions (Na^+) on the inside relative to the outside of cells, resulting in an electric charge that is transmitted across the surface of the muscle, leading to muscular contraction.

Most vertebrates have multiple copies of the genes encoding the several proteins that make up the sodium channel. These copies arose in the distant past through a series of gene duplications. Such duplications allowed the differentiation and specialization of protein function, making it possible for different sodium channels to exist in different types of tissue. In the case of electric fishes, one of the sodium channel genes expressed in muscle diverged, and a new functional protein evolved. Changes in a relatively small number of nucleotide positions in the gene resulted in modified sodium channels, allowing the development of a new organ with a unique function—the generation of externally transmitted electric energy.

The repeated evolution of electric organs from muscle tissue has been facilitated by the relative simplicity of the molecular changes required. Gene duplication has expedited the process, since redundant genes allow for such specialization in protein function. Finally, interspecific differences in sodium channel function have arisen from additional changes in the nucleotide sequences of the genes. These small differences allow different species to use different communication signals, which improves intraspecific communication while reducing interspecific interference.

How do evolutionary studies of sodium channel genes help us understand some human genetic disorders?

See answer on p. 502.

24.1 How Are Genomes Used to Study Evolution?

An organism's **genome** is the full set of genes it contains, as well as any noncoding regions of the DNA (or, in the case of some viruses, RNA). Most of the genes of eukaryotic organisms are found on chromosomes in the nucleus, but genes are also present in chloroplasts and mitochondria. In organisms that reproduce sexually, both males and females contribute nuclear genes, but mitochondrial and chloroplast genes are usually transmitted only via the cytoplasm of one of the two gametes (usually from the female parent).

Genomes must be replicated to be transmitted from parents to offspring. DNA replication does not occur without error, however. Mistakes in DNA replication—mutations—provide much of the raw material for evolutionary change. Mutations are essential for the long-term survival of life because they are the initial source of the genetic variation that permits species to evolve in response to changes in their environment.

A particular allele of a gene will not be passed on to successive generations unless an individual carrying that allele survives and reproduces. The allele must function in combination with many other genes in the genome or it will quickly be selected against. Moreover, the degree and timing of a gene's expression are affected by its location in the genome. For these reasons, the genes of an individual organism can be viewed as interacting members of a group, among which there are divisions of labor but also strong interdependencies.

A genome, then, is not simply a random collection of genes in a random order along chromosomes. Rather, it is a complex set of integrated genes, regulatory sequences, and structural elements, interspersed with vast stretches of noncoding DNA that may have little direct function. Both the positions of genes and their sequences are subject to evolutionary change, as are the extent and location of noncoding DNA. All of these changes can affect the phenotype of an organism.

Biologists have now sequenced the complete genomes of a large number of organisms, including humans. The information in these sequences is helping us understand how and why organisms differ, how they function, and how they have evolved.

Evolution of genomes results in biological diversity

The field of **molecular evolution** investigates the mechanisms and consequences of the evolution of macromolecules—particularly nucleic acids (DNA and RNA) and proteins. Molecular evolutionists study relationships between the structures of genes and proteins and the functions of organisms. They also examine molecular variation to reconstruct evolutionary history and to study the mechanisms and consequences of evolution. Students of this field ask questions such as: What does molecular variation tell us about a gene's function? Why do the genomes of different organisms vary in size? What evolutionary forces shape patterns of variation among genomes? And a crucial question from an evolutionary perspective: How do genomes acquire new functions? Investigations into the evolution of particular nucleic acids and proteins are instrumental in reconstructing the evolutionary histories of genes. Ultimately,

molecular evolutionary biologists hope to explain the molecular basis of biological diversity.

The evolution of nucleic acids and proteins depends on genetic variation introduced by mutations. One of several ways in which genes evolve is by means of nucleotide substitutions (the incorporation of point mutations in populations). In genes that encode proteins, nucleotide substitutions sometimes result in amino acid replacements that can change the charge, the structure, and other chemical and physical properties of the encoded protein. Phenotypic changes in a protein molecule often affect the way that protein functions in the organism.

Evolutionary changes in genes and proteins can be identified by comparing nucleotide or amino acid sequences from different organisms. The longer two sequences have been evolving separately, the more differences they accumulate (bearing in mind that different genes in the same species evolve at different rates). Determining how long ago changes in nucleotide or amino acid sequences occurred is a useful step toward inferring their causes. Knowledge of the pattern and rate of evolutionary change in a given macromolecule is useful in reconstructing the evolutionary history of groups of organisms.

To compare genes or proteins from Tdifferent organisms, biologists need a way to identify homologous parts of macromolecules. (Recall from Section 22.1 that homologous features are those shared by two or more species that have been inherited from a common ancestor.) Homologous parts of a protein can be identified by their homologous amino acid sequences. And, since nucleotide sequences encode amino acid sequences, the concept of homology extends down to the level of individual nucleotide positions. Therefore one of the first steps in studying the evolution of genes or proteins is to align homologous positions in the amino acid or nucleotide sequence of interest.

Genes and proteins are compared through sequence alignment

Once the nucleotide or amino acid sequences of molecules from different organisms have been determined, they can be compared. Homologous positions can be identified only if we first pinpoint the locations of deletions and insertions that have occurred in the molecules of interest in the time since the organisms diverged from a common ancestor. A simple hypothetical example illustrates this **sequence alignment** technique. In **Figure 24.1** we compare two amino acid sequences from homologous proteins in different organisms. The two sequences at first appear to differ in both the number and identity of their amino acids. If we insert a gap after the first amino acid in sequence 2 (after leucine), however, the similarities in the two sequences become obvious. This gap represents the occurrence of one of two evolutionary events: an insertion of an amino acid in the longer protein or a deletion of an amino acid in the shorter protein. Having adjusted for this insertion or deletion event, we can see that the two sequences differ by only one amino acid at position 6 (serine or phenylalanine).

Go to Activity 24.1 Amino Acid Sequence Alignment
Life10e.com/ac24.1

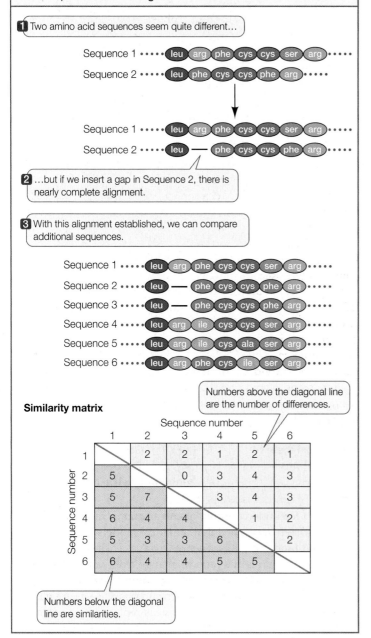

RESEARCH**TOOLS**

24.1 Amino Acid Sequence Alignment Amino acid sequence alignment is a way of arranging protein sequences to identify regions of homology between the sequences. Gaps are inserted between the amino acid residues to align similar residues in columns. Differences and similarities between each pair of aligned sequences are then summarized in a similarity matrix. Homologous DNA sequences can be aligned in a similar manner.

1 Two amino acid sequences seem quite different…

Sequence 1 ···· leu arg phe cys cys ser arg ····
Sequence 2 ···· leu phe cys cys phe arg ····

Sequence 1 ···· leu arg phe cys cys ser arg ····
Sequence 2 ···· leu — phe cys cys phe arg ····

2 …but if we insert a gap in Sequence 2, there is nearly complete alignment.

3 With this alignment established, we can compare additional sequences.

Sequence 1 ···· leu arg phe cys cys ser arg ····
Sequence 2 ···· leu — phe cys cys phe arg ····
Sequence 3 ···· leu — phe cys cys phe arg ····
Sequence 4 ···· leu arg ile cys cys ser arg ····
Sequence 5 ···· leu arg ile cys ala ser arg ····
Sequence 6 ···· leu arg phe cys ile ser arg ····

Similarity matrix

Numbers above the diagonal line are the number of differences.

Sequence number

	1	2	3	4	5	6
1		2	2	1	2	1
2	5		0	3	4	3
3	5	7		3	4	3
4	6	4	4		1	2
5	5	3	3	6		2
6	6	4	4	5	5	

Numbers below the diagonal line are similarities.

Adding a single gap—that is, identifying a deletion or an insertion—aligns the two sequences in Figure 24.1. Additional sequences can now be added to the alignment in a similar manner. Longer amino acid sequences, and those that have diverged more extensively, require more elaborate adjustments. Explicit models (incorporated into computer algorithms) have been developed to account for the relative probabilities of deletions, insertions, and particular amino acid replacements.

Having aligned the sequences, we can compare them by counting the number of nucleotides or amino acids that differ between them. Summing the numbers of the same and different amino acids in each pair of sequences allows us to construct a **similarity matrix**, which gives us a measure of the minimum number of changes that have occurred since the divergence of each pair of organisms (see Figure 24.1).

Go to Activity 24.2 Similarity Matrix Construction
Life10e.com/ac24.2

Models of sequence evolution are used to calculate evolutionary divergence

The sequence comparison procedure illustrated in Figure 24.1 gives a simple count of the number of similarities and differences between the proteins of two species. In the context of two aligned DNA sequences, we can count the number of differences at homologous nucleotide positions, and this count indicates the minimum number of nucleotide changes that must have occurred since the two sequences diverged from a common ancestral sequence.

Although it is useful in determining a *minimum* number of changes between two DNA sequences, the count provided by sequence alignment almost certainly underestimates the *actual* number of changes that have occurred since the sequences diverged. Any given change counted in a similarity matrix of DNA sequences may result from multiple substitution events that occurred at a given nucleotide position over time. As illustrated in **Figure 24.2**, any of the following events may have

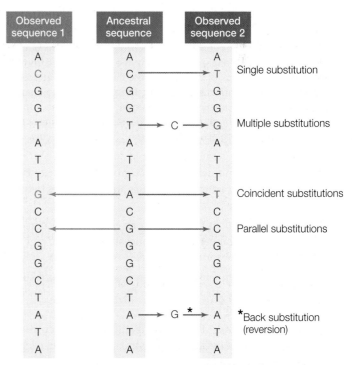

24.2 Multiple Substitutions Are Not Reflected in Pairwise Sequence Comparisons Two observed sequences descended from a common ancestral sequence (center) have undergone a series of substitutions. Although the two observed sequences differ by only three nucleotides (colored letters), these three differences result from a total of nine substitutions (arrows).

Tuna

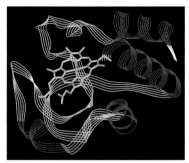

Rice

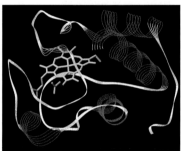

Acidic side chains

D Aspartic acid
E Glutamic acid

Basic side chains

H Histidine
K Lysine
R Arginine

Hydrophobic side chains

F Phenylalanine
I Isoleucine
L Leucine
M Methionine

V Valine
Y Tyrosine
W Tryptophan
A Alanine

Other

C Cysteine
P Proline
Q Glutamine
N Asparagine
S Serine
T Threonine
G Glycine

> The number 1 indicates an invariant position in the cytochrome *c* molecule (i.e., all the organisms have the same amino acid in this position). Such a position is probably under strong purifying selection.

> Amino acids at positions marked by red arrowheads have side chains that interact with the heme group.

Position in sequence: 1 ... 5 ... 10 ... 15 ... 20 ... 25 ... 30

Number of amino acids at the position: 1 3 5 5 5 1 3 3 4 1 4 3 2 1 3 3 1 1 2 4 3 4 2 3 4 2 1 4 1 1 2 1 5 1

Species	Sequence
Human, chimpanzee	G D V E K G K K I F I M K C S Q C H T V E K G G K H K T G P N L H G
Rhesus monkey	G D V E K G K K I F I M K C S Q C H T V E K G G K H K T G P N L H G
Horse	G D V E K G K K I F V Q K C A Q C H T V E K G G K H K T G P N L H G
Donkey	G D V E K G K K I F V Q K C A Q C H T V E K G G K H K T G P N L H G
Cow, pig, sheep	G D V E K G K K I F V Q K C A Q C H T V E K G G K H K T G P N L H G
Dog	G D V E K G K K I F V Q K C A Q C H T V E K G G K H K T G P N L H G
Rabbit	G D V E K G K K I F V Q K C A Q C H T V E K G G K H K T G P N L H G
Gray whale	G D V E K G K K I F V Q K C A Q C H T V E K G G K H K T G P N L H G
Gray kangaroo	G D V E K G K K I F V Q K C A Q C H T V E K G G K H K T G P N L N G
Chicken, turkey	G D I E K G K K I F V Q K C S Q C H T V E K G G K H K T G P N L H G
Pigeon	G D I E K G K K I F V Q K C S Q C H T V E K G G K H K T G P N L H G
Pekin duck	G D V E K G K K I F V Q K C S Q C H T V E K G G K H K T G P N L H G
Snapping turtle	G D V E K G K K I F V Q K C A Q C H T V E K G G K H K T G P N L N G
Rattlesnake	G D V E K G K K I F T M K C S Q C H T V E K G G K H K T G P N L H G
Bullfrog	G D V E K G K K I F V Q K C A Q C H T C E K G G K H K V G P N L Y G
Tuna	G D V A K G K K T F V Q K C A Q C H T V E N G G K H K V G P N L W G
Dogfish	G D V E K G K K V F V Q K C A Q C H T V E N G G K H K T G P N L S G
Samia cynthia (moth)	G N A E N G K K I F V Q R C A Q C H T V E A G G K H K V G P N L H G
Tobacco hornworm moth	G N A D N G K K I F V Q R C A Q C H T V E A G G K H K V G P N L H G
Screwworm fly	G D V E K G K K I F V Q R C A Q C H T V E A G G K H K V G P N L H G
Drosophila (fruit fly)	G D V E K G K K L F V Q R C A Q C H T V E A G G K H K V G P N L H G
Baker's yeast	G S A K K G A T L F K T R C E L C H T V E K G G P H K V G P N L H G
Candida krusei (yeast)	G S A K K G A T L F K T R C A E C H T I E A G G P H K V G P N L H G
Neurospora crassa (mold)	G D S K K G A N L F K T R C A E C H – – E – N L T Q K I G P A L H G
Wheat	G N P D A G A K I F K T K C A Q C H T V D A G A – H K Q G P N L H G
Sunflower	G D P T T G A K I F K T K C A Q C H T V E K G A – H K Q G P N L N G
Mung bean	G D S K S G E K I F K T K C A Q C H T V D K G A – H K Q G P N L N G
Rice	G N P K A G E K I F K T K C A Q C H T V D K G A – H K Q G P N L N G
Sesame	G D V K S G E K I F K T K C A Q C H T V D K G A – H K Q G P N L N G

> Gaps indicate insertion and/or deletion events.

24.3 Amino Acid Sequences of Cytochrome *c* The amino acid sequences shown in the table were obtained from analyses of the enzyme cytochrome *c* from 33 species of plants, fungi, and animals. Note the lack of variation across the sequences at positions 70–80, suggesting that this region is under strong stabilizing selection and that changing its amino acid sequence would impair the protein's function. The molecular models at the upper left are created from these sequences and show the three-dimensional structures of tuna and rice cytochrome *c*. Alpha helices are in red, and the molecule's heme group is shown in yellow.

Go to Media Clip 24.1
The Ubiquitous Protein
Life10e.com/mc24.1

occurred at a given nucleotide position that would not be revealed by a simple count of similarities and differences between two DNA sequences:

- *Multiple substitutions.* More than one change has occurred at a given position between the ancestral sequence and at least one of the observed sequences.

- *Coincident substitutions.* At a given position, different substitutions have occurred between the ancestral sequence and each observed sequence.

- *Parallel substitutions.* The same substitution has occurred independently between the ancestral sequence and each observed sequence.

- *Back substitutions* (also called reversions). In a variation on multiple substitutions, after a change at a given position, a subsequent substitution has changed the position back to the ancestral state.

To correct for undercounting of substitutions, molecular evolutionists have developed mathematical models that describe how DNA (and protein) sequences evolve. These models take into account the relative rates of change from one nucleotide to another; for example, transitions (changes between the two purines, A ↔ G, or between the pyrimidines, C ↔ T) are more frequent than transversions (a purine is replaced by a pyrimidine, or vice versa). These models also include parameters such as the different rates of substitution across different parts of a gene and the proportions of each nucleotide present in a given sequence. Once such parameters have been estimated, the model is used to correct for multiple substitutions, coincident substitutions, parallel substitutions, and back substitutions. The revised estimate accounts for the *total* number of substitutions likely to have occurred between two sequences, which is almost always greater than the observed number of differences.

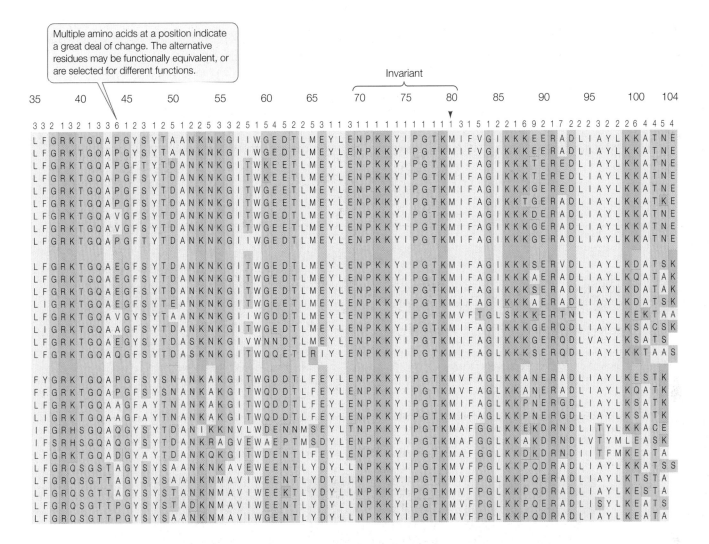

As sequence information becomes available for more and more genes in an ever-expanding database, sequence alignments can be extended across multiple homologous sequences, and the minimum number of insertions, deletions, and substitutions can be summed across homologous genes of an entire group of organisms. Similar databases have been constructed for homologous proteins. **Figure 24.3** shows aligned data for cytochrome *c* protein sequences in 33 species of animals, plants, and fungi. Such information is used extensively in determining evolutionary relationships among species.

Experimental studies examine molecular evolution directly

Although molecular evolutionists are often interested in naturally evolved genes and proteins, molecular evolution can also be observed directly in the laboratory. Increasingly, evolutionary biologists are studying evolution experimentally. Because substitution rates are related to generation time rather than to absolute time, most of these experiments use unicellular organisms or viruses with short generations. Viruses, bacteria, and unicellular eukaryotes (such as the yeasts) can be cultured in large populations in the laboratory, and many of these organisms can evolve rapidly. In the case of some RNA viruses, the

natural substitution rate may be as high as 1 substitution per 1,000 nucleotides per generation. Therefore in a virus of a few thousand nucleotides, one or more substitutions are expected (on average) every generation, and these changes can easily be determined by sequencing the entire viral genome (because of its small size). Generation time may be only tens of minutes (rather than years or decades, as in many animals), so biologists can directly observe substantial molecular evolution in a controlled population over the course of days, weeks, or months.

An example of an experimental evolutionary study is shown in **Figure 24.4**. Paul Rainey and Michael Travisano wanted to examine a potential cause of adaptive radiations, which are a major source of biological diversity (see Section 23.5). While Rainey and Travisano clearly couldn't experimentally manipulate animals over many millions of years, they could test the idea that heterogeneous environments with unoccupied niches lead to adaptive radiation by experimentally manipulating a bacterial lineage.

Rainey and Travisano inoculated several flasks containing culture medium with the same strain of the bacterium *Pseudomonas fluorescens*. They then shook some of the cultures to maintain a constantly uniform environment. They left others alone (static cultures), allowing them to develop spatially

INVESTIGATING**LIFE**

24.4 Evolution in a Heterogeneous Environment   Paul Rainey and Michael Travisano cultured the rapidly evolving bacterium *Pseudomonas fluorescens* in homogeneous and heterogeneous environments to examine the relationship among phenotypic diversity, molecular divergance, and environmental variability.[a]

HYPOTHESIS   Heterogeneous environments are more conducive to the evolution of phenotypic diversity than are homogeneous environments.

Method   One colony of *Pseudomonas fluorescens* (all of a single genotype) was used to inoculate many replicate cultures.

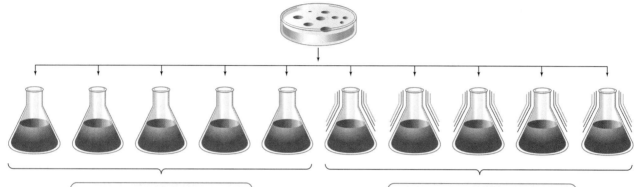

Half of replicate cultures were kept *static*, allowing many different local environments to develop.

The other half of the cultures were *shaken*, keeping the environmental conditions uniform throughout the medium.

Results   In the shaken flasks, the ancestral morph persisted; the uniform environment did not result in morphological diversification. In the static flasks, two new morphs regularly arose, each adapted to a different local environment. Molecular analysis revealed that the mutations that produce these phenotypes arose in both shaken and static cultures, but the mutations did not persist in the uniform (shaken) environment because the phenotypes they produced were selectively disadvantageous under homogeneous conditions.

Ancestral morph (smooth)

"Wrinkly spreader" morph

"Fuzzy spreader" morph

CONCLUSION   Phenotypic change and diversification are enhanced in a heterogeneous environment.

Go to **BioPortal** for discussion and relevant links for all INVESTIGATING**LIFE** figures.

[a]Rainey, P. B. and M. Travisano. 1998. *Nature* 394: 69–72.

distinct environments. In the static cultures, the environment on the surface of the medium differed from that on the walls of the flasks and from parts of the culture not touching any surfaces.

When the cultures were started, the ancestral phenotype of the bacterium produced a smooth colony phenotype, which the investigators called a "smooth morph." In just a few days, however, the static cultures consistently and independently developed two other morphs: a "wrinkly spreader" and a "fuzzy spreader." The researchers determined that the two new morphs had a genetic basis and were adaptively superior in certain environments found within the static cultures. For example, the "wrinkly spreader" cells adhered firmly to one another as well as to surfaces and thus were able to form a mat

across the surface of the medium, where they could compete successfully for oxygen.

DNA sequencing of the genomes of these morphs showed that the same phenotypes had evolved repeatedly in different static cultures and that many different substitutions could produce the same phenotypes. The homogeneous shaken cultures, in contrast, showed no changes in phenotype. The same mutations occurred in the shaken cultures, but they did not persist because the novel phenotypes they produced were selectively disadvantageous (i.e., less fit) under the "shaken" environmental conditions.

Experimental molecular evolutionary studies are used for a wide variety of purposes and have greatly expanded the ability of evolutionary biologists to test evolutionary concepts and principles. Biologists now routinely study evolution in the laboratory and, as we will see later in this chapter, they can use in vitro evolutionary techniques to produce novel molecules that perform new functions with industrial and pharmaceutical uses.

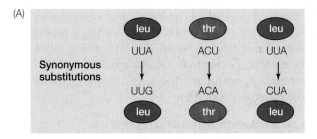

■ **RECAP 24.1**

The genomes of all organisms evolve over time. Evolutionary changes can be detected by comparing the nucleic acid and protein sequences of different species. Experimental studies of molecular evolution allow biologists to study many processes of evolution directly under controlled conditions.

- How do biologists align nucleotide and amino acid sequences they wish to compare, and how do they determine the minimum number of changes that have occurred between pairs of aligned sequences? **See pp. 486–487 and Figure 24.1**
- Explain why a simple count of nucleotide differences between two sequences underestimates the actual number of nucleotide substitutions since the sequences diverged. **See Figure 24.2**

We have seen that molecular evolutionists can directly observe the evolution of genomes over time, compare the genomes of different organisms, and reconstruct the changes that have occurred during their evolution. Let's turn now to the question of how genomes change and examine some of the consequences of those changes.

What Do Genomes Reveal about Evolutionary Processes?

A mutation, as we saw in Section 15.1, is any change in the genetic material. A nucleotide substitution is the product of one type of mutation, incorporated into a population. Many nucleotide substitutions have no effect on phenotype, even if the change occurs in a gene that encodes a protein, because most amino acids are specified by more than one codon (see Figure 14.6). A substitution that does not change the encoded amino acid is known as a **synonymous substitution** or **silent substitution** (Figure 24.5A). Synonymous substitutions do not affect the functioning of a protein (although they may have other effects, as described in Section 15.1) and are therefore less likely than other types of substitutions to be subject to natural selection.

A nucleotide substitution that *does* change the amino acid sequence encoded by a gene is known as a **nonsynonymous substitution** or **missense substitution** (Figure 24.5B). In general, nonsynonymous substitutions are likely to be deleterious to the organism. But not every amino acid replacement alters a protein's shape and charge (and hence its functional properties), so some nonsynonymous substitutions may also be selectively neutral (or nearly so). Conversely, an amino acid replacement that confers an advantage to the organism would result in positive selection for the corresponding nonsynonymous substitution.

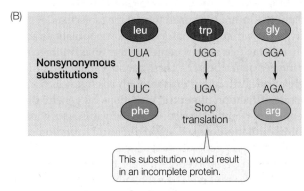

This substitution would result in an incomplete protein.

24.5 When One Nucleotide Does or Doesn't Make a Difference (A) Synonymous (silent) substitutions do not change the amino acid specified and do not affect protein function; such substitutions are unlikely to be subject to natural selection. (B) Nonsynonymous (missense) substitutions do change the amino acid sequence and are likely to have an effect (often deleterious) on protein function; such substitutions are targets for natural selection.

Investigators have measured the average rate of nonsynonymous nucleotide substitutions in several mammalian protein-coding genes at about 0.9 substitutions per position per billion years. Synonymous substitutions in these genes have occurred about five times more frequently than nonsynonymous substitutions. In other words, *substitution rates are highest at nucleotide positions that do not change the amino acid being expressed* (**Figure 24.6**). Substitution rates are even higher in **pseudogenes**, which are duplicate, nonfunctional copies of genes.

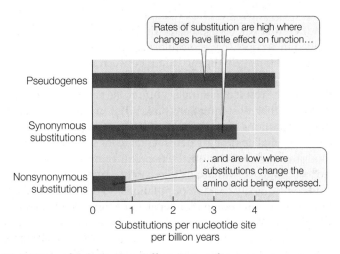

Rates of substitution are high where changes have little effect on function…

…and are low where substitutions change the amino acid being expressed.

24.6 Rates of Substitution Differ Rates of nonsynonymous substitution typically are much lower than rates of synonymous substitution and the substitution rate in pseudogenes. This pattern reflects differing levels of functional constraints.

Most natural populations harbor far more genetic variation than we would expect to find if genetic variation were influenced by natural selection alone. This discovery, combined with the knowledge that many mutations do not change molecular function, stimulated the development of the neutral theory of molecular evolution.

Much of evolution is neutral

In 1968, Motoo Kimura proposed the **neutral theory** of molecular evolution. Kimura suggested that, at the molecular level, the majority of the variants we observe in most populations are selectively neutral; that is, they confer neither an advantage nor a disadvantage on their bearers. These neutral variants accumulate through genetic drift rather than through positive selection.

The rate of fixation of neutral mutations by genetic drift is independent of population size. To see why this is so, consider a diploid population of size N and a neutral mutation rate μ (mu) per gamete per generation at a locus. The number of new mutations would be, on average, $\mu \times 2N$, because $2N$ gene copies are available to mutate. The probability that a given mutation will be fixed by drift alone is its frequency, which equals $1/(2N)$ for a newly arisen mutation. We can multiply these two terms to get the rate of fixation of neutral mutations (m) in a given population of N individuals:

$$m = 2N\mu \frac{1}{2N}$$

Therefore the rate of fixation m of neutral mutations depends only on the neutral mutation rate μ and is independent of population size. A given mutation is more likely to appear in a large population than in a small one, but any mutation that does appear is more likely to become fixed in a small population. These two influences of population size cancel each other out, so the rate of fixation of neutral mutations is equal to the mutation rate (i.e., $m = \mu$).

As as long as the underlying mutation rate is constant, macromolecules evolving in separate populations should diverge from one another in neutral changes at a constant rate. Investigators have confirmed that the rate of evolution of particular genes and proteins is often relatively constant over time and can therefore can be used as a "molecular clock." As we described in Section 22.3, molecular clocks can be used to calculate evolutionary divergence times between species.

Although much of the genetic variation we observe in populations is the result of neutral evolution, the neutral theory does not imply that most mutations have no effect on the organism. Many mutations are never observed in populations because they are lethal or strongly detrimental to the organism and are thus quickly removed from the population through natural selection. Similarly, mutations that confer a selective advantage tend to be quickly fixed in populations, so they do not result in variation at the population level either. Nonetheless, in any population, some amino acid positions will remain constant under purifying selection, others will vary through neutral genetic drift, and still others will differ between species as a result of positive selection for change. How can these evolutionary processes be distinguished?

Positive and purifying selection can be detected in the genome

As we have just seen, substitutions in a protein-coding gene can be either synonymous or nonsynonymous, depending on whether they change the resulting amino acid sequence of the protein. The relative rates of synonymous and nonsynonymous substitutions are expected to differ in regions of genes that are evolving neutrally, under positive selection for change, or staying unchanged under purifying selection.

- If a given amino acid in a protein can be one of many alternatives (without changing the protein's function), then an amino acid replacement is *neutral* with respect to the fitness of an organism. In this case, the rates of synonymous and nonsynonymous substitutions in the corresponding DNA sequences are expected to be very similar, so the ratio of the two rates should be close to 1.

- If a given amino acid position is under *positive* selection for change, the observed rate of nonsynonymous substitutions is expected to exceed the rate of synonymous substitutions in the corresponding DNA sequences.

- If a given amino acid position is under *purifying* selection, then the observed rate of synonymous substitutions is expected to be much higher than the rate of nonsynonymous substitutions in the corresponding DNA sequences.

By comparing the gene sequences that encode homologous proteins from many species, scientists can determine the history and timing of synonymous and nonsynonymous substitutions. This information can be mapped on a phylogenetic tree, as we saw in Chapter 22. Regions of genes that are evolving under neutral, purifying, or positive selection can be identified by comparing the nature and rates of substitutions across the phylogenetic tree.

A study of the evolution of lysozyme illustrates how and why particular amino acid positions might be under different modes of selection (**Figure 24.7**). The enzyme lysozyme (see Figure 3.9) is found in almost all animals. It is produced in the tears, saliva, and milk of mammals and in the albumen (whites) of bird eggs. Lysozyme digests the cell walls of bacteria, rupturing and killing them. As a result, it plays an important role as a first line of defense against invading bacteria. Most animals defend themselves against bacteria by digesting them, which is probably why most animals have lysozyme. Some animals also use lysozyme in the digestion of food.

Among mammals, a mode of digestion called foregut fermentation has evolved twice. In mammals with this mode of digestion, the foregut—the posterior esophagus or the stomach—has been converted into a chamber in which bacteria break down ingested plant matter by fermentation. Foregut fermenters can extract nutrients from the otherwise indigestible cellulose that makes up a large proportion of plant tissue. Foregut fermentation evolved independently in ruminants (a group of hoofed mammals that includes cattle) and in certain leaf-eating monkeys, such as langurs. Caro-Beth Stewart knew that these evolutionary events were independent because both langurs and ruminants have close relatives that are not foregut fermenters.

INVESTIGATINGLIFE

24.7 Convergent Molecular Evolution Langurs (a group of monkeys) and cattle are only distantly related but both have evolved foregut fermentation. They uniquely express the enzyme lysozyme in their stomachs (foreguts) to aid in breaking down bacteria that are involved in fermentation. Stewart and colleagues compared the gene sequences of lysozyme in mammals with and without foregut fermentation to see if there is convergence in the independently evolved amino acid sequences of lysozyme in langurs and cattle.[a]

HYPOTHESIS Similar selective conditions in distantly related mammals have resulted in convergence of adaptations for foregut fermentation in the amino acid sequences of lysozyme.

Method 1. Isolate and sequence lysozyme from two distantly related mammal species with foregut fermentation (langur and cattle) as well as other mammals that are more closely related to either langurs or to cattle but lack foregut fermentation.

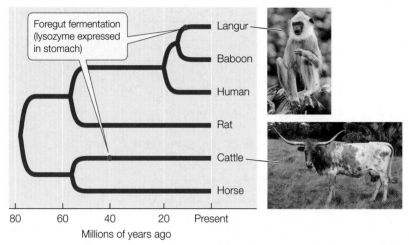

Foregut fermentation (lysozyme expressed in stomach)

Langur
Baboon
Human
Rat
Cattle
Horse

80   60   40   20   Present

Millions of years ago

2. Tabulate the pairwise differences in the amino acid sequences. Plot the amino acid changes on the phylogenetic tree and count the number of convergent similarities between each pair of species. The results can then be plotted as a matrix.

Results The matrix shows the number of *all* pairwise amino acid differences above the diagonal and the number of *convergent* similarities below the diagonal.

	Langur	Baboon	Human	Rat	Cattle	Horse
Langur		14	18	38	32	65
Baboon	0		14	33	39	65
Human	0	1		37	41	64
Rat	0	1	0		55	64
Cattle	5	0	0	0		71
Horse	0	0	0	0	1	

The lysozymes of langurs and cattle are convergent for 5 amino acids.

CONCLUSION The lysozyme sequences of the two species with foregut fermentation account for the majority of the convergent amino acid replacements observed among these species, demonstrating molecular convergence associated with the independent evolution of foregut fermentation.

Go to **BioPortal** for discussion and relevant links for all INVESTIGATINGLIFE figures.

[a]Stewart, C.-B. et al. 1987. *Nature* 330: 401–404.

metabolized by the bacteria, which the mammal then absorbs. How many changes in the lysozyme molecule were needed to allow it to perform this function amid the digestive enzymes and acidic conditions of the mammalian foregut? To answer this question, Stewart and her colleagues compared the lysozyme-coding sequences in foregut fermenters with those in several of their nonfermenting relatives. They determined which amino acids differed and which were shared among the species, as well as the rates of synonymous and nonsynonymous substitutions in lysozyme genes across the evolutionary history of the sampled species.

For many of the amino acid positions of lysozyme, the rate of synonymous substitutions in the corresponding gene sequence was much higher than the rate of nonsynonymous substitutions. This observation indicates that many of the amino acids that make up lysozyme are evolving under purifying selection. In other words, there is selection against change in the protein at these positions, and the observed amino acids must therefore be critical for lysozyme function. At other positions, several different amino acids function equally well, and the corresponding gene sequences had similar rates of synonymous and nonsynonymous substitutions. The most striking finding was that amino acid replacements in lysozyme happened at a much higher rate in the lineage leading to langurs than in any other primate lineage. The high rate of nonsynonymous substitutions in the langur lysozyme gene shows that lysozyme went through a period of rapid change in adapting to the foregut of langurs. Moreover, the lysozymes of langurs and cattle share five unique amino acid replacements, all of which lie on the surface of the lysozyme molecule, well away from the enzyme's active site. Two of these shared replacements involve changes from arginine to lysine, which makes the proteins more resistant to attack by the stomach enzyme pepsin. By understanding the functional significance of amino acid replacements, molecular evolutionists can explain the observed changes in amino acid sequences in terms of changes in the functioning of the protein.

A large body of fossil, morphological, and molecular evidence shows that langurs and ruminants do not share a recent common ancestor. However, langur and ruminant lysozymes share several amino acids that neither mammal shares with the lysozymes of its own closer relatives. The lysozymes of these two mammals

In both foregut-fermenting lineages, lysozyme has been modified to play a new, nondefensive role. This lysozyme ruptures some of the bacteria that live in the foregut, releasing nutrients

Detecting Convergence in Lysozyme Sequences

Original Paper

Stewart, C.-B., J. W. Schilling, and A. C. Wilson. 1987. Adaptive evolution in the stomach lysozymes of foregut fermenters. *Nature* 330: 401–404.

Analyze the Data

Caro-Beth Stewart and her colleagues collected lysozyme sequences from six species of mammals. A small sample of their data is shown in the table. The phylogeny of these six species is well supported from analysis of many genes and much morphological data. Using the phylogenetic tree (see Figure 24.7), plot the amino acid changes across the phylogeny of the six mammals. Assume that the ancestral state is the amino acid present at the base of the tree.

QUESTION 1

Which amino acid positions show unique convergence between the langur and cattle lineages (i.e., the derived state is found *only* in cattle and langurs)?

QUESTION 2

Which additional position is convergent between cattle and the ancestor of langurs and baboons?

QUESTION 3

Did you detect any other convergent amino acid changes between any other pair of lineages? What does this suggest about the convergent changes you observed between cattle and langurs?

Species	Amino acid position										
	2	14	17	21	50	63	75	87	117	118	130
Langur	I	K	L	K	E	Y	D	N	Q	N	V
Baboon	I	R	L	R	Q	Y	N	D	Q	N	V
Human	V	R	M	R	R	Y	N	D	Q	N	V
Rat	T	R	M	Y	Q	Y	N	D	K	N	V
Cattle	V	K	L	K	E	W	D	N	R	D	L
Horse	V	A	M	G	G	W	N	E	K	D	L
Ancestral state	V	R	M	R	Q	W	N	D	K	N	V

Go to BioPortal for all WORKING WITH**DATA** exercises

have undergone convergent evolution at some amino acid positions despite their very different ancestry. The amino acids they share give these lysozymes the ability to lyse the bacteria that ferment plant material in the foregut.

The hoatzin, an unusual leaf-eating South American bird and the only known avian foregut fermenter, offers another remarkable example of the convergent evolution of lysozyme. Many birds have an enlarged esophageal chamber called a crop. The crop of the hoatzin contains lysozyme and bacteria and acts as a fermenting chamber. Many of the amino acid replacements that occurred in the adaptation of hoatzin crop lysozyme are identical to those that evolved in ruminants and langurs. Thus even though the hoatzin and foregut-fermenting mammals have not shared a common ancestor in hundreds of millions of years, they have all evolved similar adaptations in their lysozymes that enable them to recover nutrients from their fermenting bacteria.

Genome size also evolves

We know that genome size varies tremendously among organisms. Across broad taxonomic categories, there is some correlation between genome size and organismal complexity. The genome of the tiny bacterium *Mycoplasma genitalium* has only 470 genes. *Rickettsia prowazekii*, the bacterium that causes typhus, has 634 genes. *Homo sapiens*, by contrast, has about 21,000 protein-coding genes. **Figure 24.8** shows the numbers of genes in a sample of organisms whose genomes have been fully sequenced, arranged by their evolutionary relationships.

As this figure reveals, a larger genome does not always indicate greater complexity (compare rice with the other plants, for example). It is not surprising that more complex genetic instructions are needed for building and maintaining a large multicellular organism than a small single-celled bacterium. What *is* surprising is that some multicellular organisms, such as lungfishes, some salamanders, and lilies, have about 40 times as much DNA as humans do. Structurally, a lungfish or a lily is not 40 times more complex than a human. So why does genome size vary so much?

Differences in genome size are not so great if we take into account only the portion of the DNA that actually encodes RNAs or proteins. Although the organisms with the largest total amounts of nuclear DNA (some ferns and flowering plants) have 80,000 times as much DNA as do the bacteria with the smallest genomes, no species has more than about 100 times as many protein-coding genes as a bacterium. Therefore much of the variation in genome size lies not in the number of functional genes but in the amount of noncoding DNA (**Figure 24.9**).

Why do the cells of most eukaryotic organisms have so much noncoding DNA? Does this noncoding DNA have a function, or is it "junk"? Although some of this DNA does not appear to have a direct function, it can alter the expression of the surrounding genes. The degree or timing of a gene's expression can be changed dramatically depending on the gene's position relative to noncoding sequences. Other regions of noncoding DNA consist of pseudogenes that are carried in the genome simply because the cost of doing so is very small. These

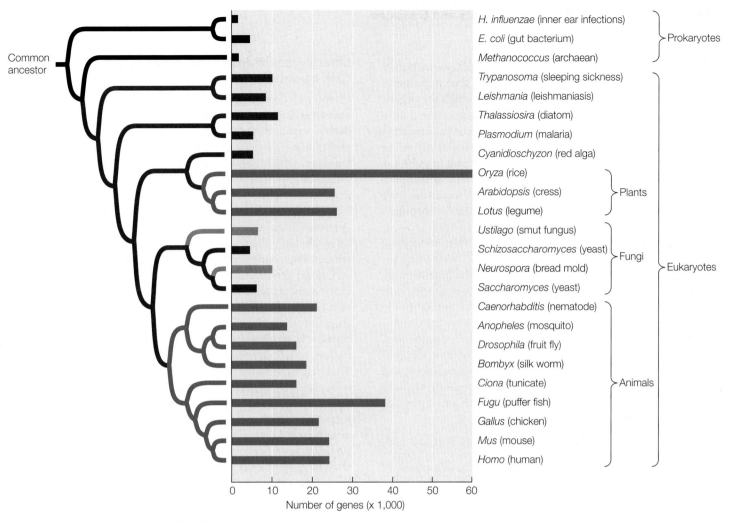

24.8 Genome Size Varies Widely This tree shows the numbers of genes from a sample of organisms whose genomes have been fully sequenced, arranged by their evolutionary relationships. Bacteria and archaea typically have fewer genes than most eukaryotes. Among eukaryotes, multicellular organisms with tissue organization (plants and animals; blue branches) have more genes than single-celled organisms (red branches) or multicellular organisms that lack pronounced tissue organization (green branches).

pseudogenes may become the raw material for the evolution of new genes with novel functions. Some noncoding DNA functions solely in maintaining chromosomal structure. Still other sequences consist of "selfish" transposable elements that proliferate because they reproduce faster than the host genome.

DNA does not just accumulate in genomes over time; noncritical nucleotide sequences are also lost from genomes. Some species differ so much in genome size because they lose noncritical sequences at very different rates. Investigators can use retrotransposons to estimate the rates at which species lose DNA. Retrotransposons are transposable elements (see Figure 17.5) that copy themselves through an RNA intermediate. The most common type of retrotransposon carries duplicated sequences at each end, called long terminal repeats, or LTRs. Occasionally, LTRs recombine in the host genome in

such a way that the DNA between them is excised. When this happens, one recombined LTR is left behind. The number of such "orphaned" LTRs in a genome is a measure of how many retrotransposons have been lost. By comparing the number of LTRs in the genomes of Hawaiian crickets (*Laupala*) and fruit flies (*Drosophila*), investigators found that *Laupala* loses DNA more than 40 times more slowly than does *Drosophila*. Therefore it is not surprising that the genome of *Laupala* is much larger than that of *Drosophila*.

24.9 A Large Proportion of DNA Is Noncoding Most of the DNA of bacteria and yeasts encodes RNAs or proteins, but a large percentage of the DNA of multicellular species is noncoding.

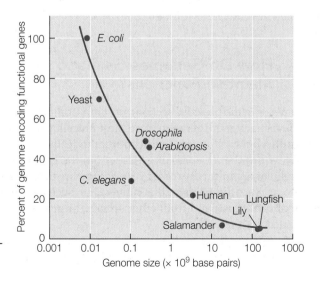

Why do species differ so greatly in the rate at which they gain or lose apparently functionless DNA? One hypothesis is that genome size is related to the rate at which the organism develops, which may be under selection pressure. Large genomes can slow down the rate of development and thus alter the relative timing of expression of particular genes. As discussed in Section 20.2, changes in the timing of gene expression—heterochrony—can produce major changes in phenotype. Thus although some noncoding DNA sequences may have no direct function, they may still affect the development of the organism.

Another hypothesis is that the proportion of noncoding DNA is related primarily to population size. Noncoding sequences that are only slightly deleterious to the organism are likely to be purged by selection most efficiently in species with large population sizes. In species with small populations, the effects of genetic drift can overwhelm selection against noncoding sequences that have small deleterious consequences. Therefore selection against the accumulation of noncoding sequences is most effective in species with large populations, and such species (such as bacteria and yeasts) have relatively little noncoding DNA compared with species with small populations (see Figure 24.9).

■ **RECAP 24.2**

The neutral theory of molecular evolution provides an explanation for the relatively constant rate of molecular change seen in many species. By examining the relative rates of synonymous and nonsynonymous substitutions in genes over time, biologists can distinguish the evolutionary mechanisms acting on individual genes.

- Describe how the ratio of synonymous to nonsynonymous substitutions can be used to determine whether a particular gene region is evolving neutrally, under positive selection, or under purifying selection. **See pp. 492–493**
- Contrast two hypotheses proposed to explain the differences in genome size among different organisms. **See pp. 494–495**

We have examined some of the ways in which organisms can lose DNA without losing gene functions. But how do organisms gain new functions through time?

24.3 How Do Genomes Gain and Maintain Functions?

As we noted in the previous section, most multicellular organisms have many more genes than do most unicellular species. But multicellular organisms evolved from unicellular ancestors. How did the numbers of genes within the genomes of multicellular organisms increase over evolutionary time? There are two primary mechanisms that can result in such increases: genes can be transferred from other species, or genes can be duplicated within species.

Lateral gene transfer can result in the gain of new functions

Chapter 23 described how, through the process of speciation, ancestral lineages divide into descendant lineages, and it is those speciation events that are captured by the branches in the tree of life. However, there are also processes of **lateral gene transfer**, which allow individual genes, organelles, or fragments of genomes to move horizontally from one lineage to another. Some species may pick up fragments of DNA directly from the environment. Other genes may be picked up in a viral genome and transferred to a new host when the virus becomes integrated into the new host's genome. Hybridization between species also results in the transfer of large numbers of genes.

Lateral gene transfer can be highly advantageous to a species that incorporates novel genes from a distant relative. Genes that confer antibiotic resistance, for example, are commonly transferred among different species of bacteria. Lateral gene transfer is another way, in addition to mutation and recombination, in which species can increase their genetic variation. That genetic variation then provides the raw material on which selection acts, resulting in evolution.

A phylogenetic tree constructed from a single laterally transferred genome fragment is likely to reflect only the evolutionary history of that fragment, rather than the overall organismal phylogeny (see Section 26.3). Most biologists prefer to build trees from large samples of genes or their products, so that the underlying species tree (as well as any lateral gene transfer events) can be reconstructed. Depictions of lateral gene transfer events on the underlying species tree are known as **reticulations**.

The degree to which lateral gene transfer events occur in various parts of the tree of life is a matter of considerable current investigation and debate. Lateral gene transfer appears to be relatively uncommon among most eukaryote lineages, although the two major endosymbioses that gave rise to mitochondria and chloroplasts can be viewed as lateral transfers of entire bacterial genomes to the eukaryote lineage. Some groups of eukaryotes, most notably some plants, are subject to relatively high levels of hybridization among closely related species. Hybridization leads to the exchange of many genes among recently separated lineages of plants. The greatest degree of lateral transfer, however, appears to occur among bacteria. Many bacterial genes have been transferred repeatedly among lineages of bacteria, to the point that relationships among bacterial species are often hard to decipher. Nonetheless, the broad relationships of the major groups of bacteria can still be determined (as we will discuss in Part Seven of this book). Lateral transfer of genes also makes it difficult to identify the boundaries of bacterial species, which is one reason why fewer bacterial species have been named than are known to exist.

Most new functions arise following gene duplication

Gene duplication is yet another way in which genomes can acquire new functions. When a gene is duplicated, one copy of that gene is potentially freed from having to perform its original function. The initially identical copies of a duplicated gene can have any one of four subsequent fates:

- Both copies of the gene may retain their original function (which can result in a change in the amount of gene product that is produced by the organism).
- Both copies of the gene may retain the ability to produce the original gene product, but the expression of the genes may diverge in different tissues or at different times in development.
- One copy of the gene may be incapacitated by the accumulation of deleterious substitutions and become a nonfunctional pseudogene, or may be eliminated from the genome altogether.
- One copy of the gene may retain its original function while the second copy accumulates enough substitutions that it can perform a different function.

How often do gene duplications arise, and which of these four outcomes is most likely? Investigators have found that rates of gene duplication are fast enough for a yeast or *Drosophila* population to acquire several hundred duplicate genes over the course of a million years. They have also found that most of the duplicated genes in these organisms are very young. Many extra genes are lost from a genome within 10 million years (which is rapid on an evolutionary time scale).

Some genes may be duplicated many times, resulting in large numbers of related pseudogenes scattered throughout the genome. In the human genome, the functional copy of the ribosomal protein gene *RPL21* is located on chromosome pair 13, but pseudogenes derived from it are found on most of the other chromosome pairs (**Figure 24.10**). Although not all genes are represented by pseudogenes, there are nearly as many known pseudogenes in the human genome as there are functional protein-coding genes.

Although many extra genes disappear rapidly, some duplication events lead to the evolution of genes with new functions. Several successive rounds of duplication and mutation may result in a **gene family**: a group of homologous genes with related functions, often arrayed in tandem along a chromosome. An example of this process is provided by the globin

gene family (see Figure 17.10). The globins were among the first proteins to be sequenced and compared. Comparisons of their amino acid sequences strongly suggest that the different globins arose via gene duplications. These comparisons also allow us to estimate how long the globins have been evolving separately because differences among these proteins have accumulated with time.

Hemoglobin, a tetramer (four-subunit molecule) consisting of two α-globin and two β-globin polypeptide chains, carries oxygen in blood. Myoglobin, a monomer, is the primary O_2 storage protein in muscle. Myoglobin's affinity for O_2 is much higher than that of hemoglobin, but hemoglobin has evolved to be more diversified in its roles. Hemoglobin binds O_2 in the lungs or gills, where the O_2 concentration is relatively high, transports it to deep body tissues, where the O_2 concentration is low, and releases it in those tissues. With its more complex tetrameric structure, hemoglobin is able to carry four molecules of O_2, as well as hydrogen ions and carbon dioxide, in the blood.

To estimate the time of the globin gene duplication that gave rise to the α- and β-globin gene clusters, we can create a **gene tree**—a phylogenetic tree that describes the evolutionary history of particular genes or gene families, in this case the gene sequences that encode the various globins (**Figure 24.11**). The rate of molecular evolution of globin genes has been estimated from other studies, using the divergence times of groups of vertebrates that are well documented in the fossil record. These studies indicate an average rate of divergence for globin genes of about 1 nucleotide substitution every 2 million years. By applying this rate to the globin gene tree, we can estimate the divergence time of the two globin gene clusters at about 450 million years ago.

Many gene duplications affect only one or a few genes at a time, but entire genomes are duplicated in polyploid organisms (which include many plants). When all the genes are duplicated, there are massive opportunities for new functions to evolve. That is exactly what appears to have happened in the evolution of vertebrates. The genomes of the jawed vertebrates appear to have four diploid sets of many major genes, which has led biologists to conclude that two genome-wide duplication events occurred in the ancestor of these species.

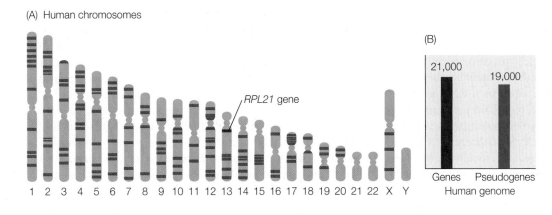

(A) Human chromosomes

RPL21 gene

(B)

21,000
19,000

Genes   Pseudogenes
Human genome

1 2 3 4 5 6 7 8 9 10 11 12 13 14 15 16 17 18 19 20 21 22 X Y

24.10 Some Functional Genes are Duplicated Many Times as Nonfunctional Pseudogenes (A) The functional gene that encodes ribosomal protein *RPL21* is located on human chromosome 13 (indicated in red). In addition, there are many nonfunctional pseudogenes of *RPL21* in the human genome, produced through repeated duplication events (indicated in blue). (B) Although *RPL21* represents a relatively extreme example of pseudogene duplication, there are almost as many known pseudogenes in the human genome as there are functional genes.

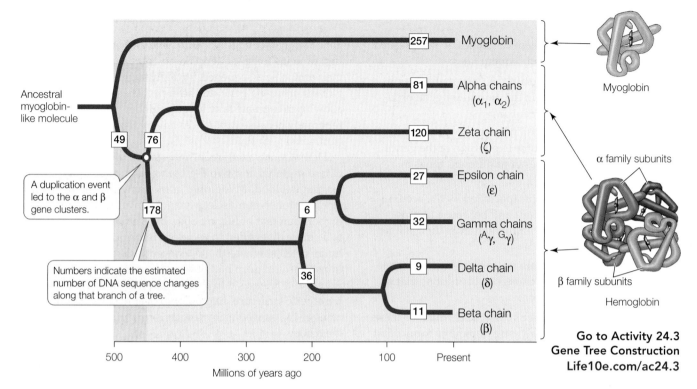

Ancestral myoglobin-like molecule

A duplication event led to the α and β gene clusters.

Numbers indicate the estimated number of DNA sequence changes along that branch of a tree.

257 Myoglobin

81 Alpha chains (α₁, α₂)

120 Zeta chain (ζ)

27 Epsilon chain (ε)

32 Gamma chains (ᴬγ, ᴳγ)

9 Delta chain (δ)

11 Beta chain (β)

Myoglobin

α family subunits

β family subunits

Hemoglobin

500 400 300 200 100 Present

Millions of years ago

Go to Activity 24.3
Gene Tree Construction
Life10e.com/ac24.3

24.11 A Globin Family Gene Tree A molecular clock analysis suggests that the α-globin (blue) and β-globin (green) gene clusters diverged about 450 million years ago, soon after the origin of the vertebrates.

These duplications have allowed considerable specialization of individual vertebrate genes, many of which are now highly tissue-specific in their expression. A good example is the duplication of sodium channel genes, which allowed the evolution of the electric organs of electric fishes described at the opening of this chapter.

Some gene families evolve through concerted evolution

Although the members of the globin gene family have diversified in form and function, the members of many other gene families do not evolve independently of one another. For instance, almost all organisms have many copies (up to thousands) of the ribosomal RNA genes. Ribosomal RNA (rRNA) is the principal structural element of ribosomes and, as such, has a primary role in protein synthesis. Every living species needs to synthesize proteins, often in large amounts (especially during early development). Having many copies of the rRNA genes ensures that organisms can rapidly produce many ribosomes and thereby maintain a high rate of protein synthesis.

Like all portions of the genome, ribosomal RNA genes evolve, and differences accumulate in the rRNA genes of different species. But within any one species, the multiple copies of the rRNA genes are very similar, both structurally and functionally. This similarity makes sense because, ideally, every ribosome in a species should synthesize proteins in the same way. In other words, within a given species, the multiple copies of these rRNA genes are evolving in concert with one another, a phenomenon called **concerted evolution**.

How does concerted evolution occur? Two different mechanisms appear to be responsible. The first of these is **unequal crossing over**. When DNA is replicated during meiosis in a diploid species, the homologous chromosome pairs align and recombine by crossing over (see Section 11.4). In the case of highly repeated genes, however, it is easy for genes to become displaced in alignment, since so many copies of the same genes are present on the chromosomes (**Figure 24.12A**). The end result is that one chromosome may gain extra copies of the repeat and the other chromosome may have fewer copies of the repeat. If a new substitution arises in one copy of the repeat, it can spread to new copies (or be eliminated) through unequal crossing over. Thus, over time, a novel substitution will either become fixed or it will be lost entirely. In either case, all copies of the repeat will remain very similar to one another.

The second mechanism that produces concerted evolution is **biased gene conversion**. This mechanism can be much faster than unequal crossing over and has been shown to be the primary mechanism for concerted evolution of rRNA genes. DNA strands are frequently broken and repaired (see Section 13.4). At many times during the cell cycle, the genes for ribosomal RNA are clustered close together. If damage occurs to one of the genes, a copy of the rRNA gene on another chromosome may be used to repair the damaged copy, and the sequence that is used as a template can thereby replace the original sequence (**Figure 24.12B**). In many cases, this repair system appears to be biased in favor of using particular sequences as templates for repair, and thus the favored sequence rapidly spreads across all copies of the gene. In this way, changes may appear in a single copy and then rapidly spread to all the other copies.

Regardless of the mechanism responsible, the net result of concerted evolution is that the copies of a highly repeated gene do not evolve independently of one another. Mutations still occur, but once they arise in one copy, they either spread rapidly

(A) Unequal crossing over

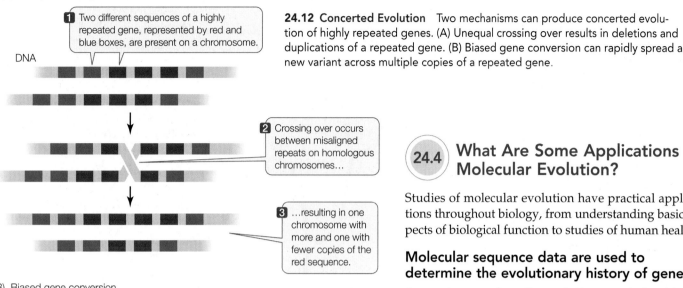

1 Two different sequences of a highly repeated gene, represented by red and blue boxes, are present on a chromosome.

DNA

2 Crossing over occurs between misaligned repeats on homologous chromosomes…

3 …resulting in one chromosome with more and one with fewer copies of the red sequence.

(B) Biased gene conversion

1 Damage occurs to the DNA of one copy of the gene.

2 Damage is repaired using the sequence indicated by red (on a homologous chromosome) as a template…

3 …resulting in one chromosome with more copies of the red sequence.

24.12 Concerted Evolution Two mechanisms can produce concerted evolution of highly repeated genes. (A) Unequal crossing over results in deletions and duplications of a repeated gene. (B) Biased gene conversion can rapidly spread a new variant across multiple copies of a repeated gene.

across all the copies or are lost from the genome completely. This process allows the products of each copy to remain similar over time in both sequence and function.

 Go to Animated Tutorial 24.1
Concerted Evolution
Life10e.com/at24.1

RECAP 24.3

Lateral gene transfer can result in the transfer of genetic functions between even distantly related species. Gene duplication can lead to the evolution of new functions. Some highly repeated genes undergo concerted evolution, which maintains uniform functionality.

- Explain the potential advantages of lateral gene transfer. **See p. 496**

- What are four possible outcomes of gene duplication? **See p. 497**

- Describe the pattern of concerted evolution among highly repeated genes and the mechanisms that lead to concerted evolution. **See p. 498 and Figure 24.12**

We have seen how the principles and methods of molecular evolution have opened new vistas in evolutionary biology. Next we will consider some of the practical applications of this field.

24.4 What Are Some Applications of Molecular Evolution?

Studies of molecular evolution have practical applications throughout biology, from understanding basic aspects of biological function to studies of human health.

Molecular sequence data are used to determine the evolutionary history of genes

A gene tree can show the evolutionary relationships of a single gene in different species, or it can trace the evolution of members of a gene family (as in Figure 24.11). The methods for constructing a gene tree are the same as those we described in Section 22.2 for building phylogenetic trees of species. The process involves identifying differences between genes and using those differences to reconstruct the evolutionary history of the genes. Gene trees are often used to construct phylogenetic trees of species, but the two types of trees are not necessarily equivalent. Processes such as gene duplication can give rise to differences between the phylogenetic trees of genes and species. From a gene tree, biologists can reconstruct the history and timing of gene duplication events and learn how gene diversification has resulted in the evolution of new protein functions.

All the genes of a particular gene family have similar sequences because they have a common ancestry. As we discussed in Section 22.1, features that are similar as a result of common ancestry are said to be homologous. When discussing gene trees, however, we usually need to distinguish between two forms of homology. Homologous genes that are found in different species and whose divergence we can trace to the speciation events that gave rise to those species are called **orthologs**. Homologous genes in the same or different species that are related through gene duplication events are called **paralogs**. When we examine a gene tree, the questions we wish to address determine whether we should compare orthologous or paralogous genes. If we wish to reconstruct the evolutionary history of the species that contain the genes, then our comparison should be restricted to orthologs (because they will reflect the history of speciation events). If we are interested in the changes in function that have resulted from gene duplication events, however, then the appropriate comparison is among paralogs (because they will reflect the history of gene duplication events). If our focus is on the diversification of a gene family through both processes, then we will want to include both paralogs and orthologs in our analysis.

Figure 24.13 depicts a gene tree for the members of a gene family called *engrailed* (its members encode transcription

24.13 Phylogeny of the *engrailed* Genes
The *engrailed* genes are homologous because they share a common ancestor. Speciation events have generated orthologous *engrailed* genes, and gene duplication events (open circles) have generated paralogous *engrailed* genes among bony vertebrates.

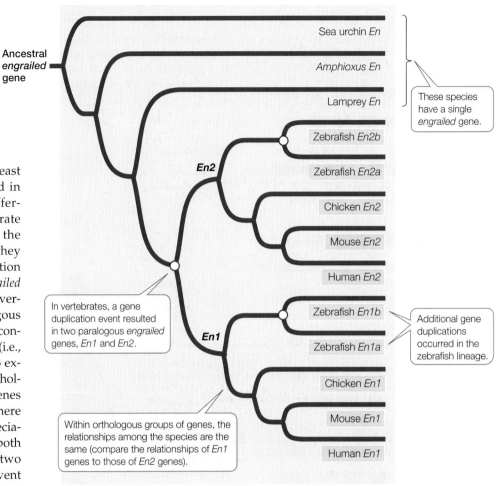

Ancestral *engrailed* gene

Sea urchin *En*

Amphioxus En

Lamprey *En*

These species have a single *engrailed* gene.

Zebrafish *En2b*

Zebrafish *En2a*

Chicken *En2*

Mouse *En2*

Human *En2*

En2

En1

Zebrafish *En1b*

Zebrafish *En1a*

Additional gene duplications occurred in the zebrafish lineage.

Chicken *En1*

Mouse *En1*

Human *En1*

In vertebrates, a gene duplication event resulted in two paralogous *engrailed* genes, *En1* and *En2*.

Within orthologous groups of genes, the relationships among the species are the same (compare the relationships of *En1* genes to those of *En2* genes).

factors that regulate development). At least three gene duplications have occurred in this family, resulting in up to four different *engrailed* genes (*En*) in some vertebrate species (such as the zebrafish). All of the *engrailed* genes are homologs because they have a common ancestor. Gene duplication events have generated paralogous *engrailed* genes (*En1* and *En2*) in some lineages of vertebrates. We could compare the orthologous sequences of the *En1* group of genes to reconstruct the history of the bony vertebrates (i.e., all the vertebrate species in Figure 24.13 except the lamprey), or we could use the orthologous sequences of the *En2* group of genes and expect the same answer (because there is only one history of the underlying speciation events). All bony vertebrates have both groups of *engrailed* genes because the two groups arose from a gene duplication event in the common ancestor of bony vertebrates. If we wanted to focus on the diversification that occurred as a result of this duplication, then the appropriate comparison would be between the paralogous genes of the *En1* versus *En2* groups.

Gene evolution is used to study protein function

Earlier in this chapter we discussed the ways in which biologists can detect regions of genes that are under positive selection for change. What are the practical uses of this information? Consider the evolution of the family of genes encoding voltage-gated sodium channels, which we introduced at the opening of this chapter. Sodium channels have many functions, including the control of nerve impulses in the nervous system. Sodium channels can be blocked by various toxins, such as tetrodotoxin (TTX), a neurotoxin present in the tissues of some puffer fishes and several other animals. A human who eats those tissues of a puffer fish that contain TTX can become paralyzed and die because the toxin blocks sodium channels and prevents nerves and muscles from functioning.

But puffer fishes have sodium channels too, so why doesn't the TTX cause paralysis in the puffer fish itself? The sodium channels of puffer fishes (and other animals that sequester TTX, such as the rough-skinned newt shown in Figure 21.20) have evolved to become resistant to the toxin. Nucleotide substitutions in the puffer fish genome have resulted in changes in the proteins that make up the sodium channels, and those changes prevent TTX from binding to the sodium channel pore.

Several different substitutions that result in TTX resistance have evolved in the various duplicated sodium channel genes of the many species of puffer fish. Many other changes that have nothing to do with the evolution of TTX resistance have occurred in these genes as well. Biologists who study the function of sodium channels can learn a great deal about how the channels work (and about neurological diseases that are caused by mutations in the sodium channel genes) by understanding which changes have been selected for TTX resistance. They do this by comparing the rates of synonymous and nonsynonymous substitutions across the genes in various lineages that have evolved TTX resistance. In a similar manner, molecular evolutionary principles are used to understand function and diversification of function in many other proteins.

In vitro evolution is used to produce new molecules

As biologists studied the relationships among selection, evolution, and function in macromolecules, they realized that molecular evolution could be used in a controlled laboratory environment to produce new molecules with novel and useful functions. Thus were born the applications of **in vitro evolution**.

Living organisms produce thousands of compounds that humans have found useful. The search for such naturally occurring compounds, which can be used for pharmaceutical, agricultural, or industrial purposes, has been termed "bioprospecting." These compounds are the result of millions of years of molecular evolution across millions of species of living organisms. Yet biologists can also imagine molecules that *could* have evolved but, lacking the right combination of selection pressures and opportunities, have not. For instance, we might want to have a molecule that binds a particular environmental contaminant so that the contaminant can be isolated and extracted from the environment. But if the contaminant is synthetic (i.e., not produced naturally), it is unlikely that any living organism will have evolved a molecule with the function we desire. This problem was the inspiration for the field of in vitro evolution.

The principles of in vitro evolution are based on the principles of molecular evolution that we have learned from the natural world. Consider the evolution of a new RNA molecule that was produced in the laboratory using the principles of mutation and selection. This molecule's intended function was to join two other RNA molecules (acting as a ribozyme with a function similar to that of the naturally occurring DNA ligase described in Section 13.3, but for RNA molecules). The process started with a large pool of random RNA sequences (10^{15} different sequences, each about 300 nucleotides long), which were then selected for any ligase activity (**Figure 24.14**). None were very effective ligases, but some were slightly better than others. The best of the ribozymes were selected and reverse-transcribed into cDNA (using the enzyme reverse transcriptase). The cDNA molecules were then amplified using the polymerase chain reaction (PCR; see Figure 13.21).

PCR amplification is not perfect, and it introduced many new mutations into the pool of sequences. These sequences were then transcribed back into RNA molecules using RNA polymerase, and the process was repeated. The ligase activity of the RNAs evolved quickly; after 10 rounds of in vitro evolution, it had increased by about 7 million times. Similar techniques have since been used to create a wide variety of molecules with novel enzymatic and binding functions.

Molecular evolution is used to study and combat diseases

Many of the most problematic human diseases are caused by living, evolving organisms that present a moving target for modern medicine. Recall the example of influenza described at the opening of Chapter 21 and that of HIV in Chapter 22. The control of these and many other human diseases depends on techniques that can track the evolution of pathogenic organisms over time.

The transportation advances of the past century have allowed humans to move around the world with unprecedented speed and frequency. Unfortunately, this mobility has allowed pathogens to be transmitted among human populations at increasing rates, which has led to the global emergence of many "new" diseases. Most of these emerging diseases are caused by viruses. Virtually all new viral diseases have been identified by evolutionary comparison of their genomes with those

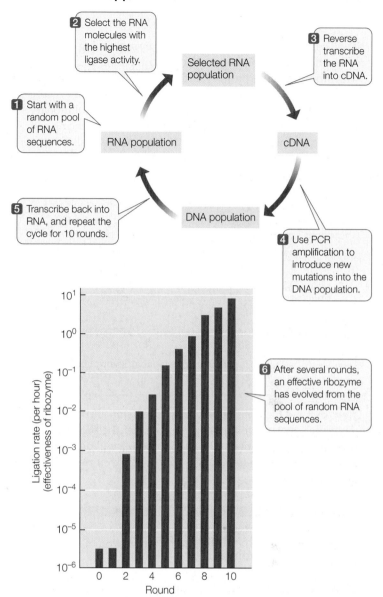

24.14 In Vitro Evolution Starting with a large pool of random RNA sequences, investigators produced a new ribozyme through rounds of mutation and selection for the ability to ligate RNA sequences.

of known viruses. In recent years, for example, rodent-borne hantaviruses have been identified as the source of widespread respiratory illnesses, and the virus (and its host) that causes Sudden Acute Respiratory Syndrome (SARS) was identified using evolutionary comparisons of genes. Studies of the origins, the timing of emergence, and the global diversity of many human pathogens depend on the principles of molecular evolution, as do the efforts to develop and use effective vaccines against these pathogens. For example, the techniques used to develop modern polio vaccines, as well as the methods used to track their effectiveness in human populations, rely on molecular evolutionary approaches.

In the future, molecular evolution will become even more critical to the identification of human (and other) diseases. Once biologists have collected data on the genomes of enough organisms, it will be possible to identify an infection by sequencing a portion of the infecting organism's genome and comparing this sequence with other sequences on an evolutionary tree. At present it is difficult to identify many common viral infections (those that cause "colds," for instance). As genomic databases and evolutionary trees increase, however, automated methods of sequencing and rapid phylogenetic comparison of the sequences will allow us to identify and treat a much wider array of human illnesses.

■ **RECAP** 24.4

Molecular evolution has provided biologists with new tools to understand the functions of macromolecules and how those functions can change over time. These tools can be used to develop synthetic molecules and to identify and combat human diseases.

- Why might a biologist limit a particular investigation to orthologous (as opposed to paralogous) genes? **See pp. 499–500**

- Explain how gene evolution can be used to study protein function. **See p. 500**

- Describe the process of in vitro evolution. **See pp. 500–501 and Figure 24.14**

Now that we have discussed how organisms and biological molecules evolve, we are ready to consider the broader evolutionary history of life on Earth. Chapter 25 will describe the long-term evolutionary changes that have given rise to all of life's diversity.

How do evolutionary studies of sodium channel genes help us understand some human genetic disorders?

ANSWER

Our understanding of how genes and proteins function is largely based on comparative evolutionary analyses of homologous genes across many species. These analyses reveal the parts of the genes that are conserved across species and evolutionary time, as well as the parts of the genes that covary with certain functions (such as the generation of electric signals, as in the case of the sodium channel genes discussed in the chapter opening story). These evolutionary patterns allow biologists to correlate sequence variation with function. These correlations, in turn, allow biologists to predict the causes (and in many cases, suggest treatments) for genetic disorders found in humans. For example, certain mutations in sodium channel genes can cause the inability to feel pain. Other mutations may result in extreme sensitivity to pain, heart disease, or sudden infant death. Understanding the relationship between sequence variation and function is the first step in understanding these disorders, as well as a necessary step in finding treatments for people who suffer from these disorders.

■ **CHAPTERSUMMARY** 24

 24.1 How Are Genomes Used to Study Evolution?

- A **genome** is an organism's full set of genes, regulatory sequences, and structural elements as well as noncoding DNA.

- The field of **molecular evolution** concerns relationships between the structures of genes and proteins and the functions of organisms.

- **Sequence alignments** of proteins or nucleic acids from different organisms allow us to compare the sequences and identify homologous positions. **Review Figure 24.1, ACTIVITY 24.1**

- The minimum number of changes between sequences can be calculated from a **similarity matrix.** Models of sequence evolution can be used to account for changes that cannot be observed directly. **Review Figure 24.2, ACTIVITY 24.2**

24.2 What Do Genomes Reveal about Evolutionary Processes?

- **Nonsynonymous substitutions** of nucleotides result in amino acid replacements in proteins, but **synonymous substitutions** do not. **Review Figure 24.5**

- The **neutral theory** of molecular evolution states that much of the molecular change in nucleotide sequences does not change genome function. The rate of fixation of neutral mutations is independent of population size and is equal to the mutation rate.

- Positive selection for change in a protein-coding gene may be detected by a higher rate of nonsynonymous than synonymous substitutions. The reverse is true of purifying selection.

- Common selective constraints can lead to convergent evolution of amino acid sequences in distantly related species. **Review Figure 24.7**

- The total size of genomes varies much more widely across multicellular species than does the number of functional genes. **Review Figures 24.8, 24.9**

 24.3 How Do Genomes Gain and Maintain Functions?

- **Lateral gene transfer** can result in the rapid acquisition of new functions from distantly related species.

- **Gene duplications** can result in increased production of the gene's product, in nonfunctional **pseudogenes**, or in new gene functions. Several rounds of gene duplication can give rise to multiple genes with related functions, collectively known as a **gene family. Review Figures 24.10, 24.11.**

- **Gene trees** describe the evolutionary history of particular genes or gene families. **See ACTIVITY 24.3**

- Some highly repeated genes undergo **concerted evolution**, in which the multiple copies within the genome maintain their similarity, even as the genes diverge among species. **Review Figure 24.12, ANIMATED TUTORIAL 24.1**

continued

24.4 What Are Some Applications of Molecular Evolution?

- **Orthologs** are genes that are related through speciation events, whereas **paralogs** are genes that are related through gene duplication events. **Review Figure 24.13**

- Protein function can be studied by examining gene evolution. Detection of positive selection can be used to identify molecular changes that have resulted in functional changes.

- **In vitro evolution** is used to produce synthetic molecules with particular desired functions. **Review Figure 24.14**

- Many diseases are identified, studied, and combated through molecular evolutionary investigations.

 Go to the Interactive Summary to review key figures, Animated Tutorials, and Activities
Life10e.com/is24

CHAPTER**REVIEW**

REMEMBERING

1. A higher rate of synonymous than nonsynonymous substitutions in a protein-coding gene is expected under
   a. purifying selection.
   b. positive selection.
   c. neutral evolution.
   d. concerted evolution.
   e. none of the above

2. Before nucleotide and amino acid sequences can be compared in an evolutionary framework, they must be aligned to account for
   a. deletions and insertions.
   b. selection and neutrality.
   c. parallelisms and convergences.
   d. gene families.
   e. all of the above

3. The rate of fixation of neutral mutations, $m = 2N\mu \dfrac{1}{2N}$, is
   a. independent of population size.
   b. higher in small populations than in large populations.
   c. higher in large populations than in small populations.
   d. slower than the rate of fixation of deleterious mutations.
   e. none of the above

4. When a gene is duplicated, which of the following may occur?
   a. Production of the gene's product may increase.
   b. The two copies may become expressed in different tissues.
   c. One copy of the gene may accumulate deleterious substitutions and become functionless.
   d. The two copies may diverge and acquire different functions.
   e. All of the above

5. Genome size varies widely among different multicellular organisms (see graph below). What is the greatest contributing cause of this variation?
   a. The number of protein-coding genes
   b. The amount of noncoding DNA
   c. The number of duplicated genes
   d. The degree of concerted evolution
   e. The amount of positive selection for change in protein-coding genes

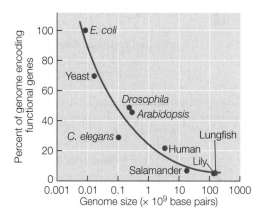

6. Paralogous genes are genes that trace back to a common
   a. speciation event.
   b. substitution event.
   c. insertion event.
   d. deletion event.
   e. duplication event.

▓▓▓▓▓ **UNDERSTANDING & APPLYING**

7. Rates of evolutionary change differ among different molecules, and different species differ widely in generation times and population sizes. How does this variation limit how and in what ways we can use the concept of a molecular clock to help us answer questions about the evolution of both molecules and organisms?

8. Based on what you have learned about evolutionary processes, what modifications could you introduce into in vitro evolution experiments (such as the one shown in Figure 24.14) to increase the diversity or rate of evolution of new molecules?

▓▓▓▓▓ **ANALYZING & EVALUATING**

9. Over evolutionary history, many groups of organisms that inhabit caves have lost the organs of sight. For instance, although surface-dwelling crayfishes have functional eyes, several crayfish species that are restricted to underground habitats lack eyes. Opsins are a group of light-sensitive proteins known to have an important function in vision (see Chapter 46), and opsin genes are expressed in eye tissues. Opsin genes are present in the genomes of eyeless, cave-dwelling crayfishes. Two alternative hypotheses that might explain the presence of opsin genes in an eyeless organism are (1) the genes are no longer experiencing purifying selection because there is no longer selection for function in vision; or (2) the genes are experiencing selection for a function other than vision. How would you investigate these two hypotheses using the sequences of the opsin genes in various species of crayfishes?

10. Analysis of synonymous and nonsynonymous substitutions in protein-coding genes can be used to distinguish neutral evolution, positive selection, and purifying selection. An investigator compared many gene sequences encoding surface proteins from influenza viruses sampled over time and collected the data shown in the table. Use the table to answer the following questions.

Amino acid position	Synonymous substitutions	Nonsynonymous substitutions
12	0	7
15	1	9
61	0	12
80	7	0
137	12	1
156	24	2
165	3	4
226	38	3

a. Which positions encode amino acids that have probably changed as a result of positive selection? Why?

b. Which positions encode amino acids that have probably changed as a result of purifying selection? Why?

(Hint: To calculate rates of each substitution type, you will need to consider the number of synonymous and nonsynonymous substitutions present *relative to the number of possible substitutions of each type*. There are approximately 3 times as many possible nonsynonymous substitutions as there are synonymous substitutions.)

Go to BioPortal at **yourBioPortal.com** for Animated Tutorials, Activities, LearningCurve Quizzes, Flashcards, and many other study and review resources.

25 The History of Life on Earth

Dragonflies as Big as Hawks *Meganeuropsis permiana* is shown here in an artist's reconstruction from fossils. Except for its size, this giant from the Permian period was similar to modern dragonflies (depicted in the inset at the same scale).

ALMOST ANYONE who has spent time around freshwater ponds is familiar with dragonflies. Their bright colors and transparent wings stimulate our visual senses on bright summer afternoons as they fly about their business of devouring mosquitoes, mating, and laying their eggs. The largest dragonflies alive today have wingspans that can be covered by a human hand. Three hundred million years ago, however, dragonflies such as *Meganeuropsis permiana* had wingspans of more than 70 centimeters—well over 2 feet, matching or exceeding the wingspans of many modern birds of prey. These dragonflies were the largest flying predators of their time.

No flying insects alive today are anywhere near this size. But during the Carboniferous and Permian geological periods, 350–250 million years ago, many groups of flying insects contained gigantic species. *Meganeuropsis* probably ate huge mayflies and other giant flying insects that shared its home in the Permian swamps. These enormous insects were themselves eaten by giant amphibians. None of these insects or amphibians would be able to survive on Earth today. The oxygen concentrations in Earth's atmosphere were about 50 percent higher then than they are now, and those high oxygen concentrations are thought to have been necessary to support giant insects and their huge amphibian predators.

Paleontologists have uncovered fossils of *Meganeuropsis permiana* in the rocks of Kansas. How do we know the age of these fossils, and how can we know how much oxygen that long-vanished atmosphere contained? The layering of rocks allows us to tell their ages relative to one another, but it does not by itself indicate a given layer's absolute age.

One of the remarkable scientific achievements of the twentieth century was the development of sophisticated techniques that use the decay rates of various radioisotopes, the ratios of certain molecules in rocks and fossils, and changes in Earth's magnetic field to infer conditions and events in the remote past and to date them accurately. It is those methods that allow us to age the fossils of *Meganeuropsis* and to calculate the concentration of oxygen in Earth's atmosphere at the time.

Earth is about 4.5 billion years old, and life has existed on it for about 3.8 billion of those years. That means human civilizations have occupied Earth for less than 0.0003 percent of the history of life. Discovering what happened before humans were around is an ongoing and exciting area of science.

Can modern experiments test hypotheses about the evolutionary impact of ancient environmental changes?

See answer on p. 522.

TABLE 25.1
Earth's Geological History

Eon	Era	Period	Onset	Major Physical Changes on Earth
Phanerozoic (~0.5 billion years long)	Cenozoic	Quaternary (Q)	2.6 mya	Cold/dry climate; repeated glaciations
		Tertiary (T)	65.5 mya	Continents near current positions; climate cools
	Mesozoic	Cretaceous (K)	145.5 mya	Laurasian continents attached to one another; Gondwana begins to drift apart; meteorite strikes near current Yucatán Peninsula at end of period
		Jurassic (J)	201.6 mya	Two large continents form: Laurasia (north) and Gondwana (south); climate warm
		Triassic (Tr)	251.0 mya	Pangaea begins to drift apart; hot/humid climate
	Paleozoic	Permian (P)	299 mya	Extensive lowland swamps; O_2 levels 50% higher than present; by end of period continents aggregate to form Pangaea, and O_2 levels drop rapidly
		Carboniferous (C)	359 mya	Climate cools; marked latitudinal climate gradients
		Devonian (D)	416 mya	Continents collide at end of period; giant meteorite probably strikes Earth
		Silurian (S)	444 mya	Sea levels rise; two large land masses emerge; hot/humid climate
		Ordovician (O)	488 mya	Massive glaciation; sea level drops 50 meters
		Cambrian (C)	542 mya	Atmospheric O_2 levels approach current levels
Proterozoic	Collectively called the Precambrian (~4 billion years long)		2.5 bya	Atmospheric O_2 levels increase from negligible to about 18%; "snowball Earth" from about 750 to 580 mya
Archean			3.8 bya	Earth accumulates more atmosphere (still almost no O_2); meteorite impacts greatly reduced
Hadean			4.5–4.6 bya	Formation of Earth; cooling of Earth's surface; atmosphere contains almost no free O_2; oceans form; Earth under almost continuous bombardment from meteorites

Note: mya, million years ago; bya, billion years ago.

25.1 How Do Scientists Date Ancient Events?

Some evolutionary changes happen rapidly enough to be studied directly and manipulated experimentally. Plant and animal breeding by agriculturalists and the evolution of surface proteins in influenza viruses are examples of rapid, short-term evolution that we have discussed in previous chapters. Other evolutionary changes, such as the appearance of new species and evolutionary lineages, usually take place over a **geological time scale** (Table 25.1).

To understand long-term patterns of evolutionary change, we must not only think in time scales spanning many millions of years, but also consider events and conditions very different from those we observe today. Earth of the distant past was so unlike our present Earth that it would seem like a foreign planet inhabited by strange organisms. The continents were not where they are now, and climates were sometimes dramatically different from those of today. We know this because much of Earth's history is recorded in its rocks.

We cannot tell the ages of rocks just by looking at them, but we can determine the ages of rocks *relative to one another*. The

first person to formally recognize this method of relative dating was the seventeenth century Danish physician Nicolaus Steno. Steno realized that in undisturbed sedimentary rocks (rocks formed by the accumulation of sediments), the oldest layers of rock, or **strata** (singular *stratum*), lie at the bottom, and successively higher strata are progressively younger.

Geologists subsequently combined Steno's insight with their observations of fossils contained in sedimentary rocks to establish the following principles of **stratigraphy**:

- Fossils of similar organisms are found in widely separated places on Earth.

- Certain fossils are always found in younger strata, and certain other fossils are always found in older strata.

- Organisms found in younger strata are more similar to modern organisms than are those found in older strata.

These patterns revealed much about the relative ages of sedimentary rocks and the fossils they contained, as well as patterns in the evolution of life. But geologists still could not tell the absolute age of these rocks. A method for *absolute* dating of rocks—that is, determining their actual age rather than just their age relative to one another)—did not become available

Major Events in the History of Life
Humans evolve; many large mammals become extinct
Diversification of birds, mammals, flowering plants, and insects
Dinosaurs continue to diversify; mass extinction at end of period (~76% of species lost)
Diverse dinosaurs; radiation of ray-finned fishes; first fossils of flowering plants
Early dinosaurs; first mammals; marine invertebrates diversify; mass extinction at end of period (~65% of species lost)
Reptiles diversify; giant amphibians and flying insects present; mass extinction at end of period (~96% of species lost)
Extensive fern/horsetail/giant club moss forests; first reptiles; insects diversify
Jawed fishes diversify; first insects and amphibians; mass extinction at end of period (~75% of marine species lost)
Jawless fishes diversify; first ray-finned fishes; plants and animals colonize land
Mass extinction at end of period (~75% of species lost)
Rapid diversification of multicellular animals; diverse photosynthetic protists
Origin of photosynthesis, multicellular organisms, and eukaryotes
Origin of life; prokaryotes flourish
Life not yet present

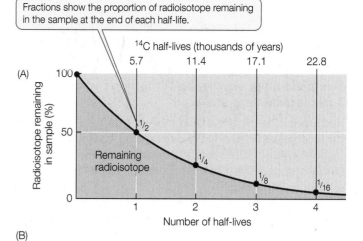

Fractions show the proportion of radioisotope remaining in the sample at the end of each half-life.

25.1 Radioactive Isotopes Allow Us to Date Ancient Rocks   The decay of radioactive isotopes into stable isotopes happens at a steady rate. A half-life is the time it takes for half of the remaining atoms to decay in this way. (A) The graph demonstrates the principle of half-life using carbon-14 (^{14}C) as an example. The half-life of ^{14}C is 5,700 years. (B) Different radioisotopes have different characteristic half-lives that allow us to estimate the ages of rocks.

until after radioactivity was discovered at the beginning of the twentieth century.

Radioisotopes provide a way to date fossils and rocks

Radioactive isotopes of elements—**radioisotopes**—decay in a predictable pattern over long periods (see Section 2.1). Over a specific time interval, known as a **half-life**, half of the atoms in a radioisotope decay to become a different, stable (nonradioactive) isotope (**Figure 25.1A**). The use of this knowledge to date fossils and rocks is known as **radiometric dating**.

To use a radioisotope to date a past event, we must know or estimate the concentration of that isotope at the time of that event, and we must know the radioisotope's half-life. For example, the production of carbon-14 (^{14}C, a radioisotope of carbon) in the upper atmosphere—by the reaction of neutrons with nitrogen-14 (^{14}N, a stable isotope of nitrogen)—just balances the natural radioactive decay of ^{14}C into ^{14}N. Therefore the ratio of ^{14}C to the more common stable isotope of carbon, carbon-12 (^{12}C), is relatively constant in living organisms and in their environment. As soon as an organism dies, however, it ceases to exchange carbon compounds with its environment.

Its decaying ^{14}C is no longer replenished, and the ratio of ^{14}C to ^{12}C in its remains decreases over time. Paleontologists can use the ratio of ^{14}C to ^{12}C in fossil material to date fossils that are less than 60,000 years old (and thus the sedimentary rocks that contain those fossils). If fossils are older than that, so little ^{14}C remains that the limits of detection using this particular isotope are reached.

Radiometric dating methods have been expanded and refined

Sedimentary rocks are formed from materials that existed for varying lengths of time before being weathered, fragmented, and transported, sometimes over long distances, to the site of their deposition. Therefore the radioisotopes in sedimentary rock do not contain reliable information about the date of its formation. Radiometric dating of rocks older than 60,000 years requires estimating radioisotope concentrations in **igneous rock**, which is formed when molten material cools. To date sedimentary strata, geologists search for places where volcanic ash or lava flows have intruded into the sedimentary rock.

A preliminary estimate of the age of an igneous rock determines which radioisotopes can be used to date it (**Figure 25.1B**). The decay of potassium-40 (which has a half-life of 1.3 billion years) to argon-40, for example, has been used to date many of the ancient events in the evolution of life. Fossils in the adjacent sedimentary rock that are similar to those in other rocks of known ages provide additional clues to the rock's age.

Scientists have used several methods to construct a geological time scale

Radiometric dating of rocks, combined with fossil analysis, is the most powerful method of determining geological age. But in places where sedimentary rocks do not contain suitable igneous intrusions and few fossils are present, paleontologists turn to other dating methods.

One method, known as paleomagnetic dating, relates the ages of rocks to patterns in Earth's magnetism, which change over time. Earth's magnetic poles move and occasionally reverse themselves. Both sedimentary and igneous rocks preserve a record of Earth's magnetic field at the time they were formed, and that record can be used to determine the ages of those rocks. Other methods of dating events in life's history use information about continental drift, information about sea level changes, and molecular clocks (a method that was described in Section 22.3).

Geologists used all of these methods to develop a geological time scale (see Table 25.1). They divided the broad history of life into four **eons**. The Hadean eon refers to the time on Earth before life evolved. The early history of life occurred in the Archean eon, which ended about the time that photosynthetic organisms first appeared on Earth. Prokaryotic life diversified rapidly in the Proterozoic eon, and the first eukaryotes in the fossil record date from this time. These three eons are sometimes referred to collectively as Precambrian time, or simply the **Precambrian**. The Precambrian lasted for approximately four billion years and thus accounts for the vast majority of geological time. It was in the Phanerozoic eon, however—a mere 542-million-year time span—that multicellular eukaryotes rapidly diversified. To emphasize the events of the Phanerozoic, Table 25.1 shows the subdivision of this eon into eras and periods. The boundaries between these divisions of time are based largely on the striking differences geologists observe in the assemblages of fossil organisms contained in successive strata. This geological record of life reveals a remarkable story of a world in which the continents and biological communities are constantly changing.

RECAP 25.1

Fossils in sedimentary rock strata enabled geologists to determine the relative ages of organisms, but absolute dating was not possible until the discovery of radioactivity. Geologists divide the history of life into eons, eras, and periods, based on assemblages of fossil organisms found in successive layers of rocks.

- What observations suggested that fossils could be used to determine the relative ages of rocks? **See p. 506**

- How is the rate of decay of radioisotopes used to estimate the absolute ages of rocks? **See p. 507 and Figure 25.1**

As geologists began to develop accurate ways to measure the age of Earth, they began to understand that Earth is far older than anyone had previously understood. During its 4.5-billion-year history, Earth has undergone massive physical changes. These changes have influenced the evolution of life, and life, in its turn, has influenced Earth's physical environment.

25.2 How Have Earth's Continents and Climates Changed over Time?

As we saw in the previous section, the Phanerozoic eon has been notable for the rapid diversification of multicellular eukaryotes. But the diversity of multicellular organisms has not simply increased steadily through time. New species have arisen, and species have gone extinct, throughout the history of life. But there have been times during which extinction rates have increased dramatically over the background levels (**Figure 25.2**). These mass extinction events are the cause of some of the striking differences in fossil assemblages that geologists use to divide the Phanerozoic eon into eras and periods. After each mass extinction, the diversity of life rebounded, although recovery took millions of years. In this section we will discuss some of the physical changes on Earth that have resulted in such dramatic changes in life's diversity.

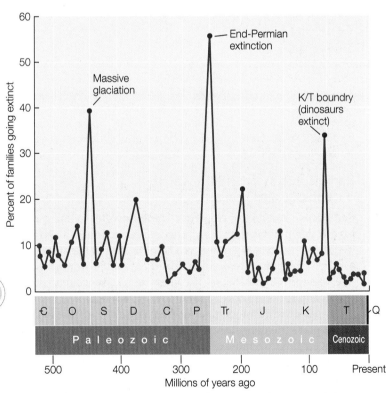

25.2 Periodic Mass Extinctions Mark Many Geologic Boundaries Five sharp rises (marked by red dots) above the background extinction rate have occurred throughout the Phanerozoic. The most sweeping of these events, the end-Permian extinction, was associated with dramatic drops in sea level (see Figure 25.4), global temperature, and atmospheric oxygen level (see Figure 25.8).

25.3 Plate Tectonics and Continental Drift (A) The heat of Earth's core generates convection currents in the viscous mantle material of the asthenosphere. These currents push the continental plates, along with the land masses they carry, together or apart. Where plates collide, one may slide under the other, creating mountain ranges and often volcanoes. (B) The Cascade Range of the Pacific Northwest of North America is an example of a mountain chain produced by subduction of an oceanic plate under a continental plate.

Go to Media Clip 25.1
Lava Flows and Magma Explosions
Life10e.com/mc25.1

(A)

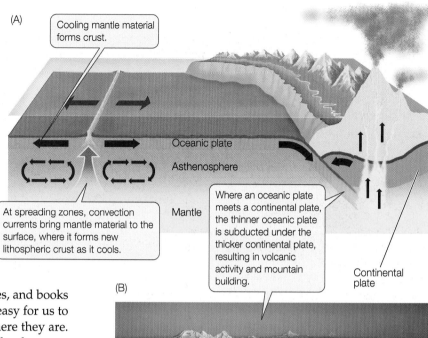

Cooling mantle material forms crust.

At spreading zones, convection currents bring mantle material to the surface, where it forms new lithospheric crust as it cools.

Oceanic plate

Asthenosphere

Mantle

Where an oceanic plate meets a continental plate, the thinner oceanic plate is subducted under the thicker continental plate, resulting in volcanic activity and mountain building.

Continental plate

(B)

The continents have not always been where they are today

The globes and maps that adorn our walls, shelves, and books give an impression of a static Earth. It would be easy for us to assume that the continents have always been where they are. But we would be wrong. The idea that Earth's land masses have changed their positions over the millennia, and that they continue to do so, was put forth in 1912 by the German meteorologist and geophysicist Alfred Wegener. His idea, known as **continental drift**, was initially met with skepticism and resistance. By the 1960s, however, physical evidence and increased understanding of **plate tectonics**—the geophysics of the movement of major land masses—had convinced virtually all geologists of the reality of Wegener's vision. Plate tectonics provided the geological mechanism that explained Wegener's hypothesis of continental drift.

Earth's crust consists of several solid plates, which collectively make up the solid **lithosphere** ("stone sphere"). Thick continental and thinner oceanic lithospheric plates overlie a viscous, malleable layer of Earth's mantle, known as the **asthenosphere** ("weak sphere"). Heat produced by radioactive decay deep in Earth's core sets up large-scale convection currents in the mantle. New crust is formed as mantle material rises between diverging plates, pushing them apart.

Where oceanic plates and continental plates converge, the thinner oceanic plate is forced underneath the thicker continental plate, a process known as **subduction**. Subduction results in volcanism and mountain building on the continental boundary (**Figure 25.3A**). For example, in the Pacific Northwest of North America, a series of volcanoes formed the Cascade mountain range as the Juan de Fuca oceanic plate has been subducted beneath a portion of the continental North American Plate (**Figure 25.3B**). When two oceanic plates collide, one is also subducted below the other, producing a deep oceanic trench and associated volcanic activity. The deepest part of the world's oceans—the Mariana Trench in the western Pacific—formed where two oceanic plates collided. Volcanic activity associated with the subduction at the Mariana Trench produced the nearby Mariana Islands.

When two thick continental plates collide, neither plate is subducted. Instead, the plates push up against one another, forming high mountain chains. The highest mountain chain in the world, the Himalayas, was formed this way when the

Indian Plate collided with the Eurasian Plate. When continental plates diverge, new crust forms in the intervening spaces, resulting in deep clefts called rift valleys in which large freshwater lakes typically form. The Great Rift Valley lakes of eastern Africa, including Lake Malawi (discussed at the opening of Chapter 23), were formed in this way. Two plates can also slide past one another, forming a transform fault boundary (such as the San Andreas Fault that produces violent seismic activity in parts of California).

Many physical conditions on Earth have oscillated in response to plate tectonic processes. We now know that the movement of the plates has sometimes brought continents together and at other times has pushed them apart, as seen in the maps across the top of Figure 25.14. The positions and sizes of the continents influence oceanic circulation patterns, global climates, and sea levels. Sea level is influenced directly by plate tectonic processes (which can influence the depth of ocean basins) and indirectly by oceanic circulation patterns, which affect patterns of glaciation. As climates cool, glaciers form and tie up water over land masses; as climates warm, glaciers melt and release water.

Go to Animated Tutorial 25.1
Movement of the Continents
Life10e.com/at25.1

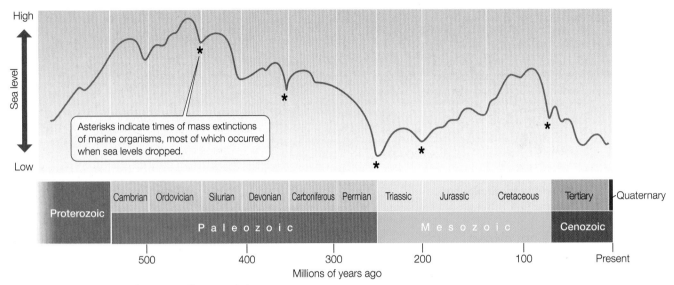

25.4 Sea Levels Have Changed Repeatedly Rapid drops in sea level are associated with periods of globally cooler temperatures and increased glaciation. Most mass extinctions of marine organisms have coincided with low sea levels.

Earth's climate has shifted between hot and cold conditions

Through much of its history, Earth's climate was considerably warmer than it is today, and temperatures decreased more gradually toward the poles. At other times, Earth was colder than it is today. Rapid drops in sea level near the ends of the Ordovician, Devonian, Permian, Triassic, and Cretaceous periods, and most recently in the Quaternary period, resulted mainly from increased global glaciation (**Figure 25.4**). Many of these drops in sea level were accompanied by mass extinctions—particularly of marine organisms, which could not survive the disappearance of the shallow seas that covered vast areas of the continental shelves.

Earth's cold periods were separated by long periods of milder climates. Because we are living in one of the colder periods, it is difficult for us to imagine the mild climates that were found at high latitudes during much of the history of life. The Quaternary period has been marked by a series of glacial advances, interspersed with warmer interglacial intervals during which the glaciers retreated.

"Weather" refers to the daily events at a given location, such as individual storms and the high and low temperatures on a given day. "Climate" refers to long-term average expectations over the various seasons at a given location. Weather often changes rapidly; climates typically change slowly. However, major climate shifts have taken place over periods as short as 5,000 to 10,000 years, primarily as a result of changes in Earth's orbit around the sun. A few climate shifts have been even more rapid: during one Quaternary interglacial period, the ice-locked Antarctic Ocean became nearly ice-free in less than 100 years. Some climate changes have been so rapid that the extinctions caused by them appear to be nearly instantaneous in the fossil record. Such rapid changes are usually caused by sudden shifts in ocean currents.

We are currently living in a time of rapid climate change thought to be caused by a buildup of atmospheric CO_2, primarily from the burning of fossil fuels by human populations. We are reversing the energy transformations that occurred with the massive burial and decomposition of organic material that occurred (especially) in the Carboniferous, Permian, and Triassic, which gave rise to the fossil fuels we are using today. But we are burning these fuels over a few hundred years, rather than the many millions of years over which those deposits accumulated. The current rate of increase in atmospheric CO_2 is unprecedented in Earth's history. A doubling of the atmospheric CO_2 concentration—which may happen during the current century—is expected to increase the average temperature of Earth, change rainfall patterns, melt glaciers and ice caps, and raise sea levels.

Volcanoes have occasionally changed the history of life

Most volcanic eruptions produce only local or short-lived effects, but a few large volcanic eruptions have had major consequences for life. When Krakatau (a volcanic island in the Sunda Strait off Indonesia) erupted in 1883, it ejected more than 25 cubic kilometers of ash and rock as well as large quantities of sulfur dioxide gas (SO_2). The SO_2 was ejected into the stratosphere and carried by high-altitude winds around the planet. Its presence led to high concentrations of sulfurous acid (H_2SO_3) in high-altitude clouds, creating a "parasol effect" that reduced the amount of sunlight reaching Earth's surface. Global temperatures dropped by 1.2°C in the year following the eruption, and global weather patterns showed strong effects for another 5 years. More recently, the eruption of Mount Pinatubo in the Philippines in 1991 (**Figure 25.5**) temporarily reduced global temperatures by about 0.5°C.

Although these individual volcanoes had only relatively short-term effects on global temperatures, they suggest that the simultaneous eruption of many volcanoes could have a much stronger effect on Earth's climate. What would cause many volcanoes to erupt at the same time? The collision of

25.5 Volcanic Eruptions Can Cool Global Temperatures When Mount Pinatubo erupted in 1991, it increased the concentrations of sulfurous acid in high-altitude clouds, which temporarily lowered global temperatures by about 0.5°C.

Iridium-rich layer at the Cretaceous-Tertiary (K/T) boundary

25.6 Evidence of a Meteorite Impact The white layers of rock are Cretaceous in age; the layers at the upper left were deposited in the Tertiary. Between the two is a thin, dark layer of clay that contains large amounts of iridium, a metal common in some meteorites but rare on Earth. Its high concentration in this sediment layer, deposited about 65.5 million years ago, suggests the impact of a large meteorite at that time.

continents during the Permian period, about 275 million years ago (mya), formed a single, gigantic land mass and caused a multitude of massive volcanic eruptions. Emissions from these eruptions blocked considerable sunlight, contributing to the advance of glaciers and a consequent drop in sea level (see Figure 25.4). Thus volcanic eruptions were probably at least part of the explanation for the greatest mass extinction in Earth's history.

Extraterrestrial events have triggered changes on Earth

At least 30 meteorites of sizes between tennis balls and soccer balls strike Earth each year. Collisions with larger meteorites or comets are rare, but such collisions have probably been responsible for several mass extinctions. Several types of evidence tell us about these collisions. Their craters, and the dramatically disfigured rocks that result from their impact, are found in many places. Geologists have discovered compounds in these rocks that contain helium and argon with isotope ratios characteristic of meteorites, which are very different from the ratios found elsewhere on Earth.

A meteorite caused or contributed to a mass extinction at the end of the Cretaceous period (about 65.5 mya). The first clue that a meteorite was responsible came from the abnormally high concentrations of the element iridium found in a thin layer separating rocks deposited during the Cretaceous from rocks deposited during the Tertiary (**Figure 25.6**). Iridium is abundant in some meteorites, but it is exceedingly rare on Earth's surface. When scientists discovered a circular crater 180 km in diameter buried beneath the northern coast of Mexico's Yucatán Peninsula, they constructed the following

scenario. When the meteorite that formed that crater collided with Earth, it released energy equivalent to that of 100 million megatons of high explosives, creating great tsunamis. A massive plume of debris rose into the atmosphere, spread around Earth, and descended. The descending debris heated the atmosphere to several hundred degrees and ignited massive fires. It also blocked the sun, preventing plants from photosynthesizing. The settling debris formed the iridium-rich layer. About a billion tons of soot with a composition matching that of smoke from forest fires were also deposited. These events had devastating effects on biodiversity. Many fossil species (including non-avian dinosaurs) that are found in Cretaceous rocks are not found in the Tertiary rocks of the next stratum.

Oxygen concentrations in Earth's atmosphere have changed over time

As the continents have moved over Earth's surface, the world has experienced other physical changes, including large increases and decreases in atmospheric oxygen concentrations. The atmosphere of early Earth probably contained little or no free oxygen gas (O_2). The increase in atmospheric O_2 came in two big steps more than a billion years apart.

The first step occurred about 2.5 billion years ago (bya), when certain bacteria gained the ability to use water as the source of hydrogen ions for photosynthesis. By chemically splitting H_2O, these bacteria generated O_2 as a waste product. They also made electrons available for reducing CO_2 to form the carbohydrate end-products of photosynthesis (see Section 10.3). The O_2 they produced dissolved in water and reacted with dissolved iron. The reaction product then precipitated

25.7 Banded Iron Formations Indicate Early Photosynthesis The alternating red and dark layers in this 2.25-billion-year-old sedimentary rock formation from Lake Superior resulted from a reaction between dissolved iron and the atmospheric oxygen produced by Earth's first photosynthetic organisms. The chemical reaction produced nearly pure iron oxide, or hematite, which forms the gray, metallic layers in this sample. The red bands are jasper tinged with much smaller amounts of iron oxide.

as iron oxide, which accumulated in alternating layers of red and dark rock known as banded iron formations (**Figure 25.7**). These formations provide evidence for the earliest photosynthetic organisms. As photosynthetic organisms continued to release O_2, oxygen gas began to accumulate in the atmosphere.

The second step occurred about a billion years later, when some of these photosynthetic bacteria became endosymbionts within eukaryotic cells, leading to the eventual evolution of chloroplasts in plants and other photosynthetic eukaryotes. This change resulted in continued accumulation of O_2 in Earth's atmosphere (**Figure 25.8**).

One group of photosynthetic bacteria, the cyanobacteria, formed rocklike structures called stromatolites, which are abundantly preserved in the fossil record. To this day, cyanobacteria still form stromatolites in a few very salty bodies of water (**Figure 25.9**). Cyanobacteria liberated enough O_2 to open the way for the evolution of oxidation reactions as the energy source for the synthesis of ATP.

Thus the evolution of life irrevocably changed the physical nature of Earth. Those physical changes, in turn, influenced the evolution of life. When it first appeared in the atmosphere, O_2 was toxic to most of the anaerobic prokaryotes that inhabited Earth at the time. Over millennia, however, prokaryotes that evolved the ability to tolerate and use O_2 not only survived but gained the advantage. Aerobic metabolism proceeds more rapidly, and harvests energy more efficiently, than anaerobic metabolism. Organisms with aerobic metabolism replaced anaerobes in most of Earth's environments.

An atmosphere rich in O_2 also made possible larger and more complex organisms. Small single-celled aquatic organisms can obtain enough oxygen by simple diffusion even when dissolved oxygen concentrations in the water are very low. Larger single-celled organisms, however, have lower surface area-to-volume ratios; to obtain enough oxygen by simple diffusion, they must live in an environment with a relatively high oxygen concentration. Bacteria can thrive at 1 percent of the current oxygen concentration; eukaryotic cells require levels that are at least 2–3 percent of the current concentration. For concentrations of dissolved oxygen in the oceans to have reached these levels, much higher atmospheric concentrations were needed.

Probably because it took many millions of years for Earth to develop an oxygenated atmosphere, only single-celled prokaryotes lived on Earth for more than 2 billion years. About 1.5 bya, atmospheric O_2 concentrations became high enough for larger eukaryotic cells to flourish. Further increases in atmospheric O_2 concentrations in the late Precambrian enabled several groups of multicellular organisms to evolve (see Figure 25.8).

Oxygen concentrations increased again during the Carboniferous and Permian periods because of the evolution of large vascular plants. These plants lived in the expansive lowland swamps that existed at the time (see Table 25.1). Massive amounts of organic material were buried in these swamps as the plants died, leading to the formation of Earth's vast coal deposits. Because the buried organic material was not subject to oxidation as it decomposed, and because the living plants were producing large quantities of O_2, atmospheric O_2 increased to

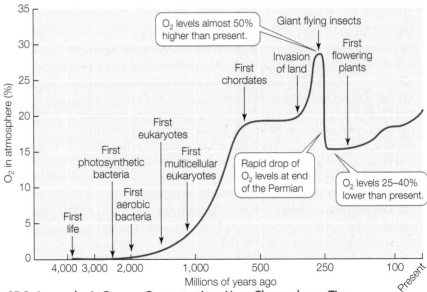

25.8 Atmospheric Oxygen Concentrations Have Changed over Time Changes in atmospheric oxygen concentrations have strongly influenced, and have been influenced by, the evolution of life. (Note that the horizontal axis of the graph is on a logarithmic scale.)

(A)

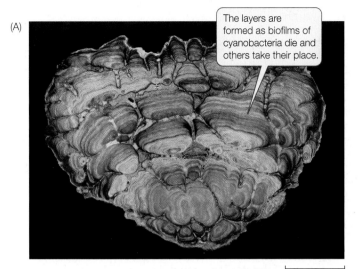

> The layers are formed as biofilms of cyanobacteria die and others take their place.

⊢――――⊣ 15 cm

(B)

> Living cyanobacteria are found in the upper parts of these stromatolites.

⊢――――⊣ 30 cm

25.9 Stromatolites (A) A vertical section through a fossil stromatolite. (B) These rocklike structures are living stromatolites that thrive in the very salty waters of Shark Bay in Western Australia.

INVESTIGATING**LIFE**

25.10 Atmospheric Oxygen Concentrations and Body Size in Insects C. Jaco Klok and his colleagues asked whether insects raised in hyperoxic conditions would evolve to be larger than their counterparts raised under today's atmospheric conditions. They raised strains of fruit flies (*Drosophila melanogaster*) under both conditions to test the effects of increased O_2 concentrations on the evolution of body size.[a]

HYPOTHESIS In hyperoxic conditions, increased partial pressure of oxygen results in evolution of increased body size in flying insects.

Method
1. Separate a population of fruit flies into multiple lineages.
2. Raise half the lineages in current atmospheric (control) conditions; raise the other lineages in hyperoxic (experimental) conditions. Continue all lineages for seven generations.
3. Raise the F_8 individuals of all lineages under identical (current) atmospheric conditions.
4. Weigh 50 flies from each of the replicate lines and test for statistical differences in body weight.

Results The average body mass of F_8 individuals of both sexes raised under hyperoxic conditions was significantly ($P < 0.001$) greater than that of individuals in the control lineages.

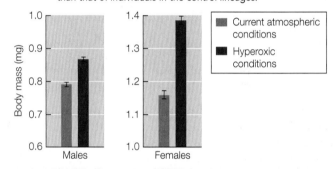

CONCLUSION Increased O_2 concentrations led to evolution of larger body size in fruit flies, consistent with the trends seen among other flying insects in the fossil record.

Go to **BioPortal** for discussion and relevant links for all INVESTIGATING**LIFE** figures.

[a]Klok, C. J. et al. 2009. *Journal of Evolutionary Biology* 22: 2496–2504.

concentrations that have not been reached again in Earth's history (see Figure 25.8). As mentioned at the opening of this chapter, these high concentrations of atmospheric O_2 allowed the evolution of giant flying insects and large amphibians that could not survive in today's atmosphere.

The drying of the lowland swamps at the end of the Permian reduced burial of organic matter as well as the production of O_2, so atmospheric O_2 concentrations dropped rapidly. Over the past 200 million years, with the diversification of flowering plants, O_2 concentrations have again increased, but not to the levels that characterized the Carboniferous and Permian periods.

Biologists have conducted experiments that demonstrate the changing selection pressures that can accompany changes in atmospheric O_2 concentrations. When fruit flies (*Drosophila*) were raised in hyperoxic conditions (i.e., with artificially increased atmospheric concentrations of O_2), they evolved larger body sizes in just a few generations (**Figure 25.10**). The present atmospheric O_2 concentrations appear to constrain body size in these flying insects; increases in O_2 appear to relax those constraints. This experiment demonstrates that the stabilizing

selection on body size at present O_2 concentrations can quickly switch to directional selection when environmental conditions change (see Section 21.3).

▋▋▋▋▋▋▋▋▋▋▋▋▋▋ **RECAP 25.2**

Physical conditions on Earth have changed dramatically over time. Changes in Earth's climate and sea levels have had major effects on biological evolution. Continental drift, volcanic eruptions, and large meteorite strikes have contributed to major climate changes during Earth's history, and many of these climate shifts have resulted in mass extinction events. Changes in atmospheric concentrations of O_2 have also influenced the evolution of life.

- How have plate tectonic processes, volcanic eruptions, and meteorite strikes resulted in mass extinction events? **See pp. 510–511 and Figure 25.4**

- Describe how increases in atmospheric concentrations of O_2 affected the evolution of multicellular organisms. **See pp. 511–512 and Figure 25.8**

WORKING WITH**DATA:**

The Effects of Oxygen Concentration on Insect Body Size

Original Papers

Harrison, J. F. and G. G. Haddad. 2011. Effects of oxygen on growth and size: Synthesis of molecular, organismal and evolutionary studies with *Drosophila melanogaster*. *Annual Review of Physiology* 73: 13.1–13.9.

Harrison, J. F., A. Kaiser and J. M. VandenBrooks. 2010. Atmospheric oxygen level and the evolution of insect body size. *Proceedings of the Royal Society of London B* 277: 1937–1946.

Klok, C. J., A. J. Hubb and J. F. Harrison. 2009. Single and multigenerational responses of body mass to atmospheric oxygen concentration in *Drosophila melanogaster*: Evidence for roles of plasticity and evolution. *Journal of Evolutionary Biology* 22: 2496–2504.

Analyze the Data

In the data shown in Figure 25.10, the body mass of individuals in the experimental population of *Drosophila* increased, on average, about 2 percent per generation under hyperoxic conditions (although the rate of increase was not constant over the experiment). Here you will extrapolate from Harrison et al.'s study to determine whether the observed rate of increase in body mass per generation these researchers observed in *Drosophila* is sufficient to account for the giant dragonflies of the Permian period.

QUESTION 1

Assume that the average rate of increase in dragonfly size during the Permian was much slower than the rate observed in the experiment in Figure 25.10. We'll assume that the actual rate of increase for dragonflies was just 0.01 percent per generation, rather than the 2 percent observed over a few generations for *Drosophila*. We'll assume further that dragonflies complete only one generation per year (as opposed to 40 or more generations for *Drosophila*). Starting with an average body mass of 1 gram, calculate the projected increase in dragonfly body mass over 50,000 years.*

QUESTION 2

What percent of the Permian period does 50,000 years represent? Use Table 25.1 for your calculation.

QUESTION 3

Given your calculations, do you think that increased O_2 concentrations during the Permian were sufficient to account for the evolution of giant dragonflies? Why or why not?

*This calculation is similar to computing compound interest for a savings account. Use the formula $W = S(1 + R)^N$ where W = the final mass, S = the starting mass, R = the rate of increase per generation (0.0001 in this case), and N = the number of generations.)

Go to BioPortal for all WORKING WITH**DATA** exercises

The many dramatic physical events in Earth's history have influenced the nature and timing of evolutionary changes among Earth's living organisms. We now will look more closely at some of the major events that characterize the history of life on Earth.

What Are the Major Events in Life's History?

How do we know about the effects of the physical changes described in the previous section on the evolution of life? To reconstruct life's history, scientists rely heavily on the fossil record. As we have seen, geologists divided Earth's history into eons, eras, and periods based on distinct fossil assemblages (see Table 25.1). Biologists refer to the assemblage of all organisms of all kinds living at a particular time or place as a **biota**. All of the plants living at a particular time or place are its **flora**; all of the animals are its **fauna**.

About 300,000 species of fossil organisms have been described and named, and the number steadily grows. The number of named species, however, is only a tiny fraction of the species that have ever lived. We do not know how many species lived in the past, but we have ways of making reasonable estimates. Of the present-day biota, about 1.8 million species have been named. The actual number of living species is estimated to be in the tens of millions, and possibly much higher, because many species have not yet been discovered and

described by biologists. So the number of described fossil species is only about 3 percent of the estimated minimum number of living species. Life has existed on Earth for about 3.8 billion years. Many species last only a few million years before undergoing speciation or going extinct; therefore Earth's biota must have turned over many times during geological history. So the total number of species that have lived over evolutionary time must vastly exceed the number living today. Why have only about 300,000 of these tens of millions of species been described from fossils to date?

Several processes contribute to the paucity of fossils

Only a tiny fraction of organisms ever become fossils, and only a tiny fraction of fossils are ever discovered by paleontologists. Most organisms live and die in oxygen-rich environments, in which they quickly decompose. Organisms are not likely to become fossils unless they are transported by wind or water to sites that lack oxygen, where decomposition proceeds slowly or not at all. Furthermore, geological processes transform many rocks, destroying the fossils they contain, and many fossil-bearing rocks are deeply buried and inaccessible. Paleontologists have studied only a tiny fraction of the sites that contain fossils, although they find and describe many new ones every year.

The fossil record is most complete for marine animals that had hard skeletons (which resist decomposition). Among the nine major animal groups with hard-shelled members, approximately 200,000 species have been described from

25.11 Insect Fossils Chunks of amber—fossilized tree resin—often contain insects such as this fly. Insects were preserved when they became trapped in the sticky resin.

fossils—roughly twice the number of living marine species in these same groups. Paleontologists lean heavily on these groups in their interpretations of the evolution of life. Insects and spiders are also relatively well represented in the fossil record because they are numerically abundant and have hard exoskeletons (**Figure 25.11**). The fossil record, though incomplete, is good enough to document clearly the factual history of the evolution of life.

By combining evidence of physical changes during Earth's history with evidence from the fossil record, scientists have composed portraits of what Earth and its inhabitants may have looked like at different times. We know in general where the continents were and how life changed over time, but many of the details are poorly known, especially for events in the more remote past.

Precambrian life was small and aquatic

Life first appeared on Earth about 3.8 bya (**Figure 25.12**). The fossil record of organisms that lived prior to the Phanerozoic is fragmentary, but it is good enough to establish that the total number of species and individuals increased dramatically in the late Precambrian.

For most of its history, life was confined to the oceans, and all organisms were small. For more than 3 billion years, all organisms lived in shallow seas. These seas slowly began to teem with microscopic prokaryotes. After the first eukaryotes appeared about 1.5 billion years ago, during the Proterozoic, unicellular eukaryotes and small multicellular animals fed on the microorganisms. Small floating organisms, known collectively as **plankton**, were strained from the water and eaten by slightly larger filter-feeding animals. Other animals ingested sediments on the seafloor and digested the remains of organisms within them. But it still took nearly a billion years before eukaryotes began to diversify rapidly into the many different morphological forms that we know today.

What limited the diversity of multicellular eukaryotes (in terms of their size and shape) for much of their early existence? It is likely that a combination of factors was responsible. We have already noted that O_2 levels increased throughout the Proterozoic, and it is likely that high atmospheric and dissolved O_2 concentrations were needed to support large multicellular organisms. In addition, geologic evidence points to a series of intensely cold periods during the late Proterozoic, which would have resulted in seas that were largely covered by

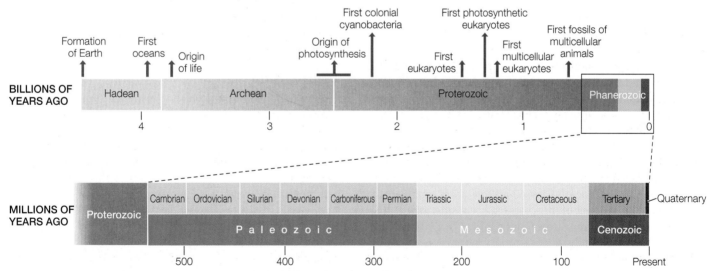

25.12 A Sense of Life's Time The top timeline shows the 4.5-billion-year history of Earth. Most of this history is accounted for by the Precambrian, a 3.4-billion-year time span that saw the origin of life and the evolution of cells, photosynthesis, and multicellularity. The final 600 million years are expanded in the bottom timeline and detailed in Figure 25.14.

25.13 Diversification of Multicellular Organisms: The "Cambrian Explosion" Shortly after the end of Proterozoic glaciations (about 580 mya), several major radiations of multicellular organisms appear in the fossil record. (A) These microscopic fossils from the Doushantuo rock formation of China are the remains of tiny one-, two-, four-, and eight-celled stages of multicellular organisms. (B) Unusual soft-bodied marine invertebrates, unlike any animals alive at present, characterize the fossilized fauna preserved at Ediacara in southern Australia. (C) By the early Phanerozoic, fossilized faunas such as those preserved in Canada's Burgess Shale include extinct representatives of some of the major animal groups alive today.

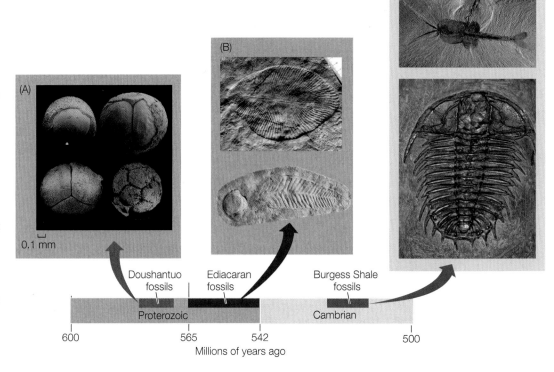

ice and continents that were covered by glaciers. The "snowball Earth" hypothesis suggests that cold conditions confined life to warm places such as hot springs, deep thermal vents, and perhaps a few equatorial oceans that avoided ice cover. The last of these Proterozoic glaciations ended about 580 million years ago, just before several major radiations of multicellular eukaryotes appear in the fossil record (**Figure 25.13**). Many of the multicellular organisms known from the late Proterozoic and early Phanerozoic were very different from any animals living today and may be members of groups that left no living descendants.

Life expanded rapidly during the Cambrian period

The Cambrian period (542–488 mya) marks the beginning of the Paleozoic, the first era of the Phanerozoic. The O_2 concentration in the Cambrian atmosphere was approaching the current level, and the glaciations of the late Proterozoic had ended nearly 40 million years earlier. Earth's land masses had come together to form several large continents. A rapid diversification of life took place that is called the **Cambrian explosion**. This name is somewhat misleading, as the series of radiations it refers to actually began before the start of the Cambrian and continued for about 60 million years into the early Cambrian (see Figure 25.13). Nonetheless, 60 million years represents a relatively short amount of time, especially considering that the first eukaryotes had appeared about a billion (= 1,000 million) years earlier. Many of the major animal groups represented by species alive today first appeared during these evolutionary radiations. **Figure 25.14** provides an overview of the numerous continental and biotic innovations that have characterized the Phanerozoic.

For the most part, fossils tell us only about the hard parts of organisms, but in three known Cambrian fossil beds—the

Burgess Shale in British Columbia, Sirius Passet in northern Greenland, and the Chengjiang site in southern China—the soft parts of many animals were preserved. Crustacean arthropods (crabs, shrimps, and their relatives) are the most diverse group in the Chinese fauna; some of them were large carnivores. Multicellular life was largely or completely aquatic during the Cambrian. If there was life on land at this time, it was probably restricted to microorganisms.

Many groups of organisms that arose during the Cambrian later diversified

Geologists divide the remainder of the Paleozoic era into the Ordovician, Silurian, Devonian, Carboniferous, and Permian periods. Each period is characterized by the diversification of specific groups of organisms. Mass extinctions marked the ends of the Ordovician, Devonian, and Permian.

THE ORDOVICIAN (488–444 MYA) During the Ordovician period, the continents, which were located primarily in the Southern Hemisphere, still lacked multicellular life. Evolutionary radiation of marine organisms was spectacular during the early Ordovician, especially among animals, such as brachiopods and mollusks, that lived on the seafloor and filtered small prey

25.14 A Brief History of Multicellular Life on Earth The geologically rapid "explosion" of life shortly before and during the Cambrian saw the rise of many major animal groups that have representatives surviving today. The following three pages depict life's history through the Phanerozoic. The movements of the major continents during the past half-billion years are shown in the maps of Earth, and associated biotas for each time period are depicted. The artists' reconstructions are based on fossils such as those shown in the photographs.

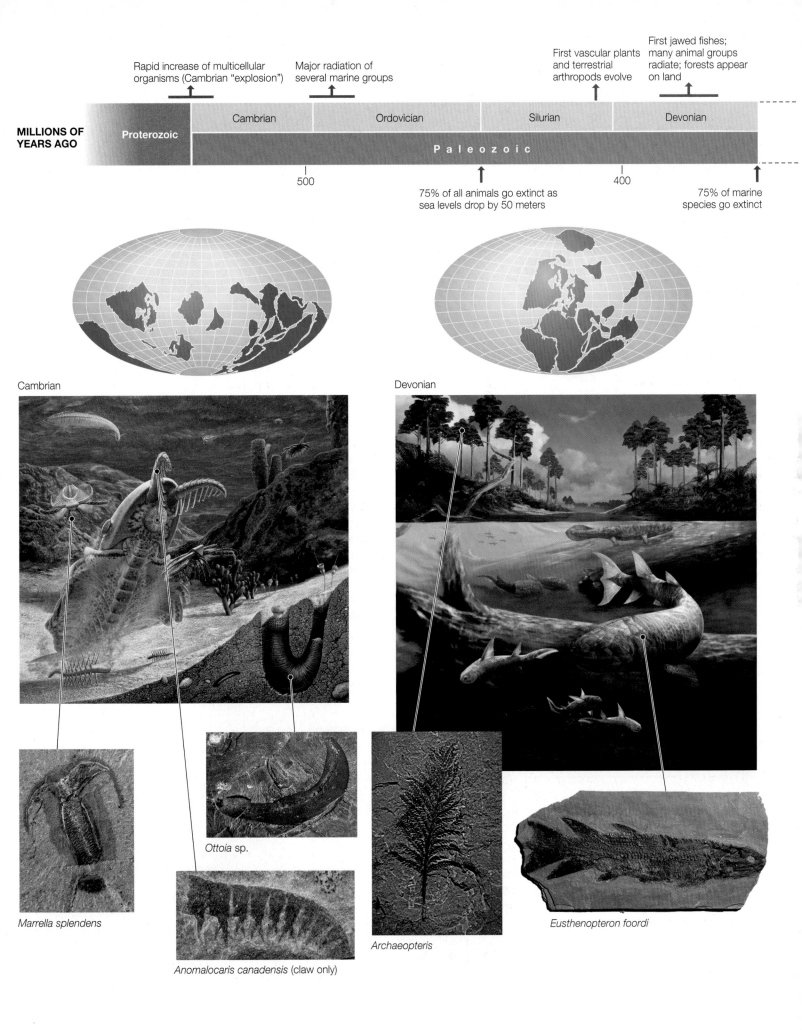

Rapid increase of multicellular organisms (Cambrian "explosion")

Major radiation of several marine groups

First vascular plants and terrestrial arthropods evolve

First jawed fishes; many animal groups radiate; forests appear on land

MILLIONS OF YEARS AGO

Proterozoic | Cambrian | Ordovician | Silurian | Devonian

P a l e o z o i c

500

400

75% of all animals go extinct as sea levels drop by 50 meters

75% of marine species go extinct

Cambrian

Devonian

Marrella splendens

Ottoia sp.

Anomalocaris canadensis (claw only)

Archaeopteris

Eusthenopteron foordi

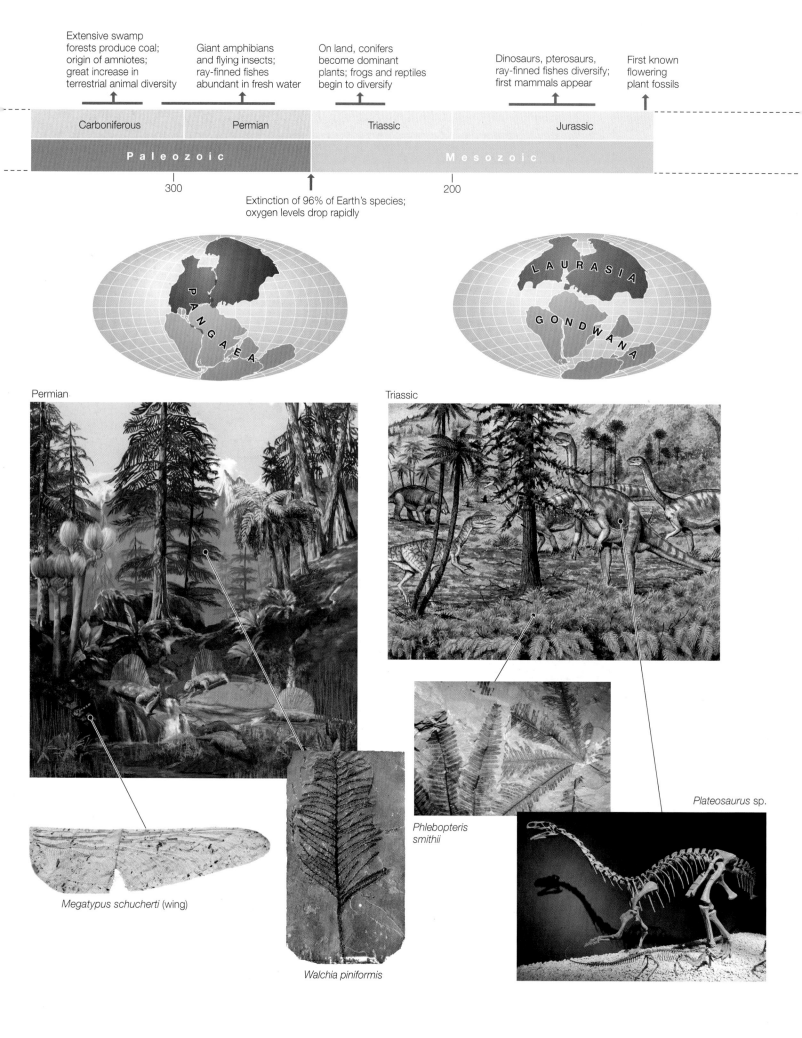

Extensive swamp forests produce coal; origin of amniotes; great increase in terrestrial animal diversity

Giant amphibians and flying insects; ray-finned fishes abundant in fresh water

On land, conifers become dominant plants; frogs and reptiles begin to diversify

Dinosaurs, pterosaurs, ray-finned fishes diversify; first mammals appear

First known flowering plant fossils

| Carboniferous | Permian | Triassic | Jurassic |

| Paleozoic | Mesozoic |

300

200

Extinction of 96% of Earth's species; oxygen levels drop rapidly

PANGAEA

LAURASIA

GONDWANA

Permian

Triassic

Megatypus schucherti (wing)

Walchia piniformis

Phlebopteris smithii

Plateosaurus sp.

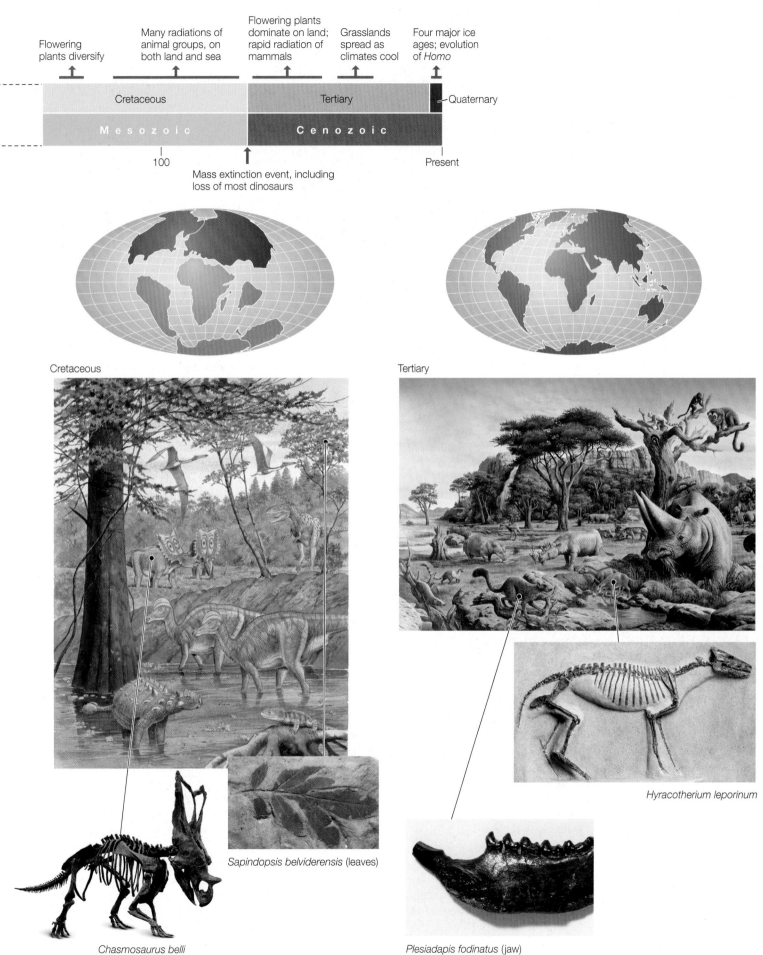

Flowering plants diversify

Many radiations of animal groups, on both land and sea

Flowering plants dominate on land; rapid radiation of mammals

Grasslands spread as climates cool

Four major ice ages; evolution of *Homo*

Cretaceous

Tertiary

Quaternary

Mesozoic

Cenozoic

100

Mass extinction event, including loss of most dinosaurs

Present

Cretaceous

Tertiary

Chasmosaurus belli

Sapindopsis belviderensis (leaves)

Plesiadapis fodinatus (jaw)

Hyracotherium leporinum

25.15 Evidence of Insect Diversification   The margins of this fossil fern leaf from the Carboniferous have been chewed by insects.

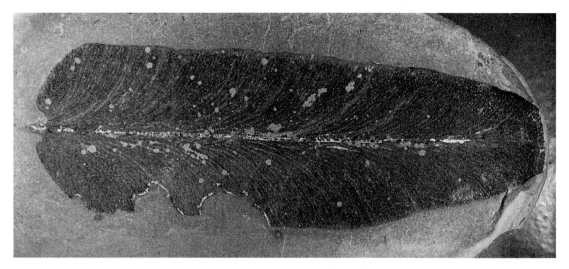

from the water. At the end of the Ordovician, as massive glaciers formed over the southern continents, sea levels dropped about 50 meters, and ocean temperatures dropped. About 75 percent of all animal species became extinct, probably because of these environmental changes.

THE SILURIAN (444–416 MYA)   During the Silurian period, the continents began to cluster together. Marine life rebounded from the mass extinction at the end of the Ordovician. Animals able to swim in open water and feed above the ocean floor appeared for the first time. Jawless fishes diversified, and the first fishes with supporting rays in their fins appeared. The tropical sea was uninterrupted by land barriers, and most marine organisms were widely distributed. On land, the first vascular plants evolved late in the Silurian (about 420 mya). The first terrestrial arthropods—scorpions and millipedes—evolved at about the same time.

THE DEVONIAN (416–359 MYA)   Rates of evolutionary change accelerated in many groups of organisms during the Devonian period. The major land masses continued to move slowly toward each other. In the oceans there were great evolutionary radiations of corals and of shelled, squidlike cephalopod mollusks. Fishes diversified as jawed forms replaced jawless ones and as bony armor gave way to the less rigid scales of modern fishes.

Terrestrial communities changed dramatically during the Devonian. Club mosses, horsetails, and ferns became common; some attained the size of large trees. Their roots accelerated the weathering of rocks, resulting in the development of the first forest soils. The first plants to produce seeds appeared in the Devonian. The earliest fossil centipedes, spiders, mites, and insects date to this period, as do the earliest terrestrial vertebrates.

A mass extinction of about 75 percent of all marine species marked the end of the Devonian. Paleontologists are uncertain about its cause, but two large meteorites that collided with Earth at about that time (one in present-day Nevada, the other in Western Australia) may have been responsible, or at least a contributing factor. The continued coalescence of the continents, with the corresponding reduction in the area of continental shelves, may have also contributed to this mass extinction.

THE CARBONIFEROUS (359–299 MYA)   Large glaciers formed over high-latitude portions of the southern land masses during the Carboniferous period, but extensive swamp forests grew on the tropical continents. These forests were dominated by giant tree ferns and horsetails with small leaves. Their fossilized remains formed the coal we now mine for energy. In the seas, crinoids (sea lilies and feather stars) reached their greatest diversity, forming "meadows" on the seafloor.

The diversity of terrestrial animals increased greatly during the Carboniferous. Snails, scorpions, centipedes, and insects were abundant and diverse. Insects evolved wings, becoming the first animals to fly. Flight gave herbivorous insects easy access to tall plants; plant fossils from this period show evidence of chewing by insects (**Figure 25.15**). The terrestrial vertebrates split into two lineages. The amphibians became larger and better adapted to terrestrial existence, while the sister lineage led to the amniotes: vertebrates with well-protected eggs that can be laid in dry places.

THE PERMIAN (299–251 MYA)   During the Permian period, the continents coalesced into a single supercontinent called **Pangaea**. Permian rocks contain representatives of many of the major groups of insects we know today. By the end of the period the amniotes had split into two lineages: the reptiles, and a second lineage that would lead to the mammals. Ray-finned fishes became common in the fresh waters of Pangaea.

Toward the end of the Permian, conditions for life deteriorated. Massive volcanic eruptions resulted in outpourings of lava that covered large areas of Earth. The ash and gases produced by the volcanoes blocked sunlight and cooled the climate. The death and decay of the massive Permian forests rapidly used up atmospheric oxygen, and the loss of photosynthetic organisms meant that relatively little new atmospheric oxygen was produced. In addition, much of Pangaea was located close to the South Pole by the end of the Permian. All of these factors combined to produce the most extensive continental glaciers since the "snowball Earth" times of the late Proterozoic. Atmospheric oxygen concentrations gradually dropped from about 30 percent to 15 percent. At such low concentrations, most animals would have been unable to survive at elevations above 500 meters; thus about half of the land area would have been uninhabitable at the end of the Permian. The combination of these changes resulted in the most drastic mass extinction in Earth's history. Scientists estimate that about 96 percent of all multicellular species became extinct at the end of the Permian.

Geographic differentiation increased during the Mesozoic era

The few organisms that survived the Permian mass extinction found themselves in a relatively empty world at the start of the Mesozoic era (251 mya). As Pangaea slowly began to break apart, the biotas of the newly separated continents began to diverge. The oceans rose and once again flooded the continental shelves, forming huge, shallow inland seas. Atmospheric oxygen concentrations gradually rose. Life once again proliferated and diversified, but different groups of organisms came to the fore. The three groups of phytoplankton (photosynthetic floating organisms) that dominate today's oceans—dinoflagellates, coccolithophores, and diatoms—became ecologically important at this time, and their remains are the primary origin of the world's oil deposits. Seed-bearing plants replaced the trees that had ruled the Permian forests.

The Mesozoic era is divided into three periods: the Triassic, Jurassic, and Cretaceous. The Triassic and Cretaceous were terminated by mass extinctions, probably caused by meteorite impacts.

THE TRIASSIC (251–201.6 MYA) Pangaea began to break apart during the Triassic period. Many invertebrate groups diversified, and many burrowing animals evolved from groups living on the surfaces of seafloor sediments. On land, conifers and seed ferns were the dominant trees. The first frogs and turtles appeared. A great radiation of reptiles began, which eventually gave rise to crocodilians, dinosaurs, and birds. The end of the Triassic was marked by a mass extinction that eliminated about 65 percent of the species on Earth.

THE JURASSIC (201.6–145.5 MYA) During the Jurassic period, Pangaea became completely divided into two large continents: **Laurasia**, which drifted northward, and **Gondwana** in the south. Ray-finned fishes rapidly diversified in the oceans. The first lizards appeared, and flying reptiles (pterosaurs) evolved. Most of the large terrestrial predators and herbivores of the period were dinosaurs. Several groups of mammals made their first appearance, and the earliest known fossils of flowering plants are from late in this period.

THE CRETACEOUS (145.5–65.5 MYA) By the early Cretaceous period, Laurasia and Gondwana had begun to break apart into the continents we know today. A continuous sea encircled the tropics. Sea levels were high, and Earth was warm and humid. Life proliferated both on land and in the oceans. Marine invertebrates increased in diversity. On land, the reptile radiation continued as dinosaurs diversified further and the first snakes appeared. Early in the Cretaceous, flowering plants began the radiation that led to their current dominance of the land. By the end of the period, many groups of mammals had appeared.

As described in Section 25.2, another meteorite-caused mass extinction took place at the end of the Cretaceous. In the seas, many planktonic organisms and bottom-dwelling invertebrates became extinct. On land, almost all animals larger than about 25 kg in body weight became extinct. Many species of insects died out, perhaps because the growth of their food plants was greatly reduced following the impact. Some species in northern North America and Eurasia survived in areas that were not subjected to the devastating fires that engulfed most low-latitude regions.

Modern biotas evolved during the Cenozoic era

By the early Cenozoic era (65.5 mya), the positions of the continents resembled those of today, but Australia was still attached to Antarctica, and the Atlantic Ocean was much narrower. The Cenozoic was characterized by an extensive radiation of mammals, but other groups were also undergoing important changes.

Flowering plants diversified extensively and came to dominate world forests except in the coolest regions, where the forests were composed primarily of gymnosperms. Mutations of two genes in one group of plants (the legumes) allowed these plants to use atmospheric nitrogen directly by forming symbioses with a few species of nitrogen-fixing bacteria. The evolution of this symbiosis, which can be thought of as the first "green revolution," dramatically increased the amount of nitrogen available for terrestrial plant growth; this symbiosis remains fundamental to the ecological base of life as we know it today.

The Cenozoic era is divided into the Tertiary and the Quaternary periods, which are commonly subdivided into **epochs** (Table 25.2).

TABLE**25.2**
Subdivisions of the Cenozoic Era

Period	Epoch	Onset (mya)
Quaternary	Holocene (Recent)	0.01 (~10,000 years ago)
	Pleistocene	2.6
Tertiary	Pliocene	5.3
	Miocene	23
	Oligocene	34
	Eocene	55.8
	Paleocene	65

THE TERTIARY (65.5–2.6 MYA) During the Tertiary period, Australia began its northward drift. By 20 mya it had nearly reached its current position. The early Tertiary was a hot and humid time, and the ranges of many plants shifted into higher latitudes. The tropics were probably too hot to support rainforest vegetation and instead were clothed in low-lying vegetation. In the middle of the Tertiary, however, Earth's climate became considerably cooler and drier. Many lineages of flowering plants evolved herbaceous (nonwoody) forms, and grasslands spread over much of Earth.

By the start of the Cenozoic era, invertebrate faunas had already come to resemble those of today. It is among the terrestrial vertebrates that evolutionary changes during the Tertiary were most rapid. Frogs, snakes, lizards, birds, and mammals all underwent extensive radiations during this period. Three waves of mammals dispersed from Asia to North America across one of the several land bridges that have intermittently connected the two continents during the past 55 million years. Rodents, marsupials, primates, and hoofed mammals appeared in North America for the first time.

THE QUATERNARY (2.6 MYA TO PRESENT) We are living in the Quaternary period. It is commonly subdivided into two epochs, the Pleistocene and the Holocene (also known as the Recent).

The Pleistocene was a time of drastic cooling and climatic fluctuations. During 4 major and about 20 minor "ice ages," massive glaciers spread across the continents, and the ranges of animal and plant populations shifted toward the equator. The last of these glaciers retreated from temperate latitudes less than 15,000 years ago. Organisms are still adjusting to this change. Many high-latitude ecological communities have occupied their current locations for no more than a few thousand years.

It was during the Pleistocene epoch that divergence within one group of mammals, the primates, resulted in the evolution of the hominoid lineage. Subsequent hominoid radiation eventually led to the species *Homo sapiens*—modern humans. Many large bird and mammal species became extinct in Australia and in the Americas when *H. sapiens* arrived on those continents about 45,000 and 15,000 years ago, respectively. Many paleontologists believe these extinctions were the result of hunting and other influences of *Homo sapiens.*

The tree of life is used to reconstruct evolutionary events

The fossil record reveals broad patterns in life's evolution. To reconstruct major events in the history of life, biologists also rely on the phylogenetic information in the tree of life. We can use phylogeny, in combination with the fossil record, to reconstruct the timing of such major events as the acquisition of mitochondria in the ancestral eukaryotic cell, the several independent origins of multicellularity, and the movement of life onto dry land. We can also follow major changes in the genomes of organisms, and we can even reconstruct many gene sequences of species that are long extinct, as described in Section 22.3.

Changes in Earth's physical environment have clearly influenced the diversity of organisms we see on the planet today. To study the evolution of that diversity, biologists examine the evolutionary relationships among species. Deciphering phylogenetic relationships is an important step in understanding how life has diversified on Earth. The next part of this book will explore the major groups of life and the different solutions these groups have evolved to meet major challenges such as reproduction, energy acquisition, dispersal, and escape from predation.

RECAP 25.3

Life evolved in the oceans about 3.8 billion years ago. It diversified as atmospheric oxygen approached its current level. Numerous climate changes and rearrangements of the continents, as well as meteorite impacts, contributed to five major mass extinctions.

- Why have so few of the multitudes of organisms that have existed over millennia become fossilized? **See pp. 514–515**
- What do we mean when we refer to the "Cambrian explosion"? **See p. 516**
- What are the major changes that have occurred in terrestrial biotic communities over the course of the Phanerozoic? **See pp. 516, 520–522 and Figure 25.14**

Can modern experiments test hypotheses about the evolutionary impact of ancient environmental changes?

ANSWER

Several experiments have been conducted to test the link between O_2 concentrations and evolution of body size in flying insects (one of these is discussed in Figure 25.10). Results of these experiments are consistent with the evolution of larger body size in flying insects in hyperoxic (high-oxygen) environments.

Experiments have also been conducted under hypoxic (low-oxygen) conditions, such as existed at the end of the Permian. Results of these experiments suggest that the evolution of body size is constrained under hypoxic conditions, even under strong artificial selection for larger body size. These latter results are consistent with the extinction of many of the large flying insects at the end of the Permian, the result of rapidly decreasing O_2 concentrations. Giant flying insects simply could not have survived the lower O_2 concentrations that existed at that time. The mass extinction at the end of the Permian is the only known mass extinction that involved considerable loss of insect diversity.

 25.1 How Do Scientists Date Ancient Events?

- The relative ages of organisms can be determined by the **strata** of **sedimentary rocks** in which their fossils are found.

- Paleontologists use a variety of **radioisotopes** with different **half-lives** to date events at different times in the remote past. **Review Figure 25.1**

- Geologists divide the history of life into **eons**, **eras**, and **periods**. These divisions are based largely on major differences in the fossil assemblages found in successive layers of rocks. **Review Table 25.1**

 25.2 How Have Earth's Continents and Climates Changed over Time?

- **Plate tectonic** processes result in **continental drift** as well as volcanism and mountain building. Changes in the positions and sizes of the continents affect oceanic circulation patterns, climate, and sea levels. **Review Figure 25.3, ANIMATED TUTORIAL 25.1**

- Major physical events on Earth, such as the collision of continents that formed the single gigantic land mass **Pangaea**, have affected Earth's surface, climate, and atmosphere. In addition, extraterrestrial events such as meteorite strikes have created sudden and dramatic climate shifts. Some dramatic changes in physical conditions on Earth have caused mass extinctions. **Review Figure 25.4**

- Oxygen-generating cyanobacteria released enough O_2 to open the door to oxidation reactions in metabolic pathways. Aerobic prokaryotes were able to harvest more energy than anaerobic organisms and began to predominate. Increases in atmospheric

O_2 levels also supported the evolution of large eukaryotic cells. **Review Figure 25.8**

 25.3 What Are the Major Events in Life's History?

- Paleontologists use the fossil record and evidence of geological changes to determine what Earth and its **biota** may have looked like at different times. **Review Figure 25.12**

- During most of its history, life was confined to the oceans. Multicellular life diversified extensively during the **Cambrian explosion**. **Review Figure 25.13**

- The remaining periods of the Paleozoic era were each characterized by the diversification of specific groups of organisms. The Paleozoic ended with the most drastic mass extinction in Earth's history, at the end of the Permian. **Review Figure 25.14**

- During the Mesozoic era, distinct terrestrial biotas evolved on each continent. Dinosaurs diversified to become the dominant large predators and herbivores. The era ended with a mass extinction event caused by the collision of a giant meteorite with Earth.

- The Cenozoic era is divided into the Tertiary and the Quaternary periods, which in turn are subdivided into **epochs**. This era saw the emergence of the modern biota as mammals radiated extensively and the angiosperms (flowering plants) became dominant. **Review Table 25.2**

See ACTIVITY 25.1 for a concept review of this chapter

 Go to the Interactive Summary to review key figures, Animated Tutorials, and Activities
Life10e.com/is25

CHAPTER**REVIEW**

▬▬▬ REMEMBERING

1. In undisturbed strata of sedimentary rock, the oldest rocks
   a. lie at the top.
   b. lie at the bottom.
   c. are in the middle.
   d. are distributed among the strata of younger rocks.
   e. None of the above

2. The concentration of oxygen in Earth's atmosphere
   a. has increased steadily over time.
   b. has decreased steadily over time.
   c. has been both higher and lower in the past than at present.
   d. was lower during most of the Permian than at present.
   e. was at its highest levels in the Cambrian.

3. Many of the coal beds we now mine for energy are largely the remains of
   a. plants that grew in swamps during the Carboniferous period.
   b. algae that grew in marshes during the Devonian period.
   c. giant insects and amphibians of the Permian period.
   d. plants that grew in the oceans during the Carboniferous period.
   e. None of the above

4. The mass extinction at the end of the Ordovician period was probably caused by
   a. the collision of Earth with a large meteorite.
   b. massive volcanic eruptions.
   c. massive glaciation on the southern continents and associated climate changes.
   d. the coming together of the continents to form Pangaea.
   e. changes in Earth's orbit.

5. The cause of the mass extinction at the end of the Mesozoic era probably was
   a. continental drift.
   b. the collision of Earth with a large meteorite.
   c. changes in Earth's orbit.
   d. massive glaciation.
   e. changes in the salt concentration of the oceans.

6. Which of the following times was marked by the largest mass extinction of life in the history of Earth?
   a. The end of the Cretaceous
   b. The end of the Devonian
   c. The end of the Permian
   d. The end of the Triassic
   e. The end of the Silurian

▓ UNDERSTANDING & APPLYING

7. Scientists date ancient events using a variety of methods, but nobody was present to witness or record those events. Accepting those dates requires us to understand the accuracy and appropriateness of indirect measurement techniques. What other basic scientific concepts are also based on the results of indirect measurement techniques?

8. Why is it useful to be able to date past events absolutely as well as relatively?

9. What conditions may have favored the evolution of multicellular groups of organisms near the end of the Precambrian?

▓ ANALYZING & EVALUATING

10. The experiment in Figure 25.10 showed that the body size of insects may evolve quickly following changes in atmospheric oxygen concentrations. What other experiments could you devise to test the effects of changing atmospheric oxygen concentrations?

Go to BioPortal at **yourBioPortal.com** for Animated Tutorials, Activities, LearningCurve Quizzes, Flashcards, and many other study and review resources.

26

Bacteria, Archaea, and Viruses

CHAPTER**OUTLINE**

ONE NIGHT IN JANUARY of 1995, the British merchant vessel *Lima* was off the coast of Somalia, near the Horn of Africa. This area is infamous for bands of pirates, so the crew was keeping a watchful eye on the seas when they spotted an eerie, whitish glow on the horizon. It was directly in their path, and there was no way to avoid it. Was this strange sight the result of some strange trick of piracy?

Within 15 minutes of first sighting the glow, the *Lima* was surrounded by shining waters for as far as her crew could see. The ship's log recorded that "it appeared as though the ship was sailing over a field of snow or gliding over the clouds." Fortunately for the crew, the glow had nothing to do with pirates.

For centuries, mariners in this part of the world had reported occasional "milky seas" in which the sea surface produced a strange glow at night, extending from horizon to horizon. Scientists up to that point had never been able to confirm the reality or the cause of such phenomena. It was well established, however, that many organisms emit light by bioluminescence, a complex, enzyme-catalyzed biochemical reaction that emits light but not heat.

What kind of organism could cause the vast expanse of bioluminescence observed by the *Lima*? Some marine organisms emit flashes of light when they are disturbed, but they could not have produced the sustained and uniform glow seen in milky seas. The only organisms known to produce the level of sustained bioluminescence consistent with milky seas are certain prokaryotes, such as bacteria of the genus *Vibrio*. Using information supplied by the *Lima*, biologists scanned satellite images of the Indian Ocean for the specific light wavelengths emitted by *Vibrio*. The satellite images clearly identified thousands of square kilometers of *Vibrio*-produced milky seas.

Vibrio's bioluminescence requires a critical concentration of a specific chemical signal produced by the bacteria, so at low densities, free-living *Vibrio* populations do not glow. But as a colony establishes itself on phytoplankton, the bacteria's population density increases and concentrations of the luminescence signal build up. Eventually bacterial density—and concentrations of the signal—become high enough for the huge colony to produce light at a rate of about 10^3 photons per second per *Vibrio* cell. Such chemical-induced action among bacterial cells is referred to as quorum sensing.

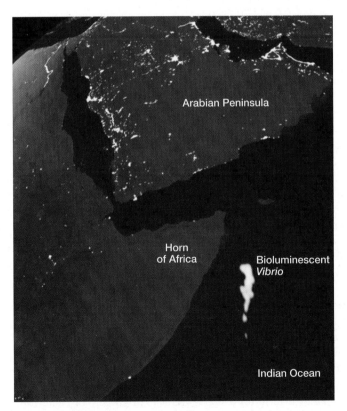

Bacteria Seen from Space A satellite image reveals vast expanses of bioluminescence in the Indian Ocean. Spreading over thousands of square kilometers, this glowing "milky sea" is produced by dense populations of *Vibrio* bacteria.

Arabian Peninsula

Horn of Africa

Bioluminescent *Vibrio*

Indian Ocean

What adaptive advantage does bioluminescence provide to *Vibrio* bacteria?

See answer on p. 546.

26.1 Where Do Prokaryotes Fit into the Tree of Life?

You may think that you have little in common with a bacterium. But all multicellular eukaryotes—including you—share many attributes with bacteria and archaea, together called **prokaryotes**. For example, all organisms, whether eukaryotes or prokaryotes,

- have plasma membranes and ribosomes (see Chapters 5 and 6).
- have a common set of metabolic pathways (see Chapters 8 and 9).
- replicate DNA semiconservatively (see Chapter 13).
- use DNA as the genetic material to encode proteins, and use similar genetic codes to produce those proteins by transcription and translation (see Chapter 14).

These shared features support the conclusion that all living organisms share a common ancestor. If life had multiple origins, there would be little reason to expect all organisms to use overwhelmingly similar genetic codes or to share structures as unique as ribosomes. Furthermore, similarities in the DNA sequences of universal genes (such as those that encode the structural components of ribosomes) confirm the monophyly of life.

Despite these commonalities, major differences have also evolved across the diversity of life. Based on the differences in cell structure and biochemical functioning that they have observed, many biologists now recognize three domains (primary divisions) of life, two prokaryotic and one eukaryotic (**Figure 26.1**).

All prokaryotic organisms are unicellular, although they may form large, coordinated colonies or communities consisting of many individuals. The domain Eukarya, by contrast, encompasses both unicellular and multicellular life forms. As we saw in Chapter 5, prokaryotic cells differ from eukaryotic cells in some important ways:

- *Prokaryotic cells do not divide by mitosis.* Instead, after replicating their DNA, prokaryotic cells divide by their own method, binary fission (see Section 11.1).

- *The organization of the genetic material differs.* The DNA of the prokaryotic cell is not organized within a membrane-enclosed nucleus. DNA molecules in prokaryotes are often circular. Many (but not all) prokaryotes have only one main chromosome and are effectively haploid, although many have additional smaller DNA molecules, called plasmids (see Section 12.6).

- *Prokaryotes have none of the membrane-enclosed cytoplasmic organelles that are found in most eukaryotes.* However, the cytoplasm of a prokaryotic cell may contain a variety of infoldings of the plasma membrane and photosynthetic membrane systems that are not found in eukaryotes.

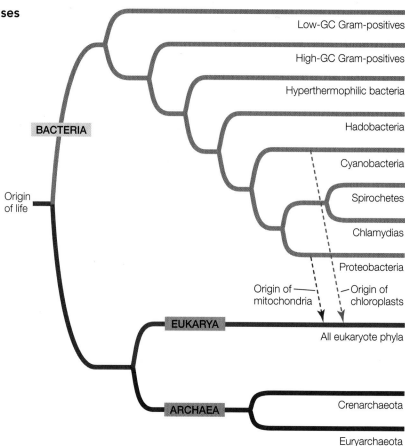

26.1 The Three Domains of the Living World This summary classification of the domains Bacteria and Archaea shows their relationships to each other and to Eukarya. The relationships among the many clades of bacteria, not all of which are listed here, are incompletely resolved at this time.

 Go to Animated Tutorial 26.1
The Evolution of the Three Domains
Life10e.com/at26.1

Although the study and classification of eukaryotic organisms goes back centuries, much of our knowledge of the evolutionarily ancient prokaryotic domains is extremely recent. Not until the final quarter of the twentieth century did advances in molecular genetics and biochemistry enable the research that revealed the deep-seated distinctions between the domains Bacteria and Archaea.

The two prokaryotic domains differ in significant ways

A glance at **Table 26.1** will show you that there are major differences between the two prokaryotic domains (most of which cannot be seen even under an electron microscope). In some ways archaea are more like eukaryotes; in other ways they are more like bacteria. (Note that we use lowercase when referring to members of these domains and initial capitals when referring to the domains themselves.) The basic unit of an archaeon (the term for a single archaeal organism) or bacterium (a single bacterial organism) is the prokaryotic cell. Each single-celled prokaryote contains a full complement of genetic and protein-synthesizing systems, including DNA, RNA, and all

TABLE**26.1** ■
The Three Domains of Life

Characteristic	Domain		
	Bacteria	Archaea	Eukarya
Membrane-enclosed nucleus	Absent	Absent	Present
Membrane-enclosed organelles	Few	Absent	Many
Peptidoglycan in cell wall	Present	Absent	Absent
Membrane lipids	Ester-linked	Ether-linked	Ester-linked
	Unbranched	Branched	Unbranched
Ribosomes[a]	70S	70S	80S
Initiator tRNA	Formylmethionine	Methionine	Methionine
Operons	Yes	Yes	Rare
Plasmids	Yes	Yes	Rare
Number of RNA polymerases[b]	One	One	Three
Ribosomes sensitive to chloramphenicol and streptomycin	Yes	No	No
Ribosomes sensitive to diphtheria toxin	No	Yes	Yes

[a]70S ribosomes are smaller than 80S ribosomes.
[b]The structure of archaeal RNA polymerase is similar to that of eukaryotic polymerases.

the enzymes needed to transcribe and translate genetic information into proteins. The prokaryotic cell also contains at least one system for generating the ATP it needs.

Genetic studies clearly indicate that all three domains of life had a single common ancestor. Across a large portion of their genome, eukaryotes share a more recent common ancestor with archaea than they do with bacteria (see Figure 26.1). However, the mitochondria of eukaryotes (as well as the chloroplasts of photosynthetic eukaryotes, such as plants) originated through endosymbiosis with bacteria (see Section 5.5). Some biologists prefer to view the origin of eukaryotes as a fusion of two equal partners (one ancestor that was related to modern archaea and another that was more closely related to modern bacteria). Others view the divergence of the early eukaryotes from the archaea as an event separate from and earlier than the later endosymbioses. In either case, some eukaryote genes are most closely related to those of archaea, whereas others are most closely related to those of bacteria. The tree of life therefore contains some merging of lineages as well as the predominant divergence of lineages.

The last common ancestor of the three domains probably lived about 3 billion years ago. It probably had DNA as its genetic material, as well as machinery for transcription and translation that produced RNAs and proteins, respectively. This ancestor almost certainly had a circular chromosome. Archaea, Bacteria, and Eukarya are all the products of billions of years of mutation, natural selection, and genetic drift, and they are all well adapted to present-day environments. The earliest prokaryote fossils, which date back at least 3.5 billion years, indicate that there was considerable diversity among the prokaryotes even during those earliest days of life.

The small size of prokaryotes has hindered our study of their evolutionary relationships

Until about 300 years ago, nobody had even *seen* an individual prokaryote; these organisms remained invisible to humans until the invention of the first simple microscope. Prokaryotes are so small, however, that even the best light microscopes don't reveal much about them. It took advanced microscopic equipment and modern molecular techniques to open up the microbial world. (Microscopic organisms—both prokaryotes and eukaryotes—are often collectively referred to as "microbes.")

Before DNA sequencing became practical, taxonomists based prokaryote classification on observable phenotypic characters such as shape, color, motility, nutritional requirements, and sensitivity to antibiotics. One of the characters most widely used to classify prokaryotes is the structure of their cell walls.

The cell walls of almost all bacteria contain **peptidoglycan**, a cross-linked polymer of amino sugars that produces a firm, protective, meshlike structure around the cell. Peptidoglycan is a substance unique to bacteria; its absence from the cell walls of archaea is a key difference between the two prokaryotic domains. Peptidoglycan is also an excellent target for combating pathogenic (disease-causing) bacteria because it has no counterpart in eukaryotic cells. Antibiotics such as penicillin and ampicillin, as well as other agents that specifically interfere with the synthesis of peptidoglycan-containing cell walls, tend to have little, if any, effect on the cells of humans and other eukaryotes.

The **Gram stain** is a technique that can be used to separate most types of bacteria into two distinct groups. A smear of bacterial cells on a microscope slide is soaked in a violet dye and treated with iodine; it is then washed with alcohol and counterstained with a red dye called safranin. **Gram-positive bacteria** retain the violet dye and appear blue to purple (**Figure

(A)

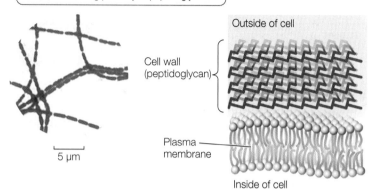

Gram-positive bacteria have a uniformly dense cell wall consisting primarily of peptidoglycan.

Outside of cell

Cell wall (peptidoglycan)

Plasma membrane

5 µm

Inside of cell

(B)

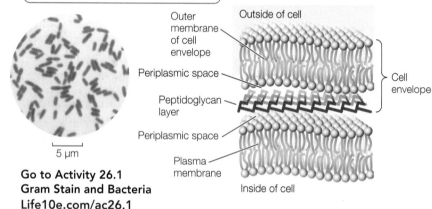

Gram-negative bacteria have a very thin peptidoglycan layer and an outer membrane, which together make up the cell envelope.

Outer membrane of cell envelope

Outside of cell

Periplasmic space

Peptidoglycan layer

Cell envelope

Periplasmic space

Plasma membrane

5 µm

Inside of cell

Go to Activity 26.1
Gram Stain and Bacteria
Life10e.com/ac26.1

26.2 The Gram Stain and the Bacterial Cell Wall   When treated with Gram-staining reagents, the cell walls of bacteria react in one of two ways. (A) Gram-positive bacteria have a thick peptidoglycan cell wall that retains the violet dye and appears deep blue or purple. (B) Gram-negative bacteria have a thin peptidoglycan layer that does not retain the violet dye, but picks up the counterstain and appears pink to red.

26.2A). The alcohol washes the violet stain out of **Gram-negative bacteria**, which then pick up the safranin counterstain and appear pink to red (**Figure 26.2B**). For most bacteria, the effect of the Gram stain is determined by the chemical structure of the cell wall:

- A *Gram-negative cell wall* usually has a thin peptidoglycan layer, which is surrounded by a second, outer membrane quite distinct in chemical makeup from the plasma membrane. Together the cell wall and the outer membrane are called the cell envelope. Between the plasma membrane and the outer membrane is a **periplasmic space**. This space contains proteins that are important in digesting some materials, transporting others, and detecting chemical gradients in the environment.

- A *Gram-positive cell wall* usually has about five times as much peptidoglycan as a Gram-negative cell wall. Its thick

peptidoglycan layer is a meshwork that may serve some of the same purposes as the periplasmic space of the Gram-negative cell envelope.

Shape is another phenotypic characteristic that is useful for the basic identification of bacteria. The three most common shapes are spheres, rods, and spiral forms (**Figure 26.3**). Many bacterial names are based on these shapes. A spherical bacterium is called a **coccus** (plural *cocci*). Cocci may live singly or may associate in two- or three-dimensional arrays such as chains, plates, blocks, or clusters of cells. A rod-shaped bacterium is called a **bacillus** (plural *bacilli*). A spiral bacterium (shaped like a corkscrew) is called a **spirillum** (plural *spirilla*). Bacilli and spirilla may be single, form chains, or gather in regular clusters. Among the other bacterial shapes are long filaments and branched filaments.

Less is known about the shapes of archaea because many of these organisms have never been seen. Many archaea are known only from samples of DNA from the environment. However, the species whose morphologies are known include cocci, bacilli, and even triangular and square species; the last grow on surfaces, arranged like sheets of postage stamps.

The nucleotide sequences of prokaryotes reveal their evolutionary relationships

Analyses of the nucleotide sequences of ribosomal RNA (rRNA) genes provided the first comprehensive evidence of evolutionary relationships among prokaryotes. Comparisons of rRNA genes from a great many organisms have revealed probable phylogenetic relationships throughout the tree of life. Databases such as GenBank contain rRNA gene sequences from hundreds of thousands of species—more than any other type of gene sequence.

For several reasons, rRNA is particularly useful for phylogenetic studies of living organisms:

- rRNA was present in the common ancestor of all life and is therefore evolutionarily ancient.

- No free-living (i.e., not parasitic) organism lacks rRNA, so rRNA genes can be compared across the tree of life.

- rRNA plays a critical role in translation in all organisms, so lateral transfer of rRNA genes among distantly related species is unlikely.

- rRNA has evolved slowly enough that gene sequences from even distantly related species can be aligned and analyzed.

Although studies of rRNA genes reveal much about the evolutionary relationships of prokaryotes, they don't always reveal the entire evolutionary history of these organisms. In some groups of prokaryotes, analyses of multiple gene sequences

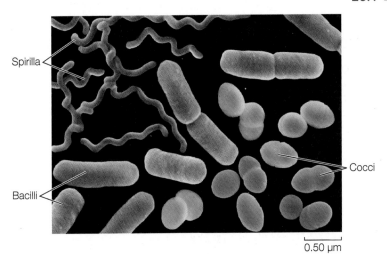

26.3 Bacterial Cell Shapes This composite, colorized micrograph shows the three most common bacterial shapes. Spherical cells are called cocci; those pictured are a species of *Enterococcus* from the mammalian gut. Rod-shaped cells are called bacilli; these *Escherichia coli* also reside in the gut. The helix-shaped spirilla are *Leptospira interrogans*, a human pathogen.

have suggested several different phylogenetic patterns. How could such differences among gene sequences arise in the same organisms? Studies of whole prokaryotic genomes have revealed that even distantly related prokaryotes sometimes exchange genetic material.

Lateral gene transfer can lead to discordant gene trees

As noted earlier, prokaryotes reproduce by binary fission. If we could follow these divisions back through evolutionary time, we would be tracing the complete tree of life for bacteria and archaea. This underlying tree of evolutionary relationships

(represented in highly abbreviated form in Appendix A) is called the organismal (or species) tree. Because binary fission is an asexual process that replicates whole genomes, we would expect phylogenetic trees constructed from most gene sequences (see Chapter 22) to reflect these same relationships.

There are other processes, however—including transformation, conjugation, and transduction—that allow the exchange of genetic information between some prokaryotes without reproduction. Thus prokaryotes can exchange and recombine their DNA with that of other individuals (this is sex in the genetic sense of the word), but this genetic exchange is not directly linked to reproduction, as it is in most eukaryotes.

From early in evolution to the present day, some genes have been moving "sideways" from one prokaryote species to another, a phenomenon known as **lateral gene transfer**. Lateral gene transfers are well documented among closely related species, and some have been documented even across the domains of life. Consider, for example, the genome of *Thermotoga maritima*, a bacterium that can survive extremely high temperatures. By comparing the 1,869 gene sequences of *T. maritima* with sequences encoding the same proteins in other species, investigators found that some of this bacterium's genes have their closest relationships not with the genes of other bacterial species, but with the genes of archaea that live in similar extreme environments.

When genes involved in lateral transfer events are sequenced and analyzed, the resulting *gene trees* will not match the organismal tree in every respect (**Figure 26.4**). The gene trees will vary because the history of lateral transfer events may be different for different genes. Biologists can reconstruct the underlying organismal phylogeny by comparing multiple genes to produce a consensus tree, or by concentrating only on genes that are unlikely to be involved in lateral gene transfer

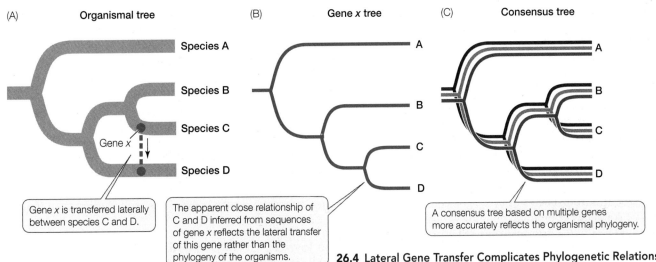

(A) Organismal tree

Species A
Species B
Species C
Species D

Gene *x*

Gene *x* is transferred laterally between species C and D.

The apparent close relationship of C and D inferred from sequences of gene *x* reflects the lateral transfer of this gene rather than the phylogeny of the organisms.

(B) Gene *x* tree

A
B
C
D

(C) Consensus tree

A
B
C
D

A consensus tree based on multiple genes more accurately reflects the organismal phylogeny.

26.4 Lateral Gene Transfer Complicates Phylogenetic Relationships
(A) The phylogeny of four hypothetical prokaryote species, two of which have been involved in a lateral transfer of gene *x*. (B) A tree based only on gene *x* shows the phylogeny of the laterally transferred gene, rather than the organismal phylogeny. (C) A consensus tree based on multiple genes is more likely to reflect the true organismal phylogeny, especially if those genes come from a stable core of genes involved in fundamental processes.

events. For example, genes that are involved in fundamental cellular processes (such as the rRNA genes discussed above) are unlikely to be replaced by the same genes from other species because functional, locally adapted copies of these genes are already present.

What kinds of genes are most likely to be involved in lateral gene transfer? Genes that result in a new adaptation that confers higher fitness on a recipient species are most likely to be retained. For example, genes that produce antibiotic resistance are often transferred among bacterial species on plasmids, especially under strong selection pressure such as that imposed by modern antibiotic medications. Improper or overly frequent use of antibiotics can select for resistant strains of pathogenic bacteria that are much harder to treat. This phenomenon explains why informed physicians have become more careful in prescribing antibiotics.

It is debatable whether lateral gene transfer has seriously complicated our attempts to resolve the tree of prokaryotic life. Recent work suggests that it has not; although lateral gene transfer complicates studies in some individual species (and makes the boundaries of "species" of bacteria difficult to determine), it need not present problems at higher taxonomic levels. It is now possible to make nucleotide sequence comparisons of entire genomes, and these studies are revealing a stable core of fundamental genes whose phylogenies are uncomplicated by lateral gene transfer. Gene trees based on this stable core more accurately reveal the organismal phylogeny. The problem that remains is that only a very small proportion of the prokaryotic world has been described and studied.

The great majority of prokaryote species have never been studied

Most prokaryotes have defied all attempts to grow them in pure culture, causing biologists to wonder how many species, and possibly even large groups of species, we might be missing. A window onto this problem was opened with the introduction of a new way of examining nucleic acid sequences. When biologists are unable to work with the whole genome of a single prokaryote species, they can instead examine individual genes collected from a random sample of the environment. This technique is known as **environmental genomics**.

Biologists now routinely isolate gene sequences, or even whole genomes, from environmental samples such as soil and seawater. Comparing such sequences with previously known ones has revealed that an extraordinary number of the sequences represent new, previously unrecognized species. Biologists have described only about 10,000 species of bacteria and only a few hundred species of archaea (see Figure 1.7). The results of some environmental genomic studies suggest that there may be millions—perhaps hundreds of millions—of prokaryote species. Other biologists put the estimate much lower, arguing that the high dispersal ability of many bacterial species greatly reduces endemism (i.e., the number of species restricted to a small geographic area). Only the magnitude of these estimates differs, however; all sides agree that we have just begun to uncover Earth's bacterial and archaeal diversity.

 RECAP 26.1

Bacteria and Archaea are distinct prokaryotic domains of the tree of life. The small size of prokaryotes, combined with their potential for lateral gene transfer, has hindered our ability to understand their evolutionary relationships. Environmental genomic studies have suggested a much higher diversity of prokaryotes than previous studies had revealed.

- What findings led to the establishment of Bacteria and Archaea as separate domains? **See pp. 526–527 and Table 26.1**
- How did biologists classify bacteria before it became possible to determine their nucleotide sequences? **See pp. 527–528 and Figures 26.2 and 26.3**
- Why are nucleotide sequences of rRNA genes particularly useful for phylogenetic studies of prokaryotes? **See p. 528**
- How does lateral gene transfer complicate phylogenetic studies of prokaryotes? **See pp. 528–529 and Figure 26.4**

Despite the challenges of reconstructing prokaryote phylogeny, taxonomists are starting to establish evolutionary classification systems for these organisms. With the understanding that new information will necessitate periodic revision of these classifications, we next apply one current system to our discussion of prokaryote diversity.

26.2 Why Are Prokaryotes So Diverse and Abundant?

The prokaryotes were alone on Earth for a very long time, adapting to new and changing environments. They are still with us, in massive numbers and incredible diversity, and they are found everywhere. If success is measured by numbers of individuals, the prokaryotes are the most successful organisms on Earth. Individual bacteria and archaea in the oceans have been estimated to number more than 3×10^{28}—perhaps 100 million times more than the number of stars in the visible universe. Closer to home, the individual bacteria living in your intestinal tract outnumber all the humans who have ever lived.

Given our still-fragmentary knowledge of prokaryote diversity, it is not surprising that there are several different hypotheses about the relationships of the major groups of prokaryotes. In this book we use a widely accepted classification system that has considerable support from nucleotide sequence data. In this section we will discuss the eight bacterial groups that have the broadest phylogenetic support and have received the most study (see Figure 26.1). We will then describe the archaea, whose diversity is even less well studied than that of the bacteria.

The low-GC Gram-positives include some of the smallest cellular organisms

The **low-GC Gram-positives**, also known as Firmicutes, derive the first part of their name from the relatively low ratio of G-C to A-T nucleotide base pairs in their DNA. The second part of their name is less accurate: some of the low-GC Gram-positives

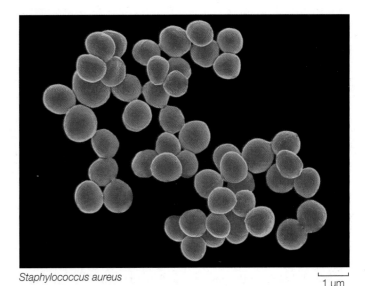

Endospore

Clostridium difficile

0.3 µm

26.5 A Structure for Waiting Out Bad Times Under harsh conditions, some low-GC Gram-positive bacteria can replicate their DNA and encase it in an endospore. The parent cell then breaks down, and the endospore survives in a dormant state until conditions improve.

are in fact Gram-negative, and some have no cell wall at all. Despite these differences, phylogenetic analyses of DNA sequences support the monophyly of this bacterial group.

One group of low-GC Gram-positives can produce resting structures called **endospores** (Figure 26.5). When a key nutrient such as nitrogen or carbon becomes scarce, the bacterium replicates its DNA and encapsulates one copy, along with some of its cytoplasm, in a tough cell wall heavily thickened with peptidoglycan and surrounded by a spore coat. The parent cell then breaks down, releasing the endospore. Endospore production is not a reproductive process; the endospore merely replaces the parent cell. The endospore, however, can survive harsh environmental conditions that would kill the parent cell, such as high or low temperatures or drought, because it is dormant—its normal metabolic activity is suspended. Later, if it encounters favorable conditions, the endospore becomes metabolically active and divides, forming new cells that are descendants of the parent cell. Members of this endospore-forming group of low-GC Gram-positives include the many species of *Clostridium* and *Bacillus*. Some of their endospores can be reactivated after more than a thousand years of dormancy. There are even credible claims of reactivation of *Bacillus* endospores millions of years old.

Endospores of *Bacillus anthracis* are the cause of anthrax. Anthrax is primarily a disease of cattle and sheep, but it can be fatal in humans. When the endospores sense macrophages in mammalian blood, they reactivate and release toxins into the bloodstream. *Bacillus anthracis* has been used as a bioterrorism agent because it is relatively easy to transport large quantities of its endospores and release them among human populations, where they may be inhaled or ingested.

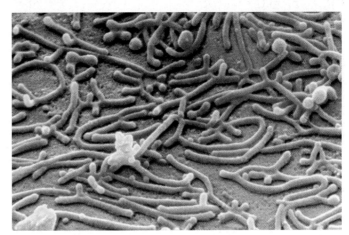

Staphylococcus aureus

1 µm

26.6 Staphylococci "Grape clusters" are the usual arrangement of these low-GC Gram-positive coccal bacteria, often the cause of skin or wound infections.

Low-GC Gram-positives of the genus *Staphylococcus*—the **staphylococci** (Figure 26.6)—are abundant on the human body surface; they are responsible for boils and many other skin problems. *Staphylococcus aureus* is the best-known human pathogen in this genus; it is present in 20 to 40 percent of normal adults (and in 50 to 70 percent of hospitalized adults). In addition to skin diseases, *S. aureus* can cause respiratory, intestinal, and wound infections.

Another interesting group of low-GC Gram-positives, the **mycoplasmas**, lack cell walls, although some have a stiffening material outside the plasma membrane. The mycoplasmas are among the smallest cellular organisms known (Figure 26.7). The smallest mycoplasmas have a diameter of about 0.2 µm. They are small in another crucial sense as well: they have less than half as much DNA as most other prokaryotes. It has been speculated

Mycoplasma sp.

0.7 µm

26.7 Tiny Cells With about one-fifth as much DNA as *E. coli*, mycoplasmas are among the smallest known bacteria.

that the DNA in a mycoplasma, which codes for fewer than 500 proteins, may be close to the minimum amount required to encode the essential properties of a living cell (see Figure 17.6).

Some high-GC Gram-positives are valuable sources of antibiotics

High-GC Gram-positives, also known as actinobacteria, have a higher ratio of G–C to A–T nucleotide base pairs than do the low-GC Gram-positives. These bacteria develop an elaborately branched system of filaments (**Figure 26.8**) that resembles the filamentous growth habit of fungi, albeit at a smaller scale. Some high-GC Gram-positives reproduce by forming chains of spores at the tips of the filaments. In species that do not form spores, the branched, filamentous growth ceases and the structure breaks up into typical cocci or bacilli, which then reproduce by binary fission.

The high-GC Gram-positives include several medically important bacteria. *Mycobacterium tuberculosis* causes tuberculosis, which kills 3 million people each year. Genetic data suggest that this bacterium arose 3 million years ago in East Africa, making it the oldest known human bacterial pathogen. The genus *Streptomyces* produces streptomycin as well as hundreds of other antibiotics. We derive most of our antibiotics from members of the high-GC Gram-positives.

Hyperthermophilic bacteria live at very high temperatures

Several lineages of bacteria and archaea are **extremophiles**: they thrive under extreme conditions that would kill most other organisms. The hyperthermophilic bacteria, for example, are thermophiles (Greek, "heat-lovers"). Genera such as *Aquifex* live near volcanic vents and in hot springs, sometimes

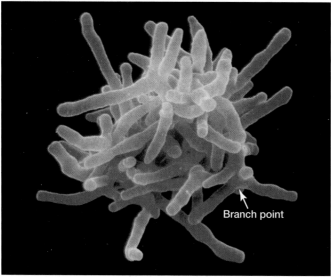

Actinomyces sp.

⊢————⊣
2 μm

26.8 Actinobacteria Are High-GC Gram-Positives The tangled, branching filaments seen in this scanning electron micrograph are typical of this medically important bacterial group.

at temperatures near the boiling point of water. Some species of *Aquifex* need only hydrogen, oxygen, carbon dioxide, and mineral salts to live and grow. Species of the genus *Thermotoga* live deep underground in oil reservoirs as well as in other high-temperature environments.

Biologists have hypothesized that high temperatures characterized the ancestral conditions for life, given that most environments on early Earth were much hotter than those of today. Reconstructions of ancestral bacterial genes have supported this hypothesis by showing that the ancestral sequences functioned best at elevated temperatures. The monophyly of the hyperthermophilic bacteria, however, is not well established.

Hadobacteria live in extreme environments

The **hadobacteria**, including such genera as *Deinococcus* and *Thermus*, are another group of thermophilic extremophiles. The group's name is derived from Hades, the ancient Greek name for the underworld. *Deinococcus* are resistant to radiation and can consume nuclear waste and other toxic materials. They can also survive extremes of cold as well as hot temperatures. Another member of this group, *Thermus aquaticus*, was the source of the thermally stable DNA polymerase that was critical for the development of the polymerase chain reaction. *Thermus aquaticus* was originally isolated from a hot spring, but it can be found wherever hot water occurs (including many residential hot water heaters).

Cyanobacteria were the first photosynthesizers

Cyanobacteria, sometimes called *blue-green bacteria* because of their pigmentation, are photosynthetic. They use chlorophyll *a* for photosynthesis and release oxygen gas (O_2); many species also fix nitrogen. The production of oxygen by these bacteria transformed the atmosphere of early Earth (see Section 25.2).

Cyanobacteria carry out the same type of photosynthesis that is characteristic of eukaryotic photosynthesizers. They contain elaborate and highly organized internal membrane systems called **photosynthetic lamellae**. As mentioned in Section 26.1, the chloroplasts of photosynthetic eukaryotes are derived from an endosymbiotic cyanobacterium.

Cyanobacteria may live free as single cells or associate in multicellular colonies. Depending on the species and on growth conditions, these colonies may range from flat sheets one cell thick to filaments to spherical balls of cells. Some filamentous colonies of cyanobacteria differentiate into three specialized cell types: vegetative cells, spores, and heterocysts (**Figure 26.9**). **Vegetative cells** photosynthesize, **spores** are resting stages that can survive harsh environmental conditions and eventually develop into new filaments, and **heterocysts** are cells specialized for nitrogen fixation. All of the known cyanobacteria with heterocysts fix nitrogen. Heterocysts also have a role in reproduction: when filaments break apart to reproduce, the heterocyst may serve as a breaking point.

 Go to Media Clip 26.1
Cyanobacteria
Life10e.com/mc26.1

(A) *Anabaena* sp.

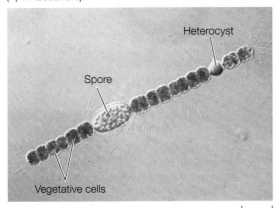

Heterocyst

Spore

Vegetative cells

(B) *Nostoc punctiforme*

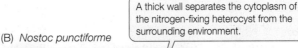
A thick wall separates the cytoplasm of the nitrogen-fixing heterocyst from the surrounding environment.

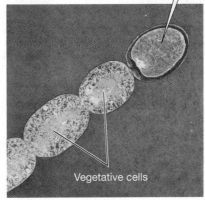

Vegetative cells

(C)

4 µm

0.4 µm

26.9 Cyanobacteria (A) Some cyanobacteria form filamentous colonies containing three cell types. (B) Heterocysts are specialized for nitrogen fixation and may serve as a breaking point when filaments reproduce. (C) This pond in Finland has experienced eutrophication: phosphorus and other nutrients generated by human activity have accumulated, feeding an immense green mat (commonly referred to as "pond scum") that is made up of several species of free-living cyanobacteria.

Spirochetes move by means of axial filaments

Spirochetes are Gram-negative, motile bacteria characterized by unique structures called axial filaments (**Figure 26.10A**), which are modified flagella running through the periplasmic space. The cell body is a long cylinder coiled into a helix (**Figure 26.10B**). The axial filaments begin at either end of the cell and overlap in the middle. Motor proteins connect the axial filaments to the cell wall, enabling the corkscrew-like movement of the bacterium. Many spirochetes are parasites of humans; a few are pathogens, including those that cause syphilis and Lyme disease. Others live free in mud or water.

Chlamydias are extremely small parasites

Chlamydias are among the smallest bacteria (0.2–1.5 µm in diameter). These tiny, Gram-negative cocci can live only as parasites in the cells of other organisms. It was once believed that their obligate parasitism resulted from an inability to produce ATP—that chlamydias were "energy parasites." However, genome sequencing indicates that chlamydias have the genetic capacity to produce at least some ATP. They can augment this capacity by using an enzyme called a translocase, which allows them to take up ATP from the cytoplasm of their host in exchange for ADP from their own cells.

Chlamydias are unique among prokaryotes because of their complex life cycle, which involves two different forms of cells, elementary bodies and reticulate bodies (**Figure 26.11**). Various strains of chlamydias cause eye infections, sexually transmitted diseases, and some forms of pneumonia in humans.

(A)

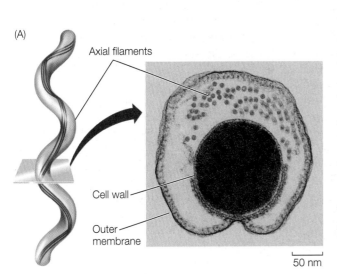

Axial filaments

Cell wall

Outer membrane

50 nm

(B)

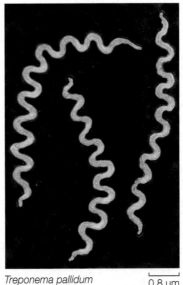

Treponema pallidum

0.8 µm

26.10 Spirochetes Get Their Shape from Axial Filaments (A) A spirochete from the gut of a termite, seen in cross section, shows the axial filaments these helical prokaryotes use to produce a corkscrew-like movement. (B) This spirochete species causes syphilis in humans.

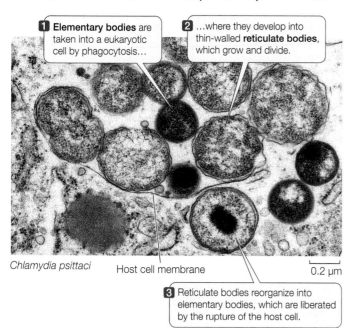

1 Elementary bodies are taken into a eukaryotic cell by phagocytosis...

2 ...where they develop into thin-walled **reticulate bodies**, which grow and divide.

Chlamydia psittaci   Host cell membrane   0.2 μm

3 Reticulate bodies reorganize into elementary bodies, which are liberated by the rupture of the host cell.

26.11 Chlamydias Change Form Elementary bodies and reticulate bodies are the two cell forms of the chlamydia life cycle.

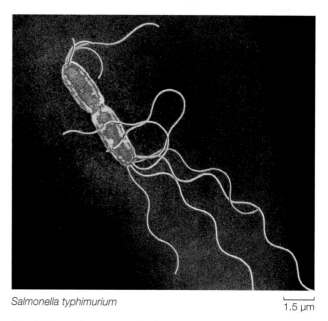

Salmonella typhimurium   1.5 μm

26.12 Proteobacteria Include Human Pathogens These conjugating cells of *Salmonella typhimurium* are exchanging genetic material. This pathogen causes a wide range of gastrointestinal illnesses in humans.

The proteobacteria are a large and diverse group

By far the largest bacterial group, in terms of numbers of described species, is the **proteobacteria**. The proteobacteria include many species of Gram-negative photoautotrophs that use light-driven reactions to metabolize sulfur, as well as dramatically diverse bacteria that bear no phenotypic resemblance to the photoautotrophic species. Genetic and morphological evidence indicates that the mitochondria of eukaryotes were derived from a proteobacterium by endosymbiosis.

Among the proteobacteria are some nitrogen-fixing genera such as *Rhizobium*, and other bacteria that contribute to the global nitrogen and sulfur cycles. *Escherichia coli,* one of the most studied organisms on Earth, is a proteobacterium. So too are many of the most famous human pathogens, such as *Yersinia pestis* (the cause of bubonic plague), *Vibrio cholerae* (cholera), and *Salmonella typhimurium* (gastrointestinal disease; **Figure 26.12**).

 Go to Media Clip 26.2
A Swarm of *Salmonella*
Life10e.com/mc26.2

The bioluminescent *Vibrio* we discussed at the opening of this chapter are also proteobacteria. There are many potential applications of the genes that encode bioluminescent proteins in bacteria; these genes are already being inserted into the genomes of other species, with the resulting bioluminescence used as a marker of gene expression. Futuristic proposals for making use of bioluminescence in bioengineered organisms include crop plants that glow when they become water-stressed and need to be irrigated; and glowing trees that could illuminate highways, replacing electric lights.

Although most plant diseases are caused by fungi and viruses, about 200 known plant diseases are of bacterial origin. *Crown gall*, with its characteristic tumors, is one of the most striking (**Figure 26.13**). The causal agent of crown gall

is *Agrobacterium tumefaciens*, a proteobacterium that harbors a plasmid often used in recombinant DNA studies as a vehicle for inserting genes into new plant hosts.

Gene sequencing enabled biologists to differentiate the domain Archaea

The original identification of Archaea as a domain separate from Bacteria and Eukarya was based on phylogenetic relationships determined from rRNA gene sequences. This separation was supported when biologists sequenced the first complete archaeal genome, which consisted of 1,738 genes—more than half of which were unlike any genes ever found in the other two domains.

26.13 Crown Gall Crown gall, a type of tumor shown here growing on the trunk of a white oak, is caused by the proteobacterium *Agrobacterium tumefaciens*.

INVESTIGATING LIFE

26.14 What Is the Highest Temperature Compatible with Life?

Can any organism thrive at temperatures above 120°C? This is the temperature used for sterilization, known to destroy all previously described organisms. Kazem Kashefi and Derek Lovley isolated an unidentified prokaryote from water samples taken near a hydrothermal vent and found it survived and even multiplied at 121°C. The organism was dubbed "Strain 121," and its gene sequencing results indicate that it is an archaeal species.[a]

HYPOTHESIS Some prokaryotes can survive at temperatures above the 120°C threshold of sterilization.

Method
1. Seal samples of unidentified, iron-reducing, thermal vent prokaryotes in tubes with a medium containing Fe^{3+} as an electron acceptor. Control tubes contain Fe^{3+} but no organisms.
2. Hold both tubes in a sterilizer at 121°C for 10 hours. If the iron-reducing organisms are metabolically active, they will reduce the Fe^{3+} to Fe^{2+} (as magnetite, which can be detected with a magnet).

Results

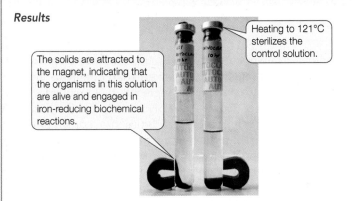

Heating to 121°C sterilizes the control solution.

The solids are attracted to the magnet, indicating that the organisms in this solution are alive and engaged in iron-reducing biochemical reactions.

CONCLUSION Archaea of "Strain 121" can survive at temperatures above the previously defined sterilization limit.

Go to **BioPortal** for discussion and relevant links for all INVESTIGATING LIFE figures.

[a]Kashefi, K. and D. R. Lovley. 2003. *Science* 301: 934.

WORKING WITH DATA:

A Relationship between Temperature and Growth in an Archaean

Original Paper

Kashefi, K. and D. R. Lovley. 2003. Extending the upper temperature limit for life. *Science* 301: 934.

Analyze the Data

After Strain 121 was isolated, its growth was examined at various temperatures. The table below shows generation time (time between cell divisions) at nine temperatures.

Temperature (°C)	Generation time (hr)
85	10
90	4
95	3
100	2.5
105	2
110	4
115	6
120	20
130	No growth, but cells not killed

QUESTION 1

Make a graph from these data showing time as a function of temperature.

QUESTION 2

Which temperature appears to be closest to optimum for the growth of Strain 121?

QUESTION 3

Note that no growth occurred at 130°C, but that the cells were not killed. How would you demonstrate that these cells were still alive?

Go to BioPortal for all WORKING WITH DATA **exercises**

Archaea are known for living in extreme habitats such as those with high salinity (salt content), low oxygen concentrations, high temperatures, or high or low pH (**Figure 26.14**). Many archaea are not extremophiles, however—they are common in soil, for example. Perhaps the largest numbers of archaea live in the ocean depths.

One current classification scheme divides Archaea into two principal groups, **Crenarchaeota** and **Euryarchaeota**. Less is known about two more recently discovered groups, **Korarchaeota** and **Nanoarchaeota**. In fact, we know relatively little about the phylogeny of archaea, in part because the study of these prokaryotes is still in its early stages.

Two characteristics shared by all archaea are the absence of peptidoglycan in their cell walls and the presence of lipids of distinctive composition in their cell membranes (see Table 26.1). The unusual lipids in the membranes of archaea are found in all archaea and in no bacteria or eukaryotes. Most lipids in bacterial and eukaryotic membranes contain

unbranched long-chain fatty acids connected to glycerol molecules by **ester linkages**:

$$-\overset{O}{\overset{\|}{C}}-O-\overset{H}{\overset{|}{\underset{|}{\underset{H}{C}}}}-$$

In contrast, some lipids in archaeal membranes contain long-chain hydrocarbons connected to glycerol molecules by **ether linkages**:

$$-\overset{H}{\overset{|}{\underset{|}{\underset{H}{C}}}}-O-\overset{H}{\overset{|}{\underset{|}{\underset{H}{C}}}}-$$

These ether linkages are a synapomorphy of archaea. In addition, the long hydrocarbon chains in the lipids of archaea are branched. One class of archaeal lipids, with hydrocarbon chains 40 carbon atoms in length, contains glycerol at *both* ends of the hydrocarbons (**Figure 26.15**). These lipids form a *lipid monolayer* structure that is unique to archaea. They still fit into a biological membrane because they are twice as long as the typical lipids in the bilayers of other membranes. Lipid monolayers

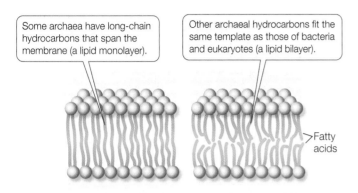

Some archaea have long-chain hydrocarbons that span the membrane (a lipid monolayer).

Other archaeal hydrocarbons fit the same template as those of bacteria and eukaryotes (a lipid bilayer).

Fatty acids

26.15 Membrane Architecture in Archaea The long-chain hydrocarbons of many archaeal lipids have glycerol molecules at both ends, so that the membranes they form consist of a lipid monolayer. In contrast, the membranes of other archaea, bacteria, and eukaryotes consist of a lipid bilayer.

and bilayers are both found among the archaea. The effects, if any, of these structural features on membrane performance are unknown. In spite of this striking difference in their lipids, the membranes of all three domains have similar overall structures, dimensions, and functions.

Most crenarchaeotes live in hot or acidic places

Most known crenarchaeotes are either thermophilic, acidophilic (acid loving), or both. Members of the genus *Sulfolobus* live in hot sulfur springs at temperatures of 70°C to 75°C. They become metabolically inactive at 55°C (131°F). Hot sulfur springs are also extremely acidic. *Sulfolobus* grows best in the range from pH 2 to pH 3, but some members of this genus readily tolerate pH values as low as 0.9. Most acidophilic thermophiles maintain an internal pH of 5.5 to 7 (close to neutral) in spite of their acidic environment. These and other crenarchaeotes thrive where very few other organisms can even survive (**Figure 26.16**).

Euryarchaeotes are found in surprising places

Some species of Euryarchaeota are **methanogens**: they produce methane (CH_4) by reducing carbon dioxide as the key step in their energy metabolism. All of the methanogens are obligate anaerobes (see Section 26.3). Comparison of their rRNA gene sequences has revealed a close evolutionary relationship among these methanogenic species, which were previously assigned to several different groups of bacteria.

Methanogenic euryarchaeotes release approximately 2 billion tons of methane gas into Earth's atmosphere each year, accounting for 80 to 90 percent of the methane that enters the atmosphere, including that produced in some mammalian digestive systems. Approximately a third of this methane comes from methanogens living in the guts of ruminants such as cattle, sheep, and deer, and another large fraction comes from methanogens living in the guts of termites and cockroaches. Methane is increasing in Earth's atmosphere by about 1 percent per year and contributes to the greenhouse effect (see Section 58.3). Part of that increase is due to increases in cattle and rice farming and the methanogens associated with both.

26.16 Some Crenarchaeotes Like It Hot Thermophilic crenarchaeotes can thrive in the intense heat of volcanic hot sulfur springs such as these in Nevada's Black Rock Desert.

Another group of euryarchaeotes, the **extreme halophiles** (salt lovers), live exclusively in very salty environments. Because they contain pink carotenoid pigments, these archaea are sometimes easy to see (**Figure 26.17**). Extreme halophiles grow in the Dead Sea and in brines of all types; the reddish pink spots that can occur on pickled fish are colonies of halophilic archaea. Few other organisms can live in the saltiest homes that the extreme halophiles occupy; most would "dry" to death, losing too much water to the hypertonic environment. Extreme halophiles have been found in lakes with pH values as high as 11.5—the most alkaline environment inhabited by living organisms, and almost as alkaline as household ammonia.

Some of the extreme halophiles have a unique system for trapping light energy and using it to form ATP—without using any form of chlorophyll—when oxygen is in short supply.

26.17 Extreme Halophiles Highly saline environments such as these commercial seawater evaporating ponds in San Francisco Bay are home to extreme halophiles. The archaea are easily visible here because of the rich red coloration of their carotenoid pigments.

26.18 A Nanoarchaeote Growing in Mixed Culture with a Crenarchaeote *Nanoarchaeum equitans* (red), discovered living near deep-sea hydrothermal vents, is the only representative of the nano-archaeote group so far identified. This tiny organism lives attached to cells of the crenarchaeote *Ignicoccus* (green). For this confocal laser micrograph, the two species were visually differentiated by fluorescent dye "tags" that are specific to their distinct gene sequences.

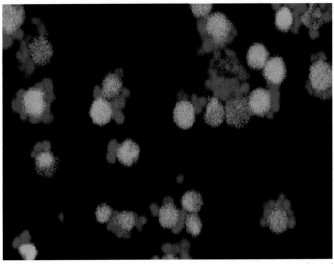

1 μm

They use the pigment retinal (also found in the vertebrate eye) combined with a protein to form a light-absorbing molecule called microbial rhodopsin.

Another member of the Euryarchaeota, *Thermoplasma*, has no cell wall. It is thermophilic and acidophilic, its metabolism is aerobic, and it lives in coal deposits. Its genome of 1,100,000 base pairs is among the smallest (along with that of the mycoplasmas) found in any free-living organism, although some parasitic organisms have even smaller genomes.

Korarchaeotes and nanoarchaeotes are less well known

The korarchaeotes are known only from DNA isolated directly from marine hydrothermal vents and freshwater hot springs. No korarchaeote has been successfully grown in pure culture.

Another distinctive archaeal lineage has been discovered at a deep-sea hydrothermal vent off the coast of Iceland. It is the first representative of the group christened Nanoarchaeota because of their minute size (Greek *nanos*, "dwarf"). This organism is a parasite that lives on cells of *Ignicoccus*, a crenarchaeote. Because of their association, the two species can be grown together in culture (**Figure 26.18**).

 RECAP 26.2

Bacteria and Archaea are highly diverse groups that survive in almost every imaginable habitat on Earth. Many can survive and even thrive in habitats where no eukaryotes can live, including extremely hot, acidic, or saline conditions.

- How does the diversity of environments occupied by prokaryotes compare with the diversity of environments occupied by multicellular organisms with which you are familiar?

- Explain why Gram staining is of limited use in understanding the evolutionary relationships of bacteria. **See pp. 530–531**

- What makes the membranes of archaea unique? **See pp. 535–536 and Figure 26.15**

Prokaryotes are found almost everywhere on Earth and live in a wide variety of ecosystems. In the next section we will examine the contributions of prokaryotes to the functioning of those ecosystems.

26.3 How Do Prokaryotes Affect Their Environments?

Many people think of prokaryotes primarily as disease-causing organisms, but only a small percentage of bacteria, and no archaea, are known to be pathogens. Prokaryotes play roles in ecosystems that reach far beyond human sickness and health.

Prokaryotes have diverse metabolic pathways

Bacteria and archaea outdo the eukaryotes in terms of metabolic diversity. Although they are much more diverse in size and shape, eukaryotes draw on fewer metabolic mechanisms for their energy needs. In fact, much of eukaryotes' energy metabolism is carried out in organelles—mitochondria and chloroplasts—that are endosymbiotic descendants of bacteria. The long evolutionary history of bacteria and archaea, during which they have had time to adapt to a wide variety of habitats, has led to the extraordinary diversity of their metabolic "lifestyles"—their use or nonuse of oxygen, their varied sources of energy and carbon atoms, and the materials they release as waste products. This diversity of metabolic pathways, some of them unique to prokaryotes, makes prokaryotes a key component of the cycling of materials through ecosystems.

ANAEROBIC VERSUS AEROBIC METABOLISM The presence of oxygen is poisonous to some prokaryotes, so these **obligate anaerobes** can live only by anaerobic metabolism. At the other extreme from the obligate anaerobes, some prokaryotes are **obligate aerobes**. They require oxygen for cellular respiration and are unable to survive for extended periods in the *absence* of oxygen. Other prokaryotes can shift their metabolism between anaerobic and aerobic modes and thus are called **facultative anaerobes**. Many facultative anaerobes alternate between anaerobic metabolic processes (such as fermentation) and cellular respiration, as conditions dictate. **Aerotolerant anaerobes** are not damaged by oxygen when it is present, but they cannot conduct cellular respiration. By definition, an anaerobe does not use oxygen as an electron acceptor for respiration.

NUTRITIONAL CATEGORIES All organisms face the same two nutritional challenges: they must synthesize energy-rich compounds such as ATP to power their life-sustaining metabolic reactions, and they must obtain carbon atoms to build their

TABLE26.2 ■
How Organisms Obtain Their Energy and Carbon

Nutritional Category	Energy Source	Carbon Source
Photoautotrophs (some bacteria, some eukaryotes)	Light	Carbon dioxide
Photoheterotrophs (some bacteria)	Light	Organic compounds
Chemoautotrophs (some bacteria, many archaea)	Inorganic substances	Carbon dioxide
Chemoheterotrophs (found in all three domains)	Usually organic compounds; sometimes inorganic substances	Organic compounds

own organic molecules. Biologists recognize four broad nutritional categories of organisms: photoautotrophs, photoheterotrophs, chemoautotrophs, and chemoheterotrophs. Prokaryotes are represented in all four groups (**Table 26.2**).

Photoautotrophs perform photosynthesis. They use light as their energy source and carbon dioxide (CO_2) as their carbon source. The cyanobacteria, like green plants and other photosynthetic eukaryotes, use chlorophyll *a* as their key photosynthetic pigment and produce oxygen gas (O_2) as a by-product of noncyclic electron transport.

There are other photoautotrophs among the bacteria, but these organisms use bacteriochlorophyll rather than chlorophyll *a* as their key photosynthetic pigment, and they do not produce O_2. Instead, some of these photosynthesizers produce particles of pure sulfur because hydrogen sulfide (H_2S), rather than H_2O, is their electron donor for photophosphorylation. Many proteobacteria fit into this category. Bacteriochlorophyll molecules absorb light of longer wavelengths than do the chlorophyll molecules used by other photosynthesizing organisms. As a result, bacteria using this pigment can grow in water under fairly dense layers of algae, using light of wavelengths that are not absorbed by the algae (**Figure 26.19**).

Photoheterotrophs use light as their energy source but must obtain their carbon atoms from organic compounds made by other organisms. Their "food" consists of organic compounds such as carbohydrates, fatty acids, and alcohols.

For example, some photoheterotrophs take up compounds released from plant roots (as in rice paddies) or from decomposing photosynthetic bacteria in hot springs and metabolize them to form building blocks for other compounds. Sunlight provides them with the energy necessary for ATP formation through photophosphorylation.

Chemoautotrophs obtain their energy by oxidizing inorganic substances, and they use some of that energy to fix carbon. Some chemoautotrophs use reactions identical to those of the typical photosynthetic cycle, but others use alternative pathways for carbon fixation. Some bacteria oxidize ammonia or nitrite ions to form nitrate ions. Others oxidize hydrogen gas, hydrogen sulfide, sulfur, and other materials. Many archaea are chemoautotrophs.

Finally, **chemoheterotrophs** obtain carbon atoms from one or more complex organic compounds that have been synthesized by other organisms, and usually obtain energy from breaking down these organic compounds as well. Most known bacteria and archaea are chemoheterotrophs—as are all animals and fungi and many protists.

Although most chemoheterotrophs rely on the breakdown of organic compounds for energy, some chemoheterotrophic prokaryotes obtain their energy by breaking down inorganic substances. Organisms that obtain energy from oxidizing inorganic substances (both chemoautotrophs as well as some chemoheterotrophs) are also known as **lithotrophs** (Greek, "rock consumers").

Prokaryotes play important roles in element cycling

The metabolic diversity of the prokaryotes makes them key players in the cycles that keep chemical elements moving through ecosystems. Many prokaryotes are **decomposers**: organisms that metabolize organic compounds in dead organic material and return the products to the environment as inorganic substances. Prokaryotes, along with fungi, return tremendous quantities of carbon to the atmosphere as carbon dioxide, thus carrying out a key step in the carbon cycle.

26.19 Bacteriochlorophyll Absorbs Long-Wavelength Light The green alga *Ulva* contains chlorophyll, which absorbs no light of wavelengths longer than 750 nm. Purple sulfur bacteria, which contain bacteriochlorophyll, can conduct photosynthesis using longer infrared wavelengths. As a result, these bacteria can grow under layers of algae.

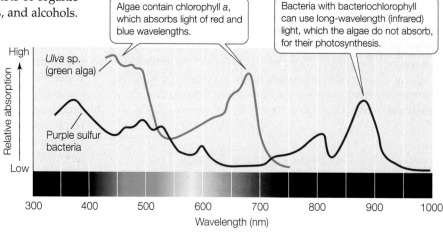

Algae contain chlorophyll *a*, which absorbs light of red and blue wavelengths.

Bacteria with bacteriochlorophyll can use long-wavelength (infrared) light, which the algae do not absorb, for their photosynthesis.

Ulva sp. (green alga)

Purple sulfur bacteria

Relative absorption

High

Low

Wavelength (nm)

300 400 500 600 700 800 900 1000

The key metabolic reactions of many prokaryotes involve nitrogen or sulfur. For example, some bacteria carry out respiratory electron transport without using oxygen as an electron acceptor. These organisms use oxidized inorganic ions such as nitrate, nitrite, or sulfate as electron acceptors. Examples include the **denitrifiers**, which release nitrogen to the atmosphere as nitrogen gas (N_2). These normally aerobic bacteria, mostly species of the genera *Bacillus* and *Pseudomonas*, use nitrate (NO_3^-) as an electron acceptor in place of oxygen under anaerobic conditions:

$$2\,NO_3^- + 10\,e^- + 12\,H^+ \rightarrow N_2 + 6\,H_2O$$

Denitrifiers play a key role in the cycling of nitrogen through ecosystems. Without denitrifiers, which convert nitrate ions into nitrogen gas, all forms of nitrogen would leach from the soil and end up in lakes and oceans; the resulting deficit of nitrogen on land would make terrestrial life much more difficult.

Nitrogen fixers convert atmospheric nitrogen gas into ammonia (NH_3), a chemical form that is usable by the nitrogen fixers themselves as well as by other organisms:

$$N_2 + 6\,H \rightarrow 2\,NH_3$$

All organisms require nitrogen in order to build proteins, nucleic acids, and other important compounds. Nitrogen fixation is thus vital to life as we know it. This all-important biochemical process is carried out by a wide variety of archaea and bacteria (including cyanobacteria) but by no other organisms, so we depend on these prokaryotes for our very existence.

Ammonia is oxidized to nitrate in soil and in seawater by chemoautotrophic bacteria and archaea called **nitrifiers**. Bacteria of two genera, *Nitrosomonas* and *Nitrosococcus*, convert ammonia (NH_3) to nitrite ions (NO_2^-), and *Nitrobacter* oxidize nitrite to nitrate (NO_3^-), the form of nitrogen most easily used by many plants. What do the nitrifiers get out of these reactions? Their metabolism is powered by the energy released by the oxidation of ammonia or nitrite. For example, by passing the electrons from nitrite through an electron transport system, *Nitrobacter* can make ATP, and using some of this ATP, can also make NADH. With this ATP and NADH, the bacterium can convert CO_2 and H_2O into glucose.

We have already seen the importance of the cyanobacteria in the cycling of oxygen: in ancient times, the oxygen generated by their photosynthesis converted Earth's atmosphere from an anaerobic to an aerobic environment. Other prokaryotes—both bacteria and archaea—contribute to the cycling of sulfur. Deep-sea hydrothermal vent ecosystems depend on chemoautotrophic prokaryotes that are incorporated into large communities of crabs, mollusks, and giant worms, all living at a depth of 2,500 meters—below any hint of sunlight. These bacteria obtain energy by oxidizing hydrogen sulfide and other substances released in the near-boiling water flowing from volcanic vents in the ocean floor.

Many prokaryotes form complex communities

Prokaryotes do not usually live in isolation. Rather, they live in communities of many different microbial species, which often include microscopic eukaryotes as well as myriad prokaryotic species. While some microbial communities are harmful to humans, others provide important services. For example, microbial communities help us digest our food, break down municipal waste, and recycle organic matter and chemical elements in the environment.

Some microbial communities form layers in sediments; others form clumps a meter or more in diameter. Many microbial communities form dense **biofilms**. Upon contacting a solid surface, the cells bind to that surface and secrete a sticky, gel-like, polysaccharide matrix that traps other cells (**Figure 26.20**). Once a biofilm forms, the cells become more difficult to kill.

Biofilms are found in many places, and in some of those places they cause problems for humans. The material on our teeth that we call dental plaque is a biofilm. Pathogenic bacteria are difficult for the immune system—and modern medicine—to combat once they form a biofilm, which may be impermeable to antibiotics. Worse, some drugs stimulate the bacteria in a biofilm to lay down more matrix, making the film even more impermeable. Biofilms may form on just about any available surface, including contact lenses and artificial joint replacements. They foul metal pipes and cause corrosion, a major problem in steam-driven electricity generation plants. Fossil stromatolites—large, rocky structures made up of alternating layers of fossilized biofilm and calcium carbonate—are among the oldest remnants of life on Earth (see Figure 25.9).

Some biologists are studying the chemical signals that prokaryotes use to communicate with one another and trigger density-linked activities such as biofilm formation. We saw one example of this type of communication—called **quorum sensing**—in the chapter-opening discussion of bioluminescent *Vibrio*. In the case of health-threatening bacteria, researchers hope to be able to block the quorum-sensing signals that lead to the production of the matrix polysaccharides, thus preventing pathological biofilms from forming.

Prokaryotes live on and in other organisms

Prokaryotes engage in many kinds of mutually beneficial relationships with eukaryotes. As we have seen, the mitochondria and chloroplasts of eukaryotes are descended from what were once free-living bacteria. Much later in evolutionary history, some plants became associated with bacteria to form cooperative nitrogen-fixing nodules on their roots.

Many animals harbor a variety of bacteria and archaea in their digestive tracts. Cattle depend on prokaryotes to perform important steps in digestion. Like most animals, cattle cannot produce cellulase, the enzyme needed to start the digestion of the cellulose that makes up the bulk of their plant food. However, bacteria living in a special section of the gut, called the rumen, produce enough cellulase to process the daily diet for the cattle. Human health also depends on many hundreds of species of symbiotic bacteria.

Microbiomes are critical to human health

Although only a few bacterial species are pathogens, popular notions of bacteria as "germs" and fear of the consequences

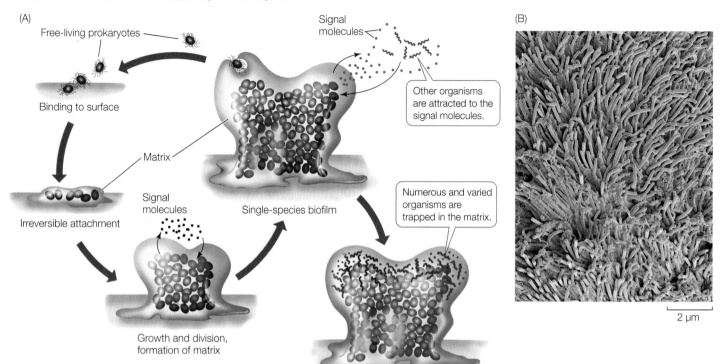

(A)

Free-living prokaryotes

Binding to surface

Matrix

Irreversible attachment

Signal molecules

Growth and division, formation of matrix

Signal molecules

Other organisms are attracted to the signal molecules.

Single-species biofilm

Numerous and varied organisms are trapped in the matrix.

Mature biofilm

(B)

2 μm

26.20 Forming a Biofilm (A) Free-living prokaryotes readily attach themselves to surfaces and form films that are stabilized and protected by a surrounding matrix. Once the population is large enough, the developing biofilm can send out chemical signals that attract other microorganisms. (B) Scanning electron micrography reveals a biofilm of dental plaque. The bacteria (red) are embedded in a matrix consisting of proteins from both bacterial secretions and saliva.

of infection cause many people to assume that most bacteria are harmful. Increasingly, however, biologists are discovering that human health depends in many ways on the health of our **microbiomes**: communities of bacteria that live in and on our bodies (**Figure 26.21**).

Every surface of your body is covered with diverse communities of bacteria. A 2009 study identified more than 1,000 species of bacteria that live on human skin. Inside your body, your digestive system teems with bacteria. If you count up all the cells in a human body, only about 10 percent of them are human cells. The rest are microbes—mostly bacteria, along with some archaea and microscopic eukaryotes.

Biologists are discovering that many complex health problems are linked to the disruption of our microbiomes. These diverse microbial communities affect the expression of our genes and play a critical role in the development and maintenance of a healthy immune system. When our microbiomes contain an appropriate community of beneficial bacteria, our bodies function normally. But these communities are strongly affected by our life experiences, by the food we eat, by the medicines we take, and by our exposure to various environmental toxins. When our microbial communities are disrupted, they must be restored before the body can function normally. The recent rapid increase in occurrences of autoimmune diseases—diseases in which our immune systems begin to attack our bodies (see Chapter 42)—has been linked to the changing diversity and composition of our microbiomes.

The early acquisition of an appropriate microbiome is critical for lifelong health. Normally, a human infant acquires much of its microbiome at birth, from the microbiome in its mother's vagina. Other components of the microbiome are also acquired from the mother, especially through breast feeding. Recent studies have shown that babies born by cesarean section, as well as babies that are bottle-fed on artificial milk formula, typically acquire microbes from a wider variety of sources. Many of the bacteria acquired in this way are not well suited for human health. Biologists have discovered that the incidence of many autoimmune diseases is much higher in people who were born by cesarean section and in those who were fed on formula as infants, compared with individuals who were born vaginally and breast-fed. The difference appears to be related to the composition of the individual's original microbiome.

Our microbiomes may be related to many other health concerns. For example, physicians have long noted a connection between autism and gastrointestinal disorders. In 2012, microbiologists discovered that children with autism have high levels of bacteria of the genus *Sutterella* adhering to their intestinal walls. These bacteria are absent or rare in children without autism. It is not yet known if *Sutterella* bacteria cause the gastrointestinal problems or if they are merely its symptoms, but it appears clear that the intestinal microbiomes of children with autism are distinctive.

Humans use some of the metabolic products—especially vitamins B_{12} and K—produced by the microorganisms living in the large intestine. Communities of bacteria line our intestines with a dense biofilm that is in intimate contact with the mucosal lining of the gut. This biofilm facilitates nutrient transfer from the intestine into the body, functioning like a specialized "tissue" that is essential to our health. This biofilm has a complex ecology that scientists have just begun to explore in

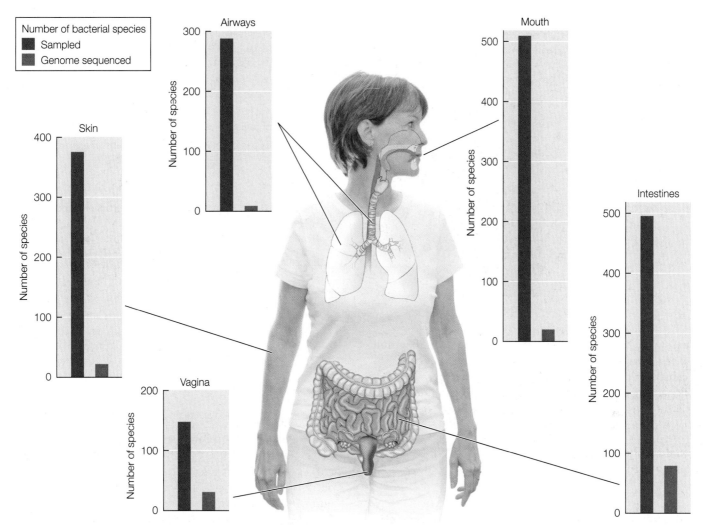

26.21 The Body's Microbiome Is Critical to the Maintenance of Health Surveys of the human microbiome have shown that this community includes thousands of diverse bacterial species that are adapted to grow in or on various parts of the body. Although we now know that the composition of this microbiome is closely associated with many aspects of human health, most of the component species are poorly characterized and remain largely unstudied by biologists. What has become clear is that, although the "subcommunities" in different parts of the body share similarities, each is a site-specific assemblage of many distinctive species.

detail—including the possibility that the species composition of an individual's gut microbiome may contribute to obesity (or the resistance to it).

A small minority of bacteria are pathogens

The tiny percentage of prokaryotes that are pathogens all fall into the domain Bacteria. Many archaea live in association with humans and other eukaryotes but are not known to cause any diseases. Why are there no pathogenic archaea? One clue is that different groups of viruses and plasmids infect bacteria than infect archaea, and pathogenesis in bacteria is largely coupled with traits carried by these plasmids and viruses. Differences in the external structures of bacteria and archaea make cross-domain infection by viruses and plasmids highly unlikely. Pathogenic traits may evolve relatively rarely, and they do not

appear to have evolved in the viruses or plasmids that infect archaea.

How can we know if a particular microbe is responsible for a disease? The late nineteenth century was a productive era in the history of medicine—a time when bacteriologists, chemists, and physicians proved that many diseases are caused by microbial agents. During this time, the German physician Robert Koch laid down a set of four rules for establishing that a particular microorganism causes a particular disease:

1. The microorganism is always found in individuals with the disease.

2. The microorganism can be taken from the host and grown in pure culture.

3. A sample of the culture produces the same disease when injected into a new, healthy host.

4. The newly infected host yields a new, pure culture of microorganisms identical to those obtained in the second step.

These rules, called **Koch's postulates**, were important tools in a time when it was not widely understood that microorganisms cause disease. Although modern medical science has more powerful diagnostic tools, Koch's postulates remain useful. For example, physicians were taken aback in the 1990s

26.22 Satisfying Koch's Postulates Robin Warren and Barry Marshall of the University of Western Australia won the 2005 Nobel Prize in Medicine for showing that ulcers are caused not by the action of stomach acid but by infection with the bacterium *Helicobacter pylori*.

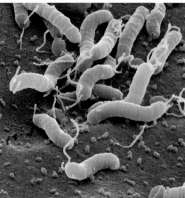

Helicobacter pylori   1.5 μm

Marshall and Warren set out to satisfy Koch's postulates:

Test 1

The microorganism must be present in every case of the disease.

Results: Biopsies from the stomachs of many patients revealed that the bacterium was always present if the stomach was inflamed or ulcerated.

Test 2

The microorganism must be cultured from a sick host.

Results: The bacterium was isolated from biopsy material and eventually grown in culture media in the laboratory.

Test 3

The isolated and cultured bacteria must be able to induce the disease.

Results: Marshall was examined and found to be free of bacteria and inflammation in his stomach. After drinking a pure culture of the bacterium, he developed stomach inflammation (gastritis).

Test 4

The bacteria must be recoverable from newly infected individuals.

Results: Biopsy of Marshall's stomach 2 weeks after he ingested the bacteria revealed the presence of the bacterium, now christened *Helicobacter pylori*, in the inflamed tissue.

Conclusion

Antibiotic treatment eliminated the bacteria and the inflammation in Marshall's stomach. The experiment was repeated on healthy volunteers, and many patients with gastric ulcers were cured with antibiotics. Thus Marshall and Warren demonstrated that the stomach inflammation leading to ulcers is caused by *H. pylori* infections in the stomach.

when stomach ulcers—long accepted and treated as the result of excess stomach acid—were shown by Koch's postulates to be caused by the bacterium *Helicobacter pylori* (**Figure 26.22**).

For an organism to be a successful pathogen, it must:

- arrive at the body surface of a potential host;
- enter the host's body;
- evade the host's defenses;
- reproduce inside the host; and
- infect a new host.

Failure to complete any of these steps ends the reproductive career of a pathogenic organism. Yet in spite of the many defenses available to potential hosts (see Chapters 39 and 42), pathogenic bacteria are often surprisingly difficult to combat, even with today's arsenal of antibiotics. One source of this difficulty, as we have seen, is their ability to form biofilms.

For the host, the consequences of a bacterial infection depend on several factors. One is the **invasiveness** of the pathogen: its ability to multiply in the host's body. Another is its **toxigenicity**, or ability to produce toxins (chemical substances that are harmful to the host's tissues). *Corynebacterium diphtheriae*, the agent that causes diphtheria, has low invasiveness and multiplies only in the throat, but its toxigenicity is so great that the entire body is affected. In contrast, *Bacillus anthracis*, which causes anthrax, has low toxigenicity, but is so invasive that the entire bloodstream ultimately teems with the bacteria.

There are two general types of bacterial toxins, exotoxins and endotoxins. **Endotoxins** are lipopolysaccharides (complexes consisting of a polysaccharide and a lipid component) that form part of the outer membrane of certain Gram-negative bacteria. They are released when these bacteria lyse (burst). Endotoxins are rarely fatal to the host; they normally cause fever, vomiting, and diarrhea. Among the endotoxin producers are some strains of the proteobacteria *Salmonella* and *Escherichia*.

Exotoxins are soluble proteins released by living, multiplying bacteria. They are highly toxic—often fatal—to the host. Human diseases induced by bacterial exotoxins include tetanus

(*Clostridium tetani*), cholera (*Vibrio cholerae*), and bubonic plague (*Yersinia pestis*). Anthrax is caused by three exotoxins produced by *Bacillus anthracis*. Botulism is caused by exotoxins produced by the obligate anaerobe *Clostridium botulinum* that are among the most poisonous ever discovered. The lethal dose for humans of one exotoxin of *C. botulinum* is about one-millionth of a gram. Nonetheless, much smaller doses of this exotoxin are marketed under various trade names (the best known being Botox®) and are used to treat muscle spasms as well as for cosmetic purposes (temporary wrinkle reduction in the skin).

▬▬▬▬▬▬▬▬▬▬▬▬▬▬▬▬▬ **RECAP** **26.3**

Many prokaryotes are beneficial and even necessary to other forms of life. Most animals, including humans, depend on a complex community of prokaryotes—a microbiome—to maintain health, especially of the immune and digestive systems. Pathogenic bacteria are the direct causes of diseases.

- How are the four nutritional categories of prokaryotes distinguished? **See p. 539 and Table 26.2**
- Why is nitrogen metabolism in the prokaryotes vital to other organisms? **See p. 538**
- How do biofilms form, and why are they of special interest to researchers? **See p. 539 and Figure 26.20**

Before moving on to discuss the diversity of eukaryotic life, it is appropriate to consider another category of life that includes some pathogens: the viruses.

(A)

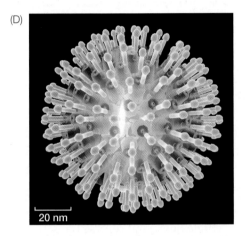

50 nm

A negative-sense single-stranded RNA virus: Influenza virus H5N1, the "bird flu" virus. Surface view.

(B)

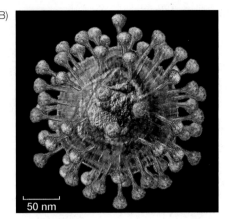

50 nm

A positive-sense single-stranded RNA virus: Coronavirus of a type thought to be responsible for severe acute respiratory syndrome (SARS). Surface view.

(C)

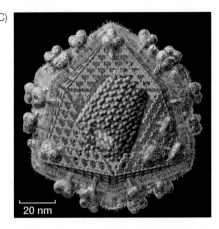

20 nm

An RNA retrovirus: One of the human immunodeficiency viruses (HIV) that causes AIDS. Cutaway view.

(D)

20 nm

A double-stranded DNA virus: One of the many herpes viruses (Herpesviridae). Different herpes viruses are responsible for many human infections, including chicken pox, shingles, cold sores. and genital herpes (HSV1/2). Surface view.

(E)

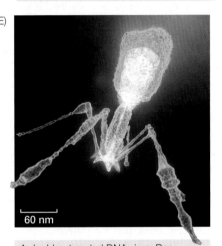

60 nm

A double-stranded DNA virus: Bacteriophage T4. Viruses that infect bacteria are referred to as bacteriophage (or simply phage). T4 attaches leglike fibers to the outside of its host cell and injects its DNA into the cytoplasm through its "tail" (pink structure in this rendition).

(F)

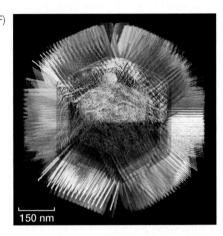

150 nm

A double-stranded DNA mimivirus: This *Acanthamoeba polyphaga* mimivirus (APMV) has the largest diameter of all known viruses and a genome larger than some prokaryote genomes. It is named for its host, an amoeba. Cutaway view.

26.23 Viruses Are Diverse Relatively small genomes and rapid evolutionary rates make it difficult to reconstruct phylogenetic relationships among some classes of viruses. Instead, viruses are classified largely by general characteristics of their genomes. The images here are computer artists' reconstructions based on cryo-electron micrographs.

26.4 How Do Viruses Relate to Life's Diversity and Ecology?

Although they are not cellular, viruses are numerically among the most abundant forms of life on Earth, and their effects on other organisms are enormous. Where did viruses come from, and how do they fit into the tree of life? Biologists are still working to answer these questions.

Some biologists do not think of viruses as living organisms, primarily because they are not cellular and must depend on cellular organisms for basic life functions such as replication and metabolism (see Section 16.1). But viruses are derived from the cells of living organisms. They use the same forms of genetic information storage and transmission as do cellular organisms. Viruses infect all cellular forms of life—bacteria, archaea, and eukaryotes. They replicate, mutate, evolve, and interact with other organisms, often causing serious diseases in their hosts. Finally, viruses clearly evolve independently of other organisms, so it is almost impossible not to treat them as a part of life.

Several factors make viral phylogeny difficult to resolve. First, the tiny size of many viral genomes restricts the phylogenetic analyses that can be conducted to relate them to the genomes of cellular organisms. Second, their rapid mutation rate, which results in rapid evolution of viral genomes, tends to cloud their evolutionary relationships over long periods. Third, there are no known fossil viruses—viruses are too small and delicate to fossilize—so the paleontological record offers no clues to virus origins. Finally, viruses are highly diverse (**Figure 26.23**). Several lines of evidence support the hypothesis that viruses have evolved repeatedly within each of the major groups of life. The difficulty in resolving their deep evolutionary relationships makes a phylogeny-based classification of viruses difficult. Instead, viruses are placed in one of several groups on the basis of the structure of their genomes, although most of these defined groups are not thought to represent monophyletic taxa.

Many RNA viruses probably represent escaped genomic components of cellular life

Although viruses are obligate parasites of cellular species, many viruses may once have been cellular components involved in basic cellular functions—that is, they may be "escaped" components of cellular life that now evolve independently of their hosts.

NEGATIVE-SENSE SINGLE-STRANDED RNA VIRUSES A case in point is a class of viruses whose genome is composed of single-stranded **negative-sense RNA**: RNA that is the complement of the mRNA needed for protein translation. Many of these negative-sense single-stranded RNA viruses have only a few genes, including one for an RNA-dependent RNA polymerase that allows them to make complementary mRNA from their negative-sense RNA genome. Modern cellular organisms cannot generate mRNA in this manner (at least in the absence of viral infections), but scientists speculate that single-stranded RNA genomes may have been common in the distant past, before DNA became the primary molecule for genetic information storage.

A self-replicating RNA polymerase gene that began to replicate independently of a cellular genome could conceivably acquire a few additional protein-coding genes through recombination with its host's DNA. If one or more of these genes were to foster the development of a protein coat, the virus might then survive outside the host and infect new hosts. It is believed that this scenario has been repeated many times independently across the tree of life, given that many of the negative-sense single-stranded RNA viruses that infect organisms from bacteria to humans are not closely related to one another. In other words, negative-sense single-stranded RNA viruses do not represent a distinct taxonomic group, but rather exemplify a particular process of cellular escape that probably happened many different times.

Familiar examples of negative-sense single-stranded RNA viruses include the viruses that cause measles, mumps, rabies, and influenza (see Figure 26.23A).

POSITIVE-SENSE SINGLE-STRANDED RNA VIRUSES The genome of another type of single-stranded RNA virus is composed of positive-sense RNA. Positive-sense genomes are already set for translation; no replication of the genome to form a complement strand is needed before protein translation can take place. Positive-sense single-stranded RNA viruses (see Figure 26.23B) are the most abundant and diverse class of viruses. Most of the viruses that cause diseases in crop plants are members of this group. These viruses kill patches of cells in the leaves or stems of plants, leaving live cells amid a patchwork of discolored dead tissue (giving them the name of mosaic or mottle viruses; **Figure 26.24**). Other viruses in this group infect bacteria, fungi, and animals. Human diseases caused by positive-sense single-stranded RNA viruses include polio, hepatitis C, and the common cold. As is true of the other functionally defined groups of viruses, these

Yellow areas are dead leaf cells, killed by the mosaic virus.

26.24 Mosaic Viruses Are a Problem for Agriculture Mosaic, or "mottle," viruses are the most diverse class of viruses. This leaf is from an apple tree infected with a mosaic virus.

viruses appear to have evolved multiple times across the tree of life from different groups of cellular ancestors.

RNA RETROVIRUSES The RNA retroviruses are best known as the group that includes the human immunodeficiency viruses (HIV; see Figure 26.23C). Like the previous two categories of viruses, RNA retroviruses have genomes composed of single-stranded RNA and probably evolved as escaped cellular components.

Retroviruses are so named because they regenerate themselves by reverse transcription. When the retrovirus enters the nucleus of its vertebrate host, viral reverse transcriptase produces complementary DNA (cDNA) from the viral RNA genome, then replicates that single-stranded cDNA to produce double-stranded DNA. Another virally encoded enzyme, called integrase, catalyzes the integration of the new piece of double-stranded viral DNA into the host's genome. The viral genome is then replicated along with the host cell's DNA; the integrated retroviral DNA is known as a **provirus**.

Retroviruses are only known to infect vertebrates, although genomic elements that resemble portions of these viruses are a component of the genomes of a wide variety of organisms, including bacteria, plants, and many animals. Several retroviruses are associated with various forms of cancer, as cells infected with these viruses are likely to undergo uncontrolled replication.

DOUBLE-STRANDED RNA VIRUSES Double-stranded RNA viruses may have evolved repeatedly from single-stranded RNA ancestors—or perhaps vice versa. These viruses, which are not closely related to one another, infect organisms from throughout the tree of life. Many plant diseases are caused by double-stranded RNA viruses. Other viruses of this type cause many cases of infant diarrhea in humans.

Some DNA viruses may have evolved from reduced cellular organisms

Another class of viruses is composed of viruses that have a double-stranded DNA genome (see Figure 26.23D–F). This group is also almost certainly polyphyletic (with many independent origins). Many of the common phage that infect bacteria are double-stranded DNA viruses, as are the viruses that cause smallpox and herpes in humans.

Some biologists think that at least some of the DNA viruses may represent highly reduced parasitic organisms that have lost their cellular structure as well as their ability to survive as free-living species. For example, the mimiviruses, which are some of the largest DNA viruses (see Figure 26.23F), have a genome in excess of a million base pairs of DNA that encodes more than 900 proteins. This genome is similar in size to the genomes of many parasitic bacteria and about twice as large as the genomes of the smallest bacteria (**Figure 26.25**). Phylogenetic analyses of these DNA viruses suggest that they have evolved repeatedly from cellular organisms. Furthermore, recombination among different viruses may have allowed the exchange of various genetic modules, further complicating the history and origins of these viruses.

Vertebrate genomes contain endogenous retroviruses

In introducing the RNA retroviruses, we noted that they insert their genomes into the genomes of their vertebrate hosts. As these incorporated retroviral genomes evolve over time, many become nonfunctional copies that are no longer expressed as functional viruses. These sequences may provide a record of ancient viral infections that plagued our ancestors. Humans, for example, carry about 100,000 fragments of endogenous retroviruses in our genome. These fragments make up about 8 percent of our DNA—a considerably larger fraction of our genome than the fraction that comprises all our protein-coding genes (about 1.2 percent of our genome).

In 2006 a French biologist named Thierry Heidmann examined the variants of one such endogenous retrovirus in a number of different people. He reasoned that each of these sequences might contain small numbers of changes from the original sequence. If so, then a consensus of the variants might resemble the original, functional sequence of a retrovirus. Heidmann constructed the consensus sequence in the lab and inserted it into human cells in culture. The cultured cells produced functional retroviruses, which could then infect other cells. Heidmann named this reconstructed virus "Phoenix" after the mythical bird that rose from the dead. Since then, biologists have resurrected other retroviruses from our genomes, and they have then identified and studied the genes that our cells use to disable these retroviral sequences and keep them from producing functional viruses. In this way, we are beginning to understand how our bodies fight retroviral infections.

Although many endogenous retroviral fragments in our genomes represent nonfunctional "ghost sequences," in some cases we appear to have derived new functions from captured retroviral sequences. For example, when a developing fetus forms a placenta, the cells in the outer layer of the placenta fuse together. This fusion is accomplished by the expression of a protein on the cell's surface, which allows the cells to attach and merge with one another. The protein that is responsible for this fusion is encoded in an endogenous retroviral sequence. Thus vertebrates appear to have co-opted some critical

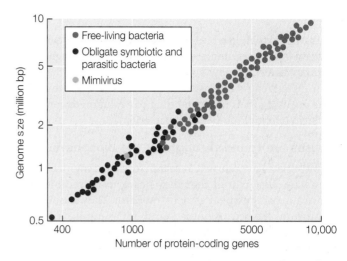

26.25 Mimiviruses Have Genomes Similar in Size to Those of Many Parasitic Bacteria The genome of *Acanthamoeba polyphaga* mimivirus contains 1,181,404 base pairs and encompasses 911 protein-encoding genes. This observation is consistent with the hypothesis that this virus evolved from a parasitic bacterium.

functions from retroviral genomes that were inserted during ancient infections.

Viruses can be used to fight bacterial infections

Although some viruses cause devastating diseases, other viruses have been used to fight disease. Most bacterial diseases are treated today with antibiotics. But antibiotics were first discovered in the 1930s, and they were not widely used to treat bacterial diseases until the 1940s; antibiotics were not available during World War I, when bacterial infections plagued the battlefields. Battlefield wounds were often infected by bacteria, and in the absence of antibiotics, these infections often led to the loss of limbs and lives. While trying to find a way to combat this problem, a physician named Felix d'Herelle discovered the first evidence of viruses that attack bacteria. He named these viruses **bacteriophages**, or "eaters of bacteria." d'Herelle extracted bacteriophages from the stool of infected patients. He then used these extracts to treat patients with deadly bacterial infections, including dysentery, cholera, and bubonic plague. This practice became known as phage therapy. After the war, phage therapy was widely used among the general public to treat bacterial infections of the skin and intestines.

 Go to Media Clip 26.3
Bacteriophages Attack *E. coli*
Life10e.com/mc26.3

Phage therapy was mostly replaced by the use of antibiotics in the 1930s and 1940s as physicians grew concerned about treating patients with live viruses. Phage therapy continued to be used in the Soviet Union but largely disappeared from Western medical practice. Today, however, many antibiotics are losing their effectiveness as bacterial pathogens evolve resistance to these drugs. Phage therapy is once again an active area of research, and it is likely that bacteriophages will become increasingly important as weapons against bacterial diseases. One advantage that bacteriophages may have over antibiotics is that, like bacteria, bacteriophages can evolve. As bacteria evolve resistance to a strain of bacteriophages, biologists can select for new strains of bacteriophages that retain their effectiveness against the pathogens. In this way, biologists are using their understanding of evolution to combat the problem of antibiotic-resistant bacteria.

Viruses are found throughout the biosphere

As biologists have learned to search for and recognize viruses, they have discovered that they occur in incredible abundance almost anywhere we look on Earth. They are abundant throughout the oceans of the world, for example, with an estimated 10^{30} individual viruses in marine environments. This means there are about 15 times more viruses in the oceans than there are cellular organisms, including all bacteria, archaea, and eukaryotes combined. Many of these viruses are bacteriophages, and they have an enormous effect on the ecology of the oceans. Every day, about half of the bacteria in the oceans are killed by viruses. Huge marine blooms of bacteria, such as the *Vibrio* bloom that produced the milky seas described at the opening of this chapter, do not last long because viral blooms soon follow the initial bacterial bloom. As the bacteriophages increase, they begin to kill bacteria faster than the bacteria can reproduce. Another species of *Vibrio* causes cholera in humans, and cholera epidemics also fade as bacteriophages increase and control these *Vibrio* population booms.

Many nutrient cycles, such as the carbon cycle (see Section 58.3), are strongly influenced by viral populations. Photosynthetic bacteria and algae produce about half of the oxygen in Earth's atmosphere. During photosynthesis, these organisms also take up carbon dioxide, which they use to produce organic carbon compounds. As these photosynthetic organisms are killed by viruses, their remains settle to the ocean bottom. These remains are a primary source of the vast oil reserves that occur under oceanic sediments. The constant growth and death of these photosynthetic organisms controlled largely by viruses. Because of their importance in controlling these populations, viruses have an enormous impact on Earth's climate and ecosystems.

RECAP 26.4

Viruses are highly diverse and appear to have evolved independently from many different cellular organisms within each of the major groups of life. Some viruses appear to have evolved from escaped components of cellular organisms, whereas other viruses may have evolved from parasitic cellular ancestors.

- Why is it difficult to place viruses precisely within the tree of life? See p. 543

- What are the two main hypotheses of viral origins? See pp. 544–545

- How can viruses be used to treat some human diseases? See pp. 545–546

- What are some of the ways that viruses can affect Earth's ecosystems? See p. 546

It may be best to view viruses as "spin-offs" from the various branches on the tree of life—sometimes evolving independently of cellular genomes, sometimes recombining with them. One way to think of viruses is as the "bark" on the tree of life: an important component all across the tree, but not quite like the main branches. The third of those main branches—the eukaryotes—will be our focus in the rest of Part 7.

What adaptive advantage does bioluminescence provide to *Vibrio* bacteria?

ANSWER

Although these marine *Vibrio* are able to live independently, they truly thrive inside the guts of fish and other marine animals. Inside a fish, *Vibrio* cells attach themselves to food particles, including phytoplankton, and are often expelled into the ocean as waste. How can they get back into their preferred environment? The bioluminescent glow produced by a dense colony of free-living *Vibrio* growing on phytoplankton attracts fish, which consume the phytoplankton and thus ingest the bacteria—which gets the bacteria into a new host fish.

Margaret McFall-Ngai and her colleagues have studied one species of bioluminescent *Vibrio*, *V. fischeri*, in which a symbiotic relationship with the Hawaiian bobtail squid (*Euprymna scolopes*) has evolved. Chemicals produced by a developing squid embryo specifically "recruit" *V. fischeri* from the surrounding seawater. The bacteria then preferentially migrate to specific tissues that develop into a bioluminescent "light organ" in the belly of the adult squid (**Figure 26.26**). The tiny (about 3 cm long) adult squid feed while floating near the sea surface at night. The soft glow produced by the bioluminescent bacteria mimics the moonlight above, so the squid are less visible to potential predators coming at them from below.

Vibrio fischeri live symbiotically inside the squid's light organ.

26.26 Bioluminescent Bacterial Symbionts *Vibrio* bacteria within the light organ emit bioluminescence downward as the Hawaiian bobtail squid floats near the ocean surface to feed. At night, this allows the squid to blend in with moonlight or starlight rather than becoming a target for a predator from below.

CHAPTER**SUMMARY** 26

 26.1 Where Do Prokaryotes Fit into The Tree of Life?

- Two of life's three domains, Bacteria and Archaea, are prokaryotic. They are distinguished from Eukarya in several ways, including their lack of a nucleus and of membrane-enclosed organelles. **Review Table 26.1**

- Eukaryotes are related to both Archaea and Bacteria and appear to have originated through endosymbiosis between members of these two lineages. The last common ancestor of all three domains probably lived about 3 billion years ago. **Review Figure 26.1, ANIMATED TUTORIAL 26.1**

- Bacteria can be classified into two groups by the **Gram stain. Gram-negative bacteria** have a periplasmic space between the plasma membrane and a distinct outer membrane. **Gram-positive bacteria** have a thick cell wall containing about five times as much peptidoglycan as a Gram-negative wall. **Review Figure 26.2, ACTIVITY 26.1**

- The three most common bacterial shapes are **cocci** (spheres), **bacilli** (rods), and **spirilla** (helices). **Review Figure 26.3**

- Phylogenetic classification of prokaryotes is now based principally on the nucleotide sequences of rRNA and other genes involved in fundamental cellular processes.

- Although **lateral gene transfer** has occurred throughout pro-karyotic evolutionary history, elucidation of many aspects of pro-karyote phylogeny is still possible. **Review Figure 26.4**

 26.2 Why Are Prokaryotes So Diverse and Abundant?

- Prokaryotes are the most numerous organisms on Earth.

- The **low-GC Gram-positives** include the **mycoplasmas**, which are among the smallest cellular organisms ever discovered.

- Some **high-GC Gram-positives** produce important antibiotics.

- The photosynthetic **cyanobacteria** release oxygen into the atmo-sphere. Cyanobacteria may live free as single cells or associate in multicellular colonies. **Review Figure 26.9**

- **Spirochetes** have unique structures called axial filaments that allow them to move in a corkscrew-like manner. **Review Figure 26.10**

- The **proteobacteria** embrace the largest number of known spe-cies of bacteria. Smaller groups include the **hyperthermophilic bacteria, hadobacteria,** and **chlamydias.**

- Many archaea are **extremophiles. Review Figure 26.14**

- **Ether linkages** in the branched long hydrocarbon chains of the lipids that make up the cell membranes are a synapomorphy of Archaea. **Review Figure 26.15**

 26.3 How Do Prokaryotes Affect Their Environments?

- Prokaryotes form complex communities, of which **biofilms** are one example. **Review Figure 26.19**

- Prokaryote metabolism is very diverse. Some prokaryotes are anaerobic, others are aerobic, and still others can shift between these modes.

- Prokaryotes fall into four broad nutritional categories: **photoauto-trophs, photoheterotrophs, chemoautotrophs,** and **chemo-heterotrophs. Review Table 26.2**

- Prokaryotes play key roles in the cycling of elements such as nitro-gen, oxygen, sulfur, and carbon.

- Diverse communities of bacteria and archaea live on and in most animals. The composition of these **microbiomes** is often closely associated with the animal's health. **Review Figure 26.21**

- **Koch's postulates** establish the criteria by which an organism may be classified as a pathogen. Relatively few bacteria—and no archaea—are known to be pathogens. **Review Figure 26.22**

 26.4 How Do Viruses Relate to Life's Diversity and Ecology?

- Viruses have evolved many times from many different groups of cellular organisms. They are placed in groups according to the structure of their genomes, but these groups are not thought to represent monophyletic taxa. **Review Figure 26.23**

- Some viruses are probably derived from escaped components of cellular organisms; others are thought to have evolved as highly reduced parasites. **Review Figure 26.25**

- A large fraction of vertebrate (including human) genomes consists of incorporated remains of retroviral genomes.

- Bacteriophages have been used to treat bacterial infections in humans.

- Viruses are found in virtually all of Earth's environments and have a huge impact on the planet's ecosystems.

 Go to the Interactive Summary to review key figures, Animated Tutorials, and Activities
Life10e.com/is26

CHAPTER**REVIEW**

 **REMEMBERING**

1. The division of the living world into three domains
   - a. is based on the number of cells in organisms of each group.
   - b. is based mostly on major morphological differences between archaea and bacteria.
   - c. emphasizes the greater importance of eukaryotes.
   - d. was proposed by the early microscopists.
   - e. is based on phylogenetic relationships determined from nucleotide sequences of rRNA and other genes.

2. Which statement about nitrogen metabolism is *not* true?
   - a. Certain prokaryotes reduce atmospheric N_2 to ammonia.
   - b. Some nitrifiers are soil bacteria.
   - c. Denitrifiers are obligate anaerobes.
   - d. Nitrifiers obtain energy by oxidizing ammonia and nitrite.
   - e. Without nitrifiers, terrestrial organisms would lack a nitrogen supply.

3. All photosynthetic bacteria
   a. use chlorophyll *a* as their photosynthetic pigment.
   b. use bacteriochlorophyll as their photosynthetic pigment.
   c. release oxygen gas (O_2).
   d. produce particles of sulfur.
   e. are photoautotrophs.

4. Gram-negative bacteria
   a. appear blue to purple following Gram staining.
   b. appear pink to red following Gram staining.
   c. are all either bacilli or cocci.
   d. contain no peptidoglycan in their cell walls.
   e. are all photosynthetic.

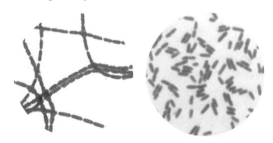

5. Archaea
   a. have cytoskeletons.
   b. have distinctive lipids in their plasma membranes.
   c. survive only at moderate temperatures and near neutral pH.
   d. all produce methane.
   e. have substantial amounts of peptidoglycan in their cell walls.

6. Genetic evidence suggests that viruses
   a. are most closely related to Bacteria.
   b. are most closely related to Archaea.
   c. are most closely related to Eukarya.
   d. have evolved multiple times from many different cellular species.
   e. evolved from the fusion of a bacterial and an archaeal species.

UNDERSTANDING & APPLYING

7. Why do systematic biologists find rRNA sequence data more useful than data on metabolism or cell structure for classifying prokaryotes?

8. The figure below shows an organismal tree in which gene *x* has undergone a lateral transfer event. Draw the phylogenetic tree you would expect based on gene *x*, as well as the phylogenetic tree you would expect based on a consensus of non-transferred genes.

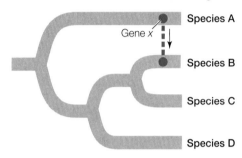

9. Do you consider viruses to be living organisms? Why or why not?

ANALYZING & EVALUATING

10. Kashefi and Lovley were able to grow an unnamed archaeal species at temperatures above 120°C only because they used Fe_3^+ as an electron acceptor—no other electron acceptor they tried allowed the archaean to grow (see Figure 26.14). How might you explore the same or other high-temperature environments for other hyperthermophilic organisms not detected by Kashefi and Lovley using Fe_3^+?

Go to BioPortal at **yourBioPortal.com** for Animated Tutorials, Activities, LearningCurve Quizzes, Flashcards, and many other study and review resources.

27

The Origin and Diversification of Eukaryotes

CHAPTEROUTLINE

27.1 How Did the Eukaryotic Cell Arise?

27.2 What Features Account for Protist Diversity?

27.3 What Is the Relationship between Sex and Reproduction in Protists?

27.4 How Do Protists Affect Their Environments?

A Toxic Sea A bloom of dinoflagellates of the genus *Noctiluca* was responsible for this red tide in Puget Sound in the U. S. state of Washington.

I N THE SUMMER of 2005, a devastating red tide crippled the shellfish industry along the Atlantic coast of North America from Canada to Massachusetts. This red tide was produced by a bloom of dinoflagellates of the genus *Alexandrium*. These protists produce a powerful toxin that accumulates in clams, mussels, and oysters. A person who eats a mollusk contaminated with the toxin can experience a syndrome known as paralytic shellfish poisoning. Many people were sickened by eating mollusks that were harvested before the problem was diagnosed, and losses to the shellfish industry in 2005 were estimated at $50 million.

Several species of dinoflagellates produce toxic red tides in many parts of the world. Along the Gulf of Mexico, red tides caused by dinoflagellates of the genus *Karenia* produce a neurotoxin that affects the central nervous systems of fish, which become paralyzed and cannot respire effectively. Huge numbers of dead fish wash up on Gulf Coast beaches during a *Karenia* red tide. In addition, wave action can produce aerosols of the *Karenia* toxin, and these aerosols often cause asthma-like symptoms in humans on shore.

After the losses that resulted from the 2005 red tide, biologists at the Woods Hole Oceanographic Institution (WHOI) on Cape Cod began to monitor and model dinoflagellate populations off the New England coast. If biologists could accurately forecast future blooms, people in the area could be made aware of the problem in advance and adjust the shellfish harvest (and their eating habits) accordingly.

Biologists from WHOI monitored counts of dinoflagellates in the water and in seafloor sediments. They also monitored river runoff, water currents, water temperature and salinity, winds, and tides. Another environmental factor they considered was the "nor'easter" storms common along the New England coast. By correlating their measurements of these environmental factors with dinoflagellate counts, biologists produced a model that predicted growth of dinoflagellate populations.

In spring 2008, the WHOI team determined that all the factors were in place to produce another red tide like the one of 2005—if a nor'easter occurred to blow the dinoflagellates toward the coast. A nor'easter did occur at just the wrong time, and another red tide materialized in summer 2008, just as predicted. But this time, people were warned. Shellfish harvesters adjusted their harvest, and many fewer people were harmed by eating toxic mollusks.

Can dinoflagellates be beneficial, as well as harmful, to marine ecosystems?

See answer on p. 566.

27.1 How Did the Eukaryotic Cell Arise?

We easily recognize trees, mushrooms, and insects as plants, fungi, and animals, respectively. But there is a dazzling assortment of other eukaryotic organisms—mostly microscopic—that do not fit into these three groups. Eukaryotes that are not plants, animals, or fungi have traditionally been called **protists**. But phylogenetic analyses reveal that many of the groups we commonly refer to as protists are not, in fact, closely related. Thus the term "protist" does not describe a formal taxonomic group, but is a convenience term for "all the eukaryotes that are not plants, animals, or fungi."

The unique characteristics of the eukaryotic cell lead scientists to conclude that the eukaryotes are monophyletic and that a single eukaryotic ancestor diversified into the many different protist lineages as well as giving rise to the plants, fungi, and animals. As we saw in Section 26.1, eukaryotes are generally thought to be more closely related to Archaea than to Bacteria. The mitochondria and chloroplasts of eukaryotes, however, are clearly derived from bacterial lineages (see Figure 26.1).

Biologists traditionally hypothesized that the split of Eukarya from Archaea was followed by endosymbioses with bacterial lineages that led to the origin of mitochondria and chloroplasts. Some biologists prefer to view the origin of eukaryotes as the *fusion* of lineages from the two prokaryote groups. This difference is largely a semantic one that hinges on the point at which we deem the eukaryote lineage to have become definitively "eukaryotic." In either case, we can make some reasonable inferences about the events that led to the evolution of a new cell type, bearing in mind that the environment underwent an enormous change—from low to high availability of atmospheric oxygen—during the course of these events.

The modern eukaryotic cell arose in several steps

At least five events were significant in the origin of the eukaryotic cell (**Figure 27.1**):

- The origin of a flexible cell surface
- The origin of a cytoskeleton
- The origin of a nuclear envelope, which enclosed a genome organized into chromosomes

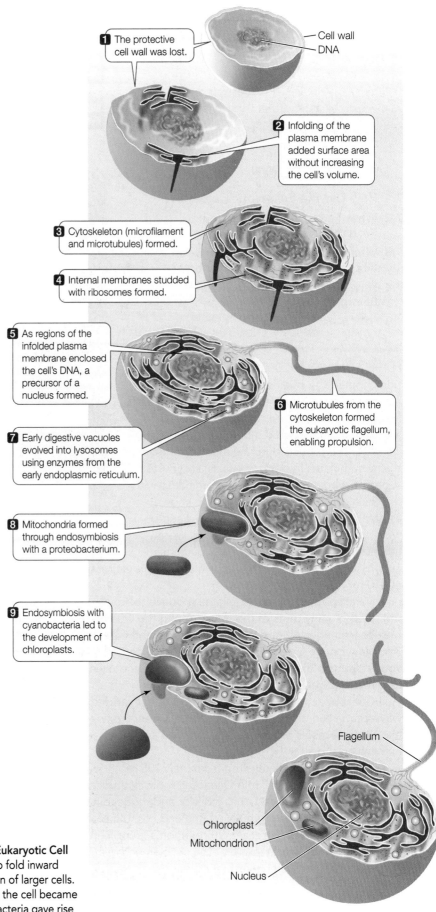

1 The protective cell wall was lost.

Cell wall
DNA

2 Infolding of the plasma membrane added surface area without increasing the cell's volume.

3 Cytoskeleton (microfilament and microtubules) formed.

4 Internal membranes studded with ribosomes formed.

5 As regions of the infolded plasma membrane enclosed the cell's DNA, a precursor of a nucleus formed.

6 Microtubules from the cytoskeleton formed the eukaryotic flagellum, enabling propulsion.

7 Early digestive vacuoles evolved into lysosomes using enzymes from the early endoplasmic reticulum.

8 Mitochondria formed through endosymbiosis with a proteobacterium.

9 Endosymbiosis with cyanobacteria led to the development of chloroplasts.

Flagellum

Chloroplast
Mitochondrion

Nucleus

27.1 A Hypothetical Sequence for the Evolution of the Eukaryotic Cell The loss of a firm cell wall allowed the plasma membrane to fold inward and create more surface area, which facilitated the evolution of larger cells. As cells grew larger, cytoskeletal complexity increased, and the cell became increasing compartmentalized. Endosymbioses involving bacteria gave rise to mitochondria and (in photosynthetic eukaryotes) to chloroplasts.

- The appearance of digestive vacuoles
- The acquisition of mitochondria and chloroplasts via endosymbiosis

FLEXIBLE CELL SURFACE We presume that ancient prokaryotic organisms, like most present-day prokaryotic cells, had firm cell walls. The first step toward the eukaryotic condition was the loss of the cell wall.

Consider the possibilities open to a flexible cell without a firm wall, starting with cell size. As a cell grows larger, its surface area-to-volume ratio decreases (see Figure 5.2). Unless the surface area can be increased, the cell's volume will reach an upper limit. If the cell's surface is flexible, however, it can fold inward, creating more surface area for gas and nutrient exchange. With a surface flexible enough to allow infolding, the cell can exchange materials with its environment rapidly enough to sustain a larger volume and more rapid metabolism (Figure 27.1, steps 1–2). Furthermore, a flexible surface can pinch off bits of the environment and bring them into the cell by endocytosis. These infoldings of the cell surface (which also exist in some modern prokaryotes) were important for the evolution of large eukaryotic cells.

CHANGES IN CELL STRUCTURE AND FUNCTION Other early steps that were important for the evolution of the eukaryotic cell are likely to have included three advances: the formation of ribosome-studded internal membranes, some of which surrounded the DNA; development of a complex cytoskeleton; and the evolution of digestive vacuoles (Figure 27.1, steps 3–7).

Until a few years ago, biologists thought that cytoskeletons were restricted to eukaryotes. Improved imaging technology and molecular analyses have now revealed homologs of many cytoskeletal proteins in prokaryotes, showing that simple cytoskeletons evolved before the origin of eukaryotes. The cytoskeleton of a eukaryote, however, is much more developed and complex than that of a prokaryote. This greater development of microfilaments and microtubules supports the eukaryotic cell and allows it to manage changes in shape, to distribute daughter chromosomes, and to move materials from one part of the large cell to other parts. In addition, the presence of microtubules in the cytoskeleton could have given rise in some cells to the characteristic eukaryotic flagellum.

The DNA of a prokaryotic cell is attached to a site on its plasma membrane. If that region of the plasma membrane were to fold into the cell, the first step would be taken toward the evolution of a nucleus, a primary feature of the eukaryotic cell.

The nuclear envelope appeared early in the eukaryote lineage. The next step was probably phagocytosis—the ability to engulf and digest other cells in digestive vacuoles. These digestive vacuoles eventually incorporated digestive enzymes to form lysosomes. Other infoldings of the plasma membrane developed into the endoplasmic reticulum and Golgi apparatus.

ENDOSYMBIOSIS At the same time the processes outlined above were taking place, cyanobacteria were generating oxygen gas as a product of photosynthesis. Increasing concentrations of O_2 in the oceans, and eventually in the atmosphere, had disastrous consequences for most organisms of the time, which were unable to tolerate the newly oxidizing environment. But some prokaryotes evolved strategies to utilize the increasing oxygen and—fortunately for us—so did some of the new phagocytic eukaryotes.

At about this time, endosymbioses began to play a role in eukaryote evolution (Figure 27.1, steps 8–9). The theory of endosymbiosis proposes that certain organelles are the descendants of prokaryotes engulfed, but not digested, by early eukaryotic cells. One crucial event in the history of eukaryotes was the incorporation of a proteobacterium that evolved into the mitochondrion. Initially, the new organelle's primary function was probably to detoxify O_2 by reducing it to water. Later, this reduction became coupled with the formation of ATP in cellular respiration. Upon completion of this step, the essential modern eukaryotic cell was complete.

Photosynthetic eukaryotes are the result of yet another endosymbiotic step: the incorporation of a prokaryote related to today's cyanobacteria, which became the chloroplast.

Chloroplasts have been transferred among eukaryotes several times

Eukaryotes in several different groups possess chloroplasts, and groups with chloroplasts appear in several distantly related eukaryote clades. Some of these groups differ in the photosynthetic pigments their chloroplasts contain. And not all chloroplasts are limited to a pair of surrounding membranes like those found in plants—in some microbial eukaryotes, they are surrounded by three or more membranes. We now view these observations as evidence of a remarkable series of endosymbioses. This conclusion is supported by extensive evidence from electron microscopy and nucleic acid sequence comparisons.

All chloroplasts trace their ancestry back to the engulfment of one cyanobacterium by a larger eukaryotic cell. This event, the step that first gave rise to the photosynthetic eukaryotes, is known as **primary endosymbiosis (Figure 27.2A)**. The cyanobacterium, a Gram-negative bacterium, had both an inner and an outer membrane (see Figure 26.2B). Thus the original chloroplasts had two surrounding membranes: the inner and outer membranes of the cyanobacterium. Remnants of the peptidoglycan-containing cell wall of the bacterium are present in the form of a bit of peptidoglycan between the chloroplast membranes of glaucophytes, the first eukaryote group to branch off following primary endosymbiosis (as we will see in Chapter 28). Primary endosymbiosis also gave rise to the chloroplasts of the red algae, green algae, and land plants. The red algal chloroplast retains certain pigments of the original cyanobacterial endosymbiont that are absent in green algal chloroplasts.

Almost all remaining photosynthetic eukaryotes are the result of additional rounds of endosymbiosis. For example, the photosynthetic euglenids derived their chloroplasts from **secondary endosymbiosis (Figure 27.2B)**. Their ancestor took up a unicellular green alga, retaining its chloroplast and eventually losing the rest of the constituents of the alga. This history explains why the photosynthetic euglenids have the same photosynthetic pigments as the green algae and land plants. It also

(A) Primary endosymbiosis

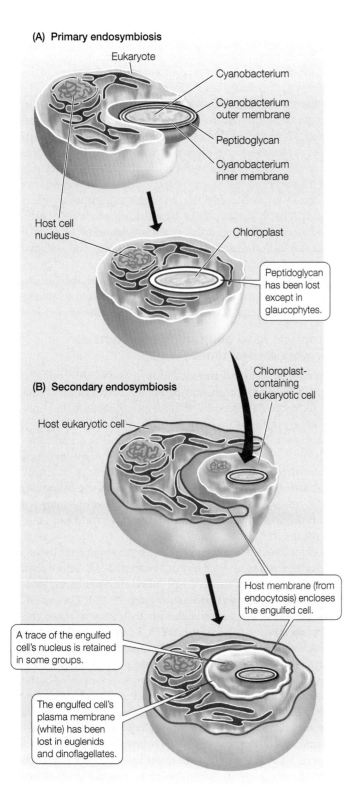

Eukaryote

Cyanobacterium

Cyanobacterium outer membrane

Peptidoglycan

Cyanobacterium inner membrane

Host cell nucleus

Chloroplast

Peptidoglycan has been lost except in glaucophytes.

Chloroplast-containing eukaryotic cell

(B) Secondary endosymbiosis

Host eukaryotic cell

Host membrane (from endocytosis) encloses the engulfed cell.

A trace of the engulfed cell's nucleus is retained in some groups.

The engulfed cell's plasma membrane (white) has been lost in euglenids and dinoflagellates.

27.2 Endosymbiotic Events in the Evolution of Chloroplasts
(A) A single instance of primary endosymbiosis ultimately gave rise to all of today's chloroplasts. (B) Secondary endosymbiosis—the uptake and retention of a chloroplast-containing cell by another eukaryotic cell—took place several times, independently.

 Go to Animated Tutorial 27.1
Family Tree of Chloroplasts
Life10e.com/at27.1

accounts for the third membrane of the euglenid chloroplast, which is derived from the euglenid's plasma membrane (as a result of endocytosis). An additional round—**tertiary endosymbiosis**—occurred when a dinoflagellate apparently lost its chloroplast and took up another protist that had acquired its chloroplast through secondary endosymbiosis.

▬▬▬▬▬▬▬▬▬▬▬▬▬▬▬▬ ▪ **RECAP 27.1**

The modern eukaryotic cell probably arose from an ancestral prokaryote in several steps, including the origin of a flexible cell surface, the enclosure of the genetic material in a nucleus, and endosymbiosis.

- What were the steps that led to the evolution of the eukaryotic cell? **See pp. 550–551 and Figure 27.1**
- Why was the development of a flexible cell surface a key event for eukaryote evolution? **See p. 551**
- Explain how increased availability of atmospheric oxygen (O_2) could have influenced the evolution of the eukaryotic cell. **See p. 551**

The features that eukaryotes gained from archaea and bacteria have allowed them to exploit many different environments. This led to the evolution of great diversity among eukaryotes, beginning with a radiation that started in the Precambrian.

27.2 **What Features Account for Protist Diversity?**

Most eukaryotes can be placed in one of eight major clades that began to diversify about 1.5 billion years ago: alveolates, stramenopiles, rhizaria, excavates, plants, amoebozoans, fungi, and animals (**Figure 27.3**). Plants, fungi, and animals each have close protist relatives (such as the choanoflagellate relatives of animals), which we will discuss along with those major multicellular eukaryote groups in Chapters 28–33.

Each of the five major groups of eukaryotes covered in this chapter consists of organisms with enormously diverse body forms and lifestyles. Some protists are motile, whereas others do not move; some are photosynthetic, others heterotrophic; most are unicellular, but some are multicellular. Most are microscopic, but a few are huge (giant kelps, for example, can grow to half the length of a football field). We refer to the unicellular species of protists as **microbial eukaryotes**, but keep in mind that there are large, multicellular protists as well.

Multicellularity has arisen dozens of times across the evolutionary history of eukaryotes. Four of the origins of multicellularity resulted in large organisms that are familiar to most people: plants, animals, fungi, and brown algae (the last are a group of stramenopiles). In addition, there are dozens of smaller and less familiar groups among the eukaryotes that include multicellular species. Recent experimental studies have shown that artificial selection for multicellularity can produce repeated, convergent evolution of multicellular forms over just a few months in some normally unicellular eukaryotic species. In addition, many unicellular species retain individual identities but nonetheless associate in large multicellular colonies.

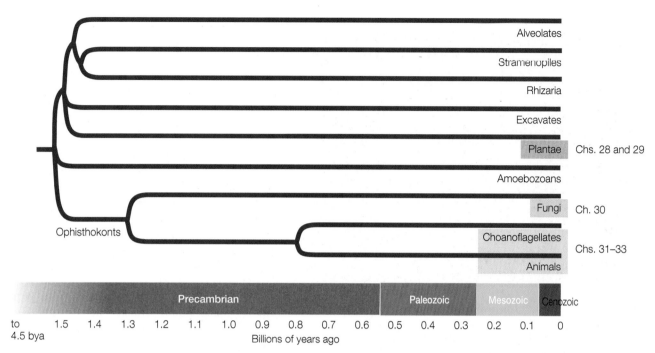

27.3 Precambrian Divergence of Major Eukaryote Groups
A phylogenetic tree shows one current hypothesis and estimated time line for the origin of the major groups of eukaryotes. The rapid divergence of major lineages between 1.5 and 1.4 billion years ago makes reconstruction of their precise relationships difficult. The major multicellular groups (tinted boxes) will be covered in subsequent chapters.

There is a near-continuum between fully integrated, multicellular organisms on the one hand and loosely integrated multicellular colonies of cells on the other. Biologists do not always agree on where to draw the line between the two.

Biologists used to classify protists largely on the basis of their life histories and reproductive features. In recent years, however, electron microscopy and gene sequencing have revealed many new patterns of evolutionary relatedness among these groups. Analyses of slowly evolving gene sequences are making it possible to explore evolutionary relationships among eukaryotes in ever greater detail and with greater confidence. Nonetheless, some substantial areas of uncertainty remain, and lateral gene transfer may complicate our efforts to reconstruct the evolutionary history of protists (as is also true for prokaryotes; see Section 26.1). Today we recognize great diversity among the many distantly related protist clades.

Alveolates have sacs under their plasma membranes

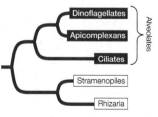

Alveolates are so named because they possess sacs, called alveoli, just beneath their plasma membranes, which may play a role in supporting the cell surface. All alveolates are unicellular, and most are photosynthetic, but they are diverse in body form. The alveolate groups we will consider in detail here are the dinoflagellates, apicomplexans, and ciliates.

DINOFLAGELLATES Most **dinoflagellates** are marine and photosynthetic; they are important primary producers of organic matter in the oceans. Although fewer *species* of dinoflagellates live in fresh water, individuals can be abundant in freshwater environments. The dinoflagellates are of great ecological, evolutionary, and morphological interest. A distinctive mixture of

photosynthetic and accessory pigments gives their chloroplasts a golden brown color. Some dinoflagellate species cause red tides, as discussed at the opening of this chapter. Other species are photosynthetic endosymbionts that live within the cells of other organisms, including invertebrate animals (such as corals; see Figure 27.21) and other marine protists (see Figure 27.12A). Still others are nonphotosynthetic and live as parasites within other marine organisms.

Dinoflagellates have a distinctive appearance. They generally have two flagella, one in an equatorial groove around the cell, the other starting near the same point as the first and passing down a longitudinal groove before extending into the surrounding medium (**Figure 27.4**). Some dinoflagellates can take on different forms, including amoeboid ones, depending on environmental conditions. It has been claimed that the dinoflagellate *Pfiesteria piscicida* can occur in at least two dozen distinct forms, although this claim is highly controversial. In any case, this remarkable dinoflagellate, when present in large enough numbers, is harmful to fish and can both stun and feed on them.

APICOMPLEXANS The exclusively parasitic **apicomplexans** derive their name from the apical complex, a mass of organelles contained in the apical end (the tip) of the cell. These organelles help the apicomplexan invade its host's tissues. For example, the apical complex enables *Plasmodium*, the causative agent of malaria, to enter its target cells in the human body after transmission by a mosquito.

Peridinium sp.

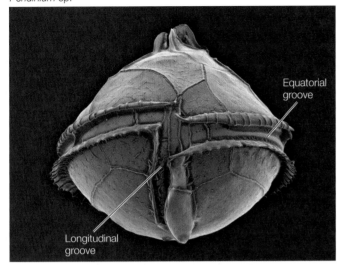

27.4 A Dinoflagellate The presence of two flagella is characteristic of many dinoflagellates, although these appendages are contained within deep grooves and thus are seldom visible. One flagellum lies within the equatorial groove and provides forward thrust and spin to the organism. The second flagellum originates in the longitudinal groove and acts like the rudder of a boat.

 Go to Media Clip 27.1
A Dinoflagellate Shows Off Its Flagellum
Life10e.com/mc27.1

Like many obligate parasites, apicomplexans have elaborate life cycles featuring asexual and sexual reproduction through a series of very dissimilar life stages (see Figure 27.20). In many species, these life stages are associated with two different types of host organisms, as is the case with *Plasmodium*. Another apicomplexan, *Toxoplasma*, alternates between cats and rats to complete its life cycle. A rat infected with *Toxoplasma* loses its fear of cats, which makes it more likely to be eaten by, and thus transfer the parasite to, a cat.

CILIATES The **ciliates** are named for their numerous hairlike cilia, which are shorter than, but otherwise identical to, eukaryotic flagella. The ciliates are much more complex in body form than are most other unicellular eukaryotes (**Figure 27.5**). Their definitive characteristic is the possession of two types of nuclei (whose roles we will describe in Section 27.3 when we discuss protist reproduction). Almost all ciliates are heterotrophic, although a few contain photosynthetic endosymbionts.

Paramecium, a frequently studied ciliate genus, exemplifies the complex structure and behavior of ciliates (**Figure 27.6**). The slipper-shaped cell is covered by an elaborate pellicle, a structure composed principally of an outer membrane and an inner layer of closely packed, membrane-enclosed sacs (the alveoli) that surround the bases of the cilia. Defensive organelles called trichocysts are also present in the pellicle. In response to a threat, a microscopic explosion expels the trichocysts in a few milliseconds, and they emerge as sharp darts, driven forward at the tip of a long, expanding filament.

The cilia provide *Paramecium* with a form of locomotion that is generally more precise than locomotion by flagella. A *Paramecium* can coordinate the beating of its cilia to propel itself either forward or backward in a spiraling manner. It can also back off swiftly when it encounters a barrier or a negative stimulus. The coordination of ciliary beating is probably the result of a differential distribution of ion channels in the plasma membrane near the two ends of the cell.

Organisms living in fresh water are hypertonic to their environment. Many freshwater protists, including *Paramecium*, address this problem by means of specialized **contractile vacuoles** that excrete the excess water the organisms constantly take in by osmosis. The excess water collects in the contractile vacuoles, which then contract and expel the water from the cell.

(A) *Paramecium* sp.

Cilia   10 µm

(B) *Didinium nasutum*

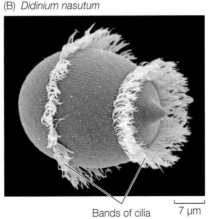

Bands of cilia   7 µm

(C) *Euplotes* sp.

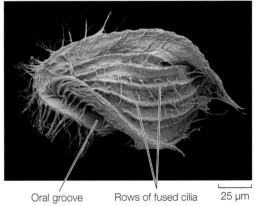

Oral groove   Rows of fused cilia   25 µm

27.5 Diversity among the Ciliates (A) A free-swimming organism, this *Paramecium* belongs to a ciliate group whose members have many cilia of uniform length. (B) The barrel-shaped *Didinium nasutum* feeds on other ciliates, including *Paramecium*. Its cilia occur in two separate bands. (C) Some of the cilia in *Euplotes* fuse into flat sheets that direct food particles into an oral groove.

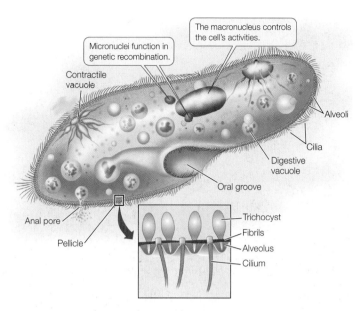

27.6 Anatomy of *Paramecium* *Paramecium*, with its many specialized organelles, exemplifies the complex body form of ciliates.

Go to Activity 27.1 Anatomy of *Paramecium*
Life10e.com/ac27.1

Paramecium and many other protists engulf solid food by endocytosis, forming a **digestive vacuole** within which the food is digested (**Figure 27.7**). Smaller vesicles containing digested food pinch away from the digestive vacuole and enter the cytoplasm. These tiny vesicles provide a large surface area across which the products of digestion can be absorbed by the rest of the cell.

Go to Animated Tutorial 27.2
Digestive Vacuoles
Life10e.com/at27.2

Stramenopiles typically have two flagella of unequal length

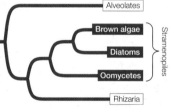

A morphological synapomorphy of most **stramenopiles** is the possession of two flagella of unequal length, with rows of tubular hairs on the longer of the two. Some stramenopiles lack flagella, but they are descended from ancestors that possessed flagella. The stramenopiles include the diatoms and the brown algae, which are photosynthetic, and the oomycetes, which are not.

DIATOMS All of the **diatoms** are unicellular, although some species associate in filaments. Many have sufficient carotenoids in their chloroplasts to give them a yellow or brownish color. All of them synthesize carbohydrates and oils as photosynthetic storage products. Diatoms lack flagella except in male gametes.

Architectural magnificence on a microscopic scale is the hallmark of the diatoms. Almost all diatoms deposit silica (hydrated silicon dioxide) in their cell walls. The cell wall of a diatom is

INVESTIGATING**LIFE**

27.7 The Role of Vacuoles in Ciliate Digestion An acidic environment is known to aid digestion in many multicellular organisms. Do ciliates also use acid to obtain nutrients?[a]

HYPOTHESIS The digestive vacuoles of *Paramecium* produce an acidic environment that allows the organism to digest food particles.

Method 1. Feed *Paramecium* yeast cells stained with Congo red, a dye that is red at neutral or basic pH but turns green at acidic pH.

2. Under a light microscope, observe the formation and degradation of digestive vacuoles within the *Paramecium*. Note time and sequence of color (i.e., acid level) changes.

Results

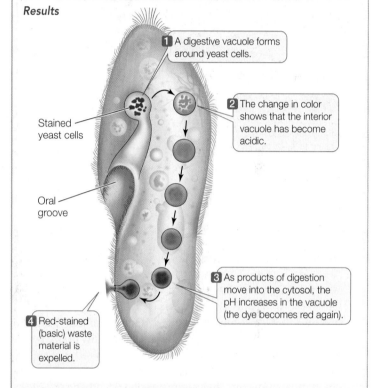

1 A digestive vacuole forms around yeast cells.

2 The change in color shows that the interior vacuole has become acidic.

Stained yeast cells

Oral groove

3 As products of digestion move into the cytosol, the pH increases in the vacuole (the dye becomes red again).

4 Red-stained (basic) waste material is expelled.

CONCLUSION Some ciliates acidify digestive vacuoles to assist in the breakdown of food.

Go to **BioPortal** for discussion and relevant links for all INVESTIGATING**LIFE** figures.

[a]Mast, S. O. 1947. *Biological Bulletin* 92: 31–72.

constructed in two pieces, with the top overlapping the bottom like the top of a petri dish. The silica-impregnated walls have intricate patterns unique to each species (**Figure 27.8**). Despite their remarkable morphological diversity, all diatoms are symmetrical—either bilaterally (with "right" and "left" halves) or radially (with the type of symmetry possessed by a circle).

Diatoms reproduce both sexually and asexually. Asexual reproduction by binary fission is somewhat constrained by the stiff cell wall. Both the top and bottom of the "petri dish" become tops of new "dishes" without changing appreciably in size; as a result, the new cell made from the former bottom is smaller than the parent cell. If this process continued indefinitely, one cell line would simply vanish, but sexual reproduction largely solves this

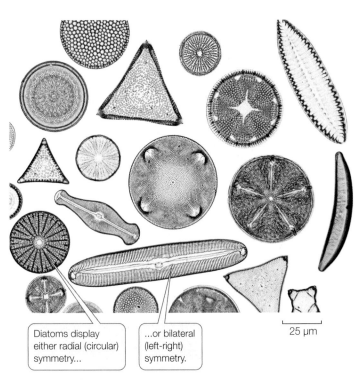

Diatoms display either radial (circular) symmetry...

...or bilateral (left-right) symmetry.

25 µm

27.8 Diatom Diversity This bright-field micrograph illustrates the variety of species-specific forms found among the diatoms.

 Go to Media Clip 27.2
Diatoms In Action
Life10e.com/mc27.2

potential problem. Gametes are formed, shed their cell walls, and fuse. The resulting zygote then grows substantially in size before a new cell wall is laid down.

Diatoms are found in all the oceans and are frequently present in great numbers. They are major photosynthetic producers in coastal waters and are among the dominant organisms in the dense "blooms" of phytoplankton that occasionally appear in the open ocean (see Section 27.4). Diatoms are also common in fresh water and even occur on the wet surfaces of terrestrial mosses.

BROWN ALGAE The **brown algae** obtain their namesake color from the carotenoid **fucoxanthin**, which is abundant in their chloroplasts. The combination of this yellow-orange pigment with the green of chlorophylls *a* and *c* yields a brownish tinge. All brown algae are multicellular, and some are extremely large. Giant kelps such as those of the genus *Macrocystis* may be up to 60 meters long (see Figure 54.9).

 Go to Media Clip 27.3
A Kelp Forest
Life10e.com/mc27.3

The brown algae are almost exclusively marine. They are composed either of branched filaments (**Figure 27.9A**) or of leaflike growths (**Figure 27.9B**). Some float in the open ocean; the most famous example is the genus *Sargassum*, which forms dense mats in the Sargasso Sea in the mid-Atlantic. Most brown

(A) *Himanthalia elongata*

(B) *Postelsia palmiformis*

Holdfasts

27.9 Brown Algae (A) This seaweed illustrates the filamentous growth form of the brown algae. (B) Sea palms exemplify the leaf-like growth form of brown algae. Sea palms and many other brown algal species are "glued" to the rocks by tough, branched structures called holdfasts that can withstand the pounding of the surf.

algae, however, attach themselves to rocks near the shore. A few thrive only where they are regularly exposed to heavy surf. All of the attached forms develop a specialized structure, called a holdfast, that literally glues them to the rocks. The "glue" of the holdfast is alginic acid, a gummy polymer found in the walls of many brown algal cells. In addition to its function in holdfasts, alginic acid cements algal cells and filaments together. It is harvested and used by humans as an emulsifier in ice cream, cosmetics, and other products.

OOMYCETES The **oomycetes** are the water molds and their terrestrial relatives. Water molds are filamentous and stationary. They are **absorptive heterotrophs**—that is, they secrete enzymes that digest large food molecules into smaller molecules that they can absorb. They are all aquatic and **saprobic**—meaning they feed on dead organic matter. If you have seen a whitish, cottony mold growing on dead fish or dead insects in water, it was probably a water mold of the common genus *Saprolegnia* (**Figure 27.10**).

Some other oomycetes, such as the downy mildews, are terrestrial. Although most of the terrestrial oomycetes are harmless or helpful decomposers of dead matter, a few are plant parasites that attack crops such as avocados, grapes, and potatoes.

Oomycetes were once classified as fungi. However, we now know that their similarity to fungi is only superficial, and that the oomycetes are more distantly related to the fungi than are many other eukaryote groups, including humans (see Figure 27.3). For example, the cell walls of oomycetes are typically made of cellulose, whereas those of fungi are made of chitin.

Saprolegnia sp.

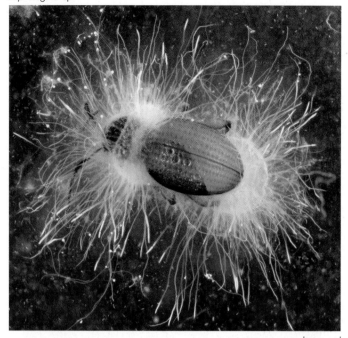

3 mm

27.10 An Oomycete The filaments of a water mold radiate from the carcass of a beetle.

Rhizaria typically have long, thin pseudopods

The three primary groups of **Rhizaria**—cercozoans, foraminiferans, and radiolarians—are unicellular and mostly aquatic. The rhiz-

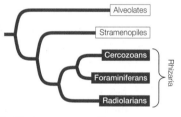

aria have contributed their shells to ocean sediments, some of which have become terrestrial features over the course of geological history.

CERCOZOANS The **cercozoans** are a diverse group with many forms and habitats. Some are aquatic; others live in soil. One group of cercozoans possesses chloroplasts derived from a green alga by secondary endosymbiosis, and those chloroplasts contain a trace of the alga's nucleus.

1 mm

27.11 Building Blocks of Limestone Some foraminiferans secrete calcium carbonate to form shells. The shells of different species have distinctive shapes. Over millions of years, the shells of foraminiferans have accumulated to form limestone deposits.

FORAMINIFERANS Some **foraminiferans** secrete external shells of calcium carbonate (**Figure 27.11**). These shells have accumulated over time to produce much of the world's limestone. Some foraminiferans live as plankton; others live on the seafloor. Living foraminiferans have been found 10,896 meters down in the western Pacific's Challenger Deep—the deepest point in the world's oceans. At that depth, however, they cannot secrete normal shells because the surrounding water is too poor in calcium carbonate.

In living planktonic foraminiferans, long, threadlike, branched pseudopods extend through numerous microscopic apertures in the shell and interconnect to create a sticky, reticulated net, which the foraminiferans use to catch smaller plankton. In some foraminiferan species, the pseudopods provide locomotion.

RADIOLARIANS **Radiolarians** are recognizable by their thin, stiff pseudopods, which are reinforced by microtubules (**Figure 27.12A**). These pseudopods greatly increase the surface area of the cell, and they help the cell stay afloat in its marine environment.

(A)

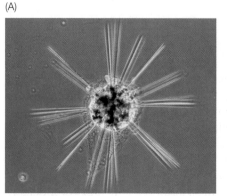

Astrolithium sp.

250 μm

(B)

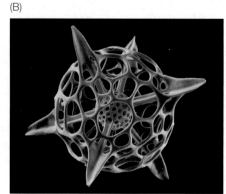

Hexacontium sp.

50 μm

27.12 Radiolarians Exhibit Distinctive Pseudopods and Radial Symmetry (A) The radiolarians are distinguished by their thin, stiff pseudopods and by their radial symmetry. The pigmentation seen at the center of this radiolarian's glassy endoskeleton is imparted by endosymbiotic dinoflagellates. (B) The endoskeleton secreted by a radiolarian.

Radiolarians also are immediately recognizable by their distinctive radial symmetry. Almost all radiolarian species secrete glassy endoskeletons (internal skeletons). The skeletons of the different species are as varied as snowflakes, and many have elaborate geometric designs (**Figure 27.12B**). A few radiolarians are among the largest of the unicellular eukaryotes, measuring several millimeters across.

Excavates began to diversify about 1.5 billion years ago

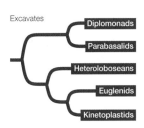

The **excavates** include a number of diverse groups that began to split from one another soon after the origin of eukaryotes. Several groups of excavates lack mitochondria, an absence that once led to the view that these groups might represent early-diverging eukaryotes that diversified before the evolution of mitochondria. However, the discovery of genes in the nucleus that are normally associated with mitochondria suggests that the absence of mitochondria is a derived condition in these organisms. In other words, ancestors of these excavate groups probably possessed mitochondria that were lost or reduced over the course of evolution. The existence of these organisms today shows that eukaryotic life is possible without mitochondria.

DIPLOMONADS AND PARABASALIDS The **diplomonads** and the **parabasalids** are unicellular and lack mitochondria (although they have reduced organelles that are derived from mitochondria). The parasitic *Giardia lamblia*, a diplomonad, causes the intestinal disease giardiasis. *Giardia* infections may result from contact with contaminated water; in the United States, such infections are most common among hikers and campers using spring or stream water in recreational areas. This tiny organism has a cytoskeleton and multiple flagella, and contains two nuclei bounded by nuclear envelopes (**Figure 27.13A**).

In addition to flagella and a cytoskeleton, the parabasalids have undulating membranes that also contribute to the cell's locomotion. *Trichomonas vaginalis* (**Figure 27.13B**) is a parabasalid responsible for a sexually transmitted disease in humans. Infection of the male urethra, where it may occur without symptoms, is less common than infection of the vagina.

HETEROLOBOSEANS The amoeboid body form appears in several protist groups that are only distantly related to one another. The body forms of **heteroloboseans**, for example, resemble those of loboseans, an amoebozoan group that is not at all closely related to heteroloboseans (see the next section). Amoebas of the free-living heterolobosean genus *Naegleria*, some of which can enter the human body and cause a fatal disease of the nervous system, usually have a two-stage life cycle, in which one stage has amoeboid cells and the other flagellated cells.

EUGLENIDS AND KINETOPLASTIDS The **euglenids** and **kinetoplastids** together constitute a clade of unicellular excavates with flagella. Their mitochondria contain distinctive disc-shaped

(A) *Giardia* sp.

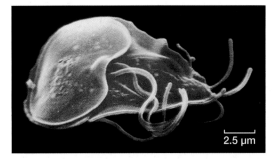

(B) *Trichomonas vaginalis*

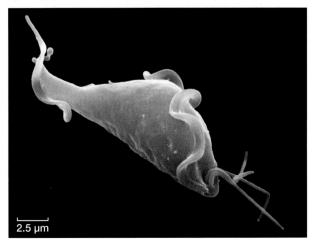

27.13 Some Excavate Groups Lack Mitochondria (A) *Giardia*, a diplomonad, has flagella and two nuclei. (B) *Trichomonas*, a parabasalid, has flagella and undulating membranes. Neither of these organisms possesses mitochondria.

cristae, and their flagella contain a crystalline rod not found in other organisms. They reproduce primarily asexually.

The flagella of euglenids arise from a pocket at the anterior end of the cell. Spiraling strips of proteins under the plasma membrane control the cell's shape. Some euglenids are photosynthetic. **Figure 27.14** depicts a typical cell of the genus *Euglena*. This common freshwater organism has a complex cell structure. It propels itself through the water with the longer of its two flagella, which may also serve as an anchor to hold the organism in place. The second flagellum is often rudimentary.

 Go to Media Clip 27.4
Euglenids
Life10e.com/mc27.4

The euglenids have diverse nutritional requirements. Many species are always heterotrophic. Other species, including species of *Euglena*, are fully autotrophic in sunlight, using chloroplasts to synthesize organic compounds through photosynthesis. When kept in the dark, these euglenids lose their photosynthetic pigment and begin to feed exclusively on dissolved organic material in the water around them. A "bleached" *Euglena* resynthesizes its photosynthetic pigment when it is returned to the light and becomes autotrophic again. But *Euglena* cells treated with certain antibiotics or mutagens lose their photosynthetic pigment completely; neither they nor

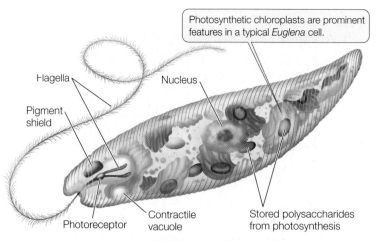

Photosynthetic chloroplasts are prominent features in a typical *Euglena* cell.

Flagella

Nucleus

Pigment shield

Photoreceptor

Contractile vacuole

Stored polysaccharides from photosynthesis

27.14 A Photosynthetic Euglenid In the *Euglena* species illustrated in this drawing, the second flagellum is rudimentary. Note that the primary flagellum originates at the anterior of the organism and trails toward its posterior.

Chaos carolinensis

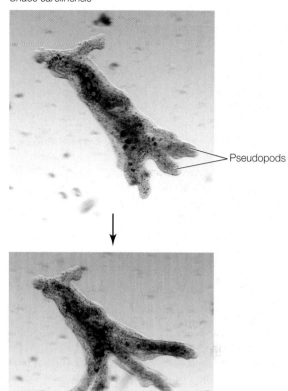

Pseudopods

120 µm

their descendants are ever autotrophs again. However, those descendants function well as heterotrophs.

The kinetoplastids are unicellular parasites with two flagella and a single, large mitochondrion. The mitochondrion contains a kinetoplast, a unique structure housing multiple circular DNA molecules and associated proteins. Some of these DNA molecules encode "guide proteins" that edit mRNA within the mitochondrion.

The kinetoplastids include several medically important species of pathogenic trypanosomes (**Table 27.1**). Some of these organisms are able to change their cell surface recognition molecules frequently. This ability allows them to evade our best attempts to kill them and thus eradicate the diseases they cause.

Amoebozoans use lobe-shaped pseudopods for locomotion

Amoebozoans appear to have diverged from other eukaryotes about 1.5 billion years ago (see Figure 27.3). It is not yet clear whether they are more closely related to opisthokonts (which include fungi and animals) or to other major groups of eukaryotes. The lobe-shaped pseudopods of amoebozoans are a hallmark of

27.15 An Amoeba in Motion The flowing pseudopods of this "chaos amoeba" (a lobosean) are constantly changing shape as it moves and feeds.

Go to Media Clip 27.5
Amoeboid Movement
Life10e.com/mc27.5

the amoeboid body form (**Figure 27.15**). Amoebozoan pseudopods differ in form and function from the slender pseudopods of rhizaria. We consider three amoebozoan groups here: the loboseans and two groups known as slime molds.

TABLE 27.1
Three Pathogenic Trypanosomes

	Trypanosoma brucei	*Trypanosoma cruzi*	*Leishmania major*
Human disease	Sleeping sickness	Chagas disease	Leishmaniasis
Insect vector	Tsetse fly	Assassin bugs (many species)	Sand fly
Vaccine or effective cure	None	None	None
Strategy for survival	Changes surface recognition molecules frequently	Causes changes in surface recognition molecules on host cell	Reduces effectiveness of macrophage hosts
Site in human body	Bloodstream; in final stages, attacks nerve tissue	Enters cells, especially muscle cells	Enters cells, primarily macrophages
Approximate number of deaths per year	50,000	45,000	60,000

Nebela collaris

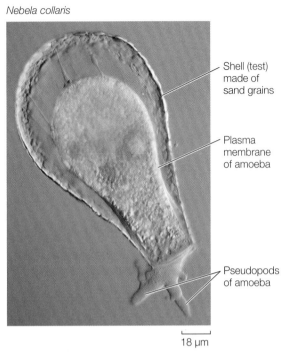

Shell (test) made of sand grains

Plasma membrane of amoeba

Pseudopods of amoeba

18 μm

27.16 Life in a Glass House This testate amoeba has built a lightbulb-shaped shell, or test, by gluing sand grains together. Its pseudopods extend through the single aperture in the test.

(A)

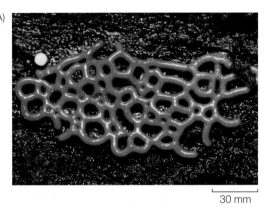

30 mm

(B)

1.5 mm

27.17 A Plasmodial Slime Mold (A) The plasmodial form of the slime mold *Hemitrichia serpula* covers rocks, decaying logs, and other objects as it engulfs bacteria and other food items; it is also responsible for the organism's common name of "pretzel mold." (B) Fruiting structures of *Hemitrichia*.

 Go to Media Clip 27.6
Plasmodial Slime Mold Growth
Life10e.com/mc27.6

LOBOSEANS **Loboseans** are small amoebozoans that feed on other small organisms and particles of organic matter by phagocytosis, engulfing them with pseudopods. Many loboseans are adapted for life on the bottoms of lakes, ponds, and other bodies of water. Their creeping locomotion and their manner of engulfing food particles fit them for life close to a relatively rich supply of sedentary organisms or organic particles. Most loboseans exist as predators, parasites, or scavengers. Members of one group of loboseans, the testate amoebas, live inside shells. Some of these amoebas produce casings by gluing sand grains together (**Figure 27.16**); other testate amoebas have shells secreted by the organism itself.

PLASMODIAL SLIME MOLDS If the nucleus of an amoeba began rapid mitotic division, accompanied by a tremendous increase in cytoplasm and organelles, but no cytokinesis, the resulting organism would resemble the multinucleate mass of a **plasmodial slime mold**. During its vegetative (feeding, nonreproductive) stage, a plasmodial slime mold is a wall-less mass of cytoplasm with numerous diploid nuclei. This mass streams very slowly over its substrate in a remarkable network of strands called a plasmodium (**Figure 27.17A**). The plasmodium is an example of a **coenocyte**: many nuclei enclosed in a single plasma membrane. The outer cytoplasm of the plasmodium (closest to the environment) is normally less fluid than the interior cytoplasm and thus provides some structural rigidity.

Plasmodial slime molds provide a dramatic example of movement by **cytoplasmic streaming**. The outer cytoplasm

of the plasmodium becomes more fluid in places, and cytoplasm rushes into those areas, stretching the plasmodium. This streaming reverses its direction every few minutes as cytoplasm rushes into a new area and drains away from an older one, moving the plasmodium over its substrate. Sometimes an entire wave of plasmodium moves across a surface, leaving strands behind. Microfilaments and a contractile protein called myxomyosin interact to produce the streaming movement. As it moves, the plasmodium engulfs food particles by endocytosis—predominantly bacteria, yeasts, spores of fungi, and other small organisms as well as decaying animal and plant remains.

A plasmodial slime mold can grow almost indefinitely in its plasmodial stage as long as the food supply is adequate and other conditions, such as moisture and pH, are favorable. If conditions become unfavorable, however, one of two things can happen. In one case, the plasmodium can form an irregular mass of hardened, cell-like components called a **sclerotium**. This resting structure rapidly becomes a plasmodium again when favorable conditions are restored.

Alternatively, the plasmodium can transform itself into spore-bearing **fruiting structures (Figure 27.17B)**. These stalked or branched structures rise from heaped masses of plasmodium. They derive their rigidity from walls that form and thicken between their nuclei. The diploid nuclei of the plasmodium divide by meiosis as the fruiting structure develops. One or more knobs, called sporangia, develop on the end of the stalk. Within a sporangium, haploid nuclei become surrounded by walls to form spores. Eventually, as the fruiting structure dries, it sheds its spores.

The spores germinate into wall-less, haploid cells called **swarm cells**, which can either divide mitotically to produce more haploid swarm cells or function as gametes. Swarm cells can live as separate individual cells that move by means of flagella or pseudopods, or they can become walled and resistant resting cysts when conditions are unfavorable; when conditions improve again, the cysts release swarm cells. Two swarm cells can also fuse to form a diploid zygote, which divides by mitosis (but without a wall forming between the nuclei) and thus forms a new coenocytic plasmodium.

CELLULAR SLIME MOLDS Whereas the plasmodium is the basic vegetative unit of the plasmodial slime molds, a single amoeboid cell is the vegetative unit of the **cellular slime molds (Figure 27.18)**. Cells called **myxamoebas**, which have single haploid nuclei, swarm together as they engulf bacteria and other food particles by endocytosis and reproduce by mitosis and fission. This simple life cycle stage, consisting of swarms of independent, isolated cells, can persist indefinitely as long as food and moisture are available.

When conditions become unfavorable, the cellular slime molds form fruiting structures, as do their plasmodial counterparts. The individual myxamoebas aggregate into a mass called a **slug** or **pseudoplasmodium**. Unlike the true plasmodium of the plasmodial slime molds, this structure is not simply a giant sheet of cytoplasm with many nuclei; the individual myxamoebas within the slug retain their plasma membranes and therefore their identity.

A slug may migrate over a substrate for several hours before becoming motionless and reorganizing to construct a delicate, stalked fruiting structure. Cells at the top of the fruiting structure develop into thick-walled spores, which are eventually released. Later, under favorable conditions, the spores germinate, releasing myxamoebas.

The cycle from myxamoebas through slug and spores to new myxamoebas is asexual. Cellular slime molds also have a sexual cycle, in which two myxamoebas fuse. The product of

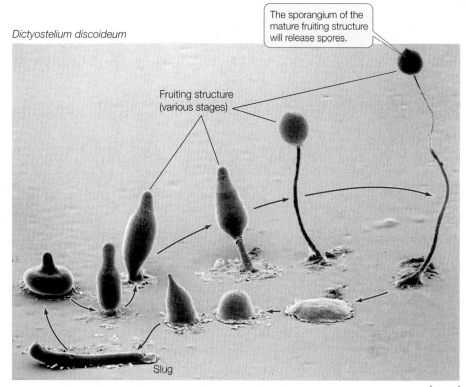

Dictyostelium discoideum

The sporangium of the mature fruiting structure will release spores.

Fruiting structure (various stages)

Slug

0.25 mm

27.18 A Cellular Slime Mold This composite micrograph shows the life cycle of the slime mold *Dictyostelium*.

Go to Media Clip 27.7
Cellular Slime Mold Aggregation
Life10e.com/mc27.7

this fusion develops into a spherical structure that ultimately germinates, releasing new haploid myxamoebas.

RECAP **27.2**

The major lineages of eukaryotes began to diverge about 1.5 billion years ago. Major groups of eukaryotes are highly diverse in their habitat, nutrition, locomotion, and body form. Many protists are photosynthetic autotrophs, but heterotrophic lineages have evolved repeatedly. Although most protists are unicellular, multicellularity has arisen independently many times.

- Contrast the major distinctive features of alveolates, excavates, stramenopiles, rhizaria, and amoebozoans.

- Give examples of alveolates, stremenoplies, and excavates that are important for medical or agricultural reasons. **See pp. 553–558 and Table 27.1.**

The ancient origins of the major eukaryote lineages and the adaptation of these lineages to a wide variety of lifestyles and environments resulted in enormous protist diversity. It is not surprising, then, that reproductive modes among protists are also highly diverse.

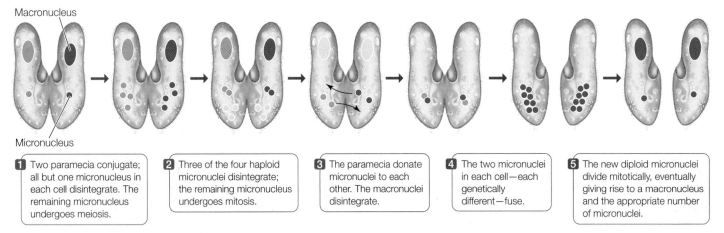

Macronucleus

Micronucleus

1 Two paramecia conjugate; all but one micronucleus in each cell disintegrate. The remaining micronucleus undergoes meiosis.

2 Three of the four haploid micronuclei disintegrate; the remaining micronucleus undergoes mitosis.

3 The paramecia donate micronuclei to each other. The macronuclei disintegrate.

4 The two micronuclei in each cell—each genetically different—fuse.

5 The new diploid micronuclei divide mitotically, eventually giving rise to a macronucleus and the appropriate number of micronuclei.

27.19 Conjugation in Paramecia The exchange of micronuclei by two conjugating *Paramecium* individuals results in genetic recombination. After conjugation, the cells separate and continue their lives as two individuals.

(27.3) What Is the Relationship between Sex and Reproduction in Protists?

Although most protists engage in both asexual and sexual reproduction, sexual reproduction has yet to be observed in some groups. In some protists, as in all prokaryotes, the acts of sex and reproduction are not directly linked. Several asexual reproductive processes have been observed among the protists:

• The equal splitting of one cell into two by mitosis followed by cytokinesis

• The splitting of one cell into multiple (i.e., more than two) cells

• The outgrowth of a new cell from the surface of an old one (known as **budding**)

• The formation of specialized cells (spores) that are capable of developing into new individuals (know as **sporulation**)

Asexual reproduction results in offspring that are genetically nearly identical to their parents (they differ only by new mutations that may arise during DNA replication). Such asexually reproduced groups of nearly identical organisms are known as **clonal lineages**, or **clones**.

Sexual reproduction among the protists takes various forms. In some protists, as in animals, the gametes are the only haploid cells. In others, the zygote is the only diploid cell. In still others, both diploid and haploid cells undergo mitosis, giving rise to alternating multicellular diploid and haploid life stages.

Some protists reproduce without sex and have sex without reproduction

As noted in Section 27.2, members of the genus *Paramecium* are ciliates, which commonly have two types of nuclei in a single cell (one macronucleus and from one to several micronuclei; see

Figure 27.6). The micronuclei are typical eukaryotic nuclei and are essential for genetic recombination. The macronucleus contains many copies of the genetic information, packaged in units containing only a few genes each. The macronuclear DNA is transcribed and translated to regulate the life of the cell.

When paramecia reproduce asexually, all of the nuclei are copied before the cell divides. Paramecia (and many other protists) also have an elaborate sexual behavior called **conjugation**, in which two individuals line up tightly against each other and fuse in the oral groove region of the body. Nuclear material is extensively reorganized and exchanged over the next several hours (**Figure 27.19**). Each cell ends up with two haploid micronuclei, one of its own and one from the other cell, which fuse to form a new diploid micronucleus. A new macronucleus develops from that micronucleus through a series of dramatic chromosomal rearrangements. The exchange of nuclei is fully reciprocal: each of the two paramecia gives and receives an equal amount of DNA. The two organisms then separate and go their own ways, each equipped with new combinations of alleles.

Conjugation in *Paramecium* is a sexual process (i.e., a process of genetic recombination), but it is not a reproductive process. Two cells begin conjugation and two cells are there at the end, so no new cells are created. As a rule, each asexual clone of paramecia must conjugate periodically. Experiments have shown that if some species are not permitted to conjugate, the clones can live through only about 350 cell divisions before dying out.

Some protist life cycles feature alternation of generations

Alternation of generations is a feature of the life cycles of many multicellular protists, all land plants, and some fungi. In these life cycles, a multicellular, diploid, spore-producing stage gives rise to a multicellular, haploid, gamete-producing stage (see Figure 28.6). When two haploid gametes fuse, a diploid organism is produced. The haploid organism, the diploid organism, or both may also reproduce asexually. Note that alternation of generations is distinct from the familiar reproductive system of animals, in which the only haploid stages are unicellular gametes produced by multicellular, diploid adults.

The two alternating (spore-producing and gamete-producing) generations differ genetically (one has diploid cells, the other haploid cells), but they may or may not differ morphologically. In **heteromorphic** alternation of generations, the two generations differ morphologically; in **isomorphic** alternation of generations, they do not. Examples of both heteromorphic and isomorphic alternation of generations are found among the brown algae.

The gamete-producing generation does not produce gametes by meiosis because the gamete-producing organism is already haploid. Instead, specialized cells of the diploid spore-producing organism, called **sporocytes**, divide meiotically to produce four haploid *spores*. The spores may eventually germinate and divide mitotically to produce the multicellular haploid generation, which then produces gametes by mitosis and cytokinesis.

Gametes, unlike spores, can produce new organisms only by fusing with other gametes. The fusion of two gametes produces a diploid zygote, which then undergoes mitotic divisions to produce a diploid organism. The diploid organism's sporocytes then undergo meiosis and produce haploid spores, starting the cycle anew.

> **RECAP 27.3**
>
> Protists reproduce both asexually and sexually, although sex occurs independently of reproduction in some species. Some multicellular protists exhibit alternation of generations, alternating between multicellular haploid and diploid life stages.
>
> - Why is conjugation between paramecia considered a sexual process but not a reproductive process? **See p. 562 and Figure 27.19**
> - Although most diploid animals have haploid stages (for example, eggs and sperm), their life cycles are not considered an example of alternation of generations. Why not? **See pp. 562–563**

Given the diversity of protists and of the environments in which they live, it is not surprising that they influence their environments in numerous ways.

27.4 How Do Protists Affect Their Environments?

As we have seen, many microbial eukaryotes are food for aquatic animals, while others poison those animals or act as pathogens. The remains of some form the sands of many modern beaches, and others are a major source of the oil that sometimes fouls those beaches.

Phytoplankton are primary producers

A single protist clade, the diatoms, performs about one-fifth of all photosynthetic carbon fixation on Earth—about the same amount as all of Earth's rainforests. These spectacular unicellular organisms (see Figure 27.8) are the predominant component of the oceanic phytoplankton, but the phytoplankton include many other protists that also contribute heavily to global photosynthesis. Like green plants on land, these "floating photosynthesizers" are the gateway for energy from the sun into the rest of the living world; in other words, they are **primary producers**. These autotrophs are eaten by heterotrophs, including animals and many other protists. Those consumers, in turn, are eaten by other consumers. Most aquatic heterotrophs (with the exception of some species in the deep sea) depend on photosynthesis performed by phytoplankton for their energy supply.

Some microbial eukaryotes are deadly

Some microbial eukaryotes are pathogens that cause serious diseases in humans and other vertebrates. The best-known pathogenic protists are members of the genus *Plasmodium*, a highly specialized group of apicomplexans that spend part of their complex life cycle as parasites in human red blood cells. *Plasmodium* parasites cause malaria, one of the world's three most serious infectious diseases; it infects more than 350 million people, and kills more than 1 million people, each year. On average, about two people die from malaria every minute of every day—most of them in sub-Saharan Africa, although malaria occurs in more than 100 countries.

Mosquitoes of the genus *Anopheles* transmit *Plasmodium* to humans when an infected female mosquito penetrates the person's skin. The parasites enter the human bloodstream and travel to cells in the liver and the lymphatic system, where they change form, multiply, and reenter the bloodstream to invade red blood cells, where they continue to multiply. Eventually the blood cells lyse (burst), releasing new swarms of *Plasmodium*. These episodes of cell lysis coincide with the primary symptoms of malaria, which include fever, shivering, vomiting, joint pain, and convulsions.

If another *Anopheles* bites the victim, the mosquito ingests *Plasmodium* cells along with blood. Some of these cells develop into gametes that unite in the mosquito, forming zygotes. The zygotes lodge in the mosquito's gut, divide several times, and move into its salivary glands, from which they can be passed on to another human host (**Figure 27.20**). Thus *Plasmodium* is an extracellular parasite in the mosquito vector and an intracellular parasite in the human host. Such an organism—that is, a parasite that requires more than one host—is said to have a **complex life cycle**.

Plasmodium is a singularly difficult pathogen for humans to combat. Its life cycle is best broken by removing stagnant water, in which mosquitoes breed. Using insecticides to reduce the *Anopheles* population can also be effective, but the benefits must be weighed against the ecological, economic, and health risks posed by the insecticides themselves.

Even some of the phytoplankton that are such important primary producers can be deadly, as described at the opening of this chapter. Some diatoms and dinoflagellates reproduce in enormous numbers when environmental conditions are favorable for their growth. In the resulting "red tides," the concentration of dinoflagellates may reach 60 million per liter of ocean water and produce potent nerve toxins that harm or kill many vertebrates, especially fish.

(A)

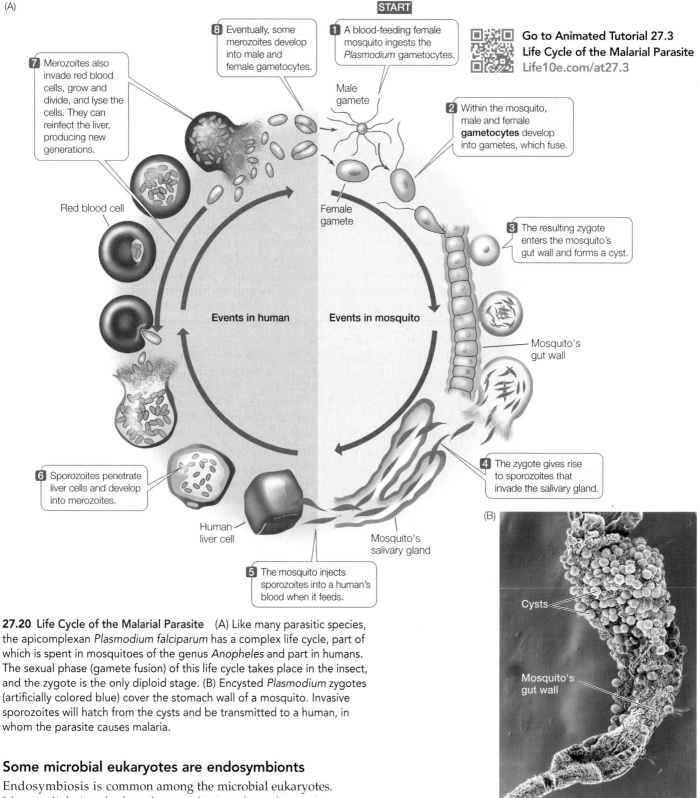

8 Eventually, some merozoites develop into male and female gametocytes.

7 Merozoites also invade red blood cells, grow and divide, and lyse the cells. They can reinfect the liver, producing new generations.

START

1 A blood-feeding female mosquito ingests the *Plasmodium* gametocytes.

Go to Animated Tutorial 27.3
Life Cycle of the Malarial Parasite
Life10e.com/at27.3

Male gamete

2 Within the mosquito, male and female **gametocytes** develop into gametes, which fuse.

Female gamete

3 The resulting zygote enters the mosquito's gut wall and forms a cyst.

Red blood cell

Events in human   **Events in mosquito**

Mosquito's gut wall

6 Sporozoites penetrate liver cells and develop into merozoites.

4 The zygote gives rise to sporozoites that invade the salivary gland.

Human liver cell

Mosquito's salivary gland

5 The mosquito injects sporozoites into a human's blood when it feeds.

(B)

Cysts

Mosquito's gut wall

170 μm

27.20 Life Cycle of the Malarial Parasite (A) Like many parasitic species, the apicomplexan *Plasmodium falciparum* has a complex life cycle, part of which is spent in mosquitoes of the genus *Anopheles* and part in humans. The sexual phase (gamete fusion) of this life cycle takes place in the insect, and the zygote is the only diploid stage. (B) Encysted *Plasmodium* zygotes (artificially colored blue) cover the stomach wall of a mosquito. Invasive sporozoites will hatch from the cysts and be transmitted to a human, in whom the parasite causes malaria.

Some microbial eukaryotes are endosymbionts

Endosymbiosis is common among the microbial eukaryotes. Many radiolarians harbor photosynthetic endosymbionts (see Figure 27.12A). As a result, these radiolarians, which are not photosynthetic themselves, appear greenish or golden, depending on the type of endosymbiont they contain. This arrangement is often mutually beneficial: the radiolarian can make use of the carbon compounds produced by its photosynthetic endosymbiont, and the endosymbiont may in turn make use of metabolites made by the host or receive physical protection. In some cases,

the endosymbiont is exploited for its photosynthetic products while receiving little or no benefit itself.

Dinoflagellates are common endosymbionts and can be found in both animals and other protists. Most, but not all, dinoflagellate endosymbionts are photosynthetic. Some dinoflagellates live endosymbiotically in the cells of corals, contributing the products of their photosynthesis to the partnership. Their importance to the corals is demonstrated when the dinoflagellates die or are expelled by the corals as a result of changing environmental conditions such as rising water temperatures or increased water turbidity. This phenomenon, known as **coral bleaching**, reduces the corals' food supply. Unless the corals can acquire new endosymbionts, they are unlikely to survive (**Figure 27.21**).

We rely on the remains of ancient marine protists

Diatoms are lovely to look at, but their importance to us goes far beyond aesthetics, and even beyond their role as primary producers. Diatoms store oil as an energy reserve and to keep themselves afloat at the correct depth in the ocean. Over millions of years, untold numbers of diatoms have died and sunk to the ocean floor, where their bodies have undergone chemical changes. In this way, diatoms have become a major source of petroleum and natural gas, two of our most important fossil fuels and political concerns.

Because the silica-containing cell walls of dead diatoms resist decomposition, some sedimentary rocks are composed almost entirely of diatom skeletons that sank to the seafloor over time. Diatomaceous earth obtained from such rocks has many industrial uses, such as insulation, filtration, toothpaste, and metal polishing. It has also been used as an "Earth-friendly" insecticide that clogs the tracheae (breathing structures) of insects.

Other ancient marine protists have also contributed to the rocks of today. Some foraminiferans, as we have seen, secrete shells of calcium carbonate. After they reproduce (by mitosis and cytokinesis), the daughter cells abandon the parent shell and make new shells of their own. The discarded shells of ancient foraminiferans form a layer hundreds to thousands of meters deep over millions of square kilometers of ocean bottom. The extensive limestone deposits seen in various parts of the world are the result of tectonic processes that have raised these layers above sea level. Foraminiferan shells also make up much of the sand of some beaches. A single gram of such sand may contain as many as 50,000 foraminiferan shells and shell fragments.

The shells of individual foraminiferans are easily preserved as fossils in marine sediments. Each geological period is characterized by a distinctive assemblage of foraminiferan species. Because the shells of foraminiferan species have such distinctive shapes (see Figure 27.11) and, because they are so abundant, the remains of foraminiferans are especially valuable in classifying and dating sedimentary rocks. In addition, analyses of the chemical makeup of foraminiferan shells can be used to estimate the global temperatures prevalent at the time when the shells were formed.

INVESTIGATING LIFE

27.21 Can Corals Reacquire Dinoflagellate Endosymbionts Lost to Bleaching? Some corals lose their chief nutritional source when their photosynthetic endosymbionts die, often as a result of changing environmental conditions. This experiment by Cynthia Lewis and Mary Alice Coffroth investigated the ability of corals to acquire new endosymbionts after bleaching.[a]

HYPOTHESIS Bleached corals can acquire new photosynthetic endosymbionts from their environment.

Method
1. Count numbers of *Symbiodinium*, a photosynthetic dinoflagellate, living symbiotically in samples of a coral (*Briareum* sp.).
2. Stimulate bleaching by maintaining all *Briareum* colonies in darkness for 12 weeks.
3. After 12 weeks of darkness, count numbers of *Symbiodinium* in the coral samples; then return all colonies to light.
4. In some of the bleached colonies (the experimental group), introduce *Symbiodinium* strain B211—dinoflagellates that contain a unique molecular marker. Do not expose the others (the control group) to strain B211. Maintain both groups in the light for 6 weeks.

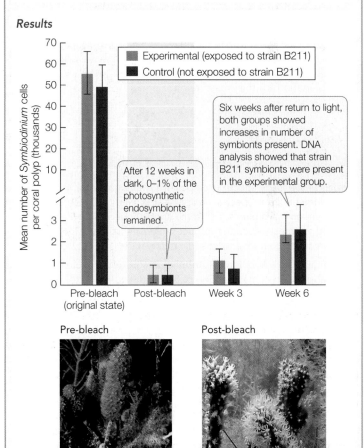

Results

Mean number of *Symbiodinium* cells per coral polyp (thousands)

■ Experimental (exposed to strain B211)
■ Control (not exposed to strain B211)

After 12 weeks in dark, 0–1% of the photosynthetic endosymbionts remained.

Six weeks after return to light, both groups showed increases in number of symbionts present. DNA analysis showed that strain B211 symbionts were present in the experimental group.

Pre-bleach (original state) Post-bleach Week 3 Week 6

Pre-bleach   Post-bleach

CONCLUSION Corals can acquire new endosymbionts from their environment following bleaching.

Go to **BioPortal** for discussion and relevant links for all INVESTIGATING LIFE figures.

[a]Lewis, C. L. and M. A. Coffroth. 2004. *Science* 304: 1490–1492.

WORKING WITH**DATA:**

Uptake of Endosymbionts After Coral Bleaching

Original Paper

Lewis, C. L. and M. A. Coffroth. 2004. The acquisition of exogenous algal symbionts by an octocoral after bleaching. *Science* 304: 1490–1492.

Analyze the Data

The data shown in the table at right come from DNA analyses of *Symbiodinium* strains found in the experimental and control colonies of corals (*Briareum*) before and after bleaching. *Symbiodinium* strain B211 (which was not present before bleaching) was introduced to the experimental colonies after bleaching. Use these data to answer the questions below.

QUESTION 1

Are new strains of *Symbiodinium* taken up only by coral colonies that have lost all their original endosymbionts?

QUESTION 2

Does the acquisition of a new *Symbiodinium* strain always result in survival of a recovering *Briareum* colony?

QUESTION 3

In week 3, only strain B211 was detected in the experimental colonies, but in week 6, non-B211 *Symbiodinium* were detected in 8 percent of the experimental colonies. Can you suggest an explanation for this observation?

	Symbiodinium strain present (% of colonies)			
	Non-B211	B211	None*	Colony died
Experimental colonies (strain B211 added)				
Pre-bleach	100	0	0	0
Post-bleach	58	0	42	0
Week 3	0	92	0	8
Week 6	8	58	8	25
Control colonies (no strain B211)				
Pre-bleach	100	0	0	0
Post-bleach	67	0	33	0
Week 3	67	0	33	0
Week 6	67	0	17	17

*Colonies remained alive but no *Symbiodinium* were detected.

Go to BioPortal for all WORKING WITH**DATA** exercises

RECAP 27.4

Protists have many effects, both positive and negative, on their environment. Some species are important primary producers, many are endosymbionts, and some are pathogens. Protists are among the most important producers of fossil fuels, and they are important components of sedimentary rocks.

- What is the role of female *Anopheles* mosquitoes in the transmission of malaria? See p. 563 and Figure 27.20

- Explain the roles of dinoflagellates in the two very different phenomena of coral bleaching and red tides. See pp. 549 and 564–565

- What are some of the ways in which diatoms are important to human society? See p. 565

The next six chapters will explore the major evolutionary radiations of multicellular eukaryotes, along with the protist ancestors from which they arose. Chapters 28 and 29 will describe the origin and diversification of plants, Chapter 30 will present the fungi, and Chapters 31–33 will provide a brief overview of the animals.

Can dinoflagellates be beneficial, as well as harmful, to marine ecosystems?

ANSWER

Not all dinoflagellate blooms produce problems for other species. Dinoflagellates are important components of many ecosystems, as we have seen throughout this chapter. Photosynthetic dinoflagellates also produce much of the atmospheric oxygen that most animals need to survive.

Corals and many other species depend on symbiotic dinoflagellates for food (see Figure 27.21). In addition, as photosynthetic organisms, free-living planktonic dinoflagellates are among the most important primary producers in aquatic food webs. They are a major component of the phytoplankton and provide an important food source for many species (see Section 27.4).

Some dinoflagellates produce a beautiful bioluminescence. Unlike the bioluminescent bacteria described at the start of Chapter 26, however, dinoflagellates cannot generate a steady bioluminescence, but produce flashes of light when disturbed, as people who swim in the ocean at night in certain regions often observe.

CHAPTER SUMMARY 27

 27.1 How Did the Eukaryotic Cell Arise?

- The term **protist** does not describe a formal taxonomic group. It is shorthand for "all eukaryotes that are not plants, animals, or fungi."

- Early events in the evolution of the eukaryotic cell probably included the loss of the firm cell wall and infolding of the plasma membrane. Such infolding probably led to segregation of the genetic material in a membrane-enclosed nucleus. **Review Figure 27.1**

- Mitochondria evolved by endosymbiosis with a proteobacterium.

- **Primary endosymbiosis** of a eukaryote and a cyanobacterium gave rise to the first chloroplasts. **Secondary endosymbiosis** and **tertiary endosymbiosis** between chloroplast-containing eukaryotes and other eukaryotes gave rise to the distinctive chloroplasts of euglenids, dinoflagellates, and other groups. **Review Figure 27.2, ANIMATED TUTORIAL 27.1**

 27.2 What Features Account for Protist Diversity?

- Most eukaryotes can be placed in one of eight major clades that diverged about 1.5 billion years ago: alveolates, stramenopiles, rhizaria, excavates, plants, amoebozoans, fungi, and animals. **Review Figure 27.3**

- Most, but not all, protists are unicellular.

- **Alveolates** are unicellular organisms with sacs (alveoli) beneath their plasma membranes. Alveolate clades include the marine **dinoflagellates**, the parasitic **apicomplexans**, and the diverse, highly motile **ciliates**. See **ACTIVITY 27.1, ANIMATED TUTORIAL 27.2**

- **Stramenopiles** typically have two flagella of unequal length, the longer one bearing rows of tubular hairs. Among the stramenopiles are the unicellular **diatoms**, the multicellular **brown algae**, and the nonphotosynthetic **oomycetes**, many of which are **saprobic**.

- **Rhizaria** are unicellular and aquatic. They include the **cercozoans**; the **foraminiferans**, which secrete shells of calcium carbonate; and the **radiolarians**, which have thin, stiff pseudopods and glassy endoskeletons.

- The **excavates** include parasitic as well as free-living species. The **diplomonads** and **parabasalids** lack typical mitochondria. **Heteroloboseans** have an amoeboid body form and a two-stage life cycle. **Euglenids** have anterior flagella; some are photosynthetic. The **kinetoplastids**, which include several human pathogens, have a single, large mitochondrion.

- The **amoebozoans** move by means of lobe-shaped pseudopods. A **lobosean** consists of a single amoeboid cell. **Plasmodial slime molds** are amoebozoans whose vegetative stage is a **coenocyte** that moves by cytoplasmic streaming. In **cellular slime molds**, the individual cells maintain their identity at all times but aggregate to form fruiting structures.

 27.3 What is the Relationship between Sex and Reproduction in Protists?

- Asexual reproduction gives rise to **clonal lineages** of organisms.

- **Conjugation** in *Paramecium* is a sexual process but not a reproductive one. **Review Figure 27.19**

- **Alternation of generations**, which includes a multicellular diploid stage and a multicellular haploid stage, is a feature of many multicellular protist life cycles (as well as those of some fungi and all land plants). The alternating generations may be **heteromorphic** or **isomorphic**.

 27.4 How Do Protists Affect Their Environments?

- The diatoms are responsible for about one-fifth of the photosynthetic carbon fixation on Earth. They and other members of the phytoplankton are important **primary producers** in the marine environment. Ancient diatoms are a major source of today's petroleum and natural gas deposits.

- Some protists are pathogens of humans and other vertebrates. **Review Figure 27.20, ANIMATED TUTORIAL 27.3**

- Endosymbiotic relationships are common among microbial protists and typically benefit both the endosymbionts and their protist or animal partners. **Review Figure 27.21**

 Go to the Interactive Summary to review key figures, Animated Tutorials, and Activities
Life10e.com/is27

CHAPTER REVIEW

 REMEMBERING

1. Which statement about eukaryotic phytoplankton is *not* true?
   a. Some are important primary producers.
   b. Some contributed to the formation of petroleum.
   c. Some form toxic "red tides."
   d. Some are food for marine animals.
   e. They constitute a clade.

2. The chloroplasts of photosynthetic protists
   a. are structurally identical.
   b. gave rise to mitochondria.
   c. are all descended from a once free-living cyanobacterium.
   d. all have exactly two surrounding membranes.
   e. are all descended from a once free-living red alga.

3. Reproduction in protists
   a. is sexual in some species
   b. is asexual in some species.
   c. can occur through both asexual and sexual processes in some species.
   d. can occur independently of sex in some species.
   e. All of the above

▓▓▓▓ UNDERSTANDING & APPLYING

4. For each pair of groups below, describe how you could recognize members of the two groups and differentiate them from each another. Then describe features that the two groups in each pair share.
   a. Foraminiferans and radiolarians
   b. Ciliates and dinoflagellates
   c. Diatoms and brown algae
   d. Plasmodial slime molds and cellular slime molds

5. Given that sex and reproduction are independent of each other in the ciliates, what does that suggest about the role of sex in maintenance of populations?

▓▓▓▓ ANALYZING & EVALUATING

Background for Questions 6–7:

In most temperate regions of the oceans, there is a spring bloom of phytoplankton. Although the red tide blooms described at the opening of this chapter are harmful, phytoplankton blooms can also be beneficial for marine communities. In fact, many species of marine life depend on these blooms for their survival. The dates of spring phytoplankton blooms near the coast of Nova Scotia, Canada, were determined by examining remote satellite images. The table below presents these dates as deviations from the mean date of the spring bloom in this region; it also gives the survival index for larval haddock (an important commercial fish) for the year after each bloom. The survival index is the ratio of the mass of juvenile fish to the mass of mature fish; higher values indicate better survival of larval fish.

Year	Deviation in bloom date* (days)	Survival index
1	+5	1.9
2	+11	2.2
3	−15	6.8
4	+5	1.9
5	−4	4.9
6	−20	10.3
7	+6	2.1
8	+14	1.9

*Negative values indicate blooms occurring earlier than the mean date; positive values indicate later blooms.

6. Plot the survival index of larval haddock against the deviation in the date of the spring phytoplankton bloom. Calculate a correlation coefficient for their relationship (see Appendix B).

7. Formulate one or more hypotheses to explain your results. Keep in mind that larval haddock include phytoplankton in their diet, and that phytoplankton blooms also provide some cover in which larval fish can hide from potential predators.

Background for Questions 8–10:

Ribosomal RNA (rRNA) genes are present in the nuclear genome of eukaryotes. There are also rRNA genes in the genomes of mitochondria and chloroplasts. Therefore photosynthetic eukaryotes have three different sets of rRNA genes, which encode the structural RNA of three separate sets of ribosomes. Translation of each genome takes place on its own set of ribosomes. The gene tree shows the evolutionary relationships among rRNA gene sequences isolated from the nuclear genomes of humans, yeast, and corn; from an archaeon (*Halobacterium*), a proteobacterium (*E. coli*), and a cyanobacterium (*Chlorobium*); and from the mitochondrial and chloroplast genomes of corn. Use the gene tree to answer the following questions.

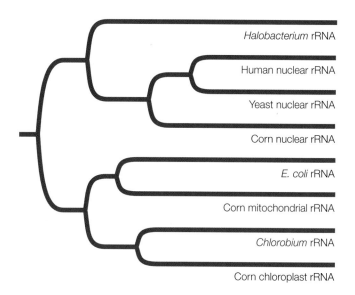

8. Why aren't the three rRNA genes of corn one another's closest relatives?

9. How would you explain the closer relationship of the mitochondrial rRNA gene of corn to the rRNA gene of *E. coli* than to the nuclear rRNA genes of other eukaryotes? Can you explain the relationship of the rRNA gene from the chloroplast of corn to the rRNA gene of the cyanobacterium?

10. If you were to sequence the rRNA genes from human and yeast mitochondrial genomes, where would you expect these two sequences to fit on the gene tree?

Go to BioPortal at **yourBioPortal.com** for Animated Tutorials, Activities, LearningCurve Quizzes, Flashcards, and many other study and review resources.

28 Plants without Seeds: From Water to Land

CHAPTEROUTLINE

Fireball on the Gulf A "blowout"—an uncontrolled release of petroleum—ignited a fireball above the drilling rig *Deepwater Horizon*. Fueled by gushing oil and natural gas, the fire could not be extinguished, and crews were forced to sink the rig without containing the blowout. Oil continued to flow from the deep-sea well for 3 months.

IN THE GULF OF MEXICO, about 60 kilometers south of the Louisiana coast, the oil rig *Deepwater Horizon* was drilling an exploratory oil well in the seafloor beneath about 1,500 meters of water when, on April 20, 2010, an explosive blowout occurred and could not be contained. Over the next 3 months, almost 5 million barrels of petroleum flowed from the well into the Gulf, making this event the worst marine oil spill in history. The spill caused massive mortality among marine life, as well as considerable damage along the coast as the oil floated to the surface and washed ashore.

Why was oil to be found so deep beneath the Gulf, and what led geologists to expect to find oil there? Most people know that petroleum is a fossil fuel, meaning that it is derived from the ancient remains of once-living organisms. Fewer people know that most petroleum is derived largely from the remains of phytoplankton, including many species of green algae (as well as other microbial groups, as discussed in Chapter 27). These algae produce complex hydrocarbons through photosynthesis. They accumulate hydrocarbons both as an energy reserve and as a way to increase their buoyancy in water. When these algae die, they drop to the bottom of the ocean, and over many millions of years, their buried remains decompose into petroleum deposits.

Today there is great interest in using solar power to help meet human energy needs. But unicellular eukaryotes first incorporated tiny solar energy converters into their cells about 1.5 billion years ago, when they formed partnerships with photosynthetic cyanobacteria. These endosymbionts—which over time would become the chloroplasts of modern plants—allowed many eukaryotes to use solar energy to drive the reactions that convert carbon dioxide into organic carbon compounds. Over many millions of years, the carbon compounds produced in the cells of marine algae accumulated in ocean sediments. Today humans are tapping that trapped solar energy in the form of petroleum and other fossil fuels.

Given that petroleum is produced naturally from green algae, can humans use green algae to produce oil commercially?

See answer on p. 585.

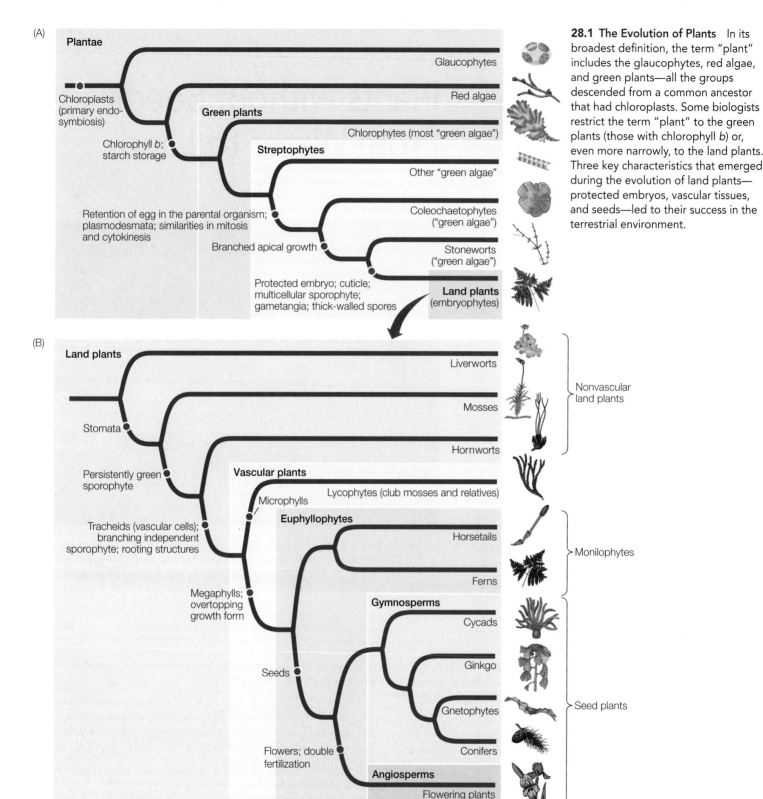

(A)

Plantae

Glaucophytes

Red algae

Chloroplasts (primary endo-symbiosis)

Green plants

Chlorophytes (most "green algae")

Chlorophyll *b*; starch storage

Streptophytes

Other "green algae"

Retention of egg in the parental organism; plasmodesmata; similarities in mitosis and cytokinesis

Coleochaetophytes ("green algae")

Branched apical growth

Stoneworts ("green algae")

Protected embryo; cuticle; multicellular sporophyte; gametangia; thick-walled spores

Land plants (embryophytes)

28.1 The Evolution of Plants In its broadest definition, the term "plant" includes the glaucophytes, red algae, and green plants—all the groups descended from a common ancestor that had chloroplasts. Some biologists restrict the term "plant" to the green plants (those with chlorophyll *b*) or, even more narrowly, to the land plants. Three key characteristics that emerged during the evolution of land plants— protected embryos, vascular tissues, and seeds—led to their success in the terrestrial environment.

(B)

Land plants

Liverworts

Nonvascular land plants

Mosses

Stomata

Hornworts

Persistently green sporophyte

Vascular plants

Lycophytes (club mosses and relatives)

Microphylls

Tracheids (vascular cells); branching independent sporophyte; rooting structures

Euphyllophytes

Horsetails

Monilophytes

Ferns

Megaphylls; overtopping growth form

Gymnosperms

Cycads

Ginkgo

Seeds

Gnetophytes

Seed plants

Conifers

Flowers; double fertilization

Angiosperms

Flowering plants (see Figure 29.18)

28.1 How Did Photosynthesis Arise in Plants?

More than a billion years ago, when a cyanobacterium was first engulfed by an early eukaryote, the history of life was altered radically. The chloroplasts that resulted from primary endosymbiosis of this cyanobacterium were obviously important for the evolution of plants and other photosynthetic eukaryotes,

but they were also critical to the evolution of all life on land. Until photosynthetic plants were able to move onto land, there was very little there to support multicellular animals or fungi, and almost all life was restricted to the oceans and fresh waters.

Primary endosymbiosis is a shared derived trait—a synapomorphy—of the group known as **Plantae** (**Figure 28.1**). Although *Plantae* is Latin for "plants," in everyday language—and throughout this book—the unmodified common name "plants"

WORKING WITH**DATA:**

The Phylogeny of Land Plants

Original Paper

Qiu, Y.-L. et al. 2006. *Proceedings of the National Academy of Sciences USA* 103: 15511–15516.

Analyze the Data

In addition to the morphological characters of land plants shown on the phylogeny in Figure 28.1, DNA sequences are widely used to study and reconstruct the evolutionary history of plants. These sequences are many tens of thousands of nucleotides long and have been collected from a large number of species. The full data set used by Yin-Long Qiu and his colleagues (available at treebase.org) includes DNA sequences from 67 genes. The table below provides sample sequences from a chloroplast gene that has been used to reconstruct the relationships of representative plant species; the table shows 27 nucleotide positions for 10 species.

QUESTION 1

Construct a phylogenetic tree of these 10 species using the parsimony method (see Section 22.2 and the examples in Table 22.1 and Figure 22.5 for instructions). Use the outgroup to root your tree. Assume that all changes among nucleotides are equally likely.

QUESTION 2

How many changes (from one nucleotide to another) occur along each branch on your tree?

QUESTION 3

Which nucleotide positions (i.e., which character states) exhibit homoplasy (convergence or reversal of the character state)?

QUESTION 4

Which group on your tree represents the streptophytes? The land plants? The vascular plants? The euphyllophytes?

Species	Nucleotide position (character state)																										
	1	2	3	4	5	6	7	8	9	10	11	12	13	14	15	16	17	18	19	20	21	22	23	24	25	26	27
Outgroup (Chlorophyte alga)	T	A	T	T	A	T	G	A	T	T	C	C	A	A	A	T	A	T	T	A	T	A	A	T	C	T	A
Stonewort	T	A	T	T	T	A	A	A	T	T	A	C	T	A	A	T	A	A	T	A	T	A	A	T	C	T	A
Liverwort	A	C	T	T	T	T	A	A	T	G	A	T	T	C	A	G	A	A	T	A	T	A	A	T	C	T	A
Moss	A	C	T	T	T	T	A	A	T	A	T	T	T	T	A	A	T	A	T	A	A	A	A	T	C	T	T
Hornwort	A	C	T	T	T	T	A	A	T	G	T	T	T	T	A	A	T	A	C	A	G	A	A	A	C	T	T
Lycophyte	A	C	T	C	C	C	G	G	T	G	T	T	C	T	G	A	T	A	C	A	A	G	G	A	C	C	T
Fern	C	C	T	C	C	G	A	G	C	G	T	T	C	T	T	A	G	A	T	A	A	G	G	A	C	C	T
Pine tree	A	C	C	C	C	G	C	G	C	G	T	T	C	T	G	A	T	G	C	G	A	G	G	A	T	C	T
Rice	A	C	C	C	C	G	C	G	C	G	T	T	C	T	G	A	T	G	C	G	A	G	G	A	T	A	T
Tobacco	A	C	C	A	C	G	C	G	C	G	T	T	C	T	G	A	T	G	C	G	A	G	G	A	T	A	T

Go to BioPortal for all WORKING WITH**DATA** exercises

is usually used to refer only to the land plants. However, the first several clades that branch off the tree of life after primary endosymbiosis are all aquatic. Most aquatic photosynthetic eukaryotes (other than those secondarily derived from land plants) are known by the common name **algae**. This name, however, is just a convenient way to refer to these groups, which are not all closely related. Many of the photosynthetic groups discussed in Chapter 27 (which acquired chloroplasts through secondary endosymbiosis) are also commonly called algae.

Several distinct clades of algae were among the first photosynthetic eukaryotes

The ancestor of Plantae was unicellular and may have been similar in general form to the modern **glaucophytes** (**Figure 28.2**). These microscopic freshwater algae are thought to be the sister group of the rest of Plantae (see Figure 28.1A). The chloroplast of glaucophytes is unique in containing a small amount of peptidoglycan between its inner and outer membranes—the same arrangement found in cyanobacteria. Peptidoglycan has been lost from the remaining photosynthetic eukaryotes.

In contrast to the glaucophytes, almost all **red algae** are multicellular (**Figure 28.3**). Their characteristic color is a result of the accessory photosynthetic pigment **phycoerythrin**, which is found in relatively large amounts in the chloroplasts of many red algae. In addition to phycoerythrin, red algal chloroplasts contain chlorophyll *a* and the accessory pigments phycocyanin and carotenoids.

The red algae include species that grow in the shallowest tide pools as well as the photosynthesizers found deepest in the ocean (as deep as 260 meters if nutrient conditions are right and the water is clear enough to permit light to penetrate). A few red algae inhabit fresh water. Most grow attached to a substrate by a holdfast.

Glaucocystis

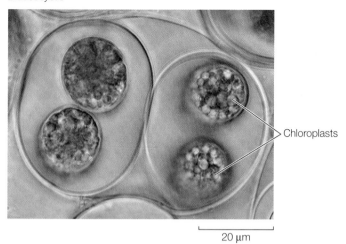

Chloroplasts

20 μm

28.2 Glaucophytes May Resemble Some of the Earliest Plantae
The large chloroplasts of unicellular glaucophytes differ from chloroplasts of other Plantae in retaining a layer of peptidoglycan. This feature is thought to have been retained from the endosymbiotic cyanobacteria that gave rise to the chloroplasts of Plantae. The photograph shows a colony of two individuals, each with two chloroplasts.

Despite their name, red algae don't always appear red in color. The ratio of two pigments—phycoerythrin (red) and chlorophyll *a* (green)—depends largely on the intensity of light that reaches the alga. In deep water, where light is dim, algae accumulate large amounts of phycoerythrin and have red coloration. But many species growing near the surface contain a higher concentration of chlorophyll *a* and are bright green.

The remaining algal groups in Plantae are the various "green algae." Like land plants, the green algae contain both chlorophylls *a* and *b* and store their reserve of photosynthetic products as starch in chloroplasts. All the groups that share these features are commonly called **green plants** because both of their photosynthetic pigments are green.

The largest clade of "green algae" is the **chlorophytes**. There are more than 17,000 species of chlorophytes, most of which are aquatic (some are marine, though more are freshwater forms), although there are a few terrestrial forms that live in moist environments. Chlorophytes range in size from microscopic unicellular forms to multicellular forms many centimeters long and display an incredible variety of shapes and body forms. Surprisingly large and well-formed colonies of cells are found in some unicellular freshwater groups, such as the genus *Volvox* (**Figure 28.4A**). Certain cells in these colonies are specialized for reproduction. The cells in these colonies are not differentiated into specialized tissues and organs, as in land plants and animals, but they show vividly how the preliminary step of this great evolutionary innovation might have been taken.

Volvox is a colonial unicellular chlorophyte, but there are also many true multicellular species of chlorophytes. Some of these are filamentous. Others, like species in the genus *Ulva* (**Figure 28.4B**), grow into thin, membranous sheets up to 30 centimeters in width.

Two groups of green algae are the closest relatives of land plants

All green algae other than the chlorophytes form a group together with the land plants known as **streptophytes** (see Figure 28.1A). Several microscopic structural features, backed by clear evidence from molecular studies, indicate that the closest relatives of the land plants are two groups of aquatic green algae, the **coleochaetophytes** (**Figure 28.5A**) and the **stoneworts** (**Figure 28.5B**). Both of these multicellular algal groups retain their eggs in the parental organism, as land plants do. As in land plants, the cytoplasm of adjacent cells in these algal groups is connected through structures called plasmodesmata; they also share similarities in the details of mitosis and cytokinesis. Of these two groups, stoneworts are thought to be the sister group of land plants (see Figure 28.1A). The growth form of stoneworts is branching and apical (new growth occurs at the tips of branches), as in most land plants. Phylogenetic

(A) *Ceramium* sp.

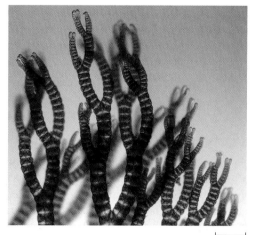

(B) *Calliarthron* sp.

28.3 Red Algae Contain a Red Accessory Photosynthetic Pigment (A) Differential contrast light microscopy reveals the rich red color of the pigment phycoerythrin. (B) Coralline red alga is named for its coral-like appearance.

1.5 mm

7.5 mm

Parent colony Somatic cells Daughter colonies produced by reproductive cells

(A) *Volvox* sp. 120 µm

(B) *Ulva rigida* 3 cm

28.4 Chlorophytes Display a Wide Diversity of Forms
(A) *Volvox* colonies are precisely spaced arrangements of individual cells. Specialized reproductive cells produce daughter colonies, which will eventually release new individuals. (B) Sea lettuce grows in marine waters and intertidal areas.

(A) *Coleochaete* sp.

(B) *Chara vulgaris* (stonewort) 1 cm

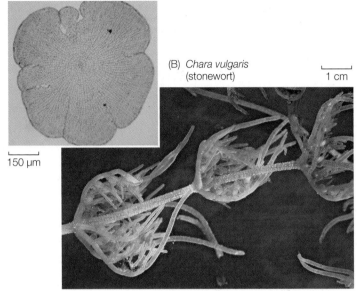

150 µm

28.5 The Closest Relatives of Land Plants (A) This species is a representative of the coleochaetophytes, the sister clade of stoneworts plus land plants. (B) The land plants probably evolved from a common ancestor shared with the stoneworts, an abundant group of multicellular green algae often found in freshwater pools and lakes (although a few species are found in marine environments). A species in the common genus *Chara* is shown here.

 Go to Media Clip 28.1
Reproductive Structures of *Chara*
Life10e.com/mc28.1

analysis of gene sequences has confirmed the close relationships of coleochaetophytes and stoneworts to the land plants.

There are ten major groups of land plants

One of the key synapomorphies of the **land plants** is development from an embryo that is protected by tissues of the parent plant. For this reason, land plants are sometimes called **embryophytes** (*phyton*, "plant"). The green plants, the streptophytes, and the land plants have each been called "the plant kingdom" by different authorities; others take an even broader view and include red algae and glaucophytes as "plants." To avoid confusion in this chapter, we will use modifying terms ("land plants" or "green plants," for example) to refer to the various clades of Plantae shown in Figure 28.1.

The land plants that exist today fall naturally into ten major clades (listed by their common names in the center column of **Table 28.1**). Members of seven of those clades possess well-developed vascular systems that transport materials throughout the plant body. We call these seven groups, collectively, the **vascular plants**, or **tracheophytes**, because they all possess fluid-conducting cells called **tracheids**. The remaining three clades (liverworts, mosses, and hornworts) lack tracheids and are referred to collectively as **nonvascular land plants**. Note,

however, that (unlike the vascular plants, which *are* a clade) *the three groups of nonvascular land plants do not form a clade.*

RECAP 28.1

Primary endosymbiosis is a synapomorphy of the Plantae. The glaucophytes, the sister clade of the other Plantae, are unicellular algae that are similar to some of the earliest photosynthetic eukaryotes. The green plants contain chlorophyll *b* in addition to the chlorophyll *a* found in all Plantae.

- Explain the different possible uses of the term "plant." **See pp. 570–571 and Figure 28.1**
- Why doesn't the name "algae" designate a formal taxonomic group? **See pp. 571–572**
- What are some of the key differences between glaucophytes, red algae, and the various clades of green algae? **See pp. 571–572**
- What evidence supports the phylogenetic relationship between land plants and the various groups of green algae? **See pp. 572–573**

The green algal ancestors of the land plants lived at the margins of ponds or marshes, ringing them with a mat of dense green. It was from such a marginal habitat, which was sometimes wet and sometimes dry, that early plants made the transition onto land.

TABLE**28.1**

Classification of Land Plants

Group	Common Name	Characteristics
Nonvascular land plants		
Hepatophyta	Liverworts	No stomata; gametophyte flat or leafy
Bryophyta	Mosses	Filamentous stage; gametophyte leafy; sporophyte grows apically (at the tip)
Anthocerophyta	Hornworts	Embedded archegonia; sporophyte grows basally (i.e., from the ground)
Vascular plants		
Lycopodiophyta	Lycophytes: Club mosses and allies	Microphylls in spirals; sporangia in leaf axils
Monilophyta	Horsetails, ferns	Simple leaves in whorls or frondlike compound leaves
SEED PLANTS		
Gymnosperms		
Cycadophyta	Cycads	Compound leaves; swimming sperm; seeds on modified leaves
Ginkgophyta	Ginkgo	Deciduous; fan-shaped leaves; swimming sperm
Gnetophyta	Gnetophytes	Vessels in vascular tissue; opposite, simple leaves
Coniferophyta	Conifers	Seeds in cones; needlelike or scalelike leaves
Angiosperms	Flowering plants	Endosperm; carpels; gametophytes much reduced; seeds contained within fruits

28.2 When and How Did Plants Colonize Land?

How did the land plants arise? To address this question, we can compare land plants with their closest relatives among the green algae. The features that differ between the two groups include the adaptations that allowed the first land plants to survive in the terrestrial environment.

Adaptations to life on land distinguish land plants from green algae

Land plants first appeared in the terrestrial environment between 450 and 500 million years ago. How did they survive in an environment that differed so dramatically from the aquatic environment of their ancestors? While the water essential for life is everywhere in the aquatic environment, water is difficult to obtain and retain in the terrestrial environment.

No longer bathed in fluid, organisms on land faced potentially lethal desiccation (drying). Large terrestrial organisms had to develop ways to transport water to body parts distant from the source of the water. And whereas water provides aquatic organisms with support against gravity, a plant living on land must either have some other support system or sprawl unsupported on the ground. A land plant must also use different mechanisms for dispersing its gametes and progeny than its aquatic relatives, which can simply release them into the water.

Survival on land was facilitated by the evolution among plants of numerous adaptations, including:

- The *cuticle*, a membrane covered in waxes to retard water loss
- *Stomata*, small openings in leaves and stems that open and close to regulate gas exchange and water loss

- *Gametangia*, multicellular organs that enclose plant gametes and prevent them from drying out
- *Embryos*, young plants contained within a protective structure
- Certain *pigments* that afford protection against the mutagenic ultraviolet radiation that bathes the terrestrial environment
- Thick *spore walls* containing a polymer that protects the spores from desiccation and resists decay
- A *mutually beneficial association with fungi* (mycorrhizae) that promotes nutrient uptake from the soil

The cuticle may be the most important—and the earliest—of these features. Composed of several unique waxy lipids that coat the leaves and stems of land plants, the cuticle has several functions, the most obvious and important of which is to keep water from evaporating from the plant body.

As ancient plants colonized the land, they not only adapted to the terrestrial environment, they also modified it by contributing to the formation of soil. Acids secreted by plants helped break down rock, and the organic compounds produced by the breakdown of dead plants contributed nutrients to the soil. Such effects are repeated today wherever plants colonize and grow in new areas.

Life cycles of land plants feature alternation of generations

A universal feature of the life cycles of land plants is alternation of generations. Recall from Section 27.3 the two hallmarks of alternation of generations:

- The life cycle includes both a multicellular diploid stage and a multicellular haploid stage.

28.6 Alternation of Generations in Land Plants A multicellular diploid sporophyte generation that produces spores by meiosis alternates with a multicellular haploid gametophyte generation that produces gametes by mitosis.

• Gametes are produced by mitosis, not by meiosis. Meiosis produces **spores** that develop into multicellular haploid organisms.

If we begin looking at the land plant life cycle at the single-cell stage—the diploid zygote—then the first phase of the cycle is the formation, by mitosis and cytokinesis, of a multicellular **embryo**, which eventually grows into a mature diploid plant. This multicellular diploid plant is called the **sporophyte** ("spore plant").

Cells contained within specialized reproductive organs of the sporophyte, called **sporangia** (singular *sporangium*), undergo meiosis to produce haploid, unicellular spores. By mitosis and cytokinesis, a spore develops into a haploid plant. This multicellular haploid plant, called the **gametophyte** ("gamete plant"), produces haploid gametes by mitosis. The fusion of two gametes (fertilization) forms a single diploid cell—the zygote—and the cycle is repeated (**Figure 28.6**).

The sporophyte generation extends from the zygote through the adult multicellular diploid plant and sporangium formation; the gametophyte generation extends from the spore through the adult multicellular haploid plant to the gametes. The transitions between the generations are accomplished by fertilization and by meiosis. In all land plants, the sporophyte and the gametophyte differ genetically: the sporophyte has diploid cells, and the gametophyte has haploid cells.

There is a trend toward reduction of the gametophyte generation in plant evolution. In the nonvascular land plants, the gametophyte is larger, longer-lived, and more self-sufficient than the sporophyte. In those groups that appeared later in plant evolution, however, the sporophyte is the larger, more conspicuous, longer-lived, and more self-sufficient generation.

Nonvascular land plants live where water is readily available

The living species of nonvascular land plants are the liverworts, mosses, and hornworts. These three groups are thought to be similar in many ways to the earliest land plants. Most of these plants grow in dense mats, usually in moist habitats. Even the largest of these species are only about half a meter tall, and most are only a few centimeters tall or long. Why have they not evolved to be taller? The probable answer is that they lack an efficient vascular system for transporting water and minerals from the soil to distant parts of the plant body.

The nonvascular land plants lack the true leaves, stems, and roots that characterize the vascular plants, although they have structures analogous to each. Their growth form allows water to move through the mats of plants by capillary action. They have leaflike structures that readily catch and hold any water that splashes onto them. They are small enough that minerals can be distributed throughout their bodies by diffusion. As in all land plants, layers of maternal tissue protect their embryos from desiccation. Many nonvascular land plants also have a cuticle, although it is often very thin (even absent in some species) and thus is not highly effective in retarding water loss.

Most nonvascular land plants live on the soil or on vascular plants, but some grow on bare rock, on dead and fallen tree trunks, and even on buildings. Their ability to grow on such marginal surfaces results from a mutualistic association with fungi. The earliest association of land plants with fungi dates back at least 460 million years. This mutualism probably facilitated the absorption of water and minerals, especially phosphorus, from the first soils.

Nonvascular land plants are widely distributed over six continents and even exist (albeit very locally) on the coast of the seventh, Antarctica. Most are terrestrial. Although a few species live in fresh water, these aquatic species are descended from terrestrial ones. None live in the oceans.

The sporophytes of nonvascular land plants are dependent on the gametophytes

In the nonvascular land plants, the conspicuous green structure visible to the naked eye is the gametophyte. The gametophyte is photosynthetic and is therefore nutritionally independent; the sporophyte may or may not be photosynthetic, but it is always nutritionally dependent on the gametophyte and remains permanently attached to it.

Figure 28.7 illustrates the life cycle of a moss, which is typical of the life cycles of nonvascular land plants. A sporophyte produces unicellular haploid spores as products of meiosis within a sporangium. When a spore germinates, it gives rise to a multicellular haploid gametophyte whose cells contain chloroplasts and are thus photosynthetic. Eventually gametes form within specialized sex organs, called the **gametangia**. The **archegonium** is a multicellular, flask-shaped female sex organ with a long neck

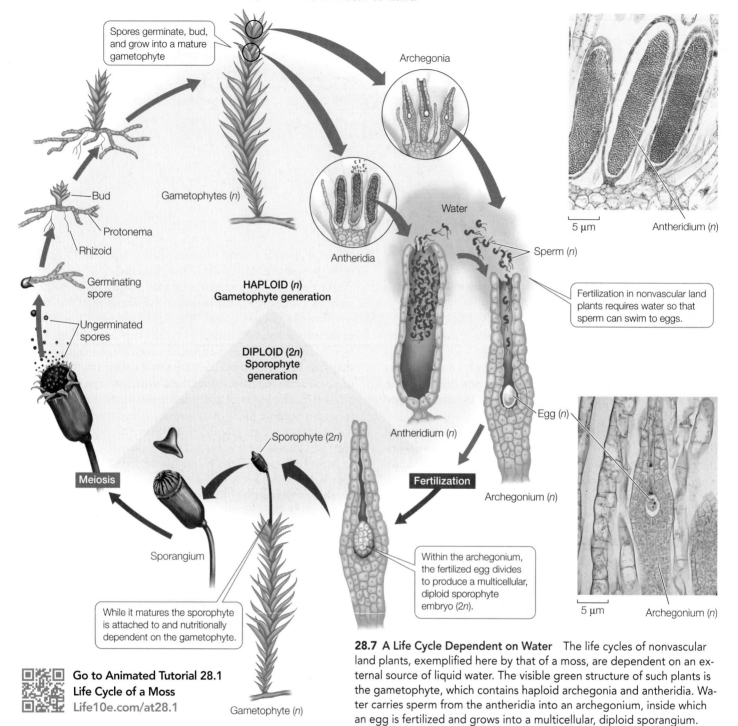

Spores germinate, bud, and grow into a mature gametophyte

Archegonia

Bud

Protonema

Rhizoid

Gametophytes (n)

Germinating spore

Ungerminated spores

Antheridia

HAPLOID (n)
Gametophyte generation

Water

Sperm (n)

5 μm

Antheridium (n)

Fertilization in nonvascular land plants requires water so that sperm can swim to eggs.

DIPLOID (2n)
Sporophyte generation

Meiosis

Sporophyte (2n)

Antheridium (n)

Fertilization

Egg (n)

Archegonium (n)

Sporangium

Within the archegonium, the fertilized egg divides to produce a multicellular, diploid sporophyte embryo (2n).

While it matures the sporophyte is attached to and nutritionally dependent on the gametophyte.

5 μm

Archegonium (n)

Go to Animated Tutorial 28.1
Life Cycle of a Moss
Life10e.com/at28.1

Gametophyte (n)

28.7 A Life Cycle Dependent on Water The life cycles of nonvascular land plants, exemplified here by that of a moss, are dependent on an external source of liquid water. The visible green structure of such plants is the gametophyte, which contains haploid archegonia and antheridia. Water carries sperm from the antheridia into an archegonium, inside which an egg is fertilized and grows into a multicellular, diploid sporangium.

 Go to Media Clip 28.2
Bryophyte Reproduction
Life10e.com/mc28.2

and a swollen base; it produces a single egg. The **antheridium** is a male sex organ in which sperm, each bearing two flagella, are produced in large numbers. Archegonia and antheridia are produced on the same individual in many species, so each individual has both male and female reproductive structures. Adjacent individuals often fertilize one another's gametes, however, which helps maintain genetic diversity in the population.

Once released from the antheridium, the sperm must swim or be splashed by raindrops to a nearby archegonium on the same or a neighboring plant—a constraint that reflects the aquatic origins of the nonvascular land plants' ancestors. The sperm are aided on their journey by chemical attractants

released by the egg or the archegonium. Before sperm can enter the archegonium, however, certain cells in the neck of the archegonium must break down, leaving a water-filled canal through which the sperm can swim to complete their journey. Notice that *all of these events require liquid water.*

Once sperm arrive at an egg, the nucleus of a sperm fuses with the egg nucleus to form a diploid zygote. Mitotic divisions of the zygote produce a multicellular, diploid sporophyte embryo. After the sporophyte grows out of the

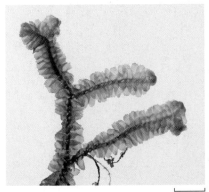

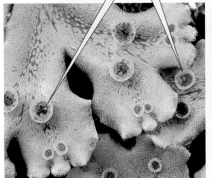

These cups contain gemmae—small, lens-shaped outgrowths of the plant body, each capable of developing into a new plant.

The banana-like structures bear archegonia.

(A) *Bazzania trilobata* 2 cm   (B) *Marchantia* sp.   0.3 cm   (C) *Marchantia polymorpha*   2.5 cm

28.8 Liverwort Diversity (A) The gametophyte of a leafy liverwort. (B) The gametophytes of the thalloid liverwort *Marchantia* lie flat to the ground. (C) *Marchantia* gametophytes bearing archegonia.

archegonium it produces a single sporangium, within which meiotic divisions produce spores and thus the next gametophyte generation.

Liverworts are the sister clade of the remaining land plants

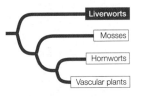

There are about 9,000 species of **liverworts**. Most liverworts have leafy gametophytes (**Figure 28.8A**). Some have thalloid gametophytes: green, leaflike layers that lie flat on the ground (**Figure 28.8B,C**). The simplest liverwort gametophytes are flat plates of cells, a centimeter or so long, that produce antheridia or archegonia on their upper surfaces and rhizoids (rootlike filaments) on their lower surfaces.

Liverwort sporophytes are shorter than those of mosses and hornworts, rarely exceeding a few millimeters. The liverwort sporophyte has a stalk that raises the sporangium above the gametophyte. In most species, the stalk elongates by expansion of cells throughout its length. This elongation raises the sporangium above ground level, allowing the spores to be dispersed more widely. The sporangia of liverworts are simple: a globular sporangium wall surrounds a mass of spores. In some species of liverworts, spores are not released by the sporophyte until the surrounding sporangium wall rots. In other liverworts, however, the spores are thrown from the sporangium by structures that shorten and compress as they dry out. When the stress becomes sufficient, the compressed structure snaps back to its resting position, throwing spores in all directions.

Among the most familiar thalloid liverworts are species of the genus *Marchantia*. *Marchantia* is easily recognized by the characteristic structures on which its male and female gametophytes bear their antheridia (Figure 28.8B) and archegonia (Figure 28.8C). Like most liverworts, *Marchantia* also reproduces asexually by simple fragmentation of the gametophyte.

In addition, *Marchantia* and some other liverworts and mosses reproduce asexually by means of gemmae (singular gemma), which are lens-shaped clumps of cells. In a few liverworts, the gemmae are held in structures called gemmae cups, which promote dispersal of the gemmae by raindrops.

Go to Media Clip 28.3
Liverwort Life Cycle
Life10e.com/mc28.3

Water and sugar transport mechanisms emerged in the mosses

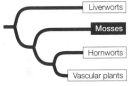

The most familiar of the nonvascular land plants are the **mosses**. These hardy little plants, of which there are about 15,000 species, are found in almost every terrestrial environment. They are often found on damp, cool ground, where they form thick mats (**Figure 28.9**). The mosses are the sister clade of the vascular plants plus the hornworts (see Figure 28.1).

The mosses, along with the hornworts and vascular plants, share an advance over the liverworts in their adaptation to life on land: they have openings called **stomata**, which allow CO_2 to enter the plant body and allow water and O_2 to leave it. Stomata are a synapomorphy of mosses and all other land plants except liverworts.

In mosses, the gametophyte begins its development following spore germination as a branched, filamentous structure called a protonema (see Figure 28.7). Although the protonema looks a bit like a filamentous green alga, this structure is unique to the mosses. Some of the filaments contain chloroplasts and are photosynthetic; others, called rhizoids, are nonphotosynthetic and anchor the protonema to the substratum. After a period of linear growth, cells close to the tips of the photosynthetic filaments divide rapidly in three dimensions to form buds. The buds eventually develop a distinct tip, or apex, and produce the familiar leafy moss shoot with leaflike structures arranged spirally. These leafy shoots produce antheridia or archegonia (see Figure 28.7).

(A)

Sporophytes

Gametophytes

(B) *Polytrichum* sp.

28.9 Mosses Often Cover the Ground in Dense Mats (A) Dense layers of moss carpet a field of solidified volcanic lava in Iceland. (B) A close-up view of moss growing on a forest floor in Michigan.

(A)

Sporophyte

Gametophyte

Sphagnum sp.

(B)

28.10 Sphagnum Moss (A) *Sphagnum* bogs are extremely dense growths of the moss shown here in a close-up view. (B) A farmer mines a bog for peat, a fossil fuel formed from decomposing *Sphagnum* mosses.

Some moss gametophytes are too large to transport enough water through their bodies solely by diffusion. Gametophytes and sporophytes of many mosses contain a type of cell called a hydroid, which dies and leaves a tiny channel through which water can travel. The hydroid is functionally similar to the tracheid, the characteristic water-conducting cell of the vascular plants, but it lacks lignin and the cell wall structure found in tracheids. The possession of hydroids and of a limited system for transport of sugar in some mosses shows that the term "nonvascular plant" is somewhat misleading when applied to these plants. Despite their simple system of internal transport, however, the mosses are not considered vascular plants because they lack tracheids or other components of xylem and phloem.

Mosses of the genus *Sphagnum* (**Figure 28.10A**) often grow in cool, swampy places, where the plants begin to decompose in water after they die. Rapidly growing upper layers of moss compress the deeper-lying, decomposing layers. Partially decomposed plant matter is called **peat**. In some parts of the world, people derive the majority of their fuel from peat bogs (**Figure 28.B**). *Sphagnum*-dominated peatlands cover a total area approximately half the size of the United States—more than 1 percent of Earth's surface. Millions of years ago, continued compression of peat composed primarily of other seedless plants gave rise to coal.

Hornworts have distinctive chloroplasts and stalkless sporophytes

The approximately 100 species of **hornworts** are so named because their sporophytes look like little horns (**Figure 28.11**). Hornworts appear at first glance to be liverworts with very simple gametophytes. Their gametophytes are flat plates of cells a few cells thick.

Hornworts have several characteristics that distinguish them from liverworts and mosses. First, the cells of hornworts each contain a single large, platelike chloroplast, whereas the cells of the other two groups contain numerous small, lens-shaped

The sporophytes of hornworts can reach 20 cm in height.

Gametophytes are flat plates a few cells thick.

Anthoceros sp.

28.11 Hornworts Get Their Name from Their Hornlike Sporophytes Unlike liverworts or mosses, the sporophytes of hornworts are persistently green. They share this trait with the vascular plants.

chloroplasts. Second, of the sporophytes in all three groups, those of the hornworts come closest to being capable of growth without a set limit. Liverwort and moss sporophytes have a stalk that stops growing as the sporangium matures, so elongation of the sporophyte is strictly limited. The hornwort sporophyte, however, has no stalk, and it is persistently green (a trait shared with vascular plants). A basal region of the sporangium remains capable of indefinite cell division, continuously producing new spore-bearing tissue above. The sporophytes of some hornworts growing in mild and continuously moist conditions can become as tall as 20 centimeters. Eventually, however, the sporophyte's growth is limited by the lack of a transport system.

Hornworts have a symbiotic relationship that promotes their growth by providing them with access to nitrogen, which is often a limiting resource. Hornworts have internal cavities filled with mucilage; these cavities are often populated by cyanobacteria that convert atmospheric nitrogen gas into a form usable by their host plant.

RECAP 28.2

The transition of plants to land required numerous adaptations, including the cuticle, stomata, gametangia, protected embryos, and mutually beneficial associations with fungi. Nonvascular land plants rely on liquid water for reproduction.

- Explain what is meant by alternation of generations. **See pp. 574–576 and Figure 28.6**

- Describe several adaptations of plants to the terrestrial environment, and describe the distribution of those adaptations among the liverworts, mosses, and hornworts.

New features appeared in plants as they continued to adapt to the terrestrial environment. One of the most important of these was vascular tissues, the characteristic that defines the vascular plants.

28.3 What Features Allowed Land Plants to Diversify in Form?

The first plants possessing vascular tissues did not arise until tens of millions of years after the earliest nonvascular plants had colonized the land. But once vascular tissues arose, their ability to transport water and food throughout the plant body allowed the vascular plants to spread to new terrestrial environments and to diversify rapidly.

Vascular tissues transport water and dissolved materials

The key synapomorphy of the vascular plants is a well-developed vascular system containing two types of tissues that are specialized for the transport of materials from one part of the plant to another. One type of vascular tissue, the **xylem**, conducts water and minerals from the soil to aerial parts of the plant. Because some of its cell walls contain a stiffening substance called lignin, xylem also provides support against gravity in the terrestrial environment. The other type of vascular tissue, the **phloem**, conducts the products of photosynthesis from sites where they are produced or released to sites where they are used or stored. (Xylem and phloem will be discussed in detail in Chapters 34 and 35.)

Although the vascular plants are an extraordinarily large and diverse group, one particular event was critical to their evolution. Sometime during the Paleozoic era, probably in the mid-Silurian (430 mya), a new cell type—the tracheid—evolved in sporophytes of the earliest vascular plants. The tracheid is the principal water-conducting element of the xylem in all vascular plants except the angiosperms (flowering plants) and gnetophytes—and tracheids persist even in these groups, along with a more specialized and efficient system derived from them.

The evolution of tracheids set the stage for the complete and permanent invasion of land by plants. First, these cells provided a pathway for the transport of water and mineral nutrients from a source of supply to regions of need in the plant body. Second, the cell walls of tracheids, stiffened by lignin, provided rigid structural support. This support is a crucial factor in a terrestrial environment because it allows plants to grow upward and thus compete for sunlight. A taller plant can intercept more direct sunlight (and thus conduct photosynthesis more readily) than a shorter plant, which may be shaded by the taller one. Increased height also improves the dispersal of spores.

The vascular plants featured another evolutionary novelty: a branching, independent sporophyte. A branching sporophyte body can produce more spores than an unbranched body, and it can develop in complex ways. The sporophyte of a vascular plant is nutritionally independent of the gametophyte at maturity. Among the vascular plants, the sporophyte is the large and obvious plant one normally notices in nature, in contrast to the relatively small, dependent sporophytes typical of most nonvascular land plants.

28.12 Artist's Reconstruction of an Ancient Forest Forests of the Carboniferous period were characterized by abundant vascular plants such as club mosses, ferns, and horsetails, some of which reached heights of 40 meters. Huge flying insects (see the opening of Chapter 25) thrived in these forests, which are the source of modern coal deposits.

Vascular plants allowed herbivores to colonize the land

The initial absence of herbivores (plant-eating animals) on land helped make the first vascular plants successful. By the late Silurian period (about 425 mya), vascular plants were being preserved as fossils that we can study today. The proliferation of these plants made the terrestrial environment more hospitable to animals. Arthropods, vertebrates, and other animals moved onto land only after vascular plants became established there.

Trees of various kinds appeared in the Devonian period and dominated the landscape of the Carboniferous period (359–299 mya). Forests of lycophytes (club mosses) up to 40 meters tall, along with horsetails and tree ferns, flourished in the tropical swamps of what would become North America and Europe (**Figure 28.12**). Plant material from those forests sank into the swamps and was gradually covered by layers of sediment. Over millions of years, as the buried plant material was subjected to intense pressure and elevated temperatures, it was transformed into coal. Today that coal provides over half of our electricity. The world's coal deposits, although huge, are not infinite, and humans are burning coal deposits at a far faster rate than they were produced.

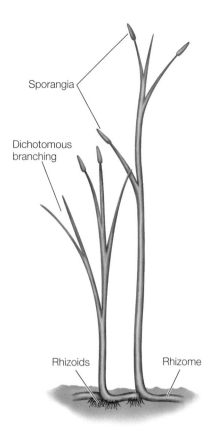

28.13 An Ancient Relative of the Vascular Plants The extinct rhyniophyte *Aglaophyton major* lacked roots and leaves. It had a central column of xylem running through its stems but no true tracheids. A horizontal underground stem called a rhizome anchored the plant. The dichotomously branching aerial stems were less than 50 centimeters tall. Some stems were topped by sporangia.

In the subsequent Permian period, when the continents came together to form Pangaea, the continental interior became warmer and drier. The 200-million-year reign of the lycophyte–fern forests came to an end as they were replaced by forests of early gymnosperms.

The closest relatives of vascular plants lacked roots

The closest relatives of living vascular plants belonged to several extinct groups called **rhyniophytes** (**Figure 28.13**). The rhyniophytes were one of a very few types of land plants in the Silurian period. The landscape at that time probably consisted mostly of bare ground, with mats of nonvascular plants and stands of rhyniophytes in low-lying moist areas. Early versions of the structural features of the vascular plant groups appeared in the rhyniophytes of that time. These shared features strengthen the case for the origin of all vascular plants from a common nonvascular land plant ancestor.

Rhyniophytes did not have roots. They were apparently anchored in the soil by horizontal portions of stem called **rhizomes**, which bore water-absorbing unicellular filaments called **rhizoids**. These plants also bore aerial branches, and sporangia—homologous to the sporangia of mosses—were found at the tips of those branches. Their branching pattern was **dichotomous**; that is, the apex (tip) of the shoot divided to produce two equivalent new branches, with each pair of branches diverging at approximately the same angle from the original stem.

The lycophytes are sister to the other vascular plants

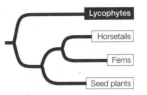

The club mosses and their relatives, the spike mosses and quillworts, are collectively called **lycophytes**. The lycophytes are the sister clade to the remaining vascular plants (see Figure 28.1B). There are relatively few (just over 1,200) surviving species of lycophytes.

The lycophytes have true roots that branch dichotomously. The arrangement of vascular tissue in their stems is simpler than that in other vascular plants. They bear simple leaflike structures called **microphylls**, which are arranged spirally on the stem. Growth in lycophytes comes entirely from apical cell division. Branching in the stems, which is also dichotomous, occurs by division of an apical cluster of dividing cells.

The sporangia of many club mosses are aggregated in cone-like structures called **strobili** (singular *strobilus*; **Figure 28.14A**), which are clusters of spore-bearing microphylls attached to the end of the stem. Other club mosses lack strobili and bear their sporangia on (or adjacent to) the upper surfaces of specialized microphylls.

Horsetails and ferns constitute a clade

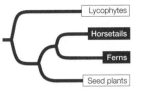

The horsetails and ferns were once thought to be only distantly related. From analysis of gene sequences we now know that they form a clade, the **monilophytes**. In the monilophytes—as in the seed plants, to which they are the sister clade (see Figure 28.1)—there is differentiation between a main stem and side branches (including the leaves derived from these branches). This pattern contrasts with the dichotomous branching characteristic of the lycophytes and rhyniophytes, in which each split gives rise to two branches of similar size.

Today there are only about 15 species of **horsetails**, all in the genus *Equisetum*. The horsetails have reduced true leaves that form in distinct whorls (circles) around the stem (**Figure 28.14B**). Horsetails are sometimes called "scouring rushes" because rough silica deposits found in their cell walls once made them useful for cleaning. They have true roots that branch irregularly. Horsetails have a large sporophyte and a small gametophyte, both independent of each other.

Strobilus

Microphylls

(A) *Lycopodium annotinum*

Leaves (in whorls)

Sporangia-bearing structure

(B) *Equisetum pratense*

(C) *Marsilea* sp.   *Salvinia* sp.

(D) *Dicksonia antarctica*

28.14 Lycophytes and Monilophytes
(A) Club mosses have microphylls arranged spirally on their stems. Strobili are visible at the tips of these stems. (B) Horsetails have a distinctive growth pattern in which the stem grows in segments above each whorl of leaves. These are fertile shoots with sporangia-bearing structures at the apex. (C) The leaves of two species of water ferns. (D) Tree ferns dominate this forest on the island of Tasmania, Australia.

28.15 Life Cycle of a Fern The most conspicuous stage in the fern life cycle is the mature diploid sporophyte, shown at the bottom of this diagram. The inset shows sori on the underside of a fern leaf. Each sorus contains many spore-producing sporangia.

Go to Activity 28.1 The Fern Life Cycle
Life10e.com/ac28.1

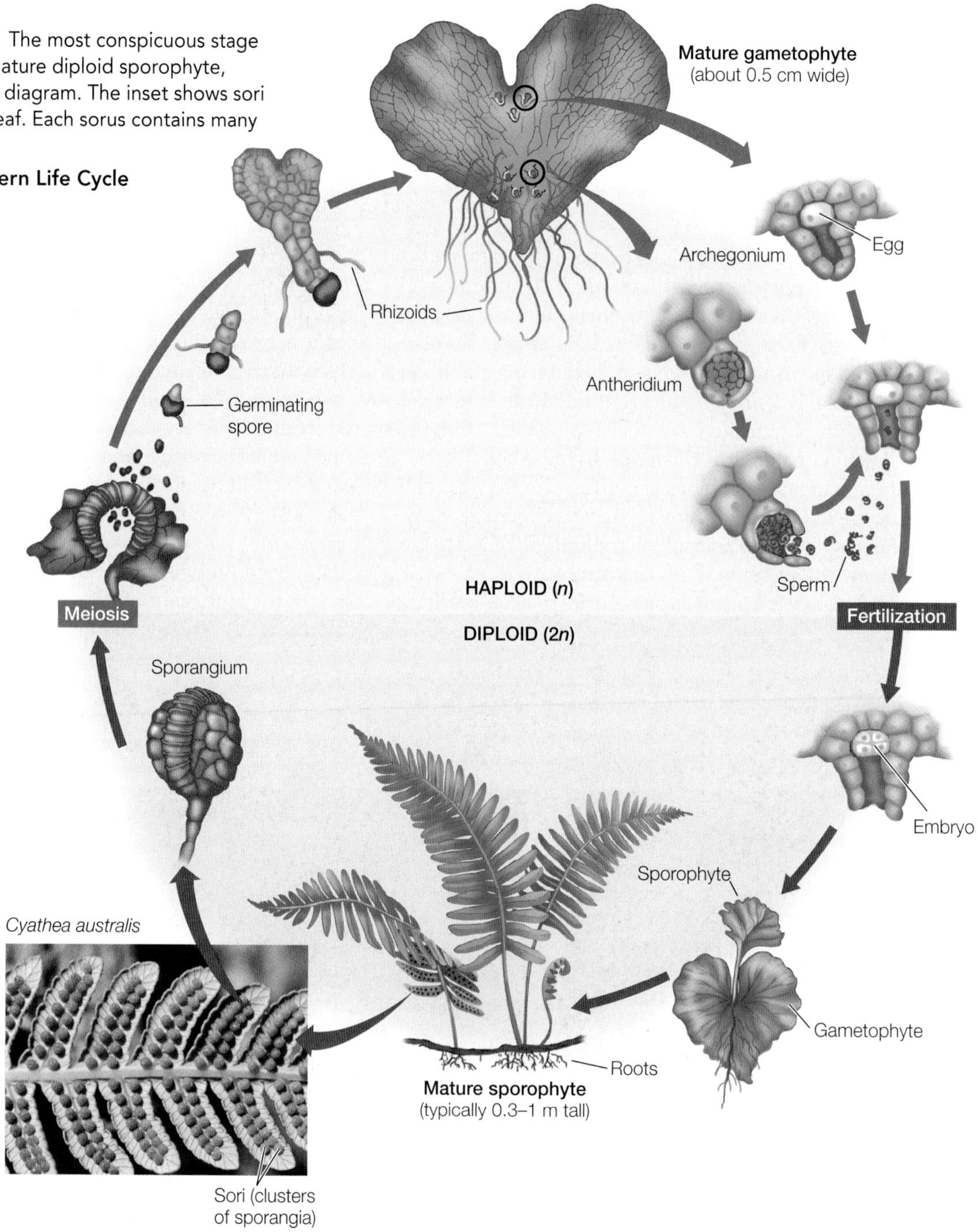

Mature gametophyte
(about 0.5 cm wide)

Archegonium

Egg

Antheridium

Rhizoids

Sperm

Germinating spore

HAPLOID (*n*)

DIPLOID (2*n*)

Fertilization

Meiosis

Sporangium

Embryo

Sporophyte

Cyathea australis

Gametophyte

Roots

Mature sporophyte
(typically 0.3–1 m tall)

Sori (clusters of sporangia)

The first ferns appeared during the Devonian period; today this group comprises more than 12,000 species. Analyses of gene sequences indicate that a few species traditionally allied with ferns may in fact be more closely related to horsetails than to ferns. Nonetheless, the majority of ferns form a monophyletic group.

Although most ferns are terrestrial, a few species live in shallow fresh water (**Figure 28.14C**). Terrestrial ferns are characterized by large leaves with branching vascular strands (**Figure 28.14D**). Some fern leaves become climbing organs and may grow to be as long as 30 meters.

In the alternating generations of a fern, the gametophyte is small, delicate, and short-lived, but the sporophyte can be very large and can sometimes survive for hundreds of years (**Figure 28.15**). Ferns require liquid water for the transport of the male gametes to the female gametes, so most ferns inhabit shaded, moist woodlands and swamps. The sporangia of ferns are typically borne on a stalk in clusters called **sori** (singular *sorus*). The sori are found on the undersurfaces of the leaves, sometimes covering the entire undersurface and sometimes located at the edges.

The vascular plants branched out

Several features that were new to the vascular plants evolved in lycophytes and monilophytes. Roots probably had their evolutionary origins as branches, either of a rhizome or of the

(A) Microphylls

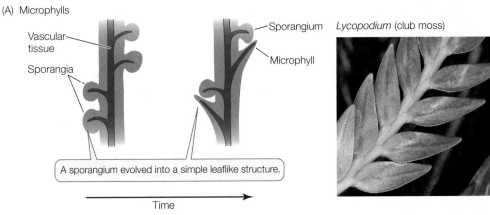

Vascular tissue

Sporangia

Sporangium

Microphyll

Lycopodium (club moss)

A sporangium evolved into a simple leaflike structure.

Time

28.16 Evolution of Leaves (A) Microphylls are thought to have evolved from sterile sporangia. (B) The megaphylls of monilo-phytes and seed plants may have arisen as photosynthetic tissue developed between branch pairs that were "left behind" as dominant branches overtopped them.

(B) Megaphylls

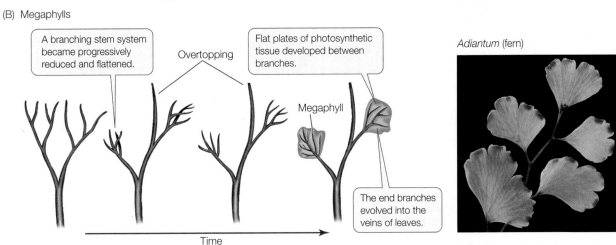

A branching stem system became progressively reduced and flattened.

Overtopping

Flat plates of photosynthetic tissue developed between branches.

Megaphyll

Adiantum (fern)

The end branches evolved into the veins of leaves.

Time

aboveground portion of a stem. These branches presumably penetrated the soil and branched further. The underground portions could anchor the plant firmly, and even in this primitive condition, they could absorb water and minerals.

The microphylls of lycophytes were probably the first leaflike structures to evolve among the vascular plants. Microphylls are usually small and only rarely have more than a single vascular strand, at least in existing species. Some biologists believe that microphylls had their evolutionary origins as sterile sporangia (**Figure 28.16A**). A typical feature of this type of leaf is a vascular strand that departs from the vascular system of the stem in such a way that the structure of the stem's vascular system is scarcely disturbed. This pattern was evident even in the lycophyte trees of the Carboniferous period, many of which had microphylls many centimeters long.

The monilophytes and seed plants constitute a clade called the **euphyllophytes** (*eu*, "true"; *phyllon*, "leaf"). An important synapomorphy of the euphyllophytes is **overtopping**, a growth pattern in which one branch differentiates from and grows beyond the others (**Figure 28.16B**). Overtopping would have given these plants an advantage in the competition for light, enabling them to shade their dichotomously branching competitors. The overtopping growth of the euphyllophytes also allowed a new type of leaflike structure to evolve. This larger, more complex leaf is called a **megaphyll**. The megaphyll

is thought to have arisen from the flattening of a portion of a branching stem system that exhibited overtopping growth. This change was followed by the development of photosynthetic tissue between the members of overtopped groups of branches, which had the advantage of increasing the photosynthetic surface area of those branches.

The first megaphylls, which were very small, appeared in the Devonian period. We might expect that evolution should have led swiftly to the appearance of more and larger megaphylls because of their greater photosynthetic capacity. However, it took some 50 million years, until the Carboniferous period, for large megaphylls to become common. Why should this have been so, especially given that other advances in plant structure were taking place during that time?

According to one theory, the high concentration of CO_2 in the atmosphere during the Devonian period reduced selection for the stomata that allow a leaf to take up CO_2 for use in photosynthesis. With more CO_2 available, fewer stomata were needed. In the Devonian, larger leaves with a limited number of stomata would have absorbed heat from sunlight, but they would have been unable to lose heat fast enough by evaporation of water through their stomata. The resulting overheating would have been lethal. Recent research has supported this hypothesis, indicating that larger megaphylls evolved only as CO_2 concentrations dropped over millions of years (**Figure 28.17**).

INVESTIGATING**LIFE**

28.17 Atmospheric CO$_2$ Concentrations and the Evolution of Megaphylls High concentrations of atmospheric CO$_2$ during the first part of the Devonian may have limited the evolution of leaf size. C. P. Osborne and colleagues compared the leaf sizes of fossil plants against estimates of CO$_2$ concentrations in the atmosphere at the time the plants were alive.[a]

HYPOTHESIS High atmospheric CO$_2$ concentrations during the early Devonian, and the resulting lack of selection for more stomata, kept leaf sizes small.

Method
1. Analyze 300 plant fossils from the Devonian and Carboniferous periods and measure the sizes of their leaves.
2. Compare the pattern of change in leaf size with that of the estimated change in atmospheric CO$_2$ concentrations over the same time span.

Results

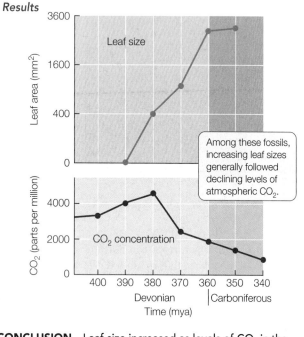

Among these fossils, increasing leaf sizes generally followed declining levels of atmospheric CO$_2$.

CONCLUSION Leaf size increased as levels of CO$_2$ in the atmosphere decreased.

Go to **BioPortal** for discussion and relevant links for all INVESTIGATING**LIFE** figures.

[a]Osborne, C. P. et al. 2004. *Proceedings of the National Academy of Sciences USA* 101: 10360–10362.

Heterospory appeared among the vascular plants

In the lineages of present-day, seedless vascular plants that are most similar to their ancestors, the gametophyte and the sporophyte are independent, and both are usually photosynthetic. The spores produced by the sporophyte are of a single type and develop into a single type of gametophyte that bears both female and male reproductive organs (see Figure 28.15). Such plants, which bear a single type of spore, are said to be **homosporous** (Figure 28.18A).

A system with two distinct types of spores evolved somewhat later. Plants of this type are said to be **heterosporous**

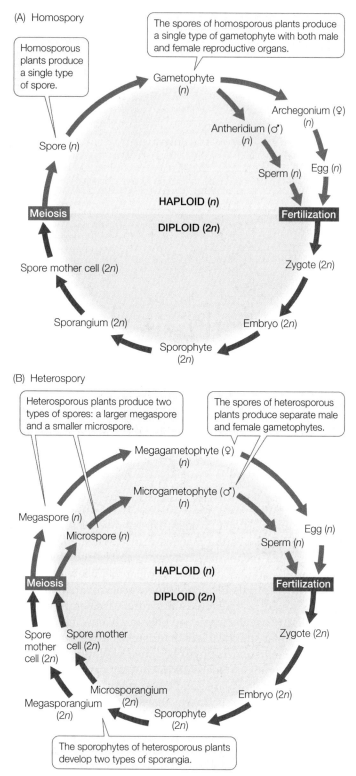

28.18 Homospory and Heterospory (A) Homosporous plants bear a single type of spore. Each gametophyte has two types of sex organs, antheridia (male) and archegonia (female). (B) Heterosporous plants bear two types of spores that develop into distinctly male and female gametophytes.

Go to Activities 28.2 Heterospory and
28.3 Homospory
Life10e.com/ac28.2
Life10e.com/ac28.3

(**Figure 28.18B**). In heterospory, one type of spore—the **mega-spore**—develops into a specifically female gametophyte (a **megagametophyte**) that produces only eggs. The other type, the **microspore**, is smaller and develops into a male gameto-phyte (a **microgametophyte**) that produces only sperm. The sporophyte produces megaspores in small numbers in **megasporangia** and microspores in large numbers in **microsporangia**. Heterospory affects not only the spores and the gametophytes but also the sporophyte plant itself, which must develop two types of sporangia.

The earliest vascular plants were all homosporous, but heterospory evidently evolved several times independently among later groups of vascular plants. The fact that heterospory evolved repeatedly suggests that it affords selective advantages. Subsequent evolution in the land plants featured ever greater specialization of the heterosporous condition.

▌ RECAP 28.3

Vascular plants are characterized by a vascular system specialized for the transport of materials from one part of the plant to another. A new type of cell, the tracheid, marked the origin of this group. Later evolutionary events included the appearance of roots, leaves, and heterospory.

- How do xylem and phloem serve the vascular plants? See p. 579

- Describe the evolution and distribution of different kinds of leaves and roots among the vascular plants. See pp. 581–583 and Figure 28.16

All of the vascular plant groups we have discussed thus far disperse by means of spores. The embryos of these seedless vascular plants develop directly into sporophytes, which either survive or die, depending on environmental conditions. The spores of some seedless plants may remain dormant and viable for long periods, but the embryos of seedless plants are relatively unprotected. Greater protection of the embryo evolved in the seed plants, which we will consider in the next chapter.

28.19 Biodiesel from Algae In this vertical-growth algal cultivation system for biofuel production, algae are grown in sheets of clear plastic, exposed to sunlight on all sides.

Given that petroleum is produced naturally from green algae, can humans use green algae to produce oil commercially?

ANSWER

Scientists are developing new methods for growing green algae for the production of biofuels (fuels produced directly from living organisms, such as biodiesel). Some species of green algae can produce up to 60 percent of their dry weight in oil. So biofuels can certainly be produced from green algae, although the process is not yet commercially viable. Like conventional fossil fuels, biofuels release carbon dioxide into the atmosphere when burned. In the production of biofuels, however, algae remove carbon dioxide from the atmosphere, so the use of these fuels is more sustainable, and results in less accumulation of CO_2 in the atmosphere over time, than the use of fossil fuels.

The primary commercial limitations to growing algae for biofuels include the need to establish efficient growing facilities, water needs, costs of fertilizers, costs and difficulties associated with harvest and refining, and labor expenses. Many new methods for growing and harvesting algae are being developed (**Figure 28.19**), and algae production in brackish water as well as in wastewater is being explored. Once some of the technical difficulties have been overcome, it is possible that commercial production of biofuels from algae will provide a significant source of energy for humans.

CHAPTER**SUMMARY** 28

 28.1 How Did Photosynthesis Arise in Plants?

- Primary endosymbiosis gave rise to chloroplasts and the subsequent diversification of the **Plantae**. The descendants of the first photosynthetic eukaryote include **glaucophytes**, **red algae**, several groups of green algae, and **land plants**, all of which contain chlorophyll *a*. Review Figure 28.1

- **Green plants**, which include the green algae and the land plants, are characterized by the presence of chlorophyll *b* (in addition to chlorophyll *a*). **Review Figure 28.1**

- Land plants, also known as **embryophytes**, arose from an aquatic green algal ancestor related to today's **stoneworts**. Land plants develop from embryos that are protected by parental tissue. **Review Figure 28.1**

28.2 When and How Did Plants Colonize Land?

- The acquisition of a **cuticle**, **stomata**, **gametangia**, a protected **embryo**, protective pigments, thick spore walls with a protective polymer, and mutualistic associations with fungi were all adaptations of land plants to terrestrial life.

- All land plant life cycles feature alternation of generations, in which a multicellular diploid **sporophyte** alternates with a multicellular haploid **gametophyte**. Review Figure 28.6

- The **nonvascular land plants** comprise the **liverworts**, **mosses**, and **hornworts**. These groups lack specialized vascular tissues for the conduction of water or nutrients through the plant body.

- The life cycles of nonvascular land plants depend on liquid water. The sporophyte is usually smaller than the gametophyte and depends on it for water and nutrition. **Review Figure 28.7, ANIMATED TUTORIAL 28.1**

- Liverworts lack stomata, but they are present in mosses, hornworts, and vascular plants. Hornworts have a persistently green sporophyte, a characteristic shared with vascular plants.

28.3 What Features Allowed Land Plants to Diversify in Form?

- The **vascular plants** have a **vascular system** consisting of **xylem** and **phloem** that conducts water, minerals, and products of photosynthesis through the plant body. The vascular system includes cells called **tracheids**.

- The **rhyniophytes**, the earliest known vascular plants, are known to us only in fossil form. They lacked true roots and leaves but possessed **rhizomes** and **rhizoids**. Review Figure 28.13

- The **lycophytes** (club mosses and relatives) have only small, simple leaflike structures (**microphylls**). **Monilophytes** (which include **horsetails** and **ferns**) have true leaves, and so together with seed plants are known as **euphyllophytes**.

- Unlike nonvascular land plants, the diploid sporophyte is the more conspicuous life stage of lycophytes and monilophytes. **Review Figure 28.15, ACTIVITY 28.1**

- Microphylls probably evolved from sterile sporangia. **Megaphylls** (true leaves) may have resulted from the flattening and reduction of a portion of a stem system with **overtopping** growth. Review Figure 28.16

- The earliest-diverging groups of vascular plants are **homosporous**, but **heterospory**—the production of distinct **megaspores** and **microspores**—has evolved several times. Megaspores develop into female **megagametophytes**; microspores develop into male **microgametophytes**. Review Figure 28.18, ACTIVITIES 28.2 and 28.3

 Go to the Interactive Summary to review key figures, Animated Tutorials, and Activities
Life10e.com/is28

CHAPTER**REVIEW**

 REMEMBERING

1. Which statement about alternation of generations in land plants is *not* true?
   a. The gametophyte and sporophyte differ in appearance.
   b. Meiosis occurs in sporangia.
   c. Gametes are always produced by meiosis.
   d. The zygote is the first cell of the sporophyte generation.
   e. The gametophyte and sporophyte differ in chromosome number.

2. Which of the following provide evidence for the close relationship between land plants and stoneworts?
   a. Phylogenetic analysis of DNA sequences
   b. Similar mechanics of mitosis and cytokinesis
   c. Similarities in chloroplast structure
   d. Similarities of their biochemistry
   e. All of the above

3. Liverworts, mosses, and hornworts
   a. lack a sporophyte generation.
   b. grow in dense masses, allowing capillary movement of water.
   c. possess xylem and phloem.
   d. possess true leaves.
   e. possess true roots.

4. Which statement about ferns is *not* true?
   a. The sporophyte is larger than the gametophyte.
   b. Most are heterosporous.
   c. The young sporophyte can grow independently of the gametophyte.
   d. The leaf is a megaphyll.
   e. The gametophytes produce archegonia and antheridia.

5. The ferns
   a. lack a gametophyte generation.
   b. have a large and prominent gametophyte but a much smaller sporophyte.
   c. are more closely related to club mosses than to horsetails.
   d. are monilophytes.
   e. produce seeds.

▌▌ UNDERSTANDING & APPLYING

6. Contrast microphylls with megaphylls in terms of structure, evolutionary origin, and occurrence among plants.

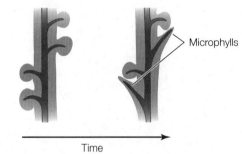

Time

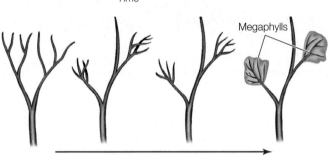

Time

7. Heterospory is thought to have evolved multiple times within plants, suggesting an advantage to the heterosporous condition. Why might heterospory provide an evolutionary advantage over homospory?

8. Contrast the life cycles of mosses (see Figure 28.7) and ferns (see Figure 28.15). How are these life cycles similar, and how do they differ?

▌▌ ANALYZING & EVALUATING

9. Are the morphological characters plotted on the phylogeny in Figure 28.1 consistent with your analysis of DNA sequences from the Working with Data exercise on p. 571? What characteristics of plants that we have discussed in this chapter appear to have evolved more than once in the history of land plants?

10. The findings of Osborne and co-workers on the evolution of megaphylls (see Figure 28.17) support the idea that large megaphylls became common only after the atmospheric CO_2 level had dropped, so that more stomata were favored, allowing evaporation from the stomata to cool larger leaves. How might you extend that work to confirm the involvement of temperature as a factor limiting leaf size?

Go to BioPortal at **yourBioPortal.com** for Animated Tutorials, Activities, LearningCurve Quizzes, Flashcards, and many other study and review resources.

29 The Evolution of Seed Plants

CHAPTER**OUTLINE**

Darwin's Ever-Fascinating Orchids On examining a specimen of the orchid *Angraecum sesquipedale*, Charles Darwin noted its exceptionally long nectar tube and predicted the existence of a pollinator with a correspondingly long proboscis. This pollinator, the sphinx moth *Xanthopan morgani*, was not discovered until after Darwin's death.

I N THE EARLY 1860s, much of middle- and upper-class England was caught up in an "orchid frenzy." Amateur gardeners and professional botanists alike were enchanted with these beautiful flowers and devoted considerable effort and money to collecting and raising them. Following the publication of *On The Origin of Species* in 1859, Charles Darwin wrote his next book on this group of plants. *On the Various Contrivances by which British and Foreign Orchids are Fertilised by Insects*, which he referred to proudly as "my little book on the fertilisation of orchids," appeared in 1862.

There are more than 25,000 species of orchids, which makes them one of the most diverse plant groups. Darwin wanted to know why orchids had experienced such rapid diversification and was particularly impressed with the role that insect pollinators might have played in this process. Seeking examples to demonstrate the power of natural selection, Darwin found such examples in abundance among the orchids.

Orchids show an impressive variety of specialized pollination mechanisms, many of which demonstrate that they have coevolved with their pollinators. For example, Darwin observed a South American orchid of the genus *Catasetum* shooting a packet of pollen at an insect that landed on its flower. When he was shown *Angraecum sesquipedale*, an orchid from Madagascar with a nectar tube more than a foot long, Darwin hypothesized that there must be a moth with a proboscis of unprecedented length that fed from and pollinated that flower. Many people scoffed at his vision, but the moth he described was eventually discovered—21 years after his death.

In 1836 the explorer Robert Schomburgk shook the botanical world with a report that he had seen flowers described as belonging to three different genera of orchids—*Catasetum*, *Monachanthus*, and *Myanthus*—growing together on a single plant. The English botanist John Lindley remarked that this observation would "shake to the foundation all our ideas of the stability of genera and species." Orchid enthusiasts were befuddled by their efforts to grow specimens of *Myanthus*, only to have them flower with the more common blooms of *Catasetum*. Darwin knew that he needed to find the explanation for these odd observations, for otherwise he would have to conclude that individual plants were able to change their specific identity, something that did not fit with his explanations of the evolution of diversity.

What was Darwin's explanation for the three distinct flowers growing on a single orchid plant?

See answer on p. 605.

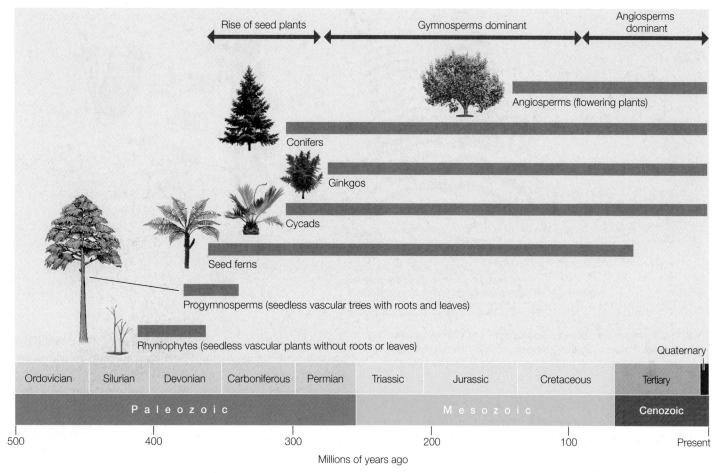

29.1 The Fossil Record of Seed Plant Evolution Woody growth evolved in the seedless progymnosperms. The now-extinct seed ferns had woody growth, fernlike foliage, and seeds attached to their leaves. New lineages of seed plants arose during the Carboniferous, but the earliest known fossils of flowering plants are from near the Jurassic–Cretaceous boundary.

29.1 How Did Seed Plants Become Today's Dominant Vegetation?

By the late Devonian period, more than 360 million years ago, Earth was home to a great variety of land plants, many of which we discussed in Chapter 28. The land plants shared the hot, humid terrestrial environment with insects, spiders, centipedes, and fishlike amphibians (early tetrapods). These plants and animals evolved together, each acting as agents of natural selection on the other. In the Devonian, a new innovation appeared when some plants developed extensively thickened woody stems. Among the first plants with this adaptation were seedless vascular plants called **progymnosperms**, all species of which are now extinct. The progymnosperms included many large trees.

Another innovation, the **seed**, arose in the seed plants. Seeds provide a secure and lasting dormant stage for the embryo. A plant embryo may safely wait within its seed (in some cases for many years, or even centuries) until conditions are right for germination.

The earliest fossil evidence of seed plants is found in late Devonian rocks. Like the progymnosperms, these now-extinct

seed ferns were woody. They possessed fernlike foliage but had seeds attached to their leaves. By the end of the Permian, other groups of seed plants had become dominant (**Figure 29.1**). Today's living seed plants fall into two major groups, the **gymnosperms** (such as pines and cycads) and the hugely diverse group known as the **angiosperms** (flowering plants).

Features of the seed plant life cycle protect gametes and embryos

In Section 28.2 we described a major trend in land plant evolution: the sporophyte became less dependent on the gametophyte, which became smaller in relation to the sporophyte. This trend continued with the seed plants, whose gametophyte generation is reduced even further than that of the ferns (**Figure 29.2**). The haploid seed plant gametophyte develops partly or entirely while attached to and nutritionally dependent on the diploid sporophyte.

Among the seed plants, only a few groups of gymnosperms (including modern cycads and ginkgos) have swimming sperm. Even in these groups, sperm is transferred via pollen grains, so fertilization does not require liquid water outside the plant body. This adaptation, along with the advent of the seed, gave seed plants the opportunity to colonize drier areas and spread over the terrestrial environment.

Seed plants are heterosporous (see Figure 28.18B)—that is, they produce two types of spores, one that becomes a

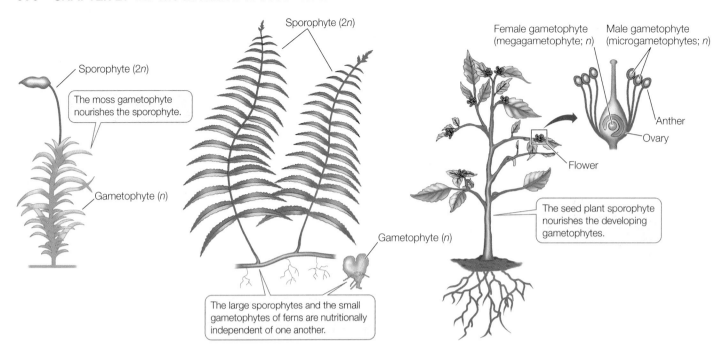

The moss gametophyte nourishes the sporophyte.

Sporophyte (2n)

Gametophyte (n)

Sporophyte (2n)

Gametophyte (n)

The large sporophytes and the small gametophytes of ferns are nutritionally independent of one another.

Female gametophyte (megagametophyte; n)

Male gametophyte (microgametophytes; n)

Anther

Ovary

Flower

The seed plant sporophyte nourishes the developing gametophytes.

29.2 The Relationship between Sporophyte and Gametophyte
In the course of plant evolution, the gametophyte (brown) has been reduced and the sporophyte (blue) has become more prominent.

microgametophyte (male gametophyte) and one that becomes a megagametophyte (female gametophyte). These plants form separate microsporangia and megasporangia on structures that are grouped on short stems.

Within the microsporangium, the meiotic products are microspores. Within its spore wall, a microspore divides mitotically one or a few times to form a multicellular male gametophyte called a **pollen grain**. Pollen grains are released from the microsporangium to be distributed by wind or by an animal pollinator (**Figure 29.3**). As in seedless land plants, the spore wall surrounding the pollen grain contains sporopollenin, the most chemically resistant biological compound known, which protects the pollen grain against dehydration and chemical damage—another advantage in terms of survival in the terrestrial environment.

In contrast to the microspores, the megaspores of seed plants are not shed. Instead, they develop into female gametophytes within the megasporangia. These megagametophytes are dependent on the sporophyte for food and water.

In most seed plant species, only one of the meiotic products in a megasporangium survives. The surviving haploid nucleus divides mitotically, and the resulting cells divide again to produce a multicellular female gametophyte. The megasporangium is surrounded by an **integument**: a layer of sporophytic tissue that protects the megasporangium and its contents. Together, the megasporangium and integument constitute the **ovule**, which will develop into a seed after fertilization.

The arrival of a pollen grain at an appropriate landing point, close to a female gametophyte on a sporophyte of the same species, is called **pollination**. A pollen grain that reaches this point produces a slender **pollen tube** that elongates and usually digests

its way toward the megagametophyte (**Figure 29.4**). When the tip of the pollen tube reaches the megagametophyte, sperm are released from the tube and fertilization occurs. The resulting diploid zygote divides repeatedly, forming an embryonic sporophyte. After a period of embryonic development, growth is temporarily suspended (the embryo enters a dormant stage). The end product at this stage is the multicellular seed.

29.3 Blown on the Wind Pollen grains are the male gametophytes of seed plants. The male flowers of this hazel tree release pollen-containing spores that are dispersed by the wind and may land near female gametophytes of the same or other hazel plants.

 Go to Media Clip 29.1
Pollen Transfer by Wind
Life10e.com/mc29.1

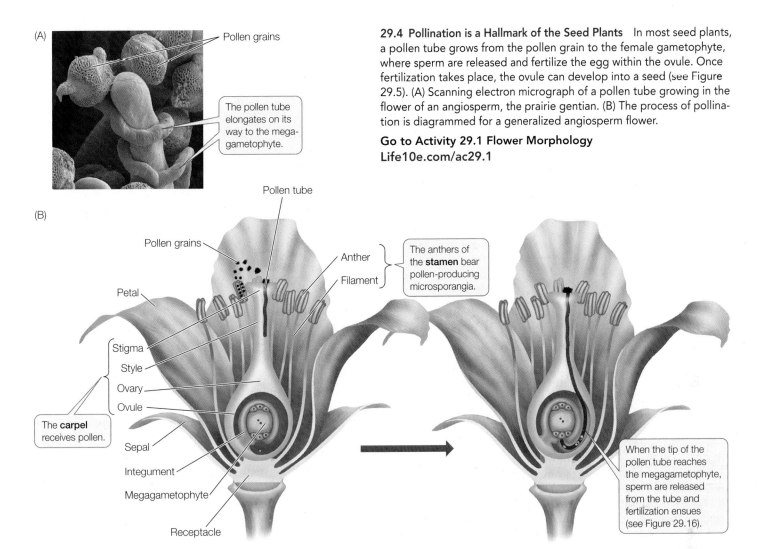

(A) Pollen grains

The pollen tube elongates on its way to the mega-gametophyte.

29.4 Pollination is a Hallmark of the Seed Plants In most seed plants, a pollen tube grows from the pollen grain to the female gametophyte, where sperm are released and fertilize the egg within the ovule. Once fertilization takes place, the ovule can develop into a seed (see Figure 29.5). (A) Scanning electron micrograph of a pollen tube growing in the flower of an angiosperm, the prairie gentian. (B) The process of pollination is diagrammed for a generalized angiosperm flower.

Go to Activity 29.1 Flower Morphology
Life10e.com/ac29.1

(B)

Pollen tube

Pollen grains

Anther

Filament

The anthers of the **stamen** bear pollen-producing microsporangia.

Petal

Stigma

Style

Ovary

Ovule

The **carpel** receives pollen.

Sepal

Integument

Megagametophyte

Receptacle

When the tip of the pollen tube reaches the megagametophyte, sperm are released from the tube and fertilization ensues (see Figure 29.16).

The seed is a complex, well-protected package

A seed contains tissues from three generations (**Figure 29.5**). A seed coat develops from the integument—the tissues of the diploid sporophyte parent that surround the megasporangium. Within the megasporangium is haploid tissue from the female gametophyte, which contains a supply of nutrients for the developing embryo. (This tissue is fairly extensive in most gymnosperm seeds. In angiosperm seeds it is greatly reduced, and nutrition for the embryo is supplied instead by a tissue called endosperm.) In the center of the seed is the third generation, the embryo of the new diploid sporophyte.

The seed is a well-protected resting stage. The seeds of some species may remain dormant but stay viable (capable of growth and development) for many years, germinating only when conditions are favorable for the growth of the sporophyte. During the dormant stage, the seed coat protects the embryo from excessive drying and may also protect it against potential predators that would otherwise consume the embryo and its nutrient reserves. Many seeds have structural adaptations that promote their dispersal by wind or, more often, by animals. When the

young sporophyte resumes growth, it draws on the food reserves in the seed. The possession of seeds is a major reason for the enormous evolutionary success of the seed plants, which are the dominant life forms of most modern terrestrial floras.

A change in stem anatomy enabled seed plants to grow to great heights

Fossils of the closest relatives of seed plants (progymnosperms) and the earliest seed plants (seed ferns) are found in late Devonian rocks (see Figure 29.1). These plants had thickened woody stems, developed through the proliferation of xylem. This type of growth, which increases the diameter of stems and roots in many modern seed plants, is called **secondary growth**. Its product is secondary xylem, or wood.

The younger portion of the wood produced by secondary growth is well adapted for water transport, but older wood becomes clogged with resins or other materials. Although no longer functional in transport, the older wood continues to provide support for the plant. This support allows woody plants to grow taller than other plants around them and thus capture more light for photosynthesis.

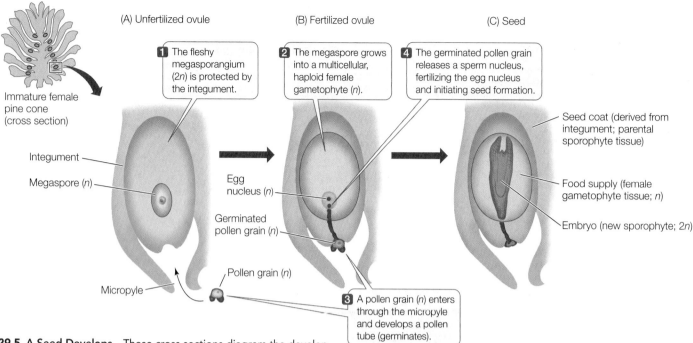

(A) Unfertilized ovule   (B) Fertilized ovule   (C) Seed

Immature female pine cone (cross section)

1 The fleshy megasporangium (*2n*) is protected by the integument.

2 The megaspore grows into a multicellular, haploid female gametophyte (*n*).

4 The germinated pollen grain releases a sperm nucleus, fertilizing the egg nucleus and initiating seed formation.

Integument

Megaspore (*n*)

Egg nucleus (*n*)

Germinated pollen grain (*n*)

Micropyle

Pollen grain (*n*)

3 A pollen grain (*n*) enters through the micropyle and develops a pollen tube (germinates).

Seed coat (derived from integument; parental sporophyte tissue)

Food supply (female gametophyte tissue; *n*)

Embryo (new sporophyte; *2n*)

29.5 A Seed Develops These cross sections diagram the development of the ovule into a seed in a gymnosperm (*Pinus* sp.). Angiosperm seed development has differences (e.g., angiosperm integuments have two layers rather than one, and the angiosperm embryo is nourished by specialized tissue called endosperm) but follows the same principle (compare Figures 29.8 and 29.16). (A) The haploid megaspore is nourished by tissues of the parental sporophyte (the diploid megasporangium). (B) The mature megaspore is fertilized by a pollen grain that penetrates the integument, germinates (grows a pollen tube; see Figure 29.4A), and releases a sperm nucleus. (C) Fertilization initiates production of a seed. A mature seed contains three generations: a diploid embryo (the new sporophyte), which is surrounded by haploid female gametophyte tissue that supplies nutrition, which is in turn surrounded by the seed coat (diploid parental sporophyte tissue).

Not all seed plants are woody. In the course of seed plant evolution, many groups lost the woody growth habit; however, other advantageous attributes helped them become established in an astonishing variety of places.

RECAP 29.1

Pollen grains, seeds, and wood are major evolutionary innovations of the seed plants. Protection of gametes and embryos is a hallmark of seed plants.

- Distinguish between the roles of the megagametophyte and the pollen grain. **See p. 590 and Figure 29.4**
- Explain the importance of pollen in freeing seed plants from dependence on liquid water. **See p. 590 and Figure 29.4**
- What are some of the advantages afforded by seeds? By wood? **See pp. 591–592 and Figure 29.5**

The seed ferns have long been extinct, but the surviving seed plants have been remarkable successes. After the seed ferns, the gymnosperms were the next group of plants to dominate terrestrial environments.

29.2 What Are the Major Groups of Gymnosperms?

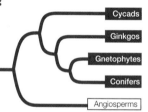

The gymnosperms are seed plants that do not form flowers or fruits. Gymnosperms (which means "naked-seeded") are so named because their ovules and seeds, unlike those of angiosperms, are not protected by ovary or fruit tissue. Although there are probably fewer than 1,200 living species of gymnosperms, these plants are second only to the angiosperms in their dominance of the terrestrial environment.

There are four major groups of living gymnosperms

The living gymnosperms can be divided into four major groups (see Figure 28.1B):

- **Cycads** are palmlike plants of the tropics and subtropics (**Figure 29.6A**). Of the present-day gymnosperms, the cycads are probably the earliest-diverging clade. There are about 300 species, some of which grow as tall as 20 meters. The tissues of many species are highly toxic to humans if ingested.

- **Ginkgos**, common during the Mesozoic era, are represented today by a single species, *Ginkgo biloba*, the maidenhair tree (**Figure 29.6B**). There are both male (microsporangiate) and female (megasporangiate) maidenhair trees. The difference is determined by X and Y sex chromosomes, as in humans; few other plants have distinct sex chromosomes.

(A) *Encephalartos sp.*

(B) *Ginkgo biloba*

(C) *Welwitschia mirabilis*

(D) *Pinus longaeva*

29.6 Diversity among the Gymnosperms
(A) Many cycads have growth forms that resemble both ferns and palms, although cycads are not closely related to either group. (B) The characteristic broad leaves of the maidenhair tree. (C) The straplike leaves of *Welwitschia*, a gnetophyte, grow throughout the life of the plant, breaking and splitting as they grow. (D) Conifers dominate many types of landscapes in the Northern Hemisphere. Bristlecone pines such as these are the longest-lived individual trees known.

- **Gnetophytes** number about 90 species in three very different genera, which share certain characteristics analogous to ones found in the angiosperms. One of the gnetophytes is *Welwitschia* (**Figure 29.6C**), a long-lived desert plant with straplike leaves that sprawl on the sand and can grow as long as 3 meters.

- **Conifers** are by far the most abundant of the gymnosperms. There are about 700 species of these cone-bearing plants, including the pines and redwoods (**Figure 29.6D**).

With the exception of the gnetophytes, the living gymnosperm groups have only tracheids as water-conducting and support cells within the xylem; they lack the vessel elements and fibers (cells specialized for water conduction and support, respectively) that are found in angiosperms. While the gymnosperm water-transport and support system may seem somewhat less efficient than that of the angiosperms, it serves some of the largest trees known. The coastal redwoods of California are the tallest gymnosperms; the largest are well over 100 meters tall.

During the Permian, as environments became warmer and dryer, the conifers and cycads flourished. Gymnosperm forests changed over time as the gymnosperm groups evolved. Gymnosperms dominated most of the Mesozoic era, during which the continents drifted apart and large dinosaurs lived. Gymnosperms were the principal trees in all forests until about 90 million years ago, and even today conifers are the dominant trees in many forests. The oldest living single organism on Earth today is a gymnosperm in California—a bristlecone pine that germinated some 4,800 years ago, at about the time the ancient Egyptians were starting to develop writing.

Conifers have cones and no swimming sperm

The great Douglas fir and cedar forests found in northwestern North America, the massive boreal forests of pine, fir, and spruce of the northern regions of the Northern Hemisphere, and the alpine forests on the upper slopes of mountain ranges everywhere rank among the great forests of the world. All these forests are dominated by trees belong to one group of gymnosperms: the conifers, or cone-bearers.

Male and female **cones** contain the reproductive structures of conifers. The female (seed-bearing) cone is known as a

(A) *Pinus contorta*

Woody scale

Seed

Central axis

Cross section of a megastrobilus

Female cones, or megastrobili

(B) *Pinus contorta*

Herbaceous scale

Pollen-containing microspores

Central axis

Cross section of a microstrobilus

Male cones, or microstrobili

29.7 Female and Male Cones (A) The woody scales of female cones (megastrobili) are modified branches. (B) The herbaceous scales of male cones (microstrobili) are modified leaves.

megastrobilus (plural *megastrobili*); an example is the familiar woody pine cone. The seeds in a megastrobilus are protected by a tight cluster of woody scales, which are modifications of branches extending from a central axis (**Figure 29.7A**). The typically much smaller male (pollen-bearing) cone is known as a **microstrobilus**. The microstrobilus is typically herbaceous rather than woody, as its scales are composed of modified leaves, beneath which are the pollen-bearing microsporangia (**Figure 29.7B**).

The life cycle of a pine illustrates reproduction in gymnosperms (**Figure 29.8**). The production of male gametophytes in the form of pollen grains frees the plant completely from dependence on liquid water for fertilization. Wind assists conifer pollen grains in their travel from the microstrobilus to near the female gametophyte inside a megastrobilus. A pollen tube provides the sperm with the means for the last stage of travel by growing through maternal sporophytic tissue. When the pollen tube reaches the female gametophyte, it releases two sperm, one of which degenerates after the other unites with an egg. Union of sperm and egg results in a zygote; mitotic divisions and further development of the zygote result in an embryo.

The megasporangium, in which the female gametophyte will form, is enclosed in a layer of sporophytic tissue—the integument—that will eventually develop into the seed coat that protects the embryo (see Figure 29.5). The integument, the megasporangium inside it, and the tissue attaching it to the maternal sporophyte constitute the ovule. The pollen grain enters the ovule through a small opening in the integument at the tip of the ovule, the **micropyle**.

Most conifer ovules, which will develop into seeds after fertilization, are borne exposed on the upper surfaces of the scales of the megastrobilus. The only protection of the ovules comes from the scales, which are tightly pressed against one another within the cone. Some pines, such as the lodgepole pine, have such tightly closed cones that only fire suffices to split them open and release the seeds. These species are said to be fire-adapted, and fire is essential to their reproduction. A fire devastated lodgepole pine forests in Yellowstone National Park in 1988, but also released large numbers of seeds from cones. As a result, large numbers of lodgepole pine seedlings are now emerging in the burn area (**Figure 29.9**).

About half of all conifer species have soft, fleshy tissues that envelop their seeds. Some of these are fleshy, fruitlike cones, as in junipers. Others are fruitlike extensions of the seeds, called arils, as in yews. These tissues, although often mistaken for "berries," are not true fruits. As we will see in the next section, true fruits are the plant's ripened ovaries, which are absent in gymnosperms. Nonetheless, the fleshy tissues that surround many conifers serve a similar purpose to that of the fruits of flowering plants, acting as an enticement for seed-dispersing animals. Animals eat these fleshy tissues and disperse the seeds in their feces, often depositing the seeds considerable distances away from the parent plant.

Go to Animated Tutorial 29.1
Life Cycle of a Conifer
Life10e.com/at29.1

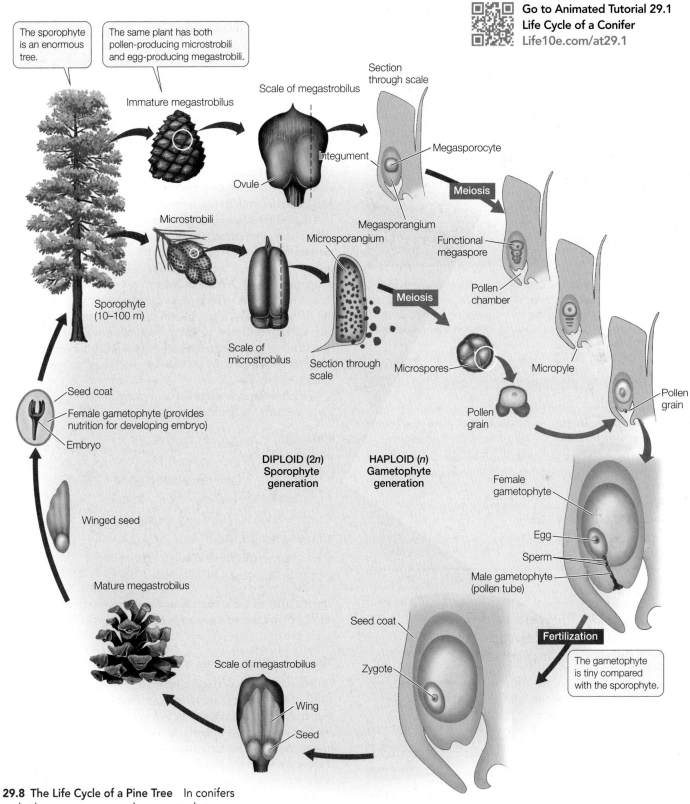

The sporophyte is an enormous tree.

The same plant has both pollen-producing microstrobili and egg-producing megastrobili.

Immature megastrobilus

Scale of megastrobilus

Section through scale

Integument

Ovule

Megasporocyte

Megasporangium

Meiosis

Microstrobili

Microsporangium

Functional megaspore

Pollen chamber

Sporophyte (10–100 m)

Scale of microstrobilus

Section through scale

Meiosis

Microspores

Micropyle

Pollen grain

Pollen grain

Seed coat

Female gametophyte (provides nutrition for developing embryo)

Embryo

DIPLOID (2n) Sporophyte generation

HAPLOID (n) Gametophyte generation

Pollen grain

Female gametophyte

Egg

Sperm

Male gametophyte (pollen tube)

Winged seed

Mature megastrobilus

Scale of megastrobilus

Seed coat

Zygote

Wing

Seed

Fertilization

The gametophyte is tiny compared with the sporophyte.

29.8 The Life Cycle of a Pine Tree In conifers and other gymnosperms, the gametophytes are small and nutritionally dependent on the sporophyte generation.

Go to Activity 29.2 Life Cycle of a Conifer
Life10e.com/ac29.2

29.9 From Devastation, New Life A stand of lodgepole pines in Yellowstone National Park. The mature trees were destroyed by a forest fire in 1988. However, the fire released large numbers of seeds from cones, and now many young lodgepole pine trees are growing in the burn area.

 RECAP **29.2**

Living gymnosperms can be divided into four major groups: cycads, ginkgos, gnetophytes, and conifers. All of these plants are woody and have seeds that are not protected by ovaries.

- Explain the differences in structure and function between a megastrobilus and a microstrobilus. **See p. 594 and Figures 29.7 and 29.8**

- What is the role of the integument in a gymnosperm seed? **See p. 594 and Figure 29.8**

- Explain how fire can be necessary for the survival of some plant species. **See p. 594**

29.3 How Do Flowers and Fruits Increase the Reproductive Success of Angiosperms?

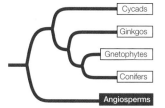

The most obvious feature defining the angiosperms is the **flower**, which is their sexual structure. Production of **fruits** is also a synapomorphy (shared derived trait) of angiosperms. After fertilization, the ovary of a flower (together with the seeds it contains) develops into a fruit that protects the seeds and can promote seed dispersal. As we will see, both flowers and fruits give angiosperms major reproductive advantages.

Angiosperms have many shared derived traits

The name *angiosperm* ("enclosed seed") is drawn from another distinctive synapomorphy of these plants that is related to

the formation of fruits: the ovules and seeds are enclosed in a modified leaf called a **carpel**. Besides protecting the ovules and seeds, the carpel often interacts with incoming pollen to prevent self-pollination, thus favoring cross-pollination and increasing genetic diversity.

The female gametophyte of the angiosperms is even more reduced than that of the gymnosperms, usually consisting of only seven cells (see Figure 29.16). Thus the angiosperms represent the current extreme of the trend we have traced throughout the evolution of the vascular plants: the sporophyte generation becomes larger and more independent of the gametophyte, while the gametophyte generation becomes smaller and more dependent on the sporophyte.

The xylem of most angiosperms is distinguished by the presence of specialized water-transporting cells called **vessel elements**. These cells are larger in diameter than tracheids and connect with one another without obstruction, allowing easy water movement. A second distinctive cell type in angiosperm xylem is the **fiber**, which plays an important role in supporting the plant body. The phloem of angiosperms possesses its own unique cell type, called a companion cell. Like the gymnosperms, woody angiosperms exhibit secondary growth, increasing in diameter by producing secondary xylem and secondary phloem.

A more comprehensive list of angiosperm synapomorphies, then, includes the following:

- Flowers
- Fruits
- Highly reduced female gametophytes
- Ovules and seeds enclosed in a carpel
- Germination of pollen on a stigma
- Double fertilization
- Endosperm (nutritive tissue for the embryo)
- Phloem with companion cells

The majority of these traits bear directly on angiosperm reproduction, which is a large factor in the success of this dominant plant group.

The sexual structures of angiosperms are flowers

Flowers come in an astonishing variety of forms—just think of a few of the flowers you recognize. Flowers may be single, or they may be grouped together to form an **inflorescence**. Different families of flowering plants have characteristic types of inflorescences, such as the compound umbels of the carrot family (**Figure 29.10A**), the heads of the aster family (**Figure 29.10B**), and the spikes of many grasses (**Figure 29.10C**).

If you examine any familiar flower, you will notice that the outer parts look somewhat like leaves. In fact, all the parts of a flower *are* modified leaves. The diagram in Figure 29.4B represents a generalized flower, for which there is no exact counterpart in nature. The structures bearing microsporangia are called **stamens**. Each stamen is composed of a **filament** bearing an **anther** that contains the pollen-producing microsporangia. The structures bearing megasporangia are called carpels. A structure

(A) *Aegopodium podagraria*

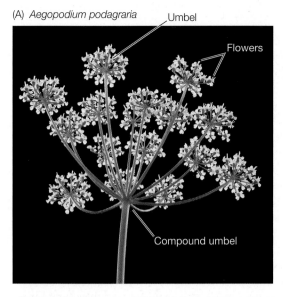

Umbel

Flowers

Compound umbel

(B) *Zinnia elegans*

Ray flowers   Disc flowers

(C) *Agropyron repens*

Spikes

Spikelets

Anthers of spikelets

29.10 Inflorescences (A) The inflorescence of bishop's gout-weed, a member of the carrot family, is a compound umbel. Each umbel bears flowers on stalks that arise from a common center. (B) Zinnias are members of the aster family; their inflorescence is a head. Within the head, each of the long, petal-like structures is a ray flower; the central portion of the head consists of dozens to hundreds of disc flowers. (C) Some grasses, such as quack grass, have inflorescences called spikes, which are composed of many smaller groups of flowers, or spikelets.

composed of one carpel or two or more fused carpels is called a **pistil**. The swollen base of the pistil, containing one or more ovules (each containing a megasporangium surrounded by two protective integuments), is called the **ovary**. The apical stalk of the pistil is the **style**, and the terminal surface that receives pollen grains is the **stigma**.

In addition, many flowers contain specialized sterile (non-spore-bearing) leaves. The inner ones are called **petals** (collectively, the corolla) and the outer ones **sepals** (collectively, the calyx). The corolla and calyx (collectively, the perianth) can be quite showy and often play roles in attracting animal pollinators to the flower. The calyx more commonly protects the immature flower in bud. From base to apex, these floral organs—sepals, petals, stamens, and carpels—are usually positioned in circular arrangements or whorls and attached to a central stalk.

The generalized flower in Figure 29.4B has both functional megasporangia and functional microsporangia; such flowers are referred to as **perfect** (or hermaphroditic). Many angiosperms produce two types of flowers, one with only megasporangia and the other with only microsporangia. Consequently, either the stamens or the carpels are nonfunctional or absent in a given flower, and the flower is referred to as **imperfect** (see Figure 38.1).

Species such as corn or birch, in which both megasporangiate (female) and microsporangiate (male) flowers occur on the same plant, are said to be **monoecious** ("one-housed"—but, it must be added, one house with

separate rooms). Complete separation of imperfect flowers occurs in some other angiosperm species, such as willows and date palms; in these species, an individual plant produces either flowers with stamens or flowers with carpels, but never both. Such species are said to be **dioecious** ("two-housed").

Flower structure has evolved over time

The flowers of the earliest-diverging clades of angiosperms have a large and variable number of tepals (undifferentiated sepals and petals), carpels, and stamens (**Figure 29.11A**). Evolutionary change within the angiosperms has included some striking modifications of this early condition: reductions in the number of each type of floral organ to a fixed number, differentiation of petals from sepals, and changes in symmetry from radial (as in a lily or magnolia) to bilateral (as in the orchid shown at the opening of this chapter), often accompanied by an extensive fusion of parts (**Figure 29.11B**).

(A) *Nymphaea* sp.

(B) *Viola tricolor*

29.11 Flower Form and Evolution (A) A water lily shows the major features of early flowers: it is radially symmetrical, and the individual tepals, stamens, and carpels are separate, numerous, and attached at their bases. (B) Violets such as this "Johnny jump-up" have a bilaterally symmetrical structure that evolved much later than radial flower symmetry.

(A) Carpel evolution

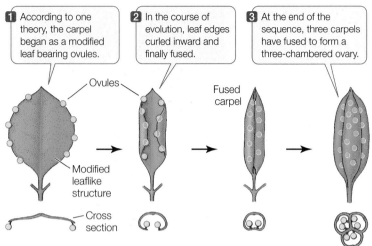

1 According to one theory, the carpel began as a modified leaf bearing ovules.

2 In the course of evolution, leaf edges curled inward and finally fused.

3 At the end of the sequence, three carpels have fused to form a three-chambered ovary.

Ovules

Fused carpel

Modified leaflike structure

Cross section

(B) Stamen evolution

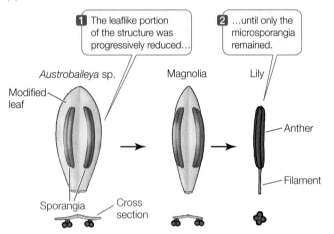

1 The leaflike portion of the structure was progressively reduced...

2 ...until only the microsporangia remained.

Austrobaileya sp.

Magnolia

Lily

Modified leaf

Anther

Filament

Sporangia

Cross section

29.12 Carpels and Stamens Evolved from Leaflike Structures (A) Possible stages in the evolution of a carpel from a more leaflike structure. (B) The stamens of three modern plants show three possible stages in the evolution of that organ. (It is *not* implied that these species evolved from one another; their structures simply illustrate the possible stages.)

According to one hypothesis, the first carpels to evolve were leaves with marginal ovules, folded but incompletely closed. Early in angiosperm evolution, carpels fused with one another, forming a single, multichambered ovary (**Figure 29.12A**). In some flowers, the other floral organs are attached at the top of the ovary rather than at the bottom as in Figure 29.4B. The stamens of the most ancient flowers may have been leaflike (**Figure 29.12B**), with little resemblance to the stamens of the generalized flower seen in Figure 29.4B.

Why do so many flowers have pistils with long styles and anthers with long filaments? Natural selection has favored length in both of these floral organs, probably because length increases the likelihood of successful pollination. Long filaments may bring the anthers into contact with insect bodies, or they may place the anthers in a better position to catch the wind. Similar arguments apply to long styles.

A perfect flower represents a compromise of sorts. On the one hand, by attracting a pollinating bird or insect, the plant is attending to both its female and male functions with a single flower type, whereas plants with imperfect flowers must create that attraction twice—once for each type of flower. On the other hand, the perfect flower can favor self-pollination, which is usually disadvantageous. Another potential problem is that the female and male functions might interfere with each other—for example, the stigma might be placed so as to make it difficult for pollinators to reach the anthers, thus reducing the export of pollen to other flowers.

Might there be a way around these problems? One solution is seen in the bush monkeyflower (*Mimulus aurantiacus*), which is pollinated by hummingbirds. Its flower has a stigma that initially serves as a screen, hiding the anthers (**Figure 29.13**). Once a hummingbird touches the stigma, however, one of the stigma's two lobes is retracted, so that subsequent hummingbird visitors pick up pollen from the previously screened anthers. Thus the first bird to visit the flower transfers pollen from another plant to the stigma. Later visitors pick up pollen from the now-accessible anthers, fulfilling the flower's male function. **Figure 29.14** describes the experiment that revealed the function of this mechanism.

Angiosperms have coevolved with animals

Whereas most gymnosperms are pollinated by wind, most angiosperms are pollinated by animals. The many different

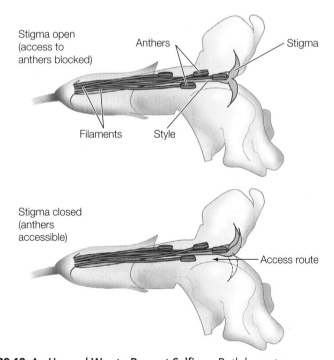

Stigma open (access to anthers blocked)

Anthers

Stigma

Filaments

Style

Stigma closed (anthers accessible)

Access route

29.13 An Unusual Way to Prevent Selfing Both long stamens and long styles facilitate cross-pollination, but if these male and female structures are too close to each other, the likelihood of (disadvantageous) self-pollination increases. In *Mimulus aurantiacus*, the stigma is initially open, blocking access to the anthers. A hummingbird's touch as it deposits pollen on the stigma causes one lobe of the stigma to retract, creating a path to the anthers and allowing pollen dispersal by subsequent hummingbird visitors.

29.14 The Effect of Stigma Retraction in Monkeyflowers

Elizabeth Fetscher's experiments showed that the unusual stigma retraction response to pollination in monkeyflowers (illustrated in Figure 29.13) enhances the dispersal of pollen to other flowers.[a]

HYPOTHESIS The stigma-retraction response in *M. aurantiacus* increases the likelihood that an individual flower's pollen will be exported to another flower once pollen from another flower has been deposited on its stigma.

Method

1. Set up three groups of monkeyflower arrays. Each array consists of one pollen-donor flower and multiple pollen-recipient flowers (with the anthers removed to prevent pollen donation).
2. In control arrays, the stigma of the pollen donor is allowed to function normally.
3. In one group of experimental arrays, the stigma of the pollen donor is permanently propped open (blocking access to the anthers).
4. In a second group of experimental arrays, the stigma of the pollen donor is artificially sealed closed (allowing access to anthers).
5. Allow hummingbirds to visit the arrays, then count the pollen grains transferred from each donor flower to the recipient flowers in the same array.

Results

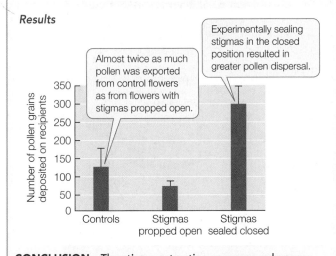

> Almost twice as much pollen was exported from control flowers as from flowers with stigmas propped open.

> Experimentally sealing stigmas in the closed position resulted in greater pollen dispersal.

Number of pollen grains deposited on recipients

Controls — Stigmas propped open — Stigmas sealed closed

CONCLUSION The stigma-retraction response enhances the male function of the flower (dispersal of pollen) once the female function (receipt of pollen) has been performed.

Go to **BioPortal** for discussion and relevant links for all INVESTIGATING**LIFE** figures.

[a]Fetscher, A. E. 2001. *Proceedings of the Royal Society B* 268: 525–529.

mutualistic pollination relationships between plants and animals are vital to both parties. We mentioned coevolution between insects and orchids at the opening of this chapter, and we'll revisit plant–pollinator coevolution in more detail when we discuss mutualisms in Chapter 56, but we'll consider a few important aspects of the plant–pollinator relationship here.

Many flowers entice animals to visit them by providing food rewards. Pollen grains themselves sometimes serve as food for animals. In addition, some flowers produce a sugary fluid called nectar, and some of these flowers have specialized structures to store and distribute it, as we saw at the opening of this chapter. In the process of visiting flowers to obtain nectar or pollen, animals often carry pollen from one flower to another or from one plant to another. Thus, in their quest for food, the animals contribute

Rubeckia fulgida

29.15 See Like a Bee To normal human vision (above), the petals of a black-eyed Susan appear solid yellow. Ultraviolet photography reveals patterns that attract bees to the central region, where pollen and nectar are located.

to the genetic diversity of the plant population. Insects, especially bees, are among the most important pollinators; birds and some species of bats are also major pollinators.

Go to Media Clip 29.2
Pollen Transfer by a Bat
Life10e.com/mc29.2

For more than 150 million years, angiosperms and their animal pollinators have coevolved in the terrestrial environment. Animals have affected the evolution of the plants, and plants have affected the evolution of the animals. Flower structure has become incredibly diverse under these selection pressures. Some of the products of coevolution are highly specific; for example, the flowers of some yucca species are pollinated by only one species of yucca moth, and that moth may exclusively pollinate just one species of yucca (see Figure 56.12). Such specific relationships provide plants with a reliable mechanism for transferring pollen only to members of their own species.

Most plant–pollinator interactions are much less specific; that is, many different animal species pollinate the same plant species, and the same animal species pollinates many different plant species. However, even these less specific interactions have developed some specialization (see Table 56.1). For example, many bird-pollinated flowers are red and odorless, whereas many insect-pollinated flowers have characteristic odors. Some bee-pollinated flowers have markings, called nectar guides, that are conspicuous only to animals, such as bees, that can see colors in the ultraviolet region of the spectrum (**Figure 29.15**).

29.16 The Life Cycle of an Angiosperm Double fertilization results in triploid endosperm in most species of angiosperms. One sperm nucleus fertilizes the egg to form the zygote, while the other combines with the two polar nuclei to form the endosperm.

Go to Animated Tutorial 29.2
Life Cycle of an Angiosperm
Life10e.com/at29.2

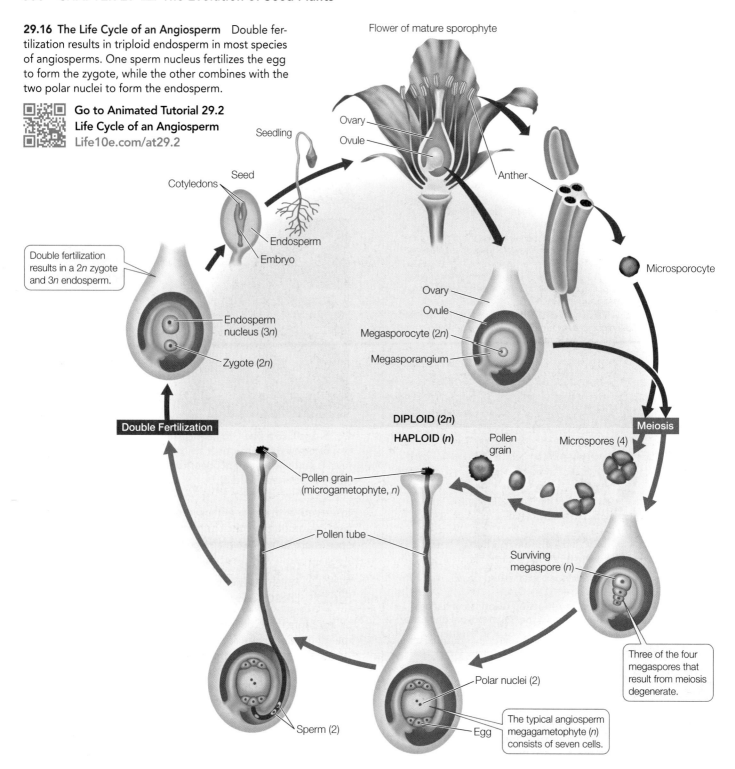

The angiosperm life cycle produces diploid zygotes nourished by triploid endosperms

Like all seed plants, angiosperms are heterosporous. As we have seen, their ovules are contained within carpels rather than being exposed on the surfaces of scales, as in most gymnosperms. The male gametophytes, as in the gymnosperms, are pollen grains.

Pollination in the angiosperms consists of the arrival of a microgametophyte—a pollen grain—on the receptive surface of a flower (the stigma). As in the gymnosperms, pollination is the first in a series of events that results in the formation of a seed. The next event is the growth of a pollen tube extending to the megagametophyte. The third event is a fertilization process that, in detail, is unique to the angiosperms (**Figure 29.16**).

In angiosperms, *two* male gametes, contained in a single microgametophyte, participate in fertilization. The nucleus of one sperm combines with that of the egg to produce a diploid zygote—the first cell of the sporophyte generation. In

29.17 Fruits Come in Many Forms (A) The single seeds inside the simple fruits of peaches are dispersed by animals. (B) Each seed of the horse chestnut is covered by a hard, woody fruit that allows it to survive drought. Although such fruits are commonly called "nuts," this is a culinary rather than a biological term. (C) The highly reduced simple fruits of dandelions are dispersed by wind. (D) A multiple fruit, the jackfruit (*Artocarpus heterophyllus*) of tropical Asia, is the largest tree-borne fruit in the world. (E) An aggregate fruit (blackberry). (F) An accessory fruit (pear).

most angiosperms, the other sperm nucleus combines with two other haploid nuclei of the female gametophyte to form a cell with a *triploid* (3*n*) nucleus. That cell in turn gives rise to a triploid tissue called the **endosperm**, which nourishes the embryonic sporophyte during its early development. This process, in which two fertilization events take place, is known as **double fertilization**. In some angiosperms, additional haploid nuclei are incorporated to form even higher ploidy levels in the endosperm, or the second sperm fuses with only one haploid nucleus, resulting in diploid endosperm.

The angiosperm zygote develops into an embryo, which consists of an embryonic axis that will become a stem and a root and one or two **cotyledons**, or "seed leaves." The cotyledons have different fates in different plants. In many, they serve as absorptive organs that take up and digest the endosperm. In others, they enlarge and become photosynthetic when the seed germinates. Often they play both roles.

The ovule develops into a seed containing the products of the double fertilization that characterizes angiosperms: the diploid zygote and a triploid endosperm (see Figure 29.16).

Fruits aid angiosperm seed dispersal

As mentioned at the start of this section, the production of seed-protecting fruits is an important synapomorphy. Fruits may attach to or be eaten by an animal. Fruits are not necessarily fleshy; they can be hard and woody, or small with modified structures that allow them to be dispersed by wind or water (**Figure 29.17**).

A fruit may consist of only the mature ovary and its seeds, or it may include other parts of the flower or structures associated with it. A **simple fruit** is one that develops from a single carpel or several fused carpels, such as a plum or peach. A raspberry is an example of an **aggregate fruit**—one that develops from several separate carpels of a single flower. Pineapples and figs are examples of **multiple fruits**, formed from a cluster of flowers (an inflorescence). Fruits derived from parts in addition to the carpel and seeds are called **accessory fruits**; examples are apples, pears, and strawberries.

 Go to Media Clip 29.3
Flower and Fruit Formation
Life10e.com/mc29.3

Recent analyses have revealed the phylogenetic relationships of angiosperms

Figure 29.18 shows the relationships among the major angiosperm clades. The two largest clades—the **monocots** and the **eudicots**—include the great majority of angiosperm species. The monocots are so called because they have a single

29.18 Evolutionary Relationships among the Angiosperms Recent analyses of many angiosperm genes have clarified the relationships among the major groups.

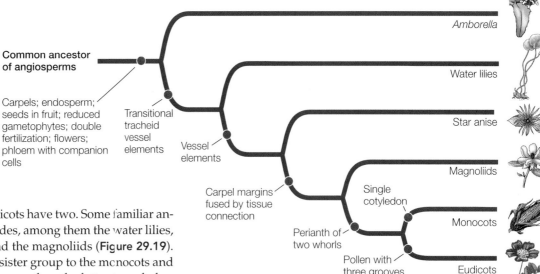

Common ancestor of angiosperms

Carpels; endosperm; seeds in fruit; reduced gametophytes; double fertilization; flowers; phloem with companion cells

Transitional tracheid vessel elements

Vessel elements

Carpel margins fused by tissue connection

Single cotyledon

Perianth of two whorls

Pollen with three grooves

Amborella

Water lilies

Star anise

Magnoliids

Monocots

Eudicots

embryonic cotyledon; the eudicots have two. Some familiar angiosperms belong to other clades, among them the water lilies, star anise and its relatives, and the magnoliids (**Figure 29.19**). The **magnoliids** are the likely sister group to the monocots and eudicots. Although less numerous than the latter two clades, the magnoliids include many familiar and useful plants, such as avocados, cinnamon, black pepper, and magnolias.

The root of the evolutionary tree of flowering plants was once a matter of great controversy. A fundamental challenge was identifying the group that is sister to the remaining angiosperms, and members of the magnoliid clade were leading candidates for the position. At the close of the twentieth century,

however, an impressive convergence of molecular and morphological evidence led to the conclusion that the sister group to the remaining flowering plants is the single species of the genus *Amborella* (see Figure 29.19A). This woody shrub, with cream-colored flowers, lives only on New Caledonia, an island in the South Pacific. Its 5 to 8 carpels have a spiral arrangement,

(A) *Amborella trichopoda*

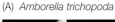

Sterile stamens

(B) *Victoria amazonica*

(C) *Illicium floridanum*

(D) *Magnolia* sp.

(E) *Aristolochia littoralis*

29.19 Monocots and Eudicots Are Not the Only Surviving Angiosperms (A) *Amborella*, a shrub, is sister to the remaining extant angiosperms. Notice the sterile stamens on this female flower, which may serve to lure insects that are searching for pollen. (B) The water lily clade was the next to diverge after *Amborella*. (C) Star

anise and its relatives belong to another early-diverging angiosperm clade. (D, E) The largest clade other than the monocots and eudicots is the magnoliid complex, which includes magnolias and the group known as "Dutchman's pipe."

(A)

(B) *Saccharum* sp.

(C) *Posidonia oceanica*

(D) *Phoenix dactylifera*

29.20 Monocots (A) Monocots include many popular garden flowers such as these hyacinths (*Muscari armeniacum*; blue), tulips (*Tulipa* sp.; red) and daffodils (*Narcissus* sp.; yellow). (B) Monocot grasses such as this sugarcane feed the world; wheat, rice, and maize (corn) are also grasses. (C) Seagrasses such as this Neptune's grass form "meadows" in the shallow, sunlit waters of the world's oceans. (D) Palms are among the few monocot trees. Date palms like these are a major food source in some areas of the world.

and it has 30 to 100 stamens. The xylem of *Amborella* lacks vessel elements, which evolved after this deepest split in the angiosperm evolutionary tree.

Representatives of the two largest angiosperm clades are everywhere. The monocots (**Figure 29.20**) include grasses, cattails, lilies, orchids, and palms. The eudicots (**Figure 29.21**) include the vast majority of familiar seed plants, including most herbs (i.e., nonwoody plants), vines, trees, and shrubs. Among the eudicots are such diverse plants as oaks, willows, beans, snapdragons, roses, and sunflowers.

(A) *Delonix regia*

(B) *Passiflora caerulea*

(C) *Echinocereus reichenbachii*

(D) *Rafflesia arnoldii*

Filaments of corona

29.21 Eudicots (A) The royal poinciana, or "flame tree," is native to Madagascar. Humans have introduced this ornamental flowering tree throughout the world's tropical regions. (B) Passionflower vines are found throughout the tropics and subtropics. Their flowers are distinguished by elaborate coronas (the purple filaments on this flower). (C) Cacti comprise a large group of eudicots, with about 1,500 species in the Americas. Many, such as this black lace cactus, bear large flowers for a brief period of each year. (D) *Rafflesia arnoldii*, found in the rainforests of Indonesia, bears the largest flower in the world. The flower lives as a parasite on tropical vines and has lost its leaf, stem, and even root structures. It smells like decaying meat, which attracts its fly pollinators.

RECAP 29.3

The synapomorphies of angiosperms include flowers, fruit, carpels, double fertilization, and endosperm. Most angiosperms also possess distinctive cells in the xylem and phloem. The largest angiosperm clades are the monocots and the eudicots.

- Explain the difference between pollination and fertilization. **See pp. 598–601**
- What are the respective roles of the two sperm in double fertilization in angiosperms? **See pp. 600–601 and Figure 29.16**
- What are the different functions of flowers, fruits, and seeds?

The remarkable diversity of the seed plants has been shaped by both biotic and abiotic components of the environments to which they have adapted. In turn, land plants—and seed plants in particular—shape their environments.

29.4 How Do Plants Benefit Human Society?

Plants make profound contributions to ecosystem services—processes by which the environment maintains resources that benefit human society. Plants produce oxygen and remove carbon dioxide from the atmosphere, and they play important roles in forming soils and renewing soil fertility. Plant roots help hold soil in place, providing a defense against erosion by wind and water (**Figure 29.22**). They also moderate local climate in various ways, such as by increasing humidity, providing shade, and blocking wind.

Seed plants have been sources of medicine since ancient times

One of the oldest human professions is that of medicine man (shaman) or "wise woman"—a person who cures others with medicines, most of which are derived from plants. It is said that in 2700 BCE, a legendary Chinese emperor understood the use of some 365 medicinal plants. Although we also use medicines derived from fungi, lichens, and actinobacteria, seed plants are the source of many of our medications; a few examples are shown in **Table 29.1**. Even in synthetic pharmaceuticals, the chemical structures of the active ingredients are often based on the biochemistry of substances isolated from plants.

How are plant-based medicines discovered? These days, many are found by systematic testing of tremendous numbers of plants from all over the world, a process that began in the 1960s. One

TABLE 29.1
Some Medicinal Plants and Their Products

Product	Plant Source	Medical Application
Atropine	Belladonna	Dilate pupils for eye examination
Bromelain	Pineapple stem	Control tissue inflammation
Digitalin	Foxglove	Strengthen heart muscle contraction
Ephedrine	*Ephedra*	Ease nasal congestion
Menthol	Japanese mint	Relieve coughing
Morphine	Opium poppy	Relieve pain
Quinine	*Cinchona* bark	Treat malaria
Taxol	Pacific yew	Treat ovarian and breast cancers
Tubocurarine	Curare plant	Muscle relaxant (used in surgery)
Vincristine	Periwinkle	Treat leukemia and lymphoma

example of a medicine discovered in this way is taxol, an important anticancer drug. Among the myriad plant samples that had been tested by 1962, extracts of the bark of the Pacific yew (*Taxus brevifolia*) showed anti-tumor activity in tests against rodent tumors. The active ingredient, taxol, was isolated in 1971 and tested against human cancers in 1977. After another 16 years, the U.S. Food and Drug Administration approved it for human use, and taxol is now widely used in treating breast and ovarian cancers as well as several other types of cancers.

Widespread screening of plant samples was eventually deemphasized in favor of a purely chemical approach. Using automation and miniaturization, pharmaceutical laboratories generate vast numbers of compounds, which are then screened just as plant materials had been screened in the search for taxol. Today,

29.22 Plants Prevent Erosion When forest vegetation was cleared on these hillsides in Malaysia, landslides and extensive soil erosion quickly followed. Adjacent forested areas did not have landslides.

however, the screening of plant materials is getting renewed interest. Both screening approaches are based on trial and error.

The other leading source of plant-based medicines is work by ethnobotanists, who study how people use and view plants in their local environments. This work proceeds all over the globe today. An older example of this approach is the discovery of quinine as a treatment for malaria. In the sixteenth century, Spanish priests in Peru became aware that the native population used the bark of local *Cinchona* trees to treat fevers. The priests successfully used the bark to treat malaria. Word of the medicine spread to Europe, where it is said to have been in use in Rome by 1631. The active ingredient of *Cinchona* bark—quinine—was identified in 1820, and quinine remained the standard malarial remedy well into the twentieth century.

Seed plants are our primary food source

Plants are primary producers; that is, they trap energy and carbon by means of photosynthesis, making those resources available not only for their own needs, but also for the herbivores and omnivores that consume them, for the carnivores and omnivores that eat the herbivores, and for the prokaryotes and fungi that complete food webs. The earliest steps in human civilization involved cultivating angiosperms to provide a reliable food supply.

Today, twelve species of seed plants stand between the human race and starvation: rice, coconut, wheat, corn (maize), potato, sweet potato, cassava (also called tapioca or manioc), sugarcane, sugar beet, soybean, common bean (*Phaseolus vulgaris*), and banana. Hundreds of other seed plants are cultivated for food,

29.23 Rice Feeds Much of the World's Human Population These rice fields, or "paddies," are on the island of Luzon in the Philippines. Rice has been cultivated in this manner for thousands of years.

but none rank with these twelve in importance. Indeed, more than half of the world's human population derives the bulk of its food energy from the seeds of a single plant, *Oryza sativa*, better known as rice. Rice is particularly important in eastern Asia, where it has been cultivated for more than 8,000 years (**Figure 29.23**). People also use rice straw in many ways, such as thatching for roofs, food and bedding for livestock, and clothing. Even rice hulls have many uses, ranging from fertilizer to fuel.

What was Darwin's explanation for the three distinct flowers growing on a single orchid plant?

ANSWER

After obtaining specimens of the plant in question and dissecting the flowers, Darwin was able to demonstrate that the orchid was a single species (*Catasetum macrocarpum*) that bore three distinct types of flowers: megasporangiate (female), microsporangiate (male), and perfect (hermaphroditic). The three types of flowers were remarkable in their morphological differences, which were great enough to have misled botanists into describing the different flower types as species in different genera. Most individual plants were either male (specimens identified as *Catasetum*) or female (specimens identified as *Monachanthus*), but some individuals that bore predominately male or female flowers also produced perfect flowers (specimens identified as *Myanthus*).

The case of *C. macrocarpum* demonstrates that some plants blur the lines between the categories we have discussed in this chapter. Most flowers on a *C. macrocarpum* individual are either male or female, but some plants can bear some perfect flowers as well.

Why do the male and female flowers of *C. macrocarpum* look so different? Part of the explanation is their different roles in pollination. Recall Darwin's observation of a *Catasetum* flower shooting a packet of pollen at an insect that landed on its flower. The pollinia (pollen packets) and associated structures in male flowers of *Catasetum* are coiled like springs and are

released suddenly when disturbed by an insect. This release forcefully propels the pollinia precisely into position on the back of the insect. The insect pollinator of *C. macrocarpum* is a specific bee species, the males of which are attracted to the odor of the flowers. The flowers produce no nectar reward, but the male bee does gather the chemical that produces the scent. The bee then moves on to another flower. When the bee visits a female flower on a different *C. macrocarpum* individual (again attracted by the same scent), no such "loaded spring" awaits. Instead, the morphology of the female flower enhances the removal of the pollinia from the insect's body. In this way, floral morphology makes cross-fertilization more likely and reduces the chances of self-pollination.

Orchids were important in forming Darwin's ideas about the mechanisms of evolution, for they showed that even aspects of the coloration and form of flowers evolved in response to natural selection. This conclusion ran counter to the thinking of the day. For example, Thomas Huxley, one of Darwin's earliest and strongest supporters, doubted that the beauty of color in plants and animals could be explained on the basis of their importance to function. Darwin showed that the beauty of flowers is indeed connected to their reproductive success and is a key element in explaining the great diversity of plants.

CHAPTER**SUMMARY** **29**

 29.1 **How Did Seed Plants Become Today's Dominant Vegetation?**

- Fossils of woody **seed ferns** are the earliest evidence of seed plants. The surviving groups of seed plants are the **gymnosperms** and **angiosperms**. Review Figure 29.1

- All seed plants are heterosporous, and their gametophytes are much smaller than (and dependent on) their sporophytes. **Review Figure 29.2**

- Seed plants do not require liquid water for fertilization. **Pollen grains**, the microgametophytes of seed plants, are carried to a megagametophyte by wind or by animals.

- An **ovule** consists of the seed plant megagametophyte and the **integument** of sporophytic tissue that protects it.

- Following **pollination**, a **pollen tube** emerges from the pollen grain, elongates, and usually delivers gametes to the megagametophyte. **Review Figure 29.4, ACTIVITY 29.1**

- The ovule develops into a **seed** that contains an embryo (the new sporophyte generation). Seeds are well protected and are often capable of long periods of dormancy, germinating only when conditions are favorable. **Review Figure 29.5**

 29.2 **What Are the Major Groups of Gymnosperms?**

- The gymnosperms produce ovules and seeds that are not protected by ovary or fruit tissues.

- The major gymnosperm groups are the **cycads**, **ginkgos**, **gnetophytes**, and **conifers**. Review Figure 29.6

- The megaspores of conifers are produced in woody **cones** called **megastrobili**; the microspores are produced in herbaceous cones called **microstrobili**. Review Figures 29.7 and 29.8, **ACTIVITY 29.2, ANIMATED TUTORIAL 29.1**

 29.3 **How Do Flowers and Fruits Increase the Reproductive Success of Angiosperms?**

- **Flowers** and **fruits** are unique to the angiosperms, distinguishing them from the gymnosperms.

- The xylem of most angiosperms is more complex than that of the gymnosperms. It contains two specialized cell types: **vessel elements**, which function in water transport, and **fibers**, which play an important role in structural support.

- The ovules and seeds of angiosperms are enclosed in and protected by **carpels**.

- The floral organs, from the base to the apex of the flower, are the **sepals**, **petals**, **stamens**, and **pistil**. Stamens bear microsporangia in **anthers**. The pistil (consisting of one or more carpels) includes an **ovary** containing ovules. The **stigma** is the receptive surface of the pistil.

- A flower with both megasporangia and microsporangia is referred to as **perfect**; a flower with only one or the other is **imperfect**.

- A **monoecious** species has megasporangiate and microsporangiate flowers on the same plant. A **dioecious** species is one in which megasporangiate and microsporangiate flowers occur on different plants.

- The carpels and stamens of flowers probably evolved from leaflike structures. **Review Figure 29.12**

- Some plants with perfect flowers have adaptations to prevent self-fertilization. **Review Figure 29.13**

- Many angiosperms have coevolved with their animal pollinators.

- Angiosperms exhibit **double fertilization**, usually resulting in the production of a diploid zygote and triploid **endosperm**. **Review Figure 29.16, ANIMATED TUTORIAL 29.2**

- The oldest evolutionary split among the angiosperms is between the clade represented by the single species in the genus *Amborella* and all the remaining flowering plants. **Review Figure 29.18**

- The most species-rich angiosperm clades are the **monocots** and the **eudicots**. The **magnoliids** are likely the sister group to the monocots and eudicots.

 29.4 **How Do Plants Benefit Human Society?**

- Plants provide ecosystem services that affect soil, water, air quality, and climate.

- Plants are primary producers and as such are the foundation of terrestrial food webs.

- Plants provide humans with many important medicinal products.

Go to the Interactive Summary to review key figures, Animated Tutorials, and Activities
Life10e.com/is29

CHAPTER**REVIEW**

▮▮▮ **REMEMBERING**

1. Which of the following statements about seed plants is *true*?
   a. Seeds are produced only by flowering plants (angiosperms).
   b. The sporophyte generation is more reduced in seed plants than in the ferns.
   c. The gametophytes of seed plants are independent of the sporophytes.
   d. All seed plant species are heterosporous.
   e. The zygote of seed plants divides repeatedly to form the gametophyte.

2. Most angiosperms
   a. have seeds enclosed in a carpel.
   b. have haploid endosperm that nourishes the developing embryo.
   c. lack secondary growth.
   d. bear two kinds of cones.
   e. lack flowers.

3. Which statement about flowers is *not* true?

   a. Pollen is produced in the anthers.

   b. Pollen is received on the stigma.

   c. An inflorescence is a cluster of flowers.

   d. A species having female and male flowers on the same plant is dioecious.

   e. A flower with both megasporangia and microsporangia is said to be perfect.

4. Which statement about the angiosperm pollen grain is *not* true?

   a. It is the male gamete.

   b. It is haploid.

   c. It produces a long tube.

   d. It is multicellular.

   e. It is produced in microsporangia.

5. The eudicots

   a. include many herbs, vines, shrubs, and trees.

   b. along with the monocots are the only extant angiosperm clades.

   c. are not a clade.

   d. include the magnolias.

   e. include orchids and palm trees.

UNDERSTANDING & APPLYING

6. Not all flowers possess all of the following floral organs: sepals, petals, stamens, and carpels. Which floral organ or organs do you think might be found in the flowers that have the smallest number of floral organ types? Discuss the possibilities, both for a single flower and for a species.

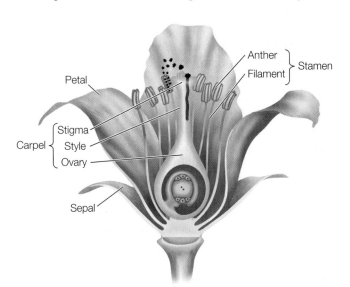

7. The origin of the angiosperms has long been "an abominable mystery," as Charles Darwin once put it. The earliest known angiosperm fossils are from the late Jurassic or early Cretaceous, but fossils of their sister group, the gymnosperms, are known from as early as the late Carboniferous (about 150 million years earlier; see Figure 29.1). Given that these two sister groups are thought to have arisen from a single split in the seed plant lineage, what might explain the lack of earlier angiosperm fossils?

ANALYZING & EVALUATING

In 1879, W. J. Beal began an experiment that he could not hope to finish in his lifetime. He prepared 20 lots of seeds for long-term storage. Each lot consisted of 50 seeds from each of 23 species. He mixed each lot of seeds with sand and placed the mixture in an uncapped bottle, then buried all the bottles on a sandy knoll. At 10-year intervals over the next century, biologists have excavated a bottle and checked the viability of its contents. The table below shows the number of germinating seeds (of the original 50) from three of the species in years 50–100 of this ongoing experiment. Use these data to answer Questions 8–10.

| | Years after burial | | | | | |
Species	50	60	70	80	90	100
Oenothera biennis (Evening primrose)	19	12	7	5	0	0
Rumex crispus (Curly dock)	26	2	7	1	0	0
Verbascum blattaria (Moth mullein)	31	34	37	35	10	21

8. Calculate the percent of viable seeds for these three species in years 50–100 and graph seed survivorship as a function of time buried.

9. No seeds of the first two species were viable after 90 years of the experiment. Assume 100 percent seed viability at the start of the experiment (year 0), and predict from your graph the approximate year when you think the last of the *Verbascum blattaria* seeds will germinate.

10. What factors do you think might influence the differences among the species in long-term seed viability?

Go to BioPortal at **yourBioPortal.com** for Animated Tutorials, Activities, LearningCurve Quizzes, Flashcards, and many other study and review resources.

30

The Evolution and Diversity of Fungi

CHAPTER**OUTLINE**

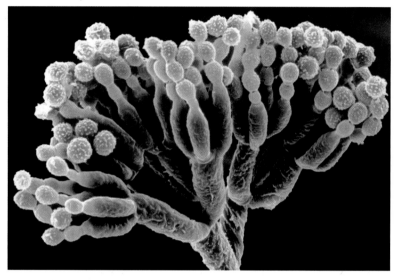

Source of a "Miracle Drug" All species of the fungus *Penicillium* are recognizable by their dense, spore-bearing structures. The derivation of the antibiotic agent penicillin from these fungi was one of the most important achievements in medical history.

A LEXANDER FLEMING WAS ALREADY A famous scientist in 1928, but his laboratory was often a mess. That year he was studying the properties of *Staphylococcus* bacteria, the agents of dangerous staph infections. In August he took a long vacation with his family. When he returned in early September, he found that some of his petri dishes of *Staphylococcus* had become infested with a fungus that killed many of the bacteria.

Many scientists would have sighed at the loss, thrown out the petri dishes, and started new cultures of bacteria. But when Fleming looked at the dishes, he saw something exciting. Around each colony of fungi was a ring within which all the bacteria were dead.

Fleming hypothesized that the bacteria-free rings around the fungal colonies were produced by a substance excreted from the fungi, which he initially called "mould juice." He identified the fungi as members of the genus *Penicillium* and eventually named the antibacterial substance they produced "penicillin." Fleming published his discovery in 1929, but initially it received very little attention.

Over the next decade, Fleming produced small quantities of penicillin for testing as an antibacterial agent. Some of the tests showed promise, but many were inconclusive, and eventually Fleming gave up on the research. But his tests had shown enough promise to attract the attention of several chemists, who worked out the practical problems of producing a stable form of the substance. Clinical trials of this stable form of penicillin were extremely successful, and by 1945 it was being produced and distributed as an antibiotic on a large scale. That same year, Fleming and two of the chemists, Howard Florey and Ernst Chain, received the Nobel Prize in Medicine for their work on penicillin.

The development of penicillin was one of the most important achievements in modern medicine. Until the introduction of modern antibiotics, the most widespread agents of human death included bacterial infections such as gangrene, tuberculosis, and syphilis. Penicillin proved to be highly effective in curing such infections, and its success led to the creation of the modern pharmaceutical industry. Soon many additional antibiotic compounds were isolated from other fungi or synthesized in the laboratory, leading to a "golden age" of human health.

Have antibiotics derived from fungi eliminated the danger of bacterial diseases in human populations?

See answer on p. 627.

30.1 What Is a Fungus?

Fungi are organisms that digest their food outside their bodies. They secrete digestive enzymes to break down large food molecules in the environment, then absorb the breakdown products through the plasma membranes of their cells in a process known as **absorptive heterotrophy**. This mode of nutrition allows them to be successful in a wide variety of environments. Many fungi are **saprobes**, which means that they absorb nutrients from dead organic matter. Others are parasites, absorbing nutrients from living hosts. Still others are mutualists, living in intimate associations with other organisms that benefit both partners.

Modern fungi are believed to have evolved from a unicellular protist ancestor that had a flagellum. The probable common ancestor of the animals was also a flagellated protist much like the living choanoflagellates (see Figure 31.2). Current evidence, including the sequences of many genes, suggests that the fungi, choanoflagellates, and animals share a common ancestor not shared by other eukaryotes. These three lineages form a group known as the **opisthokonts (Figure 30.1)**. A synapomorphy of the opisthokonts is a flagellum that, if present, is posterior, as in animal sperm. The flagella of all other eukaryotes are anterior.

Synapomorphies that distinguish the fungi as a group among the opisthokonts include absorptive heterotrophy and the presence of **chitin**, a nitrogen-containing structural polysaccharide, in their cell walls. The fungi represent one of the four independent evolutionary origins of large multicellular organisms (plants, brown algae, and animals are the other three).

Unicellular yeasts absorb nutrients directly

Most fungi are multicellular, but single-celled species are found in most fungal groups. Unicellular, free-living fungi are referred to as **yeasts (Figure 30.2)**. Some fungi have both a yeast life stage and a multicellular life stage. Thus the term "yeast" does not refer to a single taxonomic group, but rather to a lifestyle that has evolved multiple times. Yeasts live in liquid or moist environments and absorb nutrients directly across their cell surfaces.

The ease with which many yeasts can be cultured, combined with their rapid growth rates, has made them ideal model organisms for study in the laboratory. They present many of the

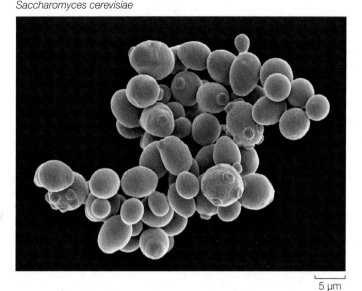

Saccharomyces cerevisiae

5 μm

30.2 Yeasts Unicellular, free-living fungi are known as yeasts. Many yeasts reproduce by budding—mitosis followed by asymmetrical cell division—as illustrated here.

same advantages to laboratory investigators as do many bacteria, but because they are eukaryotes, their genome structures and cells are much more like those of humans and other eukaryotes than are those of bacteria.

Multicellular fungi use hyphae to absorb nutrients

The body of a multicellular fungus is called a **mycelium** (plural *mycelia*). A mycelium is composed of a mass of individual tubular filaments called **hyphae** (singular *hypha*; **Figure 30.3A**), in which absorption of nutrients takes place. The cell walls of the hyphae are greatly strengthened by microscopic fibrils of chitin. In some species of fungi, the hyphae are subdivided into cell-like compartments by incomplete cross-walls called **septa** (singular *septum*); these hyphae are referred to as **septate**. Septa do not completely close off compartments in the hyphae. Pores at the centers of the septa allow organelles—sometimes even nuclei—to move in a controlled way between compartments (**Figure 30.3B**). In other species of fungi, the hyphae lack septa but may contain hundreds of nuclei; these hyphae are referred to as **coenocytic**. The coenocytic condition results from repeated nuclear divisions without cytokinesis.

Certain modified hyphae, called **rhizoids**, anchor some fungi to their substrate (i.e., the dead organism or other matter on which they are feeding). These rhizoids are not homologous to the rhizoids of plants, and they are not specialized to absorb nutrients and water.

Fungi can grow very rapidly. In some species, the total hyphal growth of a fungal mycelium (not the growth of an individual hypha) may exceed 1 kilometer a day! The hyphae may be widely dispersed to forage for nutrients over a large area, or they may clump together in a cottony mass to exploit a rich nutrient source. The familiar mushrooms you may notice in the environment are spore-producing fruiting structures (**Figure 30.3C**). In the fungal species that produce these structures, the

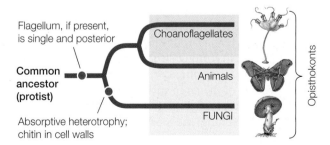

Flagellum, if present, is single and posterior

Choanoflagellates

Common ancestor (protist)

Animals

Absorptive heterotrophy; chitin in cell walls

FUNGI

Opisthokonts

30.1 Fungi in Evolutionary Context Absorptive heterotrophy and the presence of chitin in their cell walls distinguish the fungi from other opisthokonts.

Lycoperdon perlatum

(A)

Vessel in xylem   Fungal hyphae

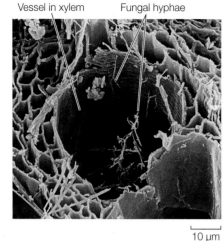

10 µm

(B)

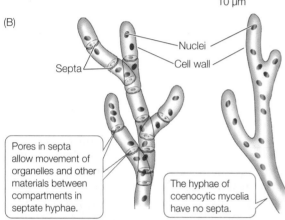

Nuclei

Cell wall

Septa

Pores in septa allow movement of organelles and other materials between compartments in septate hyphae.

The hyphae of coenocytic mycelia have no septa.

(C)

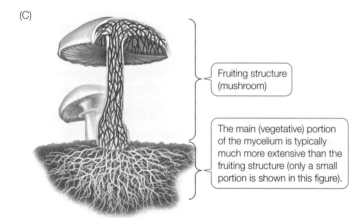

Fruiting structure (mushroom)

The main (vegetative) portion of the mycelium is typically much more extensive than the fruiting structure (only a small portion is shown in this figure).

30.3 Mycelia Are Made Up of Hyphae (A) The minute individual hyphae of fungal mycelia can penetrate small spaces. In this artificially colored micrograph, hyphae (yellow structures) of a dry-rot fungus are penetrating the xylem tissues of a log. (B) The hyphae of septate fungal species are divided into organelle-containing compartments by porous septa. The hyphae of coenocytic fungal species have no septa. (C) The fruiting structure of a club fungus is short-lived, but the filamentous, nutrient-absorbing mycelium can be long-lived and cover large areas.

mycelial mass is often far larger than the visible mushroom. The mycelium of one individual fungus discovered in Oregon covers almost 900 hectares underground and weighs considerably more than a blue whale (the largest animal). Aboveground, this individual is evident only as isolated clumps of mushrooms.

30.4 Spores Galore Puffballs (a type of club fungus) disperse trillions of spores in great bursts. Few of the spores travel very far, however; some 99 percent of them fall within 100 meters of the parent puffball.

What happens when a fungus faces a dwindling food supply? A common strategy is to reproduce rapidly and abundantly. When conditions are good, fungi produce great quantities of spores, but the rate of spore production is commonly even higher when nutrient supplies go down. The spores may then remain dormant until conditions improve, or they may be dispersed to areas where nutrient supplies are higher.

Not only are fungal spores abundant in number, but they are extremely tiny and easily spread by wind or water (**Figure 30.4**). These attributes virtually ensure that they will be scattered over great distances and that at least some of them will find conditions suitable for growth. The air we breathe contains as many as 10,000 fungal spores per cubic meter. No wonder we find fungi just about everywhere.

Fungi are in intimate contact with their environment

The filamentous hyphae of a fungus give it a unique relationship with its physical environment. The fungal mycelium has an enormous surface area-to-volume ratio compared with that of most large multicellular organisms. This large ratio is a marvelous adaptation for absorptive heterotrophy. Throughout the mycelium (except in fruiting structures), all of the hyphae are very close to their food source.

The downside of the large surface area-to-volume ratio of the mycelium is its tendency to lose water rapidly in a dry environment. Thus fungi are most common in moist environments. You have probably observed the tendency of molds, toadstools, and other fungi to appear in damp places.

Another characteristic of some fungi is a tolerance for highly hypertonic environments (those with a solute concentration higher than their own; see Section 6.3). Many fungi are more resilient than bacteria in hypertonic surroundings. Jelly in the refrigerator, for example, will not become a growth medium for bacteria because it is too hypertonic to those organisms, but it may eventually harbor mold colonies. Mold in the refrigerator illustrates yet another trait of many fungi: tolerance of temperature extremes. Many fungi grow in temperatures as low as –6°C, and some can tolerate temperatures higher than 50°C.

RECAP 30.1

Fungi, like animals, are opisthokonts. Fungi are distinguished from other opisthokonts by absorptive heterotrophy and by the presence of chitin in their cell walls. Unicellular fungi called yeasts absorb nutrients directly across their cell surfaces. The body form of multicellular fungi—a mycelium made up of rapidly growing hyphae—allows them to practice absorptive heterotrophy efficiently in a variety of moist environments.

- Describe the relationship between fungal structure and absorptive heterotrophy. **See pp. 609–610 and Figure 30.3**
- What are the advantages and disadvantages to multicellular fungi of the large surface area-to-volume ratio of the mycelium? **See p. 610**

Fungi are important components of healthy ecosystems. They interact with other organisms in many ways, some of which are harmful and some beneficial to those other organisms.

30.2 How Do Fungi Interact with Other Organisms?

Without the fungi, our planet would be very different. Picture Earth with only a few stunted plants and watery environments choked with the remains of dead organisms. Fungi do much of Earth's garbage disposal. Fungi not only help clean up the landscape and form soil, but also play key roles in the recycling of mineral nutrients. Furthermore, the colonization of the terrestrial environment was made possible in large part by associations fungi formed with land plants and other organisms.

Saprobic fungi are critical to the planetary carbon cycle

Saprobic fungi, along with bacteria, are the major decomposers on Earth. These organisms contribute to the decay of nonliving organic matter and thus to the recycling of the elements required by living things. In forests, for example, fungi digest and absorb nutrients from fallen trees, thus decomposing their wood. Fungi are the principal decomposers of cellulose and lignin, the main components of plant cell walls (which most bacteria cannot break down). Other fungi produce enzymes that decompose keratin and thus break down animal structures such as hair and nails.

Were it not for the fungal decomposers, Earth's carbon cycle would fail: great quantities of carbon atoms would remain trapped forever on forest floors and elsewhere. Instead, those carbon atoms are returned to the atmosphere in the form of

CO_2 by fungal respiration, where they are again available for photosynthesis by plants.

There was a time in Earth's history when populations of saprobic fungi declined dramatically. Vast tropical swamps existed during the Carboniferous period, as we saw in Chapter 25. When plants in these swamps died, they began to form peat. Peat formation led to acidification of the swamps; that acidity, in turn, drastically reduced the fungal population. The result? With the decomposers largely absent, large quantities of peat remained on the swamp floor and over time were converted into coal.

In contrast to their decline during the Carboniferous, fungi did very well at the end of the Permian, about 250 million years ago, when the aggregation of the continents produced volcanic eruptions that triggered a global mass extinction. The fossil record shows that even as 96 percent of all multicellular species became extinct, fungi flourished—demonstrating both their hardiness and their role in recycling the elements in dead plants and animals.

Simple sugars and the breakdown products of complex polysaccharides are the favored source of carbon for saprobic fungi. Most fungi obtain nitrogen from proteins or the products of protein breakdown. Many fungi can use nitrate (NO_3^-) or ammonium (NH_4^+) ions as their sole source of nitrogen. No known fungus can get its nitrogen directly from inorganic nitrogen gas, however, as can some bacteria and plant–bacteria associations (that is, fungi cannot fix nitrogen; see Section 26.3).

Go to Media Clip 30.1
Fungal Decomposers
Life10e.com/mc30.1

Some fungi engage in parasitic or predatory interactions

Whereas saprobic fungi obtain their energy, carbon, and nitrogen directly from dead organic matter, other species of fungi obtain their nutrition from parasitic—and even predatory—interactions.

PARASITIC FUNGI **Mycologists** (biologists who study fungi) distinguish between two classes of parasitic fungi based on their degree of dependence on their host. **Facultative parasites** can grow on living organisms but can also grow independently (including on artificial media). **Obligate parasites** can grow only on a specific, living host. The fact that their growth depends on a living host shows that obligate parasites have specialized nutritional requirements.

Plants and insects are the most common hosts of parasitic fungi. The filamentous structure of fungal hyphae is especially well suited to a life of absorbing nutrients from living plants. The slender hyphae of a parasitic fungus can invade a plant through stomata, through wounds, or in some cases, by direct penetration of epidermal cell walls (**Figure 30.5A**). Once inside the plant, the hyphae branch out to expand the mycelium. Some hyphae produce **haustoria**, branching projections that push through cell walls into living plant cells, absorbing the nutrients within those cells. The haustoria do not break

(A)

Hyphae of fungal mycelium

Leaf cells

Stoma of leaf

This hypha is penetrating the leaf's interior through a stoma.

2 μm

(B)

3 Some hyphae penetrate cells within the leaf.

Plasma membrane

The haustorium penetrates the cell wall but not the plasma membrane.

Stoma

1 Fungal spores germinate on the surface of the leaf.

Spore

2 Elongating hyphae pass through stomata into the interior of the leaf.

30.5 Invading a Leaf (A) Hyphae of the mildew *Phyllactinia guttata* growing on the surface of a hazel leaf. (B) Haustoria are fungal hyphae that push into the living cells of plants, from which they absorb nutrients.

through the plasma membranes inside the cell walls; they simply invaginate into the membranes, so that the plasma membrane fits them like a glove (**Figure 30.5B**). Fruiting structures may form, either within the plant body or on its surface.

Some parasitic fungi live in a close physical (symbiotic) relationship with a plant host that is usually not lethal to the plant. Others are *pathogenic*, sickening or even killing the host from which they derive nutrition.

Go to Media Clip 30.2
Mind-Control Killer Fungi
Life10e.com/mc30.2

PATHOGENIC FUNGI Although most human diseases are caused by bacteria or viruses, fungal pathogens are a major cause of death among people with compromised immune systems. Most people with AIDS die of fungal diseases, including the pneumonia caused by *Pneumocystis jirovecii*. Even *Candida albicans* and certain other yeasts that are normally part of a healthy microbiome can cause severe diseases, such as esophagitis (which impairs swallowing), in individuals with AIDS and in individuals taking immunosuppressive drugs. Various fungi cause other, less threatening human diseases, such as ringworm and athlete's foot. Our limited understanding of the basic biology of these fungi hampers our ability to treat the diseases they cause. As a result, fungal diseases are a growing international health problem.

The worldwide decline of amphibian species has been linked to the spread of a chytrid fungus, *Batrachochytrium dendrobatidis*. Genetic analyses indicate that the populations of this fungus that are attacking amphibian populations around the world are genetically almost identical, which suggests a recent introduction of the fungus across the globe. This chytrid appears to be native to southern Africa, and its spread around the world may have been initiated in the 1930s with exports of the African clawed frog (*Xenopus laevis*), which was once widely used in human pregnancy tests.

Fungi are by far the most important plant pathogens, much more so than bacteria and viruses. Pathogenic fungi cause crop losses amounting to billions of dollars each year. Major fungal diseases of crop plants include black stem rust of wheat and other diseases of wheat, corn, and oats. The agent of black stem rust is *Puccinia graminis*, which has a complicated life cycle that involves two plant hosts (wheat and barberry). In an epidemic in 1935, *P. graminis* was responsible for the loss of about one-fourth of the wheat crop in Canada and the United States.

PREDATORY FUNGI Some fungi have adaptations that enable them to function as active predators, trapping nearby microscopic protists or animals. The most common predatory strategy seen in fungi is to secrete sticky substances from the hyphae so that passing organisms stick to them. The hyphae then quickly invade the trapped prey, growing and branching within it, spreading through its body, absorbing nutrients, and eventually killing it.

A more dramatic adaptation for predation is the constricting ring formed by some species of soil fungi (**Figure 30.6**). When nematodes (tiny roundworms) are present in the soil, these fungi form three-celled rings with a diameter that just fits a nematode. A nematode crawling through one of these rings stimulates the fungus, causing the cells of the ring to swell and trap the worm. Fungal hyphae quickly invade and digest the unlucky victim.

Mutualistic fungi engage in relationships that benefit both partners

Certain relationships between fungi and other organisms have nutritional consequences for both partners. Two relationships of this type are **symbiotic** (the partners live in close, permanent contact with each other) as well as **mutualistic** (the relationship benefits both partners; see Section 56.3).

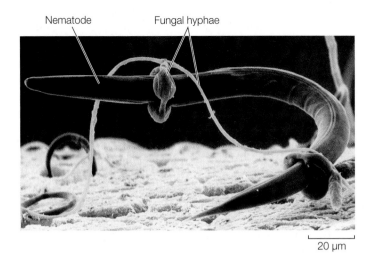

Nematode   Fungal hyphae

20 μm

30.6 Fungus as Predator A nematode is trapped by hyphal rings of the soil-dwelling fungus *Arthrobotrys dactyloides.*

LICHENS A **lichen** is not a single organism, but rather a meshwork of two radically different species: a fungus and a photosynthetic microorganism. Together, the organisms that constitute a lichen can survive some of the harshest environments on Earth. The biota of Antarctica, for example, features more than a hundred times as many species of lichens as of plants. Relatively little experimental work has focused on lichens, perhaps because they grow so slowly—typically less than 1 centimeter in a year.

There are nearly 30,000 described "species" of lichens, each of which is assigned the name of its fungal component. These fungal components may constitute as many as 20 percent of all fungal species. Most of them are sac fungi (Ascomycota). Some of them are able to grow independently without a photosynthetic partner, but most have never been observed in nature other than in a lichen association. The photosynthetic component of a lichen is most often a unicellular green alga, but it can be a cyanobacterium, or may even include both.

Lichens are found in all sorts of exposed habitats: on tree bark, on open soil, and on bare rock. Reindeer moss (not a moss at all, but the lichen *Cladonia subtenuis*) covers vast areas in Arctic, sub-Arctic, and boreal regions, where it is an important part of the diets of reindeer and other large mammals.

The body forms of lichens fall into three principal categories. **Crustose** (crustlike) lichens adhere tightly to their substrate (**Figure 30.7A**). **Foliose** (leafy) lichens are loosely attached and grow parallel to their substrate (**Figure 30.7B**) **Fruticose** lichens are highly branched and can grow upward like shrubs or hang in long strands from tree branches or rocks (**Figure 30.7C**).

A cross section of a typical foliose lichen reveals a tight upper region of fungal hyphae; a layer of photosynthetic cyanobacteria or algae; a loose hyphal layer; and finally hyphal rhizoids that attach the structure to its substrate (**Figure 30.8**). The meshwork of fungal hyphae takes up mineral nutrients needed by the photosynthetic cells and also holds water tenaciously, providing a suitably moist environment. The fungus obtains fixed carbon from the photosynthetic products of the algal or cyanobacterial cells.

Within the lichen, fungal hyphae are tightly pressed against the photosynthetic cells and sometimes invade them without breaching the plasma membrane (similar to the haustoria in parasitic fungi; see Figure 30.5). The photosynthetic cells not only survive these intrusions but continue to grow. Algal cells in a lichen "leak" photosynthetic products at a greater rate than do similar cells growing on their own, and photosynthetic cells taken from lichens grow more rapidly on their own than when associated with a fungus. On the basis of these observations, we could consider lichen fungi to be parasitic on their photosynthetic partners. In many places where lichens grow, however, the photosynthetic cells could not grow at all on their own.

Lichens can reproduce simply by fragmentation of the vegetative body (the **thallus**) or by means of specialized structures called **soredia** (singular *soredium*). Soredia consist of one or a few photosynthetic cells bound by fungal hyphae. They become detached from the lichen, are dispersed by air currents, and upon arriving at a favorable location, develop into a new

(A)   *Aspicilia* sp.   *Caloplaca* sp.

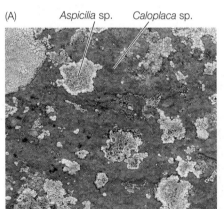

(B) *Parmotrema* sp.

(C) *Teloschistes exilis*

30.7 Lichen Body Forms The body forms of lichens fall into three principal categories. (A) Two crustose lichen species are growing together on this exposed rock surface. (B) Foliose lichens have a leafy appearance. (C) The brown and orange growth is a "shrubby" fruticose lichen.

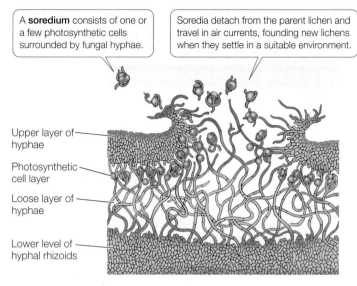

A **soredium** consists of one or a few photosynthetic cells surrounded by fungal hyphae.

Soredia detach from the parent lichen and travel in air currents, founding new lichens when they settle in a suitable environment.

Upper layer of hyphae

Photosynthetic cell layer

Loose layer of hyphae

Lower level of hyphal rhizoids

30.8 Lichen Anatomy Cross section showing the layers of a foliose lichen and the release of soredia.

lichen thallus. Alternatively, the fungal partner may go through its sexual reproductive cycle, producing haploid spores. When these spores are discharged, however, they disperse alone, unaccompanied by the photosynthetic partner.

Lichens are often the first colonists on new areas of bare rock. They get most of the mineral nutrients they need from the air and rainwater, augmented by minerals absorbed from dust. A lichen begins to grow shortly after a rain, as it begins to dry. As it grows, the lichen acidifies its environment slightly, and this acidity contributes to the slow breakdown of rocks, an early step in soil formation. With further drying, the lichen's photosynthesis ceases. The water content of the lichen may drop to less than 10 percent of its dry weight, at which point it becomes highly insensitive to extremes of temperature.

MYCORRHIZAE Many vascular plants depend on a symbiotic association with fungi. This ancient association between plants and fungi was critical to the successful exploitation of the terrestrial environment by plants. Unassisted, the root hairs of many plants often do not take up enough water or minerals to sustain their growth. However, the roots of such plants usually become infected with fungi, forming an association called a **mycorrhiza**. (We'll describe the infection process in detail in Section 36.4.) There

are two types of mycorrhizae, distinguished by whether or not the fungal hyphae penetrate the plant cell walls.

In **ectomycorrhizae**, the fungus wraps around the root, and its mass is often as great as that of the root itself (**Figure 30.9A**). The fungal hyphae wrap around individual cells in the root but do not penetrate the cell walls. An extensive web of hyphae penetrates the soil in the area around the root, so that up to 25 percent of the soil volume near the root may be fungal hyphae. The hyphae attached to the root increase its surface area for the absorption of water and minerals, and the mass of hyphae in the soil acts like a sponge to hold water in the neighborhood of the root. Infected roots are short, swollen, and club-shaped, and they lack root hairs.

The fungal hyphae of **arbuscular mycorrhizae** enter the root and penetrate the cell walls of the root cells, forming arbuscular (treelike) structures inside the cell wall but outside the plasma membrane. These structures, like the haustoria of parasitic fungi and the contact regions of fungal hyphae and photosynthetic cells in lichens, become the primary site of exchange between plant and fungus (**Figure 30.9B**). As in the ectomycorrhizae, the fungus forms a vast web of hyphae leading from the root surface into the surrounding soil.

The mycorrhizal association is important to both partners. The fungus obtains needed organic compounds, such as sugars and amino acids, from the plant. In return, the fungus, because of its high surface area-to-volume ratio and its ability to penetrate the fine structure of the soil, greatly increases the plant's ability to absorb water and minerals (especially phosphorus). The fungus may also provide the plant with certain growth hormones and may protect it against attack by disease-causing microorganisms.

Plants that have active arbuscular mycorrhizae typically are a deeper green, exhibit higher growth rates, and may resist drought and temperature extremes better than plants of the same species that have little mycorrhizal development. Attempts to introduce some plant species to new areas have failed

30.9 Mycorrhizal Associations (A) Ectomycorrhizal fungi wrap themselves around a plant root, increasing the area available for absorption of water and minerals. (B) Hyphae of arbuscular mycorrhizal fungi infect the root internally and penetrate the root cell walls, branching within the cells and forming a treelike structure, the arbuscule. (The cell cytoplasm has been removed to better visualize the arbuscule.)

(A)

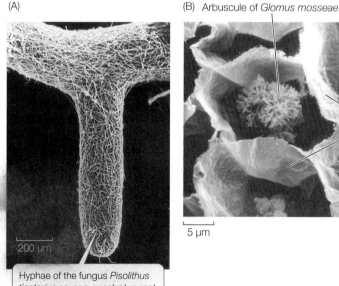

200 μm

Hyphae of the fungus *Pisolithus tinctorius* cover a eucalyptus root.

(B) Arbuscule of *Glomus mosseae*

Root cell walls

5 μm

until a bit of soil from the native area (presumably containing the fungus necessary to establish mycorrhizae) was provided. Trees without ectomycorrhizae do not grow well in the absence of abundant nutrients and water, so the health of our forests depends on the presence of ectomycorrhizal fungi. Many agricultural crops require inoculation of seeds with appropriate mycorrhizal fungi prior to planting. Without these fungi, the plants are unlikely to grow well, or in some cases at all. Certain plants that live in nitrogen-poor habitats, such as cranberry bushes and orchids, invariably have mycorrhizae. Orchid seeds will not germinate in nature unless they are already infected by the fungus that will form their mycorrhizae. Plants that lack chlorophyll always have mycorrhizae, which they often share with the roots of green, photosynthetic plants. In effect, these plants without chlorophyll are feeding on nearby green plants, using the fungus as a bridge.

Endophytic fungi protect some plants from pathogens, herbivores, and stress

In a tropical rainforest, 10,000 or more fungal spores may land on a single leaf each day. Some are plant pathogens, some do not affect the plant at all, and some invade the plant in a beneficial way. Fungi that live within aboveground parts of plants without causing obvious deleterious symptoms are called **endophytic fungi**. Recent research has shown that endophytic fungi are abundant in plants in all terrestrial environments.

Among the grasses, individual plants with endophytic fungi are more resistant to pathogens and to insect and mammalian herbivores than are plants lacking endophytes. The fungi produce alkaloids (nitrogen-containing compounds) that are toxic to animals. The alkaloids do not harm the host plant; in fact, some plants produce alkaloids (such as nicotine) themselves. The fungal alkaloids also increase the ability of grasses to resist stresses of various types, including drought (water shortage) and salty soils. Such resistance is useful in agriculture.

The role, if any, of endophytic fungi in most broad-leaved plants is unclear. They may convey protection against pathogens, or they may simply occupy space within leaves without conferring any benefit, but also without doing harm. The benefit, in fact, might be all for the fungus.

RECAP 30.2

Fungi interact with other organisms in many ways, both harmful and beneficial. Saprobic fungi play critical roles in the recycling of elements required by living organisms. Lichens are mutualistic associations of a fungus with algae or cyanobacteria. Mycorrhizae are associations of fungi and the roots of plants; they are essential for the survival of most plant species.

- What is the role of fungi in Earth's carbon cycle? **See p. 611**
- Describe the nature and benefits of the lichen association. **See pp. 613–614 and Figure 30.8**
- Why do plants grow better in the presence of mycorrhizal fungi? **See pp. 614–615 and Figure 30.9**

Before molecular techniques clarified their phylogenetic relationships, one criterion used for assigning fungi to taxonomic groups was the nature of their life cycles—including the types of fruiting structures they produced. The next section will take a closer look at life cycles in the six major groups of fungi.

30.3 How Do Major Groups of Fungi Differ in Structure and Life History?

Major fungal groups were originally distinguished by their structures and processes for sexual reproduction and, to a lesser extent, by other morphological differences. Although fungal life cycles are even more diverse than was once realized, specific types of life cycles generally characterize the six major groups of fungi: microsporidia, chytrids, zygospore fungi (Zygomycota), arbuscular mycorrhizal fungi (Glomeromycota), sac fungi (Ascomycota), and club fungi (Basidiomycota). **Figure 30.10** diagrams the evolutionary relationships of these groups as they are understood today.

The chytrids and the zygospore fungi may not represent monophyletic groups, as they each consist of several distantly related lineages that retain some ancestral features. The clades

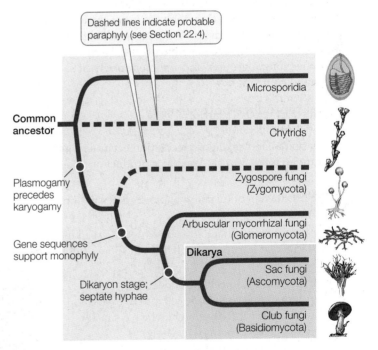

30.10 A Phylogeny of the Fungi Microsporidia are reduced, parasitic fungi whose relationships among the fungi are uncertain. They may be the sister group of most other fungi, or they may be more closely related to particular groups of chytrids or zygospore fungi. The dashed lines indicate that chytrids and zygospore fungi are thought to be paraphyletic; the relationships of the lineages within these two informal groups (see Table 30.1) are not yet well resolved. The sac fungi and club fungi together form the clade Dikarya.

Go to Activity 30.1 Fungal Phylogeny
Life10e.com/ac30.1

TABLE**30.1**

Classification of the Fungi

Group	Common Name	Features
Microsporidia	Microsporidia	Intracellular parasites of animals; greatly reduced, among smallest eukaryotes known; polar tube used to infect hosts

Chytrids (paraphyletic)[a]
  Chytridiomycota
  Neocallimastigomycota
  Blastocladiomycota
	Chytrids	Mostly aquatic and microscopic; zoospores and gametes have flagella

Zygomycota (paraphyletic)[a]
  Entomophthoromycotina
  Kickxellomycotina
  Mucoromycotina
  Zoopagomycotina
	Zygospore fungi	Reproductive structure is a unicellular zygospore with many diploid nuclei; hyphae coenocytic; no fleshy fruiting body

Glomeromycota	Arbuscular mycorrhizal fungi	Form arbuscular mycorrhizae in plant roots; only asexual reproduction is known
Ascomycota	Sac fungi	Sexual reproductive saclike structure known as an ascus, which contains haploid ascospores; hyphae septate; dikaryon
Basidiomycota	Club fungi	Sexual reproductive structure is a basidium, a swollen cell at the tip of a specialized hypha that supports haploid basidiospores; hyphae septate; dikaryon

[a]The formally named groups within the chytrids and Zygomycota are each thought to be monophyletic, but their relationships to one another (and to microsporidia) are not yet well resolved.

that are thought to be monophyletic within these two informal groupings are listed in **Table 30.1**. Recent evidence from DNA analyses has established the placement of the microsporidia among the fungi, the likely paraphyly of the chytrids and the zygospore fungi, the independence of arbuscular mycorrhizal fungi from the other fungal groups, and the monophyly of sac fungi and club fungi.

Fungi reproduce both sexually and asexually

Both asexual and sexual reproduction occur among the fungi (**Figure 30.11**). Asexual reproduction takes several forms:

- The production of (usually) haploid spores within sporangia
- The production of haploid spores (not enclosed in sporangia) at the tips of hyphae; such spores are called **conidia** (Greek *konis*, "dust")
- Cell division by unicellular fungi—either a relatively equal division of one cell into two (*fission*) or an asymmetrical division in which a smaller daughter cell is produced (*budding*)
- Simple breakage of the mycelium

Sexual reproduction is rare (or even unknown) in some groups of fungi but is common in others. Sexual reproduction may not occur in some species, or it may occur so rarely that mycologists have never observed it. Species in which no sexual stage has been observed were once placed in a separate taxonomic group because knowledge of the sexual life cycle was considered necessary for classifying fungi. Now, however, these species can be related to other species of fungi through analysis of their DNA sequences.

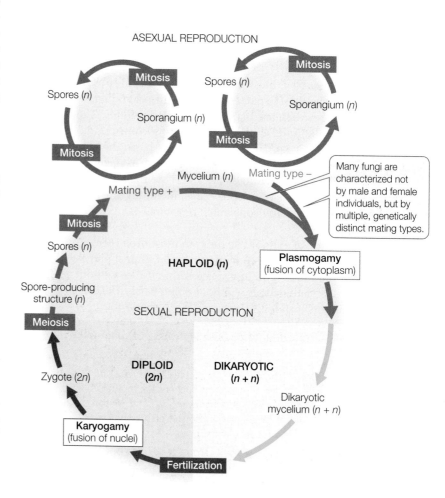

30.11 A Generalized Fungal Life Cycle   Environmental conditions may determine which mode of reproduction—sexual or asexual—takes place at a given time.

Sexual reproduction in most fungi features an interesting twist: There is no morphological distinction between female and male structures, or between female and male individuals, in most groups of fungi. Rather, there is a genetically determined distinction between two *or more* **mating types**. Individuals of the same mating type cannot mate with each other, but they can mate with individuals of another mating type within the same species, thus avoiding self-fertilization. Individuals of different mating types differ genetically but are often visually and physiologically indistinguishable.

Microsporidia are highly reduced, parasitic fungi

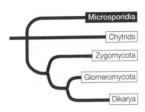

Microsporidia are unicellular parasitic fungi. They are among the smallest eukaryotes known, with infective spores that are only 1 to 40 micrometers (μm) in diameter. About 1,500 species have been described, but many more species are thought to exist. Their relationships among the eukaryotes have puzzled biologists for many decades.

Microsporidia lack true mitochondria, although they have reduced structures, known as **mitosomes**, that are derived from mitochondria. Unlike mitochondria, however, mitosomes contain no DNA; the mitochondrial genome has been completely transferred to the nucleus. Because microsporidia lack mitochondria, biologists initially suspected that they represented an early lineage of eukaryotes that diverged before the endosymbiotic event from which mitochondria evolved. The presence of mitosomes, however, indicates that this hypothesis is incorrect. DNA sequence analysis, along with the fact that their cell walls contain chitin, has confirmed that the microsporidia are in fact highly reduced, parasitic fungi, although their exact placement among the fungal lineages is still being investigated.

Microsporidia are obligate intracellular parasites of animals, especially of insects, crustaceans, and fishes. Some species are known to infect mammals, including humans. Most infections by microsporidia cause chronic disease in the host, with effects that include weight loss, reduced fertility, and a shortened life span. The host cell is penetrated by a polar tube that grows from the microsporidian spore. The function of the polar tube is to inject the contents of the spore, the sporoplasm, into the host (**Figure 30.12**). The sporoplasm replicates within the host cell and produces new infective spores. The life cycle of some species is complex and involves multiple hosts, whereas other species infect a single host. In some insects, parasitic microsporidia are transmitted vertically (i.e., from parent to offspring). Reproduction is thought to be strictly asexual in some microsporidians, but it includes poorly understood asexual and sexual cycles in other species.

Most chytrids are aquatic

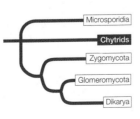

The **chytrids** (**Figure 30.13**) include several distinct lineages of aquatic microorganisms that were once classified with the protists. However, morphological evidence (cell

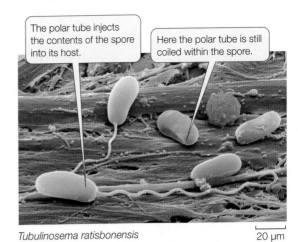

The polar tube injects the contents of the spore into its host.

Here the polar tube is still coiled within the spore.

Tubulinosema ratisbonensis   20 μm

30.12 Invasion of the Microsporidia The spores of microsporidia grow polar tubes that transfer the contents of the spores into the host's cells. The species shown here infects many animals, including humans.

walls that consist primarily of chitin) and molecular evidence support the classification of the chytrids as early-diverging fungi. In this book we use the term "chytrid" to refer to all three of the formally named clades listed as chytrids in Table 30.1, but some mycologists use this term to refer to only one of those clades, the Chytridiomycota. There are fewer than 1,000 described species among the three groups of chytrids.

Chytrids reproduce both sexually and asexually. Like the animals, chytrids that reproduce sexually possess flagellated gametes. The retention of this trait reflects the aquatic environment in which fungi first evolved. Chytrids are the only fungi that include species with flagella at any life cycle stage. Both the spores (called zoospores) and the gametes are flagellated (**Figure 30.14A**).

The alternation between multicellular haploid (n) and multicellular diploid ($2n$) generations that exists in plants and certain protist groups is seen in many fungi as well. For fungal groups other than the chytrids, this cycle differs from the usual system known as "alternation of generations" (see Figure 28.6) in that

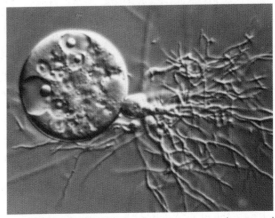

Chytriomyces hyalinus   25 μm

30.13 A Chytrid Branched rhizoids emerge from the sporangium of a mature chytrid.

(A) Chytrids

Unlike other fungi, some chytrids produce flagellated male and female gametes from a multicellular haploid stage.

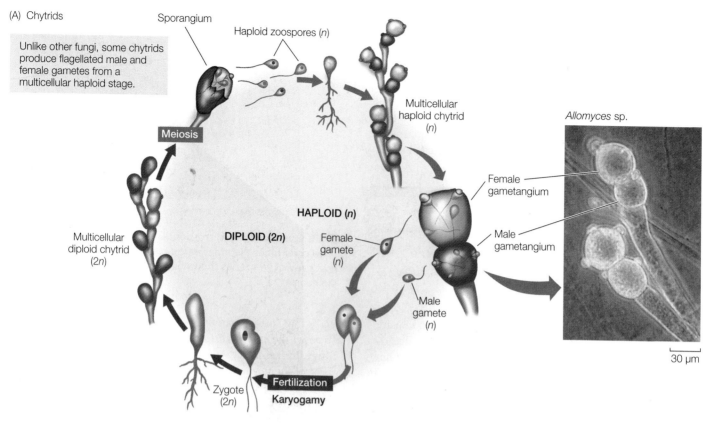

Allomyces sp.

30 μm

(B) Zygospore fungi (Zygomycota)

The single-celled zygospore is the only diploid cell in the life cycle of zygospore fungi.

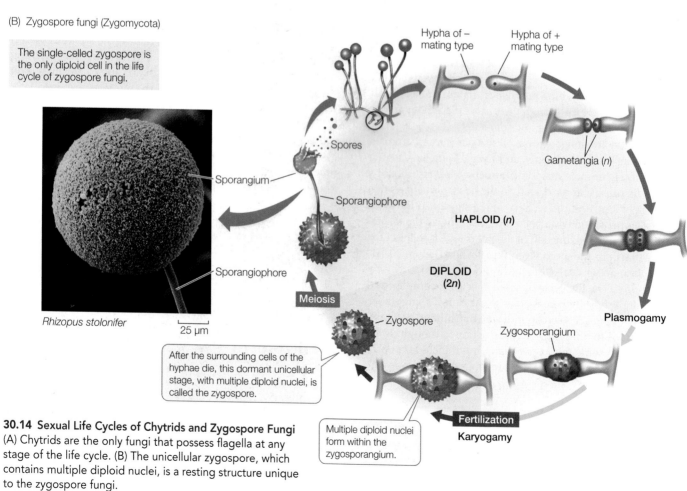

Rhizopus stolonifer

25 μm

After the surrounding cells of the hyphae die, this dormant unicellular stage, with multiple diploid nuclei, is called the zygospore.

Multiple diploid nuclei form within the zygosporangium.

30.14 Sexual Life Cycles of Chytrids and Zygospore Fungi
(A) Chytrids are the only fungi that possess flagella at any stage of the life cycle. (B) The unicellular zygospore, which contains multiple diploid nuclei, is a resting structure unique to the zygospore fungi.

the multicellular haploid stages do not produce specialized male and female gametes that are independent of the hyphae. Instead, in most fungi, cells of haploid hyphae from different mating types (of which there can be many more than two) fuse to form the diploid stage. But in sexually reproducing chytrids, the multicellular haploid stage produces independent male and female gametes, both of which are flagellated (see Figure 30.14A). The male gamete is distinguished from the female gamete by its smaller size; otherwise the two gametes are very similar in form.

The chytrids are diverse in form; some are unicellular, others have rhizoids, and still others have coenocytic hyphae. They may be parasitic (on organisms such as algae, mosquito larvae, nematodes, and amphibians) or saprobic. Some have complex mutualistic relationships with foregut-fermenting animals such as cattle and deer. Many chytrids live in freshwater habitats or in moist soil, but some are marine.

Some fungal life cycles feature separate fusion of cytoplasms and nuclei

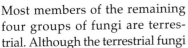

Most members of the remaining four groups of fungi are terrestrial. Although the terrestrial fungi grow in moist places, they do not have motile gametes, and they do not release gametes into the environment, so liquid water is not required for fertilization. Instead, the cytoplasms of two individuals of different mating types fuse (a process called **plasmogamy**) before their nuclei fuse (a process called **karyogamy**; see Figure 30.11). Sexual species of terrestrial fungi include some zygospore fungi, sac fungi, and club fungi.

Zygospore fungi reproduce sexually when adjacent hyphae of two different mating types release chemical signals that cause them to grow toward each other. These hyphae produce gametangia, which are specialized cells for reproduction that are retained as part of the hyphae. In the gametangia, nuclei replicate without cell division, resulting in multiple haploid nuclei in both gametangia. The two gametangia then fuse to form a zygosporangium that contains many haploid nuclei of each mating type (**Figure 30.14B**). Haploid nuclei of different mating types then pair up to form multiple diploid nuclei within the zygosporangium. A thick, multilayered cell wall forms around the zygosporangium to form a well-protected resting stage that can remain dormant for months. In harsh environmental conditions, this resting stage may be the only cell that survives as the surrounding cells of the hyphae die. At this stage the single surviving cell is known as a **zygospore**, which is the basis of the name of the zygospore fungi. When environmental conditions improve, the nuclei within the zygospore undergo meiosis and one or more stalked **sporangiophores** sprout, each bearing a sporangium. Each sporangium contains the products of meiosis: haploid nuclei that are incorporated into spores. These spores disperse and germinate to form a new generation of haploid hyphae.

The zygospore fungi include four major lineages of terrestrial fungi that live on soil as saprobes, as parasites of insects and spiders, or as mutualists of other fungi and invertebrate animals. They produce no cells with flagella, and only one diploid cell—the zygospore—appears in the entire life cycle. Their

Sporangia

Sporangiophores

Pilobolus sp.　　　150 μm

30.15 Zygospore Fungi Produce Sporangiophores These transparent structures are sporangiophores produced by a zygospore fungus growing on decomposing animal dung. The sporangiophores grow toward the light and end in tiny sporangia, which the stalked sporangiophores can eject as far as 2 meters. Animals ingest sporangia and then disseminate the spores in their feces.

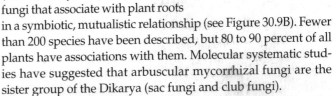

Go to Animated Tutorial 30.1
Life Cycle of a Zygospore Fungus
Life10e.com/at30.1

hyphae are coenocytic. These species do not form a fleshy fruiting structure; rather, the hyphae spread in a radial pattern from the spore, with occasional stalked sporangiophores reaching up into the air (**Figure 30.15**).

More than 1,000 species of zygospore fungi have been described. One species you may have seen is *Rhizopus stolonifer*, the black bread mold. *Rhizopus* produces many stalked sporangiophores, each bearing a single sporangium containing hundreds of minute spores (see Figure 30.14B).

Arbuscular mycorrhizal fungi form symbioses with plants

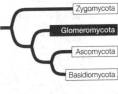

Arbuscular mycorrhizal fungi (Glomeromycota) are terrestrial fungi that associate with plant roots in a symbiotic, mutualistic relationship (see Figure 30.9B). Fewer than 200 species have been described, but 80 to 90 percent of all plants have associations with them. Molecular systematic studies have suggested that arbuscular mycorrhizal fungi are the sister group of the Dikarya (sac fungi and club fungi).

The hyphae of arbuscular mycorrhizal fungi are coenocytic. These fungi use glucose from their plant partners as their primary energy source, converting it into other, fungus-specific sugars that cannot return to the plant. Arbuscular mycorrhizal fungi reproduce asexually; there is not yet any direct evidence that they reproduce sexually.

The dikaryotic condition is a synapomorphy of sac fungi and club fungi

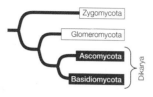

In the two remaining groups of fungi—the sac fungi and the club fungi—some stages have a nuclear configuration other than the familiar haploid or diploid states (see Figure 30.11). In these fungi, karyogamy (fusion of nuclei) occurs long after plasmogamy (fusion of cytoplasm), so that *two genetically different haploid nuclei coexist and divide within each cell of the mycelium* (**Figure 30.16**). This stage of the life cycle is called a **dikaryon** ("two nuclei"), and its ploidy is indicated as *n* + *n*. The dikaryon is a synapomorphy of the sac fungi and club fungi, which are placed together in a clade called **Dikarya**.

Eventually, specialized fruiting structures form, within which pairs of genetically dissimilar nuclei—one from each parent—fuse, giving rise to zygotes long after the original "mating." The diploid zygote nucleus then undergoes meiosis, producing four haploid nuclei. The mitotic descendants of those nuclei become spores, which germinate to give rise to the next haploid generation.

A life cycle with a dikaryon stage has several unusual features. First, there are no gamete *cells*, only gamete *nuclei*. Second, the only true diploid structure is the zygote, although for a long period the genes of both parents are present in the dikaryon and can be expressed. In effect, the dikaryon is neither diploid (2*n*) nor haploid (*n*); rather, it is dikaryotic (*n* + *n*). Therefore a harmful recessive mutation in one nucleus may be compensated for by a normal allele on the same chromosome in the other nucleus, and dikaryotic hyphae often have characteristics that are different from their *n* or 2*n* products. The dikaryotic condition is perhaps the most distinctive of the genetic peculiarities of the fungi.

The sexual reproductive structure of sac fungi is the ascus

The **sac fungi** (Ascomycota) are found in terrestrial, marine, and freshwater habitats. There are approximately 64,000 known species, nearly half of which are the fungal partners in lichens. The hyphae of sac fungi are segmented by more or less regularly spaced septa. A large pore in each septum permits movement of nuclei and organelles from one segment to the next.

Sac fungi are distinguished by the production of sacs called **asci** (singular *ascus*), which at maturity contain sexually produced haploid **ascospores** (see Figure 30.16A). The ascus is the characteristic sexual reproductive structure of the sac fungi. In the past, the sac fungi were classified on the basis of whether or not the asci are contained within a specialized fruiting structure known as an ascoma (plural *ascomata*) and on differences in the morphology of that fruiting structure. DNA sequence

30.16 Sexual Life Cycles among the Dikarya (A) In sac fungi, the products of meiosis are borne in a microscopic sac called an ascus. The fleshy fruiting structure, the ascoma, consists of both dikaryotic and haploid hyphae. (B) The basidium is the characteristic sexual reproductive structure of the club fungi. The fruiting structures, called basidiomata, consist solely of dikaryotic hyphae, and the dikaryotic phase can last a long time.

Go to Activity 30.2 Life Cycle of a Dikaryotic Fungus
Life10e.com/ac30.2

analyses have resulted in a revision of these traditional groupings, however.

SAC FUNGUS YEASTS Some species of sac fungi are unicellular yeasts. The 1,000 or so species in this group are among the most important domesticated fungi. Perhaps the best known is baker's, or brewer's, yeast (*Saccharomyces cerevisiae*; see Figure 30.2 and Section 30.4), which metabolizes glucose obtained from its environment into ethanol and carbon dioxide by fermentation. Other sac fungus yeasts live on fruits such as figs and grapes and play an important role in the making of wine. Many others are associated with insects; in the guts of some insects, they provide enzymes that break down materials that are otherwise difficult for the insects to digest, especially cellulose.

Sac fungus yeasts reproduce asexually by budding. Sexual reproduction takes place when two adjacent haploid cells of dissimilar mating types fuse. In some species, the resulting zygote buds to form a diploid cell population. In others, the zygote nucleus undergoes meiosis immediately; when this happens, the entire cell becomes an ascus. Depending on whether the products of meiosis then undergo mitosis, a yeast ascus contains either eight or four ascospores, which germinate to become haploid cells. The sac fungus yeasts have lost the dikaryon stage.

FILAMENTOUS SAC FUNGI Most sac fungi are filamentous species, such as the cup fungi (**Figure 30.17**), in which the ascomata are cup-shaped and can be as large as several centimeters across (although most are much smaller). The inner surfaces of the ascomata, which are covered with a mixture of specialized hyphae and asci, produce huge numbers of spores. The edible ascomata of some species, including morels and truffles, are regarded by humans as gourmet delicacies (and can sell at prices higher than gold). The underground ascomata of truffles have a strong odor that attracts mammals such as pigs, which then eat the fungi and disperse the spores.

The sexual reproductive cycle of filamentous sac fungi includes the formation of a dikaryon, although this stage is relatively brief compared with that in club fungi. Many filamentous sac fungi form multinucleate mating structures (see Figure 30.16A). Mating structures of two different mating types fuse and produce a dikaryotic mycelium, containing nuclei from both mating types. The dikaryotic mycelium often forms a cup-shaped ascoma, which bears the asci. Only after the formation of asci do the nuclei from the two mating types finally fuse.

(A) Sac fungi (Ascomycota)

In sac fungi, the products of meiosis are borne in a microscopic sac called an ascus. The fleshy fruiting bodies consist of both dikaryotic and haploid hyphae.

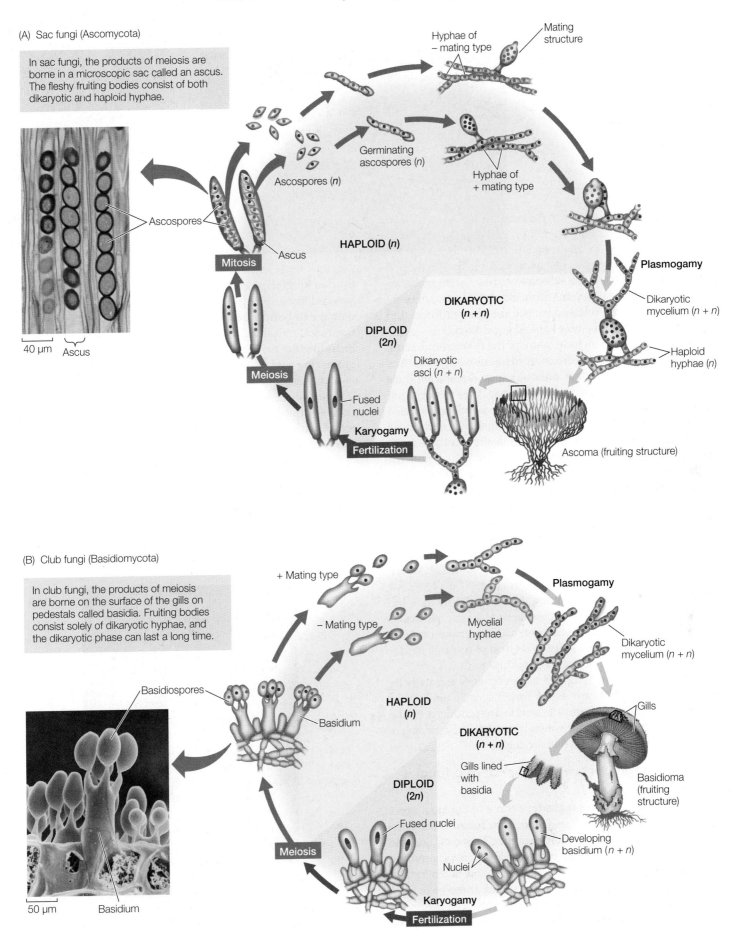

Hyphae of – mating type

Mating structure

Germinating ascospores (n)

Ascospores (n)

Hyphae of + mating type

Ascospores

Plasmogamy

HAPLOID (n)

Mitosis

Ascus

Dikaryotic mycelium (n + n)

40 μm   Ascus

DIKARYOTIC (n + n)

Haploid hyphae (n)

DIPLOID (2n)

Meiosis

Dikaryotic asci (n + n)

Fused nuclei

Karyogamy

Fertilization

Ascoma (fruiting structure)

(B) Club fungi (Basidiomycota)

In club fungi, the products of meiosis are borne on the surface of the gills on pedestals called basidia. Fruiting bodies consist solely of dikaryotic hyphae, and the dikaryotic phase can last a long time.

+ Mating type

Plasmogamy

– Mating type

Mycelial hyphae

Dikaryotic mycelium (n + n)

Basidiospores

Basidium

HAPLOID (n)

Gills

DIKARYOTIC (n + n)

Gills lined with basidia

DIPLOID (2n)

Basidioma (fruiting structure)

Fused nuclei

Meiosis

Developing basidium (n + n)

Nuclei

50 μm   Basidium

Karyogamy

Fertilization

30.17 Sac Fungi (A) These brilliant red cups are the ascomata of a cup fungus. (B) Morels, which have a spongelike ascoma and a subtle flavor, are considered a culinary delicacy by humans.

(A) *Aleuria aurantia*

(B) *Morchella esculenta*

Both nuclear fusion and the subsequent meiosis that produces haploid ascospores take place within individual asci. The ascospores are ultimately released (sometimes shot off forcefully) by the ascus to begin the new haploid generation.

The sac fungi also include many of the filamentous fungi known as molds. **Molds** consist of filamentous hyphae that do not form large ascomata, although they can still produce asci and ascospores. Many molds are parasites of flowering plants. Chestnut blight and Dutch elm disease are both caused by molds. The chestnut blight fungus, which was introduced to the United States in the 1890s, had destroyed the American chestnut as a commercial tree species by 1940. Before the blight, this species accounted for more than half the trees in eastern U.S. forests. Another familiar story is that of the American elm. Sometime before 1930, the Dutch elm disease fungus (first discovered in the Netherlands but native to Asia) was introduced into the United States on infected elm logs from Europe. Spreading rapidly—sometimes by way of connected root systems—the fungus destroyed great numbers of American elm trees.

Other plant pathogens among the sac fungi include the powdery mildews that infect cereal crops, lilacs, and roses, among many other plants. Mildews can be a serious problem to farmers and gardeners, and a great deal of research has focused on ways to control these agricultural pests.

The filamentous sac fungi can also reproduce asexually by means of conidia that form at the tips of specialized hyphae (**Figure 30.18**). Small chains of conidia are produced by the millions and can survive for weeks in nature. The conidia are what give molds their characteristic colors.

The sexual reproductive structure of club fungi is the basidium

Club fungi (Basidiomycota) produce some of the most spectacular fruiting structures found among the fungi. These fruiting structures, called **basidiomata** (singular *basidioma*), include mushrooms of all kinds, puffballs (see Figure 30.4), and the bracket fungi often encountered on trees and fallen logs in a damp forest. About 30,000 species of club fungi have been described. They include about 4,000 species of mushrooms, including both poisonous and edible species (**Figure 30.19A**).

Bracket fungi (**Figure 30.19B**) play an important role in the carbon cycle by breaking down wood; they also do great economic damage to both cut lumber and timber stands. Some of the most economically damaging plant pathogens are club fungi, including the rust fungi and smut fungi that parasitize cereal grains. In contrast, other club fungi contribute to the survival of plants as fungal partners in ectomycorrhizae.

The hyphae of club fungi characteristically have septa with small, distinctive pores. As they grow, haploid hyphae of different mating types meet and fuse, forming dikaryotic hyphae, each cell of which contains two nuclei, one from each parent hypha. The dikaryotic mycelium grows and eventually, when triggered by rain or another environmental cue, produces a basidioma. The dikaryon stage may persist for years—some club fungi live for decades or even centuries. This pattern contrasts with the life cycle of the sac fungi, in which the dikaryon is found only in the stages leading up to formation of the asci.

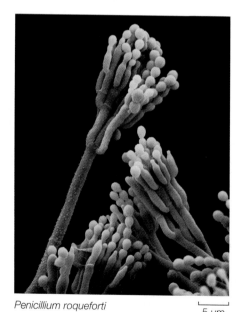

Penicillium roqueforti

5 μm

30.18 Conidia Chains of conidia (yellow) are developing at the tips of specialized hyphae arising from a *Penicillium roqueforti* mold. This species is used to produce "blue" cheese.

(A) *Armillaria* sp.

(B) *Laetiporus sulphureus*

30.19 Club Fungus Basidiomata
(A) These edible mushrooms are the fruiting structures of a honey fungus. Mycelia of this genus form connected underground masses that cover many hectares and are among the largest and longest-lived organisms in the world. (B) A bracket fungus growing parasitically on a tree. Although this particular species is edible, many similar-appearing bracket fungi are poisonous.

The **basidium** (plural *basidia*), a swollen cell at the tip of a specialized hypha, is the characteristic sexual reproductive structure of the club fungi (see Figure 30.16B). In mushroom-forming club fungi, the basidia typically form on specialized structures of the basidiomata known as gills. The basidium is the site of nuclear fusion and meiosis and thus plays the same role in the club fungi as the ascus does in the sac fungi and the zygosporangium does in the zygospore fungi.

After nuclei fuse in the basidium, the resulting diploid nucleus undergoes meiosis, and the four resulting haploid nuclei are incorporated into haploid **basidiospores**, which form on tiny stalks on the outside of the basidium. A single basidioma of the common bracket fungus *Ganoderma applanatum* can produce as many as 4.5 *trillion* basidiospores in one growing season. Basidiospores typically are forcibly discharged from their basidia and then germinate, and give rise to hyphae with haploid nuclei.

RECAP 30.3

Sexual reproduction is common in some groups of fungi but has never been observed in others. Many fungal species have two or more genetically distinct mating types. The sac fungi and club fungi share a dikaryotic condition, in which two genetically different haploid nuclei coexist in the same cell.

- Explain how microsporidia infect the cells of their animal hosts. **See p. 617 and Figure 30.12**
- What is the role of the zygospore in the life cycle of zygospore fungi? **See p. 619 and Figure 30.14B**
- Explain the dikaryon stage in terms of plasmogamy and karyogamy. **See p. 620 and Figure 30.16**
- What distinguishes the fruiting bodies of sac fungi from those of club fungi? **See pp. 620–623 and Figure 30.16**

Fungi are of special interest to biologists because of the roles they play in interactions with other organisms. But they are also useful as tools for studying many kinds of biological problems and for finding solutions to those problems.

30.4 What Are Some Applications of Fungal Biology?

We've briefly noted the important part that fungi play in the production of human foods and beverages. We have also described the diverse roles that fungi play in natural ecosystems, from decomposers to pathogens to mutualistic partners. These diverse ecological roles have led to the use of fungi in studies of environmental change and in remediation of environmental pollution. Many fungi are also important model organisms for laboratory investigations of basic biological process. Others, as we saw at the opening of this chapter, have given us treatments for human diseases.

Fungi are important in producing food and drink

Grains from grasses provide most of the world's food supply for humans. But in most cases, we do not eat these grains directly as they are produced by the plants. Instead, we use them as a source of starch. To make the starch more pleasing and digestible for human consumption, we usually convert it to more complex and tasty forms of food and drink, often with the help of fungi.

Baker's (or brewer's) yeast (*Saccharomyces cerevisiae*) converts the starch from grain into ethanol. This process also forms carbon dioxide bubbles in bread dough, causing it to rise, which gives baked bread its light texture. The ethanol and carbon dioxide are baked away in bread making (which produces the pleasant aroma of baking bread). In contrast, the ethanol and carbon dioxide are retained when yeast is used to ferment grain into beer. The carbon dioxide gives beer its fizz, and the alcohol contributes to the taste and appeal of beer to those who enjoy it. Sugars, especially from fruit such as grapes, are also converted into alcohol and carbon dioxide by yeasts in the production of wine (although the carbon dioxide is not retained in finished wine, as it is in beer). Many different strains of *S. cerevisiae* are used in wine production, which contributes to the distinctive nature of wine from different regions and wineries. Many other species of local, native yeasts are also used in producing distinctive local wines and beers. For example, fission yeast (*Schizosaccharomyces pombe*) was first isolated from African millet beer; it takes its specific name (*pombe*) from the Swahili word for beer.

Go to Media Clip 30.3
Time Lapse of Beer Formation
Life10e.com/mc30.3

Brown molds of the genus *Aspergillus* are important in some human diets. *Aspergillus tamarii* acts on soybeans in the production of soy sauce, and *A. oryzae* is used in brewing the Japanese alcoholic beverage sake from rice. *Aspergillus niger* is the source of most commercial citric acid production; citric acid gives food and soft drinks a tart taste and is also used as a food preservative. But some species of *Aspergillus* that grow on grains and on nuts such as peanuts and pecans produce extremely carcinogenic (cancer-inducing) compounds called aflatoxins. Aflatoxins can occur in high concentrations in foods such as peanut butter. In the United States and most other industrialized countries, moldy grain infected with *Aspergillus* is typically thrown out. In Africa, where food is scarcer, the grain gets eaten, moldy or not, and causes severe health problems, including high levels of certain cancers.

Penicillium is a genus of green molds, of which some species produce the antibiotic penicillin, as described in the beginning of this chapter. But several species of *Penicillium* are important for food production as well. For example, *P. camembertii* and *P. roqueforti* are the organisms responsible for the characteristic strong flavors of Camembert and Roquefort cheeses, respectively.

Many fungi serve directly as a human food source. Mushroom enthusiasts seek out the delicious fruiting structures of a wide variety of edible sac and club fungi. In the United States, relatively few species of mushrooms are grown commercially, and wild mushrooms are collected mostly for personal consumption. But in many parts of the world, a wide variety of wild mushrooms are collected for sale and consumption. Fungi used for food are not limited to fruiting bodies such as mushrooms, however. Various species of lichens are eaten in Arctic regions as well as in parts of North America and Asia. In southwestern China, for example, several species of lichens are used as a primary ingredient in cooking (**Figure 30.20**).

Fungi record and help remediate environmental pollution

Each year, biologists deposit samples of many groups of organisms in the collections of natural history museums. These museum collections serve many purposes, one of which is to document changes in the biota of our planet over time.

Collections of fungi made over many decades or centuries provide a record of the environmental pollutants that were present when the fungi were growing. Biologists can analyze these historical samples to see how different sources of pollutants were affecting our environment before anyone thought to take direct measurements. These long-term records are also useful for analyzing the effectiveness of cleanup efforts and regulatory programs for controlling environmental pollutants.

30.20 Some Lichens Are Edible In southwestern China, several species of lichens that grow on tree bark serve as a primary ingredient in the local cuisine.

We have already seen that fungi are critical to the planetary carbon cycle because of their role in breaking down dead organic matter. Fungi are also used in remediation efforts to help clean up sites that have been polluted by oil spills or contaminated with toxic petroleum-derived hydrocarbons. Many herbicides, pesticides, and other synthetic hydrocarbons are broken down primarily through the action of fungi.

Lichen diversity and abundance are indicators of air quality

Lichens can live in many harsh environments where few other species can survive, as we saw in Section 30.2. In spite of their hardiness, however, lichens are highly sensitive to air pollution because they are unable to excrete any toxic substances they absorb. This sensitivity means that lichens are good biological indicators of air pollution levels. It also explains why they are not commonly found in heavily industrialized regions or in large cities.

Sensitive biological indicators of pollution such as lichen growth allow biologists to monitor air quality without the use of specialized equipment. Monitoring the diversity and abundance of lichens growing on trees is a practical and inexpensive system for gauging air quality around cities (**Figure 30.21**). Maps of lichen diversity provide environmental biologists with a tool for tracking the spatial distribution of air pollutants and their effects. Lichens can also provide a long-term measure of the effects of air pollution across many seasons and years.

Fungi are used as model organisms in laboratory studies

Much of what we know about many basic aspects of cell and molecular biology comes from the study of model organisms. Among the eukaryotes, some fungi have numerous advantages over model plant and animals systems for laboratory investigations.

Of particular importance as model organisms are several species of sac fungi: *Aspergillus nidulans* (a brown mold), *Neurospora crassa* (a red bread mold), *Saccharomyces cerevisiae* (baker's, or brewer's, yeast), and *Schizosaccharomyces pombe* (fission yeast). These species can be cultured in large numbers in small spaces, and they have short generation times, so that genetic

WORKING WITH**DATA:**

Using Fungi to Study Environmental Contamination

Original Paper

Flegal, A. R., C. Gallon, S. Hibdon, Z. E. Kuspa, and L. F. Laporte. 2010. Declining—but persistent—atmospheric contamination in central California from the resuspension of historic leaded gasoline emissions as recorded in the lace lichen (*Ramalina menziesii* Taylor) from 1892 to 2006. *Environmental Science and Technology* 44: 5613–5618.

Analyze the Data

A. Russell Flegal and his colleagues analyzed over 100 years' worth of museum samples of lace lichens collected near San Francisco, California for evidence of lead contamination. They measured concentrations of lead (Pb) as well as the ratios of two lead isotopes, ^{206}Pb and ^{207}Pb. The isotope ratio was used to determine the source of lead contamination. Possible sources included a lead smelter that operated in the area from 1885 to 1971 (which produced emissions with a ^{206}Pb/^{207}Pb ratio of about 1.15–1.17); leaded gasoline in use from the 1930s to the early 1980s, peaking in 1970 (which produced automobile emissions with a ^{206}Pb/^{207}Pb ratio of 1.18–1.23); and resuspension of historic lead contamination as atmospheric aerosols in recent decades (with an intermediate ^{206}Pb/^{207}Pb ratio of about 1.16–1.19).

Before analyzing the data, use the information in the preceding paragraph to formulate hypotheses about the following two questions: What trends in atmospheric lead concentrations would you expect to see? What ^{206}Pb/^{207}Pb ratios would you expect to find at different times from the late 1800s to the early 2000s? After formulating your hypotheses, plot lead concentration in the lichen samples against year of sample collection. Make a second plot, this one of ^{206}Pb/^{207}Pb ratio against year of sample collection.

Sample	Year collected	Lead concentration (μg of Pb/g lichen)	^{206}Pb/^{207}Pb ratio
1	1892	11.9	1.165
2	1894	4.0	1.155
3	1906	13.7	1.154
4	1907	22.9	1.157
5	1945	49.9	1.187
6	1957	34.2	1.185
7	1978	50.9	1.221
8	1982	10.0	1.215
9	1983	4.6	1.224
10	1987	1.0	1.198
11	1988	1.3	1.199
12	1995	1.9	1.202
13	2000	0.4	1.184
14	2006	1.8	1.184

QUESTION 1

Do your analyses support the hypotheses you formulated?

QUESTION 2

Are your hypotheses consistent with your analyses of trends in both lead concentrations and ^{206}Pb/^{207}Pb ratios through time? If not, how would you modify your hypotheses, and what additional tests can you design to test your ideas?

Go to BioPortal for all WORKING WITH**DATA** exercises

30.21 More Lichens, Better Air Lichen abundance and diversity are excellent indicators of air quality. (A) In suitable environments with few pollutants in the air, many lichen species show luxuriant growth on trees. These lichens are growing on an oak tree 150 kilometers west of Austin, Texas, in an area relatively free of human habitation and industry. (B) As air quality declines, so do the number and diversity of lichens. These lichens are growing on an oak tree inside the city of Austin, where air quality is reduced (primarily by automobile exhaust).

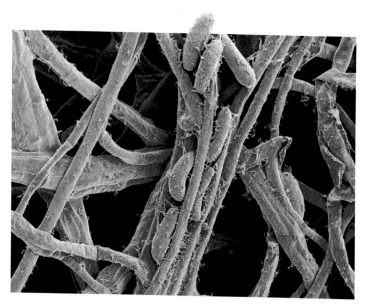

30.22 A Pathogenic Fungus Attacks a Parasitic Plant The fungus *Fusarium oxysporum* is a potent pathogen of witchweed (*Striga*), a parasitic plant that attacks crops. The fungal spores are shown in blue; fungal filaments are in tan. Both colors were added to enhance this scanning electron micrograph.

investigations can be conducted in days rather than years. Furthermore, their genomes are relatively small and encode relatively few genes compared with those of most plants and animals, so it is easier to elucidate the functions of the fungal genes responsible for basic biological functions.

Reforestation may depend on mycorrhizal fungi

When a forest is cut down, it is not just the trees that are lost. A forest is an ecosystem that depends on the interactions of many species. As we have discussed, many plants depend on close relationships with mycorrhizal fungal partners. When trees are removed from a site, the populations of mycorrhizal fungi there decline rapidly. If we wish to restore the forest on the site, we cannot simply replant it with trees and other plants and expect them to survive. The mycorrhizal fungal community must be reestablished as well. For large forest restoration projects, a planned succession of plant growth and soil improvement is often necessary before forest trees can be replanted. As the community of soil fungi gradually recovers, trees that have been inoculated with appropriate mycorrhizal fungi in tree nurseries can be planted to reintroduce greater diversity to the soil fungal community.

Fungi provide important weapons against diseases and pests

We started this chapter with the story of the discovery of penicillin. The discovery of antibiotics produced by fungi revolutionized medical treatment of bacterial diseases in humans and their domestic animals. Live strains of fungi are also used to combat various pest species of plants and animals.

In Africa, the parasitic plant witchweed (*Striga*) causes crop losses of about $7 billion every year. A group of Canadian

biologists discovered that a strain of the mold *Fusarium oxysporum* could be applied to crops to control witchweed without harming the crop plants (**Figure 30.22**). Other strains of *Fusarium* that preferentially attack coca plants, the source of cocaine, have been proposed to combat illicit drug production. Still other fungi are used to attack various animal pests, such as termites and aphids, and even malaria-carrying mosquitoes, as we'll see below.

RECAP **30.4**

Fungi are important to humans in many ways. Some species are consumed directly as food, while others are important in food production. Fungi serve as important indicators of ecosystem health and are critical in reforestation and in pollution remediation efforts. Several species are important model organisms for studies of eukaryote cell and molecular biology. Fungi are widely used to combat diseases and pests.

- What are some of the ways in which fungi or fungal products contribute to the human food supply? **See pp. 623–624**

- What are some advantages of using surveys of lichen diversity and museum collections of lichens to measure long-term changes in air quality, compared with direct measurements of atmospheric pollutants? **See pp. 624–625 and Figure 30.21**

- Why are some fungi particularly appropriate as model organisms for the study of eukaryote cell and molecular biology? **See pp. 625–626**

Whether living on their own or in symbiotic associations, fungi have spread successfully over much of Earth since their origin from a protist ancestor. An earlier ancestor of fungi also gave rise to the choanoflagellates and the animals, which we will describe in the following three chapters.

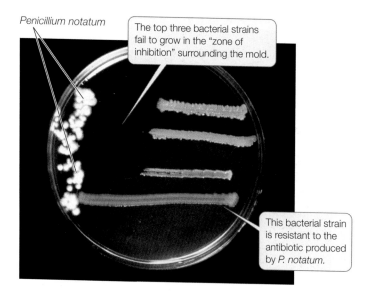

Penicillium notatum

The top three bacterial strains fail to grow in the "zone of inhibition" surrounding the mold.

This bacterial strain is resistant to the antibiotic produced by *P. notatum*.

30.23 Penicillin Resistance In a petri dish similar to those in Alexander Fleming's lab, four strains of a pathogenic bacterium have been cultured along with *Penicillium* mold. One strain is resistant to the mold's antibiotic substance, as is evidenced by its growth up to the mold.

?

Have antibiotics derived from fungi eliminated the danger
of bacterial diseases in human populations?

ANSWER

Beginning in the 1940s, antibiotics derived from fungi ushered in a "golden age" of freedom from bacterial infections. Today, however, many of these antibiotics are losing their effectiveness as pathogenic bacteria evolve resistance to them (**Figure 30.23**)

Most medical antibiotics are chemically modified forms of substances that are found naturally in fungi and other organisms. Fungi naturally produce antibiotic compounds to defend themselves against bacterial growth and to reduce competition from bacteria for nutritional resources. These naturally occurring compounds are usually chemically modified to increase their stability, to improve their effectiveness, and to facilitate synthetic production. From the late 1950s to the late 1990s,

no new major classes of antibiotics were discovered. In recent years, however, three new classes of antibiotics have been synthesized based on information learned from naturally occurring, fungally derived antibiotics, leading to improved treatment of some formerly resistant strains of bacteria.

Fungi have also been used to combat non-bacterial diseases. One of the more unusual applications of fungi is in the war against malaria. Biologists have discovered that two species of fungi, *Beauveria bassiana* and *Metarhizium anisopliae*, can kill malaria-causing mosquitoes when applied to mosquito netting. Mosquitoes have not yet shown evidence of developing resistance to these biological pathogens.

CHAPTER**SUMMARY** 30

30.1 What Is a Fungus?

- Fungi are distinguished from other **opisthokonts** by **absorptive heterotrophy** and by the presence of **chitin** in their cell walls. **Review Figure 30.1**

- Some fungi are **saprobes**, others are parasites, and some are mutualists.

- **Yeasts** are unicellular, free-living fungi.

- The body of a multicellular fungus is a **mycelium**—a meshwork of filaments called **hyphae**. Hyphae may be **septate** (having **septa**) or **coenocytic** (multinucleate). **Review Figure 30.3**

30.2 How Do Fungi Interact with Other Organisms?

- Saprobic fungi, which act as decomposers, make crucial contributions to the recycling of elements, especially carbon.

- Many fungi are parasites, harvesting nutrients from host cells by means of **haustoria**. **Review Figure 30.5**

- Certain fungi have relationships with other organisms that are **symbiotic** and **mutualistic**.

- Some fungi associate with unicellular green algae, cyanobacteria, or both to form **lichens**, which live on exposed surfaces of rocks, trees, and soil. **Review Figure 30.8**

- **Mycorrhizae** are mutualistic associations of fungi with plant roots. They improve a plant's ability to take up nutrients and water.

- **Endophytic fungi** live within plants and may provide their hosts with protection from herbivores and pathogens.

30.3 How Do Major Groups of Fungi Differ in Structure and Life History?

- The **microsporidia**, **chytrids**, and **zygospore fungi** diversified early in fungal evolution. The arbuscular mycorrhizal fungi, sac fungi, and club fungi form a monophyletic group, and the latter two groups form the clade Dikarya. **Review Figure 30.10, Table 30.1, ACTIVITY 30.1**

- Many species of fungi reproduce both sexually and asexually. In many fungi, sexual reproduction occurs between individuals of different **mating types**. **Review Figure 30.11**

- The **microsporidia** are highly reduced unicellular fungi. They are obligate intracellular parasites of animals.

- The three distinct lineages of **chytrids** all include species with flagellated gametes. **Review Figure 30.14A**

- In the sexual reproduction of terrestrial fungi, **plasmogamy** (fusion of cytoplasm) precedes **karyogamy** (fusion of nuclei).

- **Zygospore fungi** have a resting stage known as a **zygospore**, which contains many diploid nuclei. Their spores are dispersed from simple stalked **sporangiophores**. **Review Figure 30.14B, ANIMATED TUTORIAL 30.1**

- **Arbuscular mycorrhizal fungi** form symbiotic associations with plant roots. They are only known to reproduce asexually. Their hyphae are coenocytic.

- In sac fungi and club fungi, a mycelium containing two genetically different haploid nuclei, called a **dikaryon**, is formed. The **dikaryotic** ($n + n$) condition is unique to the fungi. **Review Figure 30.16, ACTIVITY 30.2**

- **Sac fungi** have septate hyphae with large pores; their sexual reproductive structures are **asci**. Some sac fungi are unicellular yeasts. Many filamentous sac fungi produce fleshy fruiting structures called **ascomata**. The dikaryon stage in the sac fungus life cycle is relatively brief. **Review Figure 30.16A**

- **Club fungi** have septate hyphae with distinctive small pores. Their fruiting structures are called **basidiomata**, and their sexual reproductive structures are **basidia**. The dikaryon stage may last for years. **Review Figure 30.16B**

30.4 What Are Some Applications of Fungal Biology?

- Some fungi are consumed as food by humans; other fungi are critical in baking, fermentation, and flavoring food.

- Fungi play important roles in cleaning up environmental pollutants such as synthetic petroleum-derived hydrocarbons.

continued

- The diversity and abundance of lichen growth on trees is a sensitive indicator of air quality.
- Reforestation projects require restoration of the mycorrhizal fungal community.

- Several species of fungi are important model organisms.
- Fungi provide important weapons against diseases and pests.

 Go to the Interactive Summary to review key figures, Animated Tutorials, and Activities
Life10e.com/is30

CHAPTER**REVIEW**

▮▮ REMEMBERING

1. The absorptive heterotrophy of fungi is aided by
   a. dikaryon formation.
   b. spore formation.
   c. the fact that they are all parasites.
   d. their large surface area-to-volume ratio.
   e. their possession of chloroplasts.

2. Which of the following is a reason that lichens can be useful indicators of environmental change?
   a. Lichens excrete the toxic substances that they absorb, and these excretions can be used to measure local contamination.
   b. Because lichens retain toxins, historical collections of lichens can be used to measure past atmospheric conditions.
   c. Atmospheric pollutants (such as ^{206}Pb) stimulate lichen growth, so heavy lichen growth is a sign of air pollution.
   d. Atmospheric pollution stimulates mycorrhizal activity, which inhibits lichen growth.
   e. None of the above

3. Which statement about the dikaryon stage is *not* true?
   a. The cytoplasm of two cells fuses before their nuclei fuse.
   b. The two haploid nuclei are genetically different.
   c. The two nuclei are of the same mating type.
   d. The dikaryon stage ends when the two nuclei fuse.
   e. Not all fungi have a dikaryon stage.

4. Microsporidia
   a. lack true mitochondria.
   b. are parasites of animals.
   c. contain mitosomes.
   d. are among the smallest eukaryotes known.
   e. All of the above

5. Which statement about lichens is *not* true?
   a. They can reproduce by fragmentation of the vegetative body.
   b. They are often the first colonists in a new area.
   c. They render their environment more basic (alkaline).
   d. They contribute to soil formation.
   e. They may contain less than 10 percent water by weight.

▮▮ UNDERSTANDING & APPLYING

6. You are shown an object that looks superficially like a pale green mushroom. Describe at least three criteria (including anatomical and chemical traits) that would enable you to tell whether the object is a piece of a plant or a piece of a fungus.

7. Many fungi are dikaryotic during part of their life cycle. Why are dikaryons described as $n + n$ instead of $2n$?

8. If all the fungi on Earth were suddenly to die, how would the surviving organisms be affected?

▮▮ ANALYZING & EVALUATING

9. Consider the following data for lichens at five survey sites:

Site number	Number of lichen species	Tree branches covered in lichens (%)
1	5	38
2	1	2
3	3	15
4	8	75
5	13	100

Predict the relative order of the sites with respect to their distance from the center of a large city. Other factors (besides distance to city center, such as prevailing wind direction) might affect your prediction. Can you think of two other major factors that might influence these results?

10. When biologists isolate DNA from whole-plant samples to study plant genomes, the investigators sometimes find that some of the genes they isolate appear to be more closely related to fungal genes than they are to the genes of close relatives of the plants they are studying. What is a likely explanation of this observation? How could you test your hypothesis?

Go to BioPortal at **yourBioPortal.com** for Animated Tutorials, Activities, LearningCurve Quizzes, Flashcards, and many other study and review resources.

Animal Origins and the Evolution of Body Plans

N 1883 THE ZOOLOGIST FRANZ SCHULZE noticed something unusual in his Austrian laboratory: transparent, flattened organisms were crawling on the sides of his saltwater aquarium. These organisms, which Schulze had collected accidentally along with the sponges that were his primary interest, appeared to be animals, but they were unlike any animals previously described—especially since they continually changed shape as they moved.

Further examination revealed that the new organisms were indeed animals. Structurally, however, they were among the simplest animals that Schulze—or anyone else—had ever observed, being made up of only four types of cells. He named the new species *Trichoplax adhaerens*, which means "sticky hairy plate," and argued that the new species had no close relationships with other major animal groups. For decades, however, most biologists dismissed Schulze's findings, insisting that the transparent organisms must be larval forms of other, well-known, animals.

In the 1960s more detailed studies confirmed the distinctive nature of *Trichoplax*. Even then, this odd animal continued to be known almost exclusively from aquariums. Only in the past decade have biologists located natural populations of *T. adhaerens* growing on hard surfaces in tropical and subtropical coastal regions. A few additional closely related species have been discovered (although most have not yet been formally named). Collectively, these species are known as placozoans (Greek, "flat animals").

The more biologists have studied *Trichoplax*, the odder this animal appears. It has the smallest genome of any animal studied to date. The mature stages lack body symmetry and have no mouth, gut, or nervous system. Is

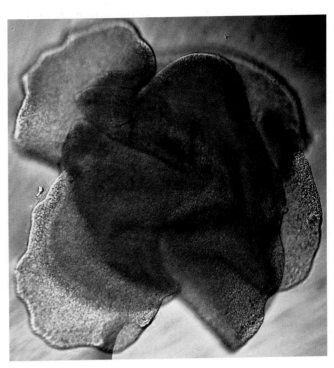

Did Placozoans Diverge at the Root of the Animal Tree? Answering that question requires an understanding of animal phylogeny. Although placozoans are morphologically simple and have the smallest genome of any animal studied to date, they may have descended from ancestors with more complex body plans.

Trichoplax a relict representative of a group of animals that appeared early in animal evolution?

Which groups of organisms are at the root of the animal tree is currently a subject of considerable investigation and debate. Several possible hypotheses of relationships are being explored with genomic analyses. The structural simplicity of *Trichoplax* is now considered by most biologists to be an evolutionary reversal from a more complex body form. Most genomic studies point to other groups of animals as forming the earliest split with the remaining species.

This chapter explores the earliest branches on the animal tree and shows how a few fundamental "body plans" have been modified to yield the remarkable variety of animal forms described in this and the following two chapters.

Which animal groups are involved in the earliest split in the animal tree?

See answer on p. 648.

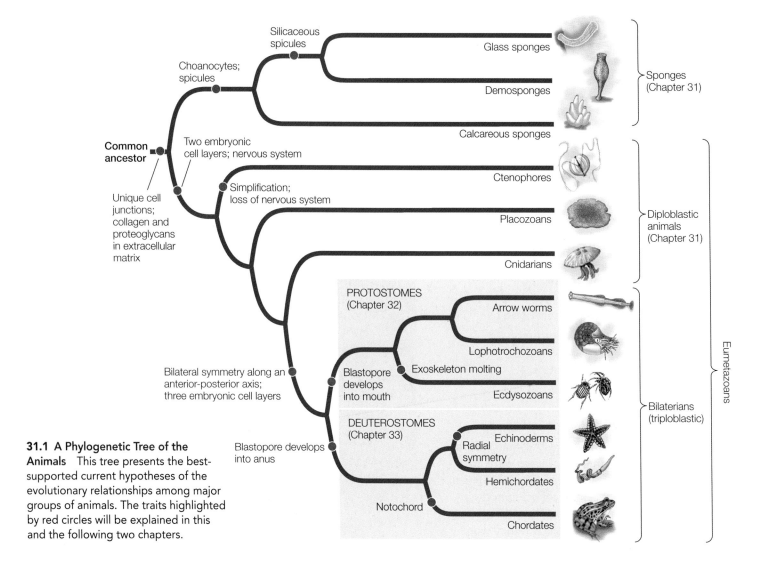

31.1 A Phylogenetic Tree of the Animals This tree presents the best-supported current hypotheses of the evolutionary relationships among major groups of animals. The traits highlighted by red circles will be explained in this and the following two chapters.

31.1 What Characteristics Distinguish the Animals?

How do we recognize an organism as an animal? That may seem obvious for many familiar animals, but less so for groups such as sponges, which were once thought to be plants. Some of the general characteristics we associate with animals include:

- *Multicellularity.* In contrast to the Bacteria, Archaea, and most protists (see Chapters 26 and 27), all animals are multicellular. Animal life cycles feature complex patterns of development from a single-celled zygote into a multicellular adult.

- *Heterotrophic metabolism.* In contrast to most plants, all animals are heterotrophs. Animals are able to synthesize very few organic molecules from inorganic chemicals, so they must obtain the necessary organic molecules from their environment.

- *Internal digestion.* Although the fungi are also heterotrophs (see Chapter 30), animals and fungi digest their food differently. While fungi digest food outside their bodies, most animals use internal processes to break down materials from their environment into the organic molecules they need. Most animals ingest food into an internal **gut** that is continuous with the outside environment and in which digestion takes place.

- *Movement and nervous systems.* In contrast to the majority of plants and fungi, most animals can move their bodies. This movement is often coordinated through a well-developed nervous system, which also typically functions as a sensory system. Animals must move to find food or bring food to them. Muscle tissue and nervous systems are unique to animals, and many animal body plans are specialized for movement and detection of prey.

Although these general features help us recognize animals, none is diagnostic for all animals. Some animals do not move, at least during certain life stages, and some plants and fungi do have limited movement. Some animals lack a centralized nervous system, and some lack a gut. Many multicellular organisms are not animals. On what basis do we group all animals together in a single clade?

Animal monophyly is supported by gene sequences and morphology

The most convincing evidence that all the organisms considered to be animals share a common ancestor comes from phylogenetic analyses of their gene sequences. Relatively few complete animal genomes are available, but more are being sequenced each year. Analyses of these genomes, as well as of many individual gene sequences, have shown that the animals are indeed monophyletic. The best-supported phylogenetic

WORKING WITH**DATA:**

Reconstructing Animal Phylogeny

Original Paper

Dunn, C. W. and 17 others 2008. Broad phylogenomic sampling improves resolution of the Animal Tree of Life. *Nature* 452: 745–749.

Analyze the Data

Several breakthroughs in our understanding of animal phylogeny have occurred in recent years as the sequences of genes and proteins have been compared across species. Casey Dunn and his colleagues compared sequences from many different proteins across a wide variety of animal groups. The table below is a sample of their data that can be used to reconstruct the relationships among these representative species. In the original paper, Dunn and colleagues reported on 11,234 amino acid positions among 77 species of animals. Twenty-seven of these amino acid positions for ten of those species are shown in the table.

QUESTION 1

As you did in Chapter 22 (see Section 22.2), construct a phylogenetic tree of these ten species using the parsimony method. Use the outgroup (a choanoflagellate) to root your tree. Assume that all changes from one amino acid to another are equally likely.

QUESTION 2

How many character state changes (i.e., changes from one amino acid to another) occur along each branch on your tree?

QUESTION 3

Which amino acid positions (i.e., which character numbers) exhibit homoplasy (convergence or reversal of the character state)?

QUESTION 4

Which group on your tree represents the bilaterian animals? The protostomes? The deuterostomes?

| Species | Character state (amino acid at position) | | | | | | | | | | | | | | | | | | | | | | | | | | |
	1	2	3	4	5	6	7	8	9	10	11	12	13	14	15	16	17	18	19	20	21	22	23	24	25	26	27
Outgroup (choano-flagellate)	Y	G	L	G	Q	D	P	N	F	P	K	S	F	S	V	A	L	T	V	I	R	Q	N	L	V	I	L
Clam	Y	S	T	G	L	H	E	N	Y	A	R	A	M	R	I	A	L	T	I	V	K	L	S	I	V	I	L
Earthworm	Y	A	T	G	L	H	E	N	Y	P	H	A	M	R	I	A	L	T	I	V	K	L	S	I	V	M	L
Tardigrade	Y	A	T	G	L	H	E	H	Y	K	R	A	M	R	V	A	T	S	I	V	R	L	N	L	V	L	L
Fruit fly	F	A	T	G	L	H	E	N	Y	K	R	A	M	R	I	A	L	S	I	V	S	L	D	L	V	L	L
Sea urchin	Y	A	T	G	L	L	E	N	Y	P	N	A	M	R	I	A	L	T	V	I	R	Q	N	L	T	V	K
Human	W	A	A	G	L	R	E	H	Y	P	K	A	I	R	I	S	V	T	V	I	R	Q	N	L	T	V	K
Chicken	W	A	A	G	L	R	E	H	Y	P	R	A	I	R	I	A	V	T	V	I	R	Q	N	L	T	V	K
Lancelet	Y	A	T	G	L	R	E	H	Y	P	K	A	M	R	I	A	V	T	V	I	R	L	N	L	T	V	K
Sponge	Y	G	L	S	L	R	P	N	F	P	K	S	M	S	V	A	L	T	V	I	R	Q	N	L	V	I	L

*Go to BioPortal for all WORKING WITH**DATA** exercises*

tree for the major animal groups is shown in **Figure 31.1. Table 31.1** summarizes the living members of those groups.

Although animals were considered to belong to a single clade long before gene sequencing became possible, surprisingly few morphological features are shared across all species of animals. Two morphological synapomorphies have been identified that distinguish the animals:

- A common set of extracellular matrix molecules, including collagen and proteoglycans (see Figure 5.22)
- Unique types of junctions between cells (tight junctions, desmosomes, and gap junctions; see Figure 6.7)

Although some animals in a few groups lack one or the other of these traits, it is believed that these traits were possessed by the ancestor of all animals and subsequently lost in those groups. Similarities among animals in the organization and function of Hox and other developmental genes (see Chapter 20) provide additional evidence of developmental mechanisms shared by a common animal ancestor.

The common ancestor of animals was probably a colonial flagellated protist similar to existing colonial choanoflagellates. Choanoflagellate colonies have clear similarities to the multicellular sponges (**Figure 31.2**). The best-supported hypothesis of animal origins postulates a choanoflagellate-like lineage in which certain cells in the colony became specialized—some for movement, others for nutrition, others for reproduction, and so on. Once this functional specialization had begun, cells could have continued to differentiate. Coordination among groups of cells could have improved by means of specific regulatory and signaling molecules that guided differentiation and migration of cells in developing embryos. Such coordinated groups of cells eventually could have evolved into the larger and more complex organisms that we call animals.

Nearly 80 percent of the 1.8 million named species of living organisms are animals, and millions of additional animal species await discovery. Evidence for the evolutionary relationships among animal groups can be found in fossils, in patterns of embryonic development, in the morphology

TABLE 31.1
Summary of Living Members of the Major Animal Groups

Group	Approximate Number of Living Species Described	Major Subgroups, Other Names, and Notes
Sponges	8,500	Demosponges, glass sponges, calcareous sponges
Ctenophores	250	Comb jellies
Placozoans	2	Additional species have been discovered but not yet formally named
Cnidarians	12,500	Anthozoans: corals, sea anemones; Hydrozoans: hydras and hydroids; Scyphozoans: jellyfish; Myxozoans: parasitic mucous animals; sometimes placed in group distinct from cnidarians
Orthonectids	45	Microscopic wormlike parasites of marine invertebrates; relationships uncertain
Rhombozoans	125	Tiny (0.5–7 mm) parasites of cephalopods; relationships uncertain
PROTOSTOMES		
Arrow worms	180	Glass worms
Lophotrochozoans		
Bryozoans	5,500	Moss animals
Entoprocts	170	Sessile aquatic animals, 0.1–7 mm long, superficially similar to bryozoans
Flatworms	30,000	Free-living flatworms; flukes and tapeworms (all parasitic); monogeneans (ectoparasites of fishes)
Gastrotrichs	800	"Hairy backs"
Rotifers and relatives	3,000	Rotifers, spiny-headed worms, and jaw worms
Ribbon worms	1,200	Proboscis worms
Brachiopods	450	Lampshells
Phoronids	10	Sessile marine filter feeders
Annelids	19,000	Polychaetes (generally marine; may not be monophyletic); Clitellates: earthworms, freshwater worms, leeches
Mollusks	117,000	Monoplacophorans; Chitons; Bivalves: clams, oysters, mussels; Gastropods: snails, slugs, limpets; Cephalopods: squids, octopuses, nautiloids
Ecdysozoans		
Kinorhynchs	180	Mud dragons
Loriciferans	30	Brush heads
Priapulids	20	Penis worms
Nematodes	25,000	Roundworms
Horsehair worms	350	Gordian worms
Tardigrades	1,200	Water bears
Onychophorans	180	Velvet worms
Arthropods		
Chelicerates	114,000	Horseshoe crabs, pycnogonids, and arachnids (scorpions, harvestmen, spiders, mites, ticks)
Myriapods	12,000	Millipedes, centipedes
Crustaceans	67,000	Crabs, shrimps, lobsters and crayfishes, barnacles, copepods
Hexapods	1,020,000	Insects and their wingless relatives
DEUTEROSTOMES		
Xenoturbellids	2	Secondarily simple marine worms; relationships uncertain
Acoels	400	Very small (mostly <2 mm) flattened marine worms; relationships uncertain
Echinoderms	7,500	Crinoids (sea lilies and feather stars); brittle stars; sea stars; sea daisies; sea urchins; sea cucumbers
Hemichordates	120	Acorn worms and pterobranchs
Tunicates	2,800	Sea squirts (ascidians), salps, and larvaceans
Lancelets	35	Cephalochordates
Vertebrates	65,000	Hagfishes, lampreys, cartilaginous fishes, ray-finned fishes, coelacanths, lungfishes, amphibians, reptiles (including birds), and mammals

(A) Choanoflagellate protists

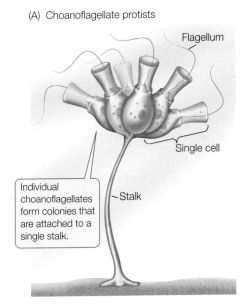

Flagellum

Single cell

Individual choanoflagellates form colonies that are attached to a single stalk.

Stalk

(B)

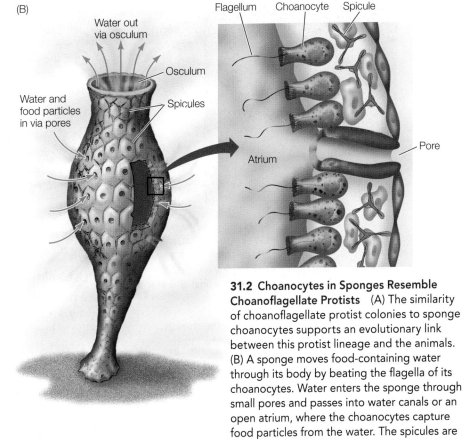

Water out via osculum

Osculum

Water and food particles in via pores

Spicules

Flagellum   Choanocyte   Spicule

Pore

Atrium

31.2 Choanocytes in Sponges Resemble Choanoflagellate Protists (A) The similarity of choanoflagellate protist colonies to sponge choanocytes supports an evolutionary link between this protist lineage and the animals. (B) A sponge moves food-containing water through its body by beating the flagella of its choanocytes. Water enters the sponge through small pores and passes into water canals or an open atrium, where the choanocytes capture food particles from the water. The spicules are supportive, skeletal structures.

and physiology of living animals, in the structure of animal proteins, and in gene sequences. Increasingly, studies of the phylogenetic relationships among major animal groups have come to depend on genomic sequence comparisons.

A few basic developmental patterns differentiate major animal groups

Differences in patterns of embryonic development have until recently provided many of the important clues to animal phylogeny. Analyses of gene sequences, however, are now showing that some developmental patterns are more evolutionarily variable than previously thought. Here we describe the basic developmental patterns that vary among the major animal clades.

The early cell divisions of an embryo are known as **cleavage**. As described in Section 44.1, several different patterns of cleavage exist among animals. Although these patterns can be useful for characterizing major animal groups, genomic analyses have shown that many changes have occurred in cleavage patterns throughout animal evolution.

Cleavage patterns are influenced by the configuration of the yolk, the acellular nutritive material that nourishes the growing embryo. The eggs of many animal groups contain a small amount of yolk that is evenly distributed throughout the egg cytoplasm. In some of these groups, the zygote and its descendant cells divide completely and evenly in a pattern known as **radial cleavage**. Radial cleavage is thought to be the ancestral condition for the animals other than sponges, as it is widely distributed among the major lineages. **Spiral cleavage**—a complicated permutation of radial cleavage—is found among many lophotrochozoans (a group that includes earthworms and clams). Lophotrochozoans with spiral cleavage are thus sometimes known as spiralians. The early branches of the ecdysozoans (molting animals, such as insects and nematodes) have radial cleavage, but most ecdysozoans have an idiosyncratic

cleavage pattern that is neither radial nor spiral in organization (see Figure 44.3C). In reptiles, the presence of a large body of yolk within the fertilized egg creates an incomplete cleavage pattern in which the dividing cells form an embryo on top of the yolk mass.

Distinct layers of cells form during the early development of most animals. These cell layers differentiate into specific organs and organ systems as development continues. The embryos of **diploblastic** animals have two cell layers: an outer **ectoderm** and an inner **endoderm**. Embryos of **triploblastic** animals have, in addition to ectoderm and endoderm, a third distinct cell layer, **mesoderm**, which lies between the ectoderm and the endoderm. The existence of three cell layers in embryos is a synapomorphy of the triploblastic animals (which form a clade), whereas the diploblastic animals (ctenophores, placozoans, and cnidarians, which are not a clade) exhibit the ancestral condition (see Figure 31.1). Some biologists consider sponges to be diploblastic, but since they do not have clearly differentiated tissue types or embryonic cell layers, the term is not usually applied to them.

During early development in many animals, in a process known as gastrulation, a hollow ball one cell thick indents to form a cup-shaped structure. The opening of the cavity formed by this indentation is called the blastopore (**Figure 31.3**). The process of gastrulation is detailed in Section 44.2; the point to remember here is that the *overall pattern* of gastrulation immediately after formation of the blastopore divides the triploblastic animals into two major groups:

Blastopore

31.3 Gastrulation Illuminates Evolutionary Relationships The blastopore is clear in this scanning electron micrograph of a sea urchin gastrula. Because sea urchins (echinoderms) are deuterostomes, this blastopore will eventually become the anal end of the animal's gut.

- In the **protostomes** (Greek, "mouth first"), the mouth arises from the blastopore, and the anus forms later.

- In the **deuterostomes** ("mouth second"), the blastopore becomes the anus, and the mouth forms later.

Although the developmental patterns of animals are more varied than suggested by this simple dichotomy, sequencing data indicate that the protostomes and deuterostomes represent distinct animal clades. Together these two groups are known as the **bilaterians** (named for their usual bilateral symmetry), and they account for the vast majority of animal species.

RECAP 31.1

The animals are thought to be monophyletic because they share several derived traits, especially among their gene sequences. Major developmental differences also provide evidence of their evolutionary relationships.

- What general features of animals distinguish this group from other living organisms? **See p. 630**

- Describe the difference between diploblastic and triploblastic embryos. **See p. 633**

- Describe the difference between protostomes and deuterostomes. **See p. 634**

We will begin our exploration of animal diversity by discussing the general features of animal body plans. Later in this chapter we will describe several groups of animals that diverged before the origin of the bilaterians. We will devote Chapter 32 to the protostomes and Chapter 33 to the deuterostomes.

31.2 What Are the Features of Animal Body Plans?

The general structure of an animal, the arrangement of its organ systems, and the integrated functioning of its parts are referred to as its **body plan**. As Chapter 20 described, the regulatory and signaling genes that govern the development of body symmetry, body cavities, segmentation, and appendages are widely shared among the different animal groups. Thus we might expect animals to share body plans. Although animal body plans vary tremendously, they can be seen as variations on five key features:

- The *symmetry* of the body
- The structure of the *body cavity*
- The *segmentation* of the body
- *External appendages* that are used for sensing, chewing, locomotion, mating, and other functions
- The development of the *nervous system*

Each of these features affects how an animal moves and interacts with its environment.

Most animals are symmetrical

The overall shape of an animal can be described by its **symmetry**. An animal is said to be symmetrical if it can be divided along at least one plane into similar halves. Animals that have no plane of symmetry are said to be asymmetrical. Placozoans and many sponges are asymmetrical, but most other animals have some kind of symmetry, which is governed by the expression of regulatory genes during development.

The simplest form of symmetry is **spherical symmetry**, in which body parts radiate out from a central point. An infinite number of planes passing through the central point can divide a spherically symmetrical organism into similar halves. Spherical symmetry is widespread among unicellular protists, but most animals possess other forms of symmetry.

In organisms with **radial symmetry**, body parts are arranged around one main axis at the body's center (**Figure 31.4A**). Ctenophores (comb jellies) are radially symmetrical, as are many cnidarians (such as sea anemones and jellyfishes) and echinoderms. A perfectly radially symmetrical animal can be divided into similar halves by any plane that contains the main axis. However, most radially symmetrical animals—including the adults of echinoderms such as sea stars—are slightly modified, so that only some planes can divide them into identical halves. Some radially symmetrical animals are sessile (they remain fixed in one place) or drift with water currents. Others move about slowly but can move equally well in any direction.

Bilateral symmetry is characteristic of animals that have a distinct front end, which typically precedes the rest of the body as the animal moves. A bilaterally symmetrical animal can be divided into mirror-image (left and right) halves by a single plane that passes through the midline of its body. This plane runs from the front, or **anterior**, end of the body, to the rear, or **posterior**, end (**Figure 31.4B**). A plane at right angles to the midline divides the body into two dissimilar sides. The back of a bilaterally symmetrical animal is its **dorsal** surface; the underside is its **ventral** surface.

Bilateral symmetry is strongly correlated with **cephalization** (Greek *kephalos*, "head"), which is the concentration of sensory organs and nervous tissues at the anterior end of the animal.

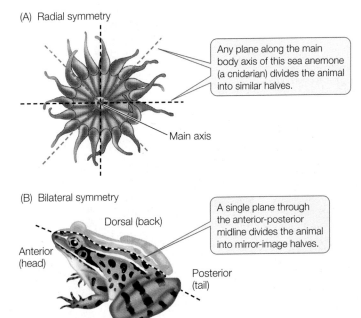

(A) Radial symmetry

> Any plane along the main body axis of this sea anemone (a cnidarian) divides the animal into similar halves.

Main axis

(B) Bilateral symmetry

Dorsal (back)

Anterior (head)

> A single plane through the anterior-posterior midline divides the animal into mirror-image halves.

Posterior (tail)

Ventral (belly)

31.4 Body Symmetry Most animals are either (A) radially or (B) bilaterally symmetrical.

Cephalization has been favored by natural selection because the anterior end of a bilaterally symmetrical animal typically encounters new environments first.

The structure of the body cavity influences movement

The body plans of triploblastic animals can be divided into three types based on the presence and structure of an internal, fluid-filled **body cavity**.

- **Acoelomate** animals such as flatworms lack an enclosed, fluid-filled body cavity. Instead, the space between the gut (derived from endoderm) and the muscular body wall (derived from mesoderm) is filled with masses of cells called **mesenchyme** (Figure 31.5A). These animals typically move by beating cilia.

- **Pseudocoelomate** animals have a body cavity called a pseudocoel, a fluid-filled space lying between the mesoderm and endoderm. Many of the internal organs are suspended in the pseudocoel, which is enclosed by muscles (mesoderm) only on its outside; there is no inner layer of mesoderm surrounding the internal organs (**Figure 31.5B**).

- **Coelomate** animals have a body cavity, the coelom, that develops within the mesoderm. The coelom is lined with a thin layer of tissue called the **peritoneum**, which also surrounds the internal organs. The coelom is thus completely enclosed by mesoderm (**Figure 31.5C**).

The structure of an animal's body cavity strongly influences the ways in which it can move. The body cavities of many animals function as **hydrostatic skeletons**. Fluids are relatively

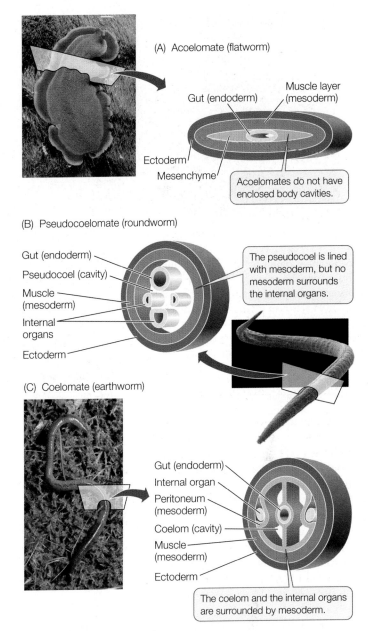

(A) Acoelomate (flatworm)

Muscle layer (mesoderm)

Gut (endoderm)

Ectoderm

Mesenchyme

> Acoelomates do not have enclosed body cavities.

(B) Pseudocoelomate (roundworm)

Gut (endoderm)

Pseudocoel (cavity)

Muscle (mesoderm)

Internal organs

Ectoderm

> The pseudocoel is lined with mesoderm, but no mesoderm surrounds the internal organs.

(C) Coelomate (earthworm)

Gut (endoderm)

Internal organ

Peritoneum (mesoderm)

Coelom (cavity)

Muscle (mesoderm)

Ectoderm

> The coelom and the internal organs are surrounded by mesoderm.

31.5 Animal Body Cavities (A) Acoelomates do not have enclosed body cavities. (B) Pseudocoelomates have a body cavity enclosed by mesoderm only on its outside. (C) Coelomates have a body cavity that is enclosed by mesoderm on both its inside and its outside.

Go to Activity 31.1 Animal Body Cavities
Life10e.com/ac31.1

incompressible, so when the muscles surrounding a fluid-filled body cavity contract, fluids shift to another part of the cavity. If the body tissues around the cavity are flexible, fluids squeezed out of one region can cause some other region to expand. The moving fluids can thus move specific body parts. (You can see how a hydrostatic skeleton works by watching a snail emerge from its shell.) A coelomate animal has better control over the movement of the fluids in its body cavity than a pseudocoelomate animal does. An animal that has longitudinal muscles (running along the length of the body) as well as

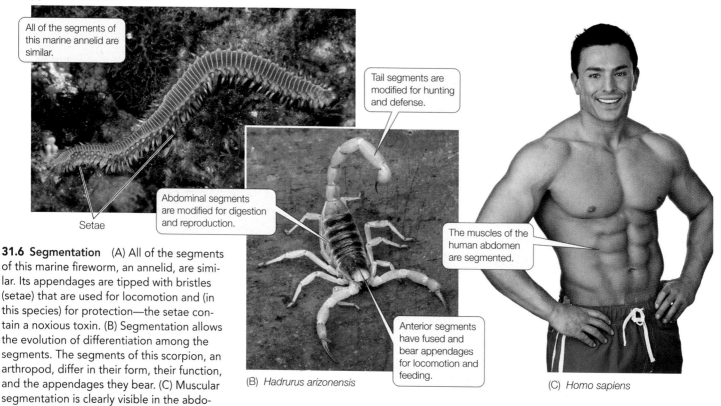

(A) *Hermodice carunculata*

All of the segments of this marine annelid are similar.

Setae

Tail segments are modified for hunting and defense.

Abdominal segments are modified for digestion and reproduction.

Anterior segments have fused and bear appendages for locomotion and feeding.

The muscles of the human abdomen are segmented.

(B) *Hadrurus arizonensis*

(C) *Homo sapiens*

31.6 Segmentation (A) All of the segments of this marine fireworm, an annelid, are similar. Its appendages are tipped with bristles (setae) that are used for locomotion and (in this species) for protection—the setae contain a noxious toxin. (B) Segmentation allows the evolution of differentiation among the segments. The segments of this scorpion, an arthropod, differ in their form, their function, and the appendages they bear. (C) Muscular segmentation is clearly visible in the abdomen of this body builder.

circular muscles (encircling the body cavity) has even greater control over its movement.

In terrestrial environments, the hydrostatic function of fluid-filled body cavities applies mostly to relatively small, soft-bodied organisms. Most larger animals (as well as many smaller ones) have hard skeletons that provide protection and facilitate movement. Muscles are attached to those firm structures, which may be inside the animal or on its outer surface (in the form of a shell or cuticle).

Segmentation improves control of movement

Segmentation—the division of the body into segments—is seen in many animal groups. Segmentation facilitates specialization of different body regions. It also allows an animal to alter the shape of its body in complex ways and to control its movements precisely. If an animal's body is segmented, muscles in each individual segment can change the shape of that segment independently of the others. In only a few segmented animals is the body cavity separated into discrete compartments, but even partly separated compartments allow better control of movement. As we will see in Chapters 32 and 33, segmentation occurs in several groups of protostomes and deuterostomes.

In many animals, such as annelids (earthworms and their relatives), similar body segments are repeated many times (**Figure 31.6A**). In other animals, including most arthropods, segments differ strikingly from one another (**Figure 31.6B**). As we'll describe in Chapter 32, the dramatic evolutionary radiation of the arthropods (including the insects, spiders, centipedes, and crustaceans) was based on modifications of a segmented body plan that features muscles attached to the inner

surface of an external skeleton, including a variety of external appendages that move these animals. In some animals, distinct body segments are not apparent externally (as with the segmented vertebrae of vertebrates, including humans). Nonetheless, muscular segmentation is clearly visible in humans with well-defined, muscular bodies (**Figure 31.6C**).

Appendages have many uses

Getting around under their own power is important to many animals. It allows them to obtain food, to avoid predators, and to find mates. Even some species that are sessile as adults, such as sea anemones, have larval stages that use cilia to swim, thus increasing the animal's chances of finding a suitable habitat.

Appendages that project from the body greatly enhance an animal's ability to move around. Many echinoderms, including sea urchins and sea stars, have myriad tube feet that allow them to move slowly across the substrate (see Figure 33.3B). Animals whose appendages have become modified into specialized limbs are capable of better controlled, more rapid movement. The presence of jointed limbs has been a prominent factor in the evolutionary success of the arthropods and the vertebrates. In four independent instances—among the arthropod insects and among the vertebrate pterosaurs, birds, and bats—body plans emerged in which limbs were modified into wings, allowing these animals to use powered flight.

Appendages also include many structures that are not used for locomotion. Many animals have antennae, which are specialized appendages used for sensing the environment. Other appendages (such as the claws and mouthparts of many arthropods) are adaptations for capturing prey or chewing food. In some species, appendages are used for reproductive purposes, such as sperm transfer or egg incubation.

Nervous systems coordinate movement and allow sensory processing

The bilaterian animals have a well-coordinated central nervous system. More diffuse nervous systems, called **nerve nets**, are present in some other animals, such as cnidarians and ctenophores. Nervous systems appear to be completely absent in a few groups, such as sponges and placozoans.

The central nervous system of bilaterians coordinates the actions of muscles, which allows coordinated movement of appendages and body parts. This coordination of muscles permits highly effective and efficient movement on land, in water, or through the air. The central nervous system is also essential for the processing of sensory information gathered from a wide variety of sensory systems. Many animals have sensory systems for detecting light, for forming images of their environment (sight), for mechanical touch, for detecting movement, for detecting sounds (hearing), for detecting electric fields, and for chemical detection (e.g., taste and smell). These sensory systems allow animals to find food, and the ability of animals to move allows them to capture or collect food from their environment. These same abilities also allow most animals to move to avoid potential predators or to search for suitable mates. Most animals can also assess the suitability of different environments and move appropriately in response to that information.

Go to Media Clip 31.1
Nervous Systems Lead to Efficient Predators
Life10e.com/mc31.1

RECAP 31.2

The body plans of animals are variations on patterns of symmetry, body cavity structure, segmentation, appendages, and nervous systems.

- Describe the main types of symmetry found in animals and explain how an animal's symmetry can influence the way it moves. **See p. 634 and Figure 31.4**

- Explain several ways in which body cavities, segmentation, and centralized nervous systems improve control over movement. **See pp. 635–637**

Many of the modifications to the body plans of animals affect their ways of finding, capturing, and processing food. Evolutionary changes in symmetry, body cavities, appendages, segmentation, and sensory systems have played key roles in enabling animals to obtain food from their environment as well as helping them avoid becoming food for other animals.

31.3 How Do Animals Get Their Food?

As noted in Section 31.1, animals are heterotrophs, or "other-feeders." Although some animals rely on photosynthetic endosymbionts to nourish them (see Figure 27.21), most animals must expend energy to obtain an outside source of nutrition, otherwise known as food.

The need to locate food has favored the evolution of sensory structures that provide animals with detailed information about their environment as well as nervous systems that can receive, process, and coordinate that information. Furthermore, in order to acquire food, animals must either move through the environment to where food is located, or move the environment and the food it contains to them. Animals that move from one place to another are **motile**; animals that stay in one place are **sessile**.

The principal feeding strategies that animals use fall into five broad categories:

- **Filter feeders** (or **suspension feeders**) strain small organisms from their environment.

- **Herbivores** eat plants or parts of plants.

- **Predators** capture and eat other animals.

- **Parasites** live in or on other, generally much larger, organisms, from which they obtain energy and nutrients.

- **Detritivores** feed on dead organic material.

Each of these strategies can be found in many different animal groups, and none of them is limited to a single group. Individuals of some species employ more than one of these feeding strategies, and some animals employ different feeding strategies at different stages of their life cycle. The constant and ongoing need to obtain food, the variety of nutrient sources available in a given environment, and the necessity of competing with other animals to obtain food means that a variety of feeding strategies can be found among all the major animal groups.

Filter feeders capture small prey

Air and water often contain small organisms and organic molecules that are potential food for animals. Moving air and water may carry those items to an animal that positions itself in a good location. Other animals can move through the environment, filtering out prey items as they move. In either case, filter feeders use some kind of straining device to filter the food from the environment.

Many sessile aquatic animals rely on water currents to bring prey to them (**Figure 31.7A**). Some sessile filter feeders (such as sponges; see Figure 31.2) expend energy to move water past their food-capturing devices. Motile filter feeders move their bodies to the nutrient source. Flamingos, for example, use their serrated beaks to filter small organisms out of the muddy mixture they pick up as they wade through shallow water (**Figure 31.7B**). Blue whales—the largest animals that have ever lived—are filter feeders that strain tiny crustaceans from the water column as they swim.

Go to Media Clip 31.2
Filter Feeders
Life10e.com/mc31.2

Herbivores eat plants

An individual plant has many different structures—leaves, wood, roots, sap, flowers, fruits, nectar, and seeds—that animals can consume. Not surprisingly, then, many different kinds of herbivores—animals that feed only on plants—may

(A) *Spirobranchus* sp.

(B) *Phoenicopterus ruber*

31.7 Filter-Feeding Strategies (A) Sessile marine filter feeders such as this Christmas tree worm, a polychaete, allow the ocean currents to bring their food—plankton—to them. (B) The greater flamingo of South America is a motile filter feeder. It uses its appendages (legs) to stir up mud as it wades through ocean lagoons and salty lakes. The bird then uses its beak (inset) to strain small organisms out of the muddy mixture.

feed on a single kind of plant, consuming different parts of the plant or feeding on the same part in different ways. Whereas an individual animal that is captured by a predator is likely to die, herbivores often feed on plants without killing them.

Animals do not need to expend energy subduing and killing plants in order to feed on them. However, plant matter can be difficult to digest and can pose special challenges to terrestrial herbivores because the dominant land plants tend to have several different kinds of tissues, many of which are tough or fibrous. Herbivorous animals typically have long, complex guts to accomplish the tasks involved in digesting plants (see Section 51.2). Animals also must expend energy to detoxify plants' defensive chemicals.

Predators and omnivores capture and subdue prey

Predators possess features that enable them to capture and subdue other animals (referred to as their **prey**). Many vertebrate predators have sensitive sensory organs that enable them to locate prey, as well as sharp teeth or claws that allow them to capture and subdue prey (**Figure 31.8**). Predators may stalk and pursue their prey or wait (often camouflaged) for their prey to come to them.

Omnivores ("all-devouring") are animals, such as raccoons and humans, that eat both plants and other animals. The diets of some omnivores differ at different life stages; many songbirds, for example, eat fruit or seeds as adults but feed insects to their young.

Parasites live in or on other organisms

Parasites obtain nutrients from another organism—a **host**—by living on or within the host. Some animal parasites consume parts of the host itself (such as ticks that suck body fluids); others highjack nutrients the host would otherwise consume (such as tapeworms that may live in the intestines of mammals).

(A) *Haliaeetus leucocephalus*

(B) *Tropidolaemus wagleri*

31.8 Active and Sit-and-Wait Predators (A) The appendages (legs and wings) of the bald eagle, along with its strong beak, are adaptations to the life of an active predatory hunter. (B) Many snakes, such as Wagler's pit viper, rely on camouflage to conceal it from potential prey. These snakes typically sit motionless, waiting in one spot for unsuspecting prey to walk within striking range.

Most animal parasites are much smaller than their hosts, and many parasites can consume parts of their host without killing it. To set up residence within a host, a parasite must first overcome the host's defenses. Parasites often have complex life cycles that rely on multiple hosts, as we will see in Section 31.4.

Parasites that live inside their hosts are called endoparasites, and these are often morphologically very simple. Endoparasites often function without a digestive system, absorbing their food directly from the host's gut or body tissues. Many flatworms are endoparasites of humans and other mammals, as we will describe in Chapter 32.

Parasites that live outside their hosts are called ectoparasites; they are generally more complex morphologically than endoparasites. Ectoparasites have digestive tracts and mouthparts that enable them to pierce the host's tissues or suck on the host's body fluids. Fleas and ticks are ectoparasitic arthropods that feed on many vertebrates, including humans.

Detritivores live on the remains of other organisms

Detritivores feed on the dead bodies or waste products of other organisms, organic matter known as **detritus**. Detritivores (sometimes called decomposers) perform an important ecosystem function by breaking down dead organic matter and returning the nutrients it contains to the environment in a form that can be used by other organisms. Detritivores are common in any soil with high organic content, as well as on the ocean floor. Well-known detritivores include earthworms and other annelids, millipedes, and many insects and crustaceans.

Charles Darwin became fascinated with earthworms and wrote a book called *The Formation of Vegetable Mould through the Action of Worms*. He was particularly impressed by the importance of earthworms in soil formation. Darwin conducted many interesting experiments to establish how quickly earthworms break down organic matter and build up rich soils.

 Go to Media Clip 31.3
Detritivores
Life10e.com/mc31.3

RECAP **31.3**

Animals are heterotrophs that must expend energy to acquire food from their environment. Most animals either move through the environment to where food is located or move the environment and the food it contains to them.

- How can you distinguish among filter feeders, predators, and parasites—all of which may feed on other animals?
  See pp. 637–639

- What adaptations are necessary for animals that eat plants? What adaptations are needed for a predatory lifestyle?
  See pp. 638–639

As an animal grows from a single-celled zygote into a larger, more complex adult, its body structure, its diet, and the environment in which it lives may all change. In the next section we will describe some animal life cycles and discuss why they are so varied.

 ## 31.4 How Do Life Cycles Differ among Animals?

The life cycle of an animal encompasses its embryonic development, birth, growth to maturity, reproduction, and death. During its life an individual animal ingests food, grows, interacts with other individuals of the same and other species, and reproduces.

Many animal life cycles feature specialized life stages

In some groups of animals, newborns look much like miniature versions of the adults (a pattern called **direct development**). Newborns of most animal species, however, differ dramatically from adults. Many animal species have a life stage called a **larva** (plural *larvae*), which is an immature form that the animal takes early in its life before assuming an adult form. Some of the most striking life cycle changes are found among insects such as beetles, flies, moths, butterflies, and bees, which undergo radical change (called **metamorphosis**) between their larval and adult stages (**Figure 31.9**). In these animals, one stage may be specialized for feeding and the other for reproduction. Adults

31.9 A Life Cycle with Complete Metamorphosis (A) The larval stage (caterpillar) of the monarch butterfly (*Danaus plexippus*) is specialized for feeding. (B) The pupa is the stage during which the transformation to the adult form occurs. (C) The adult butterfly is specialized for dispersal and reproduction.

of most moth species, for example, do not eat. In some animal species, individuals eat during all life cycle stages, but what they eat changes with the stage. For example, butterfly larvae (known as caterpillars) eat leaves and flowers, whereas most adult butterflies eat only nectar. Having different life cycle stages that are specialized for different activities may increase the efficiency with which an animal performs those activities.

Most animal life cycles have at least one dispersal stage

At some time during their lives, most animals move, or are moved, so few animals die exactly where they were born. Movement of organisms away from a parent organism or from an existing population is called **dispersal**.

Animals that are sessile as adults typically disperse as eggs or larvae. Most sessile marine animals discharge their eggs and sperm into the water, where fertilization takes place. A larva soon hatches and floats freely in the plankton, where it filters small food items from the water. Many animals that live on the seafloor, including polychaete worms and mollusks, have a radially symmetrical larval form known as a **trochophore** (**Figure 31.10A**). Other animals, such as crustaceans, have a

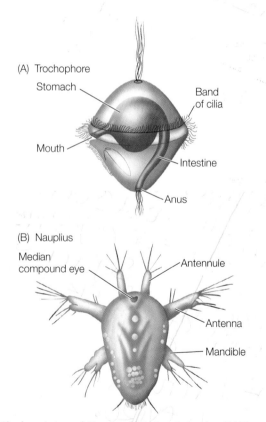

(A) Trochophore
Stomach
Band of cilia
Mouth
Intestine
Anus

(B) Nauplius
Median compound eye
Antennule
Antenna
Mandible

31.10 Planktonic Larval Forms of Marine Animals (A) The trochophore ("wheel-bearer") is a distinctive larval form found in several marine animal clades with spiral cleavage, most notably the polychaete worms and the mollusks. (B) This nauplius larva will mature into a crustacean with a segmented body and jointed appendages.

bilaterally symmetrical larval form called a **nauplius** (**Figure 31.10B**). Both types of larvae feed for some time in the plankton and may travel long distances before settling on the ocean floor and transforming into adults.

Other animal species that are motile as adults disperse when they are mature. A caterpillar, for example, may spend its entire larval stage feeding on a single plant, but after it metamorphoses into a flying adult—a butterfly—it may fly to and lay eggs on other plants located far from the one where it spent its caterpillar days. In some species, individuals disperse during several different life cycle stages.

Parasite life cycles facilitate dispersal and overcome host defenses

Animals that live as endoparasites are bathed in the nutritious tissues of their host or in the digested food that fills their host's digestive tract. Thus they may not need to exert much energy to obtain food, but to survive they must overcome the host's defenses. Furthermore, either they or their offspring must disperse to new hosts while their host is still living, because they die when their host dies.

The fertilized eggs of some parasites are voided with the host's feces and later ingested directly by other host individuals. Most parasite species, however, have complex life cycles involving one or more intermediate hosts and several larval stages (**Figure 31.11**). Some intermediate hosts transport individual parasites directly between other host species. Others house and support the parasite until another host ingests them. Complex life cycles may thus facilitate the transfer of individual parasites among hosts.

Some animals form colonies of genetically identical, physiologically integrated individuals

Most people tend to view the distinction between individuals and populations as clear-cut. In several groups of animals, however, asexual reproduction without fission leads to the formation of colonies composed of many physiologically integrated individuals. At first appearance, these colonies may look much like a single integrated organism. The individuals in the colony are clonal copies of one another, so they are genetically homogeneous.

Coloniality has arisen several times among animal groups, with widely varying levels of integration and specialization among the individuals. In some species, colonies are composed of loosely connected but integrated individuals that all function alike (**Figure 31.12**). In other colonial species, the individuals may become specialized for different functions, just as different cell types in multicellular organisms do. The Portuguese man-of-war (a cnidarian; see Figure 31.19) is an example of such a colonial animal, as it is composed of many individuals of four different specialized body forms, all integrated and functioning together. The individuals in the colony are themselves multicellular, however, unlike the cells of a single multicellular organism.

Mature
tapeworm

Final hosts
(fish-eating
mammals)

START

8 The fish is eaten by
a mammalian host; the
tapeworm matures and
reproduces in the
mammal's gut.

1 The zygote, which has developed
in a host mammal's gut, is passed
with its feces.

7 The perch is eaten
by a larger fish (third
intermediate host).

Third larval stage

2 The embryo
develops in water.

First larval stage
(free-swimming)

Second
larval stage

3 The larva
hatches.

6 The larva moves to the
muscles of the perch
(second intermediate host).

4 The free-swimming first
larval stage is ingested
by a copepod (first
intermediate host).

5 The tapeworm develops
into the second larval stage
and is passed on when a
perch eats the copepod.

31.11 Reaching a New Host by a Complex Route The
broad fish tapeworm *Diphyllobothrium latum* must pass
through the bodies of a copepod (a type of crustacean)
and at least one fish before it can reinfect its primary
host, a mammal. Such complex life cycles assist the parasite's colonization of new host individuals, but they also
provide opportunities for humans to break the cycle with
hygienic measures (such as thoroughly cooking food to
kill the parasites).

No life cycle can maximize all benefits

A common saying, "a jack-of-all-trades is master of none,"
suggests why there are constraints on the evolution of life
cycles. The characteristics an animal has in any one life cycle
stage may improve its performance in one activity but reduce
its performance in another—a situation known as a trade-off.
An animal that is good at filtering small food items from the
water, for example, is unlikely to be good at capturing large
prey. Similarly, energy devoted to building protective structures such as shells cannot be used for growth.

Some major trade-offs can be seen in animal reproduction. Some animals produce large numbers of small eggs,
each with a small energy store (**Figure 31.13A**). Other animals produce a small number of large eggs, each with a
large energy store (**Figure 31.13B**). With a fixed amount of
energy available for reproduction, a female animal can produce many small eggs or a few large eggs, but she cannot
produce many large eggs. Thus there is a trade-off between

Plumatella repens

The individual animals...

...secrete a
mineralized
shell that
connects
the colony.

1 mm

31.12 Colonial Animals This bryozoan colony consists of many
asexually reproducing, genetically homogeneous, physiologically
interacting individuals. The colony looks much like a single individual
with many parts, but in fact is a group of individuals acting together.

31.13 Many Small or Few Large Allocation of energy to eggs requires trade-offs. (A) This wood frog has divided her reproductive energy among a large number of small eggs. (B) This King penguin has invested all of her reproductive energy in one large egg.

(A) *Rana sylvatica*

(B) *Aptenodytes patagonicus*

the number of offspring produced and the energy resources each offspring receives from its mother.

The larger the energy store in an egg, the longer an offspring can develop before it must either find its own food or be fed by its parents. Birds of all species lay relatively small numbers of relatively large eggs, but incubation periods vary. In some species, eggs hatch when the young are still helpless (**Figure 31.14A**). Such **altricial** young must be fed and cared for until they can feed themselves; parents can provide for only a small number of altricial offspring. In contrast, some bird species incubate their eggs longer, and the hatchlings are developed to the point that they are able to forage for themselves almost immediately (**Figure 31.14B**). The young of such species are called **precocial**.

RECAP 31.4

Many animals have life cycle stages that differ from one another morphologically. In some animals, the larval form is a dispersal stage; in other species, the adults are more likely to disperse than are larvae. In several groups of organisms, asexual reproduction without fission leads to coloniality.

- How do trade-offs constrain the evolution of life cycles? See pp. 641–642

- Explain the difference between multicellularity and coloniality. See p. 641 and Figure 31.12

Variations in body symmetry, body cavity structure, life cycles, patterns of development, and survival strategies differentiate

(A) *Alcedo atthis*

(B) *Anser anser*

31.14 Helpless or Independent (A) The altricial young of the common kingfisher are essentially helpless when they hatch. Their parents feed and care for them for several weeks. (B) Grey goose hatchlings are precocial, ready to swim and feed independently almost immediately after hatching.

millions of animal species. In the remainder of this chapter and in Chapters 32 and 33, we will become acquainted with the major animal groups and learn how the general characteristics we have just described apply to each of them.

31.5 What Are the Major Groups of Animals?

The bilaterians make up a large monophyletic group embracing all animals other than sponges, ctenophores, placozoans, cnidarians, and a few poorly known groups of parasitic animals (see Figure 31.1). Some of the traits that support the monophyly of bilaterians are the presence of three distinct cell layers in embryos (triploblasty) and the presence of at least seven Hox genes (see Chapters 19 and 20). Although bilateral symmetry is often viewed as a synapomorphy of bilaterians (and the trait gives the group its name), some groups of cnidarians are also bilaterally symmetrical. Recent studies have shown that the genetic basis of bilateral symmetry in bilaterians and in those cnidarian groups that have this trait is similar, so this feature may have been present in the ancestor of both groups.

Bilaterian animals can be divided into the two major clades mentioned earlier, the protostomes and the deuterostomes. These two groups have been diversifying separately for more than 500 million years—ever since the "Cambrian explosion" that we described in Chapter 25. We will describe the protostomes in Chapter 32 and the deuterostomes in Chapter 33.

Go to Activity 31.2 Sponge and Diploblast Classificaton
Life10e.com/ac31.2

The remainder of this chapter describes those animal groups that are not bilaterians. The simplest animals, the sponges, have no distinct tissue types. All other animals are usually known as **eumetazoans**. Two groups treated here as eumetazoans, the ctenophores and placozoans, have weakly differentiated tissue layers (placozoans also lack a nervous system), so some biologists exclude these two groups from Eumetazoa as well. Most eumetazoans have some form of body symmetry, a gut, and tissues organized into distinct organs (although there have been secondary losses of some of these features in some eumetazoans).

Sponges are loosely organized animals

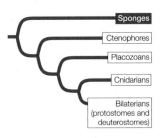

Sponges are the simplest animals. Although they have some specialized cells, they have no distinct embryonic cell layers and no true organs. Early naturalists classified sponges as plants because they were sessile and lacked body symmetry.

Sponges have hard skeletal elements called **spicules**, which may be small and simple or large and complex (see Figure 31.2B). Three major groups of sponges, which separated soon after the split between sponges and the rest of the animals, are distinguished by their spicules. Members of two groups (glass sponges and demosponges) have skeletons composed of siliceous spicules made of hydrated silicon dioxide (**Figure 31.15A,B**). These spicules are remarkable in having greater flexibility and toughness than synthetic glass rods of similar length. Members of the third group, the calcareous sponges,

(A) *Xestospongia testudinaria*

(B) *Euplectella aspergillum*

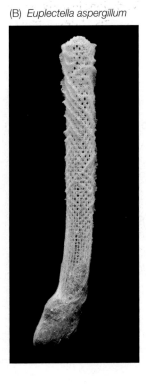

(C) *Sycon* sp.

31.15 Sponge Diversity (A) The majority of sponge species are demosponges, such as these Pacific barrel sponges. The system of pores and water canals (see Figure 31.2) that is typical of the sponge body plan is apparent in this photograph. (B) The supporting structures of both demosponges and glass sponges are siliceous spicules, seen here in the skeleton of a glass sponge. (C) The spicules of calcareous sponges are made of calcium carbonate.

take their name from their calcium carbonate spicules (**Figure 31.15C**). There is some question about the monophyly of sponges. Analyses of some gene sequences have suggested that calcareous sponges are actually more closely related to the eumetazoans than to the other groups of sponges. However, genomic analyses that combine information from many genes support the monophyly of sponges.

The body plan of sponges of all three groups—even large ones, which may reach 1 meter or more in length—is an aggregation of cells built around a water canal system. Sponges bring water into their bodies by beating the flagella of their specialized feeding cells, called **choanocytes** (see Figure 31.2B). Water, along with any food particles it contains, enters the sponge by way of small pores and passes into the water canals or a central atrium, where the choanocytes capture food particles. (You may recall from Section 31.1 that the choanocytes are similar in structure to protists known as choanoflagellates; that similarity provides evidence for the close relationship of animals to choanoflagellates.)

A skeleton of simple or branching spicules, often combined with a complex network of elastic fibers, supports the body of most sponges. Sponges also produce an extracellular matrix, composed of collagen, adhesive glycoproteins, and other molecules, that holds the cells together. Most species are filter feeders; a few species are predators that trap prey on hook-shaped spicules that protrude from the body surface.

Most of the 8,500 species of sponges are marine animals; only about 50 species live in fresh water. Sponges come in a wide variety of sizes and shapes that are adapted to different movement patterns of water. Sponges living in intertidal or shallow subtidal environments with strong wave action are firmly attached to the substrate. Most sponges that live in slowly flowing water are flattened and are oriented at right angles to the direction of current flow. They intercept water and the food items it contains as it flows past them.

Sponges reproduce both sexually and asexually. In most species, a single individual produces both eggs and sperm, but individuals do not self-fertilize. Water currents carry sperm from one individual to another. Sponges also reproduce asexually by budding and fragmentation.

Ctenophores are radially symmetrical and diploblastic

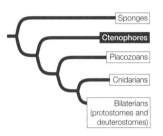

Ctenophores, also known as comb jellies, were until recently thought to be most closely related to the cnidarians (jellyfishes, corals, and their relatives). But ctenophores lack most of the Hox genes found in all other eumetazoans, and recent studies of their genomes have indicated that ctenophores were probably among the earliest lineages to split from the remaining animals. Most recent phylogenetic analyses place the ctenophores as the sister group of all other animals except sponges.

Ctenophores have a radially symmetrical, diploblastic body plan. The two cell layers are separated by an inert, gelatinous extracellular matrix called **mesoglea**. Ctenophores, unlike sponges, have a **complete gut**: food enters through a mouth, and wastes are eliminated through two anal pores.

Ctenophores move by beating cilia rather than by muscular contractions. Most of the 250 known species have eight comb-like rows of cilia-bearing plates, called **ctenes** (Figure 31.16). The feeding tentacles of ctenophores are covered with cells that discharge adhesive material when they contact prey. After capturing its prey, a ctenophore retracts its tentacles to bring

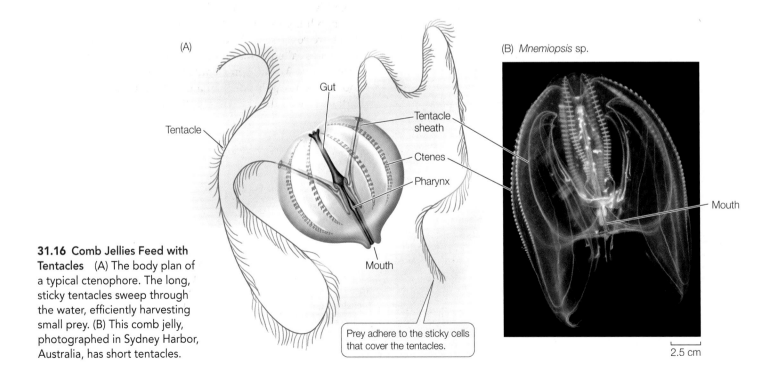

31.16 Comb Jellies Feed with Tentacles (A) The body plan of a typical ctenophore. The long, sticky tentacles sweep through the water, efficiently harvesting small prey. (B) This comb jelly, photographed in Sydney Harbor, Australia, has short tentacles.

(A)

Tentacle

Gut

Tentacle sheath

Ctenes

Pharynx

Mouth

Prey adhere to the sticky cells that cover the tentacles.

(B) *Mnemiopsis* sp.

Mouth

2.5 cm

the food to its mouth. In some species, the entire surface of the body is coated with sticky mucus that captures prey. Most ctenophores eat small planktonic organisms, although some eat other ctenophores. They are common in open seas and can become abundant in coastal bays, where large populations of ctenophores may inhibit the growth of other organisms.

Ctenophore life cycles are uncomplicated. Gametes are released into the gut and discharged through the mouth or the anal pores. Fertilization takes place in open seawater. In nearly all species, the fertilized egg develops directly into a miniature ctenophore, which gradually grows into an adult.

 Go to Media Clip 31.4
Ctenophores
Life10e.com/mc31.4

Placozoans are abundant but rarely observed

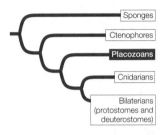

As discussed at the start of this chapter, **placozoans** are structurally very simple animals with only a few distinct cell types (**Figure 31.17A**). Individuals in the mature, asymmetrical life stage are usually observed adhering to surfaces (such as the glass of aquariums, where they were first discovered, or to rocks and other hard substrates in nature). Their structural simplicity—they have no mouth, gut, or nervous system—initially led biologists to suspect they might be the sister group of all other animals. Most phylogenetic analyses have not supported this hypothesis, however, and some aspects of the placozoans' structural simplicity may be secondarily derived. They are generally considered to have a diploblastic body plan, with upper and lower surface layers that sandwich a layer of contractile fiber cells.

Recent studies have found that placozoans have a pelagic (open-ocean) life stage that is capable of swimming (**Figure 31.17B**), but the life history of placozoans is incompletely known. Most studies have focused on the larger adherent stages that are most easily observed in aquariums. The transparent nature and small size of placozoans make them very difficult to observe in nature. Nonetheless, it is known that placozoans

can reproduce both asexually and sexually, although the details of their sexual reproduction are poorly understood. As we noted at the opening of this chapter, placozoans have been studied mainly in aquariums, where they appear after being inadvertently collected with other marine organisms, although we now know that pelagic-stage placozoans are abundant in warm seas around the world.

Cnidarians are specialized predators

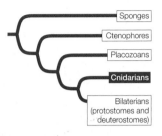

The **cnidarians** (jellyfishes, sea anemones, corals, and hydrozoans) make up the largest and most diverse group of non-bilaterian animals. The mouth of a cnidarian is connected to a blind sac called the **gastrovascular cavity** (a cnidarian thus does not have a complete gut). The gastrovascular cavity functions in digestion, circulation, and gas exchange, and it also acts as a hydrostatic skeleton. The single opening serves as both mouth and anus.

The life cycle of many cnidarians has two distinct stages, one sessile and the other motile (**Figure 31.18**), although one or the other of these stages is absent in some groups. In the sessile **polyp** stage, a cylindrical stalk attaches the animal to the substrate. The motile **medusa** (plural *medusae*) is a free-swimming stage shaped like a bell or an umbrella. It typically floats with its mouth and feeding tentacles facing downward.

Mature polyps produce medusae by asexual budding. Medusae then reproduce sexually, producing eggs or sperm by meiosis and releasing the gametes into the water. A fertilized egg develops into a free-swimming, ciliated larva called a **planula**, which eventually settles to the bottom and develops into a polyp.

Cnidarians are specialized predators adapted for capturing and subduing relatively large and complex prey. Their tentacles are covered with specialized cells that contain stinging organelles called **nematocysts**, which inject toxins into their prey (**Figure 31.19**). Some cnidarians, including many corals and anemones, gain additional nutrition from photosynthetic endosymbionts that live in their tissues.

(A)

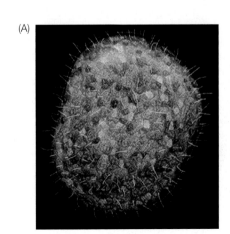

(B)

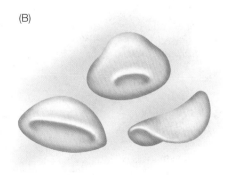

31.17 Placozoan Simplicity (A) As seen in this artist's rendition, adult placozoans are tiny (1–2 mm across), flattened, asymmetrical animals. (B) Recent studies have found a symmetrical, weakly swimming pelagic stage of placozoan to be abundant in many warm tropical and subtropical seas.

31.18 The Life Cycle of Most Cnidarians Has Two Stages The life cycle of a scyphozoan (jellyfish) exemplifies the typical cnidarian body forms: the sessile, asexual polyp and the motile, sexual medusa. Some species of cnidarians have life cycles that lack polyps or medusae.

Go to Animated Tutorial 31.1
Life Cycle of a Cnidarian
Life10e.com/at31.1

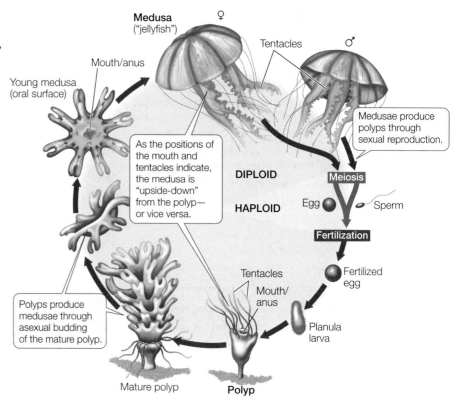

Cnidarians have cells containing muscle fibers whose contractions enable the animals to move, as well as simple nerve nets that integrate the body's activities. Their bodies also contain specialized structural molecules (collagen, actin, and myosin). Yet cnidarians, like ctenophores, are largely made up of inert mesoglea. They have low metabolic rates and can survive in environments where they encounter prey only infrequently.

Of the roughly 12,500 living cnidarian species, all but a few live in the oceans (**Figure 31.20**). The smallest cnidarians can barely be seen without a microscope. One small group, known as myxozoans, consists of tiny parasites, usually with a two-host life cycle that includes a fish and an annelid worm or a bryozoan. The largest known jellyfish is 2.5 meters in diameter, and some colonial hydrozoans (which include the Portuguese man-of-war; see Figure 31.19) can reach lengths in excess of 30 meters. Here we describe the three clades of cnidarians that contain the most species: anthozoans, scyphozoans, and hydrozoans.

ANTHOZOANS Members of the **anthozoan** clade include sea anemones, sea pens, and corals. Sea anemones (see Figure 31.20A), all of which are solitary, are widespread in both warm and cold ocean waters. Sea pens (see Figure 31.20B), by contrast, are colonial. Each colony consists of two or more different kinds of polyps. The primary polyp has a lower portion anchored in the bottom sediment and a branched upper portion that projects above the substrate. Along the upper portion, the primary polyp produces smaller secondary polyps by budding. Some of these secondary polyps differentiate into feeding polyps; in some species, other secondary polyps differentiate to circulate water through the colony.

The common names of coral groups—brain corals, staghorn corals, and organ pipe corals, among others—often describe their appearance (**Figure 31.21A**). Corals are sessile and colonial. The polyps of most species form a skeleton by secreting a matrix of organic molecules on which they then deposit calcium carbonate. As the colony grows, old polyps

31.19 Nematocysts Are Potent Weapons The tentacles of the Portuguese man-of-war, a hydrozoan, are rife with specialized cells called cnidocytes. These cells contain stinging organelles called nematocysts, which inject toxins into their prey. The Portuguese man-of-war is a colonial organism, composed of many physiologically integrated individuals with specialized functions.

die, but their calcium carbonate skeletons remain. Living corals form a layer on top of a growing bank of skeletal remains, eventually forming chains of islands and reefs (**Figure 31.21B**). The Great Barrier Reef along the northeastern

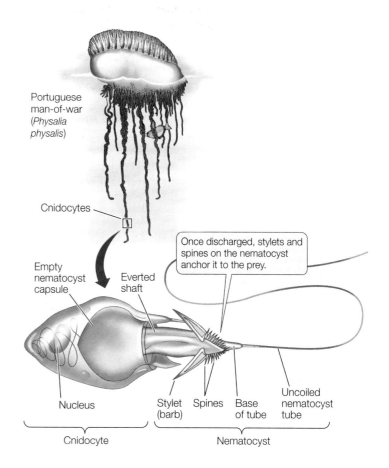

31.20 Diversity among Cnidarians (A) Sea anemones are sessile, living attached to marine substrates. Water currents carry prey into the nematocyst-studded tentacles. (B) The sea pen is a colonial cnidarian that lives in soft bottom sediments and projects polyps above the substrate. (C) This jellyfish illustrates the complexity of a scyphozoan medusa. (D) The internal structure of the medusa of a North Atlantic colonial hydrozoan is visible here.

(A) *Sagartia modesta*

(B) *Pteroeides* sp.

(C) *Gonionemus vertens*

(D) *Polyorchis penicillatus*

coast of Australia is a system of coral formations more than 2,000 kilometers long, which is about the distance from New York City to St. Louis. A single coral reef in the Red Sea has been calculated to contain more material than all the buildings in the major cities of North America combined.

Corals flourish at shallow depths in clear, nutrient-poor tropical waters. They grow well in such environments because unicellular photosynthetic dinoflagellates live endosymbiotically within their cells. These dinoflagellates provide the corals with products of photosynthesis; the corals, in turn, provide the dinoflagellates with nutrients and a place to live. This endosymbiotic relationship explains why reef-forming corals are restricted to clear surface waters, where light levels are high enough to support photosynthesis.

Coral reefs throughout the world are threatened by rising CO_2 levels (which result in increased ocean temperatures) and acidification of ocean waters. Polluted runoff from development on adjacent shorelines is an additional threat to corals. Warmer temperatures lead to the loss of coral endosymbionts (known as coral bleaching; see Figure 27.21), and acidification can cause coral skeletons to dissolve. An overabundance of nitrogen in runoff is advantageous to algae, which overgrow and eventually smother the corals.

SCYPHOZOANS The several hundred species of **scyphozoans** are all marine. The mesoglea of their medusae is thick and firm, giving rise to their common name of jellyfish (or sea jellies). The medusa rather than the polyp dominates the life cycle of scyphozoans. An individual medusa is male or female, releasing eggs or sperm into the open sea. A fertilized egg develops into a small planula larva that quickly settles on a substrate and develops into a small polyp. This polyp feeds and grows and may produce additional polyps by budding. After a period of growth, the polyp begins to bud off small medusae, which feed, grow, and transform into adult medusae (see Figures 31.18 and 31.20C).

HYDROZOANS The polyp typically dominates the life cycle of **hydrozoans**, but some species have only medusae (see Figure 31.19) and others have only polyps. Most hydrozoans are colonial. A single planula larva eventually gives rise to a colony of

(A) *Diploria labyrinthiformis*

(B)

31.21 Corals (A) The descriptive common name of this Caribbean coral is "brain coral." (B) Many different coral species form this reef in the Red Sea between Egypt and the Arabian Peninsula.

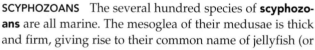

31.22 Many Hydrozoans Are Colonial The polyps in a hydrozoan colony may differentiate to perform specialized tasks. In the species whose life cycle is diagrammed here, the medusa is the sexual reproductive stage, producing eggs and sperm in organs called gonads.

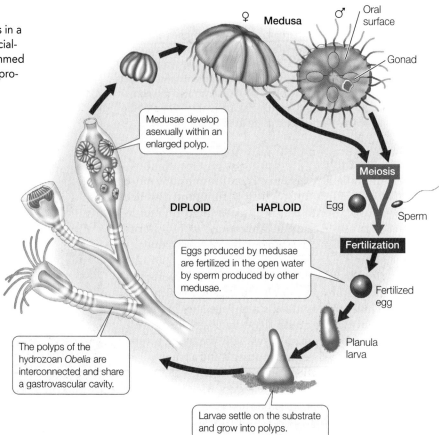

Medusae develop asexually within an enlarged polyp.

DIPLOID   HAPLOID   Egg   Sperm

Meiosis

Fertilization

Eggs produced by medusae are fertilized in the open water by sperm produced by other medusae.

Fertilized egg

Planula larva

The polyps of the hydrozoan *Obelia* are interconnected and share a gastrovascular cavity.

Larvae settle on the substrate and grow into polyps.

many polyps, all interconnected and sharing a continuous gastrovascular cavity (**Figure 31.22**). Within such a colony, some polyps have tentacles with many nematocysts; they capture prey for the colony. Other individuals lack tentacles and are unable to feed, but are specialized for the asexual production of medusae. Still others are fingerlike and defend the colony with their nematocysts.

Some small groups of parasitic animals may be the closest relatives of bilaterians

Two small groups of tiny marine parasites are listed in Table 31.1 but are not depicted in the phylogeny in Figure 31.1: the orthonectids and the rhombozoans. Recent genomic analyses suggest that these groups may be among the closest surviving relatives of the bilaterians, although their exact phylogenetic placement is uncertain. Both groups are highly reduced parasites that lack many of the structures that traditionally have been used to study animal relationships. As their genomes become more completely known, the relationships of these two groups to other animals should become clearer. Two other small groups (also listed in Table 31.1 but not shown in Figure 31.1) have been proposed as also falling just outside the bilaterians: the xenoturbellids and acoels. Recent genomic analyses, however, have suggested that these animals are actually highly specialized deuterostomes.

■ RECAP 31.5

Bilaterian animals are classified into two major clades, protostomes and deuterostomes. The non-bilaterian animals comprise the sponges, ctenophores, placozoans, cnidarians, and some small groups of parasitic animals.

- Why are sponges considered to be animals even though they lack the complex body structures found among most other animal groups? **See pp. 643–644**

- Describe some major features of the following groups: sponges, ctenophores, placozoans, and cnidarians.

Which animal groups are involved in the earliest split in the animal tree?

ANSWER

Most data support the placement of sponges as the sister group of all other animals. Morphologically, sponges are the most similar to the choanoflagellates, and most genomic analyses continue to place sponges at the base of the animal tree. The ctenophores share some superficial similarities with cnidarians and were traditionally considered their sister group, but more recent studies increasingly support the placement of the ctenophores closer to the base of the animal tree. A few studies even suggest that the ctenophores are the sister group of all other animals, including sponges, although most studies place the ctenophores as the sister group of all animals except sponges, and this is the phylogeny we show in Figure 31.1.

Although placozoans, with only four cell types and no true organs, are less structurally complex than ctenophores, this structural simplicity is now thought to represent an evolutionary reversal in the placozoan lineage. The alternative possibility, which seems less likely, is that the organ systems of ctenophores evolved independently from those of the cnidarians and bilaterians.

CHAPTER**SUMMARY** 31

 31.1 ## What Characteristics Distinguish the Animals?

- Animals share a set of derived traits not found in other groups of organisms. These traits include similarities in the sequences of many of their genes, the structure of their cell junctions, and the components of their extracellular matrix.

- Patterns of embryonic development provide clues to the evolutionary relationships among animals. **Diploblastic** animals, which include the ctenophores, placozoans, and cnidarians, develop two embryonic cell layers. **Triploblastic** animals develop three cell layers. **Review Figure 31.1**

- Differences in their patterns of early development characterize two major triploblastic clades, the **protostomes** and the **deuterostomes**.

 31.2 ## What Are the Features of Animal Body Plans?

- Animal **body plans** can be described in terms of symmetry, **body cavity** structure, **segmentation**, types of appendages, and nervous system development.

- A few animals have no symmetry, but most animals have either **radial symmetry** or **bilateral symmetry**. **Review Figure 31.4**

- Many bilaterally symmetrical animals exhibit **cephalization**: the concentration of sensory organs and nervous tissues in an anterior head.

- On the basis of their body cavity structure, animals can be described as **acoelomates**, **pseudocoelomates**, or **coelomates**. **Review Figure 31.5, ACTIVITY 31.1**

- Segmentation, which takes many forms, improves control of movement, as do appendages. **Review Figure 31.6**

- The development of a nervous system is important for the coordination of muscular movement and the processing of sensory information.

 31.3 ## How Do Animals Get Their Food?

- **Motile** animals can move to find food; **sessile** animals stay in one place, but may expend energy to move the environment and the food it contains to them

- **Filter feeders** strain small organisms and organic molecules from their environment.

- **Herbivores** consume plants, usually without killing them.

- **Predators** have morphological features such as sharp teeth, beaks, and claws that enable them to capture and subdue animal **prey**.

- **Parasites** live in or on other organisms and obtain nutrition from those **host** individuals.

- **Detritivores** consume dead organic matter and return the nutrients it contains to the ecosystem.

 31.4 ## How Do Life Cycles Differ among Animals?

- The stages of an animal's life cycle may be specialized for different activities. An immature stage whose morphology is dramatically different from that of the adult stage is called a **larva**.

- Most animal life cycles have at least one **dispersal** stage. Many sessile marine animals can be grouped by the presence of one of two distinct larval dispersal stages: **trochophore** or **nauplius**. **Review Figure 31.10**

- A characteristic of an animal or a life cycle stage may improve the animal's performance in one activity but reduce its performance in another, a situation known as a trade-off.

- Parasites have complex life cycles that may involve one or more hosts and several larval stages. **Review Figure 31.11**

- In some groups of animals, asexual reproduction without fission leads to the formation of colonies composed of many genetically homogeneous, physiologically integrated individuals.

 31.5 ## What Are the Major Groups of Animals?

- **Eumetazoans** include all animals except sponges. Animals other than sponges, ctenophores, placozoans, and cnidarians—that is, the triploblastic protostomes and deuterostomes—belong to a large monophyletic group called **bilaterians**. **Review Figure 31.1, ACTIVITY 31.2**

- **Sponges** are simple animals that lack differentiated cell layers and true organs. They have skeletons made up of siliceous or calcareous **spicules**. They create water currents and capture food with flagellated feeding cells called **choanocytes**. **Review Figure 31.2**

- **Ctenophores** are radially symmetrical and have two cell layers separated by an inert extracellular matrix called **mesoglea**. **Review Figure 31.16**

- **Placozoans** are asymmetrical as adults. They have only a few cell types and lack true organs, although their simplicity may be secondarily derived.

- The life cycle of most **cnidarians** has two distinct stages: a sessile **polyp** stage and a motile **medusa** stage that reproduces sexually. A fertilized egg develops into a free-swimming **planula** larva, which settles to the bottom and develops into a polyp. **Review Figures 31.18, 31.22, ANIMATED TUTORIAL 31.1**

 Go to the Interactive Summary to review key figures, Animated Tutorials, and Activities
Life10e.com/is31

CHAPTER**REVIEW**

▮▮▮ **REMEMBERING**

1. A bilaterally symmetrical animal can be divided into mirror-image halves by
   a. any plane through the midline of its body.
   b. any plane from its anterior to its posterior end.
   c. any plane from its dorsal to its ventral surface.
   d. a single plane through the midline of its body from its anterior to its posterior end.
   e. a single plane that divides it into dorsal and ventral halves.

2. In the common ancestor of eumetazoans, the pattern of early cleavage was
   a. spiral.
   b. radial.
   c. biradial.
   d. deterministic.
   e. haphazard.

3. A fluid-filled body cavity can function as a hydrostatic skeleton because
   a. fluids are moderately compressible.
   b. fluids are highly compressible.
   c. fluids are relatively incompressible.
   d. fluids have the same density as body tissues.
   e. fluids can be moved by ciliary action.

4. Many parasites evolved complex life cycles because
   a. they are too simple to disperse readily.
   b. they are poor at recognizing new hosts.
   c. they were driven to it by host defenses.
   d. complex life cycles increase the probability of a parasite's transfer to a new host.
   e. their nonparasitic ancestors had complex life cycles and they simply retained them.

5. The endosymbiotic dinoflagellates present in many corals
   a. provide the corals with products of photosynthesis.
   b. allow corals to flourish in clear, nutrient-poor tropical waters.
   c. can be lost when environmental conditions change.
   d. All of the above
   e. None of the above

▮▮▮ **UNDERSTANDING & APPLYING**

6. Differentiate among the members of each of the following sets of related terms:

   radial symmetry/bilateral symmetry
   protostome/deuterostome
   diploblastic/triploblastic
   coelomate/pseudocoelomate/acoelomate

7. In this chapter we listed some of the traits shared by all animals that have convinced most biologists that all animals are descendants of a single common ancestral lineage. In your opinion, which of these traits provides the most compelling evidence that animals are monophyletic?

8. Why is bilateral symmetry strongly associated with cephalization, the concentration of sensory organs in an anterior head?

9. How does a slow metabolic rate enable an animal to live in an unproductive environment?

▮▮▮ **ANALYZING & EVALUATING**

10. The discoveries that the pelagic stages of placozoans are abundant in warm seas and that the mature stages settle on smooth surfaces suggest how these organisms might be collected and surveyed. What sampling procedures might you use to discover whether placozoans occur at a particular location along a coast?

Go to BioPortal at **yourBioPortal.com** for Animated Tutorials, Activities, LearningCurve Quizzes, Flashcards, and many other study and review resources.

32

Protostome Animals

How Many Species? In an important study, entomologist Terry Erwin fogged tree canopies in a Panamanian rainforest with insecticide. Using the number of species represented among the fallen insects as a base, he extrapolated to estimate the total number of insect species on Earth.

OF THE 1.8 MILLION ANIMAL SPECIES that have been discovered and named by biologists, a large majority are protostomes. One group of protostomes, the insects, accounts for more than 1 million of these species, or more than half of all known species of living organisms. Although these numbers may seem incredibly large, they represent a relatively small fraction of the total protostome diversity that is thought to exist on Earth.

As recently as the 1980s, many biologists thought that about half of existing insect species had been described, but today they think that the number of described insect species may be a much smaller fraction of the total number of living species. Why did they change their minds?

A simple but important field study suggested that the number of existing insect species had been significantly underestimated. Knowing that the insects of tropical rainforests—the most species-rich habitat on Earth—were poorly known, entomologist Terry Erwin made a comprehensive sample of one group of insects, the beetles, in the canopies of a single species of tropical forest tree, *Luehea seemannii*, in Panama. Erwin fogged the canopies of 19 large *L. seemannii* trees with an insecticide and collected the insects that fell from the trees in collection nets. He collected about 1,200 species of beetles—many of them previously undescribed—from this one species of tree.

Erwin then used a set of assumptions to estimate the total number of insect species in tropical rainforests. His assumptions included estimates of the number of species of trees in these forests; the proportion of beetles that specialize on a specific species of host tree; the relative proportion of beetles to other insect groups; and the proportion of beetles that live in trees versus those that live in leaf litter on the ground. From this and similar studies, Erwin estimated that there may be 30 million or more species of insects on Earth. Although recent tests of Erwin's assumptions suggest that 30 million was an overestimate, it is clear that the vast majority of insect species remain to be discovered.

Erwin's pioneering study highlighted the fact that we live on a poorly known planet, most of whose species have yet to be named and described. Much of that undiscovered diversity occurs among several groups of protostomes.

Which groups of protostomes are thought to contain the most undiscovered species?

See answer on p. 675.

32.1 What Is a Protostome?

You may recall that the embryos of diploblastic animals (the ctenophores, placozoans, and cnidarians, which we discussed in Chapter 31) have two cell layers: an outer ectoderm and an inner endoderm (see Section 31.1). Sometime after the origin of the diploblastic animals, a third embryonic cell layer evolved: the mesoderm, which lies between the ectoderm and the endoderm. Mesoderm is found in the two major triploblastic animal clades, the protostomes and the deuterostomes. If we were to judge solely on the basis of numbers, both of species and of individuals, the protostomes would emerge as by far the more successful of the two groups.

As noted in Section 31.1, the name "protostome" means "mouth first." In protostomes, the embryonic blastopore becomes the mouth as the animal develops. In contrast, in deuterostomes ("mouth second"), the blastopore becomes the anal opening of the gut. The protostomes are extremely varied, but they are all bilaterally symmetrical animals whose bodies exhibit two major derived traits:

- An anterior brain that surrounds the entrance to the digestive tract
- A ventral nervous system consisting of paired or fused longitudinal nerve cords

Other aspects of protostome body organization differ widely from group to group (**Table 32.1**). Before gene sequences were available for phylogenetic analysis, biologists considered the structure of the body cavity to be a critical feature in animal classification. But the results of genetic analyses have shown that body cavity forms have undergone considerable convergence in the course of protostome evolution. Although the common ancestor of the protostomes had a coelom, subsequent modifications of the coelom distinguish many protostome lineages. In some lineages (such as the flatworms and entoprocts), the coelom has been lost (that is, these groups reverted to an acoelomate state). Some lineages are characterized by a pseudocoel, a body cavity lined with mesoderm in which the internal organs are suspended (see Figure 31.5B). In two of the most prominent protostome clades, the coelom has been highly modified:

- Arthropods lost the ancestral condition of the coelom over the course of evolution. Their internal body cavity has become a **hemocoel**, or "blood chamber," in which fluid from an open circulatory system bathes the internal organs before returning to blood vessels.

- Most mollusks have an open circulatory system with some of the attributes of the hemocoel, but they retain vestiges of an enclosed coelom around their major organs.

The protostomes can be divided into two major clades—the lophotrochozoans and the ecdysozoans—largely on the basis of DNA sequence analysis (**Figure 32.1**).

Go to Activity 32.1 Features of the Protostomes
Life10e.com/ac32.1

TABLE 32.1
Anatomical Characteristics of Some Major Protostome Groups

Group	Body Cavity	Digestive Tract	Circulatory System
Arrow worms	Coelom	Complete	None
LOPHOTROCHOZOANS			
Bryozoans	Coelom	Complete	None
Entoprocts	None	Complete	None
Flatworms	None	Blind gut	None
Rotifers	Pseudocoel	Complete	None
Gastrotrichs	Pseudocoel	Complete	None
Ribbon worms	Coelom	Complete	Closed
Brachiopods	Coelom	Complete in most	Open
Phoronids	Coelom	Complete	Closed
Annelids	Coelom	Complete	Closed or open
Mollusks	Reduced coelom	Complete	Open except in cephalopods
ECDYSOZOANS			
Nematodes	Pseudocoel	Complete	None
Horsehair worms	Pseudocoel	Greatly reduced	None
Arthropods	Hemocoel	Complete	Open

Cilia-bearing lophophores and trochophores evolved among the lophotrochozoans

Lophotrochozoans derive their name from two different ciliated features: a feeding structure known as a lophophore and a free-living larval form known as a trochophore. Neither the lophophore nor the trochophore is universal to all lophotrochozoans, however.

Several distantly related groups of lophotrochozoans (including bryozoans, entoprocts, brachiopods, and phoronids) have a **lophophore**, a circular or U-shaped ring of ciliated, hollow tentacles around the mouth (**Figure 32.2**). This complex organ is used for both food collection and gas exchange. Biologists once grouped taxa that have lophophores together as "lophophorates," but it is now clear that they are not one another's closest relatives. The lophophore appears to have evolved independently at least twice, or else it is an ancestral feature of lophotrochozoans and has been lost in many groups. Nearly all animals with a lophophore are sessile as adults. They use the tentacles and cilia of the lophophore to capture small floating organisms from the water. Other sessile lophotrochozoans have less well developed tentacles that they use for the same purpose.

 Go to Media Clip 32.1
Feeding with a Lophophore
Life10e.com/mc32.1

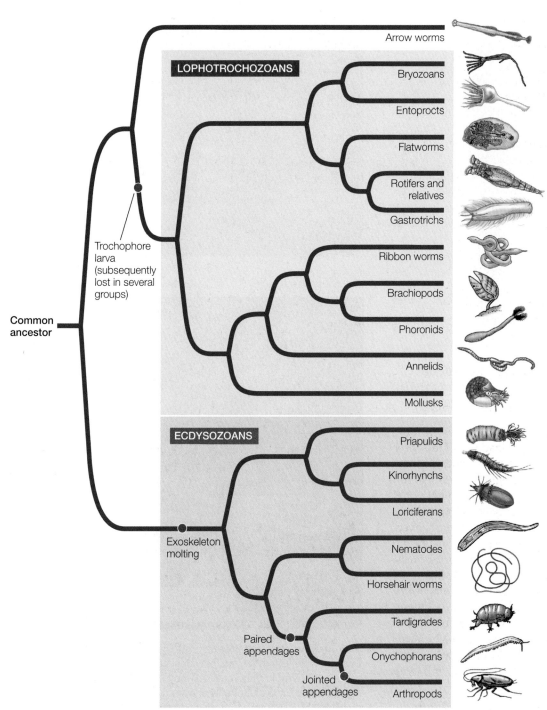

32.1 Phylogenetic Tree of Protostomes Two major lineages, the lophotrochozoans and the ecdysozoans, dominate the protostome tree. Some small groups are not included in this tree. The phylogenetic relationships shown here are supported mainly by genomic sequence data. Although genomic studies are contributing greatly to our knowledge of animal phylogeny, most species of protostomes have yet to be studied in detail.

**Go to Activity 32.2
Protosome Classification
Life10e.com/ac32.2**

Some lophotrochozoans, especially in their larval form, use cilia for locomotion. The larval form known as a **trochophore** moves by beating a band of cilia (see Figure 31.10A). This movement of cilia also brings plankton closer to the larva, where it can capture and ingest them (its cilia are therefore similar in function to the cilia of the lophophore). Trochophore larvae are found among many of the major groups of lophotrochozoans, including the mollusks, annelids, ribbon worms, entoprocts, and bryozoans. This larval form was probably present in the common ancestor of lophotrochozoans but has been subsequently lost in several lineages.

As we discussed in Chapter 31, some lophotrochozoans (including flatworms, ribbon worms, annelids, and mollusks) exhibit a form of cleavage in early development known as spiral cleavage. Some biologists group these taxa together as "spiralians," although phylogenetic analyses of gene sequences do not support a spiralian monophyly. Nonetheless, spiral cleavage may have been present in the lophotrochozoan ancestor and subsequently lost in several descendant lineages.

Many lineages of lophotrochozoans have a wormlike body form, which means that they are bilaterally symmetrical, legless, soft-bodied, and at least several times longer than they

Bryozoans can oscillate, rotate, and retract their lophophore tentacles.

Plumatella repens

100 μm

32.2 Bryozoans Use the Lophophore to Feed The extended lophophore dominates the anatomy of the colonial bryozoans. This species inhabits fresh water, although most bryozoans are marine.

(A)

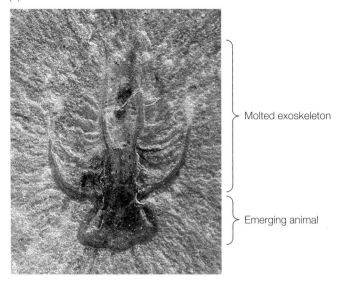

Molted exoskeleton

Emerging animal

(B) *Heterophrynus batesii*   Molted exoskeleton

The newly emerged whip scorpion's body is still soft and vulnerable.

32.3 Molting: Past and Present (A) This 500-million-year-old fossil from the Cambrian, an individual of a long-extinct arthropod species captured in the process of molting, shows that the molting process is an evolutionarily ancient trait. (B) This tailless whip scorpion has just emerged from its discarded exoskeleton. It will be highly vulnerable until its new cuticle has hardened.

are wide. A wormlike body form enables animals to burrow efficiently through marine sediment or soil. However, as we will see in Section 32.2, the mollusks—the most familiar of the lophotrochozoans to many people—have a very different body organization.

Ecdysozoans must shed their cuticles

Ecdysozoans have an external covering, or **cuticle**, that is secreted by the underlying epidermis (the outermost cell layer). The cuticle provides these animals with both protection and support. Once formed, however, the cuticle cannot grow. How, then, can ecdysozoans increase in size? They do so by shedding, or **molting**, the cuticle and replacing it with a new, larger one. This molting process gives the clade its name (Greek *ecdysis*, "to get out of").

 Go to Media Clip 32.2
Molting a Cuticle
Life10e.com/mc32.2

A fossil Cambrian arthropod preserved in the process of molting shows that molting evolved more than 500 million years ago (**Figure 32.3A**). An increasingly rich array of molecular and genetic evidence, including a set of Hox genes shared by all ecdysozoans, suggests they have a single common ancestor. Thus molting of a cuticle is a trait that may have evolved only once during animal evolution.

Before an ecdysozoan molts, a new cuticle is already forming underneath the old one. Once the old cuticle is shed, the new one expands and hardens. Until it has hardened, though,

the animal is vulnerable to its enemies, both because its outer surface is easy to penetrate and because an individual with a soft cuticle can move only slowly or not at all (**Figure 32.3B**).

In many ecdysozoans that have wormlike bodies, the cuticle is relatively thin and flexible; it offers the animal some protection but provides only modest body support. A thin cuticle allows the exchange of gases, minerals, and water across the body surface, but it restricts the animal to moist habitats. Many species of ecdysozoans with thin cuticles live in marine sediments from which they obtain food, either by ingesting sediments and extracting

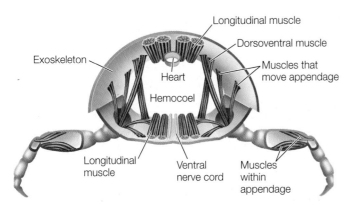

32.4 Arthropod Skeletons Are Rigid and Jointed This cross section through a thoracic segment of a generalized arthropod illustrates the arthropod body plan, which is characterized by a rigid exoskeleton with jointed appendages.

organic material from them or by capturing larger prey using a toothed **pharynx** (a muscular organ at the anterior end of the digestive tract). Some freshwater species absorb nutrients directly through their thin cuticles, as do parasitic species that live within their hosts (endoparasites). Many wormlike ecdysozoans are predators, eating protists and small animals.

The cuticles of other ecdysozoans, mainly arthropods, function as external skeletons, or **exoskeletons**. These exoskeletons are thickened by layers of protein and a strong, waterproof polysaccharide called **chitin**. An animal with a rigid, chitin-reinforced exoskeleton can neither move in a wormlike manner nor use cilia for locomotion. A hard exoskeleton also impedes the passage of oxygen and nutrients into the animal, presenting new challenges in other areas besides growth. Thus new mechanisms of locomotion and gas exchange evolved in those ecdysozoans with hard exoskeletons.

To move rapidly, an animal with a rigid exoskeleton must have body extensions that can be manipulated by muscles. Such appendages evolved in the late Precambrian, leading to the **arthropod** ("jointed foot") clade. Arthropod appendages exist in an amazing variety of forms. They serve many functions, including walking and swimming, gas exchange, food capture and manipulation, copulation, and sensory perception. Arthropods grasp food with their mouths and associated appendages and digest it internally. Their muscles are attached to the inside of the exoskeleton. Each segment has muscles that operate that segment and the appendages attached to it (**Figure 32.4**).

The arthropod exoskeleton has had a profound influence on the evolution of these animals. Encasement within a rigid body covering provides support for walking on dry land, and the waterproofing provided by chitin keeps the animal from dehydrating in dry air. Thus aquatic arthropods were, in short, excellent candidates to invade terrestrial environments. As we will see, they did so several times.

Arrow worms retain some ancestral developmental features

Nearly all triploblastic animal groups can be readily classified as either protostomes or deuterostomes, but the evolutionary relationships of one small group, the **arrow worms**, were debated for many years. Although the early development of arrow worms seems similar to that of deuterostomes, it is now

known that arrow worms merely retain developmental features that are ancestral to triploblastic animals in general. Recent studies of gene sequences clearly identify arrow worms as protostomes. There is still some question as to whether they are the closest relatives of the lophotrochozoans (as shown in Figure 32.1) or possibly the sister group of all other protostomes.

The arrow worm body is divided into three compartments: head, trunk, and tail (**Figure 32.5**). The body is transparent or translucent. Most arrow worms swim in the open sea. A few species live on the seafloor. Their abundance as fossils indicates that they were common more than 500 million years ago. The 180 or so living species of arrow worms are small enough—ranging from 3 millimeters to 12 centimeters in length—that their gas exchange and waste excretion requirements are met by diffusion through the body surface. They lack a circulatory system; wastes and nutrients are moved around the body in the coelomic fluid, which is propelled by cilia that line the coelom.

Arrow worms are hermaphroditic; that is, each individual produces both male and female gametes. Eggs are fertilized internally following elaborate courtship between two individuals. Miniature adults hatch directly from the eggs; these animals have no distinct larval stage.

Arrow worms are stabilized in the water by means of one or two pairs of lateral fins and a tail fin. They are major predators of planktonic organisms in the open ocean, ranging in size from

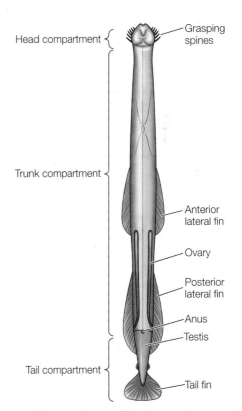

32.5 An Arrow Worm Arrow worms have a three-part body organization. Their fins and grasping spines are adaptations for a predatory lifestyle. Individuals are hermaphroditic, producing both eggs in an ovary and sperm in a testis.

small protists to young fish as large as the arrow worms themselves. An arrow worm typically lies motionless in the water until water movement signals the approach of prey. The arrow worm then darts forward and uses the stiff spines adjacent to its mouth to grasp its prey.

■ **RECAP 32.1**

The shared derived traits of protostomes include a blastopore that develops into a mouth, an anterior brain, and a ventral nervous system. Several lophotrochozoan groups are characterized by a filter-feeding structure known as a lophophore or by cilia-bearing larvae known as trochophores. Ecdysozoans, which have a body covering known as a cuticle, must molt periodically in order to grow.

- How does an animal's body covering influence the way it exchanges gases, feeds, and moves? **See pp. 654–655**
- What features made arthropods well adapted for colonizing terrestrial environments? **See p. 655**

In the next section we continue our survey of the protostomes with a more detailed look at the major groups of lophotrochozoans and the diverse body forms that are found among them.

32.2 What Features Distinguish the Major Groups of Lophotrochozoans?

Lophotrochozoans come in a variety of sizes and shapes, ranging from relatively simple animals with a blind gut (that is, a gut with only one opening) and no internal transport system to animals with a complete gut (having separate entrance and exit openings) and a complex internal transport system. They include some species-rich groups, such as flatworms, annelids, and mollusks. A number of these groups have wormlike bodies, but the lophotrochozoans encompass a wide variety of morphologies, including a few groups with external shells. Some lophotrochozoan groups have only recently been discovered by biologists.

Most bryozoans and entoprocts live in colonies

Most of the 5,500 species of **bryozoans** ("moss animals") and 170 species of **entoprocts** (meaning "anus inside") are colonial animals that live in a "house" made of material secreted by the external body wall. The colonial species are sessile, but the few solitary species can slowly move around in their environment. Almost all bryozoans and entoprocts are marine, although a few species occur in fresh or brackish water.

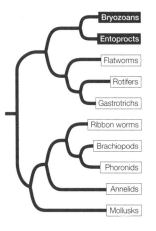

A bryozoan colony consists of many small (1–2 mm) individuals connected by strands of tissue along which nutrients

can be moved (see Figure 31.12). Bryozoan colonies can grow to contain more than 2 million individuals, all stemming from the asexual reproduction of the colony's founder. Rocks in coastal regions in many parts of the world are covered with luxuriant growths of bryozoans. Some bryozoans create miniature reefs in shallow waters. In some species, the individual colony members are differentially specialized for feeding, reproduction, defense, or support. Individual bryozoans in a colony are able to oscillate their lophophore to increase contact with prey. They can also retract it into their "house" (see Figure 32.2).

Bryozoans can reproduce sexually by releasing sperm into the water, which carries the sperm to other individuals. Eggs are fertilized internally; developing embryos are brooded before they exit as larvae to seek suitable sites for attachment to the substrate. Entoprocts can also reproduce both sexually and asexually. Some species of entoprocts release unfertilized eggs into the water for fertilization, whereas other species brood their developing young as bryozoans do.

Bryozoans and entoprocts differ in the placement of the anus. In bryozoans, the anus is located outside the ring of tentacles that make up the lophophore, whereas the anus of entoprocts is located in the center of this ring. The lophophores of the two groups also function differently: food particles are carried from the tips to the bases of the tentacles in bryozoans, but from the bases to the tips of the tentacles in entoprocts. Entoprocts lack a coelom, whereas bryozoans have a three-part coelom.

Flatworms, rotifers, and gastrotrichs are structurally diverse relatives

Flatworms, rotifers, gastrotrichs, and their close relatives are a structurally diverse group of organisms whose relationships to one another has been hypothesized only recently. If recent genomic studies prove correct, this monophyletic lophotrochozoan group includes both acoelomate subgroups (e.g., the flatworms) and pseudocoelomate subgroups (e.g., the rotifers and gastrotrichs), and yet the closest relatives of this group—the bryozoans and entoprocts—are coelomate and acoelomate, respectively. Thus this group provides an example of the evolutionary convergence in body cavity form that we described in Section 32.1.

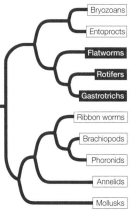

FLATWORMS Flatworms lack specialized organs for transporting oxygen to their internal tissues. In the absence of a gas transport system, each cell must be near a body surface, a requirement met by the dorsoventrally flattened body form that gives these animals their common name. The digestive tract of a flatworm consists of a mouth opening into a blind gut. The gut is often highly branched, forming intricate patterns that increase the surface area available for the absorption of nutrients. Some small free-living flatworms are cephalized, with a

(A) *Eurylepta californica*, a free-living flatworm

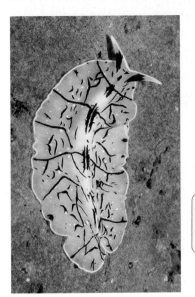

This parasitic flatworm's body is filled primarily with sex organs.

(B) Diagram of a typical parasitic flatworm

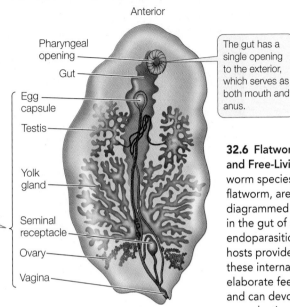

Anterior

Pharyngeal opening

Gut

The gut has a single opening to the exterior, which serves as both mouth and anus.

Egg capsule

Testis

Yolk gland

Seminal receptacle

Ovary

Vagina

Posterior

32.6 Flatworms Include Both Parasites and Free-Living Forms (A) Some flatworm species, such as this Pacific marine flatworm, are free-living. (B) The fluke diagrammed here lives endoparasitically in the gut of sea urchins and is typical of endoparasitic flatworms. Because their hosts provide all the nutrition they need, these internal parasites do not require elaborate feeding or digestive organs and can devote most of their bodies to reproduction.

head bearing chemoreceptor organs, two simple eyes, and a tiny brain composed of anterior thickenings of the longitudinal nerve cords. Free-living flatworms glide over surfaces, powered by broad bands of cilia (**Figure 32.6A**).

Although many flatworms are free-living, most flatworm species are parasites. Of the parasitic species, most are endoparasites. There are also flatworms that feed externally on animal tissues (living or dead), and some graze on plants. A likely evolutionary transition was from feeding on dead organisms to feeding on the body surfaces of dying hosts to invading and consuming parts of healthy hosts.

Most of the 30,000 species of living flatworms are tapeworms and flukes; members of these two groups are endoparasites, particularly of vertebrates (**Figure 32.6B**). Because they absorb digested food from the digestive tracts of their hosts, many endoparasitic flatworms lack digestive tracts of their own. Some cause serious human diseases, such as schistosomiasis, which is common in parts of Asia, Africa, and South America. The species that causes this devastating disease has a complex life cycle involving both freshwater snails and mammals as hosts. Members of another flatworm group, the monogeneans, are ectoparasites of fishes and other aquatic vertebrates. The turbellarians include most of the free-living species.

ROTIFERS Most species of **rotifers** are tiny (50–500 μm long)—smaller than some ciliate protists—but they have specialized internal organs (**Figure 32.7A and B**). A complete gut passes from an anterior mouth to a posterior anus; the body cavity is a pseudocoel that functions as a hydrostatic skeleton. Rotifers typically propel themselves through the water by means of rapidly beating cilia rather than by muscular contraction.

The most distinctive organ of rotifers is a conspicuous ciliated organ called the corona, which surmounts the head of many species. Coordinated beating of the cilia sweeps particles

of organic matter from the water into the animal's mouth and down to a complicated structure called the mastax, in which food is ground into small pieces. By contracting muscles around the pseudocoel, a few rotifer species that prey on protists and small animals can protrude the mastax through the mouth and seize small objects with it.

Go to Media Clip 32.3
Rotifer Feeding
Life10e.com/mc32.3

Most of the known species of rotifers live in fresh water. Some species rest on the surfaces of mosses or lichens in a desiccated, inactive state until it rains. When rain falls, they absorb water and become mobile, feeding in the films of water that temporarily cover the plants. Most rotifers live no longer than a few weeks.

Both males and females are found in some species of rotifers, but only females are known among the bdelloid rotifers (the b in "bdelloid" is silent). Biologists have concluded that the bdelloid rotifers may have existed for tens of millions of years without regular sexual reproduction. Lack of genetic recombination generally leads to the buildup of deleterious mutations, so long-term asexual reproduction typically leads to extinction (see Section 21.4). Recent studies, however, have indicated that bdelloid rotifers may avoid this problem by picking up fragments of genes from their environment during the desiccation–rehydration cycle, which allows genetic recombination among individuals in the absence of direct sexual exchange.

A few highly reduced lineages appear to have descended from the free-living rotifers. The spiny-headed worms are parasites with complex life cycles, often parasitizing several animal hosts (**Figure 32.7C**). The jaw worms are tiny marine organisms that glide between sand grains in shallow marine environments. Although spiny-headed worms and jaw worms are

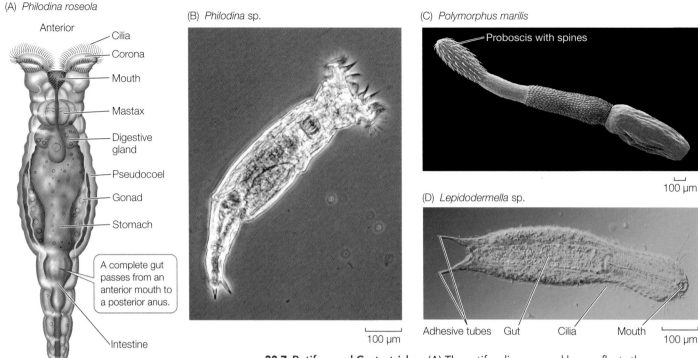

(A) *Philodina roseola*

Anterior

- Cilia
- Corona
- Mouth
- Mastax
- Digestive gland
- Pseudocoel
- Gonad
- Stomach

A complete gut passes from an anterior mouth to a posterior anus.

Intestine

"Foot" with "toes"

Anus

Posterior

(B) *Philodina* sp.

100 μm

(C) *Polymorphus marilis*

Proboscis with spines

100 μm

(D) *Lepidodermella* sp.

Adhesive tubes   Gut   Cilia   Mouth   100 μm

32.7 Rotifers and Gastrotrichs (A) The rotifer diagrammed here reflects the general structure of many rotifers. (B) A micrograph reveals the internal complexity of the microscopic rotifers. (C) Spiny-headed worms are parasitic rotifer relatives. The spines of the proboscis anchor the animal to the organs of its host. (D) Gastrotrichs superficially resemble rotifers but have flattened ventral surfaces covered with cilia, as flatworms do.

structurally quite distinct, molecular analyses have revealed that both groups are essentially highly modified rotifers.

GASTROTRICHS   The 800 species of **gastrotrichs** (also called "hairy backs") are abundant, tiny (0.05–3 mm) animals that live in marine sediments, in fresh waters, and in the water films that surround grains of soil. Their transparent bodies have a flat ventral surface that is covered with cilia (**Figure 32.7D**). Most species are simultaneous hermaphrodites, with both male and female reproductive organs, although the male organs have been greatly reduced or lost in some species that reproduce asexually.

Ribbon worms have a long, protrusible feeding organ

Ribbon worms (nemerteans) have simple nervous and excretory systems similar to those of flatworms. Unlike flatworms, however, they have a complete digestive tract with a mouth at one end and an anus at the other. Small ribbon worms move slowly by beating their cilia. Larger ones employ waves of muscle contraction to move over the surface of sediments or to burrow into them.

Bryozoans
Entoprocts
Flatworms
Rotifers
Gastrotrichs
Ribbon worms
Brachiopods
Phoronids
Annelids
Mollusks

Within the body of nearly all of the 1,200 species of ribbon worms is a fluid-filled cavity called the rhynchocoel, within which lies a hollow, muscular **proboscis**. The proboscis, which is the worm's feeding organ, may extend much of the length of the body. Contraction of the muscles surrounding the rhynchocoel causes the proboscis to evert explosively through an anterior pore (**Figure 32.8A**). The proboscis may be armed with sharp stylets that pierce prey and discharge paralysis-causing toxins into the wound.

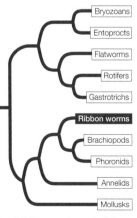

Go to Media Clip 32.4
Explosive Extrusion of Ribbon Worm Proboscis
Life10e.com/mc32.4

Most ribbon worm species are marine, although there are species that live in fresh water or on land. Most species are less than 20 centimeters long, but individuals of some species reach 20 meters or more. Some genera feature species that are conspicuous and brightly colored (**Figure 32.8B**). Recent molecular analyses suggest that ribbon worms may be most closely related to the brachiopods and phoronids.

Brachiopods and phoronids use lophophores to extract food from the water

Recall that the bryozoans and entoprocts use a lophophore to feed. Brachiopods and phoronids also feed using a lophophore, but this structure may have evolved separately in these groups. Although neither the brachiopods nor the phoronids are represented by many living species, the brachiopods (which have hard external shells and thus leave an excellent

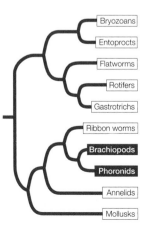

Bryozoans
Entoprocts
Flatworms
Rotifers
Gastrotrichs
Ribbon worms
Brachiopods
Phoronids
Annelids
Mollusks

(A)

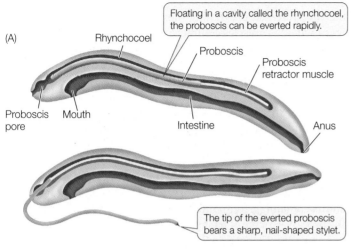

Rhynchocoel

Floating in a cavity called the rhynchocoel, the proboscis can be everted rapidly.

Proboscis

Proboscis retractor muscle

Proboscis pore   Mouth

Intestine

Anus

The tip of the everted proboscis bears a sharp, nail-shaped stylet.

(B) *Tubulanus sexlineatus*

Anterior (mouth)

Posterior (anus)

32.8 Ribbon Worms (A) The proboscis is the ribbon worm's feeding organ. (B) This large marine nemertean is found in harbors and bays along the Pacific Coast of North America. Its proboscis is not everted in this photograph.

fossil record) are known to have been much more abundant in the past.

BRACHIOPODS **Brachiopods** (lampshells) are solitary marine animals. They have a rigid shell that is divided into two parts connected by a ligament (**Figure 32.9**). The two halves can be pulled shut to protect the soft body. Brachiopods superficially resemble bivalve mollusks, but shells have evolved independently in the two groups. The two halves of the brachiopod shell are dorsal and ventral, rather than lateral as in bivalves. The lophophore is located within the shell. The beating of cilia on the lophophore draws water into the slightly opened shell. Food is trapped in the lophophore and directed to a ridge, along which it is transferred to the mouth.

Most brachiopods are 4 to 6 centimeters long. They live attached to a solid substrate or embedded in soft sediments. Most species are attached by means of a short, flexible stalk that holds the animal above the substrate. Gases are exchanged across body surfaces, especially the tentacles of the lophophore. Most brachiopods release their gametes into the water, where they are fertilized. The larvae remain among the plankton for only a few days before they settle and develop into adults.

Laqueus sp.

Lophophore ring

Tentacles

32.9 A Brachiopod's Lophophore The lophophore of this North Pacific brachiopod can be seen between the valves of its shell.

Brachiopods reached their peak abundance and diversity in Paleozoic and Mesozoic times. More than 26,000 fossil species have been described. Only about 450 species survive, but they remain common in some marine environments.

PHORONIDS The ten known species of **phoronids** are small (5–25 cm long), sessile worms that live in muddy or sandy sediments or attached to rocky substrates. Phoronids are found in marine waters from the intertidal zone to about 400 meters deep. They secrete tubes made of chitin, within which they live, and have a U-shaped gut with the anus located outside the lophophore (**Figure 32.10**). Their cilia drive water into the top of the lophophore, and the water exits through the narrow spaces between the tentacles. Suspended food particles are caught and transported to the mouth by ciliary action. Some species release eggs into the water, where they are fertilized, but other species produce large eggs that are fertilized internally and retained in the parent's body, where they are brooded until they hatch.

Annelids have segmented bodies

The wormlike bodies of **annelids** are clearly segmented. As described in Section 31.2, segmentation allows an animal to move different parts of its body independently, giving it much better control of its movement. The earliest segmented worms, preserved as fossils from the middle Cambrian, were burrowing marine annelids.

In most large annelids, the coelom in each segment is isolated from those in other segments (**Figure 32.11**). A separate nerve center called a ganglion (plural ganglia) controls each segment; nerve cords that connect the ganglia coordinate their

Bryozoans
Entoprocts
Flatworms
Rotifers
Gastrotrichs
Ribbon worms
Brachiopods
Phoronids
Annelids
Mollusks

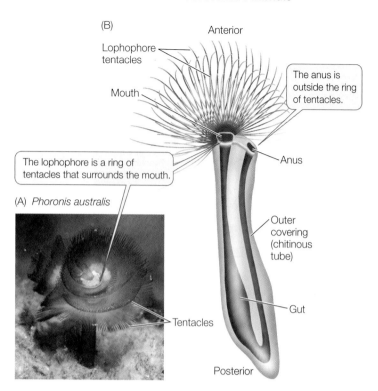

(B)

Anterior

Lophophore tentacles

Mouth

The anus is outside the ring of tentacles.

The lophophore is a ring of tentacles that surrounds the mouth.

(A) *Phoronis australis*

Anus

Tentacles

Outer covering (chitinous tube)

Gut

Posterior

32.10 Phoronids (A) The tentacles of this phoronid's lophophore form a spiral. (B) The phoronid gut is U-shaped, as seen in this generalized diagram.

the substrate and preventing the animal from slipping backward when its muscles contract.

Members of one polychaete clade, the **pogonophorans**, secrete tubes made of chitin and other substances, in which they live (**Figure 32.12B**). Pogonophorans have lost their digestive tract (they have no mouth or gut). So how do they obtain nutrition? Part of the answer is that pogonophorans can take up dissolved organic matter directly from the sediments in which they live or from the surrounding water. Much of their nutrition, however, is provided by endosymbiotic bacteria that the pogonophorans house in a specialized organ known as the trophosome. These bacteria oxidize hydrogen sulfide and other sulfur-containing compounds, fixing carbon from methane in the process. Uptake of the hydrogen sulfide, methane, and oxygen used by the bacteria is facilitated by hemoglobin in the pogonophorans' tentacles. It is this hemoglobin that gives the tentacles their red coloration.

Pogonophorans were not discovered until early in the twentieth century, when the first species were discovered on the seafloor at depths of up to a few hundred meters. In recent decades, deep-sea explorers have found them living many thousands of meters below the ocean surface. In these deep oceanic sediments, they may reach densities of many thousands per

functioning. Most annelids lack a rigid external protective covering; instead they have a thin, permeable body wall that serves as a general surface for gas exchange. Most annelids are thus restricted to moist environments because they lose body water rapidly in dry air. The approximately 19,000 described species live in marine, freshwater, and moist terrestrial environments.

POLYCHAETES More than half of all annelid species are commonly known as **polychaetes** ("many hairs"), although this is a descriptive term rather than the name of a single clade. Recent molecular studies indicate that polychaetes are paraphyletic with respect to the remaining annelids. Most polychaetes are marine, and many live in burrows in soft sediments. Most of them have one or more pairs of eyes and one or more pairs of tentacles, with which they capture prey or filter food from the surrounding water, at the anterior end of the body (**Figure 32.12A**; see also Figure 31.6A). In some species, the body wall of most segments extends laterally as a series of thin outgrowths called parapodia. The parapodia function in gas exchange, and some species use them to move. Stiff bristles called setae protrude from each parapodium, forming temporary contact with

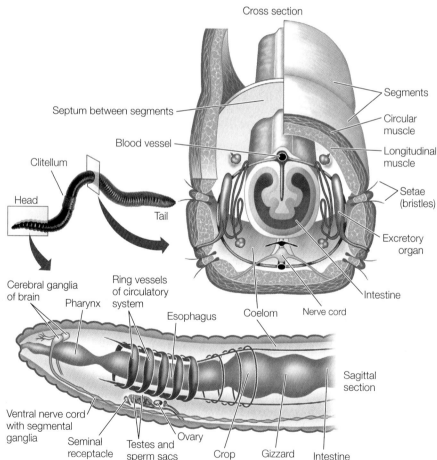

Cross section

Septum between segments

Segments

Circular muscle

Blood vessel

Longitudinal muscle

Clitellum

Setae (bristles)

Head

Tail

Excretory organ

Cerebral ganglia of brain

Ring vessels of circulatory system

Pharynx

Esophagus

Coelom

Nerve cord

Intestine

Ventral nerve cord with segmental ganglia

Seminal receptacle

Testes and sperm sacs

Ovary

Crop

Gizzard

Intestine

Sagittal section

32.11 Annelids Have Many Body Segments The segmented structure of annelids such as this earthworm is apparent both externally and internally. Many organs are repeated serially.

(A) *Spirographis spallanzanii*

(B) *Riftia* sp.

32.12 Diversity among the Annelids (A) "Fan worms" or "feather duster worms" are sessile marine polychaetes that grow in masses, filtering food from the water with their tentacles. This individual has been removed from its chitinous tube. (B) Pogonophorans live around hydrothermal vents deep in the ocean. Their tentacles can be seen protruding from their chitinous tubes. (C) Earthworms, like all oligochaetes, are hermaphroditic; when they copulate, each individual donates and receives sperm. (D) The medicinal leech has been a tool of physicians and healers for centuries. Even today, leeches have uses in clinical practice.

(C) *Lumbricus terrestris*

(D) *Hirudo medicinalis*

square meter. About 160 species have been described. The largest and most remarkable pogonophorans are 2 meters or more in length and live near deep-sea hydrothermal vents—volcanic openings in the seafloor through which hot, sulfide-rich water pours. The methane and hydrogen sulfide from these vents provide the raw materials for carbon fixation by the pogonophorans' endosymbiotic bacteria.

CLITELLATES The approximately 3,000 described species of **clitellates**, which form a well-supported clade within the annelids, are found in freshwater, marine, and terrestrial environments. The clitellates appear to be phylogenetically nested among various groups of polychaetes, although the exact relationships are not yet clear. There are two major groups of clitellates, the oligochaetes and the leeches.

Oligochaetes ("few hairs") have no parapodia, eyes, or anterior tentacles, and they have only four pairs of setae bundles per segment. Earthworms—the most familiar oligochaetes—burrow in and ingest soil, from which they extract food particles. All oligochaetes are hermaphroditic. Sperm are exchanged simultaneously between two copulating individuals (**Figure 32.12C**). Eggs and sperm are deposited in a cocoon outside the adult's body. Fertilization occurs within the cocoon after it

is shed, and when development is complete, miniature worms emerge and immediately begin independent life.

Leeches, like oligochaetes, lack parapodia and tentacles. The coelom of leeches is not divided into compartments; the coelomic space is largely filled with undifferentiated tissue. Groups of segments at each end of the body are modified to form suckers, which serve as temporary anchors that aid the leech in its movement. With its posterior sucker attached to a substrate, the leech extends its body by contracting its circular muscles. The anterior sucker is then attached, the posterior one detached, and the leech shortens itself by contracting its longitudinal muscles. Leeches live in freshwater or terrestrial habitats.

Most leeches are ectoparasites that feed by making an incision in a host, from which blood flows. A leech can ingest so much blood in a single feeding that its body may enlarge severalfold. The leech secretes an anticoagulant into the wound that keeps the host's blood flowing. For centuries, medical practitioners employed leeches to draw blood to treat diseases they believed were caused by an excess of blood or by "bad blood." Although most leeching practices (such as inserting a leech in a person's throat to alleviate swollen tonsils) have been abandoned, *Hirudo medicinalis* (the medicinal leech; **Figure 32.12D**) is used today to reduce fluid pressure and prevent blood clotting in damaged tissues, to eliminate pools of co-agulated blood, and to prevent scarring. The anticoagulants of certain other leech species also contain anesthetics and blood vessel dilators and are being studied for possible medical uses.

Go to Media Clip 32.5
Leeches Feeding on Blood
Life10e.com/mc32.5

Mollusks have undergone a dramatic evolutionary radiation

Mollusks are the most diverse group of lophotrochozoans, both in numbers of species and in the environments they occupy. Although the major groups of mollusks differ dramatically in morphology, they all share the same three major body components: a foot, a visceral mass, and a mantle (**Figure 32.13A**).

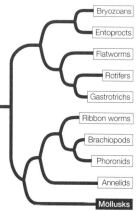

Bryozoans
Entoprocts
Flatworms
Rotifers
Gastrotrichs
Ribbon worms
Brachiopods
Phoronids
Annelids
Mollusks

- The molluscan **foot** is a large, muscular structure that originally was both an organ of locomotion and a support for the internal organs. In squids and octopuses, the foot has been modified to form arms and tentacles borne on a head with complex sensory organs. In other groups, such as clams, the foot is a burrowing organ. In some groups the foot is greatly reduced.

- The heart and the digestive, excretory, and reproductive organs are concentrated in a centralized, internal **visceral mass**.

- The **mantle** is a fold of tissue that covers the organs of the visceral mass. The mantle secretes the hard, calcareous shell that is typical of many mollusks.

In most mollusks, the mantle extends beyond the visceral mass to form a mantle cavity. Within this cavity lie gills that are used for gas exchange. When cilia on the gills beat, they create a current of water. The tissue of the gills, which is highly vascularized (contains many blood vessels), takes up oxygen from the water and releases carbon dioxide. Many mollusk species use their gills as filter-feeding devices, whereas others feed using a rasping structure known as a **radula** to scrape algae from rocks. In some mollusks, such as the marine cone snails, the radula has been modified into a drill or poison dart.

In all mollusks except cephalopods, the blood vessels do not form a closed system. Blood and other fluids empty into a large, fluid-filled hemocoel, through which fluids move and deliver oxygen to the internal organs. Eventually the fluids re-enter the blood vessels and are moved by a heart.

Monoplacophorans were the most abundant mollusks during the Cambrian period, 500 million years ago, but only a few species survive today. In monoplacophorans, in contrast to all other living mollusks, the gas exchange organs, muscles, and excretory pores are repeated over the length of the body.

The four major clades of living mollusks are the chitons, gastropods, bivalves, and cephalopods. Each of these groups is readily identifiable and distinct, even though they share variations on a common body plan.

CHITONS Eight overlapping calcareous plates, surrounded by a structure known as the girdle, protect the internal organs and muscular foot of **chitons** (**Figure 32.13B**). The chiton body is bilaterally symmetrical, and the internal organs, particularly the digestive and nervous systems, are relatively simple. Most chitons are marine omnivores that scrape algae, bryozoans, and other organisms from rocks with a sharp radula. An adult chiton spends most of its life clinging tightly to rock surfaces with its large, muscular, mucus-covered foot. It moves slowly by means of rippling waves of muscular contraction in the foot. Fertilization in most chitons takes place in the water, but in a few species fertilization is internal and embryos are brooded within the body. There are approximately 1,000 living species of chitons.

GASTROPODS **Gastropods** (**Figure 32.13C**) are the most species-rich and widely distributed mollusks, with about 85,000 living species. Snails, whelks, limpets, slugs, nudibranchs (sea slugs), and abalones are all gastropods. Most species move by gliding on the muscular foot, but in a few species—the sea butterflies and heteropods—the foot is a swimming organ with which the animal moves through open ocean waters.

Marine nudibranchs and terrestrial slugs are gastropods that have lost their protective shell over the course of evolution (**Figure 32.14**). Without a shell, these groups rely on other forms of protection from predation. The coloration of many nudibranchs is aposematic, meaning that it serves to warn potential predators of toxicity. Other nudibranch species and most terrestrial slugs exhibit camouflaged coloration.

Shelled gastropods have one-piece shells. The only mollusks that live in terrestrial environments—land snails and slugs—are gastropods. In these terrestrial species, the mantle tissue is modified into a highly vascularized lung.

BIVALVES Clams, oysters, scallops, and mussels are all familiar **bivalves**. The approximately 30,000 species are found in both marine and freshwater environments. Bivalves have a hinged, two-part shell that extends over the sides of the body as well as the top (**Figure 32.13D**). Many clams use the foot to burrow into mud and sand. Bivalves feed by taking in water through an opening called an incurrent siphon and filtering food from the water with their large gills, which are also the main sites of gas exchange. Water and gametes exit through the excurrent siphon. Fertilization takes place in open water in most species.

CEPHALOPODS The **cephalopods**—squids, cuttlefish, octopuses, and nautiluses—first appeared near the beginning of the Cambrian period. By the Ordovician period a variety of types were present. Today there are about 800 living species. In these mollusks the excurrent siphon is modified to

(A) Generalized molluscan body plan

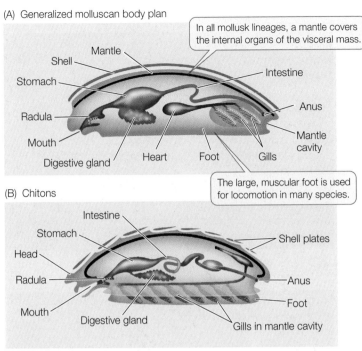

In all mollusk lineages, a mantle covers the internal organs of the visceral mass.

Mantle
Shell
Stomach
Intestine
Radula
Anus
Mouth
Mantle cavity
Digestive gland
Heart
Foot
Gills

The large, muscular foot is used for locomotion in many species.

(B) Chitons

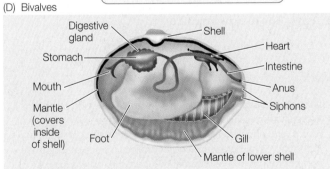

Intestine
Stomach
Head
Shell plates
Radula
Anus
Mouth
Foot
Digestive gland
Gills in mantle cavity

(C) Gastropods

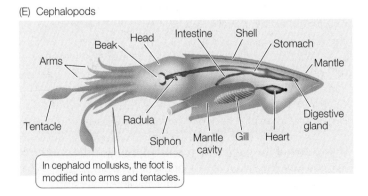

Intestine
Mantle
Mantle cavity
Gill
Shell
Anus
Heart
Siphon
Salivary gland
Cephalic tentacles
Head
Stomach
Mouth
Foot

The radula is a unique molluscan feeding structure modified for scraping.

(D) Bivalves

Digestive gland
Shell
Heart
Stomach
Intestine
Mouth
Anus
Mantle (covers inside of shell)
Siphons
Foot
Gill
Mantle of lower shell

(E) Cephalopods

Beak
Head
Intestine
Shell
Stomach
Arms
Mantle
Tentacle
Radula
Digestive gland
Siphon
Mantle cavity
Gill
Heart

In cephalod mollusks, the foot is modified into arms and tentacles.

32.13 Organization and Diversity of Molluscan Bodies (A) The major molluscan groups display different variations on a body plan that includes three major components: a foot, a visceral mass of internal organs, and a mantle. In many species, the mantle secretes a calcareous shell. (B) Chitons have eight overlapping calcareous plates surrounded by a girdle. (C) Most gastropods have a single dorsal shell, into which they can retreat for protection. (D) Bivalves get their name from their two hinged shells, which can be tightly closed. (E) Cephalopods are active predators; they use their arms and tentacles to capture prey. This cuttlefish has an internal shell but no external shell.

Chaetopleura angulata

Helix pomatia

Argopecten irradians

Sepia sp.

(A) *Flabellina iodinea*

(B) *Ariolimax columbianus*

32.14 Mollusks in Some Groups Have Lost Their Shells (A) Nudibranchs ("naked gills"), also called sea slugs, are shell-less gastropods. This species is brightly colored, alerting potential predators of its toxicity. (B) Banana slugs are terrestrial, shell-less gastropods that feed on decomposing vegetation on the damp forest floor. (C) Octopuses have neither an external nor an internal shell, which allows these cephalopods to squeeze through tight spaces.

(C) *Octopus macropus*

Go to Media Clip 32.6
Octopuses Can Pass through Small Openings
Life10e.com/mc32.6

allow the animal to control the water content of the mantle cavity (**Figure 32.13E**). The modification of the mantle into a device for forcibly ejecting water from the cavity through the siphon enables these animals to move rapidly through the water by "jet propulsion." With their greatly enhanced mobility, cephalopods became the major predators in the open waters of the Devonian oceans. They remain important marine predators today.

As is typical of active, rapidly moving predators, cephalopods have a head with complex sensory organs—most notably eyes that are comparable to those of vertebrates in their ability to resolve images (see Figure 46.14). The head is closely associated with a large, branched foot that bears the arms and/or tentacles and a siphon. Arms are distinguished by the presence of suckers along most of their length. Tentacles, in contrast, have suckers only near the tips or lack suckers altogether. Octopuses typically have eight arms and no tentacles, whereas squids and cuttlefishes have eight arms plus two tentacles. Cephalopods use their arms and tentacles to capture and subdue prey; octopuses also use their arms to move over the substrate. The large, muscular mantle provides a solid external supporting structure. The gills hang in the mantle cavity.

Many early cephalopods had an external chambered shell divided by partitions. The only surviving cephalopods with such shells are the nautiluses (genus *Nautilus*). The chambers inside nautilus shells are connected by a strand of tissue that runs through ducts in the partitions. Blood in this tissue carries water *from* the chambers and gases *into* the chambers, thus providing buoyancy. Most cephalopods retain an internal shell that functions for internal support and, in some species, is also chambered and buoyant. Octopuses have completely lost their shells, which allows them to compress their bodies through very small openings.

RECAP 32.2

Lophotrochozoans include animals with diverse body types. Wormlike forms include some flatworms, ribbon worms, phoronids, and annelids. There has been convergent evolution of lophophores (in bryozoans and entoprocts versus brachiopods and phoronids) and of external two-part shell coverings (in brachiopods versus bivalve mollusks).

- How do flatworms survive without an internal transport system? **See pp. 656–657 and Figure 32.6**

- Why are most annelids restricted to moist environments? **See p. 660**

- Briefly describe how the basic body organization of mollusks has been modified to yield a wide variety of body forms. **See pp. 662–664 and Figure 32.13**

The second of the two major protostome clades, the ecdysozoans, contains the vast majority of Earth's animal species. What evolutionary innovations led to this massive diversity?

32.3 What Features Distinguish the Major Groups of Ecdysozoans?

Many ecdysozoans are wormlike in form, although others—the arthropods, onychophorans, and tardigrades—have limbs. In this section we will look at the two clades of wormlike ecdysozoans: the priapulids, kinorhynchs, and loriciferans in one clade and the nematodes and horsehair worms in the other. Section 32.4 will be devoted to the most diverse ecdysozoans—the arthropods and their relatives—and the many forms their appendages take.

Several marine ecdysozoan groups have relatively few species

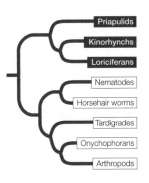

Members of several species-poor groups of wormlike marine ecdysozoans—the priapulids, kinorhynchs, and loriciferans—have relatively thin cuticles that are molted periodically as the animals grow to full size. In 2004, embryos of a fossil species related to these ecdysozoans were discovered in sediments laid down in China about 500 million years ago. This remarkable discovery shows that the ancestors of these animals developed directly from an egg to the adult form, as most of their modern descendants do.

The 20 species of **priapulids** are cylindrical, unsegmented, wormlike animals with a three-part body plan consisting of a proboscis, trunk, and caudal appendage ("tail"). It should be clear from their appearance why they were named after the Greek fertility god Priapus (**Figure 32.15A**). Priapulids range in length from 0.5 millimeters to 20 centimeters. They live in burrows in fine marine sediments and prey on soft-bodied invertebrates such as polychaetes, which they capture with a toothed, muscular pharynx that they evert through the mouth and then withdraw into the body together with the grasped prey. Fertilization is external, and most species have a larval form that also lives in the mud.

About 180 species of **kinorhynchs** have been described. They live in marine sands and muds and are virtually microscopic; no kinorhynchs are longer than 1 millimeter. Their bodies are divided into 13 segments, each covered with a separate cuticular plate (**Figure 32.15B**). These plates are periodically molted during growth. Kinorhynchs feed by ingesting sediments through a retractable proboscis (the group name means "movable snout"). They then digest the organic material found in the sediment, which may include living algae as well as dead matter. Kinorhynchs have no distinct larval stage; fertilized eggs develop directly into juveniles, which emerge from their egg cases with 11 of the 13 body segments already formed.

Loriciferans are also minute animals less than 1 millimeter long. They were not discovered until 1983. About 100 living species are known to exist, although only about 30 of these have been formally described to date. The body is divided into a head, neck, thorax, and abdomen and is covered by six plates, from which the loriciferans get their name (*Latin lorica*,

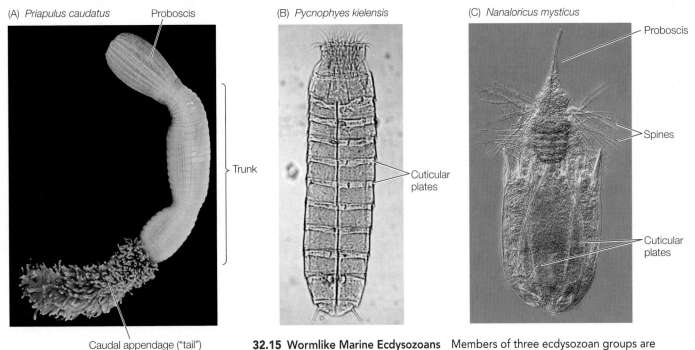

(A) *Priapulus caudatus* — Proboscis — Trunk — Caudal appendage ("tail")

(B) *Pycnophyes kielensis* — Cuticular plates

(C) *Nanaloricus mysticus* — Proboscis — Spines — Cuticular plates

32.15 Wormlike Marine Ecdysozoans Members of three ecdysozoan groups are marine bottom-dwellers. (A) Most priapulid species live in burrows on the ocean floor, extending the proboscis to feed. (B) Kinorhynchs are virtually microscopic. The cuticular plates that cover their bodies are molted periodically. (C) Six cuticular plates form a "corset" around the minute loriciferan body.

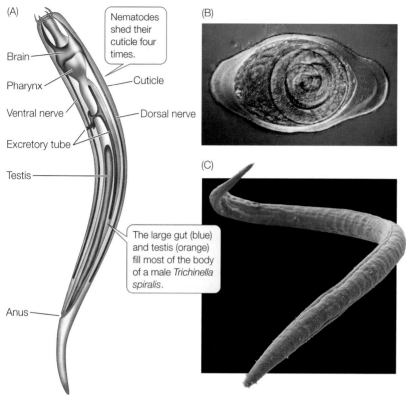

(A)

Nematodes shed their cuticle four times.

Brain
Pharynx
Cuticle
Ventral nerve
Dorsal nerve
Excretory tube
Testis
Anus

The large gut (blue) and testis (orange) fill most of the body of a male *Trichinella spiralis*.

(B)

(C)

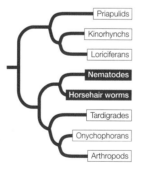

32.16 Nematodes (A) The body plan of *Trichinella spiralis*, which causes trichinosis, is typical of parasitic nematodes. (B) This polarized light micrograph shows a cyst of *T. spiralis* in the muscle tissue of a host. (C) This free-living nematode lives in freshwater environments.

"corset"). The plates around the base of the neck bear anterior-directed spines of unknown function (**Figure 32.15C**). Loriciferans live in coarse marine sediments. Little is known about what they eat, but some species apparently eat bacteria.

Nematodes and their relatives are abundant and diverse

Nematodes (roundworms) have a thick, multilayered cuticle that gives their unsegmented body its shape (**Figure 32.16**). As a nematode grows, it sheds its cuticle four times. Nematodes exchange oxygen and nutrients with their environment through both the cuticle and the gut wall, which is only one cell layer thick. Materials are moved through the gut by rhythmic contraction of a highly muscular pharynx. Nematodes move by contracting their longitudinal muscles.

Nematodes are probably the most abundant and universally distributed of all animal groups. Many nematodes are microscopic; the largest known nematode, which reaches a length of 9 meters, is a parasite in the placentas of sperm whales. About 25,000 species have been described, but the actual number of living species may be more than 1 million. Countless nematodes live as scavengers in the upper layers of the soil, on the bottoms of lakes and streams, and in marine sediments. The topsoil of rich

Priapulids
Kinorhynchs
Loriciferans
Nematodes
Horsehair worms
Tardigrades
Onychophorans
Arthropods

farmland may contain from 3 to 9 billion nematodes per acre. A single rotting apple may contain as many as 90,000 individuals.

One soil-inhabiting nematode, *Caenorhabditis elegans*, serves as a model organism in the laboratories of geneticists and developmental biologists. It is ideal for such research because it is easy to cultivate, matures in 3 days, and has a fixed number of body cells. Its genome has been completely sequenced.

Many nematodes are predators, feeding on protists and small animals (including other roundworms). Most significant to humans, however, are the many species that parasitize plants and animals. The nematodes that parasitize humans (causing serious diseases such as trichinosis and elephantiasis), domestic animals, and economically important plants have been studied intensively in an effort to find ways of controlling them.

The structure of parasitic nematodes is similar to that of free-living species, but the life cycles of many parasitic species have special stages that facilitate the transfer of individuals among hosts. *Trichinella spiralis*, the species that causes the human disease trichinosis, has a relatively simple life cycle. A person may become infected by eating the flesh of an animal (usually a pig) that has *Trichinella* larvae encysted in its muscles (see Figure 32.16B). The larvae are activated in the person's digestive tract, emerge from their cysts, and attach to the intestinal wall, where they feed. Later they bore through the intestinal wall and are carried in the bloodstream to muscles, where they form new cysts. If present in great numbers, these cysts can cause severe pain or death.

About 350 species of the unsegmented **horsehair worms** have been described. As their name implies, these animals are extremely thin in diameter; horsehair worms range from a few millimeters up to 1 meter in length. Most adult worms live in fresh water, among leaf litter and algal mats near the edges of streams and ponds. A few species live in damp soil.

Horsehair worm larvae are endoparasites of freshwater crayfishes and of terrestrial and aquatic insects (**Figure 32.17**). An adult horsehair worm has no mouth, and its gut is greatly reduced and probably nonfunctional. Some species may feed only as larvae, absorbing nutrients from their hosts across the body wall. But other species continue to shed their cuticles and grow after they have left their hosts, suggesting that adult worms may also absorb nutrients from their environment.

RECAP **32.3**

Priapulids, kinorhynchs, and loriciferans are relatively small, poorly known groups of wormlike marine ecdysozoans. Nematodes and horsehair worms have unsegmented wormlike bodies. Nematodes are probably the most abundant and universal animal group.

- Describe at least three different ways in which nematodes have a significant impact on humans. See p. 666

An adult horsehair worm leaves the body of the wood cricket it parasitized during its larval development.

32.17 Horsehair Worm Larvae Are Parasitic The larva of this horsehair worm (*Paragordius tricuspidatus*) can manipulate its host's behavior. The hatching worm causes the cricket to jump into water, where the worm emerges from the insect's body to continue its life cycle as a free-living adult. The cricket, having delivered its parasitic burden, drowns.

We will next turn to the animals that not only dominate the ecdysozoan clade but also constitute the most diverse group of animals on Earth.

32.4 Why Are Arthropods So Diverse?

Arthropods and their relatives are ecdysozoans with paired appendages. Arthropods are the most diverse group of animals in numbers of species (more than 1.2 million have been described, and many more remain to be discovered). Furthermore, the number of individual arthropods alive at any one time is estimated to be about 10^{18}, or 1 billion billion. Among the animals, only the nematodes are thought to exist in greater numbers.

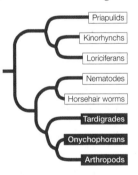

Several key features have contributed to the success of the arthropods. As we have seen, their muscles are attached to the inside of their rigid exoskeletons. Their bodies are segmented, and each segment has muscles that operate that segment and the jointed appendages attached to it (see Figure 32.4). Jointed appendages permit complex movements, and different appendages are specialized for different functions. Encasement of the body within a rigid exoskeleton provides the animal with support for walking in the water or on dry land and provides some protection against predators. The waterproofing provided by chitin keeps the animal from dehydrating in dry air.

The four major arthropod groups living today are all species-rich: chelicerates (including the arachnids—spiders, scorpions, mites, and their relatives), myriapods (millipedes and centipedes), crustaceans (including shrimps, crabs, and barnacles), and hexapods (insects and their wingless relatives). The latter three groups are together known as mandibulates.

The jointed appendages of arthropods gave the clade its name, from the Greek words *arthron*, "joint," and *podos*, "foot"

or "limb." Arthropods evolved from ancestors with simple, unjointed appendages. The exact forms of those ancestors are unknown, but some arthropod relatives with segmented bodies and unjointed appendages survive today. Before we describe the modern arthropods, we will discuss those arthropod relatives, as well as an early clade that went extinct but left an important fossil record.

Arthropod relatives have fleshy, unjointed appendages

The two living groups most closely related to the arthropods provide us with clues about the likely appearance of ancestral arthropod appendages. **Tardigrades** (water bears) have fleshy, unjointed legs and use their fluid-filled body cavities as hydrostatic skeletons (**Figure 32.18A**). Tardigrades are tiny (0.5–1.5 mm long) and lack both a circulatory system and gas exchange organs. The 1,200 known extant species live in marine sands and on temporary water films on plants. When these films dry out, the animals also lose water and shrink to small, barrel-shaped objects that can survive for at least a decade in

(A) *Echiniscus* sp.

50 μm

(B) *Macroperipatus torquatus*

1 cm

32.18 Arthropod Relatives with Unjointed Appendages
(A) Tardigrades ("water bears") can be abundant on the wet surfaces of mosses and plants and in temporary pools of water. (B) Onychophorans ("velvet worms") have unjointed legs and use the body cavity as a hydrostatic skeleton. They are sometimes referred to as "living fossils," meaning they are an ancient group that has changed very little over millennia.

a dormant state. Tardigrades have been found at densities as high as 2 million per square meter of moss.

Until fairly recently, biologists debated whether the **ony-chophorans** (velvet worms) were more closely related to annelids or to arthropods, but molecular evidence clearly links them to the latter. Indeed, with their soft, fleshy, unjointed, claw-bearing legs and elongate bodies, onychophorans may be similar in appearance to the ancestors of arthropods (**Figure 32.18B**). The 180 species of onychophorans live in leaf litter in humid tropical environments. They have soft, segmented bodies that are covered by a thin, flexible cuticle that contains chitin. Like the tardigrades, they use their fluid-filled body cavities as hydrostatic skeletons. Fertilization is internal, and the large, yolky eggs are brooded within the body of the female.

Jointed appendages appeared in the trilobites

The **trilobites** flourished in Cambrian and Ordovician seas, but they disappeared in the great Permian extinction at the close of the Paleozoic era (251 mya). Because they had heavy exoskeletons that readily fossilized, they left behind an abundant record of their existence (**Figure 32.19**). About 10,000 species have been described.

The trilobites are the earliest known arthropods to have had jointed appendages. The body segmentation and appendages of trilobites followed a relatively simple, repetitive plan, but some of their appendages were modified for different functions. This specialization of appendages is a theme in the continuing evolution of the arthropods.

Chelicerates have pointed, nonchewing mouthparts

In the **chelicerates**, the head bears two pairs of pointed appendages modified to form mouthparts, called chelicerae, that

(A) *Endeis* sp.

(B) *Limulus polyphemus*

32.20 Two Small Chelicerate Groups (A) Although they are not spiders, it is easy to see how sea spiders got their common name. (B) A spawning aggregation of horseshoe crabs. Horseshoe crabs, like the onychophorans (see Figure 32.18B), are an example of "living fossils."

are used to grasp (rather than chew) prey. Chelicerates typically have a two-part body plan, with anterior segments fused to form a cephalothorax, and rear segments fused to form an abdomen. In some groups, such as mites and ticks, there is no clear distinction between these two body parts. Most chelicerates have four pairs of walking legs. The 114,000 described species are grouped into three major clades: pycnogonids, horseshoe crabs, and arachnids.

The pycnogonids, or sea spiders, make up a poorly known group of about 1,000 marine species (**Figure 32.20A**). Most are small, with leg spans less than 1 cm, but some deep-sea species have leg spans up to 60 cm. A few pycnogonids eat algae, but most are carnivorous, eating a variety of small invertebrates.

There are only four living species of horseshoe crabs, but many close relatives are known from fossils. Horseshoe crabs, which have changed very little morphologically over their long history, have a large horseshoe-shaped covering over most of the body. They are common in shallow waters along the eastern coast of North America and the southern and eastern coasts of Asia, where they scavenge and prey on bottom-dwelling

Cheirurus ingricus

32.19 A Trilobite Fossil The relatively simple, repetitive segments of the now-extinct trilobites are illustrated by a fossil trilobite from the shallow seas of the Ordovician period, some 450 million years ago.

(A) *Poecilotheria metallica*

(B) *Pseudouroctonus minimus*

(C) *Leiobunum rotundum*

(D) *Brevipalpus phoenicis*

32.21 Arachnid Diversity (A) Tarantulas encompass several hundred species of hairy, ground-dwelling spiders, some of which can grow to the size of a dinner plate. Their venomous bite, although painful, is usually not dangerous to humans. (B) Scorpions are nocturnal predators. (C) Harvestmen, also called daddy longlegs, are scavengers. (D) Mites include many free-living species as well as blood-sucking ectoparasites.

animals. Periodically they crawl into the intertidal zone in large numbers to mate and lay eggs (**Figure 32.20B**).

Arachnids are abundant in terrestrial environments. Most arachnids have a simple life cycle in which miniature adults hatch from internally fertilized eggs and begin independent lives almost immediately. Some arachnids retain their eggs during development and give birth to live young.

The most species-rich and abundant arachnids are the spiders, scorpions, harvestmen, mites, and ticks (**Figure 32.21**). More than 60,000 described species of mites and ticks live in soil, leaf litter, mosses, and lichens, under bark, and as parasites of plants and animals. Mites are vectors for wheat and rye mosaic viruses; they cause mange in domestic animals and skin irritation in humans.

Spiders, of which about 50,000 species have been described, are important terrestrial predators with hollow chelicerae, which they use to inject venom into their prey. Some have excellent vision that enables them to chase and seize their prey.

Others spin elaborate webs made of protein threads in which they snare prey. The threads are produced by modified abdominal appendages connected to internal glands that secrete the proteins, which solidify on contact with air. The webs of different groups of spiders are strikingly varied, and this variation enables the spiders to position their snares in many different environments for many different types of prey.

Mandibles and antennae characterize the remaining arthropod groups

The remaining three arthropod groups—the myriapods, crustaceans, and hexapods—have mouthparts that are mandibles, rather than chelicerae, so they are collectively called **mandibulates**. Mandibles can be used for chewing as well as for biting and holding food. Another distinctive characteristic of the mandibulates is the presence of sensory antennae on the head.

MYRIAPODS The **myriapods** comprise the centipedes, millipedes, and their close relatives. Centipedes and millipedes have a well-formed head that bears the mandibles and antennae characteristic of mandibulates. Their distinguishing feature is a long, flexible, segmented trunk that bears many pairs of legs. Centipedes, which have one pair of legs per segment (**Figure 32.22A**), prey on insects and other small animals. In

32.22 Myriapods (A) Centipedes have modified appendages that function as poisonous fangs for capturing active prey. They have one pair of legs per segment. (B) Millipedes are scavengers and plant eaters; they have smaller jaws and legs than centipedes do, and they have two pairs of legs per segment.

millipedes, two adjacent segments are fused so that each fused segment has two pairs of legs (**Figure 32.22B**). Millipedes scavenge and eat plants. More than 3,000 species of centipedes and 9,000 species of millipedes have been described; many more species probably remain unknown. Although most myriapods are less than a few centimeters long, some tropical species are ten times that size.

CRUSTACEANS **Crustaceans** are the dominant marine arthropods today, and they are also common in fresh water and some terrestrial environments. The most familiar crustaceans are the shrimps, lobsters, crayfishes, and crabs (all decapods; **Figure 32.23A**) and the sow bugs (isopods; **Figure 32.23B**). Additional species-rich groups include the amphipods, ostracods, branchiopods (**Figure 32.23C**), and copepods (**Figure 32.23D**), all of which are found in freshwater and marine environments.

Barnacles are unusual crustaceans that are sessile as adults (**Figure 32.23E**). Adult barnacles look more like mollusks than like other crustaceans, but as the zoologist Louis Agassiz

(A) *Scolopendra hardwicki*

(B) *Sigmoria trimaculata*

(A) *Randallia ornata*

(C) *Triops longicaudatus*

(B) *Armadillidium vulgare*

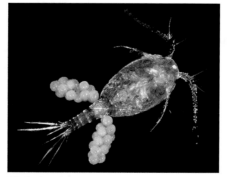

(D) Cyclopoid copepod

(E) *Lepas pectinata*

32.23 Crustacean Diversity (A) This decapod crustacean, a purple sand crab, is also called a "globe crab" based on its body shape. (B) This pillbug, a terrestrial isopod, can roll into a tight ball when threatened. (C) This tadpole shrimp, a branchiopod, is common in seasonal pools of the southwestern United States. (D) This minute copepod from a freshwater pond is brooding eggs. (E) Gooseneck barnacles attach to a substrate by their muscular stalks and feed by protruding and retracting their feeding appendages.

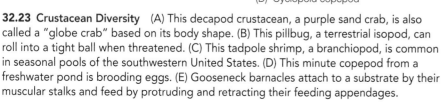

Carapace covering
head and thorax

Abdomen

Pleopods
(swimming)

Antennae
(sensing)

Maxilliped
(helps hold
food)

Pereiopods (walking and
gathering food)

Appendages are specialized for chewing,
sensing, walking, and swimming.

32.24 Crustacean Body Plan The bodies of crustaceans are divided into three regions: the head, thorax, and abdomen. Each body region bears specialized appendages. A shell-like carapace covers the head and thorax.

remarked more than a century ago, a barnacle is "nothing more than a little shrimp-like animal, standing on its head in a limestone house and kicking food into its mouth."

Go to Media Clip 32.7
Barnacles Feeding
Life10e.com/mc32.7

Most of the 67,000 described species of crustaceans have a body that is divided into three regions: head, thorax, and abdomen (**Figure 32.24**). The segments of the head are fused together, and the head bears five pairs of appendages. Each of the multiple thoracic and abdominal segments usually bears one pair of appendages. The appendages on different parts of the body are specialized for different functions, such as gas exchange, chewing, capturing food, sensing, walking, and swimming. In many species, a fold of the exoskeleton, the carapace, extends dorsally and laterally back from the head to cover and protect some of the other segments.

The fertilized eggs of most crustacean species are attached to the outside of the female's body, where they remain during their early development (see Figure 32.23D). At hatching, the young of some species are released as larvae; those of other species are released as juveniles that are similar in form to the adults. Still other species release eggs into the water or attach them to an object in the environment.

More than half of all described species are insects

During the Devonian period, more than 400 million years ago, some mandibulates colonized terrestrial environments. Of the several groups (including some crustacean isopods and decapods) that successfully colonized the land, none is more prominent today than the six-legged **hexapods**, which include the **insects** and their wingless relatives.

The wingless relatives of the insects—the springtails, two-pronged bristletails, and proturans—are probably the most similar of living forms to insect ancestors (**Figure 32.25**). These hexapods have a simple life cycle: they hatch from eggs as miniature adults. They differ from insects in having internal mouthparts. Springtails can be extremely abundant (up to 200,000 per m^2) in soil, leaf litter, and on vegetation and are the most abundant hexapods in the world in terms of number of individuals (as opposed to number of species).

As we saw at the opening of this chapter, more than 1 million of the 1.8 million described living species on Earth are insects. Like crustaceans, insects have a body with three regions: head, thorax, and abdomen. They have a single pair of antennae on the head and three pairs of legs attached to the thorax. In most groups of insects, the thorax also bears two pairs of wings. Unlike other arthropods, insects have no appendages growing from their abdominal segments (**Figure 32.26**). Insects are distinguished from springtails and other hexapods by their external mouthparts and by antennae that contain a motion-sensitive receptor called Johnston's organ. In addition, insects have a derived mechanism for gas exchange in air: a system of air sacs and tubular channels called tracheae (singular trachea) that extend from external openings called spiracles inward to tissues throughout the body (see Figure 49.4).

Table 32.2 lists the major insect groups. Two groups—the jumping bristletails and silverfish—are wingless and have simple life cycles, like the springtails and other non-insect hexapods. The remaining groups are all pterygote insects. **Pterygotes** have two pairs of wings, except in some groups in which one or both pairs of wings have been secondarily lost. These secondarily wingless groups include the parasitic lice and fleas, some beetles, and the worker individuals of many ant species.

Hatchling pterygotes do not look like adults; they undergo substantial changes at each molt. The immature stages of insects between molts are called **instars**. A substantial change that

Tomocerus minor

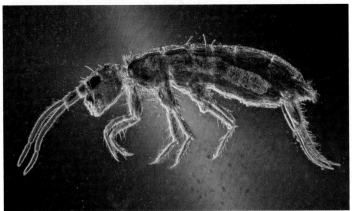

0.5 mm

32.25 Wingless Hexapods The wingless hexapods, such as this springtail, have a simple life cycle. They hatch looking like miniature adults, then grow by successive molts of the cuticle.

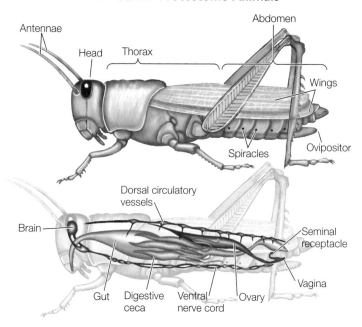

32.26 Insect Body Plan Like those of crustaceans, the bodies of insects are divided into three regions: head, thorax, and abdomen. In insects, however, the thorax bears three pairs of legs and, in most groups, two pairs of wings. Unlike other arthropods, insects have no appendages growing from their abdominal segments.

occurs between one developmental stage and another is called **metamorphosis**. If the changes between stages are gradual, an insect is said to have **incomplete metamorphosis**. If the change between at least some stages is dramatic, an insect is said to have **complete metamorphosis**. In many insects with complete metamorphosis, different stages are specialized for different functions and use different food sources. In many species the larvae are specialized for feeding and growing, whereas the adults are specialized for reproduction and dispersal.

 **Go to Media Clip 32.8**
Complete Metamorphosis
Life10e.com/mc32.8

The adults of most flying insects have two pairs of stiff, membranous wings attached to the thorax. True flies, however, have one pair of wings and a pair of stabilizers called halteres. In winged beetles, one pair of wings—the forewings—forms heavy, hardened wing covers.

Two groups of pterygotes, the mayflies and dragonflies (**Figure 32.27A**), cannot fold their wings against their bodies. This is the ancestral condition for pterygote insects, and the mayflies and dragonflies are not closely related to one another. Members of both groups have predatory or herbivorous aquatic larvae that transform into flying adults after they crawl out of the water. Dragonflies (and their relatives the damselflies) are active predators as adults. In contrast, adult mayflies lack functional digestive tracts. Mayflies live only about a day, just long enough to mate and lay eggs.

All other pterygote insects—the **neopterans**—can tuck their wings out of the way upon landing and crawl into crevices and other tight places. Some neopteran groups undergo incomplete metamorphosis, so hatchlings of these insects are sufficiently

TABLE 32.2
The Major Insect Groups[a]

Group	Approximate Number of Described Living Species
Jumping bristletails (Archaeognatha)	515
Silverfish (Thysanura)	560
PTERYGOTE (WINGED) INSECTS (PTERYGOTA)	
Mayflies (Ephemeroptera)	3,250
Dragonflies and damselflies (Odonata)	5,900
Neopterans (Neoptera)[b]	
Ice-crawlers (Grylloblattodea)	35
Gladiators (Mantophasmatodea)	15
Stoneflies (Plecoptera)	3,750
Webspinners (Embioptera)	465
Angel insects (Zoraptera)	40
Earwigs (Dermaptera)	2,000
Grasshoppers and crickets (Orthoptera)	24,000
Stick insects (Phasmida)	3,000
Cockroaches (Blattodea)	4,650
Termites (Isoptera)	2,700
Mantids (Mantodea)	2,400
Booklice and barklice (Psocoptera)	5,750
Thrips (Thysanoptera)	6,000
Lice (Phthiraptera)	5,100
True bugs, cicadas, aphids, leafhoppers (Hemiptera)	104,000
Holometabolous neopterans (Holometabola)[c]	
Ants, bees, wasps (Hymenoptera)	117,000
Beetles (Coleoptera)	388,000
Strepsipterans (Strepsiptera)	610
Lacewings, ant lions, and mantidflies (Neuroptera)	5,900
Dobsonflies, alderflies, and fishflies (Megaloptera)	350
Snakeflies (Raphidoptera)	250
Scorpionflies (Mecoptera)	760
Fleas (Siphonaptera)	2,100
True flies (Diptera)	155,000
Caddisflies (Trichoptera)	14,300
Butterflies and moths (Lepidoptera)	158,000

[a]The hexapod relatives of insects include the springtails (Collembola; 3,000 spp.), two-pronged bristletails (Diplura; 600 spp.), and proturans (Protura; 10 spp.). All are wingless and have internal mouthparts.
[b]Neopteran insects can tuck their wings close to their bodies.
[c]Holometabolous insects are neopterans that undergo complete metamorphosis.

WORKING WITH**DATA**:

How Many Species of Insects Exist on Earth?

Original Papers

Erwin, T. L. 1988. The tropical forest canopy: The heart of biotic diversity, In E. O. Wilson, ed., *Biodiversity*, 123–129. National Academy Press, Washington, D.C.,

Erwin, T. L. 1997. Biodiversity at its utmost: Tropical forest beetles. In M. L. Reaka-Kudla, D. E. Wilson, and E. O. Wilson, eds., *Biodiversity II*, 27–40. Joseph Henry Press, Washington, D.C.

Analyze the Data

The data in the table were used by entomologist Terry Erwin to estimate the undescribed diversity of insects. Review the design of Erwin's experiment in the opening story of this chapter. Then use Erwin's data to answer the questions.

QUESTION 1

From the data in the table, estimate the number of insect species in an average hectare of Panamanian forest. Assume that the data for beetles on *L. seemannii* are representative of the other tree species, and that all the species of beetles that are *not* host-specific were collected in the original sample. Remember to sum your estimates of the number of (a) host-specific beetle species in the forest canopy; (b) non-host-specific beetle species in the forest canopy; (c) beetle species on the forest floor; and (d) species of all insects other than beetles.

QUESTION 2

There are about 50,000 species of tropical forest trees. Assume that the data for beetles on *L. seemannii* are representative of other species of tropical trees and calculate the number of host-specific beetles found on these trees. Add an estimated 1 million species of non-host-specific beetles that are expected across different species of trees (including those in temperate regions). Estimate the number of ground-dwelling beetle species based on the percentage used in Question 1. Now use this information to estimate the number of insect species on Earth, based on the percentage of beetles among all insect species.

Approximate number of beetle species collected from *Luehea seemannii* trees	1,200
Estimated number of host-specific beetles in this sample	163
Number of tree species per hectare of Panamanian forest	70
Percent of beetle species living in tree canopy (as opposed to ground-dwelling species)	75%
Percent of beetles among all insect species	40%

Go to BioPortal for all WORKING WITH**DATA** exercises

similar in form to adults to be recognizable. Examples include the grasshoppers (**Figure 32.27B**), roaches, mantids, stick insects, termites, stoneflies, earwigs, thrips, true bugs (**Figure 32.27C**), aphids, cicadas, and leafhoppers. These groups acquire adult organ systems, such as wings and compound eyes, gradually through several juvenile instars.

More than 80 percent of all insects belong to a subgroup of the neopterans called the **holometabolous** insects (see Table 32.2), which undergo complete metamorphosis (**Figure 32.27D**). The many species of beetles account for almost half of this group (**Figure 32.27E**). Also included are lacewings and their relatives; caddisflies; butterflies and moths (**Figure 32.27F**); sawflies; true flies (**Figure 32.27G**); and bees, wasps, and ants, some species of which display unique and highly specialized social behaviors (**Figure 32.27H**).

Molecular data suggest that insects began to diversify about 450 million years ago, about the time of the appearance of the first land plants. These early hexapods evolved in a terrestrial environment that lacked any other similar organisms, which in part accounts for their remarkable success. But the success of the insects is also due to their wings. Pterygote insects were the first animals in evolutionary history to achieve the ability to fly. Homologous genes control the development of insect wings and crustacean appendages, suggesting that the insect wing evolved from a dorsal branch of a crustacean-like limb (**Figure 32.28**). The dorsal limb branch of crustaceans is used for gas exchange. Thus the insect wing probably evolved from a gill-like structure that had a gas exchange function.

Flight opened up many new lifestyles and feeding opportunities that only the insects could exploit, such as pollination of

(and coevolution with) flowering plants. Flight is almost certainly one of the reasons for the remarkable numbers of both insect species and individual insects, and for their unparalleled evolutionary success.

■ RECAP 32.4

All arthropods have segmented bodies. Muscles in each segment operate that segment and the appendages attached to it. Jointed, specialized appendages permit complex patterns of movement, including, in insects, the ability to fly. With flight, insects took advantage of new feeding and lifestyle opportunities, which contributed to the unparalleled evolutionary success of this group.

- What features have contributed to making arthropods among the most abundant animals on Earth, both in number of species and number of individuals? **See p. 667**

- Describe the difference between incomplete and complete metamorphosis. **See pp. 671–672**

An Overview of Protostome Evolution

The protostomes encompass a staggering number of different body forms and lifestyles. The following aspects of protostome evolution have contributed to this enormous diversity:

- The evolution of *segmentation* permitted some groups of protostomes to move different parts of the body independently of one another. Species in some groups gradually evolved the ability to move rapidly over and through the substrate, through water, and through air.

(A) *Libellula quadrimaculata*

(B) *Phymateus morbillosus*

(C) *Coquerelia ventralis*

(D) *Limnephilus sp.* (larva)

(E) *Eupholus magnificus*

(F) *Argema mittrei*

(G) *Lucilia caesar*

(H) *Polistes nympha*

32.27 Diverse Pterygotes (A) Unlike most flying insects, a dragonfly cannot fold its wings over its back. (B) Orthopteran insects such as grasshoppers have incomplete metamorphosis: they undergo several molts, but the juvenile instars resemble small adults (incomplete metamorphosis). (C) Hemipterans such as this Madagascan shield bug are known as "true" bugs. (D–H) Holometabolous insects undergo complete metamorphosis. (D) A larval caddisfly (right) emerges from its dark pupal case. (E) The beetles (Coleoptera) comprise the largest insect group; beetles such as this New Guinea weevil account for more than half of all holometabolous species. (F) Butterflies and moths are the lepidopterans, whose phases of complete metamorphosis are familiar to many (see Figure 31.9). (G) Blowflies are among the "true" flies, the Dipterans. Adult blowflies feed on pollen or nectar, but lay their eggs on carrion, upon which the larvae feed. (H) These paper wasps are hymenopterans, a group in which most members display social behaviors.

- *Complex life cycles* with dramatic changes in form between one stage and another allow individuals of different stages to specialize on different resources.
- *Parasitism* has evolved repeatedly, and many protostome groups parasitize plants and animals.

- The evolution of *diverse feeding structures* allowed protostomes to specialize on many different food sources. Specialization on food sources undoubtedly contributed to reproductive isolation and further diversification.
- Predation was a major selection pressure favoring the development of *hard external body coverings* (exoskeletons and

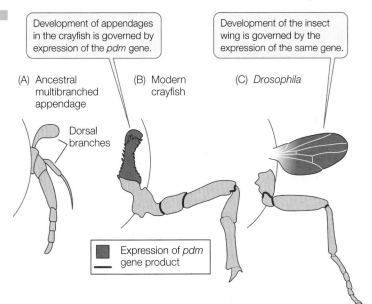

Development of appendages in the crayfish is governed by expression of the *pdm* gene.

Development of the insect wing is governed by the expression of the same gene.

(A) Ancestral multibranched appendage

Dorsal branches

(B) Modern crayfish

(C) *Drosophila*

■ Expression of *pdm* gene product

32.28 The Origin of Insect Wings? Insect wings may be derived from an ancestral appendage similar to that of modern crustaceans. (A) A diagram of the ancestral, multibranched arthropod limb. The uppermost dorsal branch may have been used for gas exchange. (B, C) The *pdm* gene, a Hox gene, is expressed throughout the dorsal limb branch and walking leg of the thoracic limb of a crayfish (B) and in the wings and legs of *Drosophila* (C).

shells). Such coverings evolved independently in many lophotrochozoan and ecdysozoan groups. In addition to providing protection, these coverings became key elements in the development of new systems of locomotion.

- *Better locomotion* permitted prey to escape from predators, but also allowed predators to pursue their prey more effectively. Thus the evolution of animals has been, and continues to be, a complex "arms race" among predators and prey.

 Go to Animated Tutorial 32.1
An Overview of the Protostomes
Life10e.com/at32.1

Many major evolutionary trends among the protostomes are shared by the deuterostomes, which include the chordates, the group to which humans belong. We turn to the deuterostomes in the next chapter.

Which groups of protostomes are thought to contain the most undiscovered species?

ANSWER

It is perhaps easier to list the groups of protostomes for which a nearly complete inventory of living species has been done than to list all the groups for which many new species remain to be described. Among the insects, the best-studied group, in terms of species, is the butterflies, which are widely collected and studied. There are still many species of other lepidopterans (such as moths), however, remaining to be discovered. Most other major insect groups contain many undescribed species.

New species discovery and description rates remain high for almost all other major groups of protostomes. Second to the insects, and perhaps even rivaling the insects in undiscovered diversity, are the nematodes. Although known nematode diversity is only about one-fortieth of known insect diversity (in terms of number of described species), the taxonomy of nematodes has been much more poorly studied than that of insects. Some biologists think there are likely to be species-specific parasitic nematodes specializing on most other species of multicellular organisms. If so, then there may be as many species of nematodes as there are of plants, fungi, and other animals combined.

Most of the other diverse groups of protostomes also contain many as yet undetected species, judging from the rate of new species descriptions. In particular, flatworms (especially the parasitic flukes and tapeworms), marine annelids, mollusks, crustaceans, myriapods, and chelicerates all contain large numbers of undescribed species.

CHAPTER**SUMMARY** 32

32.1 What Is a Protostome?

- Protostomes ("mouth first") are bilaterally symmetrical animals with an anterior brain that surrounds the entrance to the digestive tract and a ventral nervous system. The embryonic blastopore of protostomes develops into a mouth.

- There are two major clades of protostomes, the lophotrochozoans and the ecdysozoans. **Review Figure 32.1, Table 32.1, ACTIVITIES 32.1, 32.2, ANIMATED TUTORIAL 32.1**

- **Lophotrochozoans** include a wide variety of body forms. Within this group, **lophophores** (complex organs for both food collection and gas exchange), free-living **trochophore** larvae, and spiral cleavage evolved. Some of these features were subsequently lost in some lineages (or evolved convergently).

- **Ecdysozoans** have a body covering known as the **cuticle**, which they must **molt** in order to grow. Some ecdysozoans have a relatively thin cuticle. Others, especially the **arthropods**, have a rigid cuticle reinforced with **chitin** that functions as an **exoskeleton**. **Review Figure 32.4**

- **Arrow worms** may be most closely related to lophotrochozoans, or they may be the sister group of all other protostomes. **Review Figure 32.5**

32.2 What Features Distinguish the Major Groups of Lophotrochozoans?

- Lophotrochozoans range from animals with a blind gut and no internal transport system to animals with complete digestive tracts and complex internal transport systems. **Review Figure 32.6**

- Most species of **bryozoans** and **entoprocts** live in colonies produced through asexual reproduction. Individuals of both groups feed using a lophophore.

- **Flatworms**, **rotifers**, **gastrotrichs**, and their close relatives form a structurally diverse clade of ciliated lophotrochozoans. **Review Figure 32.7**

- **Ribbon worms** feed using a long, protrusible proboscis. **Review Figure 32.8**

continued

- The shelled **brachiopods** and wormlike **phoronids** use a lophophore to feed; this lophophore may have evolved independently of the lophophore in bryozoans and entoprocts. **Review Figures 32.9, 32.10**

- **Annelids** are a diverse group of segmented worms that live in moist terrestrial and aquatic environments. **Review Figure 32.11**

- **Mollusks** underwent a dramatic evolutionary radiation based on a body plan consisting of three major components: a **foot**, a **mantle**, and a **visceral mass**. The four major living molluscan clades—**chitons**, **bivalves**, **gastropods**, and **cephalopods**—demonstrate the diversity that evolved from this three-part body plan. **Review Figure 32.13**

 32.3 What Features Distinguish the Major Groups of Ecdysozoans?

- Members of several species-poor groups of wormlike marine ecdysozoans—**priapulids**, **kinorhynchs**, and **loriciferans**—have thin cuticles.

- **Nematodes** have a thick, multilayered cuticle. Nematodes are among the most abundant and universally distributed of all animal groups. **Review Figure 32.16**

- **Horsehair worms** are extremely thin; many are endoparasites as larvae.

 32.4 Why Are Arthropods So Diverse?

- One major ecdysozoan clade, the arthropods, has evolved jointed, paired appendages that have a wide diversity of functions. Collectively, arthropods are the dominant animals on Earth in number of described species, and among the most abundant in number of individuals.

- Encasement within a rigid exoskeleton provides arthropods with support for walking as well as some protection from predators. The waterproofing provided by chitin keeps arthropods from dehydrating in dry air.

- Jointed appendages permit complex movement patterns. Each arthropod segment has muscles attached to the inside of the exoskeleton that operate that segment and the appendages attached to it.

- The **onychophorans** and the **tardigrades** are arthropod relatives that have simple, unjointed appendages. **Trilobites,** the first arthropods known to have had jointed appendages, disappeared in the Permian mass extinction.

- **Chelicerates** have a two-part body and pointed mouthparts that grasp prey; most chelicerates have four pairs of walking legs.

- Mandibles and antennae are synapomorphies of the **mandibulates**, which include the myriapods, crustaceans, and hexapods.

- The bodies of **myriapods** have two regions: a head with mandibles and antennae, and a segmented trunk that bears many pairs of legs.

- **Crustaceans** have segmented bodies that are divided into three regions—head, thorax, and abdomen—with different, specialized appendages in each region. **Review Figure 32.24**

- **Hexapods**—insects and their relatives—are the dominant terrestrial arthropods. They have the same three body regions as crustaceans, but no appendages form in their abdominal segments. **Review Figure 32.26, Table 32.2**

- Wings and the ability to fly first evolved among the **pterygote** insects, allowing them to exploit new lifestyles. **Review Figure 32.28**

 Go to the Interactive Summary to review key figures, Animated Tutorials, and Activities
Life10e.com/is32

CHAPTER**REVIEW**

▨▨▨▨ **REMEMBERING**

1. Which of the following is *not* part of the molluscan body plan?
   a. Mantle
   b. Foot
   c. Radula
   d. Visceral mass
   e. Jointed appendages

2. The outer covering of ecdysozoans
   a. is always hard and rigid.
   b. is always thin and flexible.
   c. is hard and rigid in larvae but thin in adults.
   d. ranges from very thin to hard and rigid, depending on the species.
   e. grows throughout life to accommodate a growing body.

3. Which groups are arthropod relatives with unjointed legs?
   a. Trilobites and onychophorans
   b. Onychophorans and tardigrades
   c. Trilobites and tardigrades
   d. Onychophorans and chelicerates
   e. Tardigrades and chelicerates

4. The body plan of insects comprises which of the following three regions?
   a. Head, abdomen, and trachea
   b. Head, abdomen, and cephalothorax
   c. Cephalothorax, abdomen, and trachea
   d. Head, thorax, and abdomen
   e. Abdomen, trachea, and mantle

5. Insects whose hatchlings are sufficiently similar in form to adults to be recognizable are said to have
   a. instars.
   b. neopterous development.
   c. accelerated development.
   d. incomplete metamorphosis.
   e. complete metamorphosis.

6. Factors that may have contributed to the remarkable evolutionary success of insects include
   a. the lack of any other similar organisms in the terrestrial environments colonized by insects.
   b. the ability to fly.
   c. complete metamorphosis.
   d. a new mechanism for delivering oxygen to their internal tissues.
   e. All of the above

UNDERSTANDING & APPLYING

7. Segmentation either has arisen several times during animal evolution, or else arose early in animal evolution and was subsequently lost multiple times. What advantages does segmentation provide? Given these advantages, why might some animals have lost their segmentation?

8. Major structural novelties have arisen only infrequently during the course of evolution. Which of the features of protostomes do you think are major evolutionary novelties? Which of these features may have led to major evolutionary radiations?

9. There are more described and named species of insects than of all other species on Earth combined. However, only a very few insect species live in marine environments, and those species are restricted to the intertidal zone or the ocean surface. What factors may have contributed to the insects' lack of success in the oceans?

ANALYZING & EVALUATING

10. In the Working with Data exercise on page 673, you were asked to make many assumptions to estimate the number of species of insects on Earth. Do you think these assumptions are reasonable? Why or why not? Would you argue for a different set of assumptions? How do you think these changes in assumptions would affect your calculations? Can you think of ways to test these assumptions?

Go to BioPortal at **yourBioPortal.com** for Animated Tutorials, Activities, LearningCurve Quizzes, Flashcards, and many other study and review resources.

33 Deuterostome Animals

Good Parent or Cannibal? Young frogs emerge from the digestive tract of a female *Rheobatrachus silus*, one of the now-extinct gastric brooding frogs of Australia. In this unique form of gestation, eggs hatch and tadpoles develop within the protected environment of the mother's stomach.

WHY WOULD A FROG swallow its own offspring? That may not sound like a good parenting skill, but female gastric brooding frogs of Australia shut down their digestive system to brood their tadpoles in their stomach. This strategy provides the tadpoles with a safe haven from predators, making it much more likely that they will survive to metamorphosis. That was the case, at least, until gastric brooding frogs went extinct in the early 1980s, after humans introduced pathogenic fungi into their native range.

Not all adult frogs are involved in raising their young. Female bullfrogs, for example, lay thousands of eggs each year and provide no parental care for their offspring. The eggs are fertilized by a male bullfrog and left to develop on their own. The eggs hatch into tadpoles, which transform into tiny frogs—if they aren't eaten first by aquatic predators. Out of the tens of thousands of tadpoles an adult bullfrog produces in its lifetime, an average of only two offspring will survive to reproduce.

Among other species of frogs, complex behaviors associated with parental care change these long odds. Rather than producing huge numbers of offspring, each with a minimal chance of surviving, some frogs invest more energy in each individual offspring and care for the young as they grow. This strategy increases the chances that any one offspring will survive and reproduce—but it also means that far fewer offspring can be produced.

Frogs' strategies for parental care include many behaviors in addition to gastric brooding. The females of many species guard their eggs until they hatch. Other females carry their tadpoles around with them on their backs, or even in special brood pouches. Males often provide parental care as well. Males of many frog species guard egg masses or carry young, sometimes in unusual ways. In Darwin's frog of South America, for example, the tadpoles develop within the male's vocal sacs.

Parental care can extend beyond protection to feeding the young. Some female poison frogs of the tropical Americas carry each of their tadpoles to one of the many tiny pools of water that collect in bromeliad plants growing on trees. The female then returns to each bromeliad "pond" and lays unfertilized eggs as food for the single tadpole developing there.

Frogs and other vertebrates constitute one of the major groups of deuterostome animals. Deuterostomes are of particular interest to biologists not only because of their often complex behaviors, their importance in many ecosystems, and their widespread use as models in developmental biology and genetics, but because we are deuterostomes.

How has the evolution of complex behaviors affected the diversification of some major groups of deuterostomes?

See answer on p. 705.

Radial symmetry as adults,
calcified internal plates,
loss of pharyngeal slits

Ciliated
larvae

Common
ancestor
(bilaterally
symmetrical,
pharyngeal slits
present)

Notochord,
dorsal hollow
nerve cord,
post-anal tail

Vertebral column, anterior skull,
large brain, ventral heart

Echinoderms
Hemichordates
Lancelets
Tunicates
Vertebrates

Ambulacrarians
Chordates

33.1 Phylogeny of the Deuterostomes The three principal groups of deuterostomes are the echinoderms, the hemichordates, and the chordates, which include the lancelets, tunicates, and vertebrates. The echinoderms and the vertebrates contain most of the described species.

Go to Activity 33.1 Deuterostome Phylogeny
Life10e.com/ac33.1

33.1 What Is a Deuterostome?

It may surprise you to learn that both you and a sea urchin are deuterostomes. Adult sea stars, sea urchins, and sea cucumbers—the most familiar echinoderms—look so different from adult vertebrates (fishes, frogs, lizards, birds, and mammals) that it may be difficult to believe all these animals are closely related. The evidence that all deuterostomes share a common ancestor that is not shared with the protostomes includes early developmental patterns and phylogenetic analysis of gene sequences, factors that are not apparent in the forms of the adult animals.

Deuterostomes share early developmental patterns

Historically, the deuterostomes were distinguished by three early developmental patterns:

- Radial cleavage
- Development of the blastopore into the anus and formation of the mouth at the opposite end of the embryo from the blastopore (the pattern that gives the deuterostomes their name)
- Development of a coelom from mesodermal pockets that bud off from the cavity of the gastrula rather than by splitting of the mesoderm, as occurs among protostomes

These distinctions, however, are not the strongest evidence for the monophyly of the deuterostomes. Radial cleavage is not exclusive to deuterostomes, and as noted in Section 31.1, it is now thought to be the ancestral condition for all bilaterians. In fact, some of the groups now known to be protostomes were once thought to be deuterostomes because their developmental patterns are similar to those of echinoderms and chordates. The development of the blastopore into an anus does characterize the deuterostomes, but it may be the ancestral condition for bilaterians rather than a derived feature of deuterostomes. Today the strongest support for the shared evolutionary relationships of the deuterostomes comes from phylogenetic analyses of DNA sequences of many different genes.

There are three major deuterostome clades

All deuterostomes are triploblastic, coelomate animals (see Figure 31.5C). Skeletal elements, where present, are internal rather than external. Some species have segmented bodies, but the segments are less obvious than those of annelids and arthropods. Although there are far fewer species of deuterostomes than of protostomes (see Table 31.1), we have a special interest in deuterostomes because we are members of that clade. The deuterostomes are also of interest because they include many large animals that strongly influence the characteristics of ecosystems. Many deuterostome species have been intensively studied in all fields of biology. Complex behaviors, such as the parenting behaviors described at the opening of this chapter, are especially well developed among some deuterostomes.

The major groups of living deuterostomes comprise three distinct clades (**Figure 33.1**):

- **Echinoderms**: sea stars (starfish), sea urchins, and their relatives
- **Hemichordates**: acorn worms and pterobranchs
- **Chordates**: sea squirts, lancelets, and vertebrates

In addition, some recent genomic analyses suggest that two poorly known groups, the xenoturbellids and the acoels, may also be deuterostomes.

Go to Animated Tutorial 33.1
An Overview of the Deuterostomes
Life10e.com/at33.1

Fossils shed light on deuterostome ancestors

Scientists are learning much about the ancestors of modern deuterostomes from fossils recently discovered in 520-million-year-old rocks in China. Some of these early deuterostomes had skeletons similar to those of echinoderms but, unlike modern adult echinoderms, they had bilateral symmetry and a pharynx with slits through which water flowed. Another early deuterostome group, the yunnanozoans, was discovered in China's Yunnan Province. The well-preserved fossils reveal animals that had large mouths, six pairs of external gills, and a segmented posterior body section bearing a light cuticle (**Figure 33.2**). The features of these fossil animals, together with

Yunnanozoon lividum

Mouth   Esophagus   External gills   Segments

33.2 Ancestral Deuterostomes Had External Gills   The extinct yunnanozoans may be ancestral deuterostomes. This fossil, which dates from the Cambrian, shows the six pairs of external gills and segmented posterior body that characterized these animals.

findings from phylogenetic analyses of living species, show that the earliest deuterostomes were bilaterally symmetrical, segmented animals with pharyngeal slits. The adult forms of the living echinoderms with their unique symmetry (in which the body parts are arranged along five radial axes) evolved much later. Other deuterostomes retained the ancestral bilateral symmetry.

███████████████████████████████ ▌RECAP 33.1

The three major clades of deuterostomes are the echinoderms, the hemichordates, and the chordates. The common ancestry of these groups is supported by early developmental similarities and by phylogenetic analyses of DNA sequences.

- Why is radial cleavage no longer considered to be evidence for the monophyly of deuterostomes? **See p. 679**
- What three traits did the earliest deuterostomes have in common? **See pp. 679–680**

We will begin our survey of the deuterostomes with the echinoderms and hemichordates, the most distant of our relatives within that clade.

 ## What Features Distinguish the Echinoderms, Hemichordates, and Their Relatives?

About 13,000 species of echinoderms in 23 major groups have been described from their fossil remains. They are probably only a small fraction of the echinoderm species that have ever lived. Only 6 of the 23 major groups known from fossils are represented by species that survive today; many clades were lost during the periodic mass extinctions that have occurred throughout Earth's history. Nearly all of the 7,500 extant species of echinoderms live in marine environments. There are

far fewer species of living hemichordates (about 120 known species).

The echinoderms and hemichordates (together known as **ambulacrarians**) have a bilaterally symmetrical, ciliated larva (**Figure 33.3A**). Adult hemichordates are also bilaterally symmetrical. Echinoderms, however, undergo a radical change in form as they develop into adults (**Figure 33.3B**), changing from a bilaterally symmetrical larva into an adult with **pentaradial symmetry** (symmetry in five or multiples of five). As is typical of animals with radial symmetry, echinoderms have no head, and they move slowly and equally well in many directions. Rather than having an anterior–posterior (head–tail) and dorsal–ventral (back–belly) body organization, most echinoderms have an **oral** side containing the mouth and an opposite **aboral** side containing the anus.

Recent genomic analyses suggest that two groups of small, highly reduced, soft-bodied marine organisms, the xenoturbellids and the acoels, are the sister group of the ambulacrarians. The two known species of **xenoturbellids** are wormlike organisms up to 4 cm long that feed on or parasitize mollusks in the northern Atlantic Ocean. They have a very simple body plan, with almost no well-defined organ systems. The mostly tiny (<2 mm) **acoels** are also highly reduced, wormlike organisms that live as plankton, between grains of sediment, or on other organisms such as corals (**Figure 33.4**). They are among the simplest of bilaterian animals, with no gut, circulatory system, respiratory system, gonads, or excretory system. They feed by forming a vacuole around tiny food items. They are hermaphrodites, and their gametes fill the body between the epidermis and digestive vacuole. There are about 400 known species of acoels.

Echinoderms have unique structural features

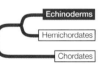

In addition to having pentaradial symmetry, adult echinoderms have two unique structural features. One is a system of calcified internal plates covered by thin layers of skin and some muscles. The calcified plates of most echinoderms are thick, and they fuse inside the entire body, forming an **internal skeleton**. The other unique feature of this group is a **water vascular system**, a network of water-filled canals leading to extensions called **tube feet** (see Figure 33.3B). This system functions in gas exchange, locomotion, and feeding. Seawater enters the system through a perforated structure called a madreporite. A calcified canal leads from the madreporite to the ring canal, which surrounds the esophagus (the tube leading from the mouth to the stomach). Radial canals branch off from the ring canal, extending through the arms (in species that have arms) and connecting with the tube feet. These structural innovations have been modified in many ways, resulting in a striking array of very different animals.

Members of one major extant echinoderm clade, the crinoids (sea lilies and feather stars), were more abundant and species-rich 300 to 500 million years ago than they are today. There are some 80 described living sea lily species, most of which are

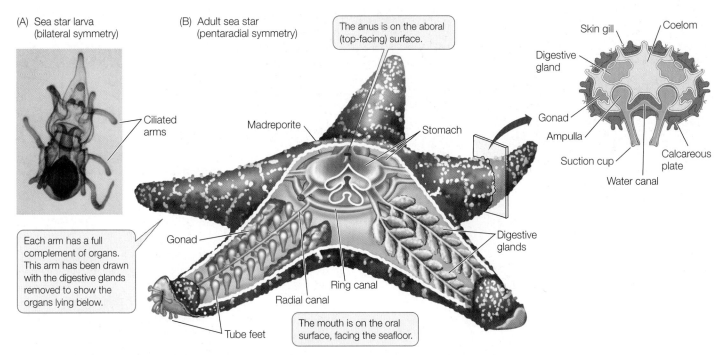

(A) Sea star larva (bilateral symmetry)

Ciliated arms

(B) Adult sea star (pentaradial symmetry)

The anus is on the aboral (top-facing) surface.

Madreporite

Stomach

Each arm has a full complement of organs. This arm has been drawn with the digestive glands removed to show the organs lying below.

Gonad

Ring canal

Radial canal

Digestive glands

Tube feet

The mouth is on the oral surface, facing the seafloor.

Skin gill

Coelom

Digestive gland

Gonad

Ampulla

Suction cup

Calcareous plate

Water canal

33.3 Echinoderms Are Bilaterally Symmetrical as Larvae but Radially Symmetrical as Adults (A) The ciliated larva of a sea star has bilateral symmetry. Hemichordates have a similar larval form. (B) An adult sea star displays the pentaradial symmetry of adult echinoderms. The canals and tube feet of the water vascular system, as well as the calcified internal skeleton, are shown in this diagram. The body's orientation is oral–aboral rather than anterior–posterior.

Wamionoa sp.

33.4 Highly Reduced Acoels Are Probably Relatives of the Ambulacrarians Acoels (yellow) are seen here living on bubble coral (white). Acoels ("without coelom") feed by enveloping food particles in a vacuole within which nutrients are digested. These hermaphroditic animals can reproduce rapidly and may become problematic in saltwater aquariums.

sessile organisms attached to the substrate by a stalk. Feather stars (**Figure 33.5A**) grasp the substrate with specialized flexible appendages that allow for limited movement. About 600 living species of feather stars have been described.

Unlike the mostly sessile crinoids, most surviving echinoderms are motile. The two main groups of motile echinoderms are the echinozoans (sea urchins and sea cucumbers) and the asterozoans (sea stars and brittle stars). Sea urchins are hemispherical in shape and lack arms (**Figure 33.5B**). They are covered with spines that are attached to the underlying skeleton with ball-and-socket joints. These joints enable the spines to be moved so they can converge toward a point that has been touched. The spines, which vary among species in size and shape, can be used for locomotion; a few produce toxic substances. They provide effective protection for the urchin, as many a scuba diver has found out the hard way. Sand dollars are flattened, disc-shaped relatives of sea urchins.

Sea cucumbers also lack arms, and their bodies are oriented in an atypical manner for an echinoderm (**Figure 33.5C**). The mouth is anterior and the anus is posterior (front and rear), in contrast to the oral–aboral (top and bottom) orientation of other echinoderms. Sea cucumbers can use most of their tube feet to move, but they use them primarily for attaching to the substrate.

Sea stars, popularly called starfish, are the most familiar echinoderms (**Figure 33.5D**). Their gonads and digestive organs are located in the arms, as seen in Figure 33.3B. Their tube feet serve as organs of locomotion, gas exchange, and attachment. Each tube foot of a sea star consists of an internal ampulla connected by a muscular tube to an external suction cup that can stick to the substrate. The tube foot is moved by expansion and contraction of the circular and longitudinal muscles of the tube. Brittle stars are similar in structure to sea stars, but their flexible arms are composed of jointed, hard plates (**Figure 33.5E**).

(A) *Oxycomanthus bennetti*

(B) *Sphaerechinus granularis*

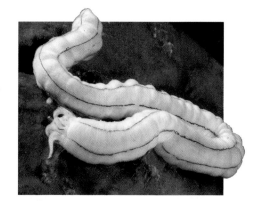

(C) *Synaptula* sp.

33.5 Echinoderm Diversity (A) The flexible arms of this golden feather star (a crinoid) are clearly visible. (B) Sea urchins are important grazers on algae in the intertidal zones of the world's oceans. (C) Sea cucumbers are unique among echinoderms in having an anterior–posterior, rather than an oral–aboral, orientation of the mouth and anus. (D) Sea stars are important predators on bivalve mollusks such as mussels and clams. Suction tips on its tube feet allow a sea star to grasp both shells of the bivalve and pull them open. (E) The arms of the brittle star are composed of hard but jointed plates.

(D) *Marthasterias glacialis*

(E) *Ophiopholis aculeata*

Echinoderms use their tube feet in a great variety of ways to capture prey. Sea lilies, for example, feed by orienting their arms in passing water currents. Food particles then strike and stick to the tube feet, which are covered with mucus-secreting glands. The tube feet transfer these particles to grooves in the arms, where ciliary action carries the food to the mouth.

Most sea urchins capture phytoplankton with their tube feet or scrape algae from rocks with a complex rasping structure. Sea cucumbers capture food with their anterior tube feet, which are modified into large, feathery, sticky tentacles that can be protruded from the mouth. Periodically, a sea cucumber withdraws the tentacles, wipes off the material that has adhered to them, and digests it.

Many sea stars use their tube feet to capture large prey such as polychaetes, gastropod and bivalve mollusks, small crustaceans such as crabs, and fishes. With hundreds of tube feet acting simultaneously, a sea star can grasp a bivalve in its arms, anchor the arms with its tube feet, and by steady contraction of the muscles in its arms, gradually exhaust the muscles the bivalve uses to keep its shell closed (see Figure 33.5D). To feed on the bivalve, the sea star can push its stomach out through its mouth and then through the narrow space between the two halves of the bivalve's shell. The sea star's stomach then secretes enzymes that digest the prey.

 Go to Media Clip 33.1
Sea Star Feeding on a Bivalve
Life10e.com/mc33.1

Most of the 2,000 species of brittle stars ingest particles from the upper layers of sediments and assimilate the organic material from them, although some species filter suspended food particles from the water, and others capture small animals.

Hemichordates are wormlike marine deuterostomes

Hemichordates—acorn worms and pterobranchs—have a bilaterally symmetrical, wormlike body organized in three major parts: a proboscis, a collar (which bears the mouth), and a trunk (which contains the other body parts). The 90 known species of acorn worms range up to 2 meters long (**Figure 33.6A**). They live in burrows in muddy and sandy marine sediments. The digestive tract of an acorn worm consists of a mouth behind which are a muscular pharynx and an intestine. The pharynx opens to the outside through several pharyngeal slits through which water can exit. Highly vascularized tissue surrounding the pharyngeal slits serves as a gas exchange apparatus. Acorn worms respire by pumping water into the mouth and out through the pharyngeal slits. They capture prey with the large proboscis, which is coated with sticky mucus to which small organisms in the sediment stick. The mucus and its attached prey are conveyed by cilia to the mouth. In the esophagus, the food-laden mucus is compacted into a ropelike mass that is moved through the digestive tract by ciliary action.

The 30 living species of pterobranchs are sedentary marine animals up to 12 millimeters long that live in a tube secreted

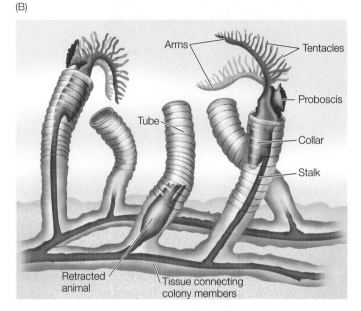

(A)

Trunk   Collar   Proboscis

Saccoglossus kowalevskii

(B)

Arms

Tentacles

Proboscis

Tube

Collar

Stalk

Retracted animal

Tissue connecting colony members

33.6 Hemichordates (A) The proboscis of an acorn worm is modified for burrowing. (B) Some pterobranch species form colonies.

by the proboscis. Some species are solitary; others form colonies of individuals joined together (**Figure 33.6B**). Behind the proboscis is a collar with anywhere from one to nine pairs of arms. The arms bear long tentacles that capture prey and function in gas exchange.

RECAP 33.2

Echinoderms are characterized by pentaradial symmetry, an internal skeleton of calcified plates, and a unique water vascular system. Hemichordates have a bilaterally symmetrical body divided into three parts: proboscis, collar, and trunk.

- How does the body form of echinoderm larvae differ from that of adults? **See p. 681 and Figure 33.3**
- Describe some of the ways that echinoderms use their tube feet to obtain food. **See p. 682**
- Explain how hemichordates obtain food. **See p. 682**

Having described the deuterostome groups that are most distantly related to us, we will next turn our attention to the unique features that evolved in the chordates, a clade dominated by the vertebrates.

33.3 What New Features Evolved in the Chordates?

As we have seen, it is not obvious from examining adult animals that echinoderms and chordates share a common ancestor. The evolutionary relationships among some chordate groups are not immediately apparent either. The features that reveal all of these evolutionary relationships are seen primarily in the larvae—in other words, it is during the early developmental stages that these evolutionary relationships are evident.

There are three principal chordate clades: the **lancelets** (also called cephalochordates), the **tunicates** (also called urochordates), and the **vertebrates** (see Figure 33.1). Adult chordates vary greatly in form, but all chordates display the following derived structures at some stage in their development (**Figure 33.7**):

- A *dorsal hollow nerve cord*
- A *tail* that extends beyond the anus
- A dorsal supporting rod, the *notochord*

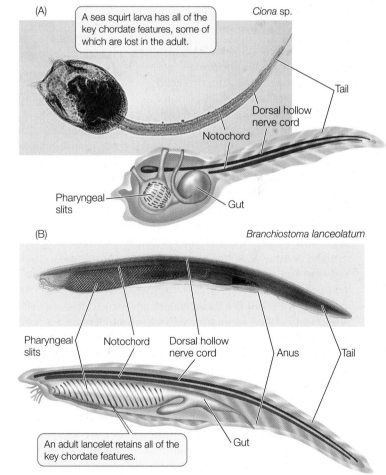

(A)

Ciona sp.

A sea squirt larva has all of the key chordate features, some of which are lost in the adult.

Tail

Dorsal hollow nerve cord

Notochord

Pharyngeal slits

Gut

(B)

Branchiostoma lanceolatum

Pharyngeal slits   Notochord   Dorsal hollow nerve cord   Anus   Tail

Gut

An adult lancelet retains all of the key chordate features.

33.7 Key Features May Be Most Apparent in Early Development (A) The sea squirt larva (but not the adult) has all three key features of chordates: a dorsal hollow nerve cord, a post-anal tail, and a notochord. (B) All three chordate synapomorphies are retained in the adult lancelet.

The **notochord** is the most distinctive derived chordate trait. It is composed of a core of large cells with turgid fluid-filled vacuoles, which make it rigid but flexible. In the tunicates the notochord is lost during metamorphosis to the adult stage. In most vertebrate species it is replaced during development by skeletal structures that provide support for the body.

The pharyngeal slits found in the common ancestor of deuterostomes are present at some developmental stage in all chordates but are often lost or greatly modified in adults. In chordates, the pharyngeal slits are separated and supported by structural elements called pharyngeal arches. In tunicates and lancelets, the pharynx functions as a straining device to filter small food particles. In fishes and larval amphibians, some of the pharyngeal arches develop into gill arches, which support the respiratory gills and are often used as feeding structures as well. Developmentally, some pharyngeal arches also develop into elements of the vertebrate jaw, as well as parts of the tongue, larynx, trachea, and middle ear of tetrapods (four-legged vertebrates). Some of the pharyngeal slits are modified in tetrapods to form the eustachian tube and middle ear chamber.

Adults of most lancelets and tunicates are sedentary

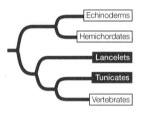

The 35 species of lancelets, also known as cephalochordates, are small animals that rarely exceed 5 centimeters in length. The notochord, which provides body support, extends the entire length of the body throughout the animal's life (see Figure 33.7B). Lancelets are found in shallow marine and brackish waters worldwide. Most of the time they lie covered in sand with their head protruding above the sediment, but they can swim. The pharynx has been enlarged and modified to form a structure called a pharyngeal basket, with which the lancelet filters prey from the water. During the reproductive season, the gonads of males and females enlarge greatly. At spawning, the walls of the gonads rupture, releasing eggs and sperm into the water column, where fertilization takes place.

All members of the three major tunicate groups—the sea squirts (also called ascidians), thaliaceans, and larvaceans—are marine animals. More than 90 percent of the 2,800 known species of tunicates are sea squirts. Individual sea squirts range in length from less than 1 millimeter to 60 centimeters. Some sea squirts form colonies by asexual budding from a single founder. Colonies may measure several meters across. The baglike body of an adult sea squirt is enclosed in a tough tunic, which is the basis for the name "tunicate" (**Figure 33.8A**). The tunic is composed of proteins and a complex polysaccharide secreted by epidermal cells. The sea squirt pharynx is enlarged into a pharyngeal basket that filters prey from the water passing through it.

In addition to its pharyngeal slits, a sea squirt larva has a dorsal hollow nerve cord and a notochord that is restricted mostly to the tail region (see Figure 33.7A). Bands of muscle that surround the notochord provide support for the body.

(A) *Clavelina dellavallei*   1 cm

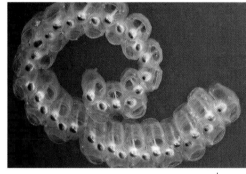

(B) *Pegea* sp. 1 cm

33.8 Adult Tunicates (A) The transparent tunic and the pharyngeal basket are clearly visible in this sea squirt. (B) A chainlike colony of thaliaceans (salps) floats in tropical waters.

After a short time swimming in the plankton, the larvae of most species settle on the seafloor and transform into sessile adults. The swimming, tadpolelike larvae suggest a close evolutionary relationship between tunicates and vertebrates (see Figure 22.6).

Thaliaceans (salps and their relatives) are tunicates that can live singly or in chainlike colonies up to several meters long (**Figure 33.8B**). They float in tropical and subtropical oceans at depths down to 1,500 meters. Larvaceans are solitary planktonic animals that retain the notochord and dorsal hollow nerve cord throughout their lives. Most larvaceans are less than 5 millimeters long, but some species that live near the bottom of deep ocean waters build delicate casings of mucus that may be more than a meter wide. They snare sinking organic particles (their primary food source) with elaborate filters built into their mucus "houses." When the old "house" gets clogged with excess debris, the animal builds a new one.

A dorsal supporting structure replaces the notochord in vertebrates

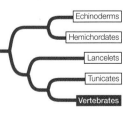

In one chordate group, the vertebrates, a new dorsal supporting structure evolved. This group takes its name from the jointed, dorsal **vertebral column** that replaces the notochord during early development as the primary supporting structure. The individual elements in the vertebral column are called vertebrae. Four other key features characterize the vertebrates as well (**Figure 33.9**):

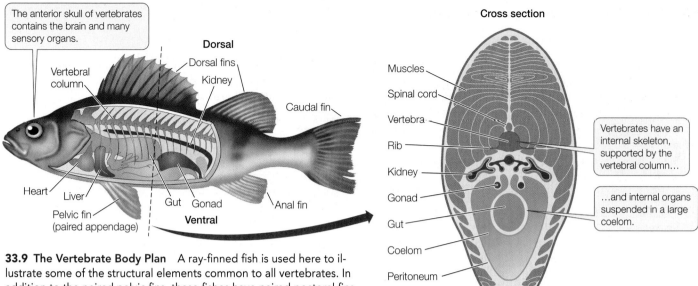

The anterior skull of vertebrates contains the brain and many sensory organs.

Dorsal

Vertebral column
Dorsal fins
Kidney
Caudal fin

Heart
Liver
Pelvic fin (paired appendage)
Gut   Gonad
Anal fin

Ventral

Cross section

Muscles
Spinal cord
Vertebra
Rib
Kidney
Gonad
Gut
Coelom
Peritoneum

Vertebrates have an internal skeleton, supported by the vertebral column…

…and internal organs suspended in a large coelom.

33.9 The Vertebrate Body Plan A ray-finned fish is used here to illustrate some of the structural elements common to all vertebrates. In addition to the paired pelvic fins, these fishes have paired pectoral fins on the sides of their bodies (not seen in this cutaway view).

- An anterior *skull* enclosing a large brain
- A rigid internal *skeleton* supported by the vertebral column
- Internal organs *suspended in a coelom*
- A well-developed *circulatory system*, driven by contractions of a ventral *heart*

These structural features can support large, active animals. The internal skeleton provides support for an extensive muscular system, which receives oxygen from the circulatory system and is controlled by the central nervous system. The evolution of these features allowed many vertebrates to become large, active predators, which in turn allowed the vertebrates to diversify widely (**Figure 33.10**).

All of the nonvertebrate deuterostomes (acoels and xenoturbellids, echinoderms, hemichordates, lancelets, and

tunicates) live in marine environments. The lineage that led to the vertebrates is also thought to have evolved in the oceans, although probably in an estuarine environment (where fresh water meets salt water). The first vertebrates appeared in the Cambrian; since then they have radiated into marine, freshwater, terrestrial, and aerial environments worldwide. There are about 65,000 species of living vertebrates.

The phylogenetic relationships of jawless fishes are uncertain

The **hagfishes** are thought by many to be the sister group to the remaining vertebrates (see Figure 33.10). Hagfishes (**Figure 33.11A**) have a weak circulatory system

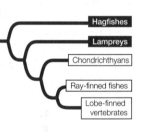

Hagfishes
Lampreys
Chondrichthyans
Ray-finned fishes
Lobe-finned vertebrates

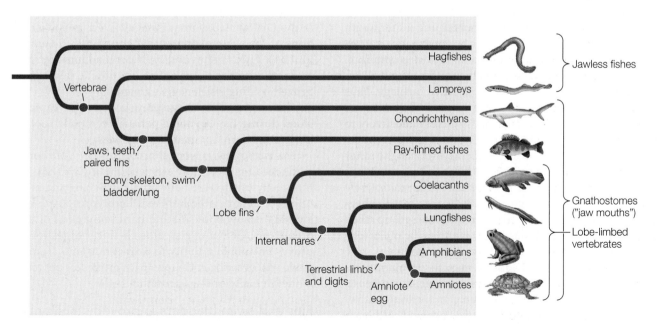

Vertebrae

Jaws, teeth, paired fins

Bony skeleton, swim bladder/lung

Lobe fins

Internal nares

Terrestrial limbs and digits

Amniote egg

Hagfishes
Lampreys
Chondrichthyans
Ray-finned fishes
Coelacanths
Lungfishes
Amphibians
Amniotes

Jawless fishes

Gnathostomes ("jaw mouths")

Lobe-limbed vertebrates

33.10 Phylogeny of the Living Vertebrates This phylogenetic tree shows the evolution of some of the key innovations among the major groups of vertebrates.

(A)

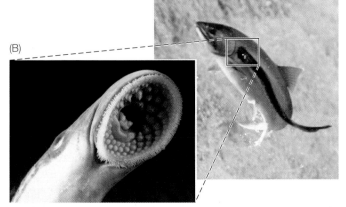

Eptatretus stoutii

(B)

Petromyzon marinus

33.11 Modern Jawless Fishes (A) Hagfishes burrow in the ocean mud, from which they extract small prey. They also scavenge on dead or dying fish. Hagfishes have degenerate eyes, which has led to their being called (inaccurately) "blind eels." (B) Sea lampreys are ectoparasites that attach to the bodies of living fish and use their large, jawless mouths to suck blood and flesh. They can survive in both fresh and salt water, as this individual attached to a salmon returning to its spawning ground will do.

with three small accessory hearts (rather than a single, large heart); a partial cranium, or skull (containing a brain with no cerebrum or cerebellum, two main regions that characterize the brains of other vertebrates); and no jaws or stomach. They also lack separate, jointed vertebrae and have a skeleton composed of a firm but pliable material called cartilage. Thus some biologists do not consider hagfishes to be vertebrates and instead use the term "craniates" to refer collectively to the hagfishes and the vertebrates. Some analyses of gene sequences suggest, however, that hagfishes may be the sister group of the **lampreys** (**Figure 33.11B**); in this phylogenetic arrangement, the hagfishes and the lampreys are collectively called the cyclostomes ("circle mouths"). If in fact the hagfishes and lampreys do form a monophyletic group, then hagfishes must have secondarily lost many of the major vertebrate morphological features during their evolution.

The 80 known species of hagfishes are unusual marine animals that produce copious quantities of mucus as a defense. They are virtually blind and rely largely on the four pairs of sensory tentacles around the mouth to detect food. Although they have no jaws, hagfishes have a tonguelike structure equipped with toothlike rasps that they use to tear apart dead organisms and to capture their principal prey, polychaete

worms. Hagfishes have direct development (no larvae), and individuals may actually change sex from year to year (from male to female and vice versa).

 Go to Media Clip 33.2
Hagfish Slime
Life10e.com/mc33.2

Although the lampreys and hagfishes may look superficially similar (with elongate eel-like bodies and no paired fins), they differ greatly in their biology. Lampreys have a complete skull and distinct and separate (although rudimentary) vertebrae, all cartilaginous rather than bony. Lampreys undergo a complete metamorphosis from filter-feeding larvae, known as ammocoetes, which are morphologically similar to adult lancelets. The adults of many species of lampreys are parasitic, although several lineages of lampreys evolved to become nonfeeding as adults. These nonfeeding adults survive for only a few weeks after metamorphosis—just long enough to breed. In the species that are parasitic as adults, the round mouth is a rasping and sucking organ that is used to attach to prey and rasp at the flesh (see Figure 33.11B).

The nearly 50 species of lampreys either live permanently in fresh water or are anadromous—meaning they live in coastal salt water and move into fresh water to breed. Some species of lampreys are critically endangered because of recent habitat changes and losses.

Jaws and teeth improved feeding efficiency

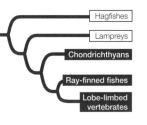

Many kinds of jawless fishes were found in the seas, estuaries, and fresh waters of the Ordovician, Silurian, and Devonian periods, but hagfishes and lampreys are the only jawless fishes that survived beyond the Devonian. Late in the Ordovician, some fishes evolved jaws via modifications of the skeletal arches that supported the gills (**Figure 33.12A**). Those fishes and their descendants are referred to as **gnathostomes** (Greek, "jaw mouths"). Jaws greatly improved feeding efficiency, as an animal with jaws can grasp, subdue, and swallow large prey. Jawed fishes rapidly diversified during the Devonian period, eventually replacing the jawless fishes in dominance of the seas.

The earliest jaws were simple, but the evolution of teeth made feeding even more efficient (**Figure 33.12B**). In predators, teeth function crucially both in grasping and in breaking up prey. In both predators and herbivores, teeth enable an animal to chew both soft and hard body parts of their food organisms. Chewing also aids chemical digestion and improves an animal's ability to extract nutrients from its food, as we will describe in Chapter 51. Vertebrates are remarkable in the diversity of their jaws and teeth.

Fins and swim bladders improved stability and control over locomotion

Most jawed fishes have a pair of pectoral fins just behind the gill slits and a pair of pelvic fins anterior to the anus (see

(A)

Jawless fishes

Skull (cartilage)

Gill arches made of cartilage supported the gills.

Gill arches

Gill slits

Early jawed fishes
(placoderms, now extinct)

Some anterior gill arches became modified to form jaws, which at first had no teeth.

Modern jawed fishes
(cartilaginous and ray-finned fishes)

Additional gill arches help support heavier, more efficient jaws, which in turn, support teeth.

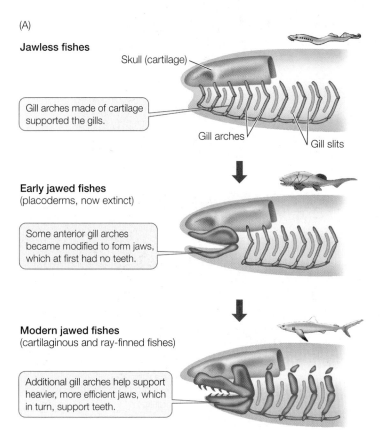

(B)

33.12 Jaws and Teeth Increased Feeding Efficiency (A) These diagrams illustrate one probable scenario for the evolution of jaws from the anterior gill arches of jawless fishes. (B) Jaws of the extinct giant shark *Carcharodon megalodon* display the teeth that indicate an extreme predatory lifestyle.

Figures 33.9 and 33.13A). These paired fins stabilize the fish's position in water (and in some cases, help propel it). Median dorsal and anal fins also stabilize the fish, or may be used for propulsion in some species. In many fishes, the caudal (tail) fin helps propel the animal and enables it to turn rapidly.

Several groups of jawed fishes became abundant during the Devonian. Among them were the **chondrichthyans**—sharks, skates, and rays (about 1,000 living species) and chimaeras (40 living species). Like hagfishes and lampreys, these fishes have a skeleton composed entirely of cartilage. Their skin is flexible and leathery, sometimes bearing scales that give it the consistency of sandpaper. Sharks move forward by means of lateral (side-to-side) undulations of the body and caudal fin (**Figure 33.13A**). Skates and rays propel themselves by means of vertical undulating movements of their greatly enlarged pectoral fins (**Figure 33.13B**).

Most sharks are predators, but some feed by straining plankton from the water. Most skates and rays live on the ocean floor, where they feed on mollusks and other animals buried in the sediments. Nearly all chondrichthyans live in the oceans, but a few are estuarine or migrate into lakes and rivers. One group of stingrays is found in river systems of South America. The less familiar chimaeras (**Figure 33.13C**) live in deep-sea or cold waters.

One lineage of aquatic gnathostomes gave rise to the bony vertebrates, which soon split into two main lineages—the **ray-finned fishes** and the **lobe-limbed vertebrates**. Bony vertebrates have internal skeletons of calcified, rigid bone rather than flexible cartilage. In early bony vertebrates, gas-filled sacs supplemented the gas exchange function of the gills by giving the animals access to atmospheric oxygen. These features enabled these fishes to live where oxygen was periodically in short supply, as it often is in freshwater environments. In the ray-finned fishes, these lunglike sacs evolved into **swim bladders**, which are organs of buoyancy. By adjusting the amount of gas in its swim bladder, a ray-finned fish can control the depth at which it remains suspended in the water while expending very little energy to maintain its position.

The outer body surface of most species of ray-finned fishes is covered with thin, flat, lightweight scales that provide protection or enhance movement through the water. The gills open into a single chamber covered by a hard flap, called an operculum. Movement of the operculum increases the flow of water over the gills, where gas exchange takes place.

Ray-finned fishes began to diversify during the Mesozoic era and continued to radiate extensively throughout the Tertiary period. Today there are about 32,000 known living species, encompassing a remarkable variety of sizes, shapes, and lifestyles (**Figure 33.14**). The smallest are less than 1 centimeter long; the largest weigh as much as 900 kilograms. Ray-finned fishes exploit nearly all types of aquatic food sources. In the oceans they filter plankton from the water, rasp algae from rocks, eat corals and other soft-bodied colonial animals, dig animals from soft sediments, and prey on virtually all kinds of other fishes. In fresh water they eat plankton, devour insects, eat fruits that fall into the water, and prey on other aquatic vertebrates and, occasionally,

(A) *Charcharodon charcharis*   Dorsal fin

Caudal fin   Pelvic fin   Pectoral fin

(B) *Myliobatis australis*   Pectoral fins

Dorsal fin

(C) *Hydrolagus colliei*   Pectoral fin   Pelvic fin

33.13 Chondrichthyans (A) Most sharks are active marine predators, as epitomized by the great white shark seen here. (B) Skates and rays, represented here by an eagle ray, feed on the ocean bottom. Their modified pectoral fins are used for propulsion; their other fins are greatly reduced. (C) A chimaera, or ratfish. Many of these deep-sea fishes possess modified dorsal fins that contain toxins.

33.14 Diversity among the Ray-Finned Fishes (A) Eels such as this moray have the large teeth and powerful jaws typical of predatory fishes. (B) There are more than 500 described species of wrasses. Many species, such as this flame fairy wrasse, inhabit coral reefs. (C) Another large ray-fin clade, the serranids, includes the sea basses and groupers. Panther groupers such as this one are endangered by the loss of Pacific coral reef habitat. (D) A unique structure that resembles a fishing lure has evolved among the anglerfishes. Deep-sea anglerfishes such as this one live below the level of light penetration; their lures are bioluminescent.

(A) *Gymnothorax meleagris*

(B) *Cirrhilabrus jordani*

(C) *Cromileptes altivelis*

(D) *Gigantactis vanhoeffeni* luring prey

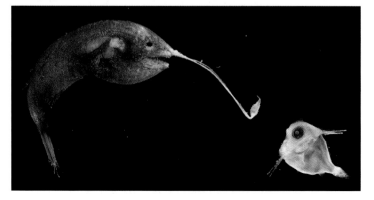

terrestrial vertebrates. Many ray-finned fishes are solitary, but in open water others form large aggregations called schools. Many species perform complicated behaviors to maintain schools, build nests, court mates, and care for their young.

Although ray-finned fishes can readily control their position in open water using their fins and swim bladder, their eggs tend to sink. Some species produce small eggs that are buoyant enough to complete their development in open water, but many marine fishes move to food-rich shallow waters to lay their eggs. That is why coastal waters and estuaries are so important in the life cycles of many marine fishes. Some ray-finned fishes, such as salmon, are anadromous, moving from the ocean to the fresh waters in which they breed.

RECAP 33.3

Chordates are characterized by a dorsal hollow nerve chord, a post-anal tail, and a dorsal supporting rod called a notochord at some point during the life cycle. Specialized structures for support (a vertebral column), locomotion (such as fins), and feeding (jaws and teeth) evolved among the vertebrates.

- What synapomorphies characterize the chordates and the vertebrates, respectively? **See pp. 683–685 and Figures 33.7 and 33.9**
- How do the hagfishes differ from the lampreys in morphology and life history? Why do some biologists contend that hagfishes are not vertebrates? **See pp. 685–686**

In the lobe-limbed vertebrates, the gas-filled sacs that gave rise to swim bladders in ray-finned fishes became specialized for another purpose: breathing air. That adaptation set the stage for the vertebrates to move onto the land.

How Did Vertebrates Colonize the Land?

The evolution of lunglike sacs in fishes set the stage for the vertebrate invasion of the land. Some early ray-finned fishes probably used those sacs to supplement their gills when oxygen levels in the water were low, as many groups of ray-finned fishes do today. But with their unjointed fins, those fishes could only flop around when out of water. Changes in the structure of the fins first allowed lobe-limbed vertebrates to support themselves better in shallow water and, later, to move better on land.

Jointed limbs enhanced support and locomotion on land

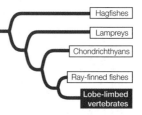

In the lobe-limbed vertebrates, the paired pelvic and pectoral fins developed into more muscular fins that were joined to the body by a single enlarged bone. The modern representatives of these lobe-limbed vertebrates include the coelacanths, lungfishes, and tetrapods.

The coelacanths flourished from the Devonian until about 65 million years ago, when they were thought to have become extinct. But in 1938 a commercial fisherman caught a living coelacanth off South Africa. Since that time, hundreds of individuals of this extraordinary fish, *Latimeria chalumnae*, have been collected. A second species, *L. menadoensis*, was discovered in 1998 off the Indonesian island of Sulawesi. *Latimeria*, a predator of other fishes, reaches a length of about 1.8 meters and weighs up to 82 kilograms (**Figure 33.15A**). Its skeleton is composed mostly of cartilage, not bone. The

(A) *Latimeria chalumnae*

33.15 The Closest Relatives of Tetrapods (A) The coelacanth, discovered in deep waters of the Indian Ocean off the South African coast, represents one of two surviving species of a group that was once thought to be extinct. (B) All surviving lungfish species, such as this African lungfish, live in the Southern Hemisphere. (C) *Tiktaalik*, a fossil lobe-limbed vertebrate from the Devonian, is believed to represent a transitional species intermediate between the finned fishes and the limbed tetrapods.

(B) *Protopterus annectens*

Tiktaalik's pectoral fins show some of the skeletal structures of tetrapod limbs.

(C) *Tiktaalik roseae*

cartilaginous skeleton is a derived feature in this clade because it had bony ancestors.

Go to Media Clip 33.3
Coelacanths in the Deep Seas
Life10e.com/mc33.3

Lungfishes were important predators in shallow-water habitats in the Devonian, but most lineages died out. The six surviving species live in stagnant swamps and muddy waters in South America, Africa, and Australia (**Figure 33.15B**). Lungfishes have lungs derived from the lunglike sacs of their ancestors as well as gills. When ponds dry up, individuals of most species can burrow deep into the mud and survive for many months in an inactive state while breathing air.

It is believed that some early aquatic lobe-limbed vertebrates began to use terrestrial food sources, became more fully adapted to life on land, and eventually evolved to become ancestral **tetrapods** ("four legs"). How was this transition from an animal that swam in water to one that walked on land accomplished? Early in 2006, scientists reported the discovery of a Devonian fossil lobe-limbed vertebrate, since then named *Tiktaalik*, which possessed intermediate appendages between the fins of fishes and the limbs of terrestrial tetrapods (**Figure 33.15C**). It appears that limbs able to prop up a large fish and make the front-to-rear movements necessary for walking evolved while these animals still lived in water. These limbs appear to have functioned in holding the animals upright in shallow water, perhaps even allowing them to hold their head above the water's surface. These same structures were then co-opted for movement on land, at first probably for foraging on brief trips out of water.

Among the lobe-limbed vertebrates, limbs capable of movement on land evolved from the short, muscular fins of aquatic ancestors (**Figure 33.16**). The resulting four terrestrial limbs give the tetrapods their name. The basic skeletal elements of those limbs can be traced through major changes in limb form and function among the terrestrial vertebrates.

An early split in the tetrapod tree led to two main groups of terrestrial vertebrates: **amphibians**, most of which remained tied to moist environments; and the **amniotes**, many of which adapted to much drier conditions.

Amphibians usually require moist environments

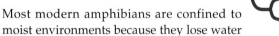

Most modern amphibians are confined to moist environments because they lose water rapidly through the skin when exposed to dry air. In addition, their eggs are enclosed within delicate membranous envelopes that cannot prevent water loss in dry conditions. In some amphibian species, adults live mostly on land but return to fresh water to lay and fertilize their eggs (**Figure 33.17**). The fertilized eggs give rise to larvae that live in water until they undergo metamorphosis to become terrestrial adults. However, many amphibians (especially those in tropical and subtropical areas) have evolved a wide variety of additional reproductive modes and types of parental care, as described in the opening of this chapter. Internal fertilization, for example, evolved

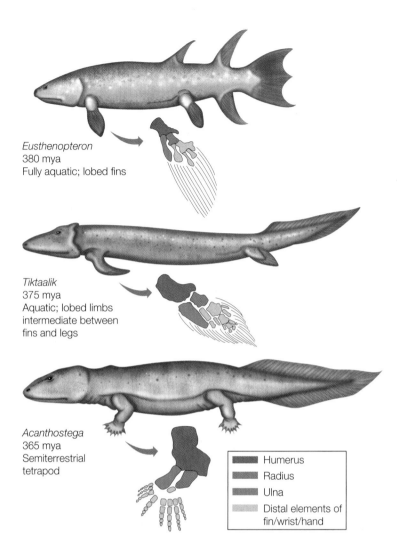

Eusthenopteron
380 mya
Fully aquatic; lobed fins

Tiktaalik
375 mya
Aquatic; lobed limbs intermediate between fins and legs

Acanthostega
365 mya
Semiterrestrial tetrapod

■ Humerus
■ Radius
■ Ulna
□ Distal elements of fin/wrist/hand

33.16 Tetrapod Limbs Are Modified Fins The major skeletal elements of the tetrapod limb were already present in aquatic lobe-limbed fishes some 380 million years ago. The relative sizes and positions of these elements changed as lobe-limbed vertebrates moved to a terrestrial environment, where limbs were needed to support and move the animal's body on land.

many times among amphibian species. Many species develop directly into adultlike forms from fertilized eggs laid on land or carried by the parents. Other species of amphibians are entirely aquatic, never leaving the water at any stage of their lives, and many of these species retain a larval-like morphology.

The more than 7,000 known species of amphibians living today belong to three major groups: the wormlike, limbless, tropical, burrowing, or aquatic caecilians (**Figure 33.18A**), the tailless frogs and toads (collectively called anurans; **Figure 33.18B**), and the tailed salamanders (**Figure 33.18C and D**).

Anurans are most diverse in wet tropical and warm temperate regions, although a few are found at very high latitudes. There are far more anurans than any other amphibians, with well over 6,000 described species and more being discovered every year. Some anurans have tough skins and other adaptations that enable them to live for long periods in deserts, whereas others live in moist terrestrial and arboreal environments. Some species are completely aquatic as adults. All anurans have a short vertebral column and a pelvic region that is

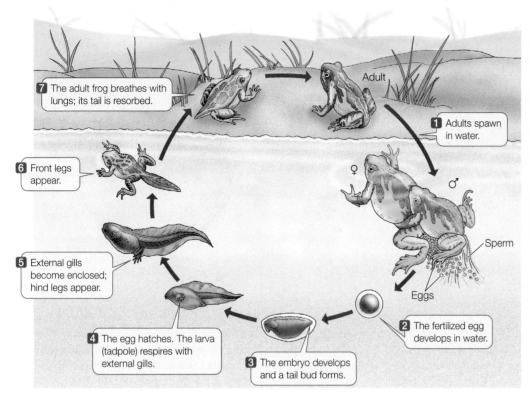

33.17 In and Out of the Water Many amphibian species have life cycles like the one diagrammed here, in which the early stages take place in water and the aquatic tadpole transforms into a terrestrial adult through metamorphosis. Some species of amphibians, however, have direct development (with no aquatic larval stage), and others are aquatic throughout life.

Go to Animated Tutorial 33.2
Life Cycle of a Frog
Life10e.com/at33.2

7 The adult frog breathes with lungs; its tail is resorbed.

1 Adults spawn in water.

Adult

6 Front legs appear.

♀ ♂

Sperm

5 External gills become enclosed; hind legs appear.

Eggs

4 The egg hatches. The larva (tadpole) respires with external gills.

2 The fertilized egg develops in water.

3 The embryo develops and a tail bud forms.

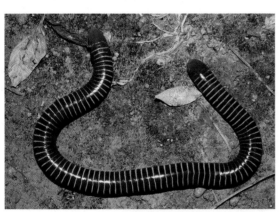

(A) *Siphonops annulatus*

(B) *Bufo periglenes*

(C) *Ambystoma mavortium*

(D) *Eurycea waterlooensis*

33.18 Diversity among the Amphibians (A) Burrowing caecilians superficially look more like worms than like amphibians. (B) Male golden toads in the cloud forest of Monteverde, Costa Rica. This species has recently become extinct, one of many amphibian species to do so in the past few decades. (C) An adult barred tiger salamander. (D) This Austin blind salamander's life cycle is completely aquatic; it has no adult terrestrial stage. The eyes of this cave dweller have become greatly reduced.

33.19 The Amniote Egg (A) The evolution of the amniote egg, with its water-retaining shell, four extraembryonic membranes, and embryo-nourishing yolk, was a major step in adaptation to the terrestrial environment. A chick egg is shown here. (B) In mammals, the developing embryo is retained inside the mother's body, with which it exchanges nutrients and wastes via the placenta. Note the correspondence between the various membranes in (A) and (B).

Go to Activity 33.2 The Amniote Egg
Life10e.com/ac33.2

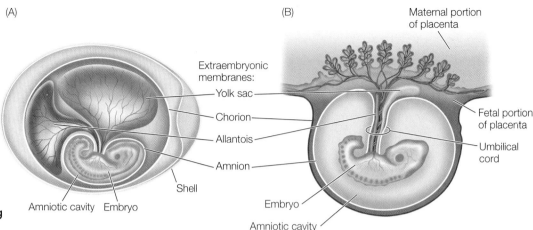

modified for leaping, hopping, or propelling the body through water by kicking the hind legs.

The more than 600 described species of salamanders are most diverse in temperate regions of the Northern Hemisphere and in cool, moist environments in the mountains of Central America, although a few species penetrate into tropical regions. Many salamanders live in rotting logs or moist soil. One major group has lost the lungs, and these species exchange gases entirely through the skin and mouth lining—body parts that all amphibians use, in addition to their lungs, for gas exchange. A completely aquatic lifestyle has evolved several times among the salamanders (see Figure 33.18D). These aquatic species have arisen through a developmental process known as **neoteny**, or the retention of juvenile traits (in this case, gills) by delayed somatic development. Most species of salamanders have internal fertilization, which is usually achieved through the transfer of a small, jellylike, sperm-embedded capsule called a spermatophore.

Many amphibians have complex social behaviors. Most male anurans utter loud, species-specific calls to attract females of their own species (and sometimes to defend breeding territories), and they compete for access to females that arrive at the breeding sites. Many amphibians lay large numbers of eggs, which they abandon once they are deposited and fertilized. As described in the opening of this chapter, however, other amphibians lay only a few eggs, which are fertilized and then cared for. A few species of frogs, salamanders, and caecilians are viviparous, meaning that they give birth to well-developed young that have received nutrition from the female during gestation.

 Go to Media Clip 33.4
Answering a Mating Call
Life10e.com/mc33.4

Amphibians are the focus of much attention today because populations of many species are declining rapidly, especially in mountainous regions of western North America, Central and South America, and northeastern Australia. Worldwide, about one-third of amphibian species are now threatened with extinction or have disappeared completely in the last few decades (as happened with the gastric brooding frogs described at the opening of this chapter). Scientists are investigating several hypotheses to account for these population declines, as described in Chapters 1 and 30.

Amniotes colonized dry environments

Several key innovations for conserving water contributed to the ability of the amniotes to exploit a wide range of terrestrial habitats. The **amniote egg** (which gives the group its name) is relatively impermeable to water and allows the embryo to develop in a contained aqueous environment (**Figure 33.19A**). Its leathery or brittle, calcium-impregnated shell retards evaporation of the fluids inside but permits passage of oxygen and carbon dioxide. The amniote egg also stores large quantities of food in the form of **yolk**, allowing the embryo to attain a relatively advanced state of development before it hatches. Within the shell are **extraembryonic membranes** that protect the embryo from desiccation and assist it with gas exchange and excretion of nitrogenous waste products of metabolism.

In several different groups of amniotes, modifications of the amniote egg allowed the embryo to develop inside (and exchange nutrients and wastes with) its mother's body (**Figure 33.19B**). In most mammals the egg lost its shell entirely while the functions of the extraembryonic membranes were retained and expanded; we will examine the roles of these membranes in detail in Section 44.5.

Other innovations evolved in the organs of terrestrial adults. A tough, impermeable skin, covered with scales or modifications of scales such as hair and feathers, greatly reduced water loss. Adaptations of the vertebrate excretory organs, called kidneys, allowed amniotes to excrete concentrated urine, ridding the body of nitrogenous wastes without losing a large amount of water in the process (see Chapter 52).

During the Carboniferous, the amniotes split into two major groups: the **reptiles** and the lineage that eventually led to the **mammals** (Figure 33.20).

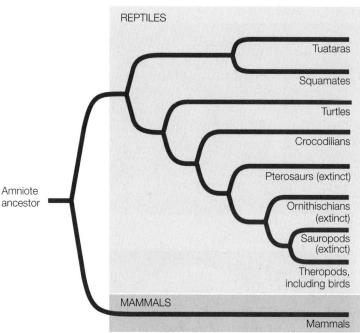

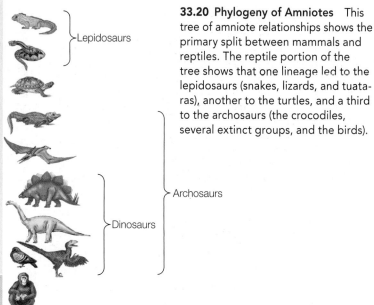

33.20 Phylogeny of Amniotes This tree of amniote relationships shows the primary split between mammals and reptiles. The reptile portion of the tree shows that one lineage led to the lepidosaurs (snakes, lizards, and tuataras), another to the turtles, and a third to the archosaurs (the crocodiles, several extinct groups, and the birds).

Reptiles adapted to life in many habitats

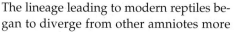

The lineage leading to modern reptiles began to diverge from other amniotes more than 300 million years ago. More than 19,000 species of reptiles exist today, more than half of which are birds. Birds are the only living representatives of the otherwise extinct dinosaurs, the dominant terrestrial predators of the Mesozoic.

The **lepidosaurs** constitute the second most species-rich clade of living reptiles. This group is composed of the **squamates** (lizards, snakes, and amphisbaenians—the last a group of mostly legless, wormlike, burrowing reptiles with greatly reduced eyes) and the **tuataras**, which superficially resemble lizards but differ from them in tooth attachment and several internal anatomical features. Many species related to the tuataras lived during the Mesozoic era, but today only two species survive (**Figure 33.21A**), and these are restricted to a few islands off New Zealand.

The skin of a lepidosaur is covered with horny scales that greatly reduce loss of water from the body surface. These scales, however, make the skin unavailable as an organ of gas exchange. Gases are exchanged almost entirely via the lungs, which are proportionally much larger in surface area than those of amphibians. A lepidosaur forces air into and out of its lungs by bellows-like movements of its ribs. The three-chambered lepidosaur heart partially separates oxygenated blood from the lungs from deoxygenated blood returning from the body. With this type of heart, lepidosaurs can generate high blood pressure and can sustain a relatively high metabolism.

Most lizards are insectivores, although some are herbivores and a few prey on other vertebrates. Most lizards walk on four limbs (**Figure 33.21B**), although limblessness has evolved repeatedly among the lizards, especially in burrowing and grassland species. The largest lizard is the predaceous Komodo

dragon of the East Indies, which grows as long as 3 meters and can weigh more than 150 kilograms.

 Go to Media Clip 33.5
Komodo Dragons Bring Down Prey
Life10e.com/mc33.5

The major group of limbless squamates is the snakes (**Figure 33.21C**). All snakes are carnivores, and many can swallow objects much larger than themselves. Several snake groups evolved venom glands and the ability to inject venom rapidly into their prey.

The **turtles** comprise a reptilian group that has changed relatively little since the early Mesozoic. In these reptiles, dorsal and ventral bony plates form a shell into which the head and limbs can be withdrawn in many species (**Figure 33.21D**). The dorsal shell is a modification of the ribs. It is a mystery how the pectoral girdles evolved to be inside the ribs of turtles, making them unlike any other vertebrates. Most turtles live in aquatic environments, but several groups, such as tortoises and box turtles, are terrestrial. Sea turtles spend their entire lives at sea except when they come ashore to lay eggs. Human exploitation of sea turtles and their eggs has resulted in worldwide declines of these species, all of which are now endangered. A few species of turtles are strict herbivores or carnivores, but most species are omnivores that eat a variety of aquatic and terrestrial plants and animals.

Crocodilians and birds share their ancestry with the dinosaurs

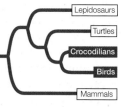

Another reptilian clade, the **archosaurs**, includes the crocodilians, pterosaurs, dinosaurs, and birds. Only the crocodilians and birds are represented by living species today. Modern **crocodilians**—crocodiles, caimans, gharials, and alligators—are

(A) *Sphenodon punctatus*

(B) *Eublepharis macularius*

(C) *Diadophis punctatus*

(D) *Chelonoidis nigra abingdonii*

33.21 Reptilian Diversity (A) This tuatara represents one of only two surviving species in its lineage. (B) The leopard gecko, a desert-dwelling lizard native to Afghanistan, Pakistan, and northwestern India. (C) The ringneck snake of North America is nonvenomous. It coils its tail to reveal a bright orange underbelly, which distracts potential predators from the vital head region. (D) Galápagos tortoises are the largest turtles and among the largest reptiles. They have been documented to live for more than 100 years in the wild.

confined to tropical and warm temperate environments (**Figure 33.22A**). All crocodilians are carnivorous; they eat vertebrates of all kinds, including large mammals. Crocodilians spend much of their time in water, but they lay their eggs in nests they build on land or on floating piles of vegetation. The eggs are warmed by heat generated by decaying organic matter that the female places in the nest. The female provides other forms of parental care as well: typically she guards the eggs until they hatch, and in some species she continues to guard and communicate with her offspring after they hatch.

Dinosaurs rose to prominence about 215 million years ago and dominated terrestrial environments for about 150 million years. However, only one group of dinosaurs, the **birds**, survived the mass extinction at the end of the Cretaceous. During the Mesozoic, most terrestrial animals more than a meter long were dinosaurs. Many were agile and could run rapidly; they had special muscles that enabled the lungs to be filled and emptied while the limbs moved. We can infer the existence of such muscles in dinosaurs from the structure of the vertebral column in fossils. Some of the largest dinosaurs weighed as much as 70,000 kilograms.

Biologists have long accepted the phylogenetic position of birds among the reptiles, although birds clearly have many unique, derived morphological features. In addition to the strong morphological evidence for this placement, fossil and molecular data emerging over the last few decades have provided definitive supporting evidence. Birds are a specialized group of **theropods**, a clade of predatory dinosaurs that shared such traits as a bipedal stance, hollow bones, a furcula ("wishbone"), elongated metatarsals with three-fingered feet, elongated forelimbs with three fingers, and a pelvis that points backward. Modern birds are endothermic, meaning that they regulate their body temperatures by producing and retaining metabolic heat, rather than by absorbing heat from their external environment (see Section 40.5). Although we cannot directly assess this physiological trait in extinct species, many fossil theropods share morphological traits that suggest they may have been endothermic as well.

The living bird species fall into two major groups that diverged about 80 to 90 million years ago from a flying ancestor. The few modern descendants of one lineage include a group of secondarily flightless and weakly flying birds, some of which are very large. This group, called the palaeognaths, includes the South and Central American tinamous and several large flightless birds of the southern continents—the rheas, emu, kiwis, cassowaries, and the world's largest bird, the ostrich (**Figure 33.22B**). The

(A) *Crocodylus porosus*

(B) *Struthio camelus*

33.22 Archosaurs The two surviving groups of archosaurs are very different. (A) The crocodilians live in tropical and warm temperate climates. This crocodile species lives in saltwater and estuarine environments along Australia's coast. (B) Birds are the only other living archosaur group, represented here by the winged but flightless ostrich.

second lineage, the neognaths, has left a much larger number of descendants, most of which have retained the ability to fly.

Feathers allowed birds to fly

Fossil theropods discovered recently in early Cretaceous deposits in Liaoning Province, in northeastern China, show that the scales of some small predatory dinosaurs were highly modified to form **feathers**. Initially these feathers were simply a body covering that probably provided insulation and enhanced coloration. But the feathers of some later dinosaurs, such as *Microraptor gui*, were structurally similar to those of modern birds (**Figure 33.23A**).

Another theropod that was even more closely related to modern birds, *Archaeopteryx*, lived about 150 million years ago. *Archaeopteryx* had teeth (unlike modern birds), but it was covered with feathers that are virtually identical to those of birds (**Figure 33.23B**). It also had well-developed wings, a long tail, and a furcula to which some of the flight muscles were probably attached. *Archaeopteryx* had clawed fingers on its forelimbs, but it also had typical perching bird claws on its hindlimbs. It probably lived in trees and shrubs and used the fingers to assist it in clambering over branches. It probably glided or flew weakly. The descendants of *Archaeopteryx* and similar Mesozoic theropods were the modern birds, most of which are accomplished fliers.

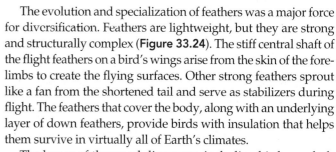

 Go to Media Clip 33.6
Falcons in Flight
Life10e.com/mc33.6

The evolution and specialization of feathers was a major force for diversification. Feathers are lightweight, but they are strong and structurally complex (**Figure 33.24**). The stiff central shaft of the flight feathers on a bird's wings arise from the skin of the forelimbs to create the flying surfaces. Other strong feathers sprout like a fan from the shortened tail and serve as stabilizers during flight. The feathers that cover the body, along with an underlying layer of down feathers, provide birds with insulation that helps them survive in virtually all of Earth's climates.

The bones of theropod dinosaurs, including birds, are hollow with internal struts that increase their strength. Hollow bones would have made early theropods lighter and more mobile; later they facilitated the evolution of flight. The sternum (breastbone) of flying birds forms a large, vertical keel to which the flight muscles are attached.

Flight is metabolically expensive. A flying bird consumes energy at a rate about 15 to 20 times faster than a running lizard of the same weight. Because birds have such high metabolic rates, they generate large amounts of heat. They control the rate of heat loss using their feathers, which may be held close to the body or elevated to alter the amount of insulation they provide. The lungs of birds allow air to flow through unidirectionally rather than pumping air in and out (see Section 49.2). This

(A)

Faint impressions of feathers can be seen around the fossilized skeletons.

(B)

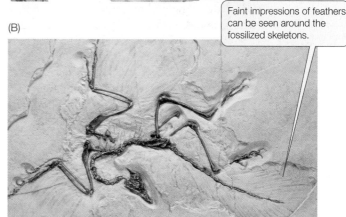

33.23 Mesozoic Bird Relatives Fossils support the evolution of birds from other theropods. (A) *Microraptor gui* was a feathered theropod from the early Cretaceous (about 140 mya). (B) *Archaeopteryx* was even more closely related to modern birds.

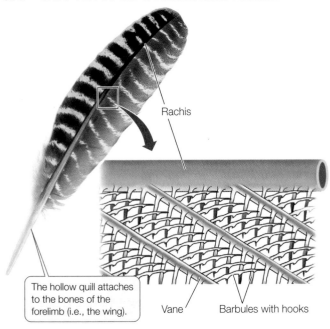

Rachis

The hollow quill attaches to the bones of the forelimb (i.e., the wing).

Vane / Barbules with hooks

33.24 Feather Anatomy The flight feathers of birds are attached to the wing's skin by the hollow portion, or quill, of a stiff central shaft. The rachis is the solid portion of the shaft from which radiate fine branches (vanes) with interlocking hooks and barbules. Overall, this structure represents a major evolutionary innovation: a strong, lightweight surface that enables flight.

flow-through structure of the lungs increases the efficiency of gas exchange and thereby supports a high metabolic rate.

There are about 10,000 species of living birds, which range in size from the 150-kilogram ostrich to a tiny hummingbird weighing only 2 grams (**Figure 33.25**). The teeth so prominent among other dinosaurs were secondarily lost in the ancestral birds, but birds nonetheless eat almost all types of animal and plant material. Insects and fruits are the most important dietary items for terrestrial species. Birds also eat seeds, nectar and pollen, leaves and buds, carrion, and other vertebrates. By eating the fruits and seeds of plants, birds serve as major agents of seed dispersal.

Mammals radiated after the extinction of non-avian dinosaurs

Reptiles
Prototherians
Marsupials
Eutherians

Small and medium-sized mammals coexisted with the large dinosaurs throughout most of the Mesozoic era, and most of the major groups of mammals that are alive today arose in the Cretaceous. After the non-avian dinosaurs disappeared during the mass extinction at the end of the Cretaceous, mammals increased dramatically in numbers, diversity, and size. Today mammals range in size from tiny shrews and bats weighing only about 2 grams to the blue whale, the largest animal on Earth, which measures up to 33

(A) *Bombycilla cedrorum*

(B) *Trichoglossus haematodus*

(C) *Tyto alba*

33.25 Diversity among the Birds (A) Perching, or passeriform, birds such as this cedar waxwing constitute the most species-rich of all bird groups. (B) Some 375 species of parrots, macaws, parakeets, and lorikeets such as the one shown here are another large bird group. (C) The barn owl is a nighttime predator that can find prey using its sensitive auditory "sonar" system. (D) The scarlet ibis of South America and the Caribbean is one of many species of wading birds.

(D) *Eudocimus ruber*

meters long and can weigh as much as 160,000 kilograms. Mammals have far fewer, but more highly differentiated, teeth than do fishes, amphibians, or reptiles. Differences among mammals in the number, type, and arrangement of teeth reflect their varied diets (see Figure 51.6).

Four key features distinguish the mammals:

- *Sweat glands,* which secrete sweat that evaporates and thereby cools an animal
- *Mammary glands*, which in females secrete a nutritive fluid (milk) on which newborn individuals feed
- *Hair*, which provides a protective and insulating covering
- A *four-chambered heart* that completely separates the oxygenated blood coming from the lungs from the deoxygenated blood returning from the body (this last characteristic is convergent with the archosaurs, including modern birds and crocodiles)

Mammalian eggs are fertilized within the female's body, and in nearly all mammalian groups the resulting embryos undergo a period of development inside the female's body in an organ called the uterus. In the uterus, the embryo is contained in an amniotic sac that is homologous to one of the three membranes found in the amniote egg (see Figure 33.19). The embryo is connected to the wall of the uterus by an organ called a placenta. The placenta allows for nutrient and gas exchange, as well as waste elimination from the developing embryo, via the female's circulatory system. Most mammals develop a covering of hair (fur), which is luxuriant in some species but has been greatly reduced in others, including cetaceans (whales and dolphins) and humans. Thick layers of insulating fat (blubber) replace hair as a heat-retention mechanism in the cetaceans; humans learned to use clothing for this purpose when they dispersed from warm tropical areas.

The approximately 5,700 species of living mammals are divided into two primary groups: the **prototherians** and the **therians** (Table 33.1). Members of the therian clade are further divided into the **marsupials** and the **eutherians**.

PROTOTHERIANS Only five species of prototherians are known, and they are found only in Australia and New Guinea. These mammals, the duck-billed platypus and four species of echidnas, differ from other mammals in laying shelled eggs and having sprawling legs (**Figure 33.26**). Prototherians supply milk for their young, but they have no nipples on their mammary glands; the milk simply oozes out and is lapped off the fur by the offspring.

MARSUPIALS Females of most marsupial species have a ventral pouch in which they carry and feed their offspring (see Figure 33.27A). Gestation (pregnancy) in marsupials is brief; the young are born tiny but with well-developed forelimbs, with which they climb to the pouch. They attach to a nipple but cannot suck. The mother ejects milk into the tiny offspring until it grows large enough to suckle. Once her offspring have left the uterus, a female marsupial may become sexually receptive again. She can then carry fertilized eggs that are capable

(A) *Tachyglossus aculeatus*

(B) *Ornithorhynchus anatinus*

33.26 Prototherians (A) The short-beaked echidna is one of four surviving species of echidnas. (B) The duck-billed platypus lives in freshwater streams in eastern Australia.

of initiating development to replace the offspring in her pouch should something happen to them.

At one time marsupials were found on all continents, but the approximately 350 living species are now restricted to Australasia (**Figure 33.27A and B**) and the Americas (especially South America; **Figure 33.27C**). Of the seven major groups of marsupials shown in Table 33.1, only the New World opossums, the shrew opossums, and the diminutive monito del monte are found in the Americas. Only one species, the Virginia opossum, is found in North America north of Mexico. Marsupials radiated to become herbivores, insectivores, and carnivores, but no marsupials live in the oceans. None can fly, although some arboreal (tree-dwelling) marsupials are gliders. The largest living marsupials are the kangaroos of Australia, which can weigh up to 90 kilograms. Much larger marsupials existed in Australia until humans exterminated them soon after reaching that continent about 40,000 years ago.

EUTHERIANS The majority of mammals are eutherians ("true" therians). Eutherians are sometimes called placental mammals, but this name is inappropriate because some marsupials also have placentas. Eutherians are more developed at birth than are marsupials; no external pouch houses them after they are born.

The more than 5,300 species of living eutherians are divided into 20 major groups (see Table 33.1). The relationships of these groups to one another have been difficult to determine because most of the major groups diverged within a short time during an explosive adaptive radiation. Modern genomic analyses have elucidated these relationships, however (**Figure**

TABLE 33.1

Major Groups of Living Mammals

Group	Number of Described Species	Examples
PROTOTHERIANS		
Monotremes (Monotremata)	5	Echidnas, duck-billed platypus
THERIANS		
Marsupials		
Diprotodonts (Diprotodontia)	146	Kangaroos, wallabies, possums, koala, wombats
New World opossums (Didelphimorphia)	93	Opossums
Carnivorous marsupials (Dasyuromorphia)	72	Quolls, dunnarts, numbat, Tasmanian devil
Omnivorous marsupials (Peramelemorphia)	22	Bandicoots and bilbies
Shrew opossums (Paucituberculata)	7	Andean rat opossums
Marsupial moles (Notoryctemorphia)	2	Southern and northern marsupial moles
Microbiothere (Microbiotherea)	1	Monito del monte
Eutherians		
Rodents (Rodentia)	2,337	Rats, mice, squirrels, woodchucks, ground squirrels, beaver, capybara
Bats (Chiroptera)	1,171	Fruit bats, echo-locating bats
Even-toed hoofed mammals and cetaceans (Cetartiodactyla)	469	Deer, sheep, goats, cattle, antelopes, giraffes, camels, swine, hippopotamus, whales, dolphins
Shrews, moles, and relatives (Soricomorpha)	428	Shrews, moles, solenodons
Primates (Primates)	396	Lemurs, monkeys, apes, humans
Carnivores (Carnivora)	284	Wolves, dogs, bears, cats, weasels, pinnipeds (seals, sea lions, walruses)
Rabbits and relatives (Lagomorpha)	92	Rabbits, hares, pikas
African insectivores (Afrosoricida)	50	Tenrecs, golden moles
Hedgehogs (Erinaceomorpha)	24	European hedgehog
Armadillos (Cingulata)	21	Giant armadillo, nine-banded armadillo
Tree shrews (Scandentia)	20	Pygmy tree shrew, pen-tailed tree shrew
Odd-toed hoofed mammals (Perissodactyla)	16	Horses, zebras, tapirs, rhinoceroses
Elephant shrews (Macroscelidea)	15	Elephant shrews, jumping shrews, sengis
Anteaters, sloths (Pilosa)	10	Anteaters, tamanduas, two- and three-toed sloths
Pangolins (Pholidota)	8	Asian and African pangolins
Hyraxes and relatives (Hyracoidea)	5	Hyraxes, dassies
Sirenians (Sirenia)	4	Manatees, dugongs
Elephants (Proboscidea)	3	African and Indian elephants
Colugos (Dermoptera)	2	Flying lemurs
Aardvark (Tubulidentata)	1	Aardvark

33.28). These studies have revealed that the major early splits in eutherian lineages are closely associated with the breakup of the continents during the Mesozoic (see Figure 25.14), after which the major groups of mammals radiated independently in Laurasia, Africa, and South America. The reconnection of South America and North America via the Panamanian land bridge about 3 million years ago resulted in a huge faunal exchange between those continents, which is particularly evident among the mammals. South American groups such as armadillos moved north into North America, and Laurasian groups such as carnivores and odd- and even-toed hoofed animals moved south into South America.

(B) *Macrotis lagotis*

(A) *Macropus giganteus*

(C) *Didelphis virginiana*

33.27 Diversity among the Marsupials
(A) Australia's eastern gray kangaroo is among the largest living marsupials. This female carries her young offspring in the characteristic marsupial pouch. (B) The greater bilby of Australia's arid regions is a nocturnal omnivore The bilby does not require drinking water, obtaining the moisture it needs from its various food items. (C) The North American opossum is the only marsupial found north of Mexico.

Eutherians are extremely varied in their form and ecology (**Figure 33.29**). The extinction of the non-avian dinosaurs at the end of the Cretaceous may have made it possible for them to diversify and radiate into a large range of ecological niches. Many eutherian species grew large, and some assumed the roles of dominant terrestrial predators previously occupied by the large dinosaurs. Among these predators, social hunting behavior evolved in several species, including members of the carnivore and primate clades.

The two most diverse groups of eutherians are the rodents and the bats, which together comprise about two-thirds of the species. Rodents are traditionally defined by the unique morphology of their teeth, which are adapted for gnawing through substances such as wood. The bats probably owe much of their

33.28 Major Groups of Eutherians Diversified as the Continents Drifted Apart This phylogenetic tree shows the relationships among most of the major terrestrial eutherian groups, the location of the earliest fossils found for each group (also indicated in the color distinctions of the various branches), and the current distributions of the groups. The major splits in eutherian evolution correspond in large degree to the tectonic history of the major continents (see Figures 25.14 and 54.12).

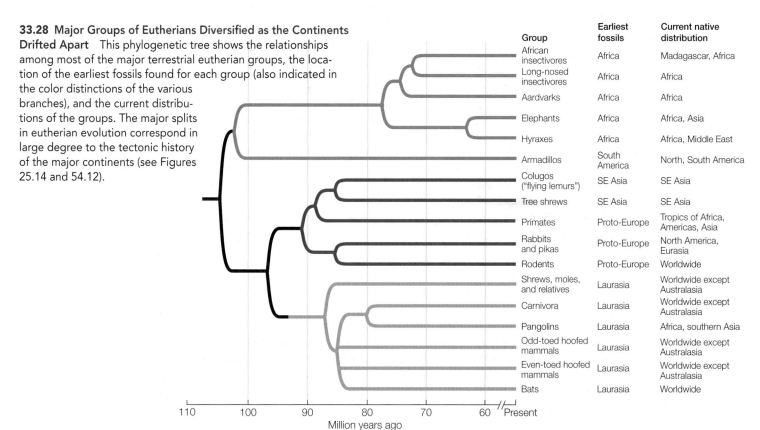

Group	Earliest fossils	Current native distribution
African insectivores	Africa	Madagascar, Africa
Long-nosed insectivores	Africa	Africa
Aardvarks	Africa	Africa
Elephants	Africa	Africa, Asia
Hyraxes	Africa	Africa, Middle East
Armadillos	South America	North, South America
Colugos ("flying lemurs")	SE Asia	SE Asia
Tree shrews	SE Asia	SE Asia
Primates	Proto-Europe	Tropics of Africa, Americas, Asia
Rabbits and pikas	Proto-Europe	North America, Eurasia
Rodents	Proto-Europe	Worldwide
Shrews, moles, and relatives	Laurasia	Worldwide except Australasia
Carnivora	Laurasia	Worldwide except Australasia
Pangolins	Laurasia	Africa, southern Asia
Odd-toed hoofed mammals	Laurasia	Worldwide except Australasia
Even-toed hoofed mammals	Laurasia	Worldwide except Australasia
Bats	Laurasia	Worldwide

110   100   90   80   70   60   Present
Million years ago

(A) *Castor canadensis*

(B) *Macroderma gigas*

(D) *Stenella longirostris*

33.29 Diversity among the Eutherians (A) The North American beaver exhibits the gnawing teeth that characterize rodents. Almost half of all eutherians are rodents. (B) Flight evolved in the ancestor of bats. This ghost bat is endemic to Australia; bats and rodents are the only eutherian groups native to that continent. (C) Large hoofed mammals such as reindeer are important herbivores in terrestrial environments. Although this bull is grazing by himself, reindeer are usually found in huge herds. (D) Spinner dolphins are cetaceans, a cetartiodactyl group that returned to the marine environment.

(C) *Rangifer tarandus*

success to the evolution of flight, which allows them to exploit a variety of food sources and colonize remote locations with relative ease.

 Go to Media Clip 33.7
Bats Feeding in Flight
Life10e.com/mc33.7

Grazing and browsing by members of several eutherian groups helped transform the terrestrial landscape. Herds of grazing herbivores fed on open grasslands, whereas browsers fed on shrubs and trees. The effects of these herbivores on plant life favored the evolution of the spines, tough leaves, and difficult-to-eat growth forms found in many plants. In turn, adaptations in the teeth and digestive systems of many herbivore lineages allowed these species to consume many plants despite such defenses—a striking example of coevolution. A large animal can survive on food of lower quality than a small animal can, and large size evolved in several groups of grazing and browsing mammals (see Figure 33.29C). The evolution of large herbivores, in turn, favored the evolution of large carnivores able to attack and overpower them.

Several lineages of terrestrial eutherians subsequently returned to the aquatic environments their ancestors had left behind (see Figure 33.29D). The completely aquatic cetaceans—whales and dolphins—evolved from even-toed hoofed ancestors (whales are closely related to the hippopotamuses). The seals, sea lions, and walruses also returned to the marine environment, and their limbs became modified into flippers. Weasel-like otters retain their limbs but have also returned to aquatic environments, colonizing both fresh and salt water. The manatees and dugongs colonized estuaries and shallow seas.

■ RECAP 33.4

The initial vertebrate colonization of dry land was facilitated by the evolution of lunglike sacs and jointed limbs. The amniotes evolved impermeable body coverings, efficient kidneys, and the amniote egg, which resists desiccation.

- What are the similarities and differences between a shelled amniote egg laid by a reptile and the retained amniotic sac of a mammal? **See p. 692 and Figure 33.19**
- How has the diversification of mammals been influenced by mass extinction events and continental drift? **See pp. 697–698 and Figure 33.28**

The biology of one eutherian group—the primates—has been the subject of extensive research. The behavior, ecology, physiology, and molecular biology of the primates are of special interest to us because this lineage includes humans.

33.5 What Traits Characterize the Primates?

One lineage of small, arboreal, insectivorous eutherians underwent extensive evolutionary radiation to become the **primates** (Figure 33.30). A nearly complete fossil of an early primate, *Carpolestes*, found in Wyoming, was dated at 56 million years ago; it had grasping feet with an opposable big toe that had a nail rather than a claw. Grasping limbs with opposable digits, an adaptation to arboreal life, are one of the major features that distinguish primates from other mammals.

Two major lineages of primates split late in the Cretaceous

About 90 million years ago, late in the Cretaceous period, the primates split into two clades: the prosimians and the anthropoids. **Prosimians**—lemurs, lorises, and galagos—once lived on all continents, but today they are restricted to Africa, Madagascar, and tropical Asia. All mainland prosimian species are arboreal and nocturnal. On the island of Madagascar, however, the site of a remarkable radiation of lemurs, there are also diurnal and terrestrial species (**Figure 33.31**). Tarsiers were once considered prosimians as well, although today we know that they are more closely related to monkeys and apes than to lemurs, lorises, and galagos.

The second primate lineage, the **anthropoids**—tarsiers, New World monkeys, Old World monkeys, and apes—began to diversify shortly after the mass extinction event at the end of the Cretaceous, in Africa or Asia. New World monkeys diverged from Old World monkeys and apes slightly later, but early enough that they may have originated in Africa and reached South America when those two continents were still close to each other. New World monkeys now live only in South and Central America, and all of them are arboreal (**Figure 33.32A**). Many of them have a long, prehensile tail with which they can grasp branches. Many Old World monkeys are arboreal as well, but several species are terrestrial (**Figure 33.32B**). No Old World monkey has a prehensile tail.

About 35 million years ago, a lineage that led to the modern apes separated from the Old World monkeys. Between 22 and 5.5 million years ago, dozens of species of apes lived in Europe, Asia, and Africa. The

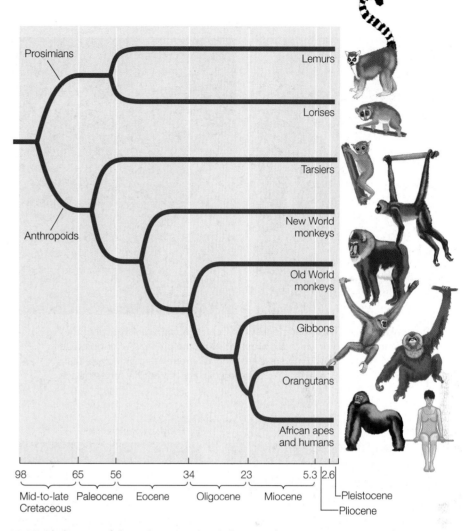

33.30 Phylogeny of the Primates The phylogeny of primates is among the best studied of any major group of mammals. This tree is based on evidence from many genes, morphology, and fossils.

Propithecus diadema

33.31 A Prosimian The diademed sifaka is one of the many lemur species found in Madagascar, where it is part of a unique assemblage of endemic plants and animals. Sifakas live in groups of up to a dozen animals and defend their territories.

(A) *Ateles geoffroyi*

(B) *Mandrillus sphinx*

33.32 Monkeys (A) The spider monkeys of Central America are typical of the New World monkeys, all of which are arboreal. Note the prehensile (gripping) tail. (B) Although many Old World monkeys are arboreal, none has a prehensile tail. Many Old World monkey species, like this mandrill, are thoroughly terrestrial.

Asian apes—gibbons and orangutans (**Figure 33.33A and B**)—descended from two of these ape lineages. Orangutans are the closest living sister group of the modern African apes: gorillas (**Figure 33.33C**), chimpanzees (**Figure 33.33D**), and humans.

Bipedal locomotion evolved in human ancestors

About 6 million years ago in Africa, a lineage split occurred that would lead to the chimpanzees on the one hand and to the **hominin** clade, which includes modern humans and their extinct close relatives, on the other.

The earliest protohominins, known as ardipithecines, had distinct morphological adaptations for **bipedal locomotion** (walking on two legs). Bipedal locomotion frees the forelimbs to manipulate objects and to carry them while walking. It also elevates the eyes, enabling the animal to see over tall vegetation to spot predators and prey. Bipedal locomotion is also energetically more economical than

(A) *Hylobates lar*

(B) *Pongo pygmaeus*

(C) *Gorilla gorilla*

33.33 Apes (A) The several genera of gibbons are all smaller in size than the other apes. Gibbons are found throughout southeastern Asia. (B) Orangutans are also native to Asia, living in the forests of Sumatra and Borneo. (C) Gorillas—the largest apes—are restricted to humid African forests. This male is a lowland gorilla. (D) Chimpanzees, our closest relatives, are found in forested regions of Africa.

(D) *Pan troglodytes*

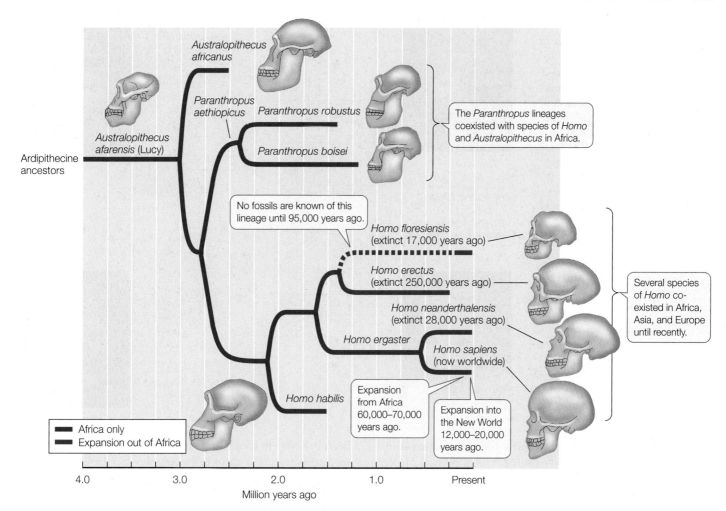

33.34 A Phylogenetic Tree of Hominins At times in the past, more than one hominin species lived on Earth at the same time. Originating in Africa, hominins spread to Europe and Asia multiple times. All but one of those species are now extinct, but that one species, modern *Homo sapiens*, has colonized nearly every corner of the planet.

quadrupedal locomotion (walking on four legs). All three advantages were probably important for the ardipithecines and their descendants, the australopithecines (**Figure 33.34**).

The first australopithecine skull was found in South Africa in 1924. Since then australopithecine fossils have been found at many sites in Africa. The most complete fossil skeleton yet found was discovered in Ethiopia in 1974. The skeleton, approximately 3.5 million years old, was that of a young female who has since become known to the world as "Lucy." Lucy was assigned to the species *Australopithecus afarensis*. Fossil remains of more than 100 *A. afarensis* individuals have since been discovered, and there have been recent discoveries of fossils of other australopithecine species that lived in Africa 4 to 5 million years ago.

Experts disagree over how many species are represented by australopithecine fossils, but it is clear that multiple species of hominins lived together over much of eastern Africa several million years ago. A lineage of larger species (weighing about 40 kilograms) is represented by *Paranthropus robustus* and *P. boisei*, both of which died out between 1 and 1.5 million years ago. A lineage of smaller australopithecines gave rise to the genus *Homo*.

Early members of the genus *Homo* lived contemporaneously with *Paranthropus* in Africa for about a million years. Some 2-million-year-old fossils of an extinct species called *H. habilis* were discovered in the Olduvai Gorge, Tanzania. Other fossils of *H. habilis* have been found in Kenya and Ethiopia. Associated with these fossils were tools that these early hominins used to obtain food.

Another extinct hominin species, *Homo erectus*, arose in Africa about 1.6 million years ago. Soon thereafter it had spread as far as eastern Asia, becoming the first hominin to leave Africa. Members of *H. erectus* were nearly as large as modern people, but their brains were smaller and they had comparatively thick skulls. The cranium, which had thick, bony walls, may have been an adaptation to protect the brain, ears, and eyes from impacts caused by a fall or a blow from a blunt object. What would have been the source of such blows? Fighting with other *H. erectus* individuals is a possible answer.

Homo erectus used fire for cooking and for hunting large animals, and made characteristic stone tools that have been found in many parts of Africa and Asia. Populations of *H. erectus* survived until at least 250,000 years ago, although more recent fossils may also be attributable to this species. In 2004 some 18,000-year-old fossil remains of a small *Homo* were found on

33.35 Neoteny in the Evolution of Humans The skulls and heads of juvenile chimpanzees and humans start out relatively similar in shape. The grid over the skulls shows how the various bony elements change in their relative proportions over the course of maturation. The adult human skull retains a shape closer to its juvenile shape, resulting in a brain that is much larger relative to other parts of the skull (notably the jaw).

the island of Flores in Indonesia. Since then, numerous additional fossils of this diminutive hominin have been found on Flores, dating from 95,000 to 17,000 years ago. Many anthropologists think that this small species, named *H. floresiensis*, was most closely related to *H. erectus*.

Human brains became larger as jaws became smaller

In another hominin lineage that diverged from *H. erectus* and *H. floresiensis*, the brain increased rapidly in size, and the jaw muscles, which were large and powerful in earlier hominins, dramatically decreased in size. These two changes were simultaneous, which suggests that they might have been developmentally linked. These changes are another example of evolution by neoteny (which, you may recall from our discussion of amphibians, is the retention of juvenile traits through delayed somatic development). Human and chimpanzee skulls are similar in shape at birth, but chimpanzee skulls undergo a dramatic change in shape as the animals mature (**Figure 33.35**). In particular, the jaw grows considerably in relation to the brain case. As human skulls grow, relative proportions much closer to those of the juvenile skull are retained, which results in a large brain case and small jaw compared with those of chimpanzees. A mutation in a regulatory gene that is expressed only in the head may have removed a barrier that had previously prevented this remodeling of the human cranium.

The striking enlargement of the brain relative to body size in the hominin lineage was probably favored by an increasingly complex social life. Any features that allowed group members to communicate more effectively with one another would have

been valuable in cooperative hunting and gathering as well as for improving one's status in the complex social interactions that must have characterized early human societies, just as they do ours today.

Several *Homo* species coexisted during the mid-Pleistocene, from about 1.5 million to about 250,000 years ago. All were skilled hunters of large mammals, but plants were important components of their diets as well. During this period another distinctly human trait emerged: rituals and a concept of life after death. Deceased individuals were buried with tools and clothing, supplies for their presumed existence in the next world.

One species, *Homo neanderthalensis*, was widespread in Europe and Asia between about 500,000 and 28,000 years ago. Neanderthals were short, stocky, and powerfully built. Their massive skull housed a brain somewhat larger than our own. They manufactured a variety of tools and hunted large mammals, which they probably ambushed and subdued in close combat. Early modern humans (*H. sapiens*) expanded out of Africa between 70,000 and 60,000 years ago. Then, about 35,000 years ago, *H. sapiens* moved into the range of *H. neanderthalensis* in Europe and western Asia, so the two species must have interacted with each other. Neanderthals abruptly disappeared about 28,000 years ago. Many anthropologists believe that Neanderthals were exterminated by those early modern humans. Scientists have been able to isolate large parts of the genome of *H. neanderthalensis* from recent fossils and compare it with our own. These studies suggest that there was some limited interbreeding between the two species while they occupied the same range. As a result, in humans with Eurasian ancestry, 1 to 4 percent of the genes in their genomes may be derived from Neanderthal ancestors.

Early modern humans made and used a variety of sophisticated tools. They created the remarkable paintings of large mammals, many of them showing scenes of hunting, found in European caves. The animals they depicted were characteristic of the cold steppes and grasslands that occupied much of Europe during periods of glacial expansion. Early modern humans also spread across Asia, reaching North America perhaps as early as 20,000 years ago, although the date of their arrival in the Americas is still uncertain. Within a few thousand years, they had spread southward through North America to the southern tip of South America.

Humans developed complex language and culture

As our ancestors evolved larger brains, their behavioral capabilities increased, especially the capacity for language. Most animal communication consists of a limited number of signals, which refer mostly to immediate circumstances and are associated with charged emotional states induced by those circumstances. Human language is far richer in its symbolic character than other animal vocalizations. Our words can refer to past and future times and to distant places. We are capable of learning thousands of words, many of them referring to abstract concepts. We can rearrange words to form sentences with complex meanings.

The expanded mental abilities of humans enabled the development of a complex culture, in which knowledge and traditions are passed along from one generation to the next by teaching and observation. Cultures can change rapidly because genetic changes are not necessary for a cultural trait to spread through a population. Cultural norms, however, are not transferred automatically and must be deliberately taught to each generation.

Go to Media Clip 33.8
Humans Develop Complex Social Behaviors
Life10e.com/mc33.8

Cultural transmission greatly facilitated the domestication of plants and animals and the resultant conversion of most human societies from ones in which food was obtained by hunting and gathering to ones in which pastoralism (herding large animals) and agriculture provided most of the food. The development of agriculture led to an increasingly sedentary life, the growth of cities, greatly expanded food supplies, rapid increases in the human population, and the appearance of occupational specializations, such as artisans, shamans, and teachers.

RECAP 33.5

Grasping limbs with opposable digits distinguish primates from other mammals. Bipedal locomotion and large brains evolved in the primate ancestors that led to humans.

- Describe the differences between Old World and New World monkeys. **See p. 701 and Figures 33.32 and 33.33**
- Explain how neoteny resulted in the development of humans with relatively large brains and small jaws. **See p. 704 and Figure 33.35**

How has the evolution of complex behaviors affected the diversification of some major groups of deuterostomes?

ANSWER

Speciation in many groups of vertebrates is closely associated with the diversification of behaviors that are used for mate recognition or mate selection. These behavioral displays can be visual, olfactory, or auditory. Many vertebrate species have elaborate and stereotypical courtship displays, usually performed by males. In some cases these displays involve male-to-male combat, and the winning male mates with any attending females. In other cases a female may watch the displays of several different males and select a mate based on her perception of the quality of the display.

Some vertebrates, including most mammals, rely on olfactory cues to determine mate receptivity and to select mates. Humans have greatly reduced olfactory senses compared with most other mammals, although research has shown that we use more olfactory cues than we realize in selecting potential mates. This is the basis for the $27.5-billion-per-year perfume industry, which attempts to create smells that other humans will find sexually attractive.

In other vertebrate groups, such as frogs, mating calls play a primary role in mate selection. The calls are species-specific, but there is often some within-species variation in the calls as well. In many frog species, females can use the pitch of the calls to select larger, older males as mates. Males that are older have demonstrated that they have successful combinations of genes that have allowed them to survive and thrive in the local environment.

Whether mate selection is based on auditory, visual, or olfactory cues, divergence in the signaling system can lead to rapid speciation. Closely related species of frogs may look very similar but have distinctly different calls. Since the females use the calls to select appropriate mates of their own species, evolution of call differences among different populations can lead to rapid reproductive isolation and speciation.

33.1 **What Is a Deuterostome?**

- Deuterostomes vary greatly in adult form, but based on the distinctive patterns of early development they share and on phylogenetic analyses of their gene sequences, they are judged to be monophyletic.

- There are far fewer species of deuterostomes than of protostomes, but many deuterostomes are large and ecologically important.

- The deuterostomes comprise three major clades: the **echinoderms**, **hemichordates**, and **chordates**. Review Figure 33.1, ACTIVITY 33.1, ANIMATED TUTORIAL 33.1

 33.2 **What Features Distinguish the Echinoderms, Hemichordates, and Their Relatives?**

- Echinoderms and hemichordates, together called **ambulacrarians**, have bilaterally symmetrical, ciliated larvae. Adult echinoderms have **pentaradial symmetry** and an **oral–aboral** body orientation. Review Figure 33.3

- The **xenoturbellids** and **acoels** are reduced, soft-bodied worm-like marine animals with few distinct organ systems. Their relationships are uncertain, but recent analyses suggest that they may be the sister group of the ambulacrarians.

- Echinoderms have an **internal skeleton** of calcified plates and a unique **water vascular system** connected to extensions called **tube feet**. Review Figure 33.3

- Hemichordates are bilaterally symmetrical and have a three-part body divided into a proboscis, collar, and trunk. **Review Figure 33.6**

 33.3 **What New Features Evolved in the Chordates?**

- Chordates fall into three principal clades: **lancelets**, **tunicates**, and **vertebrates**.

- At some stage in their development, all chordates have a dorsal hollow nerve cord, a post-anal tail, and a **notochord**. Lancelets have all three key chordate features as adults. Tunicates have these features as larvae but lose them as adults. **Review Figure 33.7**

- The vertebrate body is characterized by an internal skeleton, which is supported by a **vertebral column** that replaces the notochord. It is also characterized by internal organs suspended in a coelom, a ventral heart, and an anterior skull enclosing a large brain. **Review Figure 33.9**

- From estuarine ancestors, vertebrates diversified into many lineages of marine and freshwater fishes. One of these lineages, the lobe-limbed vertebrates, later radiated into terrestrial environments. **Review Figure 33.10**

- In the **gnathostomes**, jaws evolved from gill arches. Jaws enabled these vertebrates to grasp large prey and, together with teeth, allowed them to cut food into small pieces. **Review Figure 33.12**

- **Chondrichthyans** have skeletons of cartilage; almost all species are marine. The skeletons of **ray-finned fishes** are made of bone; these fishes have colonized all aquatic environments.

 33.4 **How Did Vertebrates Colonize the Land?**

- Lungs and jointed appendages enabled one lineage of **lobe-limbed vertebrates** to colonize the land. This lineage gave rise to the **tetrapods**. Review Figure 33.16

- The earliest split in the tetrapod tree is between the **amphibians** and the **amniotes** (reptiles and mammals).

- Most modern amphibians are confined to moist environments because their bodies and their eggs lose water rapidly. Review Figure 33.17, ANIMATED TUTORIAL 33.2

- An impermeable skin, efficient kidneys, and an **amniote egg** that could resist desiccation evolved in the amniotes. Review Figure 33.19, ACTIVITY 33.2

- The major living reptile groups are the **lepidosaurs** (tuataras, along with the **squamates**, which include lizards, snakes, and amphisbaenians), the **turtles**, and the **archosaurs** (crocodilians and **birds**). Review Figure 33.20

- Birds evolved from a group of active, predatory dinosaurs known as theropods. Feathers arose among the theropods, originally for insulation and to enhance coloration, but eventually developed into adaptations for flight in birds. Review Figures 33.23, 33.24

- Mammals are unique among animals in supplying their young with a nutritive fluid (milk) secreted by mammary glands. There are two primary mammalian clades: the **prototherians** (of which there are only five species) and the species-rich **therians**. The therian clade is further subdivided into the **marsupials** and the **eutherians**. Review Table 33.1

- Mammalian phylogeny is strongly associated with the breakup of the major continents during the Mesozoic. Major lineages of eutherians diversified in Laurasia, Africa, and South America. Review Figure 33.28

 33.5 **What Traits Characterize the Primates?**

- Grasping limbs with opposable digits distinguish **primates** from other mammals. The **prosimian** clade includes the lemurs, lorises, and galagos; the **anthropoid** clade includes tarsiers, monkeys, and apes. **Review Figure 33.30**

- The ancestors of **hominins** were terrestrial apes that developed efficient **bipedal locomotion**. Review Figure 33.34

- In the lineage leading to *Homo*, brains became larger as jaws became smaller; the two events appear to be developmentally linked and are an example of evolution via **neoteny**. Review Figure 33.35

See ACTIVITY 33.3 for a concept review of this chapter.

 Go to the Interactive Summary to review key figures, Animated Tutorials, and Activities
Life10e.com/is33

CHAPTER**REVIEW**

REMEMBERING

1. Which of the following are *not* deuterostomes?
   a. Acorn worms
   b. Sea stars
   c. Tunicates
   d. Brachiopods
   e. Lancelets

2. The structure used by adult tunicates to capture food is a
   a. pharyngeal basket.
   b. proboscis.
   c. lophophore.
   d. mucus net.
   e. radula.

3. In most ray-finned fishes, lunglike sacs evolved into
   a. pharyngeal gill slits.
   b. true lungs.
   c. coelomic cavities.
   d. swim bladders.
   e. none of the above

4. The horny scales that cover the skin of reptiles prevent them from
   a. using their skin as an organ of gas exchange.
   b. sustaining high levels of metabolic activity.
   c. laying their eggs in water.
   d. flying.
   e. crawling into small spaces.

5. Which statement about bird feathers is *not* true?
   a. They are highly modified reptilian scales.
   b. They provide insulation for the body.
   c. They exist in two layers.
   d. They help birds fly.
   e. They are important sites of gas exchange.

6. The relatively large brain case and small jaw of humans relative to chimpanzees is an example of
   a. genetic drift.
   b. neoteny.
   c. concerted evolution.
   d. cultural transfer.
   e. none of the above

UNDERSTANDING & APPLYING

7. The body plan of most vertebrates is based on four appendages. What are the varied forms that these appendages take, and how are they used? In which lineages have two or more of these appendages been lost?

8. Amphibians have survived and prospered for millions of years, but today many species are disappearing, and populations of others are declining seriously. What features of their life histories might make amphibians especially vulnerable to the kinds of environmental changes now happening on Earth?

9. In the not-too-distant past, the idea that birds were reptiles met with skepticism. Explain how fossils, morphology, and molecular evidence now support the position of birds among the reptiles.

ANALYZING & EVALUATING

10. On the phylogenetic tree of amniotes (below), map the evolutionary origin of the following traits: endothermy, hair, and feathers. Which of these traits appears to have evolved convergently in more than one lineage? What is a likely functional relationship among these traits? What are some reasons why many paleontologists think many extinct theropods were endothermic?

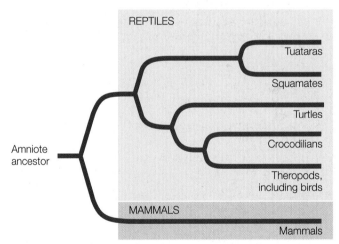

Go to BioPortal at **yourBioPortal.com** for Animated Tutorials, Activities, LearningCurve Quizzes, Flashcards, and many other study and review resources.

54 Ecology and the Distribution of Life

CHAPTER**OUTLINE**

Fynbos in the Spring   This unique vegetation community is restricted to a small region on the western coast of the South African Cape.

FYNBOS IS AN AFRIKAANS WORD derived from the Dutch *fijnbosch*, meaning "fine bush." It is the name given to an unusual assemblage of plant species found in the western Cape region of South Africa. The Dutch colonists who settled there were probably disappointed by the absence of large, sturdy trees; most of the region's shrubby vegetation lacks the heft desirable for construction. Most of the plants also have small, slender leaves that are too leathery and tough for any purpose useful to humans.

Fynbos plants may not be useful for construction, but they include some of the world's most beautiful and popular garden flowers, including geraniums, gladioli, and proteas. Although it covers less than 90,000 square kilometers (smaller than the state of Maine), the fynbos is home to 6,000 endemic species—that is, species found nowhere else in the world.

The fynbos is prone to sweeping wildfires every 15 to 30 years. These fires kill mature plants and trees, which are replaced by vigorous new plants that germinate from seeds buried in the soil or stored in fire-safe cones. If fires are too infrequent, forest trees become established and displace the unique fynbos species. About one-third of these plants depend on ants for dispersal and germination, and their seeds are equipped with a fleshy, lipid-rich structure (an elaiosome). Ants pick a seed up, nibble off the elaiosome, and then store the seed for later consumption, burying it deep enough to avoid injury by fire.

Today more than 1,700 fynbos species are threatened with extinction. For a long time, conservation efforts were stymied by the inability of fynbos seeds to germinate in the absence of extensive wildfires. In 1990 two South African botanists discovered that seeds of the rare fynbos plant *Audounia capitata* germinated in response to smoke from burning wood. Even the "liquid smoke" produced by the food-flavoring industry turned out to be enough to stimulate germination. Investigators in Australia and California then found that the chemicals in smoke stimulate germination in plants in other, entirely different families that also depend on intermittent fires for maintenance.

Fynbos is found only in a small area on the western coast of the South African Cape. Yet this *type* of vegetation—tough, shrubby, and fire-adapted—is characteristic of many areas around the world that experience hot, dry summers and cool, wet winters. This weather pattern is known as a Mediterranean climate, and similar plant communities are found not only along the Mediterranean coasts of Europe and the Near East but also along the western coasts of Australia, North America, and South America.

What is it about the western edges of continents that promotes tough, shrubby plant communities such as fynbos?

See answer on p. 1146.

54.1 What Is Ecology?

Ecology is the scientific investigation of interactions among organisms and between organisms and their physical environment. Ecology is a relatively new branch of the biological sciences; in fact, it did not even have a formal name until 1866. In 1859 Charles Darwin described the focus of *On the Origin of Species* as being "the coadaptation of organic beings to each other and to their physical conditions of life." Ernst Haeckel, a German biologist who was profoundly influenced by Darwin, constructed a new word for this new enterprise: "ecology," from the Greek root *oikos*, "household," where "household" embraces all of an organism's environment. As Haeckel put it, "Ecology is the study of all those complex interrelations referred to by Darwin as 'the conditions of the struggle for existence.'"

Ecology provides explanations of the perceptible, palpable world. Although the *consequences* of enzyme–substrate interactions may be visible, the interactions themselves are not readily visible to the casual observer. In contrast, interactions among organisms, and between organisms and their environment, can often be observed. Analysis of those observations, however, often requires persistence, ingenuity, and additional investigation. That butterflies visit flowers has been observed and admired for centuries; that butterflies perform an essential service to the flowers they visit by transporting pollen is an ecological insight less than 250 years old.

The need for sound science in making decisions about our own species' interactions with the environment is one key reason for studying ecology. Humans are part of the biotic environment, and our activities have profound effects on a tremendous variety of other organisms, as well as having incalculable effects on abiotic energy flow and nutrient cycling through the physical environment. An understanding of ecology greatly improves our ability to grow food for ourselves reliably and sustainably, to manage pests and diseases safely and effectively, and to deal with natural disasters such as floods and fires. The greater our understanding of ecological interactions, the greater the likelihood that we can accomplish these things without causing a cascade of unanticipated consequences for ourselves and other organisms.

Ecology is not the same as environmentalism

In defining what ecology is, it is important to emphasize what ecology is *not*. "Ecology" is sometimes equated with "environmentalism," but the two terms are not equivalent. Ecology is a science that generates knowledge about interactions in the natural world; as a field of inquiry, it is not inherently focused on human concerns. **Environmentalism** is the use of ecological knowledge, along with economics, ethics, and many other considerations, to inform both personal decisions and public policy relating to stewardship of natural resources and ecosystems.

Ecologists study biotic and abiotic components of ecosystems

From its beginnings, ecology has encompassed both the living, or **biotic**, components and the physical and chemical, or **abiotic**, components of ecosystems. The abiotic, physical characteristics of Earth's atmosphere, for example, determine surface temperatures and precipitation patterns, which in turn limit where organisms can live. The biotic components of an organism's environment are other organisms, so ecology includes the study of interactions within species and between species. Ecology also encompasses the study of the movement of energy and nutrients through **ecosystems**—the networks of interacting organisms in an area and the physical environment they occupy.

In Haeckel's time, a main concern of ecology was understanding the distribution and abundance of organisms. That remains true today, but the field has advanced and diversified considerably. A continuous influx of new tools—mathematical models, molecular techniques, and satellite imaging, to name just a few—as well as new or enhanced connections to other fields (particularly the physical sciences) have dramatically changed ecological research. The ultimate goal of ecology, however, remains the same: to provide objective data on the interactions of the different components of the biotic and abiotic environments and, through analysis of the data, to understand these interactions and their various results.

RECAP 54.1

Ecology is the scientific investigation of interactions among organisms and between organisms and their physical environment.

- What are some reasons to consider ecology a useful scientific enterprise? **See p. 1122**
- How does ecology differ from environmentalism? **See p. 1122**

We will begin our study of ecology as the discipline began in Haeckel's time, focusing on factors that determine the distribution and abundance of organisms. First we will look at the physical forces that result in climate variation.

54.2 Why Do Climates Vary Geographically?

The terms "weather" and "climate" both refer to atmospheric conditions—temperature, humidity, precipitation, and wind direction and velocity—but they refer to different time scales. **Weather** is the short-term state of atmospheric conditions at a particular place and time, whereas **climate** refers to the average atmospheric conditions, and the extent of their variation, at a particular place over a longer time. In other words, climate is what you expect; weather is what you get. The responses of organisms to weather are usually short-term—seeking shelter from a sudden rainstorm, for example, or shivering to keep warm when the temperature drops. Responses to climate, on the other hand, tend to be evolutionary adaptations that arise within populations over time and affect physiology, morphology, and behavior. These adaptations are among the forces driving speciation. If organisms cannot adapt to the climate of a particular place, they will not be found there.

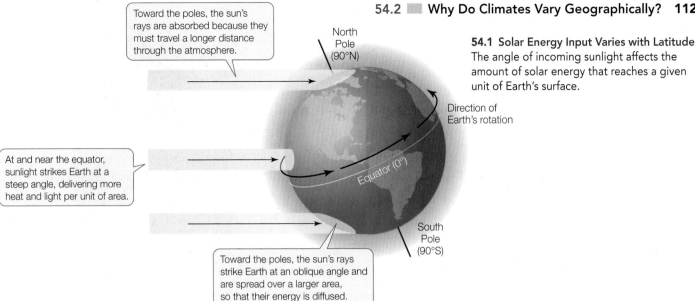

54.1 Solar Energy Input Varies with Latitude The angle of incoming sunlight affects the amount of solar energy that reaches a given unit of Earth's surface.

Solar radiation varies over Earth's surface

Solar energy input determines the air's temperature and is the major determinant of climate. The intensity of solar radiation varies over the course of a year and from place to place due to the shape of Earth, its orbit around the sun, and the tilt of its axis. The amount of solar energy reaching a given point on Earth's surface depends primarily on the angle of the sun's rays. At high latitudes (i.e., areas toward the North and South poles), sunlight strikes Earth's surface at an angle, so the incoming solar energy is distributed over a larger area (and thus is less intense) than at the equator, where sunlight strikes the surface perpendicularly (**Figure 54.1**). Moreover, when coming in at an angle, the sun's

radiation must pass through more of Earth's atmosphere, so more of its energy is absorbed or reflected before reaching the surface. Because of this difference in solar energy input, air at the poles is colder than air at the equator. The average air temperature over the course of a year decreases about 0.76°C for every degree of latitude (about 110 km) at sea level. Air temperatures also decrease with elevation, so temperatures at sea level are warmer than temperatures on mountaintops at the same latitude.

Because Earth's axis is tilted at an angle of 23.5 degrees, the amount of sunlight a particular region of Earth receives varies over the course of a year as Earth orbits the sun (**Figure 54.2**). This tilt causes seasonal variation in temperature and day length. Higher latitudes experience greater seasonal variation than lower latitudes do. Around the equator, day length and seasonal temperatures change only slightly over the course of the year, although there are seasonal shifts in precipitation patterns.

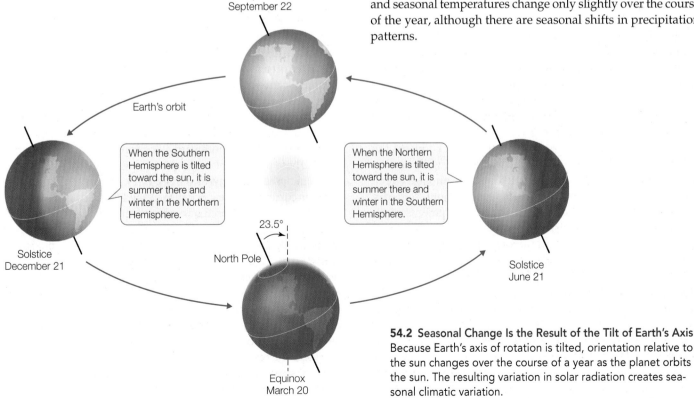

54.2 Seasonal Change Is the Result of the Tilt of Earth's Axis Because Earth's axis of rotation is tilted, orientation relative to the sun changes over the course of a year as the planet orbits the sun. The resulting variation in solar radiation creates seasonal climatic variation.

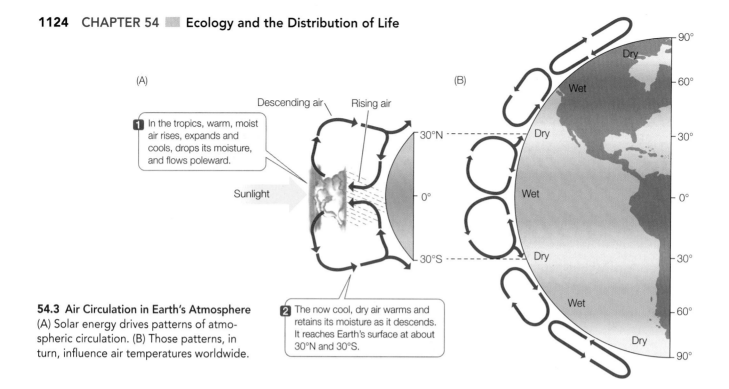

(A)

1 In the tropics, warm, moist air rises, expands and cools, drops its moisture, and flows poleward.

Sunlight

Descending air   Rising air

30°N

0°

30°S

(B)

90°
Dry
60°
Wet
30°
Dry
0°
Wet
30°
Dry
60°
Wet
Dry
90°

90°
60°
30°
0°
30°
60°
90°

54.3 Air Circulation in Earth's Atmosphere (A) Solar energy drives patterns of atmospheric circulation. (B) Those patterns, in turn, influence air temperatures worldwide.

2 The now cool, dry air warms and retains its moisture as it descends. It reaches Earth's surface at about 30°N and 30°S.

Solar energy input determines atmospheric circulation patterns

Air in the region surrounding the equator receives the greatest input of solar energy. When a parcel of air is warmed, it expands, becomes less dense, and rises. As it rises, however, it cools. Cool air cannot hold as much water vapor as warm air, so the expanding, cooling air releases moisture in the form of precipitation. Thus as the sun warms air in the tropics, that air rises into the atmosphere, cools, and releases large amounts of rainfall. As it rises, it is replaced by surface air flowing in from the north and south (**Figure 54.3A**).

High in the atmosphere, the tropical air is pushed to the north and south as newly warmed air rises to replace it. As it reaches latitudes around 30°N and 30°S, it cools and sinks. This cool, dry air, which lost its moisture as it rose over the equator, now *takes up* moisture from the ground rather than releasing it. Earth's great deserts—including the Sahara of Africa, the Gobi of China, and the deserts of Australia and the American Southwest—are located at these latitudes.

While some of the descending air flows back toward the equator, some of it flows toward the poles, setting up further cyclic movements of air. At about 60° latitude, air rises again. At the poles, where there is little solar energy input, cold, dry air descends (**Figure 54.3B**). These cyclic movements of air masses are largely responsible for determining air temperatures and precipitation patterns across Earth's surface.

Atmospheric circulation and Earth's rotation result in prevailing winds

The velocity of Earth's rotation around its axis is fastest at the equator, where Earth's diameter is greatest, and slowest close to the poles. An air mass that is not moving either to the north or the south has the same rotational velocity as Earth does at the same latitude. However, as an air mass moves toward the

equator (driven by the circulation patterns described above and in Figure 54.3), its rotational movement becomes slower than that of the planet beneath it and it is deflected to the west. Conversely, the rotational movement of an air mass moving toward either pole is faster than that of the surface beneath it and is deflected to the east. This interaction of Earth's rotation and north–south air mass movement sets up a pattern of circulating surface air referred to as the **prevailing winds** (Figure 54.4). Prevailing winds blow from east to west in the tropics (the **trade winds**); from west to east in mid-latitudes (the **westerlies**); and from east to west again above 60°N or 60°S latitude (the **easterlies**).

Prevailing winds drive ocean currents

Wind moves the water it blows over by means of frictional drag. Thus global air circulation patterns drive the circulation patterns of surface ocean waters, known as **currents**. The trade winds, for example, cause currents to converge at the equator and move westward until they encounter a continental land mass. At that point, the strong Equatorial Countercurrent brings some of the water back eastward. The remaining water divides, some moving northward and some southward along continental shores (**Figure 54.5**). These patterns of water movement set up rotating circulation patterns called **gyres** (Greek *gyros*, "spiral"). These great circular currents rotate clockwise in the Northern Hemisphere and counterclockwise in the Southern Hemisphere.

Because the ocean currents transport heat, they have a tremendous effect on Earth's climates. The poleward movement of warm water from the tropics transfers large amounts of heat to high latitudes. The Gulf Stream, for example, carries warm water from the tropical Atlantic Ocean (including the Gulf of Mexico) north across the Atlantic to northern Europe, making the European climate considerably milder than that of corresponding latitudes in North America. Similarly, currents flowing toward the equator from high latitudes bring cool, wet winters to some western coastal regions that are otherwise warm and dry.

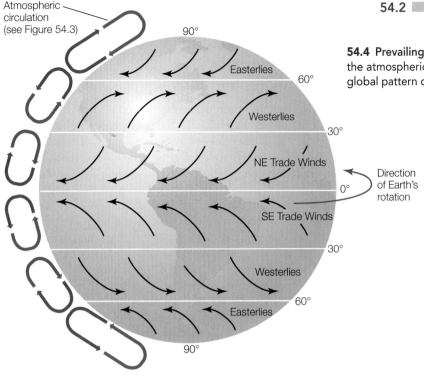

54.4 Prevailing Winds The speed of Earth's rotation combines with the atmospheric circulation of air masses (see Figure 54.3) to create a global pattern of prevailing surface winds across the planet.

Metabolic specializations help many organisms cope with an environment that periodically becomes too hot, too cold, too dry, or short of food. Organisms that must deal with extremely low temperatures, for example, are capable of surviving by a variety of strategies, chief among which is producing antifreezes. Microbes, sponges, Arctic fishes, and many temperate-zone insects are among the organisms that produce antifreezes to lower the freezing point of their cell contents or body fluids (see Section 40.3). One of only two flowering plant species native to Antarctica, the freeze-tolerant hairgrass (*Deschampsia antarctica*), survives by producing proteins that inhibit the formation of damaging large ice crystals. In Alaska the wood frog (*Rana sylvatica*), the only amphibian found north of the Arctic Circle, can survive temperatures of –6°C for more than a month with up to 65 percent of its body fluid frozen solid (**Figure 54.6**). The frog avoids damage to its cells by allowing fluids to freeze in extracellular spaces.

Adaptation to climatic conditions is often reflected in differences in morphology. For example, some endotherms living in cold climates have proportionally rounder shapes and shorter appendages than their relatives adapted to warmer climates, which gives them a smaller surface area relative to their volume and allows them to conserve heat more easily (see Figure 40.18).

Organisms adapt to climatic challenges

The patterns we have just described give rise to a mosaic of climatic conditions across Earth. The climatic conditions in a region—especially temperature and precipitation—act as selective agents on the organisms that live there. As a consequence, many organisms display adaptations to climatic conditions, which can involve physiological, morphological, or behavioral specializations.

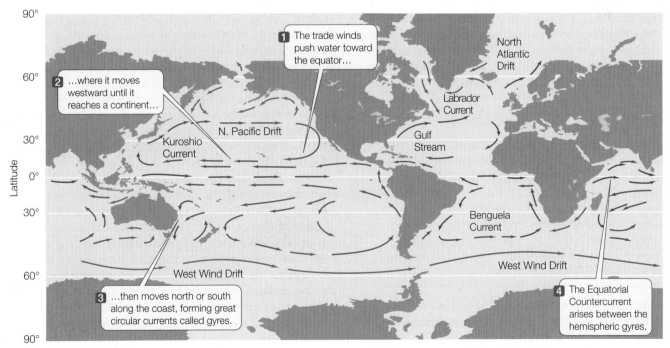

54.5 Oceanic Circulation The surface currents of the ocean are driven primarily by the prevailing winds shown in Figure 54.4.

Rana sylvatica

54.6 An Evolutionary Adaptation to Climate Wood frogs survive frigid Alaskan winters by allowing up to 65 percent of their body fluids to freeze.

Behavioral mechanisms for temperature regulation often complement physiological and morphological adaptations, particularly among ectotherms. An important behavioral adaptation to changing climatic conditions is changing one's location to find a more suitable microclimate. A **microclimate** is a subset of climatic conditions in a small specific area that generally differ from those in the environment at large. For example, some desert lizards maintain their body temperature by spending time in an underground burrow at night (see Figure 40.11). While surface temperatures may fluctuate wildly, the microclimate of the burrow is buffered against such changes.

In addition to such localized movements, long-distance movements can be key adaptations to climatic challenges. Many organisms seek new places to live when local conditions deteriorate. If repeated seasonal changes alter an environment in predictable ways, organisms may evolve life cycles that appear to anticipate those changes. Migration, one response to such cyclic environmental changes, was discussed in Section 53.5.

RECAP 54.2

Latitudinal differences in solar energy input create patterns of atmospheric circulation, which in turn drive oceanic circulation. These air and oceanic circulation patterns determine Earth's climates.

- How does latitudinal variation in solar energy input drive global air circulation patterns? **See p. 1124 and Figure 54.3**
- How do global air circulation patterns drive ocean currents? **See pp. 1124–1125 and Figure 54.5**

The tremendous variation in Earth's climates has given rise to many different environments, all home to assemblages of organisms that are adapted to the local abiotic conditions. Ecologists have found it useful to classify and name these environments based on their ecological similarities.

54.3 How Is Life Distributed in Terrestrial Environments?

A **biome** is an environment that is shaped by its climatic and geographic attributes and characterized by ecologically similar organisms. Ecologists classify terrestrial biomes principally by their dominant plants. By providing three-dimensional structure and by modifying physical conditions near the ground, the dominant plants of a terrestrial environment strongly influence the existence of the other organisms living there.

The distribution of biomes is determined largely by annual patterns of temperature and precipitation. The same biome may be found in several widely separated places, depending in large part on the presence of suitable climatic conditions (**Figure 54.7**). Different assemblages of species may be found in geographically separate regions, but due to convergent evolution in response to similar selective forces (see Section 22.1), organisms in the same biome are likely to share many physiological, morphological, and behavioral adaptations. The shrubby stature and tough leaves of fynbos vegetation described at the opening of this chapter, for example, are features shared by the dominant plants of other regions that have a Mediterranean climate.

In some biomes, such as temperate deciduous forest, precipitation is relatively constant throughout the year, but temperature varies strikingly between summer and winter. In other biomes, both temperature and precipitation change seasonally. In the tropics, seasonal temperature fluctuations are small and annual cycles are dominated by wet and dry seasons. Tropical biome types are determined primarily by the length of the dry season.

Other abiotic factors—particularly soil characteristics and wildfires—also influence the structure and life cycles of the dominant vegetation in an area and, consequently, the ecological attributes of the other organisms living there. For example, Australian desert soils are extremely nutrient-poor, and plants there have difficulty growing new foliage. Such plants often protect their leaves against consumers by producing large quantities of chemicals that reduce the leaves' digestibility. Since these leaves are not eaten, they senesce and drop to the ground, providing an abundance of highly flammable litter that feeds intense periodic fires that sweep across the landscape. As a result, succulent plants—which are easily killed by fire—are not found in Australia, although they are common in deserts on other continents.

Sometimes biomes occur in close geographic proximity to one another but differ because certain geological features alter local temperature and precipitation patterns. Major topographic features such as mountains or large lakes have regional effects on temperature and precipitation. When prevailing winds bring air masses into contact with a mountain range, for example, the air must rise to pass over the mountains, expanding and cooling as it does so. Thus clouds frequently form on the windward side of a mountain range (the side facing into the winds) and release moisture there as rain or snow. On the leeward side (that is, opposite from the direction of the winds), the now-dry air descends, warms, and once again picks up moisture. This pattern often results in a dry area called a

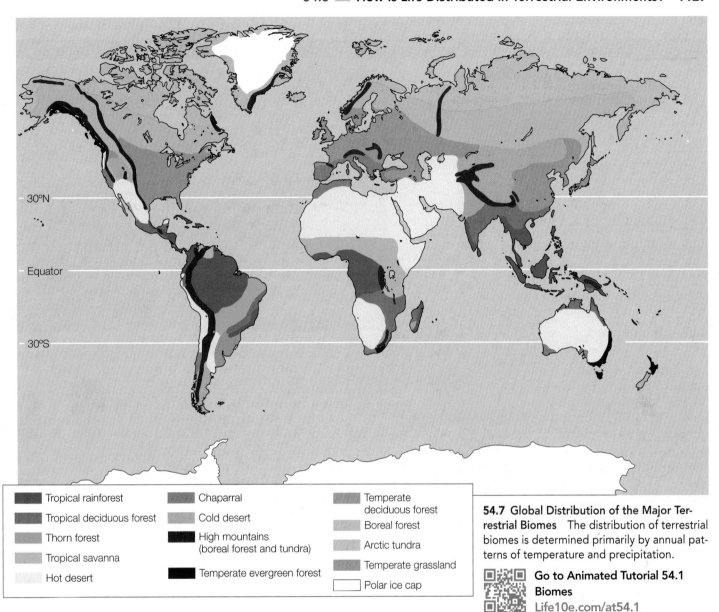

Tropical rainforest
Tropical deciduous forest
Thorn forest
Tropical savanna
Hot desert

Chaparral
Cold desert
High mountains (boreal forest and tundra)
Temperate evergreen forest

Temperate deciduous forest
Boreal forest
Arctic tundra
Temperate grassland
Polar ice cap

54.7 Global Distribution of the Major Terrestrial Biomes The distribution of terrestrial biomes is determined primarily by annual patterns of temperature and precipitation.

Go to Animated Tutorial 54.1
Biomes
Life10e.com/at54.1

rain shadow on the leeward side of the mountain range (**Figure 54.8**). The Atacama Desert, on the leeward side of the Chilean Coast Range, is one such area.

The descriptions of biomes presented here are very general and fall far short of encompassing the variation that can be found in each biome. Moreover, the boundaries ecologists draw between biomes tend to be arbitrary. Although sometimes an abrupt change is apparent in a landscape, more often one biome gradually merges into another. Despite these uncertainties, recognizing the major biomes of the world is useful because these environments share certain ecological attributes irrespective of their locations.

54.8 Rain Shadow Mean annual precipitation tends to be lower on the leeward side of a mountain range than on the windward side.

Go to Animated Tutorial 54.2
Rain Shadow
Life10e.com/at54.2

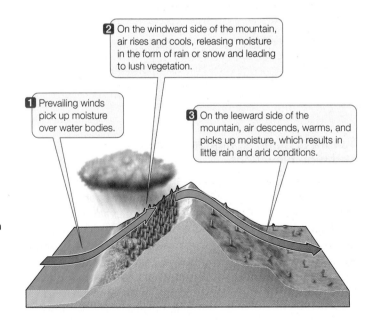

2 On the windward side of the mountain, air rises and cools, releasing moisture in the form of rain or snow and leading to lush vegetation.

1 Prevailing winds pick up moisture over water bodies.

3 On the leeward side of the mountain, air descends, warms, and picks up moisture, which results in little rain and arid conditions.

TUNDRA

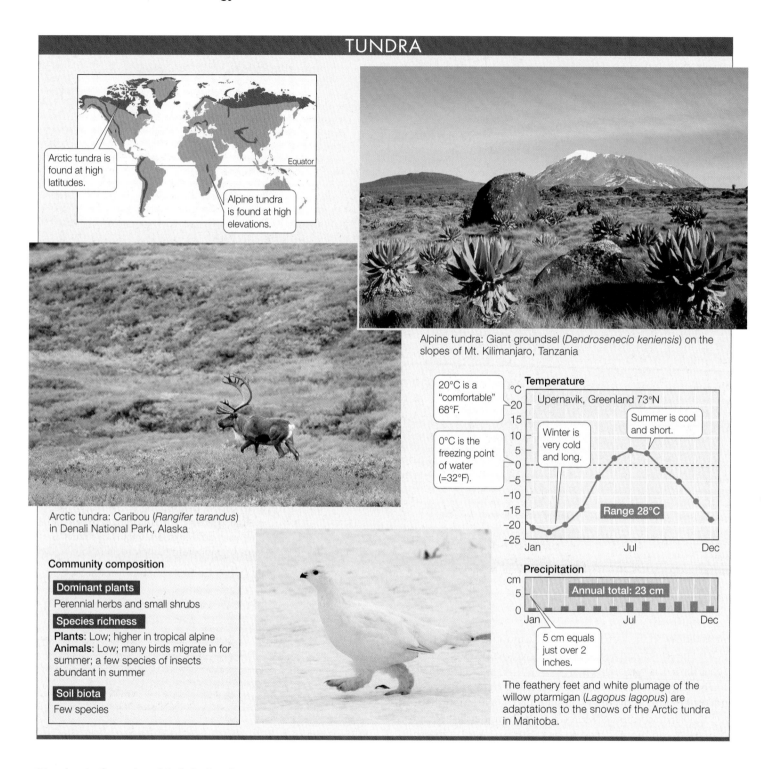

Arctic tundra is found at high latitudes.

Alpine tundra is found at high elevations.

Equator

Alpine tundra: Giant groundsel (*Dendrosenecio keniensis*) on the slopes of Mt. Kilimanjaro, Tanzania

Arctic tundra: Caribou (*Rangifer tarandus*) in Denali National Park, Alaska

20°C is a "comfortable" 68°F.

0°C is the freezing point of water (=32°F).

Temperature

°C

Upernavik, Greenland 73°N

Winter is very cold and long.

Summer is cool and short.

Range 28°C

Jan   Jul   Dec

Precipitation

cm

Annual total: 23 cm

Jan   Jul   Dec

5 cm equals just over 2 inches.

The feathery feet and white plumage of the willow ptarmigan (*Lagopus lagopus*) are adaptations to the snows of the Arctic tundra in Manitoba.

Community composition

Dominant plants

Perennial herbs and small shrubs

Species richness

Plants: Low; higher in tropical alpine
Animals: Low; many birds migrate in for summer; a few species of insects abundant in summer

Soil biota

Few species

Tundra is found at high latitudes and high elevations

The tundra biome is characterized by low temperatures and a short growing season. These conditions prevail not only at the high latitudes of the Arctic but also at high elevations in mountains at all latitudes. In Arctic tundra, the vegetation consists of low-growing perennial plants and is underlain by permafrost—soil permeated with permanently frozen water. The top few centimeters of the soil thaw during the short summers, when the sun may be above the horizon 24 hours a day. Thus even though there is little precipitation near the poles, the soil

in lowland Arctic tundra is wet because water cannot drain through the permafrost.

Tundra found at high elevations outside polar regions is called alpine tundra. Tropical alpine tundra is not underlain by permafrost, so photosynthesis and most other biological activities continue (albeit slowly) throughout the year. A variety of plant growth forms are present in tropical alpine tundra, including low-growing shrubs, perennials, and grasses.

Many tundra plants have hairy leaves that trap heat. The flowers of some species move over the course of the day, tracking the sun's warmth. Most animals are either summer

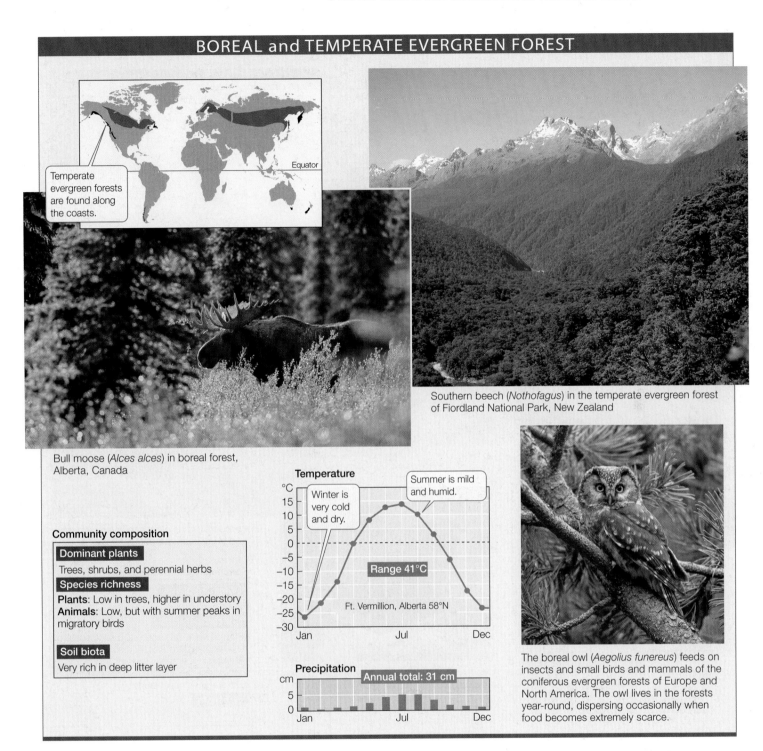

BOREAL and TEMPERATE EVERGREEN FOREST

Temperate evergreen forests are found along the coasts.

Equator

Southern beech (*Nothofagus*) in the temperate evergreen forest of Fiordland National Park, New Zealand

Bull moose (*Alces alces*) in boreal forest, Alberta, Canada

Community composition

Dominant plants
Trees, shrubs, and perennial herbs

Species richness
Plants: Low in trees, higher in understory
Animals: Low, but with summer peaks in migratory birds

Soil biota
Very rich in deep litter layer

Temperature

Winter is very cold and dry.

Summer is mild and humid.

Range 41°C

Ft. Vermillion, Alberta 58°N

Precipitation

Annual total: 31 cm

The boreal owl (*Aegolius funereus*) feeds on insects and small birds and mammals of the coniferous evergreen forests of Europe and North America. The owl lives in the forests year-round, dispersing occasionally when food becomes extremely scarce.

migrants or are dormant for much of the year. Resident birds and mammals, such as the willow ptarmigan (*Lagopus lagopus*) and Arctic fox (*Vulpes lagopus*), have thick fur or feathers that may change color with the seasons, from brown in summer to white in winter.

Evergreen trees dominate boreal and temperate evergreen forests

The boreal forest biome (also known as taiga) occurs at latitudes below Arctic tundra and at elevations below alpine tundra on temperate-zone mountains. Winters in the boreal forest are long

and very cold; summers are short, although often relatively warm. The boreal forests of the Northern Hemisphere are dominated by coniferous gymnosperm species such as spruces and firs. The evergreen leaves of conifers are needlelike rather than flat; their reduced surface area cuts down on evaporative water loss. The short summers favor evergreen leaves, which are ready to photosynthesize as soon as temperatures warm. In winter, downward-drooping limbs allow the trees to shed snow easily.

The dominant mammals of the boreal forest, such as moose and hares, eat leaves, but the seeds in conifer cones support a variety of rodents, birds, and insects. Many small mammals

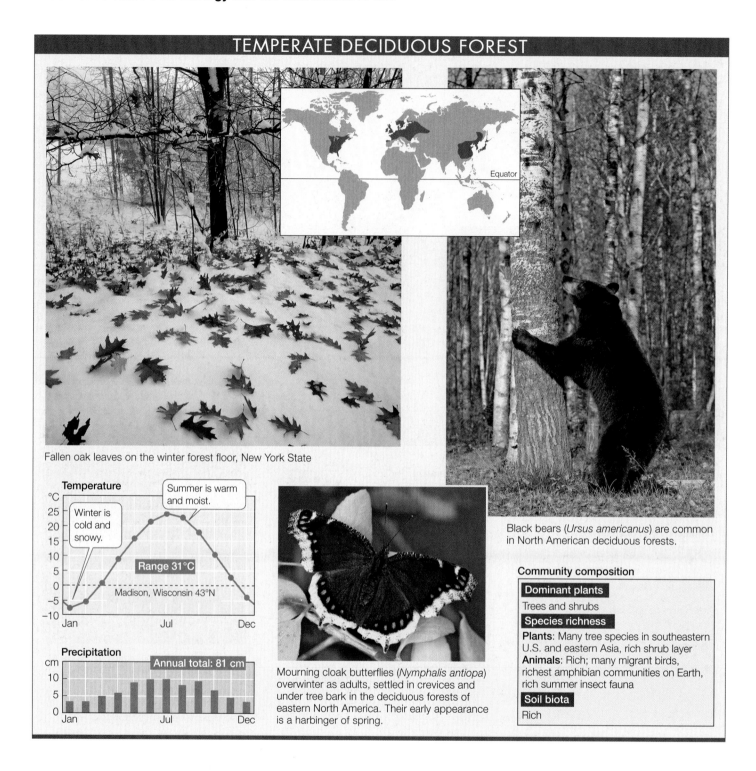

TEMPERATE DECIDUOUS FOREST

Fallen oak leaves on the winter forest floor, New York State

Temperature
°C
Winter is cold and snowy.
Summer is warm and moist.
Range 31°C
Madison, Wisconsin 43°N

Precipitation
cm
Annual total: 81 cm

Mourning cloak butterflies (*Nymphalis antiopa*) overwinter as adults, settled in crevices and under tree bark in the deciduous forests of eastern North America. Their early appearance is a harbinger of spring.

Black bears (*Ursus americanus*) are common in North American deciduous forests.

Community composition

Dominant plants
Trees and shrubs

Species richness
Plants: Many tree species in southeastern U.S. and eastern Asia, rich shrub layer
Animals: Rich; many migrant birds, richest amphibian communities on Earth, rich summer insect fauna

Soil biota
Rich

hibernate in winter, but voles, lemmings, and mice remain active under the snowpack, serving as food for predators such as foxes and owls.

Temperate evergreen forests grow along the coasts of continents in both hemispheres at middle to high latitudes, where winters are mild and wet and summers are cool and dry. In the Northern Hemisphere, the dominant trees in temperate evergreen forests are conifers, some of which are the world's most massive tree species (including the giant sequoia and coast redwood). In the Southern Hemisphere, the dominant trees are southern beeches (*Nothofagus*), some of which are evergreen.

Temperate deciduous forests change with the seasons

The temperate deciduous forest biome is found in eastern North America, eastern Asia, and Europe. Temperatures in these regions fluctuate dramatically between summer and winter, although precipitation is fairly evenly distributed throughout the year. Deciduous trees, which dominate these forests, lose their leaves during the cold winters and produce new leaves that photosynthesize rapidly during the warm, moist summers.

Many more tree species live here than in boreal forests. The temperate forests richest in species are those of the southern

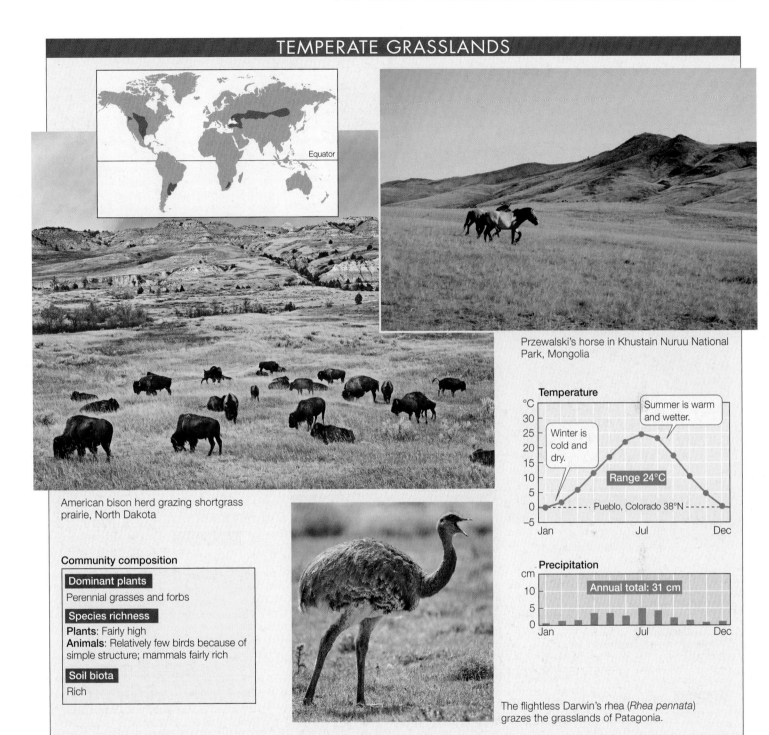

TEMPERATE GRASSLANDS

Equator

Przewalski's horse in Khustain Nuruu National Park, Mongolia

American bison herd grazing shortgrass prairie, North Dakota

Temperature

°C

Summer is warm and wetter.

Winter is cold and dry.

Range 24°C

Pueblo, Colorado 38°N

Jan   Jul   Dec

Precipitation

cm

Annual total: 31 cm

Jan   Jul   Dec

Community composition

| Dominant plants |
Perennial grasses and forbs

| Species richness |
Plants: Fairly high
Animals: Relatively few birds because of simple structure; mammals fairly rich

| Soil biota |
Rich

The flightless Darwin's rhea (*Rhea pennata*) grazes the grasslands of Patagonia.

Appalachian Mountains of the United States and those found in eastern China and Japan—areas that were not covered by glaciers during the Pleistocene. Many plant genera are shared among the three geographically separate regions where this biome is found.

Although many animals are permanent residents of deciduous forests, some (including many birds) migrate to escape the winter cold. Others that remain through the winter acquire massive fat stores in fall and hibernate (see Section 40.5), often in underground burrows. Many insects pass the winter in a state of diapause (suspended development), the onset of which is triggered by the decreasing hours of daylight—a reliable predictor of winter.

Temperate grasslands are widespread

Temperate grasslands are found in many parts of the world, all of which are relatively dry for much of the year. Most grasslands, such as the pampas of Argentina, the veldt of South Africa, and the Great Plains of North America, have hot summers and relatively cold winters. In some grasslands, most of the precipitation falls in winter (as in California grasslands); in others, the majority falls in summer (as in the Great Plains and the Russian steppe).

Grassland vegetation is structurally simple but rich in species of perennial grasses and forbs (herbaceous plants other than grasses). This abundant plant biomass supports herds

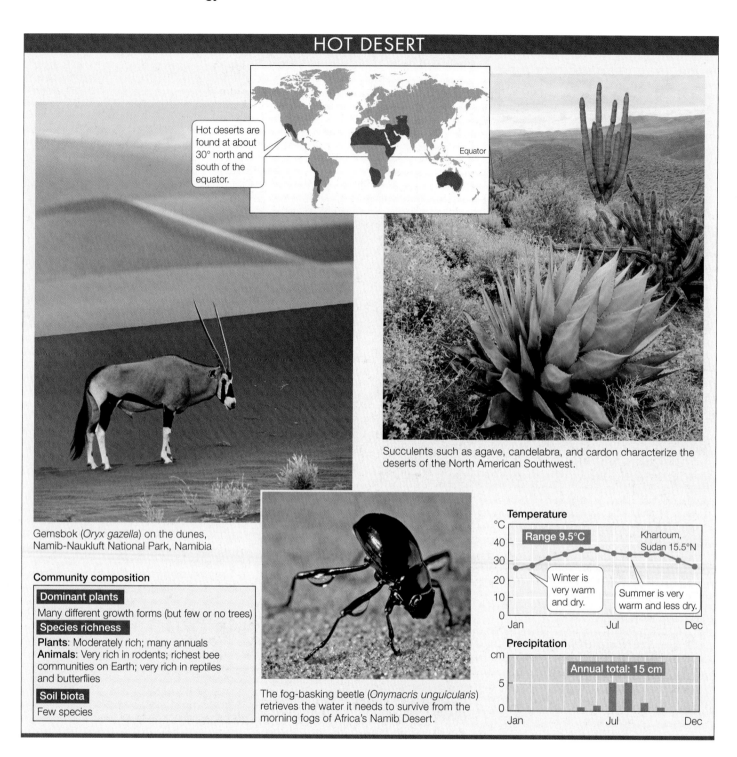

HOT DESERT

Hot deserts are found at about 30° north and south of the equator.

Equator

Succulents such as agave, candelabra, and cardon characterize the deserts of the North American Southwest.

Gemsbok (*Oryx gazella*) on the dunes, Namib-Naukluft National Park, Namibia

Community composition

Dominant plants
Many different growth forms (but few or no trees)

Species richness
Plants: Moderately rich; many annuals
Animals: Very rich in rodents; richest bee communities on Earth; very rich in reptiles and butterflies

Soil biota
Few species

The fog-basking beetle (*Onymacris unguicularis*) retrieves the water it needs to survive from the morning fogs of Africa's Namib Desert.

Temperature
°C
Range 9.5°C
Khartoum, Sudan 15.5°N
40
30
Winter is very warm and dry.
20
10
Summer is very warm and less dry.
0
Jan   Jul   Dec

Precipitation
cm
Annual total: 15 cm
5
0
Jan   Jul   Dec

of large grazing mammals. Grassland plants are adapted to grazing and to fire. They store much of their energy underground and resprout quickly after being burned or grazed. There are comparatively few trees in temperate grasslands because trees cannot survive the periodic fires. If grasslands do not burn periodically, many of the species that typify this biome will be replaced by fire-intolerant species that are superior competitors.

The topsoil of grasslands is usually rich and deep, and thus exceptionally well suited to growing crops such as corn and wheat. As a consequence, most of the world's temperate grasslands have been turned over to agriculture and no longer exist in their natural state.

Hot deserts form around 30° latitude

The hot desert biome is concentrated in two belts, centered around 30°N and 30°S latitude (where warm, dry air descends and picks up moisture; see Figure 54.3). The driest of these regions, where rains rarely penetrate, are far from the oceans, as in the center of Australia and the middle of the Sahara in Africa.

Desert plants have several structural and physiological adaptations that help them conserve water, as described in

COLD DESERT

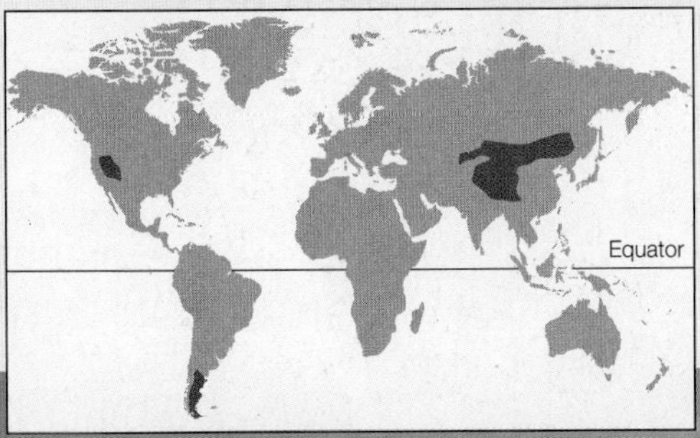

Equator

Vicuña (*Vicugna vicugna*) family herd in Andean desert, Lauca National Park, Chile

A cold desert in northern New Mexico, at about 2,000 meters in elevation, is dominated by juniper shrubs (*Juniperus* sp.).

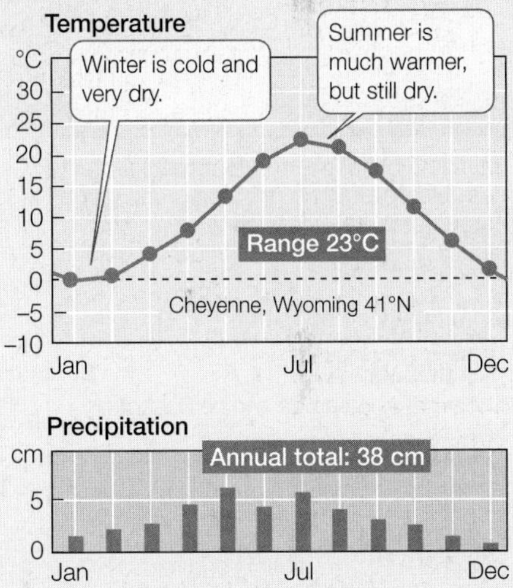

Temperature

Winter is cold and very dry.

Summer is much warmer, but still dry.

Range 23°C

Cheyenne, Wyoming 41°N

Precipitation

Annual total: 38 cm

Community composition

| Dominant plants |
Low-growing shrubs and herbaceous plants

| Species richness |
Plants: Few species
Animals: Rich in seed-eating birds, ants, and rodents; low in all other taxa

| Soil biota |
Poor in species

Living in a cold desert presents special challenges to a poikilotherm. This collared lizard (*Crotaphytus* sp.) in Bluff, Utah, is warming itself by basking on a rock.

Section 39.3. Many desert plants are xerophytes: species with adaptations for reducing water loss or storing water. The aboveground parts of many desert plants are covered with a waxy cuticle to prevent water loss; leaves may be reduced to spines to minimize surface area, as in the Cactaceae of the Western Hemisphere and the Euphorbiaceae in much of the rest of the world. Other desert plants are succulents that store water in fleshy leaves or stems. Most perennial plants go dormant during dry seasons, and then grow rapidly as soon as rains return. Their seeds tend to be heat- and drought-resistant and accumulate in a dormant state in the soil.

Small desert animals are inactive during the hottest part of the day, remaining in underground burrows. Desert mammals have physiological adaptations for conserving water, including a reduced number of sweat glands and kidneys that produce highly concentrated urine. Many desert animals require no water beyond what they can extract from the carbohydrates in their food.

Cold deserts are high and dry

The cold desert biome is found in dry regions at mid- to high latitudes, especially in the interiors of continents where mountain ranges block moisture-rich air (see Figure 54.8). Blocked by

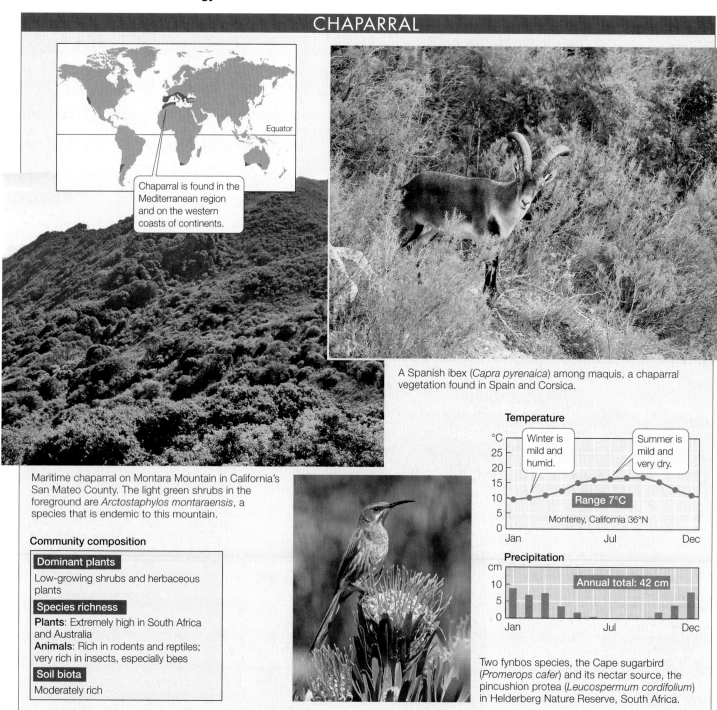

CHAPARRAL

Chaparral is found in the Mediterranean region and on the western coasts of continents.

A Spanish ibex (*Capra pyrenaica*) among maquis, a chaparral vegetation found in Spain and Corsica.

Maritime chaparral on Montara Mountain in California's San Mateo County. The light green shrubs in the foreground are *Arctostaphylos montaraensis*, a species that is endemic to this mountain.

Community composition

Dominant plants

Low-growing shrubs and herbaceous plants

Species richness

Plants: Extremely high in South Africa and Australia
Animals: Rich in rodents and reptiles; very rich in insects, especially bees

Soil biota

Moderately rich

Temperature

Winter is mild and humid.

Summer is mild and very dry.

Range 7°C

Monterey, California 36°N

Precipitation

Annual total: 42 cm

Two fynbos species, the Cape sugarbird (*Promerops cafer*) and its nectar source, the pincushion protea (*Leucospermum cordifolium*) in Helderberg Nature Reserve, South Africa.

two mountain ranges (the Andes and the Chilean Coast range), the Atacama Desert is the driest place on Earth; average yearly rainfall is less than 1 millimeter.

Cold deserts are dominated by a few species of low-growing shrubs. The surface layers of the soil are recharged with moisture in winter, and plant growth is concentrated in spring. Cold deserts are relatively species-poor, but the plants that do grow there tend to produce large numbers of seeds, supporting many species of seed-eating birds, ants, and rodents. Burrowing behavior is widespread among cold desert dwellers but—in contrast to hot desert animals—they burrow to escape cold temperatures, not excessive heat.

Chaparral has hot, dry summers and wet, cool winters

The chaparral biome is found on the western coasts of continents at mid-latitudes (around 40°). Winters in this biome are cool and wet; summers are warm and dry. Such climates are found in the Mediterranean region of Europe (for which the Mediterranean climate is named), coastal California, central Chile, extreme southern Africa, and southwestern Australia. The fynbos of the Cape region of South Africa, described at the opening of this chapter, is part of the chaparral biome.

The dominant plants of chaparral vegetation are low-growing shrubs and trees with tough evergreen leaves that conserve water.

THORN FOREST and TROPICAL SAVANNA

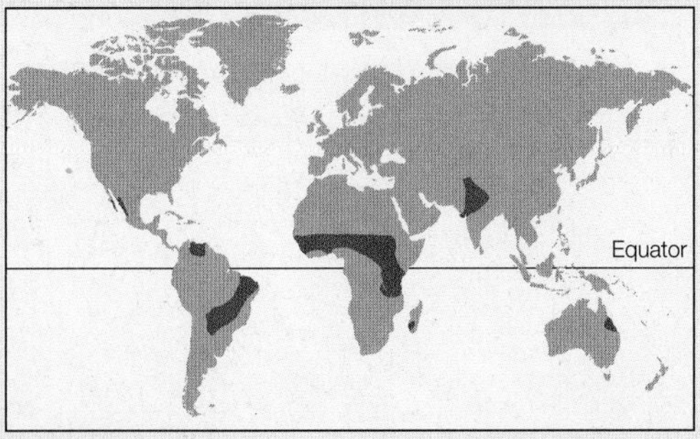
Equator

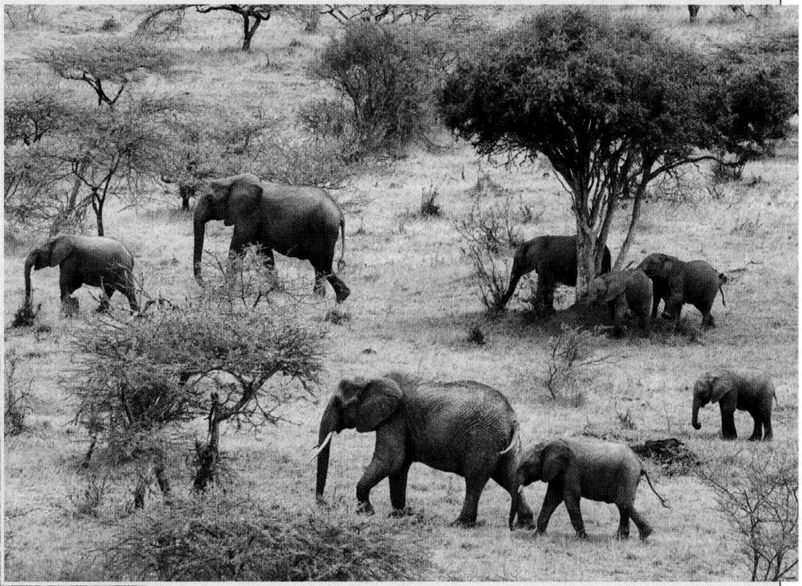

Family groups of African elephants (*Loxodonta africana*) in Kenya's Ol Malo Wildlife Sanctuary converge into a large herd for migration.

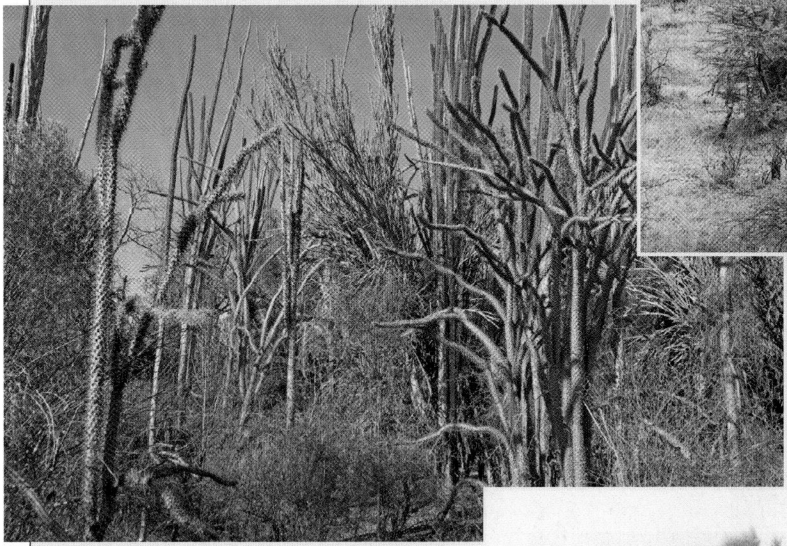

Madagascar ocotillo (*Alluadia procera*) dominate this thorn forest.

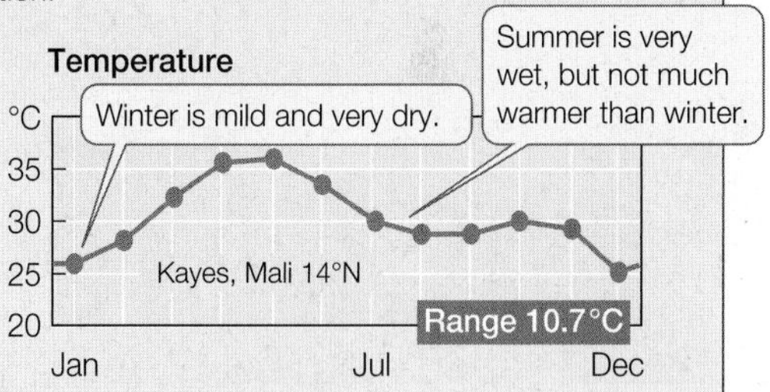
Temperature
°C
Winter is mild and very dry.
Summer is very wet, but not much warmer than winter.
35
30
25
20
Kayes, Mali 14°N
Range 10.7°C
Jan
Jul
Dec

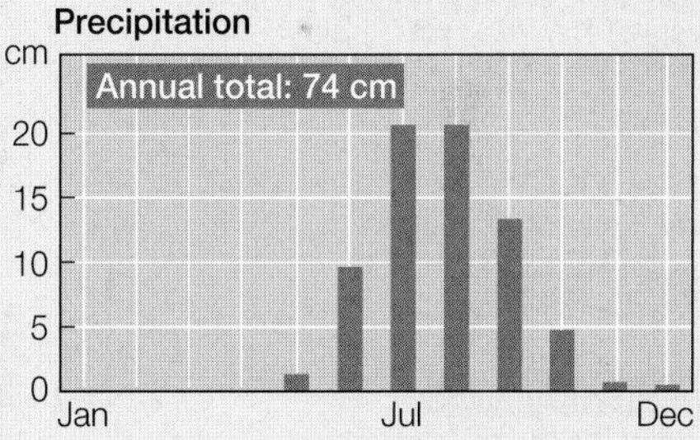
Precipitation
cm
Annual total: 74 cm
20
15
10
5
0
Jan
Jul
Dec

Community composition

Dominant plants
Shrubs and small trees; grasses

Species richness
Plants: Moderate in thorn forest; low in savanna
Animals: Rich mammal faunas; moderately rich in birds, reptiles, and insects

Soil biota
Rich

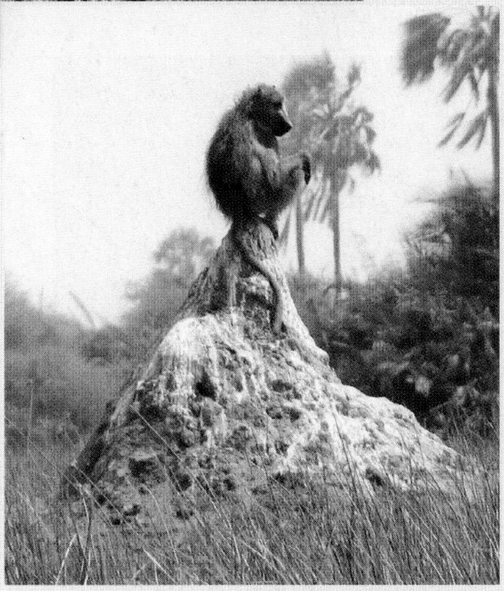

Termite colonies build huge mounds on the African savanna, providing a food source for mammals such as this baboon (*Papio* sp.).

The shrubs carry out most of their growth and photosynthesis in early spring, when insects are active and birds breed. Many chaparral species produce strong-smelling defensive chemicals to reduce losses of their hard-to-replace foliage to herbivores. Annual plants are abundant and produce large quantities of seeds that fall onto the soil, supporting many small rodents, most of which store seeds in underground burrows. Burrowing to avoid midday heat and nocturnal foraging are strategies used by many chaparral animals. Chaparral vegetation is adapted to periodic fires; the seeds of some species do not germinate until after they have survived a fire. Many shrubs of Northern Hemisphere chaparral produce bird-dispersed fruits that ripen in late fall, when large numbers of migrant birds arrive from the north. In the fynbos of South Africa, seeds equipped with elaiosomes are transported by ants, which bury them deep enough to survive the periodic fires.

Thorn forests and tropical savannas have similar climates

The thorn forest and tropical savanna biomes are found primarily at latitudes below the hot deserts of Africa, South America, and Australia. Little or no rain falls in these biomes in winter, but rainfall may be heavy during summer. Thorn forests contain many plants similar to those found in hot deserts,

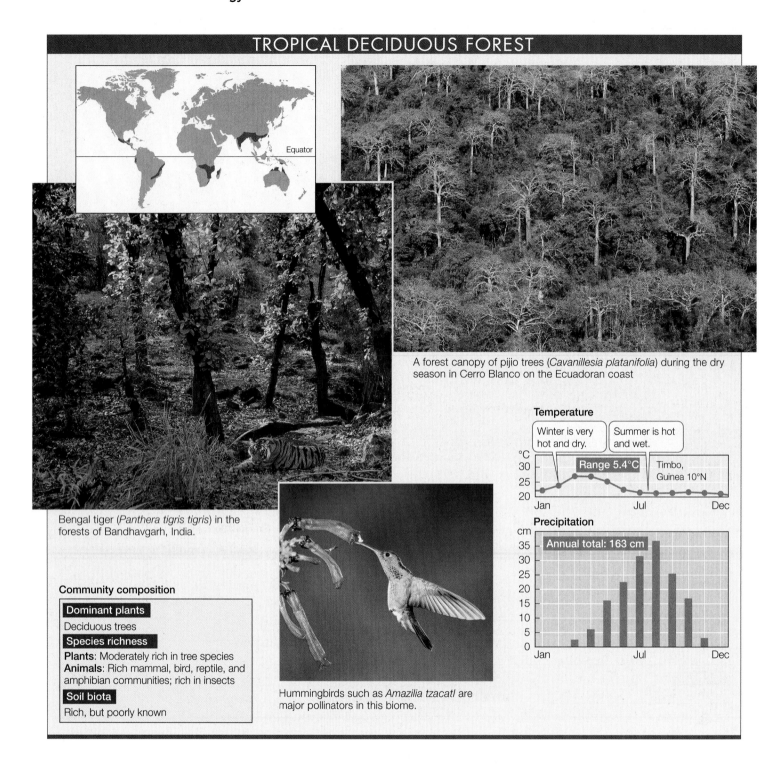

TROPICAL DECIDUOUS FOREST

Equator

A forest canopy of pijio trees (*Cavanillesia platanifolia*) during the dry season in Cerro Blanco on the Ecuadoran coast

Bengal tiger (*Panthera tigris tigris*) in the forests of Bandhavgarh, India.

Temperature

Winter is very hot and dry.

Summer is hot and wet.

°C
30
25
20

Range 5.4°C

Timbo, Guinea 10°N

Jan   Jul   Dec

Precipitation

cm
35
30
25
20
15
10
5
0

Annual total: 163 cm

Jan   Jul   Dec

Community composition

Dominant plants

Deciduous trees

Species richness

Plants: Moderately rich in tree species
Animals: Rich mammal, bird, reptile, and amphibian communities; rich in insects

Soil biota

Rich, but poorly known

Hummingbirds such as *Amazilia tzacatl* are major pollinators in this biome.

including succulents. The dominant plants are spiny shrubs and small trees, many of which drop their leaves during the long, dry winter. Trees of the genus *Acacia* are common in thorn forests and savannas worldwide. In Africa, *Andansonia* (baobab) trees are also a hallmark of these biomes.

Savanna is characterized by expanses of grasses and grass-like plants surrounding scattered individual trees. The largest tropical savannas are found in central and eastern Africa, where they are populated by herds of grazing and browsing mammals and the large carnivores that prey on them. The migration of vast herds of herbivores in search of "greener

pastures" during the dry season is another characteristic of this impressive region. Grazers and browsers maintain the savannas by disproportionately damaging shrubs and trees, which cannot withstand as much tissue loss as can the grasses. If savanna vegetation is not grazed, browsed, or burned, it typically reverts to dense thorn forest.

Tropical deciduous forests occur in hot lowlands

As the temperature and the length of the rainy season increase toward the equator, the tropical deciduous forest biome replaces thorn forest. Tropical deciduous forests have taller trees

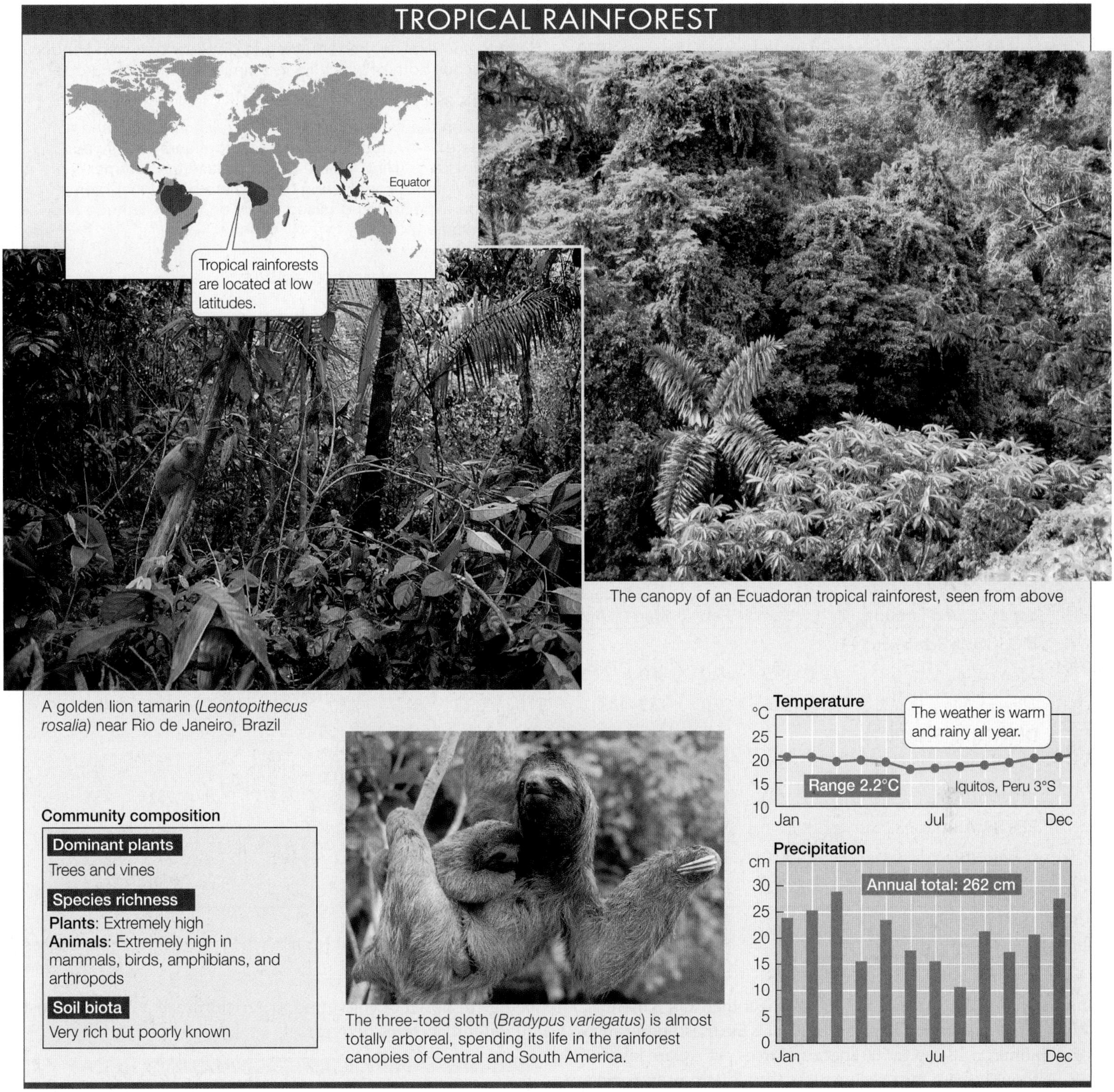

TROPICAL RAINFOREST

Tropical rainforests are located at low latitudes.

Equator

The canopy of an Ecuadoran tropical rainforest, seen from above

A golden lion tamarin (*Leontopithecus rosalia*) near Rio de Janeiro, Brazil

Community composition

Dominant plants

Trees and vines

Species richness

Plants: Extremely high
Animals: Extremely high in mammals, birds, amphibians, and arthropods

Soil biota

Very rich but poorly known

The three-toed sloth (*Bradypus variegatus*) is almost totally arboreal, spending its life in the rainforest canopies of Central and South America.

Temperature

The weather is warm and rainy all year.

Range 2.2°C   Iquitos, Peru 3°S

Precipitation

Annual total: 262 cm

and fewer succulent plants than thorn forests or savannas, and they support a much greater number of plant and animal species. Most of the trees, except for those growing along rivers, lose their leaves during the long, hot dry season. Growth increases in the rainy season; many plants flower while they are still leafless.

Most plant species in this biome are pollinated by animals. In the Sierra Madre Occidental, a mountain range in the extreme southwestern United States extending into western Mexico, tropical deciduous forests are part of a "nectar corridor," a series of patches of flowering plants that are used

as refueling stops by long-distance migrants traveling north from overwintering sites to their breeding sites in the Rocky Mountains.

The soils of this biome are among the best in the tropics for agriculture because they contain more nutrients than the soils of wetter areas. As a result, most tropical deciduous forests worldwide have been cleared for agriculture and grazing.

Tropical rainforests are rich in species

The tropical rainforest biome, or simply the rainforest, is found in equatorial regions where total rainfall exceeds 250 centimeters

Walter Climate Diagrams

Original Source

Devised by the German biogeographer Heinrich Walter in 1979, this graphic technique plots temperature and precipitation data in a simple way that visualizes a "growing season"—those months when average temperatures are above 0°C and the average precipitation trace falls above the temperature trace.

Analyze the Data

Walter climate diagrams are predicated on the "rule of thumb" that plant growth requires temperatures above 0°C and at least 20 mm of precipitation for each 10°C above 0°. They have two y-axis scales, one for temperature and one for precipitation; these axes align 0 mm of precipitation with a temperature of 0°C. The x axis shows the 12 months, with the summer solstice placed in the center of the x axis.

The Walter diagram shown at the right is for London, England. Average yearly temperature and precipitation data for three other cities are given in the table. Using these data, create Walter diagrams for each city. Use your diagrams along with the information in the preceding sections of this chapter to answer the questions.

QUESTION 1

Based solely on your diagrams, which biome do you think is represented by each location? What physical attribute other than temperature and precipitation might significantly affect the biomes of these locations?

QUESTION 2

How do you explain the temperature disparity between London and Moscow, both of which lie in a similar latitude of Europe?

QUESTION 3

Perth lies on the western coast of Australia, in the Southern Hemisphere. How does this affect the configuration of your Walter diagram? (Hint: Where did you place the summer solstice?) Without considering this climate data, what biome would you expect to find based solely on Perth's latitude and coastal location?

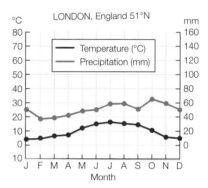

Location and latitude	Jan	Feb	Mar	Apr	May	June	July	Aug	Sept	Oct	Nov	Dec
MOSCOW, Russia 56°N												
Temp. (°C)	−10.5	−9.0	−4.0	4.5	12.0	16.5	18.5	16.5	11.0	4.0	−2.0	−7.5
Precip. (mm)	34.5	29.0	32.5	38.0	51.0	65.5	81.5	72.0	58.0	50.5	44.0	42.5
DENVER, U.S.A. 38°N												
Temp. (°C)	−1.0	0.5	4.0	9.0	14.0	19.5	23.0	22.0	17.0	11.0	4.0	−0.5
Precip. (mm)	14.0	15.5	33.5	44.5	62.5	43.0	47.0	37.5	28.5	26.0	23.0	15.0
PERTH, Australia 32°S												
Temp. (°C)	30.0	30.0	28.0	24.5	21.0	18.5	17.5	18.0	19.5	21.5	24.5	27.5
Precip. (mm)	8.5	12.5	19.0	45.0	121.5	182.0	174.0	135.5	80.0	53.5	21.0	13.5

Go to BioPortal for all WORKING WITH**DATA** exercises

---

annually and the dry season lasts no longer than 2 or 3 months. With no seasons unsuitable for growth, it is the most species-rich of all biomes, with up to 500 species of trees per square kilometer. Although these forests cover less than 2 percent of Earth's surface, they are home to more than half of all known species.

Along with the immense number of species they support, rainforests have the highest overall productivity of all terrestrial ecological communities. However, most mineral nutrients are tied up in the vegetation. The soils usually cannot support agriculture without massive applications of fertilizers. These forests are home to many epiphytes—plants that grow on other plants, deriving their nutrients and moisture from air and water rather than soil.

The rainforests provide humans with a dazzling range of products, including fruits, nuts, medicines, fuel, pulp, and furniture wood. Many more useful species undoubtedly await discovery, as only a small proportion of this biome's species have been inventoried. The rainforests, however, are currently

being cut down or converted to agriculture at a rate of almost 20 million hectares per year.

RECAP 54.3

Ecologists recognize several terrestrial environment types called biomes. The geographic distribution of biomes is determined primarily by temperature and precipitation, but is also influenced by soil characteristics and fire.

- How do temperate grasslands differ from tropical savannas? In what ways are they similar? **See pp. 1131 and 1135**

- What primary factor distinguishes a tropical biome? **See pp. 1135–1137**

About 70 percent of Earth's surface is covered by saltwater oceans and seas that support abundant life. The small percentage of the aquatic world that consists of fresh water also hosts a significant proportion of Earth's aquatic organisms.

54.4 How Is Life Distributed in Aquatic Environments?

Aquatic biomes do not depend on plants for their structure in the way that terrestrial biomes do. Salinity is the primary factor that distinguishes the aquatic biomes. The marine biome is characterized by salt water, freshwater biomes by low salinity, and estuaries by the mixing of fresh water and salt water.

The marine biome can be divided into several life zones

Earth's oceans form one large, interconnected water mass on which the atmospheric factors that distinguish terrestrial biomes have little influence. However, light penetration, water temperature, water pressure, water movement (i.e., waves and tides), and salinity all vary spatially, so most marine organisms have restricted ranges and display adaptations to particular physical conditions.

The physical and biological discontinuities within the marine biome divide it into several distinct **life zones** (Figure 54.9). These zones are identified principally by their distance from shore and from the water's surface. The depth of an ocean basin varies from the shoreline to the relatively shallow continental shelf and to the deepest part of the ocean, sometimes known as the abyssal plain.

Water depth affects how much light is available to sustain the photosynthetic organisms that form the base of the marine food chain. In both marine and freshwater environments, the layer of water reached by enough sunlight to support photosynthesis is called the **photic zone**. Approximately 90 percent of all aquatic life is found in the photic zone.

The **coastal zone** extends from the shoreline to the edge of the continental shelf; it is characterized by relatively shallow, well-oxygenated water and relatively stable temperatures and salinities. These conditions support high densities of phytoplankton (photosynthetic floating protists), which in turn support some of the world's most important fisheries. Structure in coastal-zone communities may be provided by a variety of organisms. In warm coastal waters, corals generate complex reef structures that support ecosystems rivaling the rainforest in diversity. "Forests" of multicellular algal species (seaweeds and giant kelps) grow along many coasts at higher latitudes.

The area of the coastal zone that is affected by wave action is called the **littoral zone**. The principal autotrophs in this zone—sea grasses and algae—are consumed by a variety of invertebrates as well as small fishes. The portion of the coastal zone lying between the high and low tide levels is the **intertidal zone**, where tidal movements create conditions of highly variable temperature and salinity. Intertidal organisms, including clams, barnacles, copepods, and burrowing worms, are alternately exposed to air and submerged under water.

Throughout most of the oceans, the dominant autotrophs are phytoplankton. In the open ocean, or **pelagic zone**, the principal consumers of phytoplankton are zooplankton—mainly small crustaceans and larval stages of marine animals—which

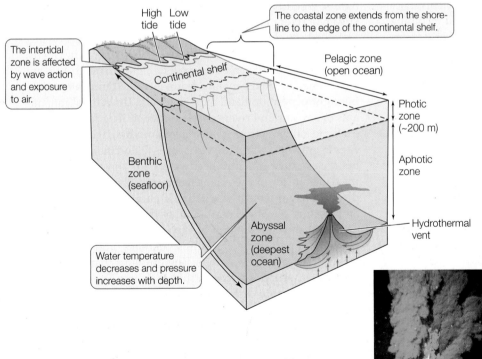

54.9 Life Zones of the Marine Biome The ocean's life zones are primarily determined by light penetration. More than 90 percent of ocean-dwelling species live in the sunlit photic zone, which comprises less than 2 percent of the volume of open water. Wave action and exposure to air affect those life zones where the ocean meets the shore.

Forests of giant kelp (*Macrocystis* sp.) dominate many coastal communities.

A large sailfish (*Istiophorus albicans*) feeds on sardines (*Sardinella aurita*) in pelagic waters of the Gulf of Mexico.

Heat and minerals from hydrothermal vents nourish unique deep-ocean communities.

54.10 Life Zones in a Freshwater Lake Like oceans, standing bodies of fresh water can be divided into life zones based on water depth and distance from shore.

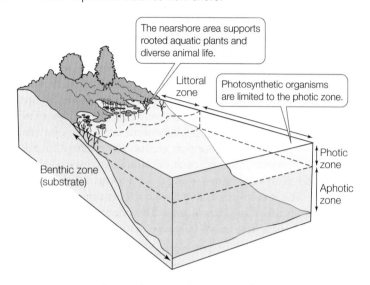

The nearshore area supports rooted aquatic plants and diverse animal life.

Littoral zone

Photosynthetic organisms are limited to the photic zone.

Benthic zone (substrate)

Photic zone

Aphotic zone

A school of bleak (*Alburnus scoranza*) among giant reeds along a lakeshore in Macedonia.

An osprey (*Pandion haliaetus*) snatches a large fish from the open waters of a Canadian lake.

in turn support many larger free-swimming vertebrate and invertebrate species.

The ocean bottom—the sediment surface—is referred to as the **benthic zone**. Many benthic organisms are adapted to life on the seafloor substrate. They include sessile animals such as sponges, bryozoans, ribbon worms, and brachiopods as well as motile bottom feeders such as crabs and sea slugs.

Where the water is too deep for light to penetrate, little photosynthesis can take place, and both plant and animal diversity are low. Depths reached by less than 1 percent of incoming sunlight constitute the **aphotic zone**. Many of the organisms inhabiting these regions subsist on decaying organic matter that sinks down from the photic zone. Some produce their own light by means of bioluminescent organs (see Figure 33.14D). Even deep-ocean trenches and rift valleys support hydrothermal vent ecosystems sustained by chemoautotrophic prokaryotes that can metabolize the nutrients in seawater without the aid of sunlight (see Section 26.3).

Freshwater biomes may be rich in species

In contrast to the vast oceans, bodies of fresh water cover less than 3 percent of Earth's surface, but they are home to about 10 percent of all aquatic species. Freshwater biomes, as the name implies, are characterized by low levels of salinity—generally below 1 percent.

Freshwater biomes are distinguished by the degree and direction of water movement. The water in streams and rivers flows (generally) in one direction, from the source to the mouth. Lakes and ponds are bodies of standing water. Wetlands constitute an intermediate biome, with water levels that fluctuate.

Lakes and ponds are found on every continent. They vary in size and persistence: some small ponds may exist for only a single season, whereas lakes thousands of square kilometers in size can persist for centuries or longer. Like the oceans, bodies of standing water can be divided into life zones based on depth and distance from shore (**Figure 54.10**). The zone along the shoreline is characterized by warm temperatures and high

species diversity. The photic upper layers of the open-water zone of open water teem with phytoplankton and the fish that feed on them; below that lies an aphotic zone where little light penetrates, oxygen levels are low, and there is little biotic diversity. As in the marine biome, the benthic zone comprises the sediments and other substrates at the lake bottom.

The physical features of a stream or river change along its length as water flows from the point of origin (the source) to its mouth, where it empties into a lake or an ocean. The source of a stream or river may be snowmelt, a spring, or a lake. The headwaters (those close to the source) tend to be cool, fast-flowing, and well oxygenated. As a river flows downstream, it widens, its rate of flow slows, and it supports a higher diversity of plant and animal life. At the mouth, sediment can accumulate, reducing light penetration and oxygen levels. The animal inhabitants of streams and rivers vary along their length as well. For example, rainbow trout (*Oncorhynchus mykiss*), which lay their eggs in gravel beds and use visual cues to find their prey, thrive in the clear, unsedimented headwaters. Certain catfish, by contrast, tolerate the oxygen-depleted, murky shallow waters near the mouths of rivers by exchanging gases through their skin and locating prey by chemical rather than visual cues.

The freshwater wetland biome is highly variable in terms of size and persistence. Swamps, marshes, and bogs are all forms of freshwater wetlands. The unifying characteristic of freshwater wetlands is intermittent flooding. The fluctuations in water level are due to inputs in the form of groundwater, surface water, and rainwater and outputs in the form of evapotranspiration, water flow below the surface, and surface runoff. Plants found in freshwater wetlands include duckweed and other floating water plants with tiny roots, emergent water plants with roots that are completely submerged, such as cattails, and trees and shrubs that grow on the margins. Although many kinds of animals are found in wetlands, frogs and other amphibians, which have a life cycle with both aquatic and terrestrial phases, fare especially well in these water-saturated terrestrial environments.

Estuaries have characteristics of both freshwater and marine environments

Estuaries form where rivers meet the ocean and salt water mixes with fresh water. Depending on local conditions, estuarine environments vary in size and species composition. In the upper part of the intertidal zone, estuaries can support salt marshes, with salt-tolerant rushes, grasses, and low-growing shrubs. Mangrove forests can be found along shorelines and in river deltas in tropical and subtropical latitudes. Dominating these forests are mangroves (*Rhizophora*). These trees display many remarkable adaptations—including aerial roots that are impervious to salt—that make them highly tolerant of high salinity, periodic anoxia, and occasional inundation. Sea grass beds can form in the subtidal zone, dominated by flowering plants such as eelgrass (*Zostera*) that can survive entirely underwater.

Diversity in estuaries tends to be very high. Having characteristics of both freshwater and marine systems, estuaries are home to many unique species and play an important role for other species as a conduit between marine and freshwater environments. Some salmon species that hatch in rivers, for example, spend many months in estuaries adjusting to higher salinities before swimming out to sea to grow into adults. The importance of estuaries as nurseries of marine life cannot be overstated.

Estuarine environments have long been a source of benefits for humans, not the least of which is their role in purifying terrestrial runoff and groundwater. In many places around the world, however, overfishing, habitat destruction, and pollution threaten the viability of estuarine ecosystems.

RECAP 54.4

The marine biome can be divided into several life zones determined by distance from the surface, which influences how much light is available to sustain photosynthesis, and distance from shore. Lakes and ponds, which are freshwater biomes, are also divided into life zones according to water depth and light penetration. Salt and fresh water mix in bodies of water known as estuaries.

- How does light penetration affect diversity in different life zones of the oceans? **See p. 1139 and Figure 54.9**
- How do estuaries link freshwater and marine systems, and why is this biome so important? **See p. 1141**

Biomes and life zones have similar physical characteristics in different parts of the world, and biome-adapted organisms share similar characteristics in widely separated regions. Yet biomes in different regions rarely have particular species in common, so climate alone cannot explain why species live where they do.

54.5 What Factors Determine the Boundaries of Biogeographic Regions?

Climate interacts with local abiotic features to influence where and how organisms live, but these are not the only factors that determine where organisms can be found. Evolutionary history—where and when groups of organisms originated and

diverged—is key to determining the distributions of organisms. Evolutionary history, in turn, is greatly influenced by geological history, which has had a profound influence on the dispersal of species.

Geological history influences the distribution of organisms

Until European naturalists traveled the globe in the nineteenth century, they had no way of knowing how organisms were distributed in other parts of the world. Alfred Russel Wallace, who along with Charles Darwin advanced the idea that natural selection could account for the evolution of life on Earth (see Section 21.1), was one of those global travelers. Wallace spent seven years in the Malay Archipelago, where he noticed some remarkable patterns in the distributions of organisms. For example, he described the dramatically different birds that inhabited the adjacent islands Bali and Lombok:

> In Bali we have barbets, fruit-thrushes and woodpeckers; on passing over to Lombock these are seen no more, but we have an abundance of cockatoos, honeysuckers, and brush-turkeys, which are equally unknown in Bali, or any island further west. The strait here is fifteen miles wide, so that we may pass in two hours from one great division of the earth to another, differing as essentially in their animal life as Europe does from America.

Wallace pointed out that these differences could not be explained by climate or by soil characteristics, because in those respects Bali and Lombok are essentially identical.

Wallace saw that, based on the distributions of plant and animal species, he could draw a line that divided the Malay Archipelago into two distinct halves (**Figure 54.11**). He correctly deduced that the dramatic differences in flora and fauna were related to the depth of the channel separating Bali and Lombok. This channel is so deep that it would have remained full of water, and thus would have been a barrier to the movement of terrestrial animals, even during the glaciations of the Pleistocene epoch, when sea level dropped more than 100 meters and Bali and the islands to the west were connected to the Asian mainland.

With these insights, Wallace established the conceptual foundations of **biogeography**, the scientific study of the patterns of distribution of populations, species, and ecological communities across Earth. In *The Geographical Distribution of Animals*, published in 1876, he detailed the factors known at the time that influence the distributions of animals, including past glaciation, land bridges, deep ocean channels, and mountain ranges. He earned some measure of scientific immortality in that the Malay discontinuity that first piqued his curiosity is known to this day as "Wallace's line."

The biotas of different parts of the world differ enough to allow us to divide Earth into several continental-scale areas called **biogeographic regions** (Figure 54.12), each containing characteristic assemblages of species occupying many different biomes. The boundaries of these biogeographic regions are drawn where assemblages of species change dramatically, often over short distances. The biotas of biogeographic regions

54.11 Wallace's Line Wallace's line corresponds to a deep-water channel between the islands of Bali and Lombok. This channel would have blocked the movement of terrestrial organisms even during the Pleistocene glaciations, when sea level was 100 meters lower than it is today.

differ because oceans, mountains, deserts, and other barriers restrict the dispersal of organisms from one region to another. Although organisms do disperse between adjacent biogeographic regions, such interchanges have not been frequent or massive enough to eliminate the striking differences between them.

Go to Activity 54.1
Major Biogeographic Regions
Life10e.com/ac54.1

Two scientific advances changed the field of biogeography

For many decades after observing that the biotas of the major biogeographic regions are strikingly different, biogeographers speculated about the causes of these differences. The field remained primarily descriptive, however, until the second half of the twentieth century when two scientific advances transformed biogeography into a dynamic, multidisciplinary field. These advances were (1) the acceptance of the theory of continental drift and (2) the development of phylogenetic taxonomy.

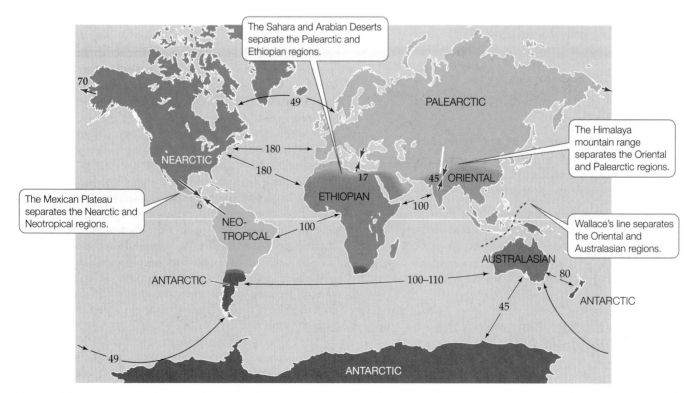

54.12 Earth's Biogeographic Regions The major biogeographic regions are separated by climatic, topographic, and/or aquatic barriers to dispersal that cause their biotas to differ strikingly from one another. The red arrows on the map show the time (in millions of years) since land masses came together. Black arrows show the time since land masses separated.

(A)

(B)

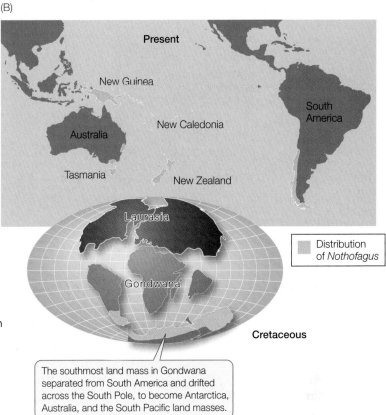

Present

New Guinea

Australia

New Caledonia

Tasmania

New Zealand

South America

Laurasia

Gondwana

Distribution of *Nothofagus*

Cretaceous

The southmost land mass in Gondwana separated from South America and drifted across the South Pole, to become Antarctica, Australia, and the South Pacific land masses.

54.13 *Nothofagus* Has a Gondwanan Distribution The modern range of southern beeches (A) is best explained by their origin in Gondwana during the Cretaceous. (B) When Gondwana broke apart, *Nothofagus* remained in South America, Australia, New Zealand, and the islands of the South Pacific.

CONTINENTAL DRIFT By the 1960s scientists knew that continents can and do move (see Section 25.2). We now know that over the course of the Triassic and Jurassic periods, the supercontinent Pangaea divided into two great land masses, Laurasia and Gondwana (see Figure 25.12), which subsequently separated into the continents we know today. Those groups of organisms that are represented on two or more continents are believed to be ancient groups whose ancestors were widely distributed over these great land masses before they broke apart. After the breakup, however, their descendants evolved independently, so groups that did not originate until after the continents separated have more discrete distributions. Thus continental drift is at least partly responsible for the existence of the biogeographic regions shown in Figure 54.12.

Continental drift explains certain biogeographic distributions that would otherwise be difficult to understand. For example, the southern beeches—trees of the genus *Nothofagus*—are found in both the Neotropical and the Australasian biogeographic regions. Their distribution suggests that the genus originated in Gondwana during the Cretaceous period and was geographically separated by the breakup of that land mass (**Figure 54.13**).

What evidence do we have that *Nothofagus* did not simply leapfrog from one biogeographic region to another? Fossilized *Nothofagus* pollen from 55 to 34 million years ago has been found in Australia, New Zealand, western Antarctica, and South America, suggesting that *Nothofagus* was once continuously distributed across a single land mass (see Figure 54.13B). Moreover, the modern distribution of aphid genera that feed exclusively on *Nothofagus* parallels the distribution of the trees.

There are no air or water currents between Chile and New Zealand that would be likely to disperse insects, indicating that the aphids arose at a time when their host plants grew on a common land mass.

PHYLOGEOGRAPHY As we saw in Chapter 22, taxonomists have developed powerful methods of reconstructing the phylogenetic relationships among organisms. Biogeographers have adapted these methods to help them understand how organisms came to occupy their present-day distributions. Biogeographers can transform phylogenetic trees into **area phylogenies** by replacing the names of the taxa on a tree with the names of the places where those taxa now live or once lived.

Suppose, for example, we wonder why zebras, which are members of the horse family (Equidae), live in Africa when the fossil record indicates that the Equidae originated in North America. An area phylogeny of living equid species suggests that the ancestors of today's horses (represented by the oldest fossils) dispersed from North America to Asia, and then from Asia to Africa, and that the subsequent speciation of zebras took place entirely in Africa (**Figure 54.14**).

 Go to Media Clip 54.1
Rafting to Madagascar
Life10e.com/mc54.1

Discontinuous distributions may result from vicariant or dispersal events

The appearance of a physical barrier to dispersal that splits the range of a species is called a **vicariant event**. A vicariant event

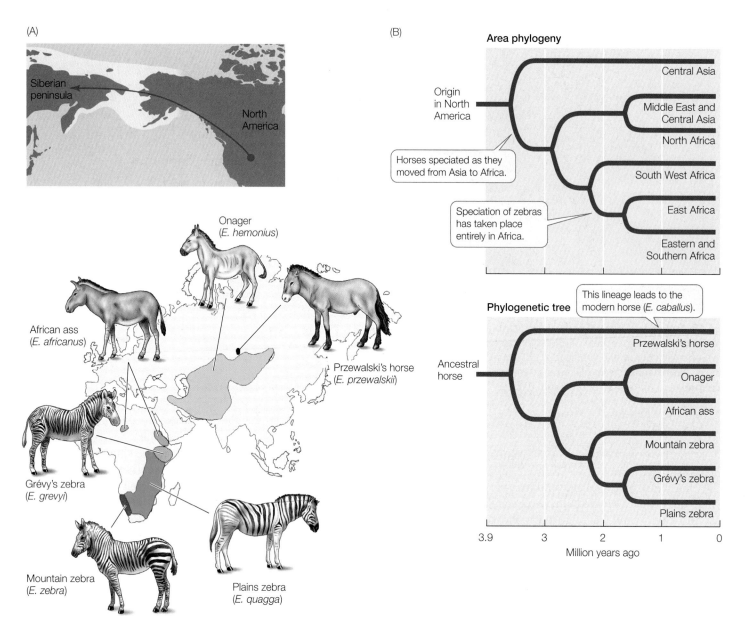

54.14 Phylogenetic Tree to Area Phylogeny The conversion of a phylogenetic tree into an area phylogeny helps biogeographers explain how the current distribution of a taxon came about. (A) The ancestor of all Asian and African equids (genus *Equus*; horses and their relatives) migrated across the Bering Strait land bridge (light green) some 10 million years ago. (B) Organismal and area phylogenies explain the Asian and African distribution of the descendants of these ancestral horses.

divides the species into two or more discontinuous populations, even though no individuals have dispersed to new areas. If, however, members of a species cross an existing barrier and establish a new population, the discontinuous range of the species is considered to be the result of dispersal.

Given that the processes of vicariance and dispersal both influence distribution patterns, how can biogeographers determine the role of each process when reconstructing the evolutionary history of a particular species? By studying area phylogenies, a biogeographer may discover evidence suggesting that the distribution of an ancestral species was influenced by a vicariant event, such as continental drift or a change in sea level. If that inference is correct, then it is reasonable to assume

that ancestral species in other lineages would have been affected by the same event and that similar distribution patterns should therefore be seen in other taxonomic groups. Differences in distribution patterns among taxonomic groups may indicate that they responded differently to the same vicariant events, that they diverged at different times, or that they had very different dispersal histories. By analyzing such similarities and differences, biogeographers seek to discover the relative roles of vicariant events and dispersal in determining today's distribution patterns.

The parsimony principle used in the reconstruction of phylogenies (see Section 22.2) can also be helpful in biogeographic studies. For example, the New Zealand flightless

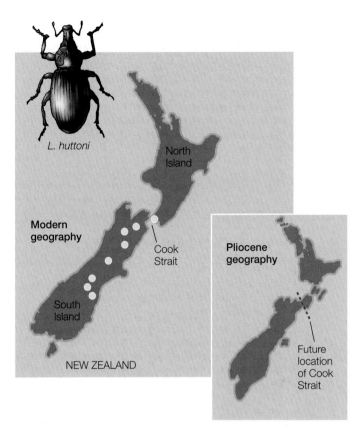

54.15 A Vicariant Distribution Yellow circles indicate the current distribution of the flightless weevil *Lyperobius huttoni*. A comparison of New Zealand's present-day geography with its geography in the Pliocene, when the southern part of today's North Island was part of South Island, suggests that a vicariant event—a physical split separating populations—explains this distribution.

weevil *Lyperobius huttoni* is found in the mountains of South Island and on sea cliffs at the extreme southwestern corner of North Island (**Figure 54.15**). At first glance, its distribution might suggest that, even though this weevil cannot fly, some individuals in the distant past managed to cross Cook Strait, the 25-km wide body of water that separates the two islands. However, more than 60 other animal and plant species, including other flightless insects, are found on both sides of Cook Strait. Irrespective of their ability to fly, wade, or swim, it is unlikely that all 60 of these species made the same ocean crossing independently at different times over the course of their evolutionary history. In fact, geological evidence indicates that the present-day southwestern tip of North Island was once united with South Island. Thus a single vicariant event—the separation of the northern tip of South Island from the remainder of the island by the newly formed Cook Strait—could have produced the distribution pattern shared by all 60 species today.

In other cases, the evidence points to dispersal. There are, for example, more than 135 species of long-horned beetles endemic to the Hawaiian Islands. The islands have never been attached to any continent (they arose by volcanism from the

deep ocean floor), so no separation of continuous populations ever took place. Therefore this distribution must have been the result of long-distance dispersal. The beetles, which are most closely related to a genus found in North and Central America, probably colonized the islands and subsequently speciated as a consequence of specializing on different host plants on different islands (see Section 23.3).

Humans exert a powerful influence on biogeographic patterns

One more force capable of generating distribution patterns that span multiple biogeographic regions is human activity. Of the insects found both in Europe and in North America, it has been estimated that more than half have been transported between the continents by humans, either by accident or deliberately. Many of these transported species have unintended consequences for other species in their new regions. Where the fynbos has been invaded by the Argentine ant, for example, seedlings often fail to appear after fires. As we saw at the beginning of this chapter, the seeds of some fynbos plants survive fire only with the help of ants. The ants must be attracted to a seed, pick it up off the ground, nibble off the lipid-rich elaiosome, and then bury the seed deep enough in the soil to avoid injury by fire. Argentine ants, which humans accidentally transported to South Africa from South America, are attracted to the seeds and eat the elaiosomes, but these tiny ants are too small to carry off large seeds and they cannot bury seeds deep enough in the ground to survive fires. In places where Argentine ants have displaced native ant species, replacement of large-seeded plants by seedlings after fires can drop by tenfold compared with areas that have not been invaded. The effects of such accidental species introductions by humans will be discussed at length in Chapter 59.

RECAP 54.5

Earth can be divided into seven biogeographic regions, each with unique assemblages of species. Vicariant events generate distribution patterns by splitting the ranges of species; distribution patterns may also change when species disperse across barriers.

- What determines the boundaries of Earth's major biogeographic regions? How are these boundaries different from those of the biomes described in Section 54.3? See p. 1126, p. 1141, and Figures 54.7 and 54.12

- Explain how the concepts of continental drift and phylogeography transformed the field of biogeography. See pp. 1142–1143

- How do vicariance and dispersal interact to generate species distribution? See pp. 1143–1145

Earth's physical environment and geological history are major factors influencing the distribution of organisms. Next we will turn to the influence of organisms and populations of organisms on one another. Abiotic, intraspecific, and interspecific forces all interact in the complex processes of population dynamics.

What is it about the western edges of continents that promotes tough, shrubby plant communities such as fynbos?

ANSWER

The fynbos is an example of the chaparral biome found in the Mediterranean regions and on the western coasts of continents at mid-latitudes (around 40°). The Mediterranean climate promotes plant communities that are nurtured by cool, damp winters and can survive the dry summer conditions and periodic fire. Continental locations of this biome are related to the proximity of the cold ocean currents that flow toward the equator offshore (see Figure 54.5). The prevailing winds set up rotating gyres of ocean water. Because of the direction of their rotation, warm water tends to move toward the poles along the east coasts of continents, whereas cool water moves toward the equator from higher latitudes along west coasts. These cool offshore currents bring cool, wet winters similar to those experienced by the Mediterranean region to large areas of continental western coastal regions.

CHAPTER**SUMMARY** 54

 54.1 What Is Ecology?

- **Ecology** is the scientific investigation of interactions among organisms, between organisms and their physical environment, and the patterns of distribution and abundance resulting from these interactions.

- **Environmentalism** is the use of ecological knowledge to inform our decisions about the stewardship of natural resources.

- An organism's environment encompasses both **abiotic** (physical and chemical) components and **biotic** components (other living organisms).

 54.2 Why Do Climates Vary Geographically?

- **Weather** refers to atmospheric conditions at a particular place and time. **Climate** is the average of atmospheric conditions, and the variation in those conditions, found in a particular place over an extended period of time.

- The solar energy that reaches a given unit of Earth's surface depends primarily on the angle of the sun's radiation, which in turn is a function of latitude. The tilt of Earth's axis results in seasonal variation in temperature and day length. **Review Figures 54.1, 54.2**

- Latitudinal variation in solar energy input drives atmospheric circulation patterns. **Review 54.3**

- Global surface wind patterns are driven by atmospheric circulation and Earth's rotation; these **prevailing winds** in turn drive ocean surface **currents**. **Review Figure 54.4, 54.5**

- Organisms respond to climatic challenges with physiological, morphological, and behavioral adaptations.

 54.3 How Is Life Distributed in Terrestrial Environments?

- A **biome** is an environment that is shaped by its climatic and geographic attributes and characterized by ecologically similar organisms. **Review Figure 54.7**

- The distribution of terrestrial biomes is determined primarily by climate, but other factors, such as soil characteristics and fire, also influence vegetation.

- Biomes include Arctic and alpine tundra, boreal forest, temperate evergreen and temperate deciduous forests, temperate grasslands, hot and cold deserts, chaparral, thorn forest and savanna, tropical deciduous forest, and tropical rainforest. **See ANIMATED TUTORIAL 54.1, 54.2**

 54.4 How Is Life Distributed in Aquatic Environments?

- Aquatic biomes do not depend on plants for their structure in the way terrestrial biomes do. Salinity is the primary factor that distinguishes aquatic biomes.

- The marine biome is characterized by high salinity. Marine **life zones** are determined by distance from the surface, which influences how much light is available to sustain photosynthetic organisms, and by distance from the shore. **Review Figure 54.9**

- Freshwater biomes are distinguished by their water movement (standing versus flowing water). Standing water (lakes and ponds), like ocean basins, can be divided into life zones distinguished by depth and distance from shore. **Review Figure 54.10**

- The physical conditions in streams and rivers change along their length as water flows from the source to the mouth.

- In freshwater wetlands, water levels fluctuate because of variation in water input and output.

- **Estuaries** are bodies of water where salt and fresh water mix. This biome supports many unique species.

 54.5 What Factors Determine the Boundaries of Biogeographic Regions?

- **Biogeography** is the scientific study of the patterns of distribution of populations, species, and ecological communities.

- The boundaries of the **biogeographic regions** are drawn where assemblages of species change dramatically over short distances. These boundaries are generally continental in scale and correspond to present or past barriers to dispersal. **Review Figures 54.11, 54.12, ACTIVITY 54.1**

- Continental drift explains some discontinuous distributions that include more than one biogeographic region. **Review Figure 54.13**

- Biogeographers can transform phylogenetic trees into **area phylogenies** to understand how organisms came to occupy their present-day distributions. **Review Figure 54.14**

- Both **vicariant events** and dispersal across barriers generate discontinuous species distributions. **Review Figure 54.15**.

 Go to the Interactive Summary to review key figures, Animated Tutorials, and Activities
Life10e.com/is54

CHAPTER**REVIEW**

REMEMBERING

1. Ecology and environmentalism are
   a. synonymous; the terms can be used interchangeably.
   b. differentiated by the emphasis placed by ecology on the biotic rather than the abiotic world.
   c. differentiated by the lack of utility of ecology in solving world problems.
   d. differentiated by the inherent focus of environmentalism on human concerns.
   e. both scientific fields of inquiry that generate knowledge about the natural world but use completely different tools.

2. Energy from the sun determines
   a. air temperature.
   b. air and wind circulation patterns.
   c. ocean surface currents.
   d. All of the above
   e. None of the above

3. The amount of solar energy that reaches a given unit of Earth's surface depends primarily on
   a. the angle of the sun's rays.
   b. the moisture content of the air.
   c. the amount of cloud cover.
   d. the strength of the winds.
   e. day length.

4. The marine biome can be divided into life zones because
   a. the rate of photosynthesis in the oceans is low.
   b. ocean currents keep organisms close to where they were born.
   c. water temperature, salinity, and food supply all vary within the ocean.
   d. trade winds keep warm and cold waters separate.
   e. continents provide barriers to movement of planktonic life forms.

5. You are choosing a location on which to grow corn (*Zea mays*) and want to minimize the amount of land you need to cultivate for maximum yield. In which biome would you locate your farm?
   a. Tropical rainforest because this biome is home to tremendous plant diversity.
   b. Temperate evergreen forest because some of the world's largest plants are found there.
   c. Temperate grassland because the topsoil is deep and rich.
   d. Arctic tundra because summer days are long and soil moisture is abundant.
   e. All of these biomes are equally well suited for efficient cultivation of corn.

UNDERSTANDING & APPLYING

6. The Chilean matorral is located on the west coast of South America between 32°S and 37°S. The region experiences wet, cool winters and hot, dry summers. Describe the appearance and life cycles of the seed plants you would expect to find there. How would they most likely disperse their seeds? Describe some specific adaptations you might expect to encounter in at least some of the plant species there.

7. In 2008 spider expert Rudy Jocqué discovered two new species of spiders in the genus *Australutica* in South Africa. Prior to this find, the only species known in the genus were all found in Australia. One of the new species was found among the oldest rock formations in southern Africa (formed about 150 million years ago). Based on what you know about continental drift, do you think undiscovered *Australutica* species might exist anywhere else in the world? Where besides Australia and southern Africa would you look for such species?

ANALYZING & EVALUATING

8. Refer to the summary chart of biome characteristics below. If mean average global temperatures increase 5°C by 2100, as predicted by some climate models, which biomes would be expected to decrease in geographic extent?

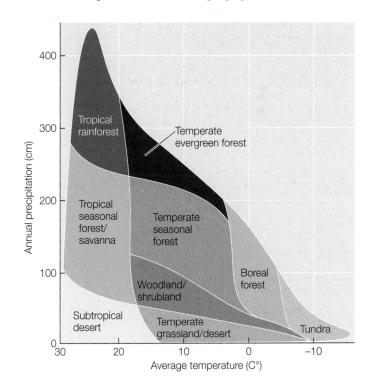

9. Today by far the greatest number of species of fruit flies (genus *Drosophila*) is found in the Hawaiian Islands. Would you conclude that the genus originated in Hawaii and spread to other regions? Under what circumstances do you think it might be accurate to conclude that a group of organisms originated in the same region where the greatest number of species live today? (Hint: Review the discussion of equid phylogeny and Figure 54.14.)

10. The map below is from a 2008 paper by H. I. McCallum and colleagues (*Ecology and Society* 13: 41–57). The paper's title is "Will Wallace's Line save Australia from avian influenza?" Based on the map and what you have learned about biogeography, how would you answer the title's question? What factor(s) do you think might be influencing the geographic spread of this disease?

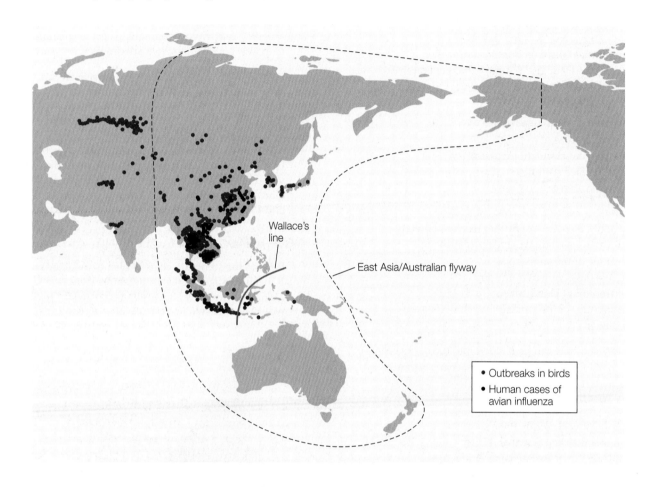

Wallace's line

East Asia/Australian flyway

- Outbreaks in birds
- Human cases of avian influenza

Go to BioPortal at **yourBioPortal.com** for Animated Tutorials, Activities, LearningCurve Quizzes, Flashcards, and many other study and review resources.

55 Population Ecology

CHAPTER**OUTLINE**

Reindeer Games Part of the St. Matthew reindeer herd is seen in this photograph taken in 1963, shortly before a particularly severe winter destroyed most of this isolated population. The herd had grown exponentially since being introduced to the island as a food source for a Coast Guard station during World War II.

DURING World War II the U.S. Coast Guard established a LORAN (Long-Range Aids to Navigation) tracking station on the tiny island of St. Matthew in Alaska, an isolated and unoccupied patch of tundra more than 300 kilometers from the nearest village. As an emergency food supply for the 19 men assigned to the island in 1944, the Coast Guard brought in 29 reindeer (*Rangifer tarandus*) by barge and released them.

The reindeer thrived on the thick, lush mat of lichens that covered the island. Other than the men, the island had no reindeer predators; the only other terrestrial vertebrate occupants were Arctic foxes, one species of vole, and a few ground-nesting birds. Then the war ended and the men left the island, leaving the reindeer behind in an environment with plentiful food and no natural predators.

In 1957 David Klein, a U.S. Fish and Wildlife biologist, visited St. Matthew. He and an assistant counted more than 1,350 reindeer, most of which appeared healthy. However, they also noticed areas of overgrazed lichen. In 1963 Klein and three colleagues hitched a ride to the island on a Coast Guard cutter. This time they counted more than 6,000 reindeer, packed in at a density of 47 per square mile. The island was covered with reindeer droppings, and the animals were smaller than the ones sighted 6 years earlier.

The winter of 1963–1964 brought punishing storms, record low temperatures, and tremendous snowfalls to St. Matthew. When Klein returned in summer 1966, the island was littered with reindeer skeletons. Klein found only 42 living reindeer, 41 of which were adult females; the lone male appeared to have deformed antlers. Lichens had disappeared, replaced almost entirely by sedges and grasses, on which reindeer cannot subsist. By 1980, the reindeer too had disappeared from the island.

Introducing large mammals to small islands is inherently risky, as the experience of reindeer on St. Matthew illustrates. But such introductions do not always end in disaster. Reindeer herds introduced to the subantarctic island of South Georgia almost a century ago have persisted, and their populations appear to be stable.

Why do populations of a particular species in one place explode and crash, but in another, seemingly similar, place remain stable? That knowledge is critical for understanding why some species become pests in some places and not in others, for managing sustainable harvests of economically important species, and for designing plans for conserving endangered species.

Why did introduced reindeer populations persist on the island of South Georgia but not on the island of St. Matthew?

See answer on p. 1166.

(A) *Loxodonta africana*

The pattern of folds and notches on each elephant's ears is as unique as a fingerprint.

(B) *Camarhyncus parvulus*

The different species of Galápagos finches are the subjects of many studies. Multiple bands identify them as individuals.

(C) *Apis mellifera*

A computer chip on a honey bee's back logs her movements between the hive and flowers.

55.1 Identifying Individuals (A) The pattern of folds and notches on the ears of an elephant is distinctive and can be used to recognize individuals in a population. (B) The idea of attaching a metal band to a bird's leg to identify an individual dates back at least to 1595, when the French king Henry IV banded the royal peregrine falcons. Scientific banding for population studies, however, did not become widely established until the early years of the twentieth century. Galápagos finches such as this small tree finch have been extensively studied and marked in this way. (C) Worker bees in a hive are individually indistinguishable to ecologists, who have come up with ingenious methods of marking. This female honey bee sports a computer chip on her back, which not only identifies her but also logs her movements between the hive and flowers.

55.1 How Do Ecologists Measure Populations?

Well before ecology became a distinct biological discipline, people engaged in population management. Whenever we grow crops or raise livestock, we are explicitly increasing the size of populations of domesticated plants and animals. Pest control strategies aim to reduce population sizes of organisms whose presence we consider undesirable. Game wardens, park managers, and conservation biologists aim to maintain stable populations of fish, wildlife, and threatened or endangered species. All of these activities require an understanding of **population dynamics**: the patterns and processes of change in populations. The study of population dynamics also allows us to understand the changes in populations we make inadvertently in the course of other human activities—as when the Coast Guard introduced reindeer to St. Matthew Island.

A **population** consists of the individuals of a species that interact with one another within a given area at a particular time. Populations are important units for study because groups of individuals that interact in time and space have ecological characteristics that individuals do not. At any given moment, an individual organism occupies only one point in space and is a particular age and size. The members of a population, however, are distributed over space, and they vary in age and size.

Population **density** is the number of individuals per unit of area or volume. Density is a function of processes that add individuals to the population (births and immigration) and processes that reduce the number of individuals in the population (deaths and emigration). Populations also have a characteristic **age structure**, or distribution of individuals across age

categories, and a characteristic **dispersion pattern**, or spatial distribution of individuals in the environment.

These properties of populations, which are constantly changing because of births, deaths, and movements of individuals, influence the stability of populations and affect the ways in which populations of one species interact with populations of other species. Thus, to study populations, ecologists need to count the individuals in a given area and determine their ages.

Ecologists use a variety of approaches to count and track individuals

How the individuals in a population are counted depends on the nature of the organism under study. Populations of animals are usually more challenging to count than populations of trees. Most animals can move, so to avoid double counting, individuals must be identified. Nevertheless, counting every tree in a forest can be logistically difficult, even though the trees are standing still.

In some species, individuals are large and distinct enough, and populations small enough, that investigators can identify all the individuals and count them. Biologists performed this type of count, called a **full census**, on the African elephant population of Samburu and Buffalo Springs National Reserves in Kenya. By monitoring the elephants for 21 months, the biologists learned to recognize each of the 760 individuals in the population, primarily by their unique and distinctive ear markings (**Figure 55.1A**).

For most species, however, recognizing individuals is impossible or impractical. If biologists are to identify individuals of such species, they must be marked in some way. No single form of artificial marking works for all species. Plants can be marked with tags tied to their branches or by stakes in the ground nearby. Birds can be marked by colored bands on their legs (**Figure 55.1B**), and butterflies and beetles with small dabs of colored paint in different patterns. Honey bees can be monitored with fully automatic radio frequency identification (RFID) technology—the same technology used for tracking supermarket purchases (**Figure 55.1C**). Individual bees are marked with a chip, and a reader is placed at the hive entrance to register movements of the marked bees. Small mammals can be marked by bleaching or dyeing their fur in strategic places.

In most species, populations are too large, and their individual members too small, too similar in appearance, or too mobile, for a full census to be conducted. Thus population sizes are often estimated from representative samples using statistical methods.

Ecologists can estimate population densities from samples

Ecologists usually measure the densities of terrestrial organisms as the number of individuals per unit of area. They may measure the densities of organisms living in soil, air, or water as the number or mass per unit of volume. Ecologists obtain these measurements from sample units, then extrapolate from these samples to estimate the total population density.

Estimating population densities is easiest for sessile organisms. Investigators need only count the individuals in a sample of representative locations and extrapolate the counts to the entire geographic range of the population. Individuals may be counted within marked and measured areas called **quadrats**. Plants are often counted along a **transect**—a line drawn across an area within the population's range (often designated by a string marked at regular intervals). Any individual that touches the line is counted. By making repeated counts with either of these methods, investigators can make reasonably good estimates of the size of a population.

Counting mobile organisms is more difficult because individuals move into and out of sampling areas. In such cases, investigators may use the **mark–recapture method** (Figure 55.2). They begin by capturing, marking, and then releasing a number of individuals. Later, after the marked individuals have had time to mix with unmarked individuals in the population (but before enough time has elapsed for births, deaths, and individual movement to affect the population size significantly), another sample of individuals is captured. This sample is then used to obtain an estimate of the total size of the population in the sampling area. This is done by applying the equation described in Figure 55.2, which assumes that the *proportion* of marked individuals in the second sample (i.e., individuals that

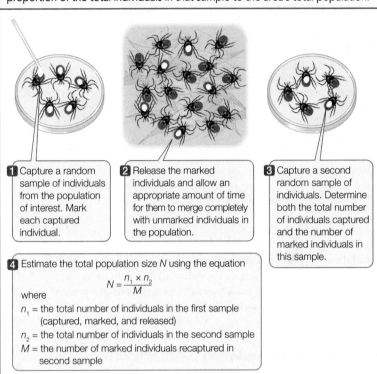

RESEARCHTOOLS

55.2 The Mark–Recapture Method   The method described here is used to estimate population sizes for animal populations in which the individuals are highly mobile (such as *Ixodes scapularis*, the black-legged tick). Once a sampling area has been determined, investigators capture, mark, and then release individuals of the organism of interest. The proportion of these marked individuals that appears in a second sample is assumed to be the same as the proportion of the total individuals in that sample to the area's total population.

1 Capture a random sample of individuals from the population of interest. Mark each captured individual.

2 Release the marked individuals and allow an appropriate amount of time for them to merge completely with unmarked individuals in the population.

3 Capture a second random sample of individuals. Determine both the total number of individuals captured and the number of marked individuals in this sample.

4 Estimate the total population size N using the equation

$$N = \frac{n_1 \times n_2}{M}$$

where

n_1 = the total number of individuals in the first sample (captured, marked, and released)

n_2 = the total number of individuals in the second sample

M = the number of marked individuals recaptured in second sample

were captured and marked in the first sample) is about the same as the proportion of individuals in the sampling area that were captured in the first sample.

For some species, however, this assumption does not apply. Some captured animals learn to avoid traps or leave the sampling area and are thus less likely to be recaptured than are unmarked individuals. Others become "trap-happy" (some mice, for example, reenter live traps repeatedly in order to snack on the peanut butter bait). In some cases the act of marking may reduce an individual's chances of survival due to the stress of handling or inadvertent alterations of appearance that make marked individuals more conspicuous to predators. Ecologists use statistical techniques to correct for these errors and improve the accuracy of population estimates.

Determining the size and density of populations is important, but these numbers are only a starting point for understanding population dynamics because not all individuals contribute equally to population growth.

A population's age structure influences its capacity to grow

The **age structure** of a population—the distribution of individuals across age categories—has a profound effect on

WORKING WITH**DATA:**

Monitoring Tick Populations

Original Paper

Falco, R. C. and O. Fish. 1988. Prevalence of *Ixodes dammini* near the homes of Lyme disease patients in Westchester County, New York. *American Journal of Epidemiolology* 127: 826–830.

Analyze the Data

Lyme disease is a chronic and debilitating condition caused by spirochete bacteria of the genus *Borrelia*, which infect humans by way of the bite of an intermediate host, the black-legged tick *Ixodes scapularis* (also known as the deer tick). The incidence of Lyme disease has increased dramatically in the past 20 years, particularly in the northeastern United States. In order to assess the risk of exposure to this disease in Westchester County, New York, investigators measured the abundance of deer ticks in suburban lawns near wooded areas using the mark–recapture method described in Figure 55.2. (Ticks are typically collected by dragging a white cloth along the ground; the ticks latch onto the cloth in much the same way they would to a passing leg.) By drag-sampling one representative lawn, they collected the data shown in the table.

QUESTION 1

Refer to Figure 55.2. Using the equation and other information described in that figure, estimate the total number of ticks in the sampled lawn from the data below.

QUESTION 2

The lawn was approximately 700 m² in size. What is the approximate density of ticks per square meter?

QUESTION 3

What do you think might be the implications of this study for residents of this neighborhood?

	Original capture event	Second capture event (3 weeks later)
Adult ticks captured	180	33
No. of marked ticks	180[a]	8

[a]All ticks captured in the first event were marked with acrylic paint and released.

Go to BioPortal for all WORKING WITH**DATA** **exercises**

---

population growth because reproductive capacity varies with age. Populations with a large proportion of individuals in their peak reproductive years have a greater potential to grow than do populations dominated by individuals that are too young or too old to reproduce.

In some species, reproduction is the province of only a tiny fraction of the population for only a short interval during the life cycle. For example, adults of the tiny insect *Clunio marinus* (the "one-hour midge") mate, lay eggs, and die within about an hour after completing their larval development. In contrast, some vertebrates, such as elephants, are capable of reproducing for years.

Consider the results of a long-term study of the age structure of the elephant population of Kidepo Valley National Park, Uganda, and how it changed over time. Relative to the population in 1970, the 2000 population had more elephants over 25 years of age (**Figure 55.3**). The change was the result of

years of differential mortality among young elephants due to drought, and among adult males due to ivory poaching. The age structure as of 2000 portends an increase in the population's growth rate, given that female African elephants become fertile around age 10 to 15 and can continue producing offspring through their fifties.

A population's dispersion pattern reflects how individuals are distributed in space

Dispersion refers to the distribution of individuals in space. Dispersion affects how individuals in a population interact with one another and thus can have important effects on population growth. In addition, ecologists must understand the dispersion patterns of a species to choose appropriate sampling areas and statistical methods for estimating population sizes.

Ecologists recognize three basic dispersion patterns:

Go to Animated Tutorial 55.1
Age Structure and Survivorship
Life10e.com/at55.1

55.3 Changes in Age Structure Influence Population Growth

The elephant population in Kidepo Valley National Park, Uganda, was monitored between 1970 and 2000. During this time, the proportion of the population in the prime reproductive age range (15–30 years) grew considerably. Such an age structure in a population is likely to result in a high rate of growth.

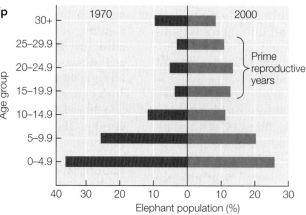

Loxodonta africana

(A) Clumped dispersion

Orcinus orca

(B) Regular dispersion

Morus bassanus

(C) Random dispersion

Taraxacum officinale

55.4 Dispersion Patterns   (A) Orcas hunting together in pods display a clumped dispersion pattern. (B) Nesting seabirds often stake out territories with a radius defined by their wingspans—an amount of space they can defend without leaving the nest. This behavior results in a regular dispersion pattern. (C) Dandelion seeds are dispersed by the wind in random fashion, so the plants that grow from those seeds show a random dispersion pattern.

- A **clumped dispersion pattern** occurs when the presence of one individual at any point in space increases the probability of others being near that point (**Figure 55.4A**).

- A **regular dispersion pattern** occurs when the presence of one individual at any point in space reduces the probability of others being near that point (**Figure 55.4B**).

- A **random dispersion pattern** occurs when the presence of one individual at any point in space does not affect the probability of other individuals being near that point (**Figure 55.4C**).

Dispersion patterns can vary among species, or among populations within species. Spatial variation in environmental conditions strongly influences dispersion patterns. Small-scale differences in temperature, humidity, or wind speed can make particular places more or less suitable for certain organisms. Aphids, for example, cluster along the protruding veins of leaves, where they are sheltered from wind. Interactions among individuals may also bring about characteristic dispersion patterns. Social life, with the cooperation it involves, tends to promote clumped dispersion patterns. By contrast, intraspecific competition for food, space, or mates tends to space individuals apart in regular dispersion patterns.

RECAP 55.1

To understand the dynamics of populations, ecologists measure their density, age structure, and dispersion patterns.

- What are some of the ways in which population density can be measured? **See p. 1151**

- How can the age structure of a population influence its growth? **See pp. 1151–1152 and Figure 55.3**

- How do environmental factors influence dispersion patterns? **See p. 1153 and Figure 55.4**

Once the sizes, densities, and other important traits of populations have been measured, these data can be used to describe various "survival strategies" found among populations and to understand and predict changes within and among populations.

55.2 How Do Ecologists Study Population Dynamics?

Quantifying population density and age structure provides useful information about populations, but those traits alone cannot explain how, when, and why populations change in size. In order to understand population growth, ecologists must measure population *processes* as well as population traits. The study of population processes is known as **demography**.

Demographic events determine the size of a population

The size of a population changes over time because of **demographic events**: births, deaths, immigration, and emigration. Over any given interval of time, the size of a population increases by the number of individuals added to the population by births and by immigration (the movement of individuals into the population from elsewhere) and decreases by the number of individuals lost from the population by deaths and by emigration (individuals leaving the population to go elsewhere). This relationship is expressed mathematically as

$$N_1 = N_0 + (B - D) + (I - E) \qquad (55.1)$$

where

N_1 = the number of individuals at time 1

N_0 = the number of individuals at time 0

B = the number of individuals born between
    time 0 and time 1

D = the number that died between time 0 and time 1

I = the number that immigrated between time 0
    and time 1

E = the number that emigrated between time 0
    and time 1

Using Equation 55.1 to estimate N_1 over multiple time intervals helps researchers estimate the *rate of change in population size over time*—that is, the growth rate of the population.

Life tables track demographic events

The study of population dynamics requires keeping track of demographic events in populations and determining the rate (number per unit of time) at which they occur. An individual is born only once and dies only once; birth rates and death rates are properties of populations. A **life table** is a tool that ecologists use for these purposes. Life insurance companies use similar tables (called actuarial tables) to determine how much to charge people of different ages for insurance policies. Data from life tables can be used to identify the principal mortality factors, or causes of death, at particular life stages, to predict future population trends, and to develop strategies for managing populations of species of commercial or ecological value.

COHORT LIFE TABLES Life tables can be constructed by a number of methods. To construct a cohort life table, investigators start with a **cohort**—a group of individuals born within the same time frame, or age class—and record their deaths until no individuals from the cohort remain alive. This type of life table is sometimes called a horizontal life table because it is based on data collected *across* the entire life span.

The age classes used in a cohort life table depend on the life cycle of the organism of interest. Age-dependent cohort life tables track demographic events as a function of calendar age. Stage-dependent cohort life tables track demographic events at various stages of the life cycle (e.g., eggs, larvae, pupae, and adults in insects). They are commonly used when survival and reproduction depend more on developmental stage than on calendar age, as is the case, for example, with insects and other animals that undergo metamorphosis.

Using the data in a cohort life table, investigators can calculate **mortality**: the proportion of individuals of each age class that die before reaching the next age class. By following a cohort, investigators can also calculate the average individual's chance of dying during a particular time interval, a value known as the per capita death rate, or d. By the same token, they can calculate **survivorship** (represented by the term l_x), which is the likelihood of an individual member of the cohort surviving to reach age x (**Table 55.1**).

ESTIMATING REPRODUCTIVE CAPACITY A cohort life table can also be used to track the degree to which individuals in different age categories contribute to reproduction (and hence population growth). Investigators can use the data in the table to calculate the number of offspring the average individual produces, or the per capita birth rate, b. Because only females produce offspring, life tables generally track the number of female offspring produced by each female during each time period—a factor called **fecundity** (indicated by the term m_x). The portion of the life table that tracks fecundity is called a fecundity schedule (see Table 55.1, rightmost column). Such data allow scientists to estimate a population's potential for growth.

The data in Table 55.1 track the survivorship (l_x) and fecundity (m_x), respectively, of a cohort of the cactus finch species *Geospiza scandens* on Isla Daphne in the Galápagos. Peter and Rosemary Grant followed a cohort of 210 birds from the time they hatched in 1978 until 1991, at which time only 3 individuals—all males—remained alive. All of the cactus finches on the island were banded so that the Grants could recognize them as individuals (see Figure 55.1B).

The *G. scandens* life table shows that mortality was high during the first year of life, then dropped dramatically for several years before increasing in later years. The fecundity data indicate that females may begin breeding as young as 1 year of age and may continue breedomg throughout their lives.

Survival and breeding success, however, are not correlated exclusively with age. Other observations of conditions on Isla Daphne revealed a correlation of fecundity with rainfall, which

TABLE 55.1

Life Table for the 1978 Cohort of *Geospiza scandens* on Isla Daphne

Age Class (years)	Number Alive	Survivorship[a]	Mortality[b]	Fecundity[c]
0–1	210	—	0.57	0.00
1–2	91	0.43	0.14	0.05
2–3	78	0.37	0.10	0.67
3–4	70	0.33	0.07	1.50
4–5 Increased rain	65	0.31	0.05	0.66
5–6	62	0.30	0.32	5.50
6–7 Drought	42	0.20	0.45	0.69
7–8	23	0.11	0.35	0.00
8–9	15	0.07	0.07	0.00
9–10	14	0.07	0.21	2.20
10–11	11	0.05	0.09	0.00
11–12	10	0.05	0.60	0.00
12–13	4	0.02	0.25	—
13	3	0.01	—	—

[a] Survivorship (l_x) = the proportion of the original cohort (here, of 210 individuals) who survive to age x.
[b] Mortality (d) = the proportion of individuals of age x who die before reaching age x + 1.
[c] Fecundity (m_x) = number of fledgling females per female per breeding season. Of the 210 birds in this cohort, 90 were females.

55.5 Survivorship Curves Ecologists recognize three general types of survivorship curves. Notice that the number of survivors has been plotted on a logarithmic scale. Three species provide real-world examples of the three types of life histories.

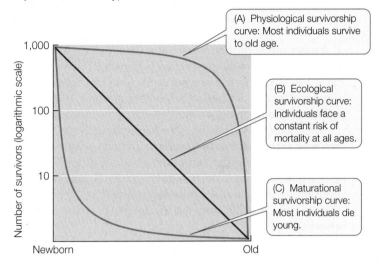

(A) Physiological survivorship curve: Most individuals survive to old age.

(B) Ecological survivorship curve: Individuals face a constant risk of mortality at all ages.

(C) Maturational survivorship curve: Most individuals die young.

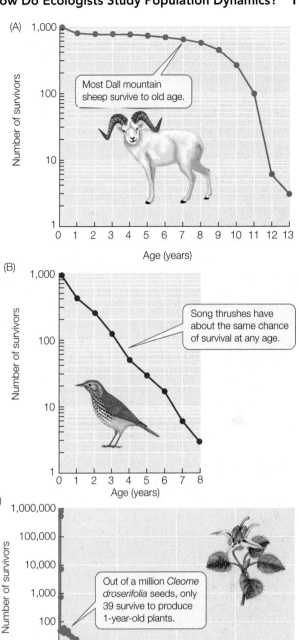

(A) Most Dall mountain sheep survive to old age.

(B) Song thrushes have about the same chance of survival at any age.

(C) Out of a million *Cleome droserifolia* seeds, only 39 survive to produce 1-year-old plants.

in the Galápagos varies dramatically from year to year. This life table and other ecological data, taken together, suggest that the survival of adult birds and the number of offspring they are able to fledge depend on food availability—that is, on cactus flower and fruit production, which is strongly correlated with rainfall (see Table 55.1). In short, life table data can be useful in identifying the ecological factors that affect population dynamics, at least over the short term.

Fecundity schedules vary greatly among species not only because organisms differ in the number of offspring they can produce, but also because they vary in the timing of reproduction. Whereas female *G. scandens* can begin breeding at the age of 1 year and in favorable conditions may fledge multiple broods each season, African elephant females do not produce offspring until they are at least 15 years old and usually produce only one calf every 5 years or so.

VERTICAL LIFE TABLES Because not all species can be easily followed over time, some life tables are constructed by sampling a population at a single time. These life tables cut across all age categories and thus are known as **vertical life tables**. One way to construct a vertical life table is to record information from a death assemblage, a collection of bodies or fossils of individuals that lived together in a particular place at a given time. Similarly, the birth and death dates on tombstones in a cemetery, for example, can be used to construct a vertical life table for a human population and to estimate its probability of reaching different ages.

Survivorship curves reflect life history strategies

The construction of life tables has allowed ecologists to observe common life history patterns, reflecting common solutions to ecological challenges, across a tremendous diversity of organisms. For example, the proportions of individuals surviving through each life stage (survivorship, l_x) can be taken from a life table and plotted graphically to construct a **survivorship curve**. Typically, a survivorship curve is constructed for a hypothetical cohort, usually of 1,000 individuals, by plotting the numbers of individuals expected to survive to reach each age category on a logarithmic scale.

Ecologists have noticed that survivorship curves tend to take one of three general shapes:

- Species with **physiological survivorship curves** experience high overall survivorship through adulthood but steep declines late in life (the graphic representation is concave; **Figure 55.5A**). Species with this type of survivorship curve (such as humans, elephants, and many other large mammals) typically have low fecundity but provide parental care to their offspring, which reduces the risk of death in early stages of development.

- Species with **ecological survivorship curves** are faced with a constant risk of mortality at all ages (the graphic representation is linear; **Figure 55.5B**). Many bird species display this pattern.

- Species with **maturational survivorship curves** experience low survivorship early in life and higher survivorship once they reach maturity (the graphic representation is convex; **Figure 55.5C**). Species with this type of survivorship curve (such as most insects, marine invertebrates, and annual plants) tend to produce many offspring but provide little or no parental care.

These different survivorship curves reflect differences in the ways in which organisms partition their time and energy among growth, maintenance, and reproduction; the way a species partitions its energetic resources is referred to as its **life history strategy**. Understanding the risks organisms face during different stages of their lives helps clarify why life histories differ among species. Although these basic strategies are to a large degree genetically and taxonomically determined, varying environmental conditions can influence life history traits. Witness how, for example, the number of offspring produced by cactus ground finches in a year depends on cactus flower and fruit availability, which in turn depends on the availability of rainfall.

▮▮▮▮▮▮▮▮▮▮▮▮▮▮▮▮▮ **RECAP 55.2**

Life tables can be constructed either by following a cohort of individuals through time or by recording age at death in a vertical life table. Survivorship curves can be constructed by plotting the likelihood of survival to different ages. Differences in the shape of these curves can shed light on differences in life history strategies.

- What kinds of information does a life table provide about a population? **See pp. 1154–1155 and Table 55.1**

- What are the differences between a vertical and a cohort life table? **See pp. 1154–1155**

- Describe the three types of survivorship curves. **See pp. 1155–1156 and Figure 55.5**

Environmental variation influences survivorship and fecundity. Comparisons across populations and species reveal different patterns in life history traits, which allow organisms to cope with different environmental challenges.

55.3 How Do Environmental Conditions Affect Life Histories?

Because resources and mortality factors vary greatly among environments, life history strategies also vary dramatically. Those variations, in turn, determine how fast populations can grow.

Survivorship and fecundity determine a population's growth rate

To see how a population is likely to grow, ecologists can use life table data to calculate the population's **per capita growth rate**, symbolized **r**. A population's growth rate is the difference between the per capita birth rate (*b*) and the per capita death rate (*d*) (leaving aside, for the moment, immigration and emigration). In other words, it is the average rate of change in population size per individual per unit of time. It is expressed by the equation

$$r = b - d \qquad (55.2)$$

If the per capita birth rate is greater than the per capita death rate, then $r > 0$ and the population is growing. If the per capita death rate is greater than the per capita birth rate, then $r < 0$ and the population is declining. The equilibrium state, $r = 0$, would indicate a stable population that is neither growing nor declining.

The maximum value of r (r_{max}) is referred to as the population's **intrinsic rate of increase**. It reflects the rate of increase that is inherent in the organism under ideal conditions—that is, independent of any external (environmental) constraints on population growth. A population can reach r_{max} only for a limited time, if at all, since environmental constraints almost always exist.

Life history traits vary with environmental conditions

Birth rates and death rates are both influenced by environmental factors, so r changes as the environment changes. The life history traits most influenced by environmental conditions include:

- age at first reproduction (generation time)

- number of broods per female (the number of times a female produces offspring)

- number of offspring per brood (the number of offspring produced each time a female reproduces)

These traits vary not only between species, but also between populations of the same species.

Opportunities for reproduction for some species are limited to certain locations or certain times of year, whereas other species and populations can breed continuously over their life span. Many desert wildflowers grow and flower only during the spring rainy season, and they may not be able to reproduce at all in years when rainfall is inadequate. In contrast, some tropical vines flower continuously in their warm, moist rainforest environment.

Species that can reproduce multiple times over the course of their adult lives are **iteroparous** (*itero*, "repeat"; *pario*, "beget"). **Semelparous** species (*semel*, "once") reproduce only once (**Figure 55.6**). Generally speaking, semelparous species produce many more offspring in a single brood than iteroparous species do over their entire lifetimes; semelparity is thus sometimes referred to as "big bang" reproduction.

Semelparity is typical of organisms that experience no great survival advantage upon reaching adulthood; it includes some fishes, many insects, and all annual plants. In contrast, iteroparity is typical of organisms whose survival chances increase once they reach maturity. For example, because environmental conditions within the nests of social insects such as honey bees and ants are remarkably stable, iteroparity is the rule among these species; some queens live 10 years or longer and reproduce over their entire adult lives.

Orgyia antiqua

55.6 Big Bang Reproduction Semelparous species reproduce only once and invest a great deal of energy in producing the maximum number of offspring. Female rusty tussock moths do not fly but remain with their empty cocoons, which are attached to the plants that are the caterpillar-stage food source. The stationary female lays a large number of eggs and then dies. When the eggs hatch the following spring, the larvae are surrounded by foliage they can eat.

Life history traits are influenced by interspecific interactions

Predation and other interactions among species can influence life history strategies in many ways. Some populations of guppies (*Poecilia reticulata*) in Trinidad, for example, live in streams where they are attacked and eaten by larger fish. But some streams have waterfalls that predatory fishes are unable to negotiate. Guppies that live in the predator-free areas upstream from those waterfalls have lower death rates than guppies below the falls. To see whether the risk of being eaten by a predator influenced the life history strategies of these guppies, David Reznick and his colleagues collected guppies from high-predation and low-predation sites and raised them in the laboratory. Some guppies from each group were provided with plentiful food, and others with limited food, to simulate the variation the fish would typically encounter in their home streams. In the laboratory, where no predators were present, guppies from high-predation sites matured earlier, reproduced more frequently, and produced more offspring in each brood than did guppies from low-predation sites, no matter how much food they received. The investigators concluded that predation had selected for early and frequent reproduction.

RECAP 55.3

The difference between birth rate and death rate provides an estimate of a population's per capita growth rate, or *r*. That rate is strongly influenced by the population's life history strategy, which in turn is highly dependent on environmental conditions.

- Give some examples of life history traits that vary with environmental conditions. **See p. 1156**
- Explain the difference between iteroparity and semelparity. **See p. 1156**
- How can predation affect the evolution of life history strategies? **See p. 1157**

In any given species, environmental factors may influence the growth of populations differently in different places and at different times.

55.4 What Factors Limit Population Densities?

What would happen if all the offspring produced by a population survived to reproduce themselves? The prospects are alarming. In 1911, L. O. Howard, then chief entomologist of the U.S. Department of Agriculture, estimated that, if all their offspring were to survive, a pair of flies beginning to reproduce on April 15 would produce a population of 5,598,720,000,000 adults by September 10 of the same year. Other entomologists took issue with Howard's calculation—they pegged the number much *higher*. Given such amazing reproductive capacities, it is clear there are forces at work that limit the growth of fly populations (and populations of every other organism).

All populations have the potential for exponential growth

As the number of individuals in a population increases, the number of reproducing individuals also increases, so the number of new individuals added per unit of time accelerates, even though the per capita rate of increase remains constant. If births and deaths occur continuously at constant rates, a graph of the population size over time forms a continuous upward curve (**Figure 55.7**). This pattern is known as **exponential growth**. Mathematically, the change in the number of individuals *N* over an interval of time *T* can be expressed as $\Delta N / \Delta T$ where Δ is the mathematical representation for "change in." The ratio between *N* and *T* can be expressed as the average contribution of each individual to population growth *r* (remember from Equation 55.2 that $r = b - d$) multiplied by the number of individuals in the population. In mathematical terms,

$$\frac{\Delta N}{\Delta T} = rN$$

Using the notation of differential calculus, which in this case simply indicates that the time interval represented by Δ is short, this equation can be expressed as

$$\frac{dN}{dT} = rN \qquad (55.3)$$

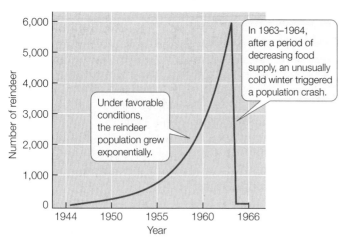

55.7 Exponential Population Growth Can Lead to a Population Crash   The reindeer herd introduced on St. Matthew Island experienced favorable conditions and grew exponentially for many years. A single catastrophically cold winter triggered a population crash that eventually resulted in the death of the entire population.

Go to Animated Tutorial 55.2
Exponential Population Growth
Life10e.com/at55.2

The term dN/dT is the rate of change in the size of the population over time, and the expression rN is sometimes called the **biotic potential** of the population.

Some populations may grow at rates close to their biotic potential, but only for short periods. During the 20 years following their introduction, the reindeer population described at the opening of this chapter grew exponentially, as seen in Figure 55.7. When the herd was first introduced, it had ample habitat, abundant food, and no predators, so there was nothing to limit the population's growth. Favorable climate conditions also allowed the population to grow exponentially. A sudden change in climate conditions—deep snow that made foraging difficult—was a major factor leading to the population's crash. Although the crash was precipitated by unusually harsh weather conditions, the relatively poor physical condition of the reindeer in the herd, caused by overcrowding and overgrazing of the lichens that were their principal food source, contributed to the massive mortality.

Go to Media Clip 55.1
The Biotic Potential of a Population
Life10e.com/mc55.1

Logistic growth occurs as a population approaches its carrying capacity

No real population can maintain exponential growth for very long. As a population increases in density, the resources it requires—such as food, nest sites, and shelter—become depleted. In the absence of adequate resources to sustain more individuals, birth rates drop and death rates rise.

Any given environment has only enough resources to support a finite number of individuals of a species indefinitely.

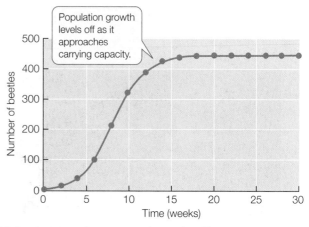

55.8 Logistic Population Growth Levels Off   In an environment with limited resources, a population typically stops growing exponentially before it reaches the environmental carrying capacity (*K*). The data here were recorded from a laboratory population of sawtoothed grain beetles maintained on a constant food supply; it is a typical logistic growth pattern, which forms an S-shaped curve.

Go to Animated Tutorial 55.3
Logistic Population Growth
Life10e.com/at55.3

That number of individuals, referred to as the environment's **carrying capacity (K)**, is a function of its resources. The growth of a population typically slows down as its density approaches the environmental carrying capacity. A population that exhibits decreasing growth as resources become more scarce displays a pattern called **logistic growth**, in which a graph of population size over time forms an S-shaped curve. **Figure 55.8** shows this growth pattern in a laboratory population of beetles maintained on a constant food supply.

To generate the S-shaped logistic growth curve from the equation for exponential growth we add a term,

$$\frac{K - N}{K}$$

This quantity represents the reduction in population growth caused by preemption of available resources and is referred to as **environmental resistance**. As long as the population size is less than the carrying capacity (i.e., $N < K$), only a fraction of the available resources are being used. As the population size approaches the carrying capacity, however, the fraction of resources available for any new individual becomes smaller. The implication is that each individual added to the population depresses population growth by an equal amount. Thus

rate of change in population size over time =
biotic potential × environmental resistance

or, in mathematical terms,

$$\frac{dN}{dT} = rN \times \frac{K - N}{K} \qquad (55.4)$$

Population growth should stop when $N = K$ because at that point, $K - N = 0$, so $(K - N)/K = 0$, and thus $dN/dT = 0$ and the population remains at a constant size.

Go to Activity 55.1 Logistic Population Growth
Life10e.com/ac55.1

r-strategists	K-strategists
HABITAT Can inhabit a broad range of habitats. High tolerance for both environmental instability and low-quality resources.	**HABITAT** Specific habitat requirements, including environmental stability. Efficient users of specific and usually high-quality resources.
PHYSIOLOGY Rapid embryonic development, rapid maturation to reproductive age, small body size.	**PHYSIOLOGY** Extended embryonic development, long maturation to reproductive age, large body size.
REPRODUCTIVE STRATEGY Random mating. Reproduce once (semelparity) resulting in a large number of offspring. Little or no parental investment in each offspring.	**REPRODUCTIVE STRATEGY** Mate choice, pair bonds. Reproduce many times (iteroparity), each event producing few offspring. Large parental investment in each offspring.
SURVIVORSHIP Short life span, density-independent mortality, typically a maturational survivorship curve (see Figure 55.4).	**SURVIVORSHIP** Long life span, density-dependent mortality, typically physiological or ecological survivorship curve (see Figure 55.4).
POPULATION FLUCTUATION Short periods of exponential population growth followed by periodic or seasonal population crashes.	**POPULATION FLUCTUATION** Slowly rising population growth that stabilizes and levels off at carrying capacity (K).
EXAMPLES Dandelions, house flies, rabbits	**EXAMPLES** Oak trees, bluebirds, polar bears

55.9 Two Life History Strategies Species whose life histories are geared to achieve the maximum possible rate of population increase are referred to as *r*-strategists; those whose population dynamics are bounded by carrying capacity are *K*-strategists. The life histories of most species combine elements of both strategies.

Different population regulation factors lead to different life history strategies

Species vary in their capacity to reproduce, as well as in the extent to which they are vulnerable to density-dependent and density-independent mortality factors. Some of this variation in life history traits appears to result from adaptation to different habitat conditions. Generally, unpredictable habitats are associated with high fecundity and correspondingly high intrinsic rates of increase as organisms make the most of rare opportunities to reproduce. Conversely, predictable habitats, where organisms have a high probability of reproductive success, are associated with low fecundity and low *r*.

Species whose life history strategies allow for high intrinsic rates of increase are called **r-strategists**, and species whose life history strategies allow them to persist at or near the carrying capacity (*K*) of their environment are called **K-strategists** (Figure 55.9). Keep in mind, however, that these categories are not absolute; many species fall along a continuum between these two strategies.

For *r*-strategists, life is uncertain. Individuals tend to reproduce only once and to produce large numbers of offspring. They can generally use a wide variety of resources and tolerate a wide range of conditions. *K*-strategists are adapted to predictable environments, are long-lived, and reproduce several times; their smaller numbers of offspring have a high probability of surviving to adulthood. *K*-strategists tend to be more specialized in their resource use and less tolerant of variation in resource quality.

That life history strategies can evolve is suggested by genetic correlations among suites of life history traits. Such genetic correlations imply either simultaneous selection on two or more life history traits or linkages among the genes that code for those traits. Across *Drosophila melanogaster* strains, for example, a high intrinsic rate of increase is correlated with the ability to reproduce under starvation conditions and with the ability to develop on a variety of media in the laboratory—both of which are consistent with the *r* strategy of tolerating a wide range of resources and conditions.

Several ecological factors explain species' characteristic population densities

Density-dependent and density-independent factors can explain how populations grow or decline, but they do not explain why

Population growth can be limited by density-dependent or density-independent factors

When resources are limited, adding more individuals to a population risks making things worse for everyone. Factors with an effect on population size that increases in proportion to population density are called **density-dependent** regulation factors. These factors include the following:

- *Food supply.* As a population increases, it may deplete its food supply, reducing the amount of food available to each individual. Poor nutrition may then increase the death rate or decrease the birth rate.

- *Predators* may be attracted to areas with high densities of their prey. If predators capture a larger proportion of the prey population than they did when that population was small, the death rate of the prey population rises.

- *Pathogens* may spread more easily in dense populations than in populations with fewer individuals per unit of area, resulting in a rise in the death rate.

Not all factors that change population size act in a density-dependent manner, however. A period of extreme cold, or an exceptionally strong hurricane, may kill a large proportion of the individuals in a population regardless of the population's density; such an event is **density-independent**. Abiotic factors tend to act in a density-independent manner, whereas biotic factors (such as food supply) tend to be density-dependent. For an ecological process to regulate population size (i.e., to maintain the population at a certain level), it must exhibit density dependence such that some sort of negative feedback is applied when populations increase.

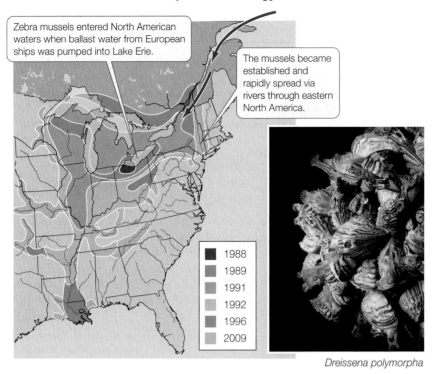

Zebra mussels entered North American waters when ballast water from European ships was pumped into Lake Erie.

The mussels became established and rapidly spread via rivers through eastern North America.

1988
1989
1991
1992
1996
2009

Dreissena polymorpha

55.10 Newly Introduced Populations Can Grow Rapidly Between 1988 and 2009, the range of zebra mussels in eastern North America increased exponentially, through rapid population growth as well as by inadvertent transport on barges moving among waterways. Female mussels can lay more than 1 million eggs in a single season, and in North America the species has few natural predators. Humans can unwittingly transport zebra mussel larvae from one body of water to another on their fishing boats and other watercraft, and in recent years the invasive pest has begun to appear in lakes and streams of the American West.

population densities much higher than those in their native ranges. Sometimes these high population densities are only temporary; these densities decline if and when new mortality factors exert an influence. In the absence of such factors, however, populations in the newly colonized habitat may remain so dense that the introduced species becomes a major problem for native species.

The population of zebra mussels (*Dreissena polymorpha*) in North America demonstrates the speed with which newly introduced populations can grow. Zebra mussels first appeared in Lake St. Clair, between Lake Erie and Lake Huron, in 1988. They were probably transported there from Europe in the ballast water of transoceanic cargo ships. They spread rapidly and today occupy most of the Great Lakes and the Mississippi River drainage (**Figure 55.10**), reaching densities as high as 400,000 individuals per square meter in some places. Because they attach to any stable underwater substrate, zebra mussels can cover the bottoms of boats and clog municipal water supply intakes and power plant pipelines. They even settle on other aquatic organisms, causing problems for native mussels. Such high densities are never found in their native Europe, where more than 100 species of predators and parasites keep zebra mussel population densities under control.

Evolutionary history may explain species abundances

The three key factors that explain variation in population densities cannot explain all differences in species abundance. For example, although Douglas firs (*Pseudotsuga menziesii*) and giant sequoias (*Sequoiadendron giganteum*) are both large trees that use the same sources of energy (sunlight) and nutrients (soil), Douglas firs are abundant throughout western North America, whereas giant sequoias are restricted to a few groves in the Sierra Nevada. Similarly, each of several species of desert pupfish (genus *Cyprinodon*) is restricted to a single spring in Death Valley, California, whereas smallmouth bass (*Micropterus dolomieu*) can be found in many of the rivers and lakes of eastern North America. To explain these differences, it is important to know not just the contemporary ecology of these organisms, but also their evolutionary history.

As Chapter 23 described, a new species can originate in several ways. Species that arise by polyploidy or by founder events inevitably begin with a very small, local population. Desert pupfish species appear to have evolved in isolation as increasing aridity in Death Valley over the past 50,000 years cut once continuous populations off from one another. Conversely, when a species is declining toward extinction (as may be happening to the giant sequoia), its range shrinks until it vanishes when the last individual dies.

some species are common whereas others are rare—that is, why the characteristic densities of species differ. Many factors explain why typical population densities vary so greatly among species, but three of these factors are especially influential:

- *Species that use abundant resources generally reach higher population densities than species that use scarce resources.* Thus, on average, the fruit fly *Drosophila melanogaster*, which feeds on yeasts and other microbes found on just about any kind of rotten fruit, reaches substantially higher population densities than do other fruit fly species that feed on the microbes found on specific fruits.

- *Species with small body sizes generally reach higher population densities than species with large body sizes.* In general, population density decreases as body size increases because, on a per capita basis, small individuals require less energy to survive than large individuals.

- *Complex social organization may facilitate high population densities.* Highly social species, including ants, termites, and humans, can achieve remarkably high population densities.

Some newly introduced species reach high population densities

Species that are introduced into a new region, where their normal predators and pathogens are absent, sometimes reach

RECAP 55.4

Population sizes are limited by the carrying capacity of the environment, which is determined by the availability of resources. Species associated with unpredictable habitats tend to be *r*-strategists, whereas species associated with predictable habitats tend to be *K*-strategists.

- Why can populations grow exponentially only for short periods? See pp. 1157–1158 and Figures 55.7 and 55.8
- What is the difference between density-dependent and density-independent factors that influence populations size? See p. 1159
- Describe the characteristics of *r*-strategists and *K*-strategists. See p. 1159 and Figure 55.9

A species is rarely found in all of the habitats that seem suitable for it. Geological history and the evolutionary histories of species supply one type of explanation for this observation (see Section 54.5). This chapter next explores another area of explanation: spatial variation in habitat suitability.

55.5 How Does Habitat Variation Affect Population Dynamics?

Most natural history field guides have maps showing the geographic range over which a species can be found. But not even the most abundant species is found everywhere within its mapped range. Every species has particular habitat requirements that determine where within its potential range it will occur.

Many populations live in separated habitat patches

Most organisms live in distinct **habitat patches**, areas of a particular habitat type surrounded by areas of less suitable habitat. For example, caterpillars of the Bay checkerspot butterfly (*Euphydryas editha bayensis*) feed on only two species of annual plants (California plantain and purple owl's clover) that grow only on outcrops of serpentine rock on hills south of San Francisco, California. The butterflies are restricted to patches of these plants and cannot establish populations in the surrounding habitats that lack them.

Some populations living in separated habitat patches are effectively divided into discrete **subpopulations** that are linked together by regular movement of individuals between patches. The larger population that includes all such subpopulations is known as a **metapopulation**. Each subpopulation has a probability of "birth" (colonization of its habitat patch) and "death" (extinction in that patch). Each subpopulation grows in the ways we have described, but because the subpopulations are much smaller than the metapopulation, local disturbances and random fluctuations in numbers of individuals are more likely to cause the extinction of a subpopulation than of the entire metapopulation, as we will explain in Chapter 59. However, if individuals move frequently between subpopulations, immigration may prevent declining subpopulations from becoming extinct, a process called the **rescue effect**.

The Bay checkerspot butterfly provides a dramatic illustration of the dynamics of metapopulations. In 1960 Paul Ehrlich and his colleagues at Stanford University began studying a population of this butterfly in the Jasper Ridge Biological Preserve near the Stanford campus. They determined that the Jasper Ridge population was actually one of several subpopulations within a large, very fragmented metapopulation. They followed the Jasper Ridge subpopulation, as well as several other subpopulations within this metapopulation, over a number of years and found that the subpopulations varied enormously and asynchronously in size. Larval survival depended on climate factors (particularly temperature), the timing of rainfall, and host plant survival.

During drought years, most host plants died early in spring, before the caterpillars had developed enough to enter their summer resting stage. A severe drought in 1975–1977 led to extinctions of some of the subpopulations. One of the empty patches was repopulated a few years later, most likely by individuals from the largest single subpopulation, Morgan Hill, which as late as 1989 contained several hundred thousand butterflies (**Figure 55.11**). In 1998, however, the Morgan Hill subpopulation, which had historically been the largest in the metapopulation, went extinct. Ehrlich and his colleagues examined 70 years of climate data for the region and concluded that increasing climate variation accounted for the extinction.

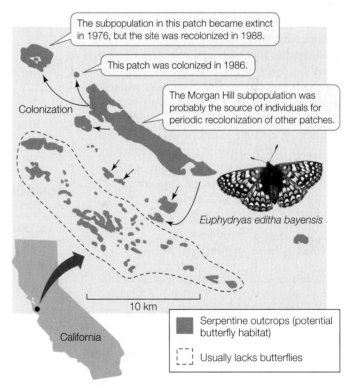

55.11 A Checkerboard of Checkerspots The Bay checkerspot butterfly metapopulation is divided into several subpopulations confined to patches of habitat (serpentine rock outcrops) that contain the species' food plants.

Without a large, stable source subpopulation to provide emigrants for recolonization, as the Morgan Hill subpopulation did during the 1970s drought, it is unlikely that any of the other subpopulations will persist without human intervention.

Corridors may allow subpopulations to persist

In any metapopulation, habitat between patches through which organisms can move, known as **corridors**, plays a critical role in facilitating dispersal to maintain subpopulations. What constitutes a corridor depends on the dispersal ability of the organism. Studying corridors is experimentally daunting because long distances may separate patches; moreover, movements of animals, particularly those that fly, can be difficult to monitor. Therefore one of the first tests of the importance of corridors was a small-scale manipulative experiment using mosses growing on rocks, which provide habitat for several small arthropod species, including springtails (minute wingless hexapods) and mites.

The investigators created patches of habitat by clearing away the mosses surrounding the patches (**Figure 55.12**). In small, completely isolated patches, the number of small arthropod species present declined about 40 percent within 6 months. The investigators also created patches that were connected by narrow corridors of moss. In some cases the corridors were left intact; in others, "pseudocorridors" were disrupted by a barrier 2 cm wide. A 2-cm barrier may seem small, but it presents a daunting obstacle to arthropods only 2 mm wide. Six months later, patches connected by unbroken corridors contained more small arthropod species than did patches connected by the disrupted pseudocorridors.

 Go to Animated Tutorial 55.4
Habitat Fragmentation
Life10e.com/at55.4

A larger scale study of the effects of corridors was conducted in Palenque National Park in Mexico. Although about one-third of the park comprises tropical rainforest, that forest is surrounded by a patchwork of cattle pasture, river habitat, and forest fragments. Investigators moved individual birds, representing a wide range of species, from one patch of forest in the park to another. Some individuals were moved between forest patches that were in close proximity but were not connected by forest corridors (i.e., they were completely surrounded by cattle pastures). The rest were moved between patches that were in close proximity and surrounded by cattle pastures, but were physically connected by narrow corridors of forest habitat. The investigators then monitored the return of the captured birds to their home forest patches. Across all species, birds were more than six times more likely to be recaptured in home forest patches connected by corridors to the patches where they were released than in home forest patches that were unconnected to the patches where they were released. Even narrow forest corridors may be beneficial to tropical forest birds, which experience an increased risk of predation and greater physiological stress when they have to fly across open areas.

INVESTIGATING**LIFE**

55.12 Corridors Can Rescue Some Populations Data from the experiments by Andrew Gonzales and Enrique Chaneton summarized here suggest that corridors between patches of habitat increase the chances of recolonization, and thus of subpopulation persistence.[a]

HYPOTHESIS Subpopulations of a fragmented metapopulation are more likely to persist if there is no barrier to recolonization.

Method 1. On replicate moss-covered boulders, scrape off the continuous cover of moss to create a "landscape" of moss "mainland" with patches surrounded by bare rock. A central 50 cm × 50 cm moss "mainland" (M) is surrounded by 12 circular patches of moss, each 10 cm^2 (subpopulations). In the "insular" treatment (I), the patches are surrounded by bare rock (which is inhospitable to moss-dwelling small arthropods, and thus a barrier to recolonization). In the "corridor" treatment (C), the patches are connected to the mainland by a 7 × 2 cm strip of live moss. In the "broken-corridor" treatment (B), the configuration is the same as the "corridor" treatment, except that the moss strip is cut by a 2-cm strip of bare rock.

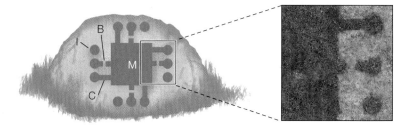

2. After 6 months, determine the number of small arthropod species present in each of the mainlands and small patches.

Results Patches connected to the mainland by corridors retained as many species as did the mainland to which they were connected. Fewer species remained in the broken-corridor and insular treatments.

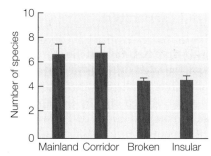

CONCLUSION Barriers to recolonization reduce the number of subpopulations that persist in a metapopulation.

Go to **BioPortal** for discussion and relevant links for all INVESTIGATING**LIFE** figures.

[a]Gonzalez, A. and E. J. Chaneton. 2002. *Journal of Animal Ecology* 71: 594–602.

A metapopulation consists of separate subpopulations living in distinct habitat patches. Corridors that facilitate movement between patches increase the chances of subpopulation persistence.

- What effects do patches of unsuitable habitat have on population structure? **See p. 1161 and Figure 55.11**
- What effects do corridors between habitat patches have on subpopulations? **See p. 1162 and Figure 55.12**

For many centuries, people have tried to reduce populations of species they consider undesirable and maintain or increase populations of desirable or useful species. Such efforts to manage populations are more likely to be successful if they are based on knowledge of how those populations grow and what determines their densities.

How Can We Use Ecological Principles to Manage Populations?

If we wish to manage other species—that is, to increase or decrease their populations—we need to understand their life histories and the dynamics of their populations. The principles of population dynamics can also help us understand the effects our own population and its activities are having on other species.

Management plans must take life history strategies into account

Knowing the life history strategy of a species can be helpful in managing populations of commercial value. The black rockfish (*Sebastes melanops*), an important game fish that lives off the Pacific coast of North America, provides one such example. Rockfish have an indeterminate growth pattern—they continue to grow throughout their lives. As in many other animals, the number of eggs a female rockfish produces is proportional to her size, so larger females produce more eggs than smaller females. In addition, older, larger females are better able to provision the eggs they produce with oil droplets, which provide energy to the newly hatched larvae, giving them a head start in life (**Figure 55.13**). Larvae from eggs with larger oil droplets, produced by larger females, grow faster and survive better than do larvae from eggs with smaller oil droplets.

These life history traits have important implications for the management of rockfish populations. Because fishermen prefer to catch big fish, intensive fishing off the Oregon coast from 1996 to 1999 reduced the average age of female rockfish from 9.5 to 6.5 years. Thus the females reproducing in 1999 were, on average, smaller than the females reproducing in 1996. This change decreased the average number of eggs produced by females and reduced the average growth rate of larvae by about 50 percent. This reduction in reproductive ability was linked to a decrease in the ability of the rockfish population to recover from intensive fishing. Maintaining productive populations of rockfish may require setting aside no-fishing zones where some females can be protected from fishing and allowed to grow to large sizes.

(A) *Sebastes melanops*

(B)

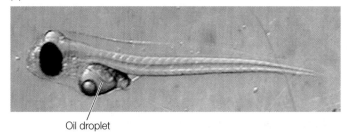

Oil droplet

55.13 Energy Stocks Give Rockfish a Head Start (A) Among rockfish, older, larger females are more reproductively successful, producing both more eggs and eggs with larger nutritive oil droplets. (B) The oil droplet attached to this rockfish larva provides it with nutrition to fuel its growth until it can feed on its own.

Management plans must be guided by the principles of population dynamics

If we look at a logistic growth curve (see Figure 55.8), we can see that the number of births tends to be highest when a population is well below its carrying capacity. Therefore if we wish to maximize the number of individuals that can be harvested from a population, we should manage the population so that it is far enough below the carrying capacity to have a high birth rate. Hunting and fishing regulations are established with this objective in mind.

Populations that have high intrinsic rates of increase can persist even if harvest rates are high. In such populations (which include many fish species), each female may produce thousands or millions of eggs. In many of these fast-reproducing populations, the growth rates of individuals are density-dependent. Therefore if prereproductive individuals are harvested at a high rate, the remaining individuals may grow faster. Some fish populations can be harvested heavily on a sustained basis because a relatively small number of females can produce sufficient numbers of eggs to maintain the population.

Fish can, however, be overharvested, as illustrated by the story of the black rockfish. Many fish populations have been greatly reduced because so many individuals were harvested that the few surviving reproductive adults could not maintain the population. Georges Bank, off the northeast coast of North America—a source of cod, haddock, and other prime food

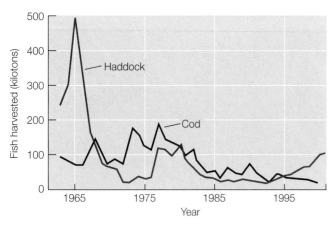

55.14 Overharvesting Can Reduce Fish Populations Populations of cod and haddock on Georges Bank—and thus harvests of these species—have crashed due to overfishing.

Bufo marinus

55.15 Biological Control Gone Awry The Central American cane toad not only failed to control destructive beetles in Australia's sugarcane fields, but increased dramatically in abundance and now threatens many native Australian species (including the native frog this individual is eating).

fishes—was exploited so heavily during the twentieth century that many fish stocks have been reduced to levels insufficient to support a commercial fishery (**Figure 55.14**). The haddock population has rebounded enough to support a fishery because commercial fishing of that species ceased and was restarted only after the population had recovered. In contrast, managers reduced fishing pressure on cod only slowly, and the cod population has failed to increase.

Many rapidly reproducing species can recover if overharvesting is stopped, but recovery is more difficult for slowly reproducing species. Twentieth-century whalers hunted the blue whale (*Balaenoptera musculus*), Earth's largest animal, nearly to extinction. These whales reproduce very slowly: they live up to 10 years before becoming reproductively mature, produce only one offspring at a time, and have long intervals between births. Not surprisingly, the population has failed to recover.

Whether we want to manage the sizes of populations of desirable species for sustainable harvesting or of undesirable species for control purposes, the same principles apply. If the dynamics of a pest population are influenced primarily by density-dependent regulation factors, then killing part of that population will only reduce it to a density at which it will grow faster. A more effective approach to reducing such a population is to remove its resources, thereby lowering the carrying capacity of its environment. For example, we can rid our cities of rats more easily by making garbage unavailable (reducing the carrying capacity of the rats' environment) than by poisoning rats (which only increases their reproductive rate).

Biological control is the use of natural enemies (predators, parasites, or pathogens) to reduce the population density of an economically damaging species. In many cases the target species is a pest only because it has been introduced to a new area. Natural enemies used for biological control are often obtained from the native region of the pest species. Biological control became popular in the nineteenth century after an outbreak of cottony-cushion scale, an Australian insect that attacks citrus, appeared in the citrus groves in California. A predaceous

ladybeetle and a parasitic fly were then introduced from Australia. Within a year of their release, these insects brought the scales under control.

Sometimes, however, introduced natural enemies not only fail to have any effect on the pest they were imported to control but also, freed of their own enemies, become pests themselves. This fact underlies the horror story of the cane toad (*Bufo marinus*) in Australia. This Central American toad (**Figure 55.15**) was introduced to control cane beetles attacking Australian sugarcane fields. But Australian cane beetles stay high on the upper stalks of the plants; the toads could not reach that high, and thus had no effect on the beetle population. Unfortunately, they had massive effects on other species.

All stages of the *B. marinus* life cycle are poisonous, and Australian reptiles (including snakes and lizards) and mammals that eat them usually die. With no enemies to limit their population growth, cane toads grow fast and outcompete native amphibian species for resources. The toads have spread from northern Australia down the east coast, where they threaten native frog species by preying on them as well as by competing with them. The Australian government is forced to spend millions of dollars in attempting to reduce their numbers.

Human population growth has been exponential

In 1798 Thomas Robert Malthus, in his *Essay on the Principle of Population*, pointed out that the human population was growing exponentially but its food supply was not, and argued that at some point, famine and death would be the ultimate fate of the human race. Malthus could not have anticipated the technological innovations over the next 200 years that would greatly enhance the capacity of humans to produce

55.16 The Human Population Is Growing Exponentially The growth rate of the human population has slowed somewhat, but its large size means that millions more people are still being added every year.

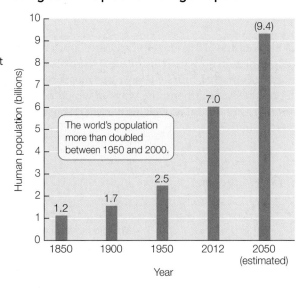

food. Today, however, the size of the human population is once again a serious concern as we confront the effects of our contributions to pollution, habitat destruction, and extinctions of other species.

For thousands of years, Earth's carrying capacity for humans was low because of the relative inefficiency with which we could obtain food and water. The development of social systems and communication, the domestication of plants and animals, ever-increasing crop and livestock yields due to ongoing technological advances, and our increasing proficiency at managing diseases all contributed to unprecedented growth of the human population. It took more than 10,000 years for the population to reach 1 billion, which happened in the early nineteenth century. Today, a mere 200 years later, the planet is home to more than 7 billion human beings (**Figure 55.16**). Population growth has slowed somewhat from its post-World War II highs—estimates place the current worldwide rate of increase to be about 1.1 percent per year—but with a base of 7 billion, even a minimal growth rate means millions more individuals each year.

Human populations are not growing at the same pace across the world. As we saw for elephants in Figure 55.2, in long-lived species the timing of births and deaths affects a population's age distribution for many years. Between 1946 and 1964, the United States experienced a "baby boom." During those years almost 75 million babies were born, and the average number of children per family grew from 2.5 to 3.8. U.S. birth rates declined during the 1960s, but in the 1970s and 1980s the baby boomers became parents, generating another demographic bulge—a "baby boom echo" (**Figure 55.17A**). Today this "echo generation" is on the threshold of becoming the dominant age class.

The age structure of the U.S. population is typical of that of many industrialized nations, but a few highly developed nations (particularly in Europe) are experiencing population declines. In the developing world, however, many countries are experiencing exponential growth rates and have populations

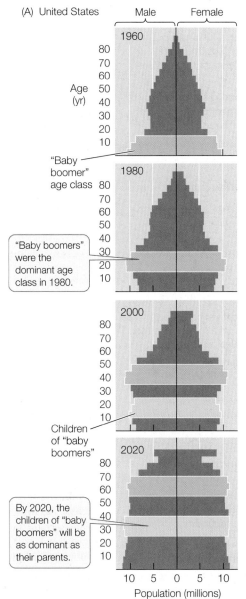

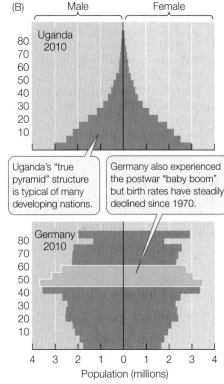

55.17 Population Pyramids (A) Observed and predicted age distributions for the human population of the United States from 1960 to 2020 show how the birth rate during the "baby boom" has influenced the age structure of the country's population over many decades. (B) In Uganda, as in many developing countries, the largest proportion of the population is found in the youngest age groups, which means a greatly increased birth rate as these individuals achieve reproductive age. Conversely, a small but increasing number of highly developed nations have the population structure seen here for Germany, which presages a population decline.

highly skewed toward younger age classes, portending high rates of population growth in the future (**Figure 55.17B**). Population growth rate, of course, is only one measure of human impact on the environment; the ways in which populations use resources are critical as well, as we will discuss further in Chapters 58 and 59.

RECAP 55.6

Efforts to manage populations are more likely to be successful if they are based on an understanding of life histories and population dynamics.

- Describe an effective strategy for reducing a pest population and explain why it is effective. **See p. 1164**
- How have humans changed the carrying capacity of Earth for our own population? **See p. 1165**

Why did introduced reindeer populations persist on the island of South Georgia but not on the island of St. Matthew?

ANSWER

The different fates of the reindeer populations on these two islands reflect differences not only in the physical conditions on the islands but also in the history and purpose of the reindeer introductions. In physical terms, average climate conditions on St. Matthew are harsher than those on South Georgia; catastrophic weather events such as the winter that essentially wiped out the St. Matthew reindeer are far less frequent on South Georgia. Stability in population size is often related to stability in environmental conditions.

In terms of history, the reindeer on South Georgia were brought there by men involved in the whaling trade with the goal of establishing a food supply for ships traveling through the area. As

a consequence, the population experienced regular harvesting (initially by whalers, who shot the reindeer for food, and later by scientists who shot them for research purposes).

That the reindeer population on South Georgia has not crashed, however, does not mean it is at a desirable size. Reindeer densities range from 40 to 85 animals per square kilometer—almost 10 times higher than densities in areas where reindeer are native. At these high densities, the reindeer are having negative effects on South Georgia's native plants and animals, such as the burrow-nesting white-chinned petrel. In February 2011, plans were made to eradicate the reindeer on South Georgia in the hope of preserving the native species.

CHAPTER**SUMMARY** 55

55.1 How Do Ecologists Measure Populations?

- A **population** consists of the individuals of a species that interact with one another within a particular area at a particular time.

- The **density** of a population is the number of individuals per unit of area or volume.

- Ecologists have developed many ways of counting individuals as well as ways of estimating population sizes from a sample, such as the **mark–recapture method**. Review Figure 55.2

- Populations have a characteristic **age structure** and pattern of **dispersion**. Review Figures 55.3, 55.4, ANIMATED TUTORIAL 55.1

55.2 How Do Ecologists Study Population Dynamics?

- **Demographic events**—births, deaths, immigration, and emigration—determine the size of a population.

- **Life tables** provide summaries of demographic events in a population. A cohort life table tracks a **cohort** of individuals born at the same time and records the **survivorship** and **fecundity** of those individuals over time. **Review Table 55.1**

- Life table data can be used to construct a **survivorship curve**. Ecologists describe three general types of survivorship curves, which reflect different life history patterns. **Review Figure 55.5**

- The **life history strategy** of an organism describes how it partitions its time and energy among growth, maintenance, and reproduction.

55.3 How Do Environmental Conditions Affect Life Histories?

- A population's **per capita growth rate** (*r*) is the difference between the per capita birth rate (*b*) and the per capita death rate (*d*).

- Life history traits within a species may vary with habitat.

- Interactions with other species and the abiotic environment can influence the evolution of a species' life history traits.

55.4 What Factors Limit Population Densities?

- Populations can exhibit **exponential growth** for short periods, but eventually their resources become depleted, causing birth rates to drop and death rates to rise. **Review Figure 55.7, ANIMATED TUTORIAL 55.2**

- **Logistic growth** is the pattern seen when the growth of a population slows as its density approaches the environmental **carrying capacity** (*K*). **Review Figure 55.8, ANIMATED TUTORIAL 55.3, ACTIVITY 55.1**

- Species that are *r*-strategists have life histories that allow for high intrinsic rates of increase. *K*-strategists persist at or near the carrying capacity (*K*) of their environment. Many species' life history strategies fall along a continuum between these two extremes. **Review Figure 55.9**

- Population densities are determined by both **density-dependent** and **density-independent** factors. Several factors—including resource abundance, body size, and social organization—influence population densities.

continued

55.5 How Does Habitat Variation Affect Population Dynamics?

- No species is found everywhere within its range. Members of most species live in distinct **habitat patches**.

- A **metapopulation** consists of separate **subpopulations** among which some individuals move on a regular basis. **Review Figure 55.11**

- Extinction of a subpopulation may be prevented by immigration of individuals from another subpopulation, a process known as the **rescue effect**. **Corridors** between patches may facilitate such movement. **Review Figure 55.12, ANIMATED TUTORIAL 55.4**

 55.6 How Can We Use Ecological Principles to Manage Populations?

- To manage populations, it is important to understand their life histories and population dynamics. To maximize the number of individuals that can be harvested from a population, the population should be kept well below carrying capacity.

- Reducing the carrying capacity of the environment for a pest species is a more effective way to reduce its population than killing its members.

- Earth's carrying capacity for humans depends on our use of resources and the effects of our activities on the environment. Human populations grow at different rates in different parts of the world. **Review Figures 55.16, 55.17**

 Go to the Interactive Summary to review key figures, Animated Tutorials, and Activities
Life10e.com/is55

CHAPTER**REVIEW**

▨ REMEMBERING

1. A group of individuals of the same species born at the same time is known as a
   a. deme.
   b. subpopulation.
   c. Mendelian population.
   d. cohort.
   e. taxon.

2. A population whose size remains constant at its carrying capacity is exhibiting
   a. exponential growth.
   b. geometric growth.
   c. logistic growth.
   d. J-shaped growth.
   e. negative growth.

3. The process by which immigrants prevent a subpopulation from becoming extinct is called the
   a. colonization effect.
   b. rescue effect.
   c. metapopulation effect.
   d. genetic drift effect.
   e. salvage effect.

4. Which of the following mortality factors is *least* likely to act in a density-dependent manner?
   a. Predation
   b. Disease
   c. Food supply
   d. Fire
   e. All of these factors act in a density-dependent manner.

5. The best way to reduce the population of an undesirable species in the long term is to
   a. reduce the carrying capacity of the environment for the species.
   b. selectively kill reproducing adults.
   c. selectively kill prereproductive individuals.
   d. attempt to kill individuals of all ages.
   e. sterilize individuals.

6. Populations that are most readily overharvested are characterized by having
   a. very long-lived adults.
   b. short prereproductive periods and many offspring.
   c. short prereproductive periods and few offspring.
   d. long prereproductive periods and few offspring.
   e. long prereproductive periods and many offspring.

▨ UNDERSTANDING & APPLYING

7. Most organisms that humans manage for higher densities are long-lived and have low reproductive rates, whereas most organisms that humans want to reduce in numbers are short-lived but have high reproductive rates. What is the significance of these differences for management strategies and the effectiveness of management practices?

8. In the mid-nineteenth century, the human population of Ireland was largely dependent on a single food crop, the potato. When a disease caused the potato crop to fail, the Irish population declined drastically for three reasons: (1) a large percent of the population emigrated to the United States and other countries; (2) the average age of a woman at marriage increased from about 20 to about 30 years; and (3) many people starved to death. None of these social changes were planned at the national level, yet they all contributed to adjusting the population size to the new carrying capacity. Discuss the ecological principles involved, using examples from other species.

ANALYZING & EVALUATING

9. One method of controlling introduced pest species is to introduce a natural enemy (a predator, parasite, or pathogen) from the pest's native habitat to reduce its population density. However, some species introduced to control a pest have become pests themselves. Some scientists argue that biological controls should not be used under any circumstances for pest management. Others argue that, provided they are properly studied and thoroughly vetted, we should continue to use biological control organisms as part of our set of tools for managing pests. Which view do you support, and why?

10. Section 55.5 described two studies of the effects of corridors on metapopulation dynamics—one on tiny arthropods with limited dispersal abilities (see Figure 55.12) and another on birds of tropical forests in Mexico (see p. 1162). Given the differences in size and mobility between tiny, wingless arthropods and forest birds, is it possible to come up with a general definition of a corridor? How could an investigator conduct a single experiment to determine the effects of corridors on multiple organisms that differ widely in size and mobility? Is it important to consider more than one group of organisms in trying to understand the effects of corridors in fragmented habitats?

Go to BioPortal at **yourBioPortal.com** for Animated Tutorials, Activities, LearningCurve Quizzes, Flashcards, and many other study and review resources.

56

Species Interactions and Coevolution

Fungus Farmers *Atta cephalotes* is one Central American species of leafcutter ant. Leafcutter ants harvest and transport leaf fragments to their nests where the vegetation will nourish a thriving crop of fungus, which the ants consume.

NOT MANY INSECTS can claim to have been the subject of a Hollywood film-making feud, but ants are an exception. In 1998 two animated films from competing studios, *Antz* and *A Bug's Life*, were released within a month of each other. Whether as animated entertainment or the subject of scientific study, the behaviors of these social insects have long fascinated humans.

The 50 or so species of leafcutter or "parasol" ants owe their name to their habit of clipping bits of leaves and holding the pieces above their heads like parasols as they cart them off to their nests. The ants don't eat the leaf matter they collect, however. The cut leaves will serve as a substrate for growing the fungi on which leafcutter ants feed.

When a new queen ant leaves her mother's nest, she takes with her a portion of the fungal mass on which she was raised. After mating, she digs into the soil to form a tunnel ending in a chamber, in which she places the pellet of fungus and lays eggs. Her offspring eat the fungus and develop into pint-size workers, which then collect leaf material to "feed" the fungus. As the fungus garden expands, the ants construct more nest chambers. In 3 years the number of workers in a nest can reach 8 million and the nest can measure more than 30 meters across. More than 2 kilograms of leaves each day are needed to maintain an average colony's fungus garden, so these ants can easily strip an area of vegetation. Fungus production on a large scale necessitates a division of labor; different individuals harvest leaf pieces, care for the fungus, defend the nest, clear trails for the leaf collectors, and guard the leaf fragments being carried back to the nest.

The fungi in leafcutter nests cannot exist without the ants, which supply the fungi with leaves to grow on and add fertilizer in the form of their fecal droplets. Leafcutter ants even evaluate leaves, avoiding those that contain fungus-killing chemicals. The fungal gardens, however, are vulnerable to invasion by undesirable microbes. To fend off one such invader, the green mold *Escovopsis*, the ants bring in another partner. They have special structures for carrying *Pseudonocardia* bacteria, which manufacture powerful antibiotics that suppress the unwelcome mold but do no harm to the cultivated fungus. Also present are *Klebsiella* bacteria, which fix atmospheric nitrogen to help fertilize the fungus garden and satisfy ant nutritional requirements. No doubt other organisms lurk in the fungus gardens awaiting discovery by curious ecologists or future filmmakers.

The fungi in leaf-cutter nests cannot survive without the ants, but can leaf-cutter ants survive without the fungus?

See answer on p. 1185.

56.1 What Types of Interactions Do Ecologists Study?

One of life's certainties is that, at some point between birth and death, every individual will encounter and interact with individuals of other species. These interactions have consequences that can affect each individual's fitness. Thus they can influence the densities of populations and the distributions of species, and, over the long term, they can lead to evolutionary change in one or more of the interacting species.

Interactions among species can be grouped into several categories

Although the kinds of interactions that take place among living things on Earth are essentially limitless, ecologists group interactions among species into a few basic categories. These categories reflect whether the outcome of the interactions is positive (+), negative (–), or neutral (0) for each of the species involved (**Figure 56.1**). We will introduce five broad categories of species interactions in this chapter.

ANTAGONISTIC INTERACTIONS **Antagonistic interactions** are those in which one species benefits and the other is harmed. Antagonistic interactions include three basic types:

- **Predation**, in which an individual of one species kills and consumes multiple individuals of other species (its **prey**).

- **Herbivory**, in which an individual of another species consumes part or (rarely) all of a plant.

- **Parasitism**, in which one species consumes only certain tissues in one or a few individuals of another species (its **host**) without necessarily killing them. Some parasites are pathogens that cause symptoms of disease in their hosts.

MUTUALISM **Mutualism** is a type of interaction between species that benefits both species. The interaction between leaf-cutter ants and fungi described at the opening of this chapter is an example of mutualism: the ants feed and cultivate the fungi, and the fungi, in turn, serve as food for the ants. Mutualisms exist between widely varied pairs of partners, including not only animals and fungi but also fungi and plants, animals and plants, animals and animals, and microbes and all other kinds of organisms.

COMPETITION **Competition** between species refers to interactions in which two or more species use the same resource. The outcomes of these interactions depend on resource availability. In some cases competitors can coexist by using the resource in different ways; if the resource is in extremely short supply,

(A)

Categories of Species Interactions

Type of interaction	Effect on species 1	Effect on species 2
Predation (predator-prey)	+	–
Herbivory (plant-herbivore)	+	–
Parasitism (parasite/ pathogen host)	+	–
Mutualism	+	+
Competition	–	–
Commensalism (commensal-host)	+	0
Amensalism	0	–

Antagonistic interactions: { Predation, Herbivory, Parasitism }

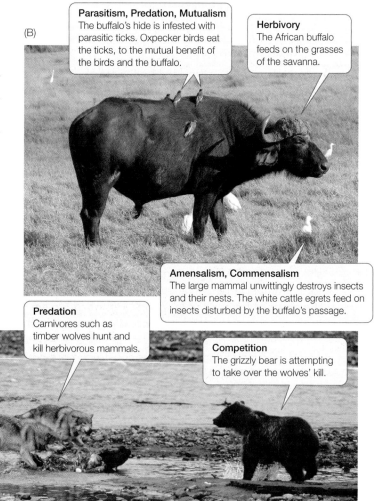

(B)

Parasitism, Predation, Mutualism
The buffalo's hide is infested with parasitic ticks. Oxpecker birds eat the ticks, to the mutual benefit of the birds and the buffalo.

Herbivory
The African buffalo feeds on the grasses of the savanna.

Amensalism, Commensalism
The large mammal unwittingly destroys insects and their nests. The white cattle egrets feed on insects disturbed by the buffalo's passage.

Predation
Carnivores such as timber wolves hunt and kill herbivorous mammals.

Competition
The grizzly bear is attempting to take over the wolves' kill.

56.1 Types of Species Interactions (A) Interactions among species can be grouped into categories based on whether their influence on each of the interacting species is positive (+), negative (–), or neutral (0). (B) Even small scenes can encompass many different species interactions.

Go to Activity 56.1 Ecological Interactions
Life10e.com/ac56.1

however, the outcome can be negative for all competing species. At some point a resource may be in such short supply that a population can no longer sustain itself; when a resource becomes limiting in this way, competition becomes intense. Competition can occur along with almost any other kind of interaction: between predators that depend on the same prey species, between herbivores that feed on the same host plant, or between pathogenic microbes attacking the same host. The limiting resource need not be food; species may compete for water, for space, for nesting sites, or even (in the case of plants) for sunlight.

COMMENSALISM AND AMENSALISM   Antagonistic interactions, mutualism, and competition all affect the fitness of both participants, but there are two other types of interactions that affect only one participant. **Commensalism** is a type of interaction in which one participant benefits but the other is unaffected. Most examples of commensalism (Latin, "at the same table") involve one species feeding in, on, or around another species. For instance, one species may associate with another species that, by virtue of its own feeding behavior, makes food more accessible. Cattle egrets, for example, feed on insects disturbed by large grazing animals, but their activities have no effect on the grazers (see Figure 56.1B).

Another form of commensalism involves association for the purpose of transport, often to reach food resources that are rare and short-lived. Piles of mammal dung, for example, are a valuable resource for some detritivores, but they can be hard to find and never last long. Many kinds of detritivores that cannot fly—mites, nematodes, and even fungi—attach themselves to the bodies of dung beetles, which not only can fly but are also very good at locating fresh dung. These hitchhikers have no known effect on the dung beetles' fitness.

Amensalism is a type of interaction in which one participant is unaffected while the other is harmed. A herd of elephants moving through a forest crushes insects and plants with each step, but the elephants are unaffected by this carnage. Amensal interactions tend to be more random, and thus less predictable, than other types of interactions.

Interaction types are not always clear-cut

Although ecologists find it useful to group interactions among species into a few basic categories, the boundaries between categories are not always clear. For example, sea anemones in the Pacific Ocean sting and eat small fish, but a select few fish species (mostly in the genus *Amphiprion*) live inside sea anemones and are unaffected by their stings. Safe from their predators, the anemonefish move freely among the stinging tentacles to scavenge the cnidarians' leavings (**Figure 56.2**).

Anemonefish must acclimate to the anemone's venom, and the anemone, in turn, must acclimate to the fish. The acclimation process appears to involve a change in the mucus coat of the fish; wiping off the mucus of an acclimated fish results in immediate stinging, whereas anemones do not sting fish with intact mucus. The benefits of this relationship to the anemonefish are clear: it escapes its own predators by hiding behind the anemone's stinging tentacles, and it has no need to forage

Amphiprion ocellaris

56.2 Interactions between Species Are Not Always Clear-Cut Ecologists long believed that the relationship between sea anemones and anemonefish was a commensalism: that the fish, by living among the anemone's stinging tentacles, gained protection from its predators. But could it also be considered a mutualism, if the fish's feces provide the anemone with beneficial nutrients?

widely for food. But does the anemone benefit from the association? By defecating while in residence, the anemonefish may provide nitrogen-rich nutrients to the anemone. On the other hand, the fish may occasionally steal the anemone's prey, which has a negative effect on the anemone's fitness.

The interaction types described in this section are in reality part of a continuum, and over evolutionary time they may shift from one type to another. Their outcomes depend on both ecological and evolutionary circumstances, including the presence and influence of other species.

Some types of interactions result in coevolution

All types of interactions have the potential to influence the population densities of the interacting species. By contributing to the differential survival or reproduction of individuals with different traits, they can also alter genotype frequencies within the interacting populations over time. Thus these interactions have both ecological consequences, as when they affect the distribution and abundance of a species, and evolutionary consequences, as when they lead to adaptations. In some cases an adaptation in one species may lead to the evolution of a reciprocal adaptation in a species it interacts with, a process known as **coevolution**.

Darwin observed that evolutionary change occurs not only in response to physical conditions, as described in Chapter 54, but also in response to interactions among species. In his introduction to *On the Origin of Species*, Darwin pointed out that woodpeckers have feet, tails, beaks, and tongues "admirably adapted to catch insects under the bark of trees" as a result of their long-standing interactions with their insect prey.

While abiotic factors also act as agents of selection, they differ in a fundamental way from biotic agents of selection in that they do not themselves undergo change as a result of the interaction. Snow and ice do not become more deadly as a result of encountering cold-resistant organisms, but predators can, over evolutionary time, become swifter, more powerful, or more efficient at capturing their prey. In response, prey species may become swifter, tougher, less conspicuous, or more poisonous, all of which decrease the likelihood of being consumed.

A series of reciprocal adaptations can lead to what has been dubbed a coevolutionary **arms race**. The arms race analogy, first used in the context of interactions between herbivores and plants, can be applied to most antagonistic interactions. The evolution of traits that increase the fitness of a predator, herbivore, or parasite species exerts selection pressure on its prey or host species to counter the consumer's adaptation. The prey or host adaptation, in turn, exerts selection pressure on the consumer to improve its fitness even more, resulting in an escalating series of reciprocal adaptations.

The types of interactions most likely to lead to coevolution are those that occur predictably and with high frequency over time and that have a strong effect on the interacting species. Thus most amensal and commensal interactions are less likely to coevolve than are many antagonistic and mutualistic interactions.

Go to Animated Tutorial 56.1
Coevolution: Strategies for Survival
Life10e.com/at56.1

RECAP **56.1**

Species interactions can be grouped into categories based on whether they benefit or harm each of the species involved. Some species interactions can lead to reciprocal adaptations and coevolution.

- Describe the categories of interspecific interactions. **See p. 1170 and Figure 56.1**
- What is meant by a coevolutionary arms race? **See p. 1172**

Sections 31.3 and 51.1 looked at a number of heterotrophic feeding strategies from the consumer's point of view. In the next section we will see how the antagonistic interactions—predation, herbivory, and parasitism—influence both consumer and resource species.

56.2 How Do Antagonistic Interactions Evolve?

Every species serves as a food resource, in one way or another, for at least one other species. Consumers can increase their fitness by acquiring food, whereas resource species can increase their fitness by avoiding being consumed. Thus the interests of consumer and resource species set up an antagonistic relationship that can lead to a coevolutionary arms race. These consumptive relationships need not, however, be fatal; organisms make meals of one another in many different ways.

Predator–prey interactions result in a range of adaptations

Predator–prey interactions are probably the most familiar, and the most dramatic, type of antagonistic interaction. Predators invariably kill the prey individuals they consume, and over its lifetime, a predator kills and consumes many prey individuals. Predators tend to be less specialized than other types of consumers.

The fitness of predators depends on balancing the cost of pursuing, subduing, and handling prey against the energetic benefit of consuming it, as we saw in Section 53.4. Thus many predators are larger than their prey, and many of them use strength or swiftness to capture prey. This is true of predators of all sizes: tigers pursuing deer and tiger beetles pursuing smaller insects are both fast, powerful predators and both are equipped with strong jaws (**Figure 56.3**). Predators that are smaller than their prey rely on other strategies that increase their efficiency. Many spiders, for example, capture their prey in webs. The tiny short-tailed shrew, among the smallest mammalian predators, produces venomous saliva that paralyzes not

(A) *Panthera tigris*

(B) *Cicindela campestris*

56.3 Predators Use Many Weapons (A) Tigers embody most people's image of a predator—a large animal that uses stealth, speed, strength, teeth, and claws to capture prey. (B) The 1.3-centimeter green tiger beetle is also formidable to its prey, including caterpillars. The beetle's huge jaws account for much of its body length, and it is one of the speediest runners among the insects.

(A) *Hyla versicolor*

(B) *Mimetica* sp.

56.4 Avoiding Consumption by Avoiding Detection (A) The gray tree frog can change its coloration to blend in with its substrate. (B) Resemblance to an inedible object can be an effective defense against visually hunting predators. Birds searching for insect prey are likely to bypass a katydid that looks like a partially eaten leaf.

only earthworms and snails but also prey much larger than itself, including mice and small birds.

Prey species have many different kinds of defenses against predators. Many animals can escape from predators simply by flying or running away. Others have morphological defenses. Tough skin, shells, spines, or hair can foil even a determined predator. In turn, however, adaptations evolve in predators that may overcome these defenses.

AVOIDING DETECTION Prey species can often escape predators by hiding. One form of hiding is camouflage, or background matching, also called **crypsis**. Some animals can even change their coloration to match the substrate they find themselves on (**Figure 56.4A**). The camouflage of some species allows them to resemble objects their predators consider inedible. The katydid in **Figure 56.4B**, for example, looks very much like a dead leaf, even down to the likeness of a spot of fungal decay.

Because the vision of many types of predators is adapted to spot moving prey, many prey species simply stop moving if they are being pursued. "Playing possum," a term that is sometimes applied to this strategy, refers to the ability of the opossum (*Didelphis virginiana*) to simulate death.

CHEMICAL DEFENSES Many animals use chemical defenses to escape or repel their predators. Chemical defenses are generally the province of animal prey that are small, weak, sessile, or otherwise unprotected. Among the mollusks, for example, the weaker a species' shell, the more likely it is to use chemical defenses; for example, the sea slug in Figure 56.5B has no shell but is highly toxic. Some vertebrates also rely on chemicals to repel their predators.

Many insects produce sprays, oozes, or froths when attacked. Bombardier beetles, for example, possess a pair of glands near the anal opening. Each gland has two compartments lined with a protective cuticle. The inner compartment contains a mix of relatively nontoxic chemicals, along with hydrogen peroxide. The outer compartment contains enzymes. When the beetle is disturbed, it discharges the contents of the inner compartment into the outer compartment, which leads to an instant, energy-releasing chemical reaction. Oxygen is one of the end products generated by this reaction, and the resulting pressure discharges the mixture with an audible "pop." Because of the energy released by the reaction, the temperature of the spray is approximately 100°C. The reaction of predators—including humans—to this hot, explosive secretion is predictable, and bombardier beetles have very few enemies.

Go to Media Clip 56.1
Bombardier Beetle Sprays Its Enemies
Life10e.com/mc56.1

But adaptions may evolve in predators that overcome their prey's chemical defenses, as we saw in the case of the rough-skinned newt and the garter snakes that have become insensitive to its protective toxin (see Figure 21.20). Some predators are not only undeterred by their prey's defensive chemicals, but ingest them and sequester them in their bodies as defenses against their own predators. Sea slugs are able to feed on a variety of well-defended prey with impunity and are masters at acquiring defenses from their food. Some species that feed on sponges concentrate toxic chemicals expropriated from their prey, whereas others, which feed on hydrozoans, incorporate the stinging cells of their prey, still active, into their own bodies.

APOSEMATISM Some prey species that defend themselves with toxic chemicals advertise that fact. This form of advertisement is called **aposematism**, or **warning coloration**. Aposematic prey species exploit the fact that predators can learn to avoid certain warning signals. Their warning signals may be visual (many toxic species are brightly colored) or acoustical (the rattlesnake's warning rattle, for example), depending on what sensory cues their predators use to find prey.

Many toxic prey sport bright colors or striking patterns to protect themselves against visually orienting predators. Such warning coloration increases the probability that a predator will learn to recognize and avoid the toxic species (**Figure 56.5**). Some vertebrate predators that rely on visual cues can learn quickly to associate certain color patterns with an unpleasant dining experience. Thus aposematic species are characteristically tough enough to survive a brief encounter with a predator. Any encounter that results in the death of the aposematic individual is unlikely to result in selection for its aposematic pattern. Sometimes field researchers find aposematic butterflies with damage inflicted by a bird beak—an indication of having survived being tasted by an uneducated avian predator.

MIMICRY SYSTEMS Even some nontoxic species benefit from warning coloration. We have seen that some prey species avoid consumption by mimicking inedible objects (see Figure 56.4B).

(A) *Danaus plexippus* larva

(B) *Chromodoris* sp.

(C) *Dendrobates reticulatus*

56.5 Some Prey Come with Warning Labels Some toxic prey warn potential predators with aposematic coloration. (A) Milkweed plants are toxic, and many of the insects that feed on them, such as this monarch butterfly larva, incorporate the plant's toxic chemicals into their systems. (B) Nudibranchs (sea slugs) are mollusks without protective shells; however, they may possess stinging nematocysts (acquired from their hydrozoan prey). (C) Poison dart frogs of Central and South American sequester highly toxic chemicals in their brightly colored skin.

Others do so by mimicking aposematic species. This strategy has led to the evolution of mimicry systems of two types. In **Batesian mimicry**, a benign, edible species (the mimic) closely resembles a dangerous, toxic species (the model) and benefits from the avoidance behavior learned by the model species' predators (**Figure 56.6A**). Mimicry may extend beyond physical appearance; many mimics also simulate distinctive behaviors of their models. In the Kalahari Desert of southern Africa, adult

Eremias lugubris lizards are cryptically colored to blend in with the sand, but juvenile lizards of this species are conspicuously black and white, resembling the dangerous oogpister beetles native to the same region. Oogpisters (Afrikaans for "piss in your eye"), like the bombardier beetles described earlier, can emit a noxious spray over a considerable distance. Young lizards will press their tails to the ground and arch their backs, thus enhancing their resemblance to an oogpister about to "fire."

(A) Batesian mimics

This harmless blenny...

...closely resembles a venomous related species.

Petroscirtes breviceps

Meiacanthus grammistes

(B) Müllerian mimics

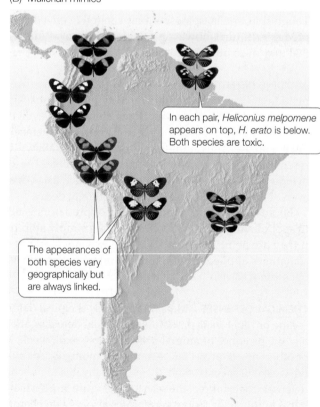

In each pair, *Heliconius melpomene* appears on top, *H. erato* is below. Both species are toxic.

The appearances of both species vary geographically but are always linked.

56.6 Truth in Labeling? (A) Batesian mimics are vulnerable species that gain protection by mimicking the aposematic signals of dangerous species. The appearance of the harmless blenny species *Petroscirtes breviceps* closely resembles that of the fanged striped blenny, which possesses a pair of grooved fangs with associated venom glands. (The male fish seen here are guarding eggs that females laid inside discarded bottles on the seafloor.) (B) The shared aposematic coloration of Müllerian mimics is an honest advertisement of their toxicity. As caterpillars, all of the longwing butterflies (genus *Heliconius*) of South America feed on toxic passionflower plants and incorporate the toxins into their adult bodies. The *Heliconius* species living together in a particular region have similar warning coloration.

In **Müllerian mimicry**, a number of aposematic species converge on a common color pattern; all benefit from providing a stronger recognition signal to predators. Many of the Neotropical longwing butterflies (*Heliconius*), which as caterpillars feed on toxic passionflower plants and incorporate the plant toxins into their bodies, are Müllerian mimics, and *Heliconius* species living together in a particular geographic region are likely to have similar coloration and share a common warning patern (**Figure 56.6B**). Genome sequencing of *Heliconius* Müllerian mimics has identified one gene, *optix*, that codes for a transcription factor that can, by changing gene expression patterns, create the same color patterns in *Heliconius* species that are not very closely related, thus leading to the evolution of mimetic color patterns within a geographic region.

Herbivory is a widespread interaction

The most common interaction among Earth's multicellular organisms is that between plants and the herbivores that eat them. Herbivores have a relatively easy time acquiring food, since plants are sessile and cannot claw, bite, or run away. Every major class of vertebrates includes at least a few herbivores. In marine systems, organisms that feed on plants and algae include mollusks, crustaceans, echinoderms, and annelids. But in terms of numbers of individuals as well as numbers of species, the vast majority of the world's herbivores are insects.

More than 90 percent of herbivorous insects are **oligophagous**, or specialists that dine on just one or a few, often taxonomically related, plant species. **Polyphagous** species, in contrast, feed on as many as hundreds of unrelated plant species. Vertebrate herbivores are generally polyphagous; a cow grazing in a pasture, for example, can consume many different plant species in a single afternoon. There are exceptions to this pattern, however. Australian koalas famously feed exclusively on the foliage of eucalyptus trees, and the diet of giant pandas is made up almost entirely of bamboo.

Herbivores, particularly insects, generally consume only parts of their food plants and usually do not kill them. In most natural ecosystems, insects rarely remove more than 20 percent of plant biomass. For that reason, some ecologists question the ability of insects to exert selection pressure on plant traits. Mortality is not, however, the only form of selection that leads to evolutionary change; herbivores can reduce plant fitness if the plants they attack produce fewer offspring.

PLANT DEFENSES AGAINST HERBIVORES The defenses of plants against their diverse consumers are necessarily highly diverse. For most plant species, chemistry is the principal defense mechanism. As we saw at the opening of this chapter, the leaves of some plants contain chemicals that prevent them from being consumed by fungi—and thus, incidentally, from being harvested by leafcutter ants. The amazing variety of secondary metabolites produced by plants to defend themselves against herbivores is the topic of Section 39.2. Many plants, however, have additional defenses.

Some plants protect themselves by being physically difficult to ingest. Thorns and spines are effective deterrents to browsing vertebrate herbivores. Smaller herbivores, including many insects, can be deterred by small hooked hairs on leaf surfaces. The soft bodies of leafhoppers can be pierced by these hairs, which fix the insect in place until it eventually dies from starvation or loss of blood. The plant's cuticle may also act as a physical barrier. Most grasses contain silica, which wears down sharp edges of herbivore teeth. Insects that feed only on grasses tend to have chisel-like mandibles that slice through leaf tissue, and their heads are enlarged to accommodate the larger jaw muscles needed to process their food.

RECIPROCAL ADAPTATIONS IN HERBIVORES AND PLANTS The concept of coevolution was first described in the context of interactions between herbivores and plants. In 1959 the entomologist Gottfried Fraenkel reached the conclusion after many years of study that all green plants are essentially nutritionally equivalent for insects. Why, then, are so many insects such picky eaters? Fraenkel proposed the novel hypothesis that ecological factors underlie the diversity of secondary metabolites that deter insect herbivores. A few years later, the entomologist Paul Ehrlich and the botanist Peter Raven proposed the following evolutionary scenario to account for patterns of host plant use among herbivorous insects (specifically, in their case, butterfly families):

- Certain plants, by mutation or recombination, evolve a novel secondary metabolite.
- If the chemical reduces the plant's appeal to herbivores, then plant genotypes producing the chemical are favored by natural selection.
- Freed from mortality associated with herbivory, plants possessing the novel chemical undergo an adaptive radiation.
- Certain herbivores, by mutation or recombination, evolve resistance to the chemical, and these resistant herbivores undergo their own adaptive radiation.
- With sufficient selection pressure, a resistant herbivore can evolve to use the chemical as a defense against its own predators.

This stepwise coevolutionary process explains not only the biochemical diversity of flowering plants but also the tremendous diversity of herbivorous insects. The ecological scenario outlined by Ehrlich and Raven is another example of a coevolutionary arms race.

A spectacular variety of adaptations to plant defenses has evolved in herbivores. Many herbivores circumvent plant defenses by behavioral means. For example, the secondary metabolites produced by a plant called St. Johnswort (*Hypericum perforatum*) require exposure to sunlight for optimal toxicity, so some insects that feed on this plant roll its leaves into a light-impervious cylinder and feed in comfort in the dark. The laticifer-cutting beetles described in Section 39.2 have a different method of detoxifying their food plant. Many large polyphagous herbivores, such as deer, horses, and the like, graze on a wide variety of plant species, minimizing their exposure to any particular defensive chemical. Long-lived and with relatively good memories, they can learn to avoid plants with an unpleasant taste.

Unlike large mammalian herbivores, caterpillars and many other insect herbivores may spend their entire lives feeding on a single individual plant. Such oligophagous diets are associated with highly specialized detoxification systems. The diamondback moth caterpillar eats plants in the cabbage family, which are rich in toxic mustard oil glycosides. In its gut is an enzyme that breaks down the glycosides into harmless by-products, allowing it to eat these plants with impunity.

Some herbivores take resistance a step further by storing, or sequestering, plant toxins in specialized organs or tissues that are insensitive to those toxins. In this way they can accumulate large quantities of toxins in their bodies with no ill effects. This strategy also makes the expropriated chemicals available for defense against the herbivores' own enemies. The caterpillar of the monarch butterfly, for example, is insensitive to the neurotoxic glycosides in its milkweed host plants, but most of its enemies, including insect-eating birds, cannot tolerate these compounds (as the caterpillar's aposematic coloration suggests; see Figure 56.5A).

Yet the plants continue their side of the coevolutionary arms race. As we have seen, longwing butterflies principally consume passionflower plants. These oligophagous butterflies lay eggs only on passionflower plants, and their larvae sequester host plant toxins in their bodies as they feed on the leaves. Some passionflower species, however, have modified leaf structures that resemble the eggs of butterflies. Some longwing butterfly species will not lay eggs on plants already containing eggs, so the egg mimics reduce the plant's probability of being consumed (**Figure 56.7**).

Spots on the leaves of *Passiflora* mimic butterfly eggs.

56.7 Using Mimicry to Avoid Herbivory The leaves of some passionflower species develop structures that resemble the eggs of their principal herbivores, longwing butterflies (*Heliconius* spp.; see Figure 56.6B). Females of many longwing butterfly species will not lay eggs on plants already containing eggs, so the egg mimics deter these females, thus protecting the plant from being eaten by hatchling caterpillars.

Parasite–host interactions may be pathogenic

Parasitism is an interaction in which one species consumes only certain tissues in one or a few host individuals of another species without necessarily killing them. Keeping the host alive is important for parasites that are highly specialized; killing the host would leave the parasite with no way to make a living.

MICROPARASITES **Microparasites** are many orders of magnitude smaller than their hosts and generally live and reproduce inside their hosts. Microparasites include in their ranks viruses, bacteria, and protists. Multiple generations may reside within a single host individual, and a host may harbor thousands or millions of them. Many microparasites, in the process of acquiring nutrients at the expense of their host, cause symptoms of disease—that is, they are pathogens. Section 39.1 describes the array of secondary metabolites that plants produce to defend themselves against pathogens, and Chapter 42 describes the immune system defenses of animals.

Infection by pathogens may in some cases result in the death of the host, but death is by no means the inevitable outcome of these interactions. If a pathogen strain is to persist in a host population, the pathogens must continually infect new host individuals. A less deadly strain that kills a smaller proportion of host individuals may be able to infect a larger number of new hosts. Thus pathogen and host may reach a state of coexistence as increased host resistance (ability to withstand the effects of a pathogen) and decreased pathogen virulence (ability to cause disease) evolve. Yet new virulent strains may also arise, reminding us that the arms race goes on.

The pathogens' hosts fall into three classes: susceptible (capable of being infected), infected, or recovered (and thus, in many cases, immune). A pathogen can readily invade a host population dominated by susceptible individuals, but as the infection spreads, fewer susceptible individuals remain to be infected. Eventually a point is reached at which most infected individuals no longer transmit the infection to susceptible individuals. Thus rates of infection typically rise, then fall, and do not rise again until a sufficiently large population of susceptible host individuals has reappeared.

MACROPARASITES While microparasites generally live and reproduce inside their hosts, larger **macroparasites** are associated with their hosts in a slightly less intimate way. Although macroparasites rarely cause the same kinds of disease symptoms that pathogenic microparasites cause, they may nevertheless affect host survival and reproduction and can thereby act as selective agents on their hosts. **Ectoparasites** are macroparasites that live outside the bodies of their hosts. **Endoparasites**, such as the tapeworms described in Section 31.3, are macroparasites that spend at least part of their life cycle inside the bodies of their hosts.

Some ectoparasites—leeches, mosquitoes, and the like—are only casually associated with their hosts, interacting with them just long enough to eat their fill and then moving on. Others spend their entire lives on their hosts; these sedentary ectoparasites have a number of attributes that keep them attached

(A) *Pthirus pubis*

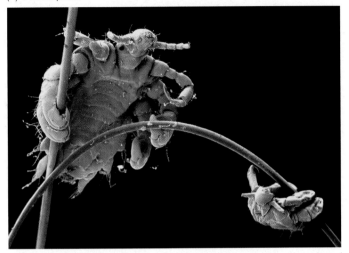

0.5 μm

(B) *Macaca fuscata*

56.8 Ectoparasites and Primates (A) Ectoparasites such as crab lice tend to be tiny, wingless, flattened, and equipped with strong claws for gripping. Humans are the only known host of this species, which infests the pubic hair. (B) Reciprocal grooming behaviors among primates is believed to have evolved in response to ectoparasites. Japanese macaques form social groups in which this behavior plays a significant role.

to their hosts. Crab lice, which are generally found in the pubic region of their human hosts, have claws on the tips of their legs that clamp around pubic hairs with great precision (**Figure 56.8A**). Pulling off a crab louse will often leave the legs behind, still firmly attached to the hair. Other adaptations that reduce the ability of irritated hosts to remove an ectoparasite include flattened bodies and a thick, tough cuticle. Most sedentary ectoparasitic insects are highly specialized, sometimes feeding on only a single host species.

Most hosts actively work to rid themselves of their ectoparasites. Grooming behavior—an important component of the social interactions of many primates—may have evolved in response to ectoparasites. The Japanese macaque (*Macaca fuscata*), for example, is prone to infestation by two species of lice, which tend to lay their multitudinous eggs on the outer surfaces of the host's back, arms, and legs. To keep louse populations in check, macaques form and maintain social bonds that ensure the consistent presence of grooming partners (**Figure 56.8B**). Some biologists believe that humans' hairlessness and bipedal posture (which freed the hands for manipulating small objects), as well as the opposable thumb, were evolutionary responses to ectoparasites.

Like antagonistic interactions, mutually beneficial interactions between species can result in coevolution. A mutually beneficial exchange of goods or services can ensure the predictability and frequency of such interactions over evolutionary time; thus many mutualistic interactions are tightly coevolved.

 56.3 How Do Mutualistic Interactions Evolve?

Mutualisms are interactions between two species that benefit both partners. There are few taxonomic limits on mutualistic interactions: many organisms have mutualistic partners from other domains and distant branches on the tree of life. Mutualistic interactions often arise in environments where resources are in short supply. Consequently, many mutualisms involve an exchange of food for housing or defense. Corals and their photosynthetic endosymbionts (see Figure 27.21) and lichens formed from fungi and photosynthetic algae (see Section 30.2) are examples of mutualistic interactions in which food is exchanged for housing. In another common type of mutualism, sessile organisms, particularly flowering plants, rely on more mobile species for mating or dispersal. In this chapter we will focus on mutualisms that involve animals, which can form mutualistic associations with other animals, with plants, and with a wide range of microorganisms.

Many mutualisms are asymmetrical—in other words, one party benefits more than the other. One or both partners may evolve adaptations that ensure that the exchange benefits both of them. Reciprocal adaptations are most likely to arise

RECAP 56.2

Predator–prey, herbivore–plant, and parasite–host interactions are all antagonistic. Consumers have adaptations for finding and using their resource species efficiently. Their resource species in turn have adaptations that reduce their probability of being discovered, captured, or eaten.

- What are some of the adaptations that help prey species avoid consumption by predators? **See pp. 1172–1175**

- How are aposematism and mimicry related? **See pp. 1173–1175 and Figures 56.4–56.6**

- Explain the scenario for coevolution between insect herbivores and their host plants proposed by Ehrlich and Raven. **See p. 1175**

in mutualistic interactions if an increase in dependence on a partner provides an increase in the benefits realized from the interaction. If increased dependence provides no selective advantage, mutualists (particularly species in asymmetrical mutualisms) may evolve into parasites, lose their partners and live independently, or even go extinct.

Some mutualistic partners exchange food for care or transport

Some organisms, such as the leafcutter ants described at the opening of this chapter, get their food by "farming" fungi. Fungus farming has been documented in a wide variety of species, including beetles, termites, and even a snail. In most cases the farmers provide housing, nutrition, and care for the fungal partner. The fungus provides food for the host, producing enzymes that degrade plant proteins and cellulose and thus converting plant materials that the insects could not have digested by themselves into an edible form.

Over the past 50 years the fungus-farming southern pine beetle (*Dendroctonus frontalis*) has destroyed huge tracts of valuable pine forests in the southeastern United States (**Figure 56.9**). The beetle owes much of its efficiency to its mutualistic partners. Masses of adult beetles attack a pine tree at once, overwhelming the tree's ability to defend itself (the tree's defense is to release large quantities of resin under pressure to force out the beetles). The beetles then excavate a series of galleries through the vascular tissue underneath the bark in which to lay their eggs. Female beetles also carry spores of their partner fungus into the galleries. The fungus grows on and breaks down the gallery walls, and the beetles feed directly on the fungus and the partially digested wood. The beetles also introduce a bacterium that produces an antibiotic to keep harmful bacteria from attacking the fungus (such as the leafcutter ants at the opening of this chapter). This insect–fungus–bacteria consortium overcomes the tree's antiherbivore defenses, to the partners' mutual benefit but to the great detriment of pine forests.

Some mutualistic partners exchange food or housing for defense

Some plants are not only food resources for insects, they are also mutualistic partners. The best known of these interactions is that between ants and acacia trees in Central America. In 1874, in Nicaragua, the naturalist Thomas Belt observed a peculiar interaction between bullhorn acacia trees (*Acacia cornigera*) and *Pseudomyrmex* ants, known as acacia ants because they are found only in association with acacias. Bullhorn acacias get their common name from the enlarged, hollow thorns, in which the ants build nests. The trees also produce rewards for the ants, both in nectar-producing extrafloral structures and modified leaflet tips that are rich in oil and protein. These structures have no apparent purpose other than providing food for ants.

Belt suggested that the notoriously aggressive acacia ants defend the plants against herbivores in exchange for food and shelter. But his idea was not tested until Daniel Janzen conducted experiments in 1966. By removing ants from some

(A) *Dendroctonus frontalis*

(B)

Galleries in vascular tissue of bark

(C)

56.9 A Mutualistic Interaction Brings Death to Pine Trees
(A) The southern pine beetle has a mutualistic relationship with a fungus, which it "farms" within the vascular tissue of pine trees. (B) The beetles excavate galleries inside the trees' vascular tissue. Here they lay eggs and farm fungus; the fungus digests wood and provides nutrition for the larvae. (C) Masses of pine beetles have overwhelmed this forest, resulting in widespread death of pine trees.

acacias with insecticide, Janzen demonstrated that trees without ants suffered a reduction in growth and an increase in mortality (**Figure 56.10**). Although this experimental design was imperfect—the insecticides removed non-ants as well as ants from the experimental acacias and may have also influenced the ability of the trees to grow—it was the first experimental demonstration that plants may benefit from an association with ants, which is now a widely accepted concept. In fact, since Janzen conducted his experiment, additional work on ants and acacias has revealed that the ants do more than simply defend the plant against herbivorous enemies; they also clip weeds from around the base of the plants, presumably reducing competition for nutrients.

Go to Animated Tutorial 56.2
Mutualism
Life10e.com/at56.2

56.10 Are Ants and Acacias Mutualists? Bullhorn acacia trees (*Acacia cornigera*) have numerous structures that provide food and shelter for ants of the genus *Pseudomyrmex* (acacia ants). Daniel Janzen's experiments demonstrated that the trees benefit greatly from their association with these ants, and that the energy expended in growing ant-attractive structures is repaid with increased growth and survival.[a]

HYPOTHESIS *Acacia cornigera* trees deprived of their *Pseudomyrmex* ant populations will survive and grow less well than trees populated by ant colonies.

Method

1. Define a population of *A. cornigera* trees; randomly designate some of them as untreated controls and the rest as experiment subjects.
2. Fumigate the experimental trees with insecticide to eliminate all *Pseudomyrmex* ants.
3. Apply Tanglefoot® (a sticky material) to the base of the experimental trees to prevent ants from recolonizing them.
4. Record the survival and growth rates of the trees in both groups over a 10-month period.

The "bull's horns" are enlarged, hollow thorns in which the ants build nests.

Results

After 10 months, control trees (with ants) had considerably higher survival and growth rates than did trees without ant populations.

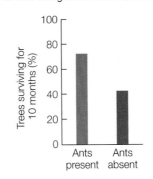
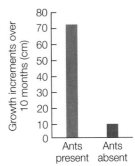

CONCLUSION *Pseudomyrmex* ants provide substantial fitness benefits to *Acacia cornigera* trees.

Go to **BioPortal** for discussion and relevant links for all INVESTIGATING**LIFE** figures.

[a]Janzen, D. H. 1966. *Evolution* 20: 249–275.

WORKING WITH**DATA:**

A Complex Species Interaction

Original Paper

Ness, J. H. 2006. A mutualism's indirect costs: The most aggressive plant bodyguards also deter pollinators. *Oikos* 113: 506–514.

Analyze the Data

As mentioned in the text, bullhorn acacias produce sugary substances on extrafloral nectaries that attract ants. Certain ant species act as pugnacious bodyguards for the plants, attacking other insects that may threaten the plants. In this way the ants and plants act as mutualists (see Figure 56.10). However, not all insects that are discouraged by ant bodyguards necessarily threaten the plant. In another such mutualism, the fishhook barrel cactus (*Ferocactus wislizeni*) of the western United States produces extrafloral nectaries that attract several species of ants. In 2003 and 2004, Joshua

Ness collected data on visitation by bee pollinators from *F. wislizeni* colonized by each of four ant species. Use the data in the table to answer the questions, explaining your reasoning in each case.

QUESTION 1

Which ant species is the best defender against herbivores?

QUESTION 2

In the presence of which ant species are bees most likely to forage?

QUESTION 3

Why do you think that the reproductive success of these plants (as measured by seed mass and number of seeds produced) varies according to which ants are guarding them?

	Ant species present[a]			
Trait	*Crematogaster opuntiae*	*Forelius* sp.	*Solenopsis aurea*	*Solenopsis xyloni*
No. ants/flower	1.3 ± 0.2	2.5 ± 0.3	4.0 ± 0.5	3.8 ± 0.3
Flowers occupied by ants (%)	10.8 ± 3.1	52.0 ± 6.8	30.9 ± 4.1	44.2 ± 4.3
Flowers occupied by pollinators (%)	23.3 ± 4.5	14.2 ± 4.9	13.6 ± 3.1	4.2 ± 1.2
Bee foraging time/flower (sec)	43 ± 12.0	44 ± 18	76 ± 19.0	15 ± 3.0
Individual seed mass (mg)	237 ± 6.0	253 ± 15	232 ± 8.0	203 ± 8
No. seeds/fruit	1017 ± 88.0	1037 ± 206	1239 ± 107	871 ± 10.7
Seed mass/fruit (g)	2.32 ± 0.16	2.43 ± 0.38	2.82 ± 0.21	1.77 ± 0.20

[a]Only one ant species was present on each cactus plant studied. Values are means ± standard error.

Go to BioPortal for all WORKING WITH**DATA** exercises

Plants and pollinators exchange food for pollen transport

For about three-fourths of the planet's 250,000 flowering plant species, reproduction requires the transport of pollen by an animal partner. The benefit from the plant's perspective is clear: the animal partner moves pollen from one sessile individual to another and thereby promotes sexual reproduction and thus genetic diversity. In this section we will focus on the benefits accruing to the animal pollinator.

A mutualistic pollination system requires several features:

- An attractant or reward that entices a pollinator to visit the plant
- Behavior by the pollinator that ensures it will visit more than one individual of the same plant species
- Anatomical features that allow the pollinator to transport the plant's pollen

Floral characteristics influence the type of pollinator that is attracted to a flower. Ultraviolet color patterns, for example, are highly attractive to bees (see Figure 29.15) but are invisible to most other pollinators. The depth and width of a flower can restrict the size and shape of the pollinator mouthparts that can gain access to its nectar (see Figure 23.13). The timing of a plant's flowering can also restrict the number of potential pollinator species and encourage pollinator fidelity.

Flowers entice pollinators in many ways. The most direct reward for pollinators is the pollen itself, which sometimes serves as food. Pollen was probably the original attractant in the evolutionary history of plant–pollinator interactions. Plant reproduction would not be served, however, if pollinators were to eat *all* of a plant's pollen; thus plants have evolved various adaptations to ensure that they benefit from the exchange. For example, some plants have two types of anthers: feeding anthers to produce pollen for pollinators, and fertilization anthers to produce pollen for reproduction. These two types of anthers are shaped and positioned differently, so that as the pollinator dines on pollen from the feeding anthers, the fertilization anthers deposit pollen on a part of its body that will transfer it to the stigma of another flower of the same species.

Compared with pollen, nectar—a sugar-rich solution produced by some angiosperms—is a relatively new evolutionary development. Of the floral rewards, nectar has the greatest appeal and is consumed by the widest range of animal pollinators, including birds (such as hummingbirds) and mammals (such as bats) as well as insects. While nectar is particularly effective for attracting potential pollinators, it is also prone to removal by "nectar thieves": animals such as ants that consume the nectar without transporting pollen. Nectar thieves lower plant fitness by depleting nectar that would otherwise attract actual pollinators.

Plants may also take advantage of their pollinators. Some orchid species have evolved flowers that resemble the females of particular wasp species (sometimes even producing the same chemical substance the female wasp uses to attract mates). The plants are pollinated by male wasps that attempt to copulate with the flower (**Figure 56.11**).

Plants not only need to attract pollinators, but must also ensure that those pollinators carry their pollen to other members

Argogorytes mystaceus (male wasp)

The flower of *Ophrys insectifera* closely resembles a female wasp.

56.11 Taking Advantage of a Pollinator Flowers of the orchid *Ophrys insectifera* look and smell like female *Argogorytes mystaceus* wasps. A male wasp of this species expends energy in a futile attempt to mate with the orchid's flower, getting pollen on his body in the process. The wasp then carries the pollen to the next flower he visits in his quest for a genuine mate.

of the same species. Repeat visits by a pollinator to different individuals of a particular plant species increase the likelihood that the pollen will end up on the appropriate stigma; thus some plants have adaptations to limit the diversity of their animal visitors. Botanists have long wondered why certain plants produce small amounts of toxic substances in their nectar. The nectar of tobacco flowers, for example, contains trace amounts of nicotine, an insecticidal neurotoxin. Many flower visitors, including hummingbirds, can ingest only tiny amounts of nicotine-laced nectar before moving on to other flowers. To other pollinators, however, nicotine may actually be addictive. Honey bees, for example, overwhelmingly prefer artificial nectar spiked with nicotine in laboratory tests. Putting small amounts of a potentially addictive substance in nectar may be one way tobacco plants improve their odds of a repeat visit by the right pollinator species.

Most flowers can be successfully pollinated by several different animal species. The evolution of broad suites of floral characteristics that attract certain groups of pollinators is an example of **diffuse coevolution**: the evolution of similar suites of traits in species experiencing similar selection pressures (**Table 56.1**). Scarlet gilia (*Ipomopsis aggregata*), a common wildflower in the Rocky Mountains, has successfully combined two different pollinator attraction strategies. Early in its growing season, it produces red flowers that attract hummingbirds; later in the season, the gilia shifts to producing white flowers because by then the most abundant pollinators are hawk moths, which cannot see red but are attracted to white.

A few plant–pollinator relationships are much more exclusive; these relationships lead to highly specific, rather than diffuse, coevolution. Yucca plants, for example, are pollinated only by a group of moths collectively known as yucca moths, whose

TABLE 56.1
Pollination Syndromes Resulting from Diffuse Coevolution

Preferred pollinator	Suite of anatomical traits			
	Flower shape	Flower color	Reward	Odor
Bees	Irregular	Many	Nectar, pollen	Sweet
Flesh flies	Irregular	Purplish	None	Carrion
Beetles	Bowl	White or pale	Pollen	Faint
Butterflies	Tubular	Many	Nectar	Faint
Moths	Often pendant	White or pale	Nectar	Heavy
Hummingbirds	Tubular	Red	Nectar	Imperceptible
Bats	Cuplike	White or pale	Copious nectar	Musty

larvae feed exclusively on yucca seeds. The stigma of the yucca flower is located deep within the pistil, and fertilization will not occur unless pollen is physically placed there. The specialized mouthparts of female yucca moths have distinctive long tentacles, which the moths use to pack masses of pollen from one yucca flower into transportable balls that they then carry to another flower. The moth pushes the pollen ball deep into the recess in which the flower's stigma is tucked, then turns around and deposits her eggs inside the flower's ovule (**Figure 56.12**). When the eggs hatch, the moth caterpillars will consume some—but not all—of the flower's developing seeds. Neither of these species can reproduce in the absence of the other.

Plants and frugivores exchange food for seed transport

Many animals that eat fruits (called **frugivores**) provide a valuable service to the plants that produce those fruits by dispersing seeds. Seed dispersal by animals not only offers plants the advantages of delivery to potential germination sites away from the parent plant (described in Section 38.1), but comes with the bonus of organic fertilizer for the seeds. Interactions between plants and frugivores, however, are not always reciprocal; in many cases, one party benefits more than the other. Whereas the frugivore is paid "in advance" for its transportation services, the seeds may never reach an appropriate destination for germination (your windshield, for example, will not do). From the plant's perspective, its partnership with frugivores requires a delicate balance between discouraging them from eating fruits before the seeds are capable of germinating and attracting them when the seeds are ready. In addition, the plant must protect the seeds from destruction in the frugivore's digestive tract and defend them against inappropriate consumers that would damage the seeds or fail to disperse them at all.

The chemical process of fruit ripening ensures that fruits are most attractive to frugivores when the seeds are mature and ready for dispersal. In many fruits, ripening is accompanied by a decrease in organic acids, which make many unripe fruits sour. Color changes, which result from loss of chlorophyll and the accumulation of other pigments (the conversion of peppers from green to red during ripening is an example), have enormous signal value to many frugivores. Green, unripe fruits are generally difficult for vertebrate frugivores to see against green foliage; red and bicolored red and black fruits contrast with foliage. Fruit softens as it ripens, allowing for gentle processing by the frugivore

Yucca filamentosa

Tegeticula yuccasella

...and pushes the pollen deep into the pistil.

This female is laying eggs in the yucca's ovule.

The female moth collects and stores pollen grains in specialized mouthparts...

56.12 Pistil-Packing Mama
Yucca flowers are pollinated only by yucca moths, and the larvae of yucca moths feed only on the seeds of yucca plants. The moth *Tegeticula yuccasella* is the exclusive pollinator of *Yucca filamentosa*.

Dicaeum hirundinaceum

56.13 A Frugivore Plants and Fertilizes a Seed at the Same Time After a mistletoebird eats the fruit of the parasitic mistletoe plant, the seeds inside the fruit pass through the bird's digestive tract intact. As the seeds are voided, their sticky outer coat makes them stick to the bird's feathers. As the bird wipes itself clean on a branch, the seed sticks to the branch, where it germinates.

and rapid passage through its gut. Another conspicuous change in ripening fruits is an increase in sugar content—the "reward" most sought by frugivores. Seed coats, fruit pulp, and epidermis may all contain secondary chemicals designed to discourage inappropriate frugivores from consuming the fruit.

Because of the often asymmetrical nature of the mutualism between frugivores and plants, relatively few highly specialized frugivores exist. One apparently reciprocal interaction is between mistletoes—parasitic plants that grow on trees—and the mistletoebird that serves as the plants' primary dispersal agent in Asia and Australia (**Figure 56.13**). This bird dines largely on the fleshy berries of mistletoe. The seeds, covered with a gluelike outer coat, experience little enzymatic or mechanical damage as they pass through the thin-walled guts of the birds that swallow them. When the seeds are voided with a bird's droppings, the sticky outer coat causes the seeds to adhere to the bird's feathers, prompting it to wipe its bottom across the tree branch on which it is perched. Once the seed is wiped on the branch, the gluey coat keeps it there—in an ideal location for a mistletoe seed to germinate.

RECAP 56.3

Mutualistic interactions involve an exchange of benefits. Most plants rely on mutualisms with animals for fertilization and seed dispersal.

- Give examples of benefits that are exchanged in at least two mutualisms between plants and animals.

- What three features are required by a plant–pollinator mutualism? **See p. 1180**

- How are plant–pollinator interactions different from plant–frugivore interactions? **See pp. 1180–1181**

- Describe some adaptations in plants that help maintain the balance in their relationships with frugivores. **See p. 1181–1182**

From our discussion of interactions between two species that benefit both, we will now move to a type of interaction that benefits neither: competition. This type of interaction is widespread because it can arise wherever two or more species require the same resources.

56.4 What Are the Outcomes of Competition?

Antagonistic interactions can be quite attention-getting; the scene of a lion stalking a gazelle is almost emblematic of the African savanna. But at the same time a predator is interacting with its prey, it may also be interacting with other predators that hunt the same prey species. Lions are not the only predators of gazelles; cheetahs, hyenas, and even crocodiles hunt and kill gazelles, potentially reducing the supply of food available for lions.

Whenever any resource is not sufficiently abundant to meet the needs of all the organisms with an interest in that resource, organisms must compete with one another to gain enough of that resource to survive. Competition not only influences the evolution of species but also plays an important role in determining the structure and composition of communities, as we will see in the next chapter.

Competition is widespread because all species share resources

Virtually no species enjoys exclusive access to any given resource. The vast majority of species must compete for at least some resources with other species. As we saw in Section 55.4, limited resources are the main reason why populations do not grow indefinitely. When resources are limited, individuals in the population compete for those resources. Such **intraspecific competition**—competition among individuals of the same species—may result in reduced growth and reproductive rates for some individuals, may exclude some individuals from better habitats, and may cause the deaths of others. **Interspecific competition**—competition among individuals of different species—affects individuals in much the same way. At some point an essential resource may be in such short supply that a population is in danger of becoming unable to sustain itself; when a resource becomes limiting in this way, competition becomes intense and can influence the persistence and evolution of species.

The principle of competitive exclusion holds that no two species can share the same limiting resource indefinitely. If one species can prevent all members of another species from using the resource, the inferior competitor may go locally extinct, a result called **competitive exclusion**. In other cases, selection pressures resulting from interspecific competition cause changes in the ways in which the competing species use the limiting resource. If those changes allow the species to coexist, the result is called **resource partitioning**.

Whether it is interspecific or intraspecific, competition occurs by two major mechanisms. **Interference competition** occurs when a competitor interferes with another competitor's access to a limiting resource. **Exploitation competition** occurs when a limiting resource is available to all competitors and the

Geospiza fuliginosa

Xylocopa darwinii on *Opuntia* flower

Nectar Use and Wingspan of *G. fuliginosa*		
Island	Time spent feeding on nectar (%)	Mean wingspan (mm)
Bees absent		
Pinta	10	59.8
Marchena	28	58.2
Bees present		
Fernandina	1	64.8
Santa Cruz	14	64.0
San Salvador	0	63.8
Española	0	64.7
Isabela	7	64.5

56.14 Competition with Bees Influences Finch Morphology On islands in the Galápagos archipelago where *Geospiza fuliginosa* is the sole pollinators of cactus flowers, a short wingspan increases these birds' ability to negotiate the flowers. On islands where carpenter bees compete with the birds for cactus nectar, *G. fuliginosa* individuals have a longer wingspan and feed more heavily on other foods.

outcome of competition depends on the relative efficiency with which the competitors use the resource.

Interference competition may restrict habitat use

Interference competition can take many forms. A graphic example involves the desert ant *Conomyrma bicolor* and the honeypot ant *Myrmecocystus mexicanus*. These two ant species occupy the same type of habitat—arid areas containing little vegetation—and they feed on similar foods—the sugary excretions of aphids and other sap-feeding insects as well as occasional arthropods, none of which is in great supply. When *C. bicolor* workers find the entrance of a honeypot ant nest, they pick up small stones in their mandibles, carry them to the rim of the nest opening, and drop them down the hole—up to 200 stones in a 5-minute interval. This activity is enough to stop the honeypot ants from going out foraging. Some honeypot ant colonies, under constant stone-dropping attack for several weeks, may be almost entirely deprived of food.

Even microorganisms interfere with one another's use of resources. In the highly structured environment of the rhizosphere, or "root-world," of the soil, competitive interactions can be locally intense. Many soil bacteria produce substances that subdue their microbial competitors. Actinobacteria, for example, produce chemicals that interfere with essentially every life process in other kinds of bacteria. Many of the chemicals that these remarkably well defended microbes produce to defeat their competitors are used as antibiotics by mutualistic partners, such as the bark beetles described in Section 56.3, as well as in human pharmacology.

Exploitation competition may lead to coexistence

Exploitation competition may lead to coexistence, provided that the species relying on the same resource evolve adaptations to divide up, or partition, that resource. For example, in many Rocky Mountain communities, at least three species of bees consume the nectar of *Agave schottii*, the shindagger agave. The three bee species differ in where and when they collect shindagger nectar. Honey bees tend to forage in places with the greatest numbers of shindagger flowers, bumblebees in places with intermediate numbers of flowers, and carpenter bees where flowers are few and far between. Honey bees also tend to be most active when nectar output is greatest. With their larger nests and greater numbers of offspring to support, honey bees require greater foraging efficiency and greater energy intake. Foraging sites that are not worth their while are left to the other bees.

In some cases individuals within a species display different behaviors or morphologies depending on whether they are competing for resources with another species. Darwin remarked in *On the Origin of Species* that "Natural Selection leads to divergence of character; for more living beings can be supported on the same area the more they diverge in structure, habits, and constitutions." This "divergence of character" is referred to today as **character displacement**. On some of the islands of the Galápagos archipelago, for example, certain cactus species are pollinated exclusively by the small ground finch *Geospiza fuliginosa*, for which cactus nectar is an important food source (see Figure 23.8). On other islands, a carpenter bee (*Xylocopa darwinii*) competes with the finches for cactus nectar; the birds consequently feed more heavily on seeds and insects. On the islands where bees are absent, the birds feed on nectar more often and have smaller wingspans than on islands where they share cacti with bees (**Figure 56.14**).

Sometimes organisms respond to competition by changing their location to avoid confrontations. The African wild dog (*Lycaon pictus*) is a carnivore that lives and forages in packs (groups of related individuals). Frequent vocalizations, called twitters, function to keep the pack together, but these acoustical signals also can alert their chief competitors for prey, African lions (*Panthera leo*). Lions hearing the dogs' twitters can use them to locate dog packs and steal their kills. The dogs avoid competing with lions by selecting areas for their dens where the likelihood of being overheard by lions is low; thus wild dog densities are inversely correlated with lion densities. The wild dog is considered a **fugitive species**—a species that leaves an otherwise suitable habitat in order to avoid competition with another species.

Species may compete indirectly for a resource

Species may compete indirectly for a resource even when they are not present in the same habitat at the same time. Sometimes a species so alters the quality of a resource that it is rendered less usable by other species that may encounter it afterward. For example, feeding by sap-sucking leafhoppers on potato plants early in the growing season can cause curled leaves and chlorosis (loss of chlorophyll). Potato beetles feed on potato plants later in the growing season. Beetles that consume leaves damaged by leafhoppers suffer reduced growth and survival rates. Even though these two herbivores do not feed at the same time, one species influences the use of the shared food resource by its competitor.

Indirect competition can also result when two species share a common predator. For example, the parasitoid wasp *Venturia canescens* is a consumer of two different species of caterpillars that infest stored food products such as flour, the Indianmeal moth caterpillar (*Plodia interpunctella*) and the Mediterranean flour moth caterpillar (*Ephestia kuehniella*). The two caterpillar species can coexist in a flour bin, but when the wasp is present, it preferentially attacks and kills the flour moth caterpillars. Thus in the presence of the wasp, the competitive balance between the two caterpillar species is altered in the meal moth's favor. This type of competition is indirect because the outcome of competition depends not on how the two competitors use the shared resource, but on how the two competitors interact with a shared predator.

Competition may determine a species' niche

Competition is important in determining where a species can be found. A species' **niche** is the set of physical and biological conditions it requires to survive, grow, and reproduce. Thus a species' niche is partly defined by the resources available in the environment. Although a species might be physiologically able to live under a wide range of conditions, competitors may restrict its use of resources in a particular location. Thus every species has a **fundamental niche**, defined by its physiological capabilities, and a **realized niche**, defined by its interactions with other species.

Two species of barnacles, the rock barnacle (*Semibalanus balanoides*) and Poll's stellate barnacle (*Chthamalus stellatus*), compete for space on the rocky shorelines of the North Atlantic Ocean. The planktonic larvae of both species settle in the intertidal zone and metamorphose into sessile adults. The smaller stellate barnacles generally live at higher levels in the intertidal zone, where they face longer periods of exposure and desiccation (drying out) than do rock barnacles, which live at a lower level. There is little overlap between the areas occupied by adults of the two species (**Figure 56.15**). What explains their distinct distributions in the intertidal zone?

In a famous study conducted more than 50 years ago, Joseph Connell experimentally removed one or the other species from

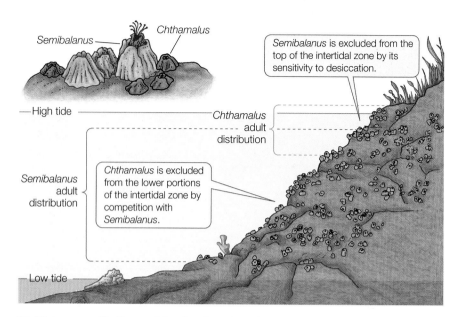

56.15 Interspecific Competition Can Restrict a Species' Range Interspecific competition with rock barnacles (*Semibalanus*) restricts stellate barnacles (*Chthamalus*) to a smaller portion of the intertidal zone than they could otherwise occupy. Larvae of both species settle throughout the intertidal, but at lower levels, rock barnacles grow much faster and eliminate the stellate barnacle larvae. In the upper reaches of the intertidal, however, the greater susceptibility of rock barnacles to desiccation (drying out) allows stellate barnacles to outcompete them. The two species can coexist in a small portion of the intertidal zone.

its characteristic zone and observed the response of the remaining species. Stellate barnacle larvae normally settle in large numbers throughout much of the intertidal zone, including the lower levels where rock barnacles are found (their fundamental niche), but they thrive at those lower levels only when rock barnacles are not present (their realized niche). Connell found that the rock barnacles grew so fast they smothered, crushed, or undercut the stellate barnacle larvae. However, removing stellate barnacles from their spots higher in the intertidal zone did not lead to their replacement by rock barnacles; the rock barnacles are less tolerant of desiccation and failed to thrive there even when stellate barnacles were absent. The result of the competitive interaction between the two species is a distinctive pattern of intertidal zonation, with stellate barnacles restricted in their distribution by competition and rock barnacles restricted in their distribution by their physiological limitations.

RECAP 56.4

Competition occurs when two or more species require a resource that is in limited supply. No two species can share the same limiting resource indefinitely. The outcome of competition may be competitive exclusion, in the form of local extinction, or coexistence, in the form of resource partitioning.

- How does exploitation competition differ from interference competition? **See pp. 1182–1183**

- How can competition lead to character displacement? **See p. 1183 and Figure 56.14**

- Explain the difference between an organism's fundamental niche and its realized niche. **See p. 1184 and Figure 56.15**

56.16 A Fungal Garden Cutaway view of a South American leaf-cutter ant nest chamber filled with fungus. Several winged ants (*Atta colombica*) can be seen in the crevices of the fungal mass.

The study of interactions among the species in a community is a large part of community ecology—the topic of the next chapter. Every kind of interaction we have studied in this chapter influences the nature and structure of communities. Competition helps determine which species persist and which go extinct, as well as dictating how many different species can be supported by a particular resource. Similarly, antagonistic interactions have important effects on the distribution and abundance of consumer and resource species, and the presence of mutualistic partners may dictate whether a particular species can exist in a particular community.

The fungi in leafcutter nests cannot survive without the ants, but can leafcutter ants survive without the fungus?

ANSWER

The relationship between the leafcutter ants and their fungi is a coevolved mutualism. The ants are so specialized behaviorally that they would starve without their fungus gardens. Recent sequencing of the genome of one leafcutter species, *Atta cephalotes*, revealed that this species has lost several genes possessed by other ants. Those genes include the ones encoding the entire pathway for biosynthesizing the amino acid argi-nine, which the leafcutter obtains from its fungal food, as well as several genes that break down plant toxins (which the fungus does for the leafcutter as it metabolizes the leaf substrate). In the case of leafcutter ants and fungi, we can assume that the increasing dependence of each species on its partner provided it with an increase in benefits from the partnership, resulting in reciprocal adaptations.

CHAPTER**SUMMARY** 56

 56.1 What Types of Interactions Do Ecologists Study?

- Species interactions can be grouped into categories. **Antagonistic interactions** include **predation**, **herbivory**, and **parasitism**, all of which benefit a consumer while harming the species that is consumed. **Mutualism** benefits both participants, whereas **competition** harms both. **Commensalism** benefits one participant with no effect on the other; **amensalism** has no effect on one participant but harms the other. **Review Figure 56.1, ACTIVITY 56.1**

- The evolution of an adaptation in one species may lead to the evolution of an adaptation in a species with which it interacts, a process known as **coevolution**. A series of reciprocal adaptations among consumers and their resource species can lead to a coevolutionary **arms race**. See **ANIMATED TUTORIAL 56.1**

56.2 How Do Antagonistic Interactions Evolve?

- Predators kill the individuals they consume (their prey). Over its lifetime, a predator kills and consumes many prey individuals.

- Some prey species avoid detection by means such as **crypsis**. Others defend themselves by physical or chemical means. Chemically defended animals often advertise their toxicity with **aposematism**, or warning coloration. **Review Figure 56.4, 56.5**

- In **Batesian mimicry**, a nontoxic species mimics a toxic species. In **Müllerian mimicry**, two or more toxic species converge to resemble one another. **Review Figure 56.6**

- Herbivores generally consume only parts of their food plants and usually do not kill them.

- Many herbivores have evolved resistance to the defensive secondary metabolites produced by plants, and some have incorporated them into their own defenses against predators.

- Parasites consume certain tissues in one or a few host individuals of another species without necessarily killing them. **Microparasites** include viruses, bacteria, and protists; large numbers of these organisms can live and reproduce within the body of the host and are often pathogenic. **Macroparasites** are less intimately associated with their hosts but can nonetheless affect host fitness.

56.3 How Do Mutualistic Interactions Evolve?

- Mutualistic interactions involve an exchange of benefits. Many mutualisms arise in environments where resources are in short supply.

- Reciprocal adaptations are most likely to arise when an increase in dependency on a partner provides an increase in the benefits realized from the interaction.

- Some animals "farm" fungal species, which provide them with food. Other mutualisms involve an exchange of food or housing for defense. **Review Figures 56.9, 56.10, ANIMATED TUTORIAL 56.2**

continued

- Many mutualisms between plants and animals involve an exchange of food for transport. In plant–pollinator interactions, animals that collect and transport pollen are rewarded with pollen or nectar.

- Broad suites of floral characteristics that are attractive to certain types of pollinators exemplify **diffuse coevolution**. Some plant–pollinator mutualisms, however, are much more specific and exclusive. **Review Figure 56.11, Table 56.1**

- Plants that depend on **frugivores** for seed dispersal must balance the need to discourage frugivores from eating fruits before the seeds are mature, attract frugivores when the seeds are mature, and protect the seeds from destruction in a frugivore's digestive tract.

 56.4 What Are the Outcomes of Competition?

- Competition occurs whenever a resource is not sufficient to meet the needs of all organisms with an interest in that resource.

- Competition may result in local extinction of the inferior competitor, an outcome called **competitive exclusion**. Alternatively, selection pressures resulting from competition may change the ways in which the competing species use a limiting resource, an outcome called **resource partitioning**. **Interference**

competition occurs when an individual interferes with a competitor's access to a limiting resource. **Exploitation competition** occurs when a limiting resource is available to all competitors and the outcome of competition depends on the relative efficiency with which competitors use the resource.

- Exploitation competition may lead to **character displacement**, in which attributes of a species vary depending on whether a competitor is present or absent. **Review Figure 56.14**

- Species may compete indirectly even when they are not present in the same place at the same time, as, for example, when they share a common predator.

- A species' **niche** is the set of physical and biological conditions it requires to persist. Although a species may be able to persist under a wide range of resource conditions (its **fundamental niche**), competitors may restrict its use of resources in a particular location (its **realized niche**). **Review Figure 56.15**

 Go to the Interactive Summary to review key figures, Animated Tutorials, and Activities
Life10e.com/is56

CHAPTER**REVIEW**

▓▓ **REMEMBERING**

1. Predation, herbivory, and parasitism are all examples of
   a. antagonistic interactions.
   b. mutualistic interactions.
   c. commensal interactions.
   d. amensal interactions.
   e. competitive interactions.

2. In a coevolutionary arms race, after a plant evolves a novel chemical defense against an herbivore,
   a. the herbivore can be expected to go extinct.
   b. the plant can be expected to undergo a range restriction because of the cost of producing the novel chemical.
   c. the herbivore can be expected to evolve resistance to the plant's defense.
   d. the plant can be expected to experience reduced fitness because of the cost of producing the novel chemical.
   e. the plant can be expected to stop producing other types of defenses.

3. Damage caused to shrubs by branches falling from overhead trees is an example of
   a. interference competition.
   b. predation.
   c. amensalism.
   d. commensalism.
   e. diffuse coevolution.

4. A hummingbird sips nectar from the flowers of a plant, pollinating those flowers in the process. This interaction is best classified as
   a. parasitism, because the hummingbird consumes the flower's nectar.
   b. predation, because the hummingbird eats the plant's seeds.
   c. commensalism, because the hummingbird benefits from consuming nectar and the plant is unaffected.
   d. mutualism, because the plant provides nectar for the hummingbird and the hummingbird transports pollen for the plant.
   e. Not enough information is provided to classify this interaction.

5. One factor that can constrain the realized niche occupied by an organism is
   a. crypsis
   b. aposematism.
   c. mimicry.
   d. commensalism.
   e. competition.

▓▓ **UNDERSTANDING & APPLYING**

6. The different types of interspecific interactions are part of a continuum, and their outcomes often depend on circumstances. Refer to the Working with Data exercise on p. 1179. How does this example exemplify the various types of interspecific interactions (mutualism, competition, predation, parasitism, commensalism)? Do you think a continuum is represented here? What aspects of the situation described by the data could change the interactions?

7. Like the southern pine beetle in Figure 56.9, the mountain pine beetle (*Dendroctonus ponderosae*) attacks pine trees with the help of a symbiotic fungus that infects the host tree. In 2009 these beetles infested almost 4 million acres of pine forest across Montana, Wyoming, Colorado, Idaho, Utah, Oregon, and Washington. How would you determine which pine trees are susceptible to mountain pine beetle attack? How could the fact that this beetle has a symbiotic partner affect approaches for managing the outbreak?

8. *Salmonella serovar typhimurium* is a bacterium that lives in the intestines of a wide variety of animals—including humans, in which it can cause gastroenteritis. In the United States, raw chicken and other poultry are frequent dietary sources of this bacterium and thus present a significant health hazard. Although warning labels now appear on packaged raw poultry, the food industry is testing new ways to reduce the likelihood of contamination by reducing *Salmonella* populations in chickens before they are slaughtered. In one such test, broiler chicks were given a culture of three species of bacteria: *Lactobacillus plantarum*, *L. acidophilus*, and *Lactococcus lactis*. These birds, along with control birds that had not ingested the cultures, were exposed to *Salmonella* gut colonization and then tested to see if they maintained populations of *Salmonella* in their guts. Chicks that had been given bacterial cultures consistently had significantly lower populations of *Salmonella* than the control group.

    a. What ecological principle is being applied by the poultry industry?

    b. What other ecological outcomes might this experiment have produced?

    c. What other problems might this ecological principle be useful in tackling?

ANALYZING & EVALUATING

9. Many ectoparasites feed on only a narrow range of host species. Until recently, investigators used close genetic relationships among bird lice as evidence of close genetic relationships among their host bird species. Some ornithologists thought that flamingoes were closely related to ducks and geese, citing as evidence the observation that three of the four genera of bird lice found on flamingoes also parasitize ducks and geese. DNA analysis, however, showed that flamingoes are not close relatives of ducks, but are more closely related to grebes, another group of waterfowl. If this analysis is accurate, what would you predict about the lice parasitizing grebes? How could you use modern methods of molecular analysis to determine relationships among bird lice and their hosts? Given that grebes, ducks, and flamingoes are all water birds, what other factors might contribute to host shifts in ectoparasitic lice?

10. Even though nectar serves no function in the life of a plant other than to attract and reward pollinators, some plants produce toxic compounds in their nectar. As we've seen, in some cases these substances are addictive and encourage pollinators to revisit the same plant species; honey bees, for example, may visit tobacco flowers repeatedly because they become "addicted" to nicotine (see Section 56.3). But another study had a different outcome. Some researchers created genetically modified tobacco plants that produced different levels of nicotine in their nectar and found that higher concentrations led to shorter visits by pollinating hummingbirds and hawk moths—and more successful pollination than in plants whose flowers hosted longer visits. Why might shorter visits increase pollination success? What other factors might influence how much nicotine a tobacco plant should produce in its nectar?

Go to BioPortal at **yourBioPortal.com** for Animated Tutorials, Activities, LearningCurve Quizzes, Flashcards, and many other study and review resources.

57 Community Ecology

Dead Reckoning Forensic entomologists use pig carcasses like this one as models to measure successional changes in corpse communities. They are able to apply these measurements to human corpses because pigs are in the same weight range as humans and, like humans, pigs are mostly hairless.

O NE DAY IN MARCH OF 1996, the body of an apparent suicide victim was discovered under the bushes near the railroad tracks in Cologne, Germany. The badly decomposed corpse contained masses of maggots—fly larvae—and the dried outer skin was peppered with pale yellow insect eggs. Examining the body, Mark Benecke, a forensic entomologist, recovered a single adult fly, which he identified as *Piophila casei*, the cheese skipper.

A diverse community of insect species colonizes human corpses, and its composition varies predictably as decomposition progresses. There are three major stages of decomposition: autolysis (degradation of proteins and lipids), putrefaction, and finally, decomposition. Each stage is characterized by a distinctive faunal community of species that use decomposing corpses in many ways. Some species, such as cheese skippers, consume dead flesh; others, such as hide beetles, eat hair and nails; still others prey on the flesh-eating insects or consume their excrement. Among these species, the cheese skippers are latecomers, arriving at corpses only when autolysis is well advanced and the body's proteins start breaking down—typically after 1 to 3 months. The abundance of cheese skipper eggs suggested to Benecke that these insects had undergone at least two generations in the corpse. Knowing that completion of the insect's life cycle required from 11 to 19 days under local weather conditions, he calculated that the first adults probably laid eggs about 90 days after death, and that 22 to 38 more days were needed to complete two generations. Thus he calculated that death must have occurred

between 112 and 128 days earlier. As it happened, a 38-year-old woman had been reported missing about 4 months before the body was found; the estimated postmortem interval helped investigators identify her as the suicide victim found by the tracks.

What species occur where and when is the concern of community ecologists. The species present in a community can change in predictable ways, at spatial scales ranging from a dead body in a patch of shrubbery to an Amazonian rainforest, and at time scales ranging from days to millennia. But the ecological processes affecting communities are similar whatever the scale. The study of the seemingly esoteric changes in the composition of carrion communities has not only allowed ecologists to help find evidence that can identify a missing person or convict a murderer, but has also added to our understanding of ecological communities.

How do the insect species in a corpse community influence one another's ability to survive?

See answer on p. 1203.

(A) *Sarracenia purpurea*

(B)

57.1 Ecological Communities Exist at Different Scales (A) The microorganisms and tiny invertebrates (such as insect larvae) existing within a single pitcher plant constitute an ecological community. (B) Lake Superior in central North America is Earth's largest freshwater lake. It encompasses a large biological community with boundaries defined by its shoreline. Despite having such a defined boundary, the lake community is subject to effects from species and activities far beyond its contained waters.

57.1 What Are Ecological Communities?

In ecological terms, a **community** is a group of species that co-exist and interact within a defined area. Although each species has unique interactions with the other species in its community (as we saw in the previous chapter), ecologists often find it useful to study the properties of the community as a whole.

Communities vary greatly in size and scope. The organisms colonizing a dead body constitute one type of definable community. Another type of easily identifiable community is that living within the purple pitcher plant (*Sarracenia purpurea*), a plant common in North American wetlands (**Figure 57.1A**). Each plant has several leaves that form rainwater-collecting "pitchers," and the plant derives nitrogen from insect prey that get trapped in the pitchers and drown (see Section 36.5). However, the pitchers are also occupied by thriving communities of living microorganisms and tiny invertebrates, including bacteria, protists, rotifers, and mosquito larvae.

The boundaries defining a community, particularly a large one, are not always so easy to recognize. The community of organisms inhabiting Lake Superior, for example, is for the most part bounded by the shores of the lake (**Figure 57.1B**). However, even though Lake Superior may appear to be self-contained, many of its components originate far away. For example, mallards and other ducks may consume seeds in one location, and in the 5 to 11 hours it takes for the seeds to move from one end of the duck digestive tract to the other, the birds may fly up to 1,400 kilometers before depositing the seeds in their excrement.

When borders become unclear, ecologists may designate boundaries somewhat arbitrarily, based on their ability to study the community. By the same token, because a community can contain thousands of different species from microscopic bacteria to towering trees, it is usually impractical or impossible to study all species within a community. Ecologists often define a community of interest taxonomically—for example, the fish community of Lake Superior.

Communities are characterized (1) by their species composition—the number and kinds of species they contain; and (2) by the relative abundances of those species. Species composition is determined by the same factors that determine the distributions of species because, as we saw in Chapter 54, a species can occur in a particular place only if it can colonize that place and if environmental conditions allow it to persist there. Thus even the same type of community may contain different numbers of species in different places. The insect community of human corpses, for example, varies with local climate and burial conditions (e.g., whether the body is buried, deposited on the soil surface, or submerged in a river or pond).

Although communities vary in size and complexity, ecologists have devised methods for quantifying basic properties of community structure and organization irrespective of their scale. These methods have revealed patterns that reflect underlying community assembly rules and general principles.

What determines how many species constitute a community in any particular place? One important factor is the amount of energy available to sustain organisms.

Energy enters communities through primary producers

Sunlight is the ultimate source of energy for most of Earth's communities. Sunlight makes photosynthesis possible, and photosynthesis, in the vast majority of communities, makes energy available to other organisms in an edible form. All nonphotosynthetic organisms (heterotrophs) consume, either directly or indirectly, the energy-rich organic molecules produced by the plants and other photosynthetic organisms that get their energy directly from sunlight (autotrophs). Photosynthetic autotrophs, along with a handful of chemoautotrophs (organisms that obtain chemical energy from inorganic molecules in their environment), are known as **primary producers**.

Gross primary productivity (**GPP**) is the rate at which all the primary producers in a particular community turn solar energy into stored chemical energy via photosynthesis. The energy that is accumulated by primary producers is called gross

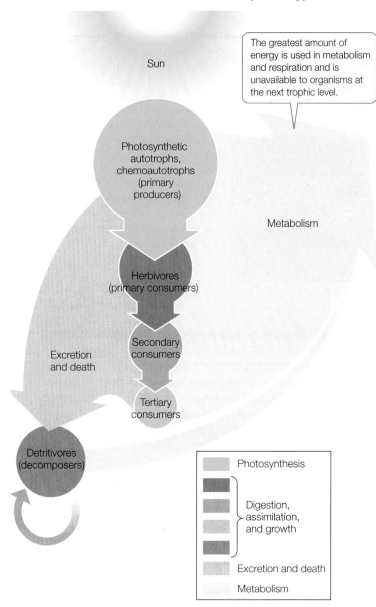

57.2 Energy Flow through Trophic Levels Much of the energy accumulated at each trophic level is lost (often as heat) to metabolism and respiration by the organisms at that level. In this diagram, the width of each arrow is roughly proportional to the amount of energy flowing through that channel. Arrows indicate directions of energy flow.

Go to Activity 57.1
Energy Flow through an Ecological Community
Life10e.com/ac57.1

primary production. (The terms "productivity" and "production" are often used interchangeably: "productivity" is the rate of energy accumulation; "production" is a measure of accumulated energy as a product.)

Not all GPP accumulated by primary producers becomes available to heterotrophs because primary producers use some of that energy for their own respiration and other metabolic processes. **Net primary productivity** (**NPP**) is the rate at which energy is incorporated into primary producers' bodies through growth and reproduction. Thus net primary

production can be measured as the amount of primary producer biomass (the weight of organic matter) that is available for consumption by heterotrophs. This relationship is described mathematically as

$$NPP = GPP - R$$

where R is the energy lost through respiration. These relationships are represented in a highly simplified form in **Figure 57.2**.

Consumers use diverse sources of energy

A **food chain** is a diagram that depicts the linear sequence of who eats whom in a given community. Food chains can be interwoven into a more realistic depiction of community feeding relationships, called a **food web** (**Figure 57.3**). Most communities contain so many species interacting in so many different ways that it is impossible to enumerate (or even identify) all of the links. Nevertheless, simplified food webs are useful in envisioning the sequence of energy flow through a community. An organism's **trophic level** indicates where in that sequence it obtains its energy (**Table 57.1**). Primary producers start the chain of trophic levels. At the next level are **primary consumers**—the herbivores that dine on the primary producers. Organisms that eat herbivores, called **secondary consumers**, are the next trophic level. Those that eat secondary consumers are tertiary consumers, and so on. The waste products and dead bodies of organisms (known as **detritus**) provide another source of energy, as we saw at the opening of this chapter. Organisms that consume such materials are called **detritivores** or **decomposers** (see Figure 57.2).

 Go to Media Clip 57.1
A Food Chain in Africa
Life10e.com/mc57.1

Most species in a community eat and are eaten by more than one other species, so food webs and trophic levels are necessarily oversimplified. Some organisms, called **omnivores**, feed on multiple trophic levels. Opossums, for example, are famously omnivorous. Investigators in Portland, Oregon, dissected the stomachs of road-killed opossums and found remains of mammals, birds, insects, earthworms, snails, fruits, bulbs, seeds, leaves, grass, pet food, and garbage, along with some items they couldn't identify—a diet that would be difficult to depict in a food web.

Fewer individuals and less biomass can be supported at higher trophic levels

One important way to characterize a community is by the distribution of energy and biomass within it. The flow of energy through a community is governed by the physical laws that regulate energy transformations, foremost among which are the first and second laws of thermodynamics. Recall from Section 8.1 that the amount of energy in the universe is constant, and that when energy is converted from one form to another, part of it becomes unavailable to do work. As Figure 57.2 showed, energy is lost to metabolism and respiration at each trophic level.

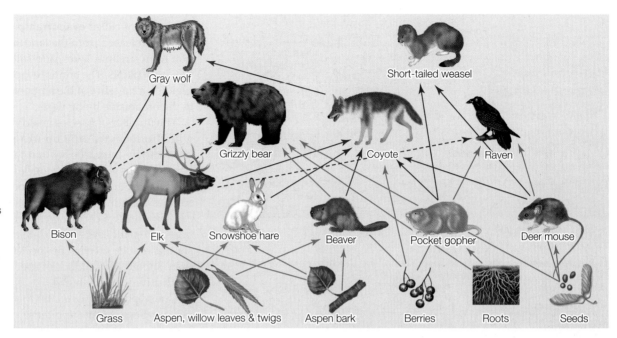

57.3 Food Webs Show Trophic Interactions in a Community
This simplified food web for the grasslands of Yellowstone National Park includes only large vertebrates and the plants on which they depend. The arrows show who eats whom. Species whose sole source of food is plants (green arrows) are primary consumers. Carnivores that kill and eat animals (red arrows) are secondary and tertiary consumers. Omnivores such as grizzly bears, coyotes, and ravens eat both plant and animal tissues; ravens and grizzlies also eat carrion (dashed red arrows), so these species are also detritivores.

Go to Activity 57.2 The Major Trophic Levels
Life10e.com/ac57.2

On average, only about 10 percent of the energy of one trophic level is transferred to the next, for a number of reasons:

- *Heat loss.* Organisms incorporate much of the energy they accumulate into biomass, but they use much more of it for respiration and other metabolic processes. That energy is dissipated as heat and is lost to the community.

- *Biomass availability.* Not all of the biomass in a community can be ingested. Grazers routinely miss blades of grass; effective plant defenses prevent herbivory; prey can escape predators or leave the community.

- *Indigestibility.* Not all of the biomass ingested can be assimilated by consumers. Tree bark, for example, contains lignin and cellulose, which cannot be digested by most herbivores.

The overall transfer of energy from one trophic level to the next (which can be expressed as the ratio of consumer production to producer production) is called **ecological efficiency**.

Pyramid diagrams such as those in **Figure 57.4A** can be used to illustrate the proportions of energy transferred from each trophic level to the next and to compare those proportions among different communities. Pyramid diagrams can also be used to illustrate the amount of biomass or numbers of individuals found at each trophic level (**Figure 57.4B**). As these diagrams show, a given environment typically supports fewer individuals, less biomass, and fewer species at higher trophic levels than at lower trophic levels. Progressive energy loss through the inefficiencies of energy transfer also limits the number of trophic levels in a food chain or food web; largely for this reason, most communities support only three to five trophic levels. One conspicuous exception to this general pattern occurs in the open oceans. The phytoplankton that are the primary producers

TABLE**57.1**

The Major Trophic Levels

Trophic Level	Source of Energy	Examples
Photosynthesizers (primary producers)	Solar energy	Green plants, photosynthetic bacteria, diatoms
Herbivores (primary consumers)	Tissues of primary producers	Copepods, grasshoppers, bark beetles, deer, geese, white-footed mice
Primary carnivores (secondary consumers)	Tissues of herbivores	Spiders, warblers, wolves, anchovies
Secondary carnivores (tertiary consumers)	Tissues of primary carnivores	Tuna, falcons, killer whales
Omnivores	Several trophic levels	Humans, opossums, crabs, robins
Decomposers (detritivores)	Dead tissues and waste products of other organisms	Many fungi, many bacteria, vultures, earthworms, termites

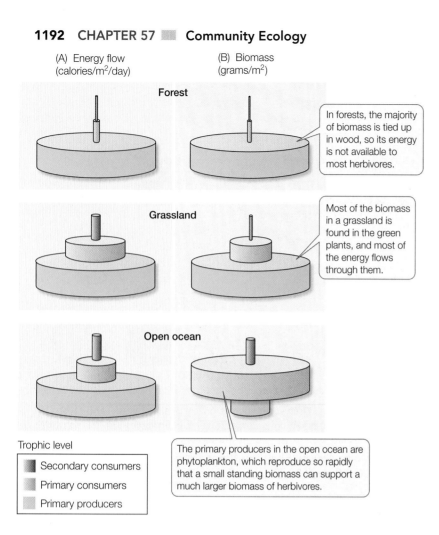

(A) Energy flow (calories/m²/day)

(B) Biomass (grams/m²)

Forest

In forests, the majority of biomass is tied up in wood, so its energy is not available to most herbivores.

Grassland

Most of the biomass in a grassland is found in the green plants, and most of the energy flows through them.

Open ocean

The primary producers in the open ocean are phytoplankton, which reproduce so rapidly that a small standing biomass can support a much larger biomass of herbivores.

Trophic level

■ Secondary consumers
■ Primary consumers
■ Primary producers

57.4 Energy and Biomass Distributions Pyramid diagrams allow ecologists to compare patterns of energy flow through trophic levels (A) and the amount of biomass present at different trophic levels in different communities (B). Biomass distribution in the open ocean differs ftrom that in most other communities because most of the biomass is not at the primary producer level.

those regions called **evapotranspiration**: the amount of water released from the land surface by evaporation from streams, lakes, and soil and by transpiration from plants. The annual evapotranspiration of a region is a measure of the amount of water available to the organisms living there.

The number of species in a community increases with productivity only up to a point, however. If productivity increases beyond that point, the number of species may actually decline (see Figure 57.5). Why should that be? As local productivity increases, so does the number of individuals the local habitat can support (its carrying capacity). Thus populations can grow larger, and larger population sizes should reduce the risk of species extinction. Why, then, should the number of species decrease when productivity is very high?

One hypothesis postulates that interspecific competition becomes more intense when productivity is very high, resulting in competitive exclusion of some species (see Section 56.4). This hypothesis is supported by the results of a long-term experiment at the Rothamsted Experiment Station in England, begun in 1857 and still going on. Fertilizer has been added regularly to selected plots of land to increase their productivity, and fertilized and unfertilized plots have been monitored continuously. Over 150 years, the number of plant species in the unfertilized plots has remained roughly constant, whereas the number of species in the fertilized plots has declined, supporting the premise that species diversity can decline when productivity rises.

in this community grow and reproduce so much faster than the zooplankton and small fish that consume them that their smaller biomass, with its rapid rate of primary production, can actually support a larger biomass of primary consumers.

Productivity and species diversity are linked

Just as the diversity of a trophic level tends to be positively correlated with the amount of energy available to it, the species diversity of a community tends to be positively correlated with its productivity, up to a point (**Figure 57.5**). A number of factors that influence productivity vary among communities. The most obvious of these factors is energy input. The amount of solar energy reaching Earth's surface varies by latitude, as we saw in Figure 54.1. The ability of plants to photosynthesize, however, depends not only on the supply of energy from the sun but also on the supply of water and nutrients. The number of tree species across different regions of North America, for example, can be best predicted not by measuring incoming solar radiation but by measuring an ecological attribute of

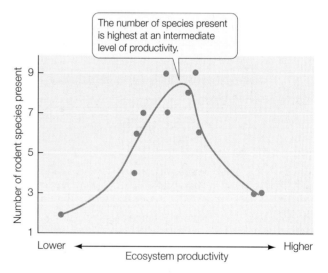

The number of species present is highest at an intermediate level of productivity.

57.5 Species Diversity Peaks at Intermediate Productivity The number of rodent species living in ecosystems of varying productivity in the Gobi Desert exemplifies a pattern in which species richness increases only up to a certain point. Beyond that point, it can actually decline with productivity.

RECAP 57.1

An ecological community is a group of species that coexist and interact within a defined area. Energy enters an ecological community through the primary producers and moves through the food web by means of trophic interactions.

- Explain the difference between gross primary productivity and net primary productivity. **See pp. 1189–1190 and Figure 57.2**
- Describe three trophic levels that might appear in a food web. **See pp. 1190–1191, Table 57.1, and Figure 57.3**
- What is a typical distribution of energy among the trophic levels of a community? **See pp. 1190–1191 and Figures 57.2 and 57.4**

Energy and biomass distributions primarily reflect the abiotic factors that influence the community. In the next section we will see how the interactions among species described in Chapter 56 shape community structure.

57.2 How Do Interactions among Species Influence Communities?

The antagonistic interactions described in Section 56.2, which transfer energy between trophic levels, have a strong influence on ecological communities. But species are not identical bags of biomass through which energy flows. Species in one part of a food web can affect many other species without necessarily eating them.

Species interactions can cause trophic cascades

The interactions of a single consumer species with other species in its community can cause a progression of successive effects throughout an entire food web, a pattern called a **trophic cascade**. The reintroduction of wolves into Lamar Valley in Yellowstone National Park initiated just such a pattern.

The food web in the grasslands of Yellowstone National Park is far more complex than the "streamlined" rendition in Figure 57.3 indicates. Gray wolves in the park feed on elk, bison, and coyotes. Although they share some of these prey species with coyotes and grizzly bears, wolves exert particularly strong effects on the park community's structure and dynamics, as demonstrated by the effects of their absence during most of the twentieth century.

By 1926, unrestricted hunting had eliminated wolves from the Yellowstone community. To prevent elk from exceeding the park's carrying capacity, the park service culled elk herds (that is, they selectively killed some members of each herd) until 1968, when, in response to public pressure, the culls were stopped and the elk population rapidly increased (**Figure 57.6A**). The elk browsed aspen trees so intensely that the number of young trees recruited (added to the population) declined precipitously, and by 1960 no new trees were recruited at all (**Figure 57.6B**). The elk also severely browsed streamside willows, with the result that beavers, which depend on willows for food, were nearly exterminated from Lamar Valley. In regions of the park where elk were absent, however, aspen and willow trees flourished. This observation suggested that the decline of the trees in Lamar Valley was

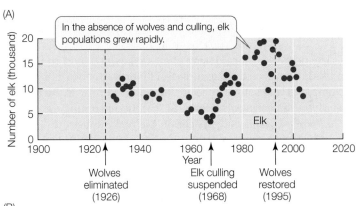

(A)

In the absence of wolves and culling, elk populations grew rapidly.

Elk

Wolves eliminated (1926) Elk culling suspended (1968) Wolves restored (1995)

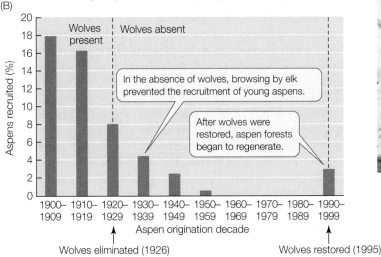

(B)

Wolves present | Wolves absent

In the absence of wolves, browsing by elk prevented the recruitment of young aspens.

After wolves were restored, aspen forests began to regenerate.

Aspen origination decade

Wolves eliminated (1926) Wolves restored (1995)

57.6 Wolves Initiated a Trophic Cascade in Yellowstone National Park (A) Number of elk in Wyoming's Yellowstone National Park. (B) Aspen recruitment (that is, the number of new trees becoming established) in the presence and absence of wolves.

indeed due to elk browsing rather than to climate conditions or some other factor.

In 1995, after wolves had been absent for 70 years, park managers reintroduced them to Yellowstone, and their population grew rapidly. The wolves preyed primarily on elk. The elk population of Lamar Valley dropped, and elk avoided the aspen groves, where they were especially vulnerable to wolf predation. Young aspen began to grow, willows regrew along streams, and the number of beaver colonies increased from one in 1996 to seven in 2003. Thus the presence or absence of a single predator influenced not only populations of its prey but also populations of its prey's food resource and of other species that depended on that resource.

Herbivores, too, can have indirect effects on other trophic levels. The savannas of central Kenya are dominated by large grazing mammals such as zebras, eland, elephants, Grant's gazelles, giraffes, buffaloes, and hartebeests. A team of investigators used exclosures (areas protected by a barrier, such as a fence, designed to keep organisms out) to investigate the influence of these grazers on the savanna community. They created six exclosures and compared community structure within the exclosures with that in control sites where large mammals could graze freely. Over 19 months of monitoring, they found that trees, beetles, and an insectivorous lizard species were all more abundant within the exclosures. The elimination of grazing increased the abundance of trees, which in turn increased the number of beetles by providing more food and habitat, which in turn increased the number of lizards, which feed preferentially on beetles.

Trophic cascades can be initiated by the behaviors of species as well as by their diets. Beavers preferentially cut down some species of trees to build their dams; by so doing, they alter the composition of the vegetation. In addition, the beavers' activities create wetlands, meadows, and ponds that provide habitat for species that would otherwise not be able to live in the area.

Organisms that build structures that alter existing habitats or create new habitats are called **ecosystem engineers**. Beavers create new habitats by cutting down (and killing) trees and using them to dam streams and create ponds. Trees in terrestrial forests, and corals and kelp in aquatic communities, are ecosystem engineers in a different way: they modify the environment by changing their size and shape over time. For example, by becoming larger and more structurally complex as it grows, the stilt palm (*Socratea exorrhiza*) of Panama can support more than 60 species of epiphytes.

Keystone species have disproportionate effects on their communities

Architects call the single wedge-shaped stone in the center of an archway the "keystone" because it holds all the other stones in place. Ecologists thus refer to any species that exerts an influence on the other members of its community that is disproportionate to its abundance as a **keystone species**. Keystone species can influence both the number of species and the number of trophic levels in a community.

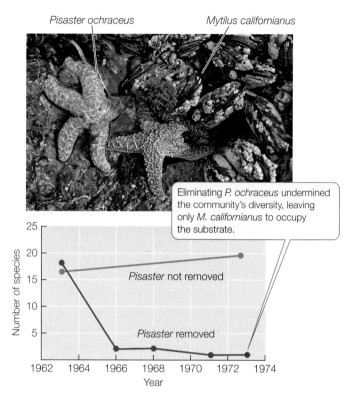

57.7 Ochre Sea Stars Are a Keystone Species   (A) Along the Pacific coast of northern North America, ochre sea stars (*Pisaster ochraceus*) consume large quantities of the mussel *Mytilus californianus*, creating bare substrate that is then inhabited by a variety of intertidal species. (B) Experiments by Robert Paine and his colleagues demonstrated that when sea stars were excluded from this intertidal community, *M. californianus* outcompeted all other species for space on the substrate and eliminated the community's diversity.

The ochre sea star (*Pisaster ochraceus*), which lives in rocky intertidal zones on the Pacific coast of North America, is a good example of a keystone species. Its preferred prey is the mussel *Mytilus californianus* (**Figure 57.7A**) By consuming mussels, *P. ochraceus* creates bare spaces on the rocky substrate. These spaces are then taken over by a variety of intertidal species.

In a classic experiment, Robert Paine of the University of Washington demonstrated the disproportionate influence of *P. ochraceus* on species diversity by removing the sea stars from selected sites repeatedly over a 5-year period. Two major changes occurred in the areas where sea stars were absent. First, the lower edge of the mussel bed extended farther down into the intertidal zone, showing that sea stars are able to eliminate mussels completely in areas that are submerged most of the time. Second, and more dramatically, 18 species of animals and algae disappeared from the sea star removal sites. Eventually only *M. californianus*—the dominant competitor for space in the community—occupied the entire substrate (**Figure 57.7B**). Through its effect on competitive relationships among its prey, predation by the sea star determines how many species can thrive in its rocky intertidal community. In the absence of sea stars, *M. californianus* crowds out other organisms in a broad belt of the intertidal zone.

Species other than consumers can be keystone species. Fig trees in tropical forests, for example, produce fruits several times every year, so their fruits are abundant at times when few, if any, other trees are fruiting. Dozens of frugivores depend on figs when no other fruits are present. Fig-eating animals include fruit bats, parrots, toucans, pigeons, flycatchers, trogons, orioles, rodents, howler monkeys, and even fish, which eat figs that fall into nearby streams. All of these animals provide prey for a diverse community of predators. Moreover, the trunks of fig trees provide habitat for several thousand species of insects, reptiles, rodents, and birds. Without fig trees, rainforest communities around the world would be profoundly diminished in terms of the number of species they contain.

RECAP 57.2

Some interactions between consumers and their resource species result in a trophic cascade of indirect effects on species at other trophic levels. A keystone species has effects on its community that are disproportionate to its abundance.

- Describe an example of a consumer whose interactions with other species cause indirect effects across trophic levels. **See p. 1194**

- What are some of the ways in which keystone species can affect other species in their communities? **See p. 1194**

We have seen how certain keystone species influence the species composition of their communities. But a simple count of species is only one measure of community diversity. The next section will look more closely at the different ways in which ecologists measure species diversity.

57.3 What Patterns of Species Diversity Have Ecologists Observed?

Communities clearly vary in their diversity, both geographically, on scales ranging from local to global, and over time, on scales ranging from a day to centuries. Comparing the diversity of two or more communities can be challenging because diversity has many different components depending on the scale at which it is measured. In some cases it is important to measure species diversity within a single community or habitat, but it can also be useful to measure diversity across an entire region, encompassing a range of communities or habitats.

Diversity comprises both the number and the relative abundance of species

The most straightforward way to quantify the diversity of a community is to count the number of species present in a sample. This number is the **species richness** of the community. The larger the area that is sampled, the greater the likelihood that rare species in the community will be found, and the more accurate the resulting assessment of its species richness will be.

To say that two communities of the same size have the same species richness tells only part of the story, however. How abundant each species is in the community also affects diversity. This aspect of diversity—the distribution of abundances of individuals across species—is called **species evenness**. Imagine that we take samples of 12 individuals in each

Community A

Community B

57.8 Species Richness and Species Evenness Both Contribute to Diversity These two hypothetical mushroom communities are the same size (12 individuals) and have the same species richness (four species), but community A has a more even distribution of species and is thus more diverse than community B.

of two different communities. Our sample from community A contains 3 individuals of each of four species (an even distribution of individuals). Our sample from community B, however, contains 9 individuals of one species and only 1 individual of each of the other three species (an uneven distribution). Even though the species richness of the two communities is the same (four), community B is less diverse because the less abundant species are encountered infrequently compared with the single most abundant species (**Figure 57.8**). Thus estimates of diversity should account for both species richness and species evenness.

A study of diversity patterns in an agricultural region of southern England demonstrates how measures of diversity can be applied at different spatial scales. Investigators sampled plants and macroinvertebrates in three types of freshwater communities—rivers, ponds, and ditches—in this area of English countryside. Rivers had high species richness, but most rivers in the area contained the same assortment of species. In contrast, the ponds displayed a wide range of variation in species richness. Thus the ponds contributed more to the overall diversity of the area than the rivers did because they contained

57.9 Latitudinal Gradients in Diversity Among swallowtail butterflies (Papilionidae), species diversity decreases with latitude both north and south of the equator. Similar latitudinal gradients of diversity have been observed in many other taxa.

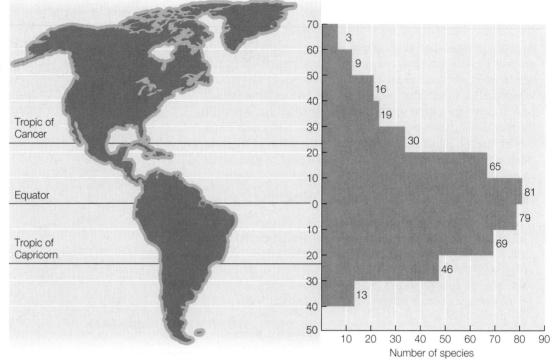

Number of species

more unique or rare species. Surprisingly, the ditches, many of which contained water for only a short time, had the lowest species richness of the three community types, but many of the species in the ditches were found nowhere else in the area, including insects that live only in temporary bodies of water (including a very rare water beetle). Ponds and ditches, despite being relatively species-poor, contributed disproportionately to regional diversity because the few species they did support were found nowhere else in the region. Partitioning diversity within a community, between communities, and across an entire region in this way can provide insights into the ecological characteristics of the species making up those communities as well as the processes by which those communities were assembled.

Ecologists have observed latitudinal gradients in diversity

Diversity can be measured at a wide range of scales, but the broader the scale at which it is measured, the more difficult it can be to understand the ecological process underlying the differences observed. Understanding diversity at a global scale has proved to be a challenge.

About 200 years ago, the German explorer and naturalist Alexander von Humboldt spent 5 years traveling around Latin America. He remarked in the account of his voyages that "the nearer we approach the tropics, the greater the increase in the variety of structure, grace of form, and mixture of colors, as also in perpetual youth and vigour of organic life." Humboldt would not have been surprised to learn that, if he had sailed toward the poles, the diversity he observed would have decreased. These latitudinal gradients in diversity have been observed repeatedly in both hemispheres and in a wide variety of taxa, including birds, mammals, flowering plants, and insects (**Figure 57.9**).

Although most ecologists agree that latitudinal gradients in diversity exist, there is less consensus as to why they exist. At least four hypotheses have been advanced to account for latitudinal gradients in diversity:

- The *time hypothesis* argues that over evolutionary time, organisms in tropical regions have had more time to diversify under relatively stable climate conditions than have those in more temperate regions.

- The *spatial heterogeneity hypothesis* suggests that tropical regions have higher spatial heterogeneity—more different types of microclimates, vegetation, soils, and so forth—and thus contain more distinct habitats and many more species.

- The *specialization hypothesis* attributes latitudinal gradients to greater interspecific competition in the tropics, which leads to narrower realized niches (see Figure 56.15).

- The *predation hypothesis* proposes that predation intensity is greater in the tropics. Where predation is high, it argues, prey populations are held to levels so low that interspecific competition never comes into play, and rare species can persist.

Corroborative evidence can be found for each of these hypotheses, varying with taxon, locality, and scale, and not all of the hypotheses are mutually exclusive. It may be that multiple factors are responsible for this widespread ecological pattern.

The theory of island biogeography suggests that immigration and extinction rates determine diversity on islands

While latitudinal diversity gradients prevail on a global scale, other factors influence species diversity at smaller spatial scales. For example, small islands tend to have fewer species

WORKING WITH**DATA:**

Latitudinal Gradients in Pitcher Plant Communities

Original Paper

Buckley, H. L., T. E. Miller, A. Ellison, and N. J. Gotelli. 2003. Reverse latitudinal trend in the species richness of an entire community at two spatial scales. *Ecology Letters* 6: 825–829.

Analyze the Data

In 2003 Hannah Buckley and several colleagues examined the diversity of species present in the community of water-filled leaves—pitchers—of *Sarracenia purpurea* (see Figure 57.1A). The researchers determined the relative abundances of all species of invertebrates (including insect larvae), heterotrophic protists (protozoa), and bacteria present in each of 20 pitcher plants collected at each of 39 sites that spanned the plant's entire north–south range. They summarized the data at two scales: as the average number of species per site (the site-wide scale) and as the average number of species per pitcher at each site (the pitcher scale). Their data are plotted in the graphs at the right.

QUESTION 1

Overall, does this community display a typical diversity gradient with latitude? Of the individual taxa, which ones depart from the typical diversity gradient?

QUESTION 2

The top predator in this system is the larva of the mosquito *Wyeomyia smithii*, which filter-feeds on bacteria and protozoans. This species is significantly more abundant at low latitudes (i.e., farther south in the Northern Hemisphere) than at high latitudes. How might the abundance of a top predator explain the pattern observed by these investigators?

QUESTION 3

Do you think *Wyeomyia smithii* is a keystone species? Why or why not?

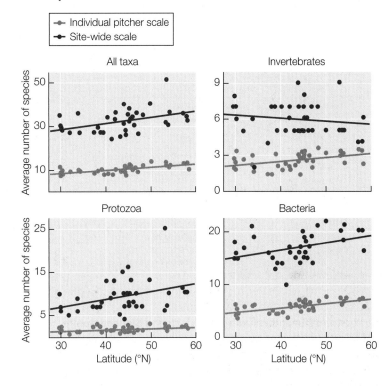

Go to BioPortal for all WORKING WITH**DATA** exercises

---

than large islands, irrespective of latitude (**Figure 57.10**). The biologist Edward O. Wilson was struck by this **species–area relationship**, which he encountered through his exhaustive collection of ant species from all over the world.

With Robert MacArthur, Wilson developed the theory of **island biogeography**. They based their theory on just two processes: the immigration of new species to an island and the extinction of species already present on that island (**Figure 57.11A**). The premise of island biogeography is that the number of species on an island represents a balance, or equilibrium, between the rate at which species immigrate to and colonize the island and the rate at which resident species go locally extinct.

The rate of immigration is determined in part by the number of species in the source area providing the immigrants, known as the **species pool**. In the case of oceanic islands, the species pool comprises all the species on the nearest land mass (usually a continent). Not all species that reach the island will persist there, however. The more species there are on an island, the greater the likelihood that there will be competition for limited resources, and the higher the likelihood that any of the species on the island will go extinct. Therefore at some point the number of species arriving from

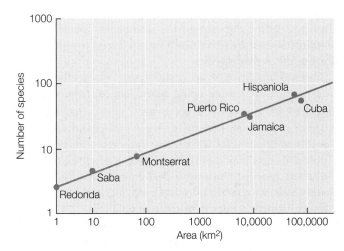

57.10 The Species–Area Relationship E. O. Wilson and Robert MacArthur plotted the number of species of reptiles and amphibians against the size of several islands in the Antilles of the Caribbean. Larger islands consistently contained more species, regardless of latitude, a fact these scientists incorporated in their theory of island biogeography.

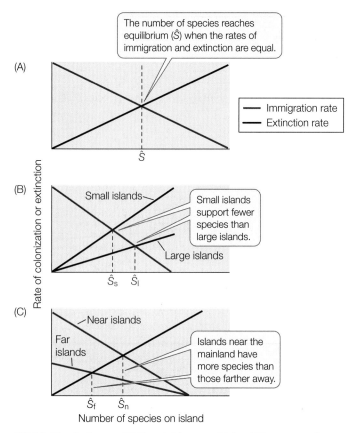

The number of species reaches equilibrium (Ŝ) when the rates of immigration and extinction are equal.

(A)

— Immigration rate
— Extinction rate

Ŝ

(B)

Small islands

Small islands support fewer species than large islands.

Large islands

Ŝ_s Ŝ_l

(C)

Near islands

Far islands

Islands near the mainland have more species than those farther away.

Ŝ_f Ŝ_n

Number of species on island

57.11 MacArthur and Wilson's Theory of Island Biogeography
(A) The rate of arrival of new species and the rate of extinction of species already present determine the equilibrium number of species on an island. These rates and the eventual equilibrium numbers are affected by the size of the island (B) and by the island's distance from the mainland (C). For simplicity these rates are depicted as linear, but in reality immigration and extinction rates tend to be curvilinear.

the source area and the number of resident species going extinct should balance, and the number of species should remain stable at this balance point, referred to as the equilibrium number of species.

Rates of immigration and extinction, and thus the equilibrium number of species on an island, are influenced by two other factors:

• *The size (area) of the island.* The smaller the island, the fewer resources it provides, the greater the potential for competition, and the higher the extinction rate will be (**Figure 57.11B**). Larger islands provide greater habitat diversity and can sustain larger populations (which tend to have lower extinction rates than small populations).

• *Distance of the island from the species pool.* The farther the island is from the source of immigrants, the lower the immigration rate—the rate at which new species arrive—will be (**Figure 57.11C**).

Between 1966 and 1969, Wilson and his student Daniel Simberloff conducted an ingenious experiment to test the theory

of island biogeography (although given concerns over environmental impacts, this experiment might not have been approved today). Simberloff and Wilson identified four small, isolated clumps of red mangrove (*Rhizophora mangle*), all approximately the same size (11–12 meters in diameter), in the Florida Keys. These mangrove islands were small enough to allow an accurate count of the arthropod species on each one. They were also small enough for the research team to enclose each island in a tent and gas it with methyl bromide to kill all the arthropods. After this defaunation, Simberloff and Wilson monitored and tracked recolonization of the islands by arthropods (**Figure 57.12**). After 2 years, species richness on all but the farthest island was close to what it was before defaunation. This observation is consistent with the idea that the number of species on the islands prior to defaunation represented an equilibrium number of species.

The theory of island biogeography can be applied equally well to **habitat islands**—isolated patches of suitable habitat surrounded by extensive areas of unsuitable habitat. Thus a pond in the English countryside or a forest surrounded by housing subdivisions may acquire an equilibrium number of species in much the same way an oceanic island does. The theory of island biogeography also has important applications for the conservation of endangered species. As habitat islands decrease in size because of human encroachment, more and more species become vulnerable to population declines, especially those that require large areas in order to live and breed successfully. Even more broadly, the processes of immigration and extinction contribute to determining community composition at continental scales.

Go to Animated Tutorial 57.1
Biogeography Simulation
Life10e.com/at57.1

■ RECAP **57.3**

Measures of species diversity encompass both species richness and species evenness. Species diversity is highest in the tropics, decreasing at higher latitudes. Island biogeography theory states that species diversity on an island or other isolated habitat represents a balance between immigration and extinction rates.

• What is the difference between species richness and species evenness? **See p. 1195 and Figure 57.8**

• Explain how scale can influence estimates of diversity. **See pp. 1195–1196**

• What factors influence the equilibrium number of species on an island, according to the theory of island biogeography? **See pp. 1197–1198 and Figure 57.11**

The composition of a community is dynamic: as we have seen, it can change over time and space. Processes of change such as immigration to and extinction on islands are generally predictable and consistent. But community composition can change dramatically in response to less predictable forces as well.

INVESTIGATINGLIFE

57.12 The Theory of Island Biogeography Can Be Tested

By experimentally removing all the arthropods on four small mangrove islands of equal size but different distance from the mainland, Simberloff and Wilson were able to observe the process of recolonization and compare the results with the predictions of island biogeography theory.

HYPOTHESIS If mangrove islands are populated by an equilibrium number of species, then the rate at which they will accumulate species after defaunation will decrease with distance from a mainland source of colonists, as will their eventual species richness.

Method

1. Census the terrestrial arthropods on 4 small mangrove islands of equal size (11–12 m diameter) but different distance from a mainland source of colonists.

2. Erect scaffolding and tent the islands. Fumigate with methyl bromide (a chemical that kills arthropods but does not harm plants).

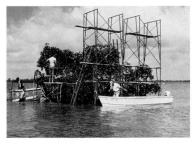

3. Remove tenting. Monitor recolonization for the following 2 years, periodically censusing terrestrial arthropod species.

Results Recolonization was fastest on the closer islands, slowest on the one farthest from the mainland. Two years after defaunation, each island had about the same number of species it had before the experiment.

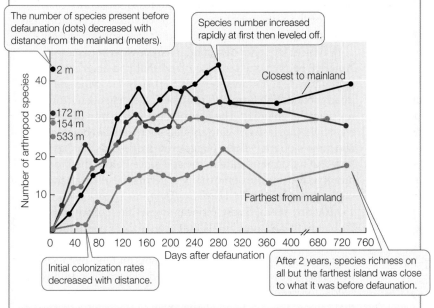

The number of species present before defaunation (dots) decreased with distance from the mainland (meters).

Species number increased rapidly at first then leveled off.

Closest to mainland

Farthest from mainland

Initial colonization rates decreased with distance.

After 2 years, species richness on all but the farthest island was close to what it was before defaunation.

CONCLUSION The data support the theory that species richness on islands represents a dynamic balance between colonization and extinction rates.

Go to **BioPortal** for discussion and relevant links for all INVESTIGATINGLIFE figures.

Simberloff, D. S. and E. O. Wilson 1970. *Ecology* 51: 934–937

How Do Disturbances Affect Ecological Communities?

57.4

An ecological **disturbance** is a disruption in a community caused by a discrete external force, often abiotic in nature. Disturbances may remove some species from a community but may open up space and resources for other species.

The magnitude of a disturbance's effects varies enormously. Some disturbances are limited to small areas—for example, a log carried by waves may crush algae and animals attached to rocks in an intertidal community. In contrast, hurricanes, forest fires, and volcanic eruptions affect communities over hundreds or thousands of hectares. Although small-scale disturbances are far more frequent, a few large-scale events may be responsible for most of the changes in a community. A single hurricane, for example, may fell more trees than several years of "normal" storms, and the movements of a glacier change community composition across millennia.

A community's history of disturbance may explain patterns of species diversity that would otherwise be puzzling. The Province Islands, located just above the Antarctic Circle in the South Indian Ocean, provide an example. The climate of these islands is cool, with average temperatures above freezing for only 6 months of the year; precipitation is high, and gale force winds are common. The vegetation is primarily tundra. South African entomologist S. L. Chown, an expert on life in the Antarctic, compared the insect faunas on the four largest of these islands and found that the two largest, Marion and Kerguelen, housed fewer arthropod species (16 and 22 species, respectively) than the significantly smaller Cochons and Possessions (26 and 38 species). Why do the species diversity patterns on these islands fail to conform to the theory of island biogeography? One possible explanation lies in their different disturbance histories. The two largest islands were once covered by glaciers and thus experienced considerably more disturbance than did the two smaller islands.

Succession is the predictable pattern of change in a community after a disturbance

How does a community recover from a disturbance, particularly one as massive as a glacier? The pattern of change in community composition following a disturbance is known as **succession**. The most common type of succession is **directional succession**, which is characterized by an orderly (or at least predictable) progression

57.13 Primary Succession As the community occupying a glacial moraine at Glacier Bay, Alaska, changes from an assemblage of pioneer plants such as *Dryas* to a spruce forest, soil depth increases and nitrogen accumulates in the soil.

 Go to Animated Tutorial 57.2
Primary Succession on a Glacial Moraine
Life10e.com/at57.2

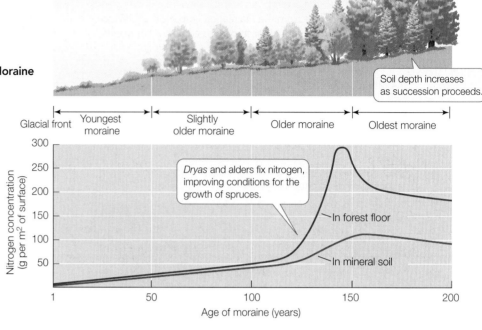

of community assemblages. Species come and go until a particular community—one that is capable of perpetuating itself under the local climate and soil conditions—persists for a relatively long time. This persistent stage is called the **climax community**.

Directional succession is easiest to observe after a disturbance strips away all preexisting living organisms and exposes a bare substrate. The type of directional change that occurs under these circumstances is known as **primary succession**. Glaciers (see Figure 1.16), volcanic activity, and in some cases, floods can initiate primary succession.

Primary succession can be seen in the successive changes in plant growth form and community composition in the wake of the retreat of the glacier in Glacier Bay, Alaska, over the last 200 years (**Figure 57.13**). The glacier scraped the landscape down to bare rock and left a series of moraines—gravel deposits dropped where the glacial front was stationary for a number of years. No human observer was present to measure changes over the entire 200-year period, but ecologists have inferred the temporal pattern of succession by studying the vegetation on moraines of different ages. The youngest moraines, closest to the current glacial front, are populated with bacteria, fungi, and photosynthetic microorganisms that can support themselves on bare rock. Slightly older moraines farther from the glacial front are home to lichens, which break down rocks and, when they die, decompose and contribute to the buildup of soil. Mosses and a few species of shallow-rooted herbs such as mountain avens (*Dryas octopetala*) become established and contribute to soil-building as they die and decompose. Still farther from the glacial front, successively older moraines have deeper soil layers that support shrubby willows, alders, and spruces.

Nitrogen is virtually absent from glacial moraines, so the plants that grow best on recently formed moraines at Glacier Bay are *Dryas* and alders (*Alnus*), both of which have nitrogen-fixing bacteria in nodules on their roots (see Figure 36.7B). Nitrogen fixation by these plants improves the soil so that spruces can grow. Spruces then outcompete and displace the early colonists. If the local climate does not change dramatically, a climax community dominated by spruce trees is likely to persist for many centuries on old moraines at Glacier Bay.

Directional succession following a disturbance that some organisms, particularly those in the soil, survive is called **secondary succession**. Secondary succession is often initiated by human activities (such as the clearing of a forest) as well as by natural disasters (such as storms and fires). Generally easier to monitor than primary succession, secondary succession also tends to occur more frequently and progress more rapidly.

A typical sequence of secondary succession in eastern North America begins when forested land that had been cleared for agriculture is abandoned (**Figure 57.14**). Because the soil is left intact, abandoned farmland provides an excellent environment for plants to colonize. The first plants to appear in old agricultural fields are fast-growing annuals such as pigweed, ragweed, and lamb's-quarter. These plants grow from seeds that have persisted in a dormant state in the soil since before the land was farmed, awaiting the opportunity to germinate. A single individual lamb's-quarter (*Chenopodium album*) can produce more than 150,000 tiny seeds. These pioneer species are quickly replaced by biennials and perennials that are stronger competitors for resources, such as milkweed, goldenrod, and thistles. The seeds of many of these plants have mechanisms for long-distance dispersal that allow them to colonize newly cleared sites. Milkweed seeds, for example, come equipped with a silky "parachute" that catches the wind and allows them to travel long distances. Eventually shrubby plants such as dogwood, eastern red cedar, and sumac become established, followed by tree species such as cottonwood, cherry, and red maple. Ultimately, shade-tolerant tree species whose seedlings can survive the shady conditions under the established trees, including beech and sugar maple, dominate the landscape. The beech–maple forest is the climax community for much of the region.

Successional stage

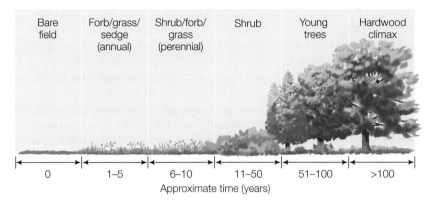

57.14 Secondary Succession Land that was once an agricultural field ultimately supports a long-lasting climax community characterized by shade-tolerant trees.

Directional succession, irrespective of where it takes place, is characterized by certain trends. In general, the pioneer species of early successional stages tend to be good dispersers with high rates of increase (*r*-strategists; see Section 55.4). Early stages of succession are characterized by high productivity and simple food webs; most nutrients are present in detritus or in the soil. As succession proceeds, nutrients accumulate in living biomass, food webs become more complex, and abiotic sources of nutrients become less important. Species typical of late successional stages tend to be good competitors with relatively low intrinsic rates of increase (*K*-strategists).

Both facilitation and inhibition influence succession

To some extent, the progress of succession depends on the activity of successive colonists, each of which modifies the environment in such a way as to facilitate colonization by other species. Predators are unlikely to colonize a habitat with no prey species, nor can primary consumers exist before plants are established. The fixation of nitrogen by *Dryas* and alders, which allows spruces to become established in Glacier Bay (see Figure 57.13), is an example of such **facilitation**.

Although secondary succession is often described in terms of changes in plant species composition, colonization by plants is actually facilitated by heterotrophs. Many of the first organisms to arrive on bare soil after a disturbance are detritivores, which process dead organic matter and release nutrients (especially nitrogen), thus facilitating the establishment of plants. In a study of intensively burned 20-year-old pine plantations in northern Germany, the first organisms to colonize these burned forests were algae, slime molds, liverworts, mosses, and mushrooms. These were followed by algae-feeding flies, fungus-feeding beetles, and moss-feeding springtails. Flowering plants such as fireweed moved in, at which point leaf-feeding insects appeared, soon followed by predaceous ground beetles and wolf spiders. Even in the corpse communities described at the opening of this chapter, early colonists can make resources available for subsequent colonizers: some spider beetles, for example, cannot feed on rotting flesh directly, but

rather feed on the excrement of the early flesh-eating colonizers.

In other cases the effect of early colonists does not facilitate but rather inhibits colonization by other species. The roots of some old-field species such as goldenrod and thistle exude chemicals that inhibit the germination and growth of potential competitors. Eventually, when these plants grow old and die, other plant species can become established. Similarly, in rocky intertidal communities, when wave action turns over boulders and clears rock surfaces, the green alga *Ulva* colonizes the cleared spaces quickly and efficiently, preventing colonization by the slower-growing perennial red algae. Certain crab species then selectively graze on *Ulva*, helping to undermine this inhibition and thus promoting the establishment of the red algae that dominate the community in later stages.

Cyclical succession requires adaptation to periodic disturbances

Some forms of disturbance recur with regularity, even if their recurrence is not always predictable. Such recurrent disturbances are associated with a pattern called **cyclical succession** because the climax community depends on the periodic disturbances in order to persist. The lodgepole pine (*Pinus contorta*) forests of southern Oregon, for example, are maintained by periodic forest fires (**Figure 57.15**). In this fire-adapted community, fires return nutrients to the soil in the form of burned organic matter and thus provide favorable conditions for seed germination. The cones of lodgepole pines are sealed shut by resins; only when they are subjected to high temperatures that melt the resins do they open and release their seeds.

57.15 Some Communities Are Adapted to Disturbance Forest communities dominated by lodgepole pine (*Pinus contorta*) are adapted to periodic fires. Fire removes mature trees weakened by pest infestation, revitalizes the soil, and provides favorable conditions for seed germination and new growth.

Lodgepole pine trees are attacked by the mountain pine beetle (*Dendroctonus ponderosae*) and are also prone to infection by the fungus *Phaeolus schweinitzii*, which causes the roots and heartwood of the trees to rot. Trees that have lived long enough to experience and be scarred by a fire are much more likely to become infected by the fungus than are trees that have not been scarred. Fungus-infected, weakened trees are preferentially attacked by beetles. After a beetle outbreak, in which many mature, fire-scarred trees are killed, the dead trees serve as potential fuel for a fire that will free up their nutrients for use by the remaining trees as well as new seedlings.

Heterotrophic succession generates distinctive communities

Plants play a vital role in most patterns of succession because, as autotrophs, they are the source of energy for the other organisms in the community. Successional changes, however, can take place without the participation of plants. Detritus-based communities—found in dung, dead plants, and carrion—undergo a series of changes known as **heterotrophic succession**. In these communities, energy resources are greatest when the habitat first becomes available to colonists and are depleted as succession takes place. There is no mechanism, such as photosynthesis, for generating more energy. Thus in contrast to most other forms of succession, biomass and species diversity decrease over time because the resource base declines. In addition, these temporary habitats are not really self-contained, so predators, which do not have to confine themselves to one dead body to live out their lives, can outnumber primary consumers (detritivores, in these communities), in apparent violation of the laws of thermodynamics.

RECAP 57.4

Ecological disturbances may remove some species from a community but may open up space and resources for other species. Ecological succession—a predictable pattern of change in community composition—typically follows a disturbance.

- How do primary succession and secondary succession differ? **See p. 1200 and Figures 57.13 and 57.14**
- Describe some ways in which early colonists facilitate or inhibit colonization by the species that follow them in a pattern of succession. **See p. 1201**

Now that we have seen how communities change in composition over time, let's look at what allows certain communities (such as climax communities) to persist over time and withstand disturbance with little change. In this context we will return to diversity as an important arbiter of community stability.

57.5 How Does Species Richness Influence Community Stability?

Up to a point, higher productivity favors higher species richness, as we saw in Section 57.1. Does species richness, in turn,

influence productivity? And how do both of those properties influence community stability? We might expect species richness to enhance productivity because no two species in a community use resources in exactly the same way (as illustrated in Section 56.4 by the principle of competitive exclusion), so a mixture of more species might result in a more complete use of the available resources. Moreover, if environmental conditions should change, a species-rich community is more likely to contain some species that can persist under the new conditions. Thus a species-rich community should be more stable—that is, it should change less over time in either productivity or species composition—than a species-poor community.

Species richness is associated with productivity and stability

To test the hypothesis that species-rich communities are more stable than species-poor communities, David Tilman and his colleagues at the University of Minnesota cleared 120 outdoor plots, in which they planted grasses in mixtures ranging from 2 to 22 grass species. At the end of each growing season, they measured total plant cover (a measure of grass biomass, and thus of net primary production) and the population densities of all the grasses in each plot. Over a period of 11 years, which included a serious drought, the plots with more species were more productive (**Figure 57.16A**), and their productivity was less variable from year to year. These findings were consistent with the hypothesis that species richness promotes productivity and keeps productivity stable. Moreover, in the plots with greater species richness, soil nitrogen was used more efficiently (**Figure 57.16B**). However, the population densities of *individual* species in the plots were not stable over the years (regardless of a plot's species richness) because different species performed better during drought years and wet years. In other words, higher species richness increased the stability of productivity in the plots, but not the stability of their species composition.

Researchers continue to debate whether species diversity is responsible for maintaining stability or is simply correlated with stability. This question is important because many of the alterations humans have made in the structure of natural communities have reduced their species richness, and many of these human-altered communities—notably agricultural communities—are notoriously unstable.

Diversity, productivity, and stability differ between natural and managed communities

Although ecologists have been debating the relationships among species richness, productivity, and stability for only a few decades, humans have been experimenting with those relationships, albeit inadvertently, for millennia—since plants were domesticated and agriculture was invented. Since the dawn of agriculture, crops have been susceptible to diseases and insect outbreaks: massive (often sudden) increases in populations of species that destroy or damage crops.

The practice of growing crops as **monocultures**—plantings of a single crop species—is one reason why managed

(A) Plant cover

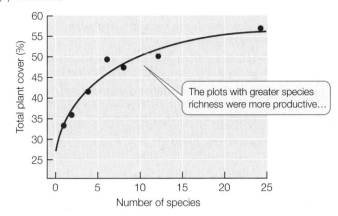

The plots with greater species richness were more productive...

(B) Efficiency of nitrogen use

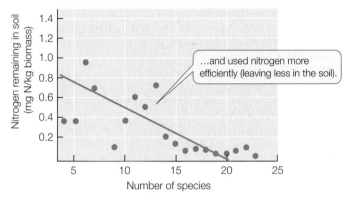

...and used nitrogen more efficiently (leaving less in the soil).

57.16 Species Richness Enhances Community Productivity
Tilman and colleagues cultivated a total of 120 grassland plots, containing from 2 to 22 grass species, for 11 years. (A) Total plant cover (a measure of grass biomass, and thus of net primary production). (B) The amount of nitrogen remaining in the soil is a measure of resource use efficiency.

agricultural communities are particularly unstable. Most farmers have little tolerance for the presence of any potential competitors for their crops and actively eliminate weeds (and the herbivore species that live with them) from their fields. Thus a typical agricultural community has very low species diversity. So the answer to the question of whether diversity causes or is merely correlated with stability may be sought in modern farming practices. The predisposition of agricultural communities to play host to outbreaks may well result from human influences on community structure.

For the last 20 years, ecologists have been using traditional subsistence agricultural plots as experimental models for testing the relationships between diversity and stability. Throughout the world, many farmers with small land holdings grow multiple crops on the same plot. In Costa Rica, for example, farmers often grow corn together with sweet potato. Such corn–sweet potato dicultures contain fewer sweet potato pests and many more parasitoid wasps (which feed on those pests)

than sweet potato monocultures do. The wasps feed on the corn pollen, and the tall corn plants act as a structural barrier, shade plant, and source of disruptive chemical signals that interfere with the ability of the sweet potato pests to find their host plants.

In recent years such applications of community ecology have been paying dividends. Although monoculture is overwhelmingly the dominant agricultural practice, polycultures are under development for agricultural production systems as varied as carp and shrimp farming, vermicomposting (raising worms for compost), and biofuel feedstock production.

RECAP 57.5

Communities with higher species diversity tend to be more productive and more stable than less diverse communities because they use resources more efficiently. The instability of modern agricultural monocultures suggests that diversity results in stability.

- What relationships have ecologists observed between species diversity, community productivity, and community stability? See p. 1202 and Figure 57.16
- Describe some agricultural practices that might result in more stable ecological communities. See pp. 1202–1203

How do the insect species in a corpse community influence one another's ability to survive?

ANSWER

The animal communities in decomposing corpses consist primarily of insects, but their exact composition varies with the factors that influence the rate and nature of decomposition: climate, season, and the condition of the body, including whether it is immersed or buried, wrapped or exposed. Typically, immediately after death blow flies, bluebottle flies, and house flies arrive to lay eggs. As a detectable odor develops, other flies, including greenbottles and flesh flies, arrive. Fat breakdown, with its accompanying release of volatile fatty acids, attracts a range of carrion-feeding beetles. As proteins decompose, cheese skippers can colonize. Other species have less interest in the corpse than in the corpse-eaters: rove beetles prey on the maggots that develop from the flies' eggs. During decay, skin beetles, hide beetles, and clothes moths (which can feed on the keratin in mammalian hair) dominate. In the final stages of decomposition, spider beetles and other scavengers arrive to feed on the excrement and shed exoskeletons of the insects that have been consuming the corpse. This succession varies tremendously with climate and geography, but within any particular region it is sufficiently predictable that it is admissible in court as evidence of the time of death.

57.1 What Are Ecological Communities?

- A **community** is a group of species that coexist and interact within a defined area.

- **Gross primary productivity (GPP)** is the rate at which the **primary producers** in a community turn solar energy into chemical energy via photosynthesis. **Net primary production** represents the energy incorporated into primary producer **biomass. Review Figure 57.2, ACTIVITY 57.1**

- A **food web** is a diagram of the feeding relationships in a community. **Review Figure 57.3**

- The organisms in a community can be divided into **trophic levels** based on the energy sources they use. Primary producers get their energy from sunlight; **primary consumers** get their energy by eating primary producers; **secondary consumers** get their energy by eating primary consumers; and so on. **Review Table 57.1, ACTIVITY 57.2**

- Organisms that consume the dead bodies of other organisms or their waste products are called **detritivores** or **decomposers. Omnivores** are organisms that feed at multiple trophic levels.

- **Ecological efficiency** is the overall transfer of energy from one trophic level to the next. Pyramid diagrams illustrate the proportions of energy or biomass that flow to each successive trophic level. **Review Figure 57.4**

- Species diversity tends to increase with productivity up to a point; however, if productivity increases beyond that point, species diversity may decline. **Review Figure 57.5**

57.2 How Do Interactions among Species Influence Communities?

- The interactions of a consumer with other species can result in a **trophic cascade**: a series of indirect effects across successive trophic levels. **Review Figure 57.6**

- Organisms that build structures that create habitat for other species are known as **ecosystem engineers.**

- **Keystone species** have an influence on their community that is disproportionate to their abundance. **Review Figure 57.7**

57.3 What Patterns of Species Diversity Have Ecologists Observed?

- Species diversity encompasses **species evenness** as well as **species richness. Review Figure 57.8**

- Species diversity can be measured at multiple spatial scales: within a single community or habitat, or over a range of communities in a geographic region.

- Latitudinal gradients in diversity, with the greatest diversity at low latitudes, have been observed in many taxa. **Review Figure 57.9**

- According to the theory of **island biogeography**, the equilibrium number of species on an island represents a balance between the rate at which species immigrate to the island from the mainland **species pool** and the rate at which resident species go extinct. **Review Figure 57.11, ANIMATED TUTORIAL 57.1**

57.4 How Do Disturbances Affect Ecological Communities?

- A **disturbance** is a disruption in a community caused by a discrete external force, often abiotic in nature.

- **Succession** is a predictable pattern of change in community composition following a disturbance. In **directional succession**, species come and go in a predictable sequence until a **climax community** forms and persists for an extended time.

- **Primary succession** begins on sites that lack living organisms. **Secondary succession** begins on sites where some organisms have survived a disturbance. **Review Figures 57.13, 57.14, ANIMATED TUTORIAL 57.2**

- In any pattern of succession, species that become established may facilitate or inhibit colonization by other species.

- In **cyclical succession**, the climax community is maintained by periodic disturbances.

- **Heterotrophic succession** in detritus-based communities does not rely on photosynthesis and therefore differs in a number of ways from other types of succession.

57.5 How Does Species Richness Influence Community Stability?

- Species-rich communities use resources more efficiently, and thus tend to vary less in productivity, than do less diverse communities. **Review Figure 57.16**

- **Monocultures** are subject to pest outbreaks, whereas agricultural communities containing greater species diversity tend to be more stable.

Go to the Interactive Summary to review key figures, Animated Tutorials, and Activities
Life10e.com/is57

CHAPTER**REVIEW**

REMEMBERING

1. An ecological community is
   a. a group of species that coexist and interact within a defined area.
   b. a group of species that coexist and interact in an area together with the abiotic environment.
   c. all the species in an area that belong to a particular trophic level.
   d. all the species that are members of a local food web.
   e. All of the above

2. A trophic level consists of those organisms
   a. whose energy has passed through the same number of steps to reach them.
   b. that use similar foraging methods to obtain food.
   c. that are eaten by a similar set of predators.
   d. that eat both plants and other animals.
   e. that compete with one another for food.

3. Net primary production is
   a. the total amount of photosynthesis in a community.
   b. the total amount of primary producer biomass available for consumption by heterotrophs.
   c. the total amount of biomass produced by all autotrophs and heterotrophs in a community.
   d. the total amount of biomass consumed by heterotrophs.
   e. gross primary productivity minus secondary productivity.

4. Pyramid diagrams of energy and biomass distribution for forests and for grasslands differ because
   a. forests are more productive than grasslands.
   b. forests are less productive than grasslands.
   c. large mammals avoid living in forests.
   d. wood presents more nutritional challenges to herbivores than grasses do.
   e. grasses grow faster than trees.

5. The theory of island biogeography
   a. predicts that the equilibrium number of species on an island is a balance between the rate of immigration of new species and the rate of extinction of resident species.
   b. predicts that the rate of immigration of new species will decline with island distance from the mainland species pool.
   c. predicts that the rate of extinction of resident species will decrease as island size increases.
   d. applies to isolated habitat patches as well as to oceanic islands.
   e. All of the above

6. Which of the following events is *not* followed by primary succession?
   a. A glacier recedes.
   b. A volcano erupts.
   c. A fire destroys a forest.
   d. A hurricane creates a bare-sand beach.
   e. All of these disturbances are followed by primary succession.

7. Early stages of succession are characterized by
   a. species that are good dispersers.
   b. species with high rates of reproduction.
   c. simple food webs.
   d. nutrients that are available primarily from detritus and abiotic sources.
   e. All of the above

UNDERSTANDING & APPLYING

8. Recent analyses of human gut flora using genomic methods have revealed tremendous microbial diversity, including many previously unknown species (see Figure 26.21). This microbial diversity has effects on human health. Describe the ecological methods you could use to investigate (a) how microbial diversity varies between individuals or across populations; and (b) which particular microbes might be keystone species or ecosystem engineers.

9. Jan Beck and Ian Kitching examined patterns of hawk moth diversity in the 113 islands of Thailand and mainland Malaysia. How do the findings of their study as summarized in the figure below relate to the theory of island biogeography?

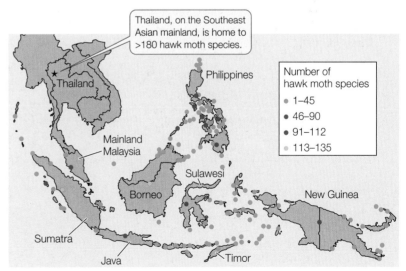

Thailand, on the Southeast Asian mainland, is home to >180 hawk moth species.

Number of hawk moth species
- 1–45
- 46–90
- 91–112
- 113–135

Philippines

Thailand

Mainland Malaysia

Sulawesi

Borneo

New Guinea

Sumatra

Java

Timor

ANALYZING & EVALUATING

10. Marek Sammul, Lauri Oksanen, and Merike Mägi investigated the effect of productivity on species richness in 16 different plant communities in western Estonia and northern Norway. When they removed one perennial species (the goldenrod *Solidago virgaurea*) from these communities, they found that its competitors, particularly the grass *Anthoxanthum odoratum*, increased in biomass, most noticeably in communities with high productivity (where living plant biomass was greater than 200 g/m²). In less productive communities, such increases could not be detected. How might interspecific competition lead to a decrease in species richness at high levels of productivity? What other hypotheses might explain this puzzling relationship, and how would you test them?

11. Sea lampreys (*Petromyzon marinus*) are parasitic fish that fasten onto the bodies of host fish with their disc-shaped mouths and remain attached for long periods, feeding on the host's blood and other body fluids (see Figure 33.11B). This invasive species was responsible for reducing populations of sport fish in the Great Lakes, and consequently has been the target of many extermination campaigns. Recent studies, however, have revealed that *P. marinus* spawns in fast-moving freshwater streams, where these fish build elaborate nests by burrowing and by moving stones around, to create nesting mounds. While they are nesting, sea lampreys do not parasitize other fish. Also, when adult lampreys die, their decomposing remains help restore nutrients to freshwater habitats. Later in the season, salmon and brook trout move from the ocean to freshwater streams; they spawn in the same habitats as the lampreys and are known to use abandoned lamprey nests with great success. Given all this, should lampreys be eliminated as damaging parasites of game fish, or should they be encouraged as ecosystem engineers? What other kinds of information might you need to decide on an ecologically sound lamprey management strategy?

Go to BioPortal at **yourBioPortal.com** for Animated Tutorials, Activities, LearningCurve Quizzes, Flashcards, and many other study and review resources.

58 Ecosystems and Global Ecology

The Gulf Dead Zone High concentrations of nitrogen and phosphorus in the runoff from agricultural lands in the U.S. interior are carried by the Mississippi River to the Gulf of Mexico. This nutrient enrichment of Gulf waters creates a "dead zone" in which many aquatic organisms cannot survive.

HOW CAN A CORNFIELD in Illinois, 1,500 kilometers from the nearest ocean, affect the price of sushi? When farmers in the Midwest apply chemical fertilizers to their fields, nitrogen and phosphorus from those fertilizers are dissolved in rainwater and washed into streams, which carry them into the Mississippi River. The river water eventually reaches the Gulf of Mexico, where the enormous inputs of dissolved nitrogen and phosphorus nourish explosive blooms of floating photosynthetic organisms (phytoplankton), including algae.

During the day, phytoplankton photosynthesize and produce oxygen. At night, however, they take up oxygen from the water to carry out cellular respiration. When these short-lived organisms die, their bodies sink to the bottom, where bacterial decomposition further depletes oxygen. This process results in a state of hypoxia (the reduction of dissolved oxygen to levels below 2 milligrams per liter of water), suffocating other organisms and creating a "dead zone." There are more than 140 coastal dead zones around the world. The one in the northern Gulf of Mexico, which has been mapped since 1985, is among the largest, spanning an area the size of New Jersey.

The Atlantic croaker (*Micropogonias undulatus*) lives along the coasts of eastern North America and the Gulf of Mexico. These bottom-feeding fish, which consume invertebrates, can reach 30 centimeters in length when mature. Their white, firm flesh is used in imitation crabmeat, or surimi, a popular ingredient in sushi. But recently croakers in the dead zone have experienced strange symptoms that may forecast a decrease in their populations. The croaker sex ratio, which is 50:50 among fish caught east of the Delta, is skewed toward males in the dead zone, and about one-fifth of the females in dead zone samples were found to have male germ cells in their ovaries. Laboratory experiments confirmed that only 10 weeks of exposure to hypoxia could induce these sexual defects. Depriving fish brains of oxygen apparently inhibits production of the neurohormones that promote normal ovary development. The croaker owes its name to its ability to make a drumming sound, but if the dead zone continues to expand and resulting sexual defects reduce their population growth rates, these fish might end up "croaking," both literally and figuratively.

Human activities such as farming change the movement patterns of mineral nutrients, and these changes affect not just croakers but many organisms that live far from where the human activities take place.

How can we determine to what extent dead zones result from human actions and to what extent they are the result of natural processes?

See answer on p. 1225.

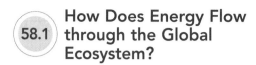

58.1 How Does Energy Flow through the Global Ecosystem?

An **ecosystem** includes all of the organisms in an ecological community as well as the physical and chemical factors that influence those organisms. In other words, ecosystems have both biotic and abiotic components. Ecosystems can occupy a wide range of spatial scales, from the entire planet to a watershed, a specific forest, a lake or pond, or even a patch of lichen on a rock. To some degree, ecologists must define the boundaries of ecosystems arbitrarily. In this chapter we will focus on the global ecosystem, noting that all smaller-scale ecosystems are linked by the global flows of energy and chemical elements that are considered in this chapter.

Energy flows and chemicals cycle through ecosystems

Earth is essentially a closed system with respect to chemical elements, but it is an open system with respect to energy. The sun delivers a nearly constant amount of energy to Earth every day and has done so for billions of years. When captured by primary producers such as plants and photosynthetic bacteria, that energy flows through the trophic levels of food webs in one direction (see Figure 57.3). Much of the energy that enters each trophic level is used to power the metabolism of producers and consumers; that energy is eventually dissipated as heat and is lost from the ecosystem (see Figure 57.2). Chemical elements, by contrast, are not altered when they are transferred between organisms. Furthermore, they are not lost from the global ecosystem, although they may become unavailable to organisms for long periods; instead chemical elements cycle continually between living organisms and the abiotic components of ecosystems (**Figure 58.1**).

In this first section we examine how the geographic distribution of incoming solar radiation influences the amount of energy assimilated by primary producers and how human activities are modifying energy flow through the global ecosystem.

The geographic distribution of energy flow is uneven

Nearly all energy used by organisms comes (or once came) from the sun. The only exceptions are found in those few ecosystems in which solar energy is not the main energy source (such as some caves and deep-sea hydrothermal vent ecosystems). Even the fossil fuels—coal, oil, and natural gas—on which the economy of modern human civilization is based are reserves of captured solar energy locked in the remains of organisms that lived (and died) millions of years ago (see Section 25.2).

As described in Section 57.1, solar energy enters ecosystems by way of plants and other photosynthetic organisms.

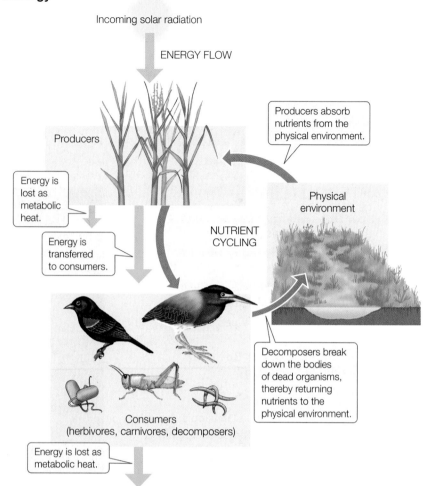

58.1 Energy Flows and Chemical Nutrients Cycle through Ecosystems Each time one organism eats another, a portion of the solar energy originally captured by a primary producer is lost as heat (gold arrows). As a result, energy flows through the ecosystem in a single direction. Chemical elements that organisms use as nutrients, however, cycle repeatedly between organisms and the physical environment (orange arrows).

These primary producers use some of the energy they assimilate for their own metabolism; the rest—net primary production (NPP)—is stored in their bodies or used for their growth and reproduction (see Figure 57.2). Because only the energy of NPP is potentially available to other organisms, which obtain it by consuming primary producers, NPP can be used as a rough measure of energy influx into an ecosystem. NPP varies among ecosystem types, but because ecosystem types also vary greatly in their geographic extent, the most productive ecosystem types do not necessarily contribute the most to Earth's net primary production (**Figure 58.2**).

The geographic distribution of NPP reflects the geographic variation in incoming solar radiation described in Section 54.2 and the climate patterns that result from it. In other words, the distribution of temperatures and moisture makes some ecosystems more productive than others (**Figure 58.3**). Close to the equator at sea level, temperatures are high throughout the year and the water supply is adequate for plant growth much of the time. In these climates, productive forests thrive. In deserts, plant growth is limited by lack of moisture and NPP is low. At higher latitudes, even though moisture is generally available, NPP is low because

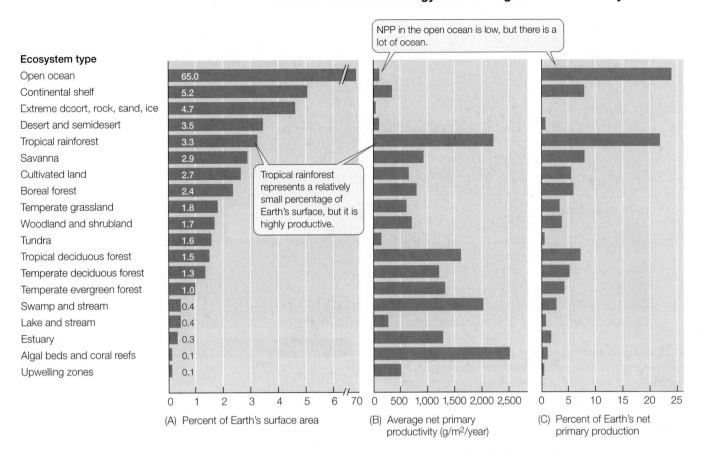

58.2 Energy Flow Contributions by Ecosystem Type The contributions of different ecosystem types to global energy flow can be measured by (A) their geographic extent and (B) their average net primary productivity. (C) Combining these two measures gives us a proportional contribution of each ecosystem type to Earth's total net primary production.

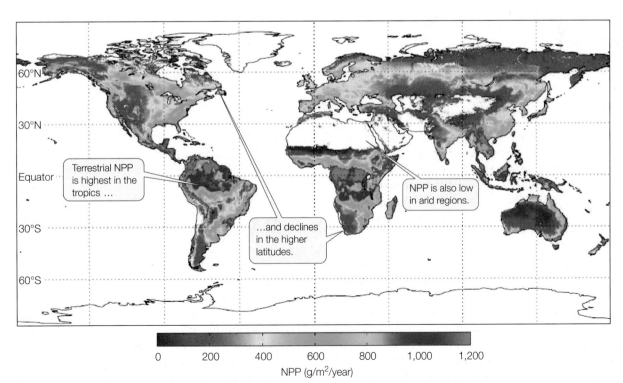

58.3 Geographic Variation in Terrestrial NPP This map of estimated terrestrial net primary production is based on satellite sensor data accumulated from 2000 through 2005. White spaces represent unvegetated areas, including deserts and ice caps.

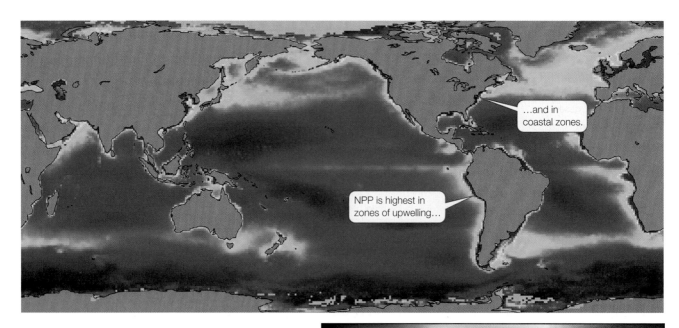

NPP (g/m²/year)

58.4 Geographic Variation in Marine NPP The availability of nutrients determines how much primary production occurs in any part of the photic zone. NPP is highest where runoff from land brings nutrients into shallow coastal waters and where upwellings bring nutrients from the seafloor to the surface.

it is relatively cold much of the year and the growing season is short (see Working with Data, p. 1138).

Production in aquatic ecosystems is limited by light, which decreases rapidly with depth (see Figure 54.9), and by temperature, which also decreases with depth, except in areas on the seafloor near hydrothermal vents. Production in aquatic ecosystems is also strongly limited by nutrient availability, as we'll see in the following section. Net primary production in the oceans tends to be highest in coastal zones, where runoff from land and upwellings from deeper waters bring nutrients into shallow waters (**Figure 58.4**).

Human activities modify the flow of energy

The effects of human activities on energy flow through the global ecosystem have accelerated markedly in the last 150 years. Some human activities decrease net primary production, as when forests are cut down and replaced by cities; some increase it, as when prairies are converted to extensive agricultural fields.

Humans consume about one-quarter of Earth's average annual net primary production. More than 50 percent of this consumption results from the croplands and rangelands that occupy more than one-third of Earth's ice-free surface; another 40 percent results from productivity changes brought about by alterations in land use; and 7 percent represents biomass consumed in fires caused by humans. The percentage varies strikingly among regions, however; urban areas consume as much as 300 times the NPP they generate, but people in sparsely inhabited parts of the Amazon Basin consume vanishingly small amounts of the NPP generated there.

▮▮▮▮▮▮▮▮▮▮▮▮▮▮▮▮▮▮▮▮▮▮▮▮▮▮▮▮ RECAP 58.1

Nearly all energy used by living organisms comes from the sun. Energy flow through ecosystems, as measured by net primary production, varies with geographic location.

- How does the distribution of temperature and moisture influence the geographic distribution of net primary production in terrestrial systems? **See p. 1208 and Figure 58.3**

- What percentage of Earth's average annual net primary production is appropriated by humans, and how does that percentage vary regionally? **See p. 1210**

Ecosystem productivity is influenced not only by energy flow but also by the availability of the nutrients and other materials required by organisms to build their bodies and to power their metabolism. The next section will describe how biochemical materials and nutrients move around the abiotic environment and become available to living organisms.

58.2 How Do Materials Move through the Global Ecosystem?

In contrast to the energy that powers biological processes, which comes from the sun, the chemical elements that make up the bodies of organisms come from within the Earth system itself. As we have seen, these elements cycle continually through the global ecosystem. But they are not always available in the right place, or in the right form, to be useful to organisms. Because nutrient availability, in addition to energy input, influences productivity, ecologists are interested in knowing where on Earth nutrients are located and how they move from one location to another.

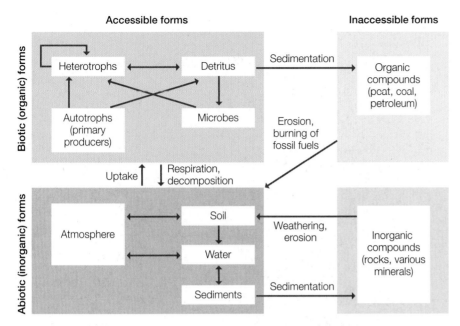

Accessible forms   **Inaccessible forms**

58.5 Chemical Elements Cycle through the Biosphere The different forms and locations of the chemical elements determine whether or not they are accessible to living organisms. Biological, geological, and chemical processes cycle matter among these biotic and abiotic components of the global ecosystem.

Energy from the sun, combined with energy from the radioactive decay in Earth's interior, drives the biological, geological, and chemical processes that transform chemical elements and move them around the planet (**Figure 58.5**). In this section we'll examine the properties of some of the abiotic and biotic components of the global ecosystem—referred to as compartments—through which elements move, as well as the processes that move them. The rate at which an element moves through a compartment is called its **flux**; the term "flux" is also applied to the movements of energy. Elements may accumulate, or "pool," in some compartments.

Elements move between biotic and abiotic compartments of ecosystems

All materials in the bodies of organisms ultimately originate from abiotic sources, but organisms acquire these materials in many different ways. Autotrophs such as plants take up certain elements directly from soil, water, and the atmosphere and incorporate them into organic molecules to build biomass. Heterotrophs generally acquire elements by consuming the biomass produced by other organisms, then reassemble those elements, via chemical reactions, in different ways. Some heterotrophs, however, acquire some elements by housing mutualistic microbes that convert those elements into forms that are usable by their hosts.

Respiration by living organisms returns certain elements to the atmosphere as gases. After organisms die, the materials in their bodies become detritus and are broken down by decomposers into simpler biochemical components. In this way the elements are freed to be taken up again by autotrophs. Elements that do not get taken up by autotrophs can accumulate in soil, water, or sediments.

At times in the remote past, great quantities of organic material were removed from active cycling when organisms died in large numbers and were buried in sediments that lacked oxygen. In such anaerobic environments, decomposers could not efficiently break down organic molecules to their inorganic forms. Instead organic molecules accumulated and were eventually transformed into deposits of oil, natural gas, coal, or peat—the **fossil fuels** that modern humans use as a combustible source of energy.

The movement of elements through food webs from uptake to decomposition—that is, through the biotic compartments of ecosystems—occurs primarily on a local scale. In contrast, abiotic processes can move elements far beyond the boundaries of the local ecosystem. The various abiotic compartments of the global ecosystem differ in fundamental ways, and the quantities of different elements in each compartment (e.g., atmosphere, ocean, soil), how long those elements remain there (their **residence time**), the forms they take, and the rates at which they enter and leave also differ. Moreover, the cycling of nutrients among compartments is influenced by the ways in which energy flows through them.

The atmosphere contains large pools of the gases required by living organisms

The outermost compartment of the global ecosystem is the atmosphere, a thin layer of gases surrounding Earth. The atmosphere is 78.08 percent nitrogen gas (N_2), 20.95 percent oxygen gas (O_2), 1 percent water vapor, 0.93 percent argon, and 0.03 percent carbon dioxide (CO_2). It also contains traces of hydrogen gas, neon, helium, krypton, xenon, ozone, and methane. It contains Earth's biggest pool of nitrogen as well as a large proportion of its oxygen.

About 80 percent of the mass of the atmosphere lies in its lowest layer, the **troposphere**. This layer extends upward from Earth's surface about 17 km in the tropics and subtropics, but only about 10 km at high latitudes. Most global air circulation takes place within the troposphere, and virtually all of the atmospheric water vapor is found there (**Figure 58.6**).

The **stratosphere**, which extends from the top of the troposphere up to about 50 km above Earth's surface, contains very little water vapor. Most materials enter the stratosphere from the region of the troposphere that encircles the equator, where air heated by the sun's energy rises to high altitudes (see Section 54.2).

The stratosphere contains a layer of ozone (O_3) that absorbs most of the biologically damaging ultraviolet radiation that enters the atmosphere. Over the last several decades, this **ozone layer** has been seriously damaged by human activities, particularly the widespread release of chlorinated fluorocarbons (CFCs), such as the refrigerant freon. CFCs remain stable as they ascend to the stratosphere, where they interact with and break

58.6 Earth's Atmosphere Has Two Layers
The troposphere and the stratosphere differ in their circulation patterns, the amount of water vapor they contain, and the amount of ultraviolet radiation they receive.

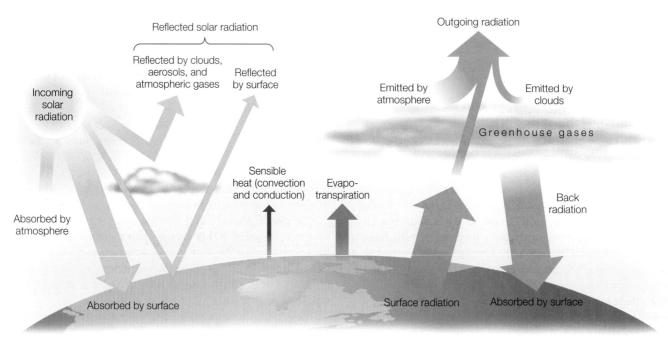

down ozone molecules in the presence of ultraviolet light. The more ozone that is lost, the more ultraviolet light can reach Earth's surface. Increases in ultraviolet radiation are associated with increased rates of skin cancer, cataract formation, and crop damage. Since 1989, when a global treaty was enacted to phase out production of CFCs, atmospheric levels of CFCs have for the most part stabilized or declined, and there is encouraging evidence from both satellite and ground station measurements that stratospheric ozone levels are slowly returning to their normal state.

The atmosphere moderates temperatures at and near Earth's surface by trapping heat energy. If Earth had no atmosphere, its average surface temperature would be about –18°C, rather than its actual +17°C. Carbon dioxide, methane (CH_4), nitrous oxide (N_2O), water vapor, and certain other gases in the atmosphere are known as **greenhouse gases** because they are transparent to sunlight but trap heat radiating from Earth's surface back toward space (**Figure 58.7**). Ozone, when it is present in the troposphere rather than the stratosphere, also

acts as a greenhouse gas. As we'll see later in this chapter, human activities are increasing the concentrations of greenhouse gases in the atmosphere, and those increases are altering the climate.

58.7 Radiant Energy Warms the Planet Solar energy input (yellow arrows) is absorbed by Earth's atmosphere and surface. Much of this energy is radiated from Earth's surface in the form of heat (orange arrows). Much of this radiation is prevented from escaping back into space by greenhouse gases in the atmosphere. The widths of the arrows here are roughly proportional to the sizes of the energy fluxes.

 Go to Animated Tutorial 58.1
Earth's Radiation Balance
Life10e.com/at58.1

The terrestrial surface is influenced by slow geological processes

About one-fourth of Earth's surface consists of land above sea level. Because the geological processes that move elements through minerals and soils are so slow, regional and local variations in the supply of particular elements greatly influence terrestrial ecosystem processes.

Nearly all the rocks that underlie the continents have been transformed at least once through a complex cycle of plate tectonic processes (see Figure 25.3). The physical and chemical processes of weathering break down surface rocks into soil. The type of soil in an area and the chemical nutrients that soil contains are determined in large part by the underlying rock from which the soil forms, although climate, topography, the local biota, and the length of time that soil-forming processes have been acting also influence the nature of soil (see Section 36.3). Chemical elements in rocks are released by weathering and by certain biological processes; these elements are then carried in solution into streams and groundwater, which transport them ultimately into the oceans. Structural features of the land surface affect how rapidly and in what direction wind and water currents can transport elements.

Water transports elements among compartments

The high heat capacity of water and its ability to change states from gas to liquid to solid at temperatures found on Earth (see Section 2.4) mean that it can move freely among all compartments, including the biosphere. Water is a powerful solvent that can carry materials in solution within and among all the compartments of the global ecosystem.

FRESH WATERS The liquid fresh waters of the global ecosystem consist of streams, lakes, and **groundwater** (water occupying pore spaces in rock, sand, and soil). Only a small fraction of Earth's water resides in lakes and streams at any given time, but because water moves so rapidly through the freshwater compartment, most of Earth's water spends some time there. Some mineral nutrients enter fresh waters from the atmosphere in rainfall, but most are released from rocks by weathering. They are dissolved and carried into streams by surface runoff or by movements of groundwater in a process called erosion. After entering streams, mineral nutrients are usually carried rapidly to lakes or to the oceans.

TURNOVER IN LAKES The nutrients in lakes are taken up by and incorporated into the bodies of the organisms living there. Those organisms eventually die and sink to the lake bottom, where decomposition of their tissues by microbes releases the nutrients but consumes most of the oxygen in the bottom water. The surface waters of deep lakes thus quickly become depleted of nutrients while deeper waters become depleted of oxygen. The waters of most deep lakes in temperate climates have an annual cycle of **turnover**, vertical movements of the water column that bring nutrients to the surface and oxygen to deeper water (**Figure 58.8**).

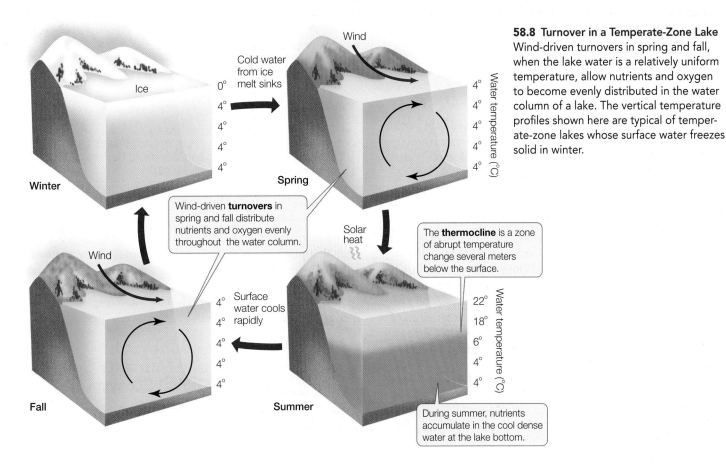

58.8 Turnover in a Temperate-Zone Lake Wind-driven turnovers in spring and fall, when the lake water is a relatively uniform temperature, allow nutrients and oxygen to become evenly distributed in the water column of a lake. The vertical temperature profiles shown here are typical of temperate-zone lakes whose surface water freezes solid in winter.

Wind-driven **turnovers** in spring and fall distribute nutrients and oxygen evenly throughout the water column.

The **thermocline** is a zone of abrupt temperature change several meters below the surface.

During summer, nutrients accumulate in the cool dense water at the lake bottom.

Turnover depends on the unique physical properties of water. Liquid water is densest at 4°C, a few degrees above its freezing point of 0°C (at which point it becomes solid and floats). Above 4°C, water expands. Thus in winter the coldest liquid water in a lake is at the surface, often just beneath a layer of ice, and the dense waters of the depth remain at 4°C. In spring, when the sun melts the ice and warms the surface water to 4°C, there is a time at which water density is uniform throughout the lake, and even modest winds will readily mix the entire water column. As spring and summer progress, the surface water becomes warmer still, and the depth of the warm water layer gradually increases. However, there is a well-defined depth, called the **thermocline**, at which the temperature changes abruptly. Only if the lake is shallow enough so that water warms up all the way to the bottom does the temperature of the deepest water rise above 4°C.

Another turnover occurs in fall as the process reverses itself. The surface of the lake cools until the water there is denser than the warmer water below it, at which point it sinks and is replaced by warmer water from below. Once again, water density becomes uniform, and winds can mix the entire water column.

A similar process contributes to the formation of the Gulf of Mexico dead zone described at the opening of this chapter, but in that case the density differences are caused by differences in salinity rather than in temperature. Fresh water is less dense than salt water, so when the nutrient-rich fresh water of the Mississippi River flows into the Gulf, it does not mix with the salt water but floats on top of it. Dying algae sink to the bottom, and bacterial decomposition depletes the available oxygen there. The difference in water density prevents oxygen from the surface water from mixing with the hypoxic salt water below.

 Go to Media Clip 58.1
Tracking Dead Zones from Space
Life10e.com/mc58.1

OCEANS Over time scales of hundreds to thousands of years, most materials that cycle through the global ecosystem end up in the oceans, which hold almost 97 percent of Earth's water. The oceans are enormous, but they exchange materials with the atmosphere only at their surfaces, so they respond very slowly to inputs from that compartment. They receive materials from land primarily in runoff from rivers.

Except on continental shelves—the shallow ocean waters surrounding large land masses—ocean waters mix slowly. Most materials that enter the oceans from the land or the atmosphere gradually sink to the seafloor, where they may remain for millions of years, until intermittent plate tectonic processes lift seafloor sediments above sea level. Thus concentrations of mineral nutrients in most ocean waters are very low (except where human activities have released materials into the water).

Near the coasts of continents, however, offshore winds may push the warmer surface waters away from shore, causing cold water from the bottom to rise and bringing nutrients back to the surface waters. An area where water from depths below 50 meters rises in this way is known as an **upwelling zone**.

Upwelling zones support high rates of primary production by phytoplankton, which in turn support dense consumer populations. Most of the world's great fisheries are concentrated in upwelling zones. For example, the upwelling zone off the coast of Peru is the source of much of the world's supply of anchovies. This rich fish population supports vast seabird communities, which in turn produce enormous quantities of guano, or excrement, that have provided Peru with an important raw material to support a major fertilizer industry.

Fire is a major mover of elements

Every year 200 to 400 million hectares of savannas, 5 to 15 million hectares of boreal forests, and smaller expanses of other biomes catch fire and burn. Lightning ignites some of these fires, but humans start most of them to manage vegetation (as when they cut down and burn forests to clear land for growing crops). Fires rapidly consume the energy stored in, and release the chemical elements from, the vegetation they burn. Some nutrients, such as nitrogen, are readily vaporized by fire. Nitrogen enters the atmosphere in smoke or is carried into groundwater by rain falling on burned ground.

Fires also release large amounts of carbon into the atmosphere. The global annual flux of carbon to the atmosphere from savanna and forest fires is estimated at 1.7 to 4.1 petagrams (1 pg = 10^{15} g or 10^{12} kg). Biomass burning (which includes combustion of wood and alcohol, wildfires, and land clearing, but not the burning of fossil fuels) is responsible for about 40 percent of Earth's annual flux of CO_2 into the atmosphere and contributes to the production of other greenhouse gases as well. Large-scale wildfires in the western and southeastern United States can release as much CO_2 into the atmosphere as motor vehicles in those regions release over an entire year.

 RECAP 58.2

Biological, geological, and chemical processes move materials within and among biotic and abiotic compartments of the global ecosystem.

- How does the atmosphere keep temperatures at and close to Earth's surface warmer than they would be in its absence? **See pp. 1211–1212 and Figure 58.7**
- Describe the process of turnover in a temperate-zone lake in fall. **See pp. 1213–1214 and Figure 58.8**

As we learned in Chapters 3 and 10, most of the chemical energy that primary producers convert from sunlight is stored in carbon-containing compounds. In the next section we will consider how carbon and other chemical elements required by living organisms cycle through the biotic and abiotic compartments of the global ecosystem.

58.3 How Do Specific Nutrients Cycle through the Global Ecosystem?

Each of the chemical elements that organisms use in large quantities cycles in a distinctive way through the biotic and

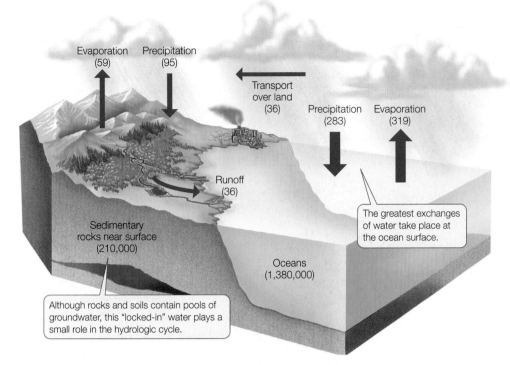

Evaporation (59) Precipitation (95)

Transport over land (36)

Precipitation (283) Evaporation (319)

Runoff (36)

Sedimentary rocks near surface (210,000)

Oceans (1,380,000)

The greatest exchanges of water take place at the ocean surface.

Although rocks and soils contain pools of groundwater, this "locked-in" water plays a small role in the hydrologic cycle.

58.9 The Global Hydrologic Cycle
As described in Section 2.4, the unique properties of water are essential to life as we know it on Earth. The numbers in parentheses show the estimated amounts of water (expressed as exagrams, one exagram equalling 10^{18} g or 10^{15} kg) held in or exchanged annually by fluxes (arrows) among compartments of the global ecosystem. The widths of the arrows are proportional to the sizes of the fluxes.

 Go to Animated Tutorial 58.2
The Global Hydrologic Cycle
Life10e.com/at58.2

abiotic compartments of the global ecosystem. Because geological, chemical, and biological processes are all important in moving materials around the planet, the pattern of movement of an element is called its **biogeochemical cycle**. The nature of each biogeochemical cycle depends on the physical and chemical properties of the element and on the ways in which it is used by organisms.

Water cycles rapidly through the ecosystem

In addition to being a compartment of the global ecosystem where nutrients are found and a medium that transports those nutrients between other compartments, water is itself a material. Water cycles through the ecosystem in the global **hydrologic cycle** (Figure 58.9).

Energy from the sun drives the hydrologic cycle, taking up water by evaporation from the vast surfaces of the oceans. The cycle operates because more water evaporates from the ocean surfaces than is returned to the oceans in the form of precipitation. On land, water evaporates from soils, lakes, and rivers and is taken up from the leaves of plants by transpiration. However, the total amount evaporated and transpired from terrestrial surfaces is less than the amount that falls on them as precipitation. Excess terrestrial precipitation eventually returns to the oceans via streams, coastal runoff, and groundwater flows. More than half of this volume of water is carried back to the oceans by Earth's four largest rivers: the Amazon in South America, the Nile in Africa, the Mississippi in North America, and the Yangtze in Asia.

Despite their relatively small volume, rivers play a disproportionate role in the hydrologic cycle because the average residence time of a water molecule in rivers is only a few years. By comparison, the average residence time of a water molecule in lakes ranges from a few years to centuries. The larger the lake,

the longer the residence time; the residence time for water in the top portion of Lake Superior, for example, is 1,500 to 2,000 years, and the water at the bottom of this massive lake never cycles. In the oceans the average residence time of a water molecule is about 3,000 years. Other pools of water include glaciers (with residence times of 20–100 years), seasonal snow cover (a few months), and soil moisture (1–2 months). The average residence time of water in the bodies of organisms is particularly brief, averaging just under a week.

Although large amounts of groundwater are present in underground pools called **aquifers**, this water has a long residence time underground and plays only a small role in the hydrologic cycle. In some places, however, aquifers are being depleted because humans are using groundwater more rapidly than it can be replaced, primarily by pumping it for irrigation. Much of the groundwater being used today in the Northern Hemisphere was deposited during the most recent ice age, when regional precipitation was much greater than it is now. Using this groundwater for irrigation and other purposes has increased flows of water to the oceans and has contributed to the sea level rise of the past century.

The effects of groundwater depletion are already being felt. On the North China Plain, depletion of shallow aquifers is forcing people to sink wells more than 1,000 meters deep to reach groundwater. Worldwide, more than 1 billion people have no access to safe drinking water. If current water consumption patterns continue, by 2025 at least 48 percent of the current world population will live in areas with inadequate water supplies. However, per capita water consumption in the United States and Europe is declining as a result of increasing use of water-efficient home appliances as well as implementation of new regulations that restrict water use. If such trends continue, global water use in 2025 could be lower than it is today, despite continued population growth.

58.10 The Global Carbon Cycle Carbon is the basis of the organic molecules essential to life. The numbers in parentheses show the quantities of carbon in petagrams (1 pg = 10^{15} g or 10^{12} kg) held in or exchanged annually by fluxes (arrows) among compartments of the global ecosystem. The widths of the arrows are proportional to the sizes of the fluxes.

Go to Animated Tutorial 58.3
The Global Carbon Cycle
Life10e.com/at58.3

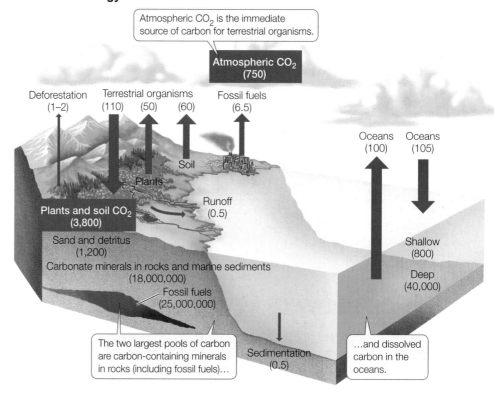

Atmospheric CO_2 is the immediate source of carbon for terrestrial organisms.

Atmospheric CO_2 (750)

Deforestation (1–2)　Terrestrial organisms (110)　(50)　(60)　Fossil fuels (6.5)

Oceans (100)　Oceans (105)

Soil

Plants

Runoff (0.5)

Plants and soil CO_2 (3,800)

Sand and detritus (1,200)

Carbonate minerals in rocks and marine sediments (18,000,000)

Fossil fuels (25,000,000)

Shallow (800)

Deep (40,000)

The two largest pools of carbon are carbon-containing minerals in rocks (including fossil fuels)…

Sedimentation (0.5)

…and dissolved carbon in the oceans.

The carbon cycle has been altered by human activities

As described in Part One of this book, all of the important macromolecules that make up living organisms contain carbon, and much of the energy that organisms use to fuel their metabolic activities is stored in carbon-containing (organic) compounds. Carbon in the atmosphere, in the form of CO_2, is taken up by autotrophs and incorporated into organic molecules by photosynthesis. All heterotrophic organisms obtain carbon by consuming autotrophs or other heterotrophs, their remains, or their waste products.

On land, biological processes move carbon directly between organisms and the atmosphere as terrestrial organisms take up carbon during photosynthesis and return it to the atmosphere through respiration and metabolism. In contrast, carbon dioxide moves into ocean waters from the atmosphere primarily by simple diffusion at the ocean surface; this dissolved CO_2 is the source of the carbon used by marine primary producers (**Figure 58.10**). Even taken together, however, the amounts of carbon in the atmosphere, in soils, and in living and dead organisms are dwarfed by the vast quantities of carbon stored in terrestrial rocks, in fossil fuels, in marine sediments, and in seawater in the form of carbonate ions (CO_3^{-2}) or bicarbonate ions (HCO_3^-).

At times in the remote past, quantities of carbon were removed from active cycling when organisms died in large numbers and were buried in sediments lacking oxygen. In such anaerobic environments, with no detritivores to reduce organic carbon to CO_2, organic molecules accumulate and are eventually transformed into deposits of oil, natural gas, coal, or peat. Humans have discovered and used these fossil fuels at ever-increasing rates during the past 150 years. As a result, CO_2, one of the final products of burning fossil fuels, is being released into today's atmosphere faster than it is dissolving in the oceans or being incorporated into terrestrial biomass (**Figure 58.11**). Based on a variety of calculations, atmospheric scientists estimate that

before the Industrial Revolution, the concentration of CO_2 in Earth's atmosphere was probably about 265 parts per million. Today it is 392 parts per million, representing a rate of increase more than 10 times faster than at any other time for millions of years.

WHERE HAS ALL THE CARBON GONE? Less than half of the CO_2 released into the atmosphere by human activities remains in

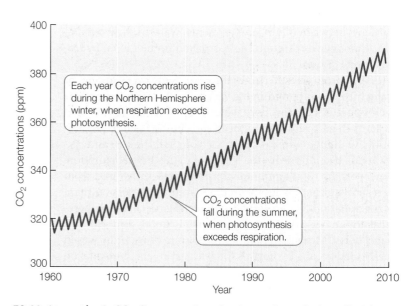

Each year CO_2 concentrations rise during the Northern Hemisphere winter, when respiration exceeds photosynthesis.

CO_2 concentrations fall during the summer, when photosynthesis exceeds respiration.

58.11 Atmospheric CO_2 Concentrations Are Increasing Carbon dioxide concentrations, expressed as parts per million by volume of dry air, have been recorded since 1960 on top of Mauna Loa, Hawaii, far from most sources of human-generated CO_2 emissions. Although concentrations vary seasonally, the trend has been consistently upward.

the atmosphere. Where does the rest of the CO_2 wind up? Much of it is dissolved in the oceans in inorganic forms. Over decades to centuries, the oceans, which contain 50 times more carbon than the atmosphere, determine atmospheric CO_2 concentrations. The rate at which CO_2 diffuses from the atmosphere into the oceans depends in part on photosynthesis by phytoplankton in the surface waters. These organisms remove dissolved CO_2 from water, thereby increasing the rate at which atmospheric CO_2 is absorbed by surface waters. In addition, many marine organisms (including clams, oysters, corals, and planktonic foraminiferans) incorporate carbon in their shells and other structures in the form of calcium carbonate ($CaCO_3$), which is synthesized by combining bicarbonate ions (HCO_3^-) and calcium ions (Ca^{2+}) dissolved in seawater. When these organisms die, those shells and their embedded carbon sink to the ocean floor.

Today's oceans absorb 20 to 25 million tons of CO_2 from the atmosphere each day—more than at any time during the past 20 million years. As a result, water near the ocean surface is becoming more acidic. As CO_2 concentrations in the atmosphere rise, more of the gas diffuses into the water at the ocean surface, where it reacts with water to form carbonic acid (H_2CO_3). As levels of carbonic acid rise, the pH of seawater drops. This increase in acidity can have negative effects on many marine organisms, particularly corals. The combination of decreasing pH and increasing water temperature to which corals are being exposed kills their symbiotic algae, "bleaching" the corals and killing them as well (see Figure 27.21). Because so many other reef species depend on corals and the structure they provide, the entire reef community can collapse if the corals fail to thrive.

Photosynthesis by terrestrial vegetation, principally in forests and savannas, typically absorbs about the same amount of carbon that is released by terrestrial metabolism—about half of it released by plants and half by microbes in the soil that break down plant detritus. The photosynthetic consumption of CO_2 currently exceeds the metabolic production of CO_2, which means Earth's terrestrial vegetation is storing carbon that would otherwise be increasing atmospheric CO_2 concentrations—but we cannot count on terrestrial vegetation to store the vast amounts of excess CO_2 that human activities produce. Furthermore, climate warming (another result of increasing atmospheric CO_2 concentrations, as we will see next) increases plant metabolism and is thus likely to increase the flux of CO_2 from vegetation into the atmosphere.

ATMOSPHERIC CO_2 AND GLOBAL CLIMATE CHANGE Carbon dioxide is a greenhouse gas, so we would expect increasing atmospheric CO_2 concentrations to raise temperatures at Earth's surface. What evidence do we have that this is occurring? Measurements of gases in air trapped in the Antarctic and Greenland

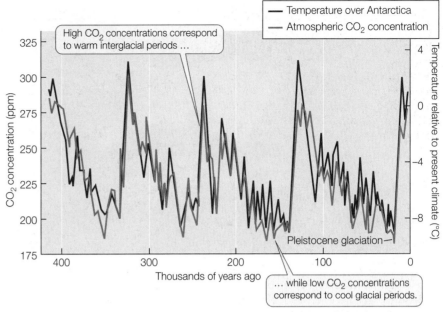

58.12 Higher Atmospheric CO_2 Concentrations Correlate with Warmer Temperatures Atmospheric concentrations of CO_2 (measured in air bubbles trapped in Antarctic ice) have varied with temperatures over Antarctica (estimated by a technique known as oxygen isotope analysis).

ice caps show that atmospheric CO_2 concentrations have been higher when Earth has been warmer and lower when it has been cooler (**Figure 58.12**). For example, the atmospheric CO_2 concentration was very low during the most recent glaciation, between 30,000 and 15,000 years ago, when temperatures were presumably much colder than they are today. In contrast, during a warm interval 5,000 years ago, atmospheric CO_2 concentration may have been slightly higher than it is today.

How global climates and ecosystems will change in response to this rapid warming is a subject of intense investigation. Complex computer models of the global ecosystem indicate that a doubling of today's atmospheric CO_2 concentration would increase mean annual temperatures worldwide and would probably result in droughts in the central regions of continents, but would increase precipitation in coastal areas. Global warming has already resulted in the shrinking of Arctic sea ice (**Figure 58.13**). The five smallest expanses of Arctic sea ice ever measured were recorded in the past six years; overall Arctic sea ice extent is declining by 3.5 to 12 percent per decade. Based on scientific climate models, if glacial ice continues to melt and temperatures continue to rise, sea level will rise (because of both thermal expansion of ocean waters and the addition of glacial meltwater), increasing the probability of flooding of coastal cities and agricultural lands.

Global climate warming is having profound effects on the distributions and abundances of species and, consequently, on species interactions. One clear example is an increase in the severity of insect infestations in certain temperate forest communities. Pine trees in some temperate forests are attacked by pine bark beetles, which carry with them a symbiotic fungus that infects the trees and helps the beetles overcome the trees' defenses (see Figure 56.9). In Colorado, cold winters have

The outline shows the mean extent of Arctic sea ice for September 16, 1979–2000.

Russia

Greenland

Alaska

Hudson Bay   Quebec

58.13 Shrinking Ice Caps A NASA satellite image shows the extent of Arctic sea ice (bright white) on September 16, 2012. The plus symbol marks the geographic North Pole. The extent of the ice cap in 2012 was about one-half that of the September 16 mean for the years 1979–2000.

historically limited the ability of these beetles to kill many trees. In 2008, however, Colorado experienced an outbreak of mountain pine beetles that destroyed more than 400,000 hectares (1 million acres) of trees. The lack of an extended period of below-zero temperatures in the previous winter had allowed large numbers of overwintering beetles to survive.

The nitrogen cycle depends on both biotic and abiotic processes

Nitrogen gas is the most abundant gas in Earth's atmosphere, but most organisms cannot use nitrogen in its gaseous form.

Only a few species of microorganisms can convert atmospheric N_2 into forms such as ammonia and nitrate that are usable by plants, a process called nitrogen fixation (see Figure 36.10). Other microorganisms carry out denitrification, the principal process that removes nitrogen from the biosphere and returns it to the atmosphere as N_2. Collectively this microbial processing of nitrogen accounts for about 95 percent of all natural nitrogen flux on Earth (**Figure 58.14**).

Abiotic weathering is an important source of nitrogen in some terrestrial ecosystems. In temperate forests growing on land underlain by nitrogen-rich sedimentary rocks, for example, the soils and foliage have 50 percent higher levels of nitrogen than in temperate forests growing on land underlain by nitrogen-poor igneous rocks.

All living organisms require nitrogen, and the inability of the vast majority of organisms to use N_2 means that usable nitrogen is often in short supply. Populations of nitrogen-fixing organisms rarely increase in abundance to the extent that nitrogen is no longer limiting because the end products of their nitrogen fixation are rapidly lost from ecosystems (ammonia by vaporization and denitrification; and nitrate, which is highly water-soluble, by leaching).

Human activities that fix nitrogen, such as the manufacture of artificial fertilizers, have had some unanticipated effects on the nitrogen cycle. The extensive use of artificial fertilizers on agricultural crops, coupled with the burning of fossil fuels (which generates nitric oxide and nitrogen dioxide), has resulted in total nitrogen fixation by humans being nearly equal

58.14 The Global Nitrogen Cycle The largest pool of nitrogen is held in the atmosphere in the form of nitrogen gas, N_2. Nitrogen cycles through the biosphere primarily via the processes of nitrogen fixation, which converts inorganic nitrogen to an organic form usable by plants, and denitrification, which returns N_2 to the atmosphere. The numbers in parentheses show the quantities of nitrogen in teragrams (1 tg = 10^{12} g or 10^9 kg) exchanged annually by compartments of the global ecosystem. The widths of the arrows are proportional to the sizes of the fluxes.

Go to Animated Tutorial 58.4
The Global Nitrogen Cycle
Life10e.com/at58.4

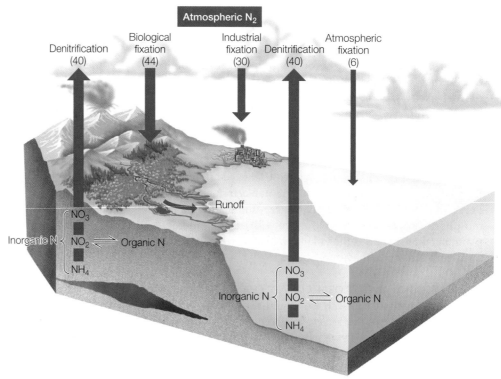

Atmospheric N_2

Denitrification (40)

Biological fixation (44)

Industrial fixation (30)

Denitrification (40)

Atmospheric fixation (6)

Runoff

Inorganic N { NO_3 / NO_2 ⇌ Organic N / NH_4 }

Inorganic N { NO_3 / NO_2 ⇌ Organic N / NH_4 }

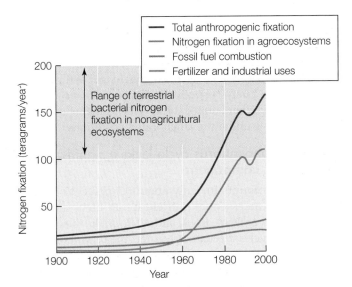

58.15 Human Activities Have Increased Nitrogen Fixation Most of the nitrogen fixed by industrial processes is used in agricultural fertilizers. Some fixation is a by-product of fossil fuel combustion. Fixation by natural processes in managed agroecosystems (e.g., by legumes grown as crops) also contributes to anthropogenic (human-caused) effects on nitrogen flux.

to global natural nitrogen fixation (**Figure 58.15**). This human-generated nitrogen flux has been increasing over the past half-century and is expected to continue to increase.

Eutrophication is an increase in biomass production in a body of water due to inputs of nutrients. Eutrophication occurs naturally as part of the aging process in lakes. As lakes become more shallow with the accumulation of sediments brought in by streams, their water warms more rapidly with the onset of summer, and exploding populations of photosynthetic cyanobacteria and single-celled algae, called blooms, can deplete oxygen levels (see Figure 26.9C). Human nutrient inputs greatly increase the likelihood and frequency of these blooms. When more nitrogen fertilizer is applied to croplands than can be taken up by the crops, the excess nitrogen moves out of the system in surface runoff, or downward into groundwater, and ultimately ends up in rivers, lakes, and oceans. The dead zone that has formed near the mouth of the Mississippi River in the Gulf of Mexico, described at the opening of this chapter, is a result of water flows from agricultural fields in the U.S. interior carrying high concentrations of nitrogen from fertilizer.

The human increase in nitrogen fixation has also increased atmospheric concentrations of the greenhouse gas nitrous oxide (N_2O), resulting in the production of tropospheric ozone—also a greenhouse gas—and smog. Some of the nitrogen that enters the atmosphere falls back to land in precipitation or as dry particles. This deposition of nitrogen from the atmosphere has increased dramatically during recent decades. Nitrogen deposition affects the composition of terrestrial vegetation by favoring those plant species that are best adapted to take advantage of high nutrient levels, which then outcompete other species. Spatial variation in nitrogen deposition rates has

allowed ecologists to determine that plant species richness in grasslands declines as the rate of nitrogen deposition increases. Rates of nitrogen deposition are high enough over much of Europe and eastern North America to cause substantial reductions in species richness in grasslands on both continents.

The burning of fossil fuels affects the sulfur cycle

As a component of proteins, sulfur is required by all organisms. Most of Earth's sulfur supply is locked up in rocks on land and as sulfate salts in deep-sea sediments, but some sulfur moves between the atmosphere and land. Emissions of the gases sulfur dioxide (SO_2) and hydrogen sulfide (H_2S) from volcanoes account for between 10 and 20 percent of the total natural abiotic flux of sulfur to the atmosphere, but they occur only intermittently (although volcanic eruptions spew great quantities of sulfur over broad areas, they are rare events). In the atmosphere, H_2S can combine with oxygen to form SO_2, which dissolves in atmospheric water and reaches the ground as sulfuric acid in precipitation and fog.

When sulfur in the soil comes in contact with atmospheric oxygen, it is converted to sulfate salts, which can be taken up by plants and incorporated into proteins. This sulfur ultimately is returned to the atmosphere via microbial decomposition. In marine systems, too, microbial decomposition is important in returning sulfur to the atmosphere. Many marine phytoplankton and seaweeds manufacture large quantities of a sulfur-containing compound (dimethylsulfoniopropionate, or DMSP) to maintain their salt and water balance. When broken down, DMSP releases dimethyl sulfide (CH_3SCH_3), the principal odorant of rotting seaweed stench. Because the quantities of phytoplankton in the oceans are enormous, dimethyl sulfide production accounts for about half of the biotic component of the global sulfur cycle.

Atmospheric sulfur plays an important role in global climate. Even if air is moist, clouds do not form readily unless there are small particles in the atmosphere around which water can condense. Dimethyl sulfide is the major component of such particles, so increases in atmospheric sulfur concentrations increase cloud cover and reduce the amount of incoming solar radiation that ultimately reaches Earth's surface.

Human use of fossil fuels alters the sulfur cycle as well as the carbon and nitrogen cycles. The combustion of fossil fuels releases sulfur in the form of SO_2, as well as nitrogen in the form of nitrogen dioxide (NO_2), into the atmosphere. Both compounds react with water molecules in the atmosphere to form sulfuric acid (H_2SO_4) and nitric acid (HNO_3), respectively. These acids can travel hundreds of kilometers in the atmosphere before they settle to Earth as dry particles or in precipitation. Rain or snow that contains enough nitric and sulfuric acid to lower its pH is called **acid precipitation**.

Acid precipitation now falls in all major industrialized countries and is particularly widespread in eastern North America and Europe. The normal pH of unpolluted precipitation is about 5.6, but precipitation in New England now averages about pH 4.5, and there have been occasional rainfalls

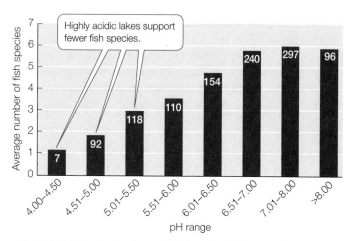

58.16 Acidification Reduces Fish Species Richness The average number of fish species found in lakes sampled in the Adirondack region of New York is directly correlated with pH. Numbers in the bars indicate the number of lakes in each pH range.

and snowfalls with a pH as low as 3.0. Precipitation with a pH of about 3.5 or lower damages the leaves of plants and reduces rates of photosynthesis. Acidification of lakes in the Adirondack region of New York State has reduced fish species richness by causing the extinction of acid-sensitive species (**Figure 58.16**). Many invertebrates that are primary consumers in aquatic communities are sensitive to pH; in particular, multiple species of mayflies and caddisflies (important food resources for fish) have experienced local population reductions in acidified streams around the world. Even when its effects are not lethal, acidification can have subtle effects on behavior that reduce the viability of aquatic organisms. Diving beetles, for example, lose their ability to regulate their underwater oxygen supply when pH levels drop significantly.

Regulations instituted by the U.S. Clean Air Act in 1990 have raised the pH of precipitation in much of the eastern United States, primarily by lowering sulfur emissions. There are indications that once emissions have been reduced, acidified aquatic systems can recover quickly. David Schindler at the University of Alberta studied the effects of acid precipitation by adding enough H_2SO_4 to two small Canadian lakes to reduce their pH from about 6.6 to a moderately acidic level of 5.2. In both lakes, nitrifying bacteria failed to survive, and nitrogen cycling within the lake was blocked. When Schindler stopped adding acid to one of the lakes, its pH returned to its original value in about a year and nitrification resumed. That said, the larger organisms

in acidified aquatic systems may be slower to bounce back. In Wales, investigators conducted a 25-year study of 14 rivers to determine the impacts of acid rain reduction. They were disappointed to find that, even after that length of time, only 4 insect species—2 mayfly species and 2 caddisfly species—had recolonized the rivers out of the 29 species that should have been able to live under the ameliorated conditions.

The global phosphorus cycle lacks a significant atmospheric component

Phosphorus accounts for only about 0.1 percent of Earth's crust, but it is an essential nutrient for all life forms. It is a key component of DNA, RNA, and ATP. Unlike the other biogeochemical cycles discussed thus far, the phosphorus cycle lacks a significant gaseous component (**Figure 58.17**). Some phosphorus is transported on dust particles, but very little of the phosphorus cycle takes place in the atmosphere. Most of Earth's phosphorus is present in the form of phosphate salts in rocks and deep-sea sediments. Abiotic cycling of phosphorus takes millions of years because the processes of sedimentary rock formation, uplift, and weathering all take a long time. In contrast, phosphorus often cycles rapidly among organisms, and it is often a limiting factor for their growth, particularly for plants. Artificial fertilizers routinely include phosphorus as well as nitrogen.

Human activity has radically accelerated some parts of the phosphorus cycle. One consequence of the massive use of artificial fertilizers described above is that between 10.5 and 15.5 teragrams (1 tg = 10^{12} g or 10^9 kg) of phosphorus accumulate in

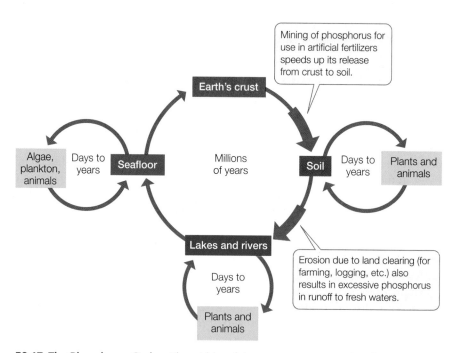

58.17 The Phosphorus Cycle The widths of the arrows are proportional to the sizes of the fluxes. Two large increases in phosphorus flux are the result of human activities.

soils each year, primarily in agricultural fields. When the concentration of phosphorus in soils exceeds the capacity of plants to take it up, the excess moves into streams and lakes. Soil erosion due to the clearing of land for purposes such as agriculture and logging also increases the amount of phosphorus and other nutrients in runoff. Phosphorus is a limiting nutrient in many lakes, so when it enters those lakes through runoff, eutrophication results, in much the same way that nitrate enrichment can cause eutrophication.

The capacity of phosphorus-charged runoff to cause extensive eutrophication was graphically illustrated almost 50 years ago. A technological innovation in the formulation of laundry detergents in the early 1960s was the use of sodium tripolyphosphate (STPP) ($Na_5P_3O_{10}$) as a water softener and dirt-breaking agent to enhance cleaning efficiency. Widespread adoption of these new detergents led to massive phosphorus enrichment of lakes and streams and resulting eutrophication and blooms of phytoplankton. Water quality declined so much across the United States that phosphate-based detergents were banned in many states. In today's detergents, phosphonates—forms of phosphorus that do not appear to promote algal growth—and aluminum silicates perform the functions STPP used to serve.

Detergents, however, are only one agent contributing to eutrophication. Human waste is rich in phosphorus, as are manure from domesticated animals and industrial wastes of various types. Two hundred years ago, Lake Erie, one of the Great Lakes on the border between the United States and Canada, had only moderate phytoplankton populations and clear, oxygenated water. With increasing industrialization in the early part of the twentieth century, nutrient concentrations in the lake increased greatly, and algae proliferated. At the water filtration plant in Cleveland, Ohio, algae increased from 81 individuals per milliliter in 1929 to 2,423 per milliliter in 1962. Populations of bacteria also increased; *Escherichia coli* levels rose so high that many of the lake's beaches were declared health hazards. Since 1972 the United States and Canada have invested more than U.S. $9 billion to improve municipal waste treatment facilities and reduce discharges of pollutants. As a result, the amount of phosphorus added to Lake Erie has decreased more than 80 percent from its highest level, and phosphorus concentrations in the lake have declined substantially. The deeper waters of Lake Erie are still oxygen-poor during the summer months, but the rate of oxygen depletion is declining.

We could greatly reduce phosphorus pollution by recovering and recycling phosphorus. The phosphorus discarded in sewage and animal wastes could supply much of the needs of the detergent and fertilizer industries. More careful application of fertilizers on agricultural lands could reduce the rate of phosphorus accumulation in soils without reducing crop yields. However, reduction of phosphorus concentrations in soils will take many decades after remedial actions are initiated, and eutrophication of lakes and streams may persist even after these actions are taken.

Other biogeochemical cycles are also important

Other elements are important to the global ecosystem because they are essential nutrients for organisms, even though they are needed only in very small amounts. One such element is iron (Fe), an essential micronutrient for almost all organisms. Iron is a key component of the enzymes involved in chlorophyll synthesis as well as an essential component of many animal enzymes. Iron confers oxygen-binding ability on hemoglobin in vertebrate blood. Members of the cytochrome P450 family of enzymes, which in most aerobic organisms play a central role in detoxifying environmental poisons, rely on iron for their catalytic activity.

Iron is readily available on land in rocks and minerals. It moves into coastal waters in streams and into the open oceans in atmospheric dust. Because iron is insoluble in oxygenated water, it rapidly sinks to the ocean floor. Therefore in most marine communities the rate of photosynthesis is limited by iron. In 1996 investigators launched an ecosystem-scale experiment in which surface waters of the equatorial Pacific Ocean were seeded with dissolved iron. The response was a tremendous phytoplankton bloom, accompanied by massive uptake of nitrate and carbon dioxide, which had apparently been underused because of iron limitation.

Iodine is an example of an element that is globally rare but is an essential micronutrient for living organisms. Endothermic vertebrates in particular require iodine in concentrations that exceed the supply in many environments. It is an essential component of the hormone thyroxine, which governs many metabolic processes (see Section 41.4). Iodine is found on land in mineral deposits and in seawater as an inorganic salt.

Biogeochemical cycles interact

The biogeochemical cycles of different elements interact with one another in complex ways, and perturbations of one cycle can have profound effects on other cycles. In recent years, human-induced perturbations have made these interactions glaringly apparent. For example, nitrate released by human activities can have profound effects on the biogeochemical cycle of arsenic. The bottom sediments of some urban lakes contain arsenic levels in excess of 2,000 parts per million. Nitrate is a powerful oxidant, so nitrate pollution can increase the oxidation of arsenic in lake sediments, releasing it into the water in a form that is carcinogenic and has negative effects on embryonic development. Every year scientists discover interactions of which they were previously unaware, and studies exploring biogeochemical interactions are increasing in number.

Changes in atmospheric CO_2 concentrations have been a particular focus of investigation in recent years because of their potential for interacting with other biogeochemical cycles through photosynthesis. A case in point is a study of the effect of elevated atmospheric CO_2 concentrations on rates of nitrogen fixation by microorganisms associated with plant roots. Bruce Hungate and his colleagues grew a nitrogen-fixing vine (i.e., a legume) called Elliott's milkpea (*Galactia elliottii*) under

INVESTIGATINGLIFE

58.18 Effects of Atmospheric CO_2 Concentration on Nitrogen Fixation   Some scientists have hypothesized that rising concentrations of CO_2 in the atmosphere could lead to increased rates of photosynthesis, increased rates of nitrogen fixation, and eventually to the fixation of large amounts of carbon in the soil—which could potentially reduce global warming. In a 7-year-long experiment, Bruce Hungate and his colleagues at Northern Arizona University expected to find that nitrogen fixation in a leguminous vine would be enhanced by increased atmospheric CO_2.[a]

HYPOTHESIS   Exposure of legumes to elevated CO_2 concentrations will enhance nitrogen fixation by their symbiotic bacteria.

Method   Grow plots of the leguminous vine *Galactia elliottii* under baseline (typical) and artificially elevated concentrations of CO_2. Measure nitrogen fixation over 7 years.

Results

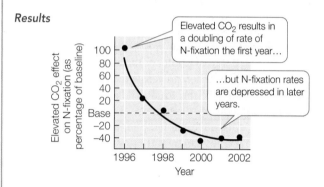

In an attempt to explain these results, (which did not support the hypothesis), the investigators measured the concentrations of iron and molybdenum—two micronutrients that are essential for nitrogen fixation—in the leaves of the 7-year-old plants.

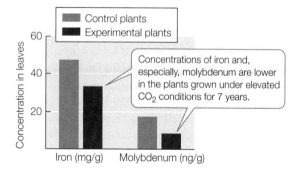

CONCLUSION   Although enhanced CO_2 levels initially increase nitrogen fixation, lowered levels of essential micronutrients in plants growing under these conditions soon leads to decreased rates of nitrogen fixation.

Go to **BioPortal** for discussion and relevant links for all INVESTIGATINGLIFE figures.

[a]Hungate, B. A. et al. 2004. *Science* 304: 1291.

artificially increased CO_2 concentrations. Higher CO_2 concentrations led to an enhancement of nitrogen fixation during the first year of the experiment, but surprisingly, the positive effect disappeared by the third year, and elevated CO_2 concentrations actually *reduced* nitrogen fixation below baseline levels during the fourth, fifth, sixth, and seventh years of the experiment (**Figure 58.18**).

WORKING WITHDATA:

How Does Molybdenum Concentration Affect Nitrogen Fixation?

Original Paper

Hungate, B. A. et al. 2004. CO_2 elicits long-term decline in nitrogen fixation. *Science* 304: 1291.

Analyze the Data

The experiments in Figure 58.18 were conducted in an oak woodland where *G. elliottii* grew naturally. The investigators used open-top chambers to produce a 350-ppm increase in the concentration of CO_2 in the air around the plants. The study site had a sandy, acidic soil known to have low concentrations of molybdenum.

Because nitrogen-fixing plants are sensitive to light availability, an alternative explanation for the results in Figure 15.18 is that increased shading resulting from greater leaf area of the CO_2-stimulated plants could have caused the subsequent decline in fixation. To test this possibility, the investigators computed the leaf-area index (LAI), a measure of the amount of leaf-surface area per unit of ground area. They found no correlation between LAI and nitrogen fixation, but they did find a positive correlation between concentrations of molybdenum in *G. elliottii* leaves and the rate of nitrogen fixation.

QUESTION 1

The authors claim that this regression analysis provides strong evidence in favor of low availability of molybdenum rather than low light availability as the reason for the decline in rate of nitrogen fixation. Do you agree? Why or why not?

QUESTION 2

What does this experiment suggest about the kinds of future studies and the range of ecosystem types and nutrient elements that should be investigated to determine the likely overall response of Earth's ecosystems to increasing atmospheric concentrations of CO_2?

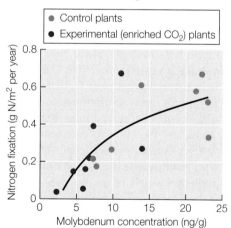

Go to BioPortal for all WORKING WITHDATA **exercises**

The investigators suspected that a lack of micronutrients such as iron and molybdenum had caused the reduction in nitrogen fixation, so they measured concentrations of those elements in the leaves of the *G. elliottii* grown under high CO_2 concentrations. They found that concentrations of molybdenum in those plants were particularly low. Hungate and his colleagues proposed a mechanism by which this could occur: elevated CO_2 concentrations could increase the acidity of water in soil by enhancing the rate of carbonic acid formation. By enhancing photosynthesis, elevated CO_2 concentrations could increase the accumulation of organic matter in the soil. Both types of change would increase the tendency of iron and molybdenum to bind to soil particles, which would reduce their availability to nitrogen-fixing bacteria and cause a decrease in nitrogen fixation rates.

■ RECAP 58.3

The pattern of movement of a chemical element through the biotic and abiotic compartments of the global ecosystem is its biogeochemical cycle. Human activities have affected many biogeochemical cycles, especially those of water, carbon, nitrogen, sulfur, and phosphorus. Increasing concentrations of carbon dioxide and other greenhouse gases in the atmosphere are implicated in global climate change.

- Describe the global hydrologic cycle and explain what drives it. **See p. 1215 and Figure 58.9**

- How do biological processes move carbon from the atmosphere to land and then return it to the atmosphere? **See p. 1216 and Figure 58.10**

- What are some results of human-induced alterations of the sulfur and nitrogen cycles? **See pp. 1219–1221**

- Name two elements that can cause eutrophication in aquatic ecosystems and describe their effects on those systems. **See pp. 1219 and 1221**

The biogeochemical cycles of chemical elements are intimately involved in ecosystem function. Just as human alterations of those cycles are having many effects on ecosystems worldwide, the resulting changes in those ecosystems are having profound effects on human lives.

58.4 What Goods and Services Do Ecosystems Provide?

Although it seems obvious today that humans depend on natural ecosystems for survival, explicit recognition of the value of those ecosystems is rather recent. Environmental writers introduced the idea of "natural capital" in the 1940s; it was in 1970 that ecosystems were first said to provide people with a variety of "goods and services." The goods include food, clean water, clean air, fiber, building materials, and fuel; the services include flood control and water quality, soil stabilization, pollination, and climate regulation. Most of these benefits either are irreplaceable, or the technology necessary to replace them is prohibitively expensive. For example, fresh drinking water can be provided by desalinating seawater, but only at great cost. The aesthetic, psychological, spiritual, and recreational benefits of ecosystems are less tangible, but no less important, and no more easily replaced.

Humans have increasingly altered Earth's ecosystems in ways that increase the systems' capacity to provide us with necessities such as food, fresh water, timber, fiber, and fuel. The benefits of these ecosystem alterations have not been equally distributed, and some human populations have been harmed by manipulations of natural ecosystems. Moreover, short-term increases in some ecosystem goods and services often comes at the cost of the long-term degradation of others.

Although humans have been altering natural ecosystems for millennia, the pace and scope of the shift to intense human use have increased considerably in the past century. More land was converted to cropland between 1950 and 1980 than in the 150 years between 1700 and 1850. Ecosystem conversions have been particularly rapid in tropical and subtropical biomes. Aquatic ecosystems have suffered losses at an increasing pace as well. In freshwater ecosystems, so much water is now impounded behind dams that artificial reservoirs hold about six times as much water today as do natural rivers. These freshwater systems are being rapidly depleted: the amount of water withdrawn from rivers—most of it for agriculture—has doubled since 1960. In terms of nutrient cycling, more than half of all the artificial nitrogen fertilizer ever used on Earth has been used since 1985.

Human alteration of ecosystems has had many positive effects on human health and prosperity, but it necessarily involves trade-offs. Agriculture, for example, feeds and employs huge numbers of people. But the spread of agriculture into marginal lands may degrade soils and compromise the ability of ecosystems to provide clean water, as when overuse of artificial fertilizers results in eutrophication. Extensive use of pesticides controls insect pests, but also reduces populations of pollinators and the services they provide to both crops and native plants.

Similarly, the loss of wetlands and other natural buffers has reduced the ability of ecosystems to regulate flooding and other natural hazards. The damage from the tsunami that hit Indonesia and other Southeast Asian countries in December 2004 was greater in many places than it would have been had the mangrove forests that protect the coast not been cut down and converted to cropland. Hurricane Katrina, which struck the U.S. Gulf Coast less than a year later, would not have caused as much flooding in New Orleans had the wetlands surrounding the city been intact. Katrina's devastating effects were due in part to a situation that had been developing for decades.

New Orleans is located on the Mississippi River delta. Much of the city lies below sea level, buffered by dams and levees constructed by the Army Corps of Engineers. The upstream dams that protect New Orleans from flooding also prevent the river from depositing the sediments that have sustained the surrounding delta wetlands for centuries. Oil and natural gas producers have cut thousands of small canals through those wetlands in order to lay pipelines and install drilling rigs, and the extraction of oil and gas from beneath the land has caused it to sink. Increased dredging of shipping lanes and rising sea levels due to global warming have contributed to a rise in salinity, killing off many of the great cypress tree swamps. These extensive alterations resulted in the loss of more than 80 percent (more than 50,000 hectares, or 1.2 million acres) of the delta wetlands between 1930 and 2005. By the time Katrina made landfall, those wetlands could no longer protect New Orleans from flooding. Storm surges raced along the paths carved by canals and shipping lanes to breach the levees, inundating much of the city.

As it flooded New Orleans, Hurricane Katrina raised awareness and appreciation of an ecosystem hitherto taken for granted by most people. Crucial not only for their flood control services, the delta wetlands provide winter habitat for some 70 percent of the migrating birds in the huge Mississippi Valley. They are also the spawning grounds for marine organisms, some of which are commercially valuable. The delta's famous shrimping industry contributes about 30 percent by weight of the total commercial fish harvest in the continental United States. The importance of coastal wetlands—and, indeed, of a wide range of other ecosystems—to human well-being mandates careful ecosystem management to guarantee a sustained flow of ecosystem goods and services.

RECAP 58.4

Ecosystems provide human society with indispensable goods and services. Altering ecosystems can compromise their ability to provide these goods and services.

- What are some of the essential goods and services that ecosystems provide to humans? **See p. 1223**
- Give an example of a human effort to increase the provision of some ecosystem goods or services that caused the degradation of others. **See p. 1223**

How can we meet the challenge of obtaining goods and services from ecosystems without compromising their ability to provide those goods and services over the long term? What options exist for sustainable management of ecosystems?

58.5 How Can Ecosystems Be Sustainably Managed?

Practices that allow us to conserve or enhance ecosystems so as to benefit from specific ecosystem goods and services over the long term without compromising others are referred to as **sustainable**. In many cases the total economic value of a sustainably managed ecosystem is higher than that of a converted or intensively exploited ecosystem (**Figure 58.19**). Furthermore, the long-term economic benefits of preventing overexploitation of ecosystems are enormous. For example, the collapse of the cod fishery on Georges Bank due to overfishing (see Figure 55.14) resulted in the loss of tens of thousands of jobs.

Impeding the establishment of policies that encourage sustainable practices is the impression that ecosystem services are "public goods" with no market value. People who do not stand to profit from the services provided by natural ecosystems have no incentive to pay for them, whereas individuals who own converted ecosystems can reap great economic benefits. Government action may be needed to create incentives encouraging sustainable ecosystem management. Examples of such action might include:

- Elimination of subsidies that promote damaging exploitation of ecosystems. For example, the billions of dollars

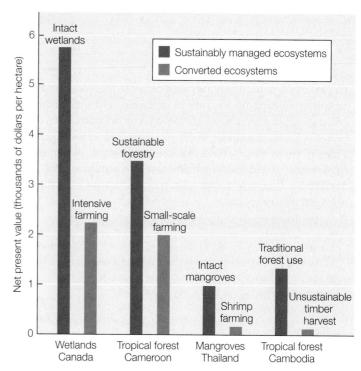

58.19 The Economic Value of Sustainably Managed Ecosystems Many types of ecosystems are able provide more goods and services when they are sustainably managed than when they are completely converted to human use and intensively exploited.

paid by governments in developed nations to subsidize domestic agriculture (with the aim of insulating farmers from economic risk) have led to greater food production than the global market warrants, promoted excessive use of fertilizers, and reduced the profitability of agriculture in developing countries.

- More sustainable use of fresh water from rivers and aquifers could be achieved by charging users the full cost of providing water, by developing methods to use water more efficiently, and by altering the allocation of water rights so that the incentives favor conservation rather than wasteful consumption.

- More sustainable use of marine fisheries could be achieved by establishing more protected marine reserves and "no-take" zones where fish can grow to reproductive age. Recent discussions of marine fisheries management center on what is colloquially known as the BOFFF ("Big, Old, Fat, Female Fish") hypothesis, which proposes that it is most important to protect the largest and oldest females in a population because they outreproduce younger fish by an enormous margin (see Section 55.6).

Raising public awareness is essential to the implementation of sustainable management programs. Most people do not realize the long-term value of ecosystem goods and services or understand how human activities affect the functioning of ecosystems. Maintaining and enhancing ecosystem goods and services that have no established market value is especially difficult. Perhaps the most difficult aspect of ecosystem function to maintain in the face of increasingly intensive human use of the global ecosystem will be biological diversity. The final chapter of this book is devoted to this important topic.

How can we determine to what extent dead zones result from human actions and to what extent they are the result of natural biogeochemical processes?

ANSWER

Hypoxia occurs naturally in aquatic systems, but throughout history it has been restricted primarily to deep-water ecosystems such as deep ocean basins, fjords, and the bottoms of the largest lakes. The appearance of dead zones in shallow coastal waters and estuaries is a twentieth-century phenomenon (the Gulf of Mexico dead zone was first identified in 1972). This timing, as well as the observation that dead zones are increasing in number and size, suggests human involvement.

The White House Office of Science and Technology Policy's Committee on Environment and Natural Resources initiated a scientific assessment of the causes and consequences of Gulf hypoxia in 1997. As part of this effort, sediment cores were taken to determine historic levels of algal deposition in sediment. These cores revealed a clear pattern of increases in the second half of the twentieth century. Sophisticated computer models demonstrated a significant association between river loads of dissolved inorganic nitrogen and rates of oxygen depletion, and the most dramatic increases coincided with historic records of changes in human activities that increased nitrate loads in the river system. Given that the Gulf of Mexico provides almost three-fourths of the shrimp and two-thirds of the oysters harvested in the U.S., as well as recreational fishing resources and ecologically vital forage fishes, further expansion of the dead zone could have devastating economic consequences.

CHAPTER**SUMMARY** 58

58.1 How Does Energy Flow through the Global Ecosystem?

- An **ecosystem** includes all of the organisms in an ecological community as well as the physical and chemical factors that influence those organisms.

- Energy flows and chemical elements cycle through ecosystems. **Review Figure 58.1**

- Terrestrial net primary production varies across the globe, reflecting differences in solar energy input and the climate patterns that result from them. **Review Figures 58.2, 58.3**

- Productivity in aquatic ecosystems is limited by light, temperature, and nutrient availability. **Review Figure 58.4**

- Humans appropriate about one-quarter of Earth's average annual net primary production, although this amount varies regionally.

58.2 How Do Materials Move through the Global Ecosystem?

- Chemical elements cycle through biotic and abiotic compartments of the global ecosystem. **Review Figure 58.5**

- The movement of elements through the biotic compartment of ecosystems, from uptake by autotrophs to decomposition, generally occurs on a local scale.

- Most global air circulation takes place in the lowest layer of the atmosphere, the **troposphere**. An **ozone layer** in the **stratosphere** absorbs ultraviolet radiation. **Review Figure 58.6**

- Carbon dioxide, water vapor, and other **greenhouse gases** in the atmosphere are transparent to sunlight but trap heat, thus warming Earth's surface. **Review Figure 58.7, ANIMATED TUTORIAL 58.1**

- Because the geological processes that move elements on land are so slow (on the scale of millions of years), there are large regional and local variations in the supply of particular elements within the terrestrial compartments.

- Some nutrients enter fresh waters from the atmosphere in rainfall, but most are released from rocks by weathering. They are usually carried rapidly to lakes or to the oceans.

- **Turnover** occurs regularly in temperate-zone lakes in both spring and fall, bringing nutrients to the surface and oxygen to the deeper waters. **Review Figure 58.8**

- Most materials that cycle through biotic and abiotic compartments end up in the oceans, where they eventually sink to the bottom.

- Fires release the chemical elements from the vegetation they burn. Those vaporized elements enter the atmosphere, where they can be carried into groundwater by rain.

58.3 How Do Specific Nutrients Cycle through the Global Ecosystem?

- The pattern of movement of a chemical element through the biotic and abiotic compartments of the global ecosystem is its **biogeochemical cycle**.

- The **hydrologic cycle** is driven by the sun, which evaporates more water from the ocean surface than it returns by precipitation. The excess precipitation that falls on land eventually returns to the oceans, primarily in rivers. **Review Figure 58.9, ANIMATED TUTORIAL 58.2**

- Groundwater plays a minor role in the hydrologic cycle, but underground **aquifers** are being seriously depleted by human activities.

- Carbon is removed from the atmosphere by photosynthesis and returned to the atmosphere by metabolism and burning. **Review Figure 58.10, ANIMATED TUTORIAL 58.3**

- The concentration of CO_2 in the atmosphere has increased greatly in the last 150 years, largely because of the burning of fossil fuels. This buildup of CO_2 is warming the global climate. **Review Figures 58.11–58.13**

- As a result of agricultural use of fertilizers and the burning of fossil fuels, total nitrogen fixation by humans is nearly equal to natural nitrogen fixation. **Review Figures 58.14. 58.15, ANIMATED TUTORIAL 58.4**

- Human alteration of the nitrogen cycle has resulted in excesses of nitrogen compounds in bodies of water, leading to **eutrophication** and dead zones.

- The burning of fossil fuels releases sulfur and nitrogen into the atmosphere, leading to **acid precipitation**. **Review Figure 58.16**

- Agricultural use of fertilizers and clearing of land have dramatically increased the input of phosphorus into soils and fresh waters. **Review Figure 58.17**

continued

What What Goods and Services Do Ecosystems Provide?

- The goods and services provided by ecosystems include food, clean water, flood control, pollination, pest control, climate regulation, spiritual fulfillment, and aesthetic enjoyment. Most ecosystem services either are irreplaceable or the technology necessary to replace them is prohibitively expensive.

- Efforts to enhance the capacity of an ecosystem to provide some goods and services often come at the cost of the system's ability to provide others.

How Can Ecosystems Be Sustainably Managed?

- The total economic value of an ecosystem managed in a **sustainable** manner often is higher than that of a converted or intensively exploited ecosystem. **Review Figure 58.19**

- Recognition of the value of ecosystem goods and services that are now perceived as "public goods" may induce government action to protect them. Public education is needed to make people aware of how much they benefit from ecosystem goods and services.

See ACTIVITY 58.1 for a concept review of this chapter

 Go to the Interactive Summary to review key figures, Animated Tutorials, and Activities
Life10e.com/is58

CHAPTER**REVIEW**

■ REMEMBERING

1. What features of Earth influence its ecosystem dynamics?
   a. Lithospheric plates that move continuously
   b. Atmospheric gases that moderate surface temperatures
   c. Large amounts of water in liquid form
   d. A diversity of living organisms
   e. All of the above

2. Marine upwelling zones are important because
   a. they help scientists measure the chemistry of deep ocean waters.
   b. they bring to the surface organisms that are difficult to observe elsewhere.
   c. ships can sail faster in these zones.
   d. they increase marine productivity by bringing nutrients back to surface ocean waters.
   e. they bring oxygenated water to the surface.

3. The hydrologic cycle operates as it does because
   a. water flows into the oceans via rivers.
   b. water evaporates from the leaves of plants.
   c. more water evaporates from the surface of the oceans than is returned to the oceans as precipitation.
   d. precipitation falls on land.
   e. more water falls on the oceans as precipitation than evaporates from its surface.

4. Carbon dioxide is called a greenhouse gas because
   a. it is used in greenhouses to increase plant growth.
   b. it is transparent to heat but traps sunlight.
   c. it is transparent to sunlight but traps heat.
   d. it is transparent to both sunlight and heat.
   e. it traps both sunlight and heat.

5. The biogeochemical cycle of phosphorus differs from the cycles of carbon and nitrogen in that
   a. phosphorus lacks an atmospheric component.
   b. phosphorus lacks a liquid phase.
   c. only phosphorus is cycled through marine organisms.
   d. living organisms do not need phosphorus.
   e. The phosphorus cycle does not differ importantly from the carbon and nitrogen cycles.

6. Maintaining the capacity of ecosystems to provide goods and services is important because
   a. most ecosystem services cannot be replicated by any other means.
   b. replacing them with technological substitutes is prohibitively expensive.
   c. technological substitutes take up valuable land.
   d. governments cannot function without taxing ecosystem services.
   e. It is not important. Humans could survive quite well even if ecosystem services declined greatly.

■ UNDERSTANDING & APPLYING

7. The waters of Lake Washington, the second largest lake in the state of Washington and lying adjacent to the city of Seattle, returned to their preindustrial condition within 10 years after sewage was diverted from the lake to Puget Sound, an arm of the Pacific Ocean. Would all lakes being polluted with sewage clean themselves up as quickly as Lake Washington if sewage inputs were stopped? What characteristics of a lake are most important to its rate of recovery following reduction of nutrient inputs?

8. A government official authorizes construction of a large coal-burning power plant in a former wilderness area. Its smokestacks discharge great quantities of combustion wastes. List and describe all the likely effects of this action on ecosystems at local, regional, and global levels. If the wastes were thoroughly scrubbed from the stack gases, which of the effects you have just outlined would still happen?

ANALYZING & EVALUATING

9. What types of experiments would you conduct to assess the likely consequences of fertilization of the oceans with iron to increase rates of photosynthesis? At what spatial and temporal scales should these experiments be conducted?

10. One mechanism proposed for reducing the anthropogenic (human-caused) flux of carbon into the atmosphere is called "cap and trade," whereby a government sets a "cap," or limit, on carbon emissions by polluters, but allows facilities that emit less than their emission allowance to sell their excess credits to polluters who would otherwise exceed their emission allowance. What benefits and drawbacks can you see to such an approach to reducing carbon emissions?

11. A string of powerful hurricanes struck the east coast of the United States over the course of a single year's hurricane season. Some people claim that this disaster was due to warming of the oceans caused by greenhouse gases in the atmosphere. Others assert that global warming is not responsible because hurricanes have occurred for many centuries. How would you evaluate these conflicting claims?

Go to BioPortal at **yourBioPortal.com** for Animated Tutorials, Activities, LearningCurve Quizzes, Flashcards, and many other study and review resources.

59

Biodiversity and Conservation Biology

A Natural Christmas Tree During the 1980s, large numbers of migrating bald eagles (*Haliaeetus leucocephalus*) turned Montana's Glacier National Park into a tourist attraction each fall as the birds stopped to feed on spawning kokanee salmon. When the salmon population fell victim to introduced lake trout, the eagles also disappeared and tourism declined.

MANY HUMAN ACTIVITIES have factored in the extinction of animal species; even good intentions can pose a threat. Flathead Lake in northwestern Montana originally had fewer than a dozen native fish species, including bull trout (*Salvelinus confluentus*) and westslope cutthroat trout (*Oncorhynchus clarkii lewisi*). To encourage sport fishing, non-native sport fish species were introduced. Most of these introductions were unsuccessful, but the kokanee salmon (*Oncorhynchus nerka*), introduced from western Canada, eventually prospered and by the mid 1980s was the dominant sport fish.

Because kokanee were popular with anglers, efforts were made to establish them in nearby lakes. Fisheries managers introduced opossum shrimp (*Mysis diluviana*) into neighboring lakes to provide a food source for juvenile kokanee, which feed on zooplankton. But kokanee are daytime feeders that use vision to find their prey; opossum shrimp remain on lake bottoms during the day, thus escaping predation by the young kokanee.

Somehow the shrimp made their way to Flathead Lake, where they proved to be a bonanza for lake trout (*Salvelinus namaycush*), another introduced species but one that had never become abundant because of limited food supply on the lake bottom, where the trout feed as juveniles. With the new food source, the lake trout population exploded. Adult lake trout are voracious consumers of other fish, and kokanee numbers plummeted as they fell prey to the lake trout. By 1992 kokanee were gone from Flathead Lake. The native bull trout may be next; this species was officially designated "vulnerable to extinction" in 1999 and its future is uncertain.

These changes have had economic impacts well beyond sport fishing. In the 1980s, flocks of migrating bald eagles gorged on the abundant kokanee spawning upstream of Flathead Lake in Glacier National Park. The sight was a tremendous tourist draw every fall. Without the salmon, fewer eagles visit the area, and without the eagles, there are fewer tourists. This example of unanticipated effects of human activities is far from unique. Given what we know about species interactions, it should not be surprising that a species introduced in the wrong place can endanger other species.

How can adverse impacts of species introductions be anticipated before lasting damage occurs?

See answer on p. 1245.

59.1 What Is Conservation Biology?

Virtually all natural ecosystems on Earth have been altered by human activities. Many habitats have disappeared completely, and many others have been greatly modified. Even Earth's climate and its global biogeochemical cycles have been altered, as we saw in Chapter 58. One consequence of these changes has been a rapid increase in the rate at which species go extinct.

Conservation biology is a scientific discipline devoted to protecting and managing Earth's biodiversity. The discipline draws heavily on the principles of ecology, ethology, evolutionary biology, and wildlife management particularly when elucidating the factors that determine whether a given population will persist.

Early conservation efforts were characterized by tensions between people whose principal goal was to conserve natural resources for their economic benefits and people who believed that nature has intrinsic value independent of human economic interests. Today conservation biologists study the full array of goods and services that humans derive from species and ecosystems, including aesthetic and psychological benefits. We now know that understanding the global ecosystem and the effects of human activities on that system is essential to the long-term well-being of *Homo sapiens*.

Conservation biology is an applied discipline, which is to say that it involves the practical application of scientific knowledge to solve problems. Workers in conservation biology are guided by three basic principles:

- *The processes of evolution unite all forms of life.* To effectively protect and manage biodiversity, we must understand the evolutionary processes that generate and maintain it.

- *The ecological world is dynamic.* Because populations and communities change continuously over time, there is no static "balance of nature" that can serve as a goal of conservation activities.

- *Humans are a part of ecosystems.* Human interests and activities must be incorporated into conservation goals and practices.

Conservation biology aims to protect and manage biodiversity

The term **biodiversity**, a contraction of "biological diversity," has multiple definitions. We may speak of biodiversity as the degree of genetic variation within a species. Genetic variation can be measured as the number of alleles at a locus, the number of polymorphic loci in a genome, or the number of individuals in a population that are polymorphic at given loci. As we have seen throughout this book, genetic variation allows organisms to adapt to environmental change. Biodiversity can also be defined in terms of species richness in a particular community. At

59.1 Extinct Megafauna The extinction of some large North American mammals during the Pleistocene may have been driven by the arrival of *Homo sapiens*. This portion of a museum re-creation shows Columbian mammoths (*Mammuthus columbi*), ancient bison (*Bison antiquus*, an extinct ancestor of the American bison) and western horses (*Equus occidentalis*).

a larger scale, biodiversity also embraces ecosystem diversity—particularly the complex interactions within and between ecosystems. While we may study these components of biodiversity separately, in life they are intimately interconnected.

One conspicuous manifestation of biodiversity loss is species extinction. Extinction is a constant theme in the history of life; most of the species that have lived on Earth over the ages are extinct today. Consider, for example, the anaerobic organisms that were lost as early photosynthetic prokaryotes and eukaryotes added oxygen to Earth's atmosphere, as described in Section 25.2. Extinctions have occurred throughout Earth's history at what is referred to as a "background" rate as changes in environmental conditions have favored some species and negatively affected others. But the rate of extinctions taking place today rivals those of the five great mass extinctions (see Table 25.1 and Figure 25.2). The past mass extinction episodes were the result of cataclysmic natural disturbances, whereas the majority of modern extinctions can be attributed to effects of human activities.

Humans have a tremendous capacity to alter ecosystems and, accordingly, to cause extinctions. When humans first arrived in North America from Siberia about 14,000 years ago, they encountered a diverse and spectacular fauna of large mammals, including saber-toothed cats, dire wolves, mammoths, mastodons, giant ground sloths, and giant beavers (Figure 59.1). Most of this megafauna went extinct within a few thousand years after humans arrived. Although several hypotheses have been advanced to account for the geologically rapid and simultaneous disappearance of so many large animals, overhunting by humans is the most likely explanation. Losses of megafauna coinciding with the arrival of humans have also been documented in Australia and Hawaii.

Over the past 400 years, increasing industrialization and urbanization have accelerated the rate of species extinctions astronomically. The renowned evolutionary biologist Edward O. Wilson estimates that Earth is losing some 30,000 species per year, putting us in the midst of a sixth mass extinction. The mass extinction events in Earth's past occurred relatively far apart in time, and each one provided ecological opportunities for other groups, which subsequently underwent adaptive radiations. Protecting Earth's biodiversity today requires maintaining the processes that generate new species as well as bringing extinction rates closer to background levels.

Biodiversity has great value to human society

Conservation biologists are concerned about the escalating loss of Earth's biodiversity for many reasons:

- *Humans depend on thousands of other species for food, fiber, and medicine.* Humans have domesticated countless plants as sources of food, and more than 2,000 plant species are used for fiber worldwide. In India alone, more than 7,000 species of plants are used in traditional medicine, and in the United States more than one-fourth of all medical prescriptions contain or are based on plant products. Hundreds of animal species also supply us with food, clothing, and medicine.

- *Losing species can threaten ecosystem functioning.* Throughout Part 10 we have described many complex interactions among species. When species are lost, entire communities and ecosystems may change or be lost completely and humans may lose the goods and services those ecosystems provide.

- *Humans derive enormous psychological benefits, including aesthetic pleasure, from interacting with other organisms.* These aesthetic benefits give biodiversity economic value. Trees growing on a residential lot, for example, can increase the lot's property value by an amount that is greater than the value of the lumber that could be made from the trees.

- *Extinctions deprive the scientific community of opportunities to study and understand ecological relationships among organisms.* The more species that are lost, the more difficult it will be to understand the structure and functioning of ecological communities and ecosystems.

- *Living in ways that cause the extinction of other species raises ethical issues.* Losses of biodiversity increasingly concern philosophers, ethicists, and religious leaders, who believe species to have intrinsic value.

All of these concerns, to varying degrees, may be integrated by conservation biologists into strategies for protecting biodiversity.

RECAP 59.1

Conservation biology is an applied scientific discipline aimed at protecting and managing biodiversity, which is rapidly decreasing due to extinctions that are the result of human activities.

- Explain the multiple meanings of the term "biodiversity." See p. 1229

- What are some of the ways in which biodiversity is valuable to humans? See p. 1230

Conservation biologists must understand biodiversity as it exists today as well as how and why it is changing. An important

goal of conservation biologists is to predict which species are most likely to go extinct and how soon extinction is likely to happen.

59.2 How Do Conservation Biologists Predict Changes in Biodiversity?

How many, and which, species will go extinct will depend both on human activities and on natural events. Conservation biologists attempt to track the extinctions that are occurring and to predict the ones that are likely to occur during the coming century.

Our knowledge of biodiversity is incomplete

Tracking and predicting extinctions is difficult for several reasons. First, we do not know how many species live on Earth today. Many species that are likely to go extinct in the near future have not even been named and described by scientists. Insects provide a case in point: although more than 1 million species have been described (see Section 32.4), estimates of the number of species yet to be discovered range from 2 million to more than 50 million. Even in the case of larger organisms, our understanding of biodiversity is far from complete. For example, in an 18-month period in 2005–2006, more than 50 species of animals and plants previously unknown to science were discovered in the rainforests of Borneo. Worldwide, an annual inventory of newly described species counted 19,232 species discovered in 2009 alone; this list included 9,738 insects, 2,184 plants, 1,360 fungi, 71 mammals, and 7 birds.

Second, we do not know where species live. The ranges of most described species, particularly those that are small, reclusive, and rare to start with, are poorly known. One tiny North American true bug, *Corixidea major* (so rare that it has no common name), had been found in only one location near Clarksville, Tennessee, until entomologists collecting insects attracted to lights at night discovered it in Virginia and Florida, extending its known range by more than 1,000 kilometers.

Third, it is difficult to determine whether a species is truly extinct. Rarely is the death of the last surviving member of a species recorded with certainty, as it was in the case of the last passenger pigeon (*Ectopistes migratorius*), a female named Martha, who died in the Cincinnati Zoo on September 1, 1914. The status of rare, reclusive species with poorly known life histories is much more difficult to determine, as has been the case with the ivory-billed woodpecker (*Campephilus principalis*) in the southeastern United States, which is thought to be extinct despite reported sightings (**Figure 59.2A**). Pygmy tarsiers (*Tarsius pumilus*), tiny primates weighing less than 60 grams, were thought to have gone extinct from their native cloud forests on the island of Sulawesi in Indonesia. In 2008—85 years after the last reported sighting of a living *T. pumilus*—a research team from Texas A&M University discovered individuals of this species living in one of the island's national parks (**Figure 59.2B**).

Fourth, we rarely know all of the connections among species. At the opening of this chapter, we saw how the introduction

(A) *Campephilus principalis*

(B) *Tarsius pumilus*

59.2 Is It Really Extinct? (A) The ivory-billed woodpecker (shown here in a nineteenth-century Audubon print) was presumed to be extinct until reports of sightings in 2004. Clear proof of the living bird has so far eluded ornithologists. (B) In 2008, pygmy tarsiers were discovered in an Indonesian national park after having been presumed extinct for 85 years.

Rarity in and of itself is not always a cause for concern. Some species that are specialized for living exclusively in rare and unusual habitats have probably never been especially abundant and are well adapted to being rare. The Cayman crab fly (*Drosophila endobranchia*), for example, has probably never been abundant. It is found only in the Cayman Islands, where it parasitizes only two species of terrestrial crabs. Small population sizes are of concern, however, especially for species whose populations shrink suddenly. These "newly rare" species are usually at high risk of extinction, as large and rapid reductions in population size can lead to genetic drift and loss of genetic variation (see Chapter 21).

Certain aspects of species' life histories are particularly important in predicting the ability of populations to recover from reductions in size (see Chapter 55). In fishes and mammals, for example, one of the best predictors of extinction risk is age at maturity, a life history trait that has a profound influence on rates of population growth. Ecological niche requirements can also influence the ability of populations to recover from rapid declines. Species with specialized habitat or dietary requirements, for

of one non-native species altered Flathead Lake's entire food web and placed other species at risk. How many species are at risk is unknown, however, because the ecological interactions among all of the lake inhabitants have never been thoroughly characterized.

 Go to Media Clip 59.1
New Species Found in the Twenty-first Century
Life10e.com/mc59.1

We can predict the effects of human activities on biodiversity

Despite these gaps in our understanding of biodiversity, methods exist for estimating probable rates of extinction resulting from human activities. To estimate the risk that a particular population will become extinct, conservation biologists develop statistical models that incorporate information about a population's size, its genetic variation, its life history traits, and the physiology and behavior of its members. The International Union for the Conservation of Nature (IUCN) has published categories that define a species' danger of extinction. Species in imminent danger of extinction in all or most of their range are classified as "endangered" or "critically endangered"; those believed to be susceptible to extinction in the near future are classified as "vulnerable." Biologists consider species in any one of these three categories to be "threatened" (**Figure 59.3**).

59.3 Species at Risk of Extinction (A) The breakdown by extinction risk category of all 47,677 species assessed by th IUCN. (B) The bars show the numbers and proportions of species in the various extinction risk categories (together termed "threatened") in several taxonomic groups that have been comprehensively assessed.

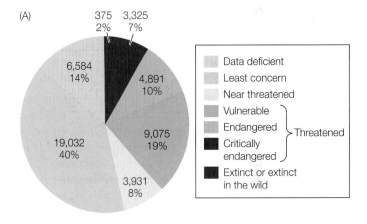

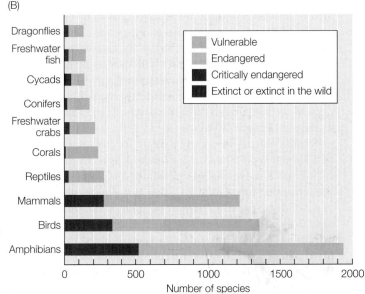

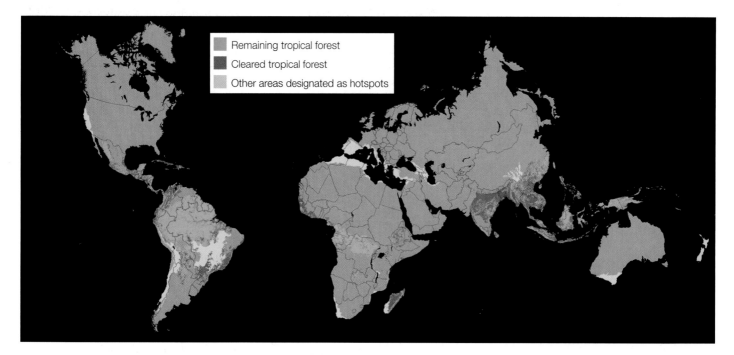

Remaining tropical forest
Cleared tropical forest
Other areas designated as hotspots

59.4 The Disappearing Rainforest Since around 1950, tropical forests have been destroyed at tremendous rates as land is cleared for agriculture, highways, timber resources, and other needs of an exploding human population. Rainforests have long been recognized as centers of biodiversity, or "hotspots," that harbor vast numbers of species (see Section 59.4 and Figure 59.10).

example, are more likely to become extinct than species with more generalized requirements.

In addition, populations reduced to a small size or confined to a small range can easily be eliminated by local disturbances. For example, populations of the Cozumel thrasher (*Toxostoma guttatum*), a member of the mockingbird family known only on the island of Cozumel off the coast of Mexico, had been declining since 1970 due to a combination of factors, including the unintentional introduction of boa constrictors to the island. Then, beginning in 1988, a series of strong hurricanes had a catastrophic effect on the remaining thrasher populations. Surveys done in 2006 failed to document any surviving individuals, and today *Toxostoma guttatum* is most likely extinct.

Conservation biologists apply the principles of the species–area relationship and the theory of island biogeography (see Section 57.3) to predict the effects on species of habitat loss—the major cause of extinction today. By measuring the rate at which species richness decreases with decreasing habitat patch size, they have found that, on average, a 90 percent loss of habitat area results in the loss of half the species that live in and depend on that habitat. We will examine a key example of such a study in Section 59.3.

Similar calculations can be made for the total global area of a habitat type. The current rate of loss of tropical rainforest—Earth's most species-rich biome—is about 2 percent of the remaining forest each year due to the increasing demands of a rapidly expanding human population not only for forest resources but also for cleared agricultural land. Most of the rainforests of Asia have already been reduced to small fragments, the only extensive tracts remaining being found on the islands of New Guinea and, to a much lesser extent, Borneo (**Figure 59.4**). Between 2000 and 2010, the highest rate of

tropical deforestation took place in Central America. If the current rate of loss continues, close to 1 million rainforest species (a conservative estimate) could become extinct before the end of this century.

RECAP 59.2

Predicting changes in biodiversity is difficult because our knowledge of biodiversity is incomplete. The species–area relationship can be used to predict rates of extinction in areas that are subject to habitat loss.

- What are some of the gaps in our current knowledge of biodiversity? **See pp. 1230–1231**
- What are some of the factors that render a species especially vulnerable to extinction? **See pp. 1231–1232**

Many factors can place species at risk of extinction, but human activities have had a disproportionate impact on the mass extinction that Earth is currently experiencing. Understanding how particular human activities present challenges to species survival is essential for developing ways to mitigate biodiversity losses.

 What Human Activities Threaten Species Persistence?

Human activities that threaten the persistence of species include habitat alteration and destruction, introductions of non-native species, overexploitation, and climate alteration. Conservation biologists determine how these activities affect species and use that information to devise strategies to protect species that are endangered or threatened.

Habitat losses endanger species

Global Biodiversity Outlook 3, a report published by the United Nations in 2010, identified five principal pressures on biodiversity. Topping the list is habitat loss and degradation, including fragmentation or outright destruction of habitat by human activities.

Many habitats—particularly freshwater habitats—are being degraded by pollution. Many toxic substances released into natural habitats by human activities have negative effects on the reproduction, development, and behavior of species, reducing both their survivorship and their competitive ability. Among the most troublesome toxic pollutants today are heavy metal waste products of mining and manufacturing, polycyclic aromatic hydrocarbons arising from fossil fuel combustion, and synthetic organic chemicals released into the environment to control pests.

Pollutants do not necessarily have to be toxic to cause problems. Nondegradable plastic trash dumped in the ocean poses a choking hazard to marine birds and mammals, which can mistake floating bits of plastic for prey that they then try to eat. Fish, corals, and other sea life can become entangled in discarded plastic, often resulting in their death.

Habitat loss can also occur through outright habitat elimination. As we saw in Section 58.4, natural ecosystems are being converted to human use at an increasing rate. Physical destruction of a particular habitat, as when tropical rainforest is cut down and the land converted to agricultural use, eliminates species that cannot survive anywhere else. Habitat loss also affects nearby habitats that are not destroyed. As portions of a habitat are lost to human activities, the remaining habitat becomes **fragmented** into habitat patches that become ever smaller and more isolated.

Small habitat patches are qualitatively different from larger patches of the same habitat in ways that affect species persistence. Small patches cannot maintain populations of species that require large areas, and they can support only small populations of those species that can survive in small patches. In addition, the fraction of a patch influenced by external factors increases disproportionately as patch size decreases (**Figure 59.5**). Close to the edges of a forest patch, winds are stronger, temperatures are higher, humidity is lower, and light levels are higher than they are farther inside the forest. Species from surrounding habitats often colonize the edges of a patch, where they compete with or prey on the species living in the patch. These influences are known as **edge effects**. A proliferation of edges can benefit some species, such as those that depend on resources in multiple habitats and must travel among them. For many other species, however, edge effects render habitats unsuitable or promote the establishment of competitors, predators, or parasites.

One effect of forest fragmentation in midwestern North America has been an increase in the abundance of the brown-headed cowbird (*Molothrus ater*). This bird is a brood parasite—that is, it lays its eggs in the nests of other bird species and its hatchlings are raised by the host parents, to the detriment of their own young (see the opening of Chapter 53). Historically,

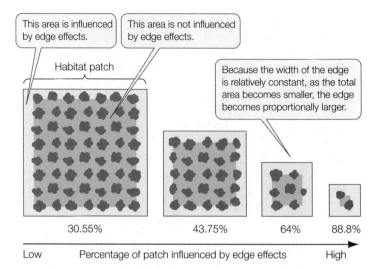

This area is influenced by edge effects.

This area is not influenced by edge effects.

Habitat patch

Because the width of the edge is relatively constant, as the total area becomes smaller, the edge becomes proportionally larger.

| 30.55% | 43.75% | 64% | 88.8% |

Low ⟶ Percentage of patch influenced by edge effects ⟶ High

59.5 Edge Effects The smaller a patch of habitat, the greater the proportion of that patch that is influenced by conditions in the surrounding environment.

Go to Animated Tutorial 59.1
Edge Effects
Life10e.com/at59.1

cowbirds followed bison and other grazing mammals, feeding on insects kicked up by the herds; thus their eggs were laid primarily in nests of grassland host species. Forest fragmentation, however, opened up new opportunities for the cowbirds, which can now lay their eggs in the nests of forest birds in forest edges. Fragmented forests, with relatively more edge than intact forests, thus favor the proliferation of cowbirds at the expense of forest species.

Because so many habitats have already undergone fragmentation by the time they are first investigated by ecologists, determining the effects of fragmentation on the original communities can be difficult. Timely surveys can provide some of this information. For example, a major research project was launched in 1979 in a tropical rainforest near Manaus, Brazil, that was slated for conversion to pasture. The landowners agreed to preserve forest plots of certain sizes and configurations laid out by biologists (**Figure 59.6A**). The biologists counted the species in the future "fragments" while they were still part of the continuous forest, then monitored these plots after the surrounding forest was cut (**Figure 59.6B**). Species soon began to disappear from isolated plots. The first species to be eliminated were monkeys that travel over large areas. Army ants and the birds that follow army ant swarms also disappeared quickly.

Species that are lost from small habitat fragments are unlikely to become reestablished there because dispersing individuals are unlikely to find the isolated fragments. As Section 55.5 pointed out, however, a species may persist in a small patch if it is connected to other patches by habitat corridors through which individuals can disperse. Among the experimental forest plots in Brazil, those that were completely isolated

(A)

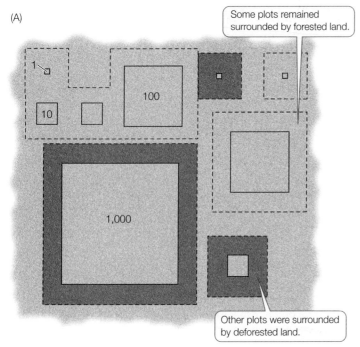

Some plots remained surrounded by forested land.

Other plots were surrounded by deforested land.

(B)

Isolated plots lost species much more quickly...

...than plots connected to unfragmented forest.

Even large plots lost some species of animals.

59.6 Species Losses in Fragmented Brazilian Forest Biologists studied plots of tropical rainforest near Manaus, Brazil, before and after they were isolated by forest clearing. (A) The landowners agreed to preserve forest plots of certain sizes and configurations according to a plan laid out by the biologists. (B) Some of the plots after clearing. The results of the study demonstrated that small, isolated habitat plots lost species more quickly than did larger plots of the same habitat.

lost species more rapidly than did those that were connected to unfragmented forest by corridors. Since the experiment began, some of the pastures that surrounded the experimental plots have been abandoned, and young forests now grow in them. Within 9 years of abandonment, army ants and some of the birds that follow them recolonized forest fragments connected to larger forest fragments by young forests that served as dispersal corridors. Other birds that forage in the forest canopy also reestablished themselves. Young forest is not a suitable permanent habitat for most of these species, but they can disperse through it to find more appropriate habitat.

Insight into the importance of corridors has led to new regional conservation initiatives, among the most notable of which is the Yellowstone to Yukon Conservation Initiative. This joint Canada–United States nonprofit organization has as its goal the sustainable preservation of the mountain ecosystem extending from Yellowstone National Park in the United States to Yukon, Canada. This stretch of land, the largest intact ecosystem of its kind on the planet, contains high-quality habitat for many of North America's most imperiled animals, including grizzly bears, gray wolves, lynx, and native fishes. The initiative works with landowners to find sustainable ways of preserving high-quality, well-connected wildlife habitat in the region. Managing the entire region in this way will not only provide habitat for these species, but will also provide room for their populations to shift in response to global climate change.

Overexploitation has driven many species to extinction

Overexploitation was once the most important cause of species extinctions. Although habitat loss now presents a greater threat to more species, many species are still threatened by overexploitation. Particularly at risk are species with life history traits that are linked to slow population growth (see Section 55.3), which make recovery from losses less likely. Elephants and rhinoceroses, for example, are slow to reach reproductive maturity and produce relatively few offspring over the course of their lives; they are at risk in much of Africa and Asia because poachers kill elephants for their valuable ivory tusks and rhinoceroses for their horns (primarily based on a long-prevalent but false belief that imbibing drinks made with powdered rhinoceros horn boosts a man's sexual potency).

The principal threat to the continued survival of tigers, whose numbers have declined by more 90 percent since 1900 (**Figure 59.7A**), is the use of their body parts in traditional medicine—bones to treat rheumatism, eyes to cure epilepsy, and penises to enhance virility. In 2009 a bowl of tiger penis soup could be obtained for $300 in Taiwan. There is some hope that the availability of inexpensive drugs for treating erectile dysfunction will reduce the incidence of poaching of these and other endangered species, but hopes are dim for eliminating poaching altogether in Asia and Africa. Demand for traditional animal-based medicines remains high, and animal aphrodisiacs provide an economic boon for impoverished hunters and a status symbol for the rich.

Massive international trade in exotic pets and aquarium fishes, ornamental plants, and tropical forest hardwoods has decimated many species. The Banggai cardinalfish (*Pterapogon kauderni*; **Figure 59.7B**) is on the brink of extinction entirely because of the pet trade; almost a million of these critically endangered fish are hauled out of the waters annually near Sulawesi, Indonesia, to satisfy the demand from saltwater aquarium enthusiasts.

(A)

Boiga irregularis

59.8 An Agent of Extinction Since it was accidentally introduced onto the tiny Pacific island of Guam, the brown tree snake (*Boiga irregularis*) has eaten 15 species of land birds to extinction.

(B) *Pterapogon kauderni*

59.7 Endangered by Exploitation (A) Skins confiscated at the China–Myanmar border illustrate the extent of poaching of endangered tigers (*Panthera tigris*). Beyond the value of their pelts, tiger bones and other body parts are highly prized in Asian traditional medicine. (B) The international pet trade has brought the Banggai cardinalfish to the brink of extinction. Each year almost a million of these critically endangered fish are hauled out of Indonesian coral reef waters to satisfy the demand from saltwater aquarium enthusiasts.

Burgeoning human populations in need of food are also placing unprecedented pressure on species harvested from the wild. Humans have captured wild fish for food for at least 40,000 years, but in recent centuries innovations in technology and increasing demand have led to removal of fish from wild populations at rates that far exceed the capacity of the remaining individuals to reproduce. An estimated 25 percent of the world's wild fisheries are currently at risk of overexploitation and collapse. Deep-sea fish that are slow to mature and produce relatively few offspring, such as the orange roughy (*Hoplostethus atlanticus*), are especially sensitive to overexploitation.

Invasive predators, competitors, and pathogens threaten many species

As people travel, they deliberately or inadvertently move species to regions outside their original ranges. Some of these non-native species become **invasive**—that is, they reproduce rapidly, spread widely, and have negative effects on the native

species of the region. As we saw in Section 55.4, species that are introduced into a region where their natural enemies are absent may reach very high population densities. Moreover, the native species in an invader's new range may not have evolved defenses against these new antagonists and competitors.

Invasive species are spread in several ways. Marine organisms have been spread throughout the oceans by ballast water, taken on by ships at the port of departure and discharged at the destination port along with its content of surviving animals and plants. The notorious zebra mussel (*Dreissena polymorpha*; see Figure 55.10) is thought to have arrived in North America in this way. The brown tree snake (*Boiga irregularis*; **Figure 59.8**) arrived on Guam in air cargo shortly after World War II. Until then, the only snake on Guam was a tiny insect-eating species. The number of *B. irregularis* on Guam remained low for some 20 years, but in the 1960s the species began to multiply and today can be found at densities up to 5,000 individuals per square kilometer. The snake has exterminated 15 species of land birds, including 3 found only on Guam.

Over the past 400 years, Europeans colonizing new continents have deliberately introduced plants and animals to their new homes in an effort to reconstruct their familiar surroundings. Many of these introductions have had disastrous effects on native flora and fauna. In Australia the introduction of European rabbits and foxes for sport hunting and of dogs and cats as pets has led to the extermination of nearly half the small- to medium-sized native marsupials over the last 100 years. Some species have been introduced deliberately to control other invasive species, and then have themselves caused even greater problems. One such example is the cane toad (*Bufo marinus*), introduced into Australia to control sugarcane pests (see Figure 55.15).

It can be difficult for people to imagine that plants that are desirable and attractive in their place of origin can "go rogue" when they escape from cultivation in a new region. Some of today's most noxious weeds were deliberately transported and

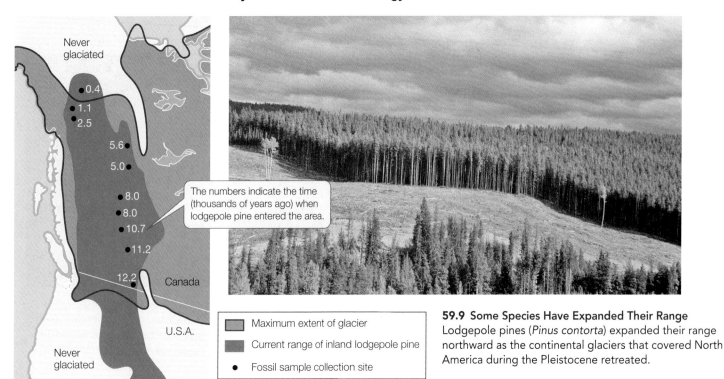

The numbers indicate the time (thousands of years ago) when lodgepole pine entered the area.

- Maximum extent of glacier
- Current range of inland lodgepole pine
- • Fossil sample collection site

59.9 Some Species Have Expanded Their Range
Lodgepole pines (*Pinus contorta*) expanded their range northward as the continental glaciers that covered North America during the Pleistocene retreated.

planted in new places for their beauty, fragrance, or culinary value. Once established in their new environments, however, these invasive plants have had profoundly negative effects. While native plants must devote considerable energy and resources to defending themselves against native herbivores, invasive plants are less prone to attack, in part because their natural enemies have been left behind in their original range. Therefore invasive plants can devote more resources to growth and reproduction and fewer to producing defensive secondary compounds. The majority of plants considered by U.S. farmers to be weeds are non-native, and controlling them, primarily with chemical herbicides, costs billions of dollars every year.

Introduced pathogens have also wreaked havoc among native species, as exemplified by the effects of avian malaria in the Hawaiian Islands. Before the arrival of Europeans, no mosquitoes existed in the islands. The first mosquito species was found there in 1827, and over the next century several others followed. At the start of the twentieth century, the microbial pathogen that causes avian malaria arrived, most likely carried by imported caged birds. Not having been exposed to malaria over the course of their evolutionary history, Hawaii's many endemic bird species were exceptionally vulnerable to infection. Today nearly all species living below 1,500 meters elevation (the current upper limit of the range of the mosquito vectors) have been eliminated, mostly by avian malaria. Species living at higher elevations have fared better, but the range of the mosquitoes appears to be expanding upward as the climate warms, placing the surviving endemic species at risk.

Rapid climate change can cause species extinctions

As we saw in Section 58.3, human-generated emissions of greenhouse gases are hastening global climate warming, and that warming is likely to become an increasingly important cause of extinctions. Across North America, for example, average annual temperatures are predicted to increase anywhere from 2°C to 5°C by the end of the twenty-first century. If the climate warms to that extent, the average temperature found at any given location in North America today will shift 500 to 800 kilometers to the north. Those species that cannot adapt to the warmer climate will have to shift their geographic ranges by that distance within less than a century if they are to persist. Some biomes, such as alpine tundra, could disappear entirely as temperate forests expand up mountain slopes.

Efforts to control or reverse global warming present a challenge to people worldwide. Conservation biologists can contribute to discussions about how to respond to climate change by predicting how it may affect organisms and looking for ways to mitigate those effects. Their research activities include analyses of past climate changes and studies of sites currently undergoing rapid climate change. It would be helpful to know, for example, how rapidly species responded to the end of the most recent ice age, about 10,000 years ago. Which species did and did not keep pace with the warming climate? How much, and in what ways, do past ecological communities differ from those of today as a result of differences in the rates at which species' ranges shifted?

Species that can disperse easily, such as birds and insects that can fly considerable distances, may be able to shift their ranges as rapidly as the climate changes, provided they can find appropriate habitats. However, the ranges of other species are likely to shift more slowly. For example, after the glaciers retreated in North America about 8,000 years ago, the ranges of some pine trees, which have lightweight seeds that can be carried great distances by wind, expanded northward, so that today they grow as far north as the current climate permits (**Figure 59.9**). Native earthworms, on the other hand, fared less well—the glaciers may well have eliminated all earthworm species in Canada, and they have not been replaced by other North American species, which have moved their ranges northward only slowly. (Many of the earthworms found in the United States today are non-native species accidentally introduced from elsewhere.)

If Earth's surface warms as predicted, entirely new climates will develop, especially at low elevations in the tropics, where a warming of even 2°C would result in conditions warmer than those found anywhere in the humid tropics today. Adaptation to those climates may prove difficult even for many tropical organisms. Since the mid-1980s, the average minimum nightly temperature at La Selva Biological Station, in the Caribbean lowlands of Costa Rica, has increased from about 20°C to 22°C. On warmer nights, trees use more of their energy reserves to maintain themselves. As a result, even this small rise in temperature has reduced the average growth rates of six different tree species by about 20 percent.

RECAP 59.3

Several human activities threaten the persistence of species, including habitat degradation, fragmentation, and destruction; overexploitation; introductions of invasive species; and activities that cause rapid climate change.

- Describe three ways in which habitat loss is occurring today. See p. 1233

- Why are rates of species loss high in small habitat patches? See pp. 1233–1234 and Figures 59.5 and 59.6

- How can dispersal ability and climate change interact to affect the probability of extinction? See pp. 1236–1237

Demonstrating that species are endangered is an empty exercise if we cannot implement a plan of action to save them. In the next section we will consider some of the positive steps that are being taken to protect biodiversity.

59.4 What Strategies Are Used to Protect Biodiversity?

Conservation biologists use scientific theory, empirical data, and tools from a variety of disciplines to help protect endangered and threatened species and ecosystems. They identify the factors that present risks to species and use that information to devise action plans. Implementing those plans, however, often requires the cooperation of many different groups of people, so conservation biologists also work with landowners, politicians, lawyers, environmental activists, and the general public. It is thus very useful to examine the actions that conservation biologists and policy makers take to protect biodiversity in order to determine which approaches have been most successful and to understand what aspects have contributed to their success.

Protected areas preserve habitat and prevent overexploitation

The establishment of **protected areas**, in which habitat alteration and exploitation are restricted or prohibited, is an important component of efforts to conserve biological diversity. Protected areas allow populations to maintain themselves in the preserved habitat and may also serve as nurseries from which individuals can disperse into exploited areas, replenishing populations that might otherwise become extinct.

Deciding which areas to protect is a challenging enterprise. Two robust criteria are species richness (the total number of species living in an area; see Section 57.3) and **endemism** (the number of species in an area that are found nowhere else—a measure of its uniqueness). Using these two criteria, biologists have identified regions of unusual richness and endemism, which they have labeled **biodiversity hotspots** (Figure 59.10). These hotspots occupy slightly less than 16 percent of Earth's land surface, but they are home to approximately 77 percent of its terrestrial vertebrate species. Most of these hotspots are also areas of high human population density, which means habitat loss is ongoing and often rapid. Developing a conservation strategy for any of these regions requires not only a detailed analysis of the distributions of species and the locations of special habitat resources (such as caves, freshwater springs, or migratory stopover areas for birds), but also an analysis of factors that threaten and factors that support biodiversity in the region.

In 2010, in an effort to pinpoint sites with threatened species that are found nowhere else, conservation biologists identified 587 "centers of imminent extinction." These sites are concentrated in tropical forests, on islands, and in mountainous regions (**Figure 59.11**). Only about half the sites are even partially protected by law, and most of them are surrounded by land that is undergoing rapid development. Unless protective actions are taken soon, species extinctions at these sites are inevitable. Identifying biodiversity hotspots and centers of imminent extinction has been helpful in prioritizing conservation efforts worldwide, encouraging international cooperation, and raising public awareness of critical threats to species persistence.

Degraded ecosystems can be restored

When a species is endangered as a consequence of habitat degradation rather than outright habitat loss, protecting the species may require restoring the habitat to a more natural state. Many degraded ecosystems recover only slowly, if at all, without human assistance. Practitioners of **restoration ecology** are developing methods aimed at just such habitat reconstitution.

Because the soil that supports them is so rich, grasslands all over the world have been converted to agriculture. By the middle of the twentieth century, for example, most North American prairies had been converted to cropland or were heavily grazed by domestic livestock. The herds of large mammals that roamed the prairies when European settlers arrived have been reduced to tiny remnant populations confined to small areas. Most of these populations are too small to maintain their genetic diversity or to function in their original ecological roles. The species *have* survived, however, so opportunities exist to reintroduce them if their habitat can be restored.

A major prairie restoration project is underway in northeastern Montana. When Lewis and Clark mapped this region 200

(A) Tropical rainforest hotspots

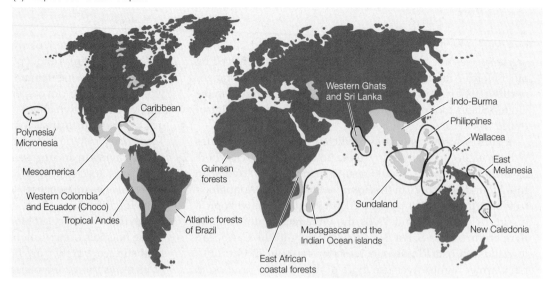

(B) Hotspots in other biomes

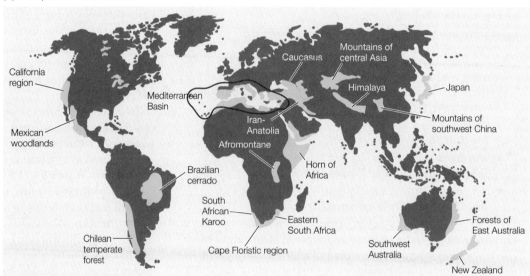

59.10 Hotspots of Biodiversity
(A) Almost half of the world's terrestrial biodiversity hotspots are regions of tropical rainforest habitat. There are only three remaining areas of extensive unbroken rainforest (Amazonia, the Congo Basin of Africa, and the island of New Guinea; see Figure 59.4). (B) Eighteen additional hotspots represent non-rainforest ecosystems.

years ago, they saw large herds of bison, elk, deer, and pronghorn as well as abundant populations of their predators. The goal of the restoration project, which is run by the World Wildlife Fund and the American Prairie Foundation in cooperation with public land managers and several other private conservation organizations, is to restore the native prairie and its fauna in a 15,000-km² area near the Missouri River (**Figure 59.12**).

This ambitious project is feasible for three reasons. First, the private land in the area is owned by a small number of ranchers, each of whom owns extensive grazing leases on public lands administered by either U.S. federal agencies or the State of Montana. Second, most of the land has never been plowed,

59.11 Centers of Imminent Extinction Areas shown in yellow include many of the world's 587 "centers of imminent extinction" (as designated in 2010 by the Alliance for Zero Extinction, a coalition of more than 80 conservation groups). Although there are scattered centers in other regions, the areas highlighted here harbor an estimated 1,000 endemic species (species found nowhere else) known to be at high risk of extinction.

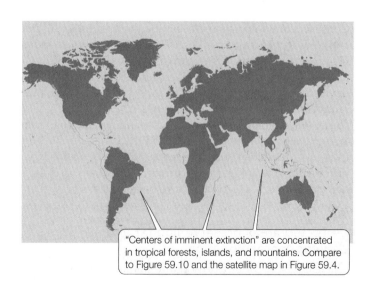

"Centers of imminent extinction" are concentrated in tropical forests, islands, and mountains. Compare to Figure 59.10 and the satellite map in Figure 59.4.

(A)

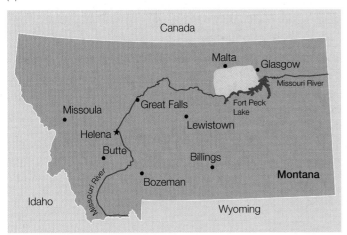

(B) *Cynomys ludovicianus*

(C) *Bison bison*

59.12 Restoring a North American Prairie   (A) A major prairie restoration project (yellow area) is under way north of the Missouri River in the state of Montana. (B) Native prairie dogs maintain the vegetation by digging extensive burrows and clipping plants. (C) The first bison were reintroduced to the area in 2005.

so native vegetation may recover rapidly when grazing pressures are reduced. Third, the area's human population is decreasing. Ranchers are aging, and their children are abandoning the hard work and uncertain profits of ranching for careers in urban settings. Once a free-ranging herd of several thousand bison and large numbers of elk—along with their predators (wolves)—has been established, nature-minded tourists are expected to flock to the area to view the wildlife spectacle. Over the long term, the restored ecosystem should deliver major economic benefits to the region.

In the United States, the sense that humans are capable of creating functioning ecosystems to replace those lost to development underlies policies that allow developers to destroy habitats. Destruction of wetlands, in particular, is often permitted because developers assert that those ecosystems can be replaced. However, creating new wetlands requires detailed ecological knowledge that generally surpasses what is currently available.

In southern California, where 90 percent of the coastal wetlands have been destroyed, wetland restoration is a high priority. Species have been lost from degraded coastal wetlands, so restoration requires species introductions. Early attempts at restoration, in which one or two common wetland species were introduced, did not succeed; other wetland-associated species failed to recolonize the "rehabilitated" wetlands. To understand why, conservation biologists established a large field experiment at the Tijuana Estuary near San Diego (**Figure 59.13**). Here they found that experimental plots planted with species-rich mixtures were covered with vegetation faster, developed a complex vegetation structure (which is important to insects and birds) more rapidly, and accumulated nitrogen (required for plant growth) faster than did species-poor plots (**Figure 59.14**). This outcome represents a practical demonstration

of the relationship between community stability and species richness (see Section 57.5).

Disturbance patterns sometimes need to be restored

Many species depend on particular patterns of disturbance—such as fires or windstorms—to maintain their populations (see Section 57.4). Recognition of the need for periodic disturbance to maintain healthy ecosystems is a relatively new dimension

59.13 A Wetlands Laboratory   The Tijuana Estuary near San Diego is a shallow-water wetland habitat. Experiments at this natural research reserve have advanced efforts to restore this valuable ecosystem.

INVESTIGATING**LIFE**

59.14 Species Richness Can Enhance Wetland Restoration

In a large-scale experiment in the Tijuana Estuary, John Callaway and other ecologists from the Southern California Wetlands Recovery Project compared different methods for restoring shallow-water wetlands. They found that several measures of ecosystem function improved more rapidly in species-rich than in species-poor plantings.[a]

HYPOTHESIS   Faster progress toward restoring a shallow-water wetland community to its original condition will be made by planting mixtures of species than by planting a single species.

Method   1. In an area of wetland denuded of vegetation, mark off replicate experimental plots, all of the same size.

2. Choose 8 native species typical of the region. Plant some plots with 1 of the 8 species by itself, others with different subsets of 3 species, and others with different subsets of 6 species. Plant the same total number of seedlings in each plot. Leave control plots unplanted.

3. Measure the percent ground cover, number of canopy layers, and soil nitrogen levels at 6-month intervals over the next 18 months.

Results   In the plots with higher species richness, more of the ground was covered by plants, the vegetation structure was more complex, and more nitrogen accumulated in the soil.

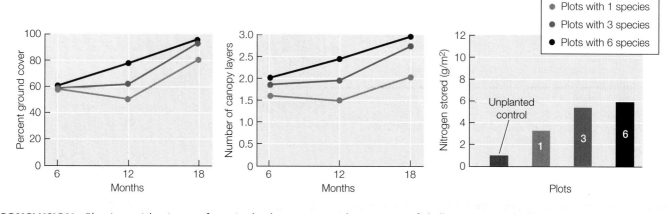

CONCLUSION   Planting a rich mixture of species leads to more rapid restoration of shallow-water wetlands.

Go to **BioPortal** for discussion and relevant links for all INVESTIGATING**LIFE** figures.

[a] Callaway, J. C., G. Sullivan, and J. B. Zedler. 2003. *Ecological Applications* 13: 1626–1639.

of conservation biology. For example, although many plant species require periodic fires for successful establishment and survival, for many years the official policy of the U.S. Forest Service, symbolized by the iconic mascot Smokey Bear, was to suppress all forest fires. Today, however, controlled burning is common, particularly in western North America. In order to use fire as an ecosystem management tool, it is important to know the historical pattern of fires in an area, which can be determined in part by studies of the annual growth rings and fire scars of trees (**Figure 59.15**). A schedule of controlled burning that recreates the historical pattern can reduce forest floor litter, avoiding a buildup of fuel that can lead to intense, tree-killing canopy fires.

Ending trade is crucial to saving some species

Most endangered species cannot survive any further reductions in their breeding populations, so it is important to prevent their exploitation. The legal mechanism for prohibiting trade in these species or their products is an international agreement called the Convention on International Trade in Endangered Species (CITES). Most nations of the world are members of CITES. CITES rules currently prohibit international trade in items such as whale meat, rhinoceros horns, and many species of parrots, orchids, and others.

The recent history of elephant poaching and trade in ivory illustrates how complex preventing exploitation of endangered species can be. CITES instituted a ban on international trade in African elephant ivory in 1989, but demand for ivory remains strong, especially in Asia. As a result, poaching of elephants continues in the forests of central and East Africa, where the animals are threatened. However, some countries, including Malawi and Zambia, have so many elephants that government officials kill them to control populations and prevent the animals from damaging crops. These countries would like to sell the ivory from culled elephants to fund conservation efforts, but other countries are worried that if restrictions are relaxed, poaching will escalate everywhere.

Control of ivory trade might be possible if scientists could determine where the ivory comes from. Conservation biologist Samuel Wasser and his colleagues identified 16 DNA markers from elephant feces collected by park rangers in Malawi and Zambia. The source of an elephant tusk could then be determined by matching DNA extracted from the ivory with the geographically based frequencies of the 16 DNA markers in the dung samples. Such safeguards were partially responsible for the controversial decision to sanction sales of ivory from Namibia, Botswana, Zimbabwe, and South Africa in 2008, the first such sales in close to a decade. More than 100 tons of

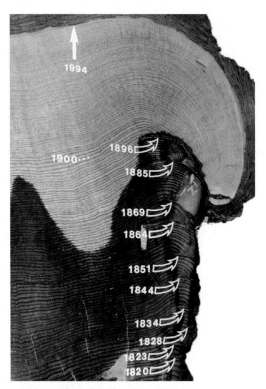

59.15 A Natural Disturbance Pattern As revealed by scars (arrows) in the growth rings of this ponderosa pine (*Pinus ponderosa*), low-intensity, nonlethal ground fires were frequent in the pine forests of the southwestern United States prior to fire suppression.

elephant tusks—the equivalent of 20,000 dead elephants—were auctioned off to authorized buyers from China and Japan, for use primarily in folk medicine. This legal sale generated some $15 million for elephant conservation efforts.

Although the 2008 sales were monitored by CITES, concerns remain that the flood of legal ivory will be intermingled with poached ivory. One promising development in curbing illegal sales was the decision by eBay, the international internet marketplace, to ban sales of ivory on its platform as of January 2009. An independent investigation by the International Fund for Animal Welfare stated that two-thirds of online sales of protected wildlife products take place on eBay, so conservationists hope eBay's actions will be effective in drying up markets. Notwithstanding such efforts, illegal poaching, smuggling, and trafficking of ivory have all increased, and representatives from 175 countries attending the 2010 meeting of CITES in Qatar voted to ban sales of stockpiled elephant ivory for at least 3 years.

Species invasions must be controlled or prevented

The best way to reduce the damage caused by invasive species is to prevent their introduction in the first place. Given the tremendous volume of global trade, it might seem impossible to curtail their spread, but some promising strategies do exist. For example, transoceanic transport of invasive species in ballast water (responsible for the devastation caused by the invasive

zebra mussel; see Figure 55.10) could be largely eliminated by the simple procedure of deoxygenating ballast water before it is pumped out. This practice not only kills most organisms in the water but also extends the life of ballast tanks—an economic benefit to shippers.

In 1996 the U.S. Congress responded to concerns about ballast water with legislative action. After years of wrangling, in 2012 the U.S. Coast Guard amended its regulations on managing ballast water to set standards for "the allowable concentration of living organisms in ballast water discharged from ships in waters of the United States." The Coast Guard relied on scientific reports issued by the National Academy of Sciences and the U.S. Environmental Protection Agency Science Advisory Board to specify the most stringent discharge standards achievable with current technology. Despite the adoption of these strict standards for protecting U.S. waterways, the challenge of achieving global uniformity remains. The transport of invasive aquatic organisms in ballast water is an international problem whose potential solutions continue to run up against political and economic barriers.

Regulating the importation and sale of non-native plant species has been more successful in reducing deliberate introductions. In 2002, members of the American horticulture industry crafted a voluntary code of conduct for their profession, stating that the invasive potential of a plant should be assessed prior to its introduction and marketing. Horticulturists were encouraged to work with biologists to determine which species are currently invasive, or are likely to become so, and to identify suitable alternative species.

Conservation biologists have developed a "decision tree" based on the traits that characterize plant species that have become invasive (**Figure 59.16**). The tree is used to help horticulturists and regulators determine whether a non-native plant species should be allowed into North America. Although the protocols stipulated by this decision tree cannot eliminate all detrimental introductions, if followed conscientiously they can greatly reduce the risk of such events.

Biodiversity has economic value

Many studies have demonstrated the market value of protecting biodiversity. Markets already exist for many products of biodiversity; to cite just one example, sales of pharmaceuticals derived from plants worldwide amount to more than $30 billion annually. Thus the argument for conservation is compelling not only from an ecological or ethical perspective but also from an economic perspective. The interdisciplinary field of **ecological economics** provides tools for assessing the economic value of biodiversity

Crucial to ecological economics is recognizing and determining the value of services that thriving ecosystems provide to human society (see Section 58.4). These services depend on biodiversity, but it is difficult to assess their value in monetary terms. Ecosystem services that depend on the maintenance of biodiversity include:

• *Provisioning services*, including the availability of food and water for human consumption.

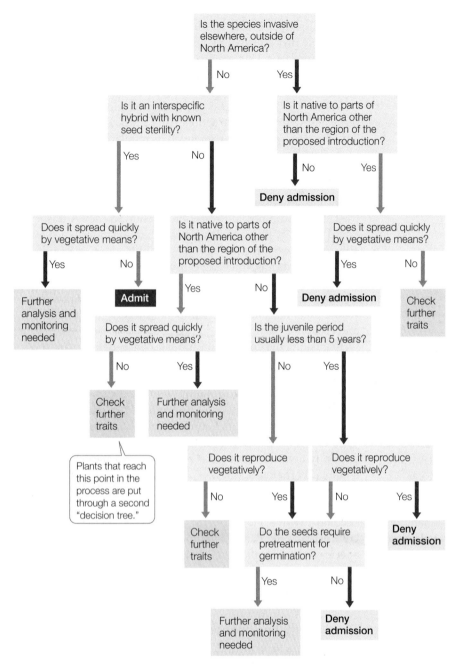

59.16 A "Decision Tree" This diagram sets criteria for evaluating proposed introductions of non-native plant species and helps regulators identify those species with the potential to become invasive.

- *Regulating services*, including the contributions of ecosystems to purification of water, flood control, pollination, and regulation of disease or pest outbreaks.

- *Supporting services*, including formation of soils and maintenance of nutrient cycles.

- *Cultural services*, including the provision of nonmaterial benefits such as recreational activities, psychological well-being, and spiritual enrichment (one example being the opportunities to fish for salmon and to watch eagles at Flathead Lake, described at the opening of this chapter).

The following three examples illustrate how biodiversity can offer a range of benefits to human populations that more than justify investing in conservation.

WILD DOGS AND ECOTOURISM Ecotourism—environmentally responsible travel to natural areas, the proceeds of which support conservation efforts and the economic well-being of the local communities—is a major source of income for many developing nations. For example, tourists visiting Africa often express interest in seeing wild dogs (*Lycaon pictus*; **Figure 59.17**). However, diseases such as rabies and canine distemper, along with habitat loss, roadkills, deliberate extermination due to a perceived threat to livestock, and many other factors have decimated wild dog populations, making this the second most endangered carnivore in Africa. (Another canid, the Ethiopian wolf *Canis simensis* is first.) South Africa is home to about 400 of Africa's remaining 5,000 dogs, most of which live in Kruger National Park. Their endangered status has piqued tourist interest in these charismatic animals; a survey of visitors to South Africa revealed that nearly three-fourths of them would be willing to pay an extra U.S. $12 for the opportunity to see wild dogs. Conservation biologists are working with lodge owners and ranchers elsewhere in South Africa and in Kenya to encourage them to reestablish wild dogs in areas from which they have disappeared.

FYNBOS Studies by a group of economists, ecologists, and land managers have attempted to calculate the value of the economic benefits provided by the spectacularly species-rich fynbos community (described at the opening of Chapter 54) to the Western Cape Province of South Africa. More than two-thirds of the 8,500 plant species in the fynbos community are endemic, thriving despite summer droughts, nutrient-poor soils, and periodic fires. Some of the endemic plants, including proteas, are harvested for cut and dried flowers. An international market has developed for rooibos, a fynbos shrub used for herbal tea. Income also comes from hundreds of thousands of ecotourists who visit the region. The fynbos also provides recreational opportunities for local residents in urban areas nearby. Perhaps most importantly, however, the highland watershed in which fynbos thrive provides about two-thirds of the Western Cape's water supply.

In recent years several trees and shrubs from other continents have invaded the fynbos. Taller and faster-growing than the endemics, they displace the native vegetation, increasing the intensity and severity of fires. Moreover, because the

59.17 A Sight for Ecotourists' Eyes Tourists visiting Africa to experience its wildlife often express a desire to see the endangered African wild dog (*Lycaon pictus*). Protecting such species (as seen here in South Africa's Mala Mala Game Reserve) can be in the economic interest of a region.

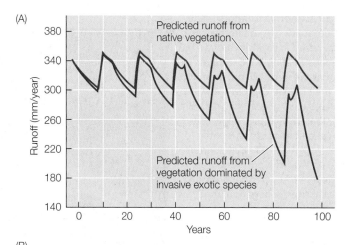

59.18 Biodiversity Maintains Ecosystem Functioning
(A) A computer simulation of change in stream flows over time from fynbos watersheds that have and have not been invaded by non-native trees and shrubs. (B) The Working for Water Programme has launched an effort to preserve the Western Cape Province's water supply by physically removing invasive species.

invaders transpire more water than the endemics, they could decrease stream flows to less than half of those from areas covered with native plants, drastically reducing the water supply for people in the region (**Figure 59.18A**).

We can get an idea of the economic value of fynbos biodiversity to the Western Cape Province by estimating the cost of maintaining or replacing the water supply services the fynbos provides. The South African government has launched an effort to maintain fynbos biodiversity by felling and digging out invasive trees and shrubs and by controlled fires (**Figure 59.18B**). This effort costs between $140 and $830 per hectare, depending on the densities of the invaders. Alternatively, the services provided by fynbos vegetation could be replaced, but at a much higher cost. A sewage purification plant that would deliver the same amount of water to the Western Cape Province as a well-managed watershed of 10,000 hectares would cost $135 million to build and $2.6 million per year to operate. Desalination of seawater would cost four times as much. Thus the available alternatives would deliver water at a cost somewhere between 1.8 and 6.7 times higher than the cost of maintaining natural vegetation in the watershed. Maintaining the fynbos is less expensive and more labor-intensive—thus generating more employment—than the technologically sophisticated methods that could substitute for the services it provides.

COFFEE AND POLLINATORS Taylor Ricketts and colleagues at Stanford University assessed the economic value of the pollination services provided by the bees that live in, and depend on, tropical forest patches adjacent to a coffee plantation in Costa Rica. They found that coffee production was highest at the sites that were closest to forest patches. They also hand-pollinated some coffee plants to show that the difference in production was a result of pollination services rather than

other environmental conditions. The investigators calculated that the value of pollination services to the plantation on which the experiments were carried out was about $60,000 per year, more than the current conservation incentive payments offered to landowners to preserve forest patches.

Changes in human-dominated landscapes can help protect biodiversity

Establishing protected areas is an essential component of efforts to maintain biodiversity, but this action alone is insufficient to stem global biodiversity loss. The extensive landscapes in which people live and extract resources must also play important roles in biodiversity conservation. The good news is that, when carefully used, these lands can contribute much more to conservation than they currently do. The practice of using ecosystems for residences, resources, or recreation in ways that sustain their biodiversity is known as **reconciliation ecology**.

59.19 California Condors Make a Comeback (A) California condors raised in captivity are fed by humans wearing hand puppets so that the birds will not imprint on their human captors and will be able to survive in the wild. (B) Numbered wing tags allow conservation biologists to identify and track released adult condors. The survival of North America's largest bird species depends on this captive propagation project.

(A)

(B) *Gymnogyps californianus*

Reconciliation ecology is based on the principle that most ecosystem services are provided locally, and that people are more motivated to work to protect their local interests than they are to work on national or global issues. The National Wildlife Federation has established a successful program in which people petition to have their backyards certified as wildlife-friendly. Criteria for certification include planting shrubs that provide food for birds and refraining from applying pesticides.

Even some industrial sites can support biodiversity. The Turkey Point power plant in southern Florida uses large amounts of water to cool its generating units. To cool the heated water before discharging it, the Florida Power & Light Company dug a system of 38 canals that covers 6,000 acres. These cooling canals are separated by low-lying berms that support a variety of native and non-native plants. Red mangroves grow along the edges of the canals. Today they support a thriving population of American crocodiles, a highly endangered species. Crocodiles living in the canals yield about 10 percent of all young crocodiles born in the United States. Having discovered the biodiversity value of its cooling system, the company employs biologists to monitor the crocodiles and works actively to ensure their continued reproductive success.

Captive breeding programs can maintain a few species

A few of the world's endangered species can be maintained in captivity while the external threats to their persistence are reduced or removed. However, captive propagation is only a temporary measure that buys time to deal with those threats. Zoos, aquariums, and botanical gardens do not have enough space to maintain adequate populations of more than a small fraction of Earth's endangered and threatened species. Nonetheless, captive propagation can play an important role by maintaining species during critical periods, providing a source of individuals for reintroduction into the wild, and raising public awareness of threatened and endangered species.

The California condor, North America's largest bird, survives today only because of captive propagation (**Figure 59.19**). Two centuries ago, condors ranged from British Columbia to northern Mexico, but by 1978 the wild population was plunging toward extinction. Many of the birds, which are scavengers, had died from ingesting animal carcasses containing lead shot or bullets. To save the condor from certain extinction, biologists captured all the remaining condors—only 22 individuals—and initiated a captive breeding program in 1983.

The first captive-bred birds were released in the mountains north of Los Angeles in 1992. Since that time, there have also been releases in northern Arizona and Baja California. Today captive-bred birds use the same roosting sites, bathing pools, and mountain ridges that their wild-born predecessors did. In 2003 a wild-born chick fledged in the wild for the first time in more than two decades. By 2012 the number of condors living in the wild had reached 226, with another 179 living in captivity.

Most of the major threats to condor survival, including power lines, pesticides, and museum collectors, have been mitigated. Lead poisoning is still a problem, but as of July 1, 2008, under the Ridley–Tree Condor Preservation Act, California hunters are required to use non-lead bullets when hunting within the condor's range. Passage of this legislation marks a change in public attitudes from the days when cattle ranchers, in the mistaken belief that the condors killed livestock, vociferously opposed their reintroduction into the wild.

Earth is not a ship, a spaceship, or an airplane

Biologists often use metaphors to convey complex concepts, and ecologists in particular are fond of them. To convey the notion that human activities may be overwhelming the capacity of Earth's ecosystems to accommodate them, Herman Daly in 1966 introduced the metaphor of the Plimsoll line.

Since the nineteenth century, the Plimsoll line, or load line, has been painted on the side of ships that travel in international waters. If the Plimsoll line of a fully loaded ship anchored in port sinks below the water, there is a significant risk that, once at sea, the ship will capsize and the crew and cargo will be lost. Another ecological metaphor (also introduced in 1966, by Kenneth Boulding) is the concept of Spaceship Earth, which views humans as passengers on a planetary spaceship with limited supplies who must learn how to reuse and recycle in a "cyclical ecological system." And finally, in their 1985 book *Extinction*, Paul and Anne Ehrlich used an airplane metaphor to frame their "rivet hypothesis." According to this metaphor, ecosystems, like well-constructed airplanes, have redundant design features that allow certain parts to fail without causing a catastrophic breakdown: "A dozen rivets, or a dozen species, might never be missed. On the other hand, a thirteenth rivet popped from a wing flap, or the extinction of a key species … could lead to a serious accident."

Even though they are remarkable human inventions, ships, spaceships, and airplanes are hopelessly inadequate to convey the enormous complexity of life. Moreover, relying on these metaphors creates a sense that we can rely on a few individuals—the captain or pilot and the "ship's" crew—to ensure that the planet doesn't capsize, run out of supplies, or crash. In fact, *every individual has a responsibility to understand how Earth provides essential resources and services and to recognize the effects that using those resources has on other living beings.*

Humans share Earth with a staggering diversity of organisms, and our continued existence depends on our interactions with them. We may be unique as a species, however, in that we have the intellectual capacity to recognize, quantify, and if need be, mitigate the effects of those interactions. Learning all we can about life on Earth is perhaps our best tool for improving the quality of our own existence on the planet for the long haul.

▓▓▓▓▓▓ **RECAP 59.4**

To conserve biodiversity, it is necessary to set aside protected areas, restore ecosystems and natural disturbance patterns, restrict trade in endangered species and transport of invasive species, increase populations of endangered species, and otherwise recognize the benefits of maintaining functioning ecosystems.

- What are some of the priorities that conservation biologists consider when establishing protected areas? **See p. 1237 and Figures 59.10 and 59.11**
- What are some of the material and non-material ecosystem services that depend on maintaining biodiversity? **See pp. 1241 and 1241–1243**
- Explain the difference between restoration ecology and reconciliation ecology. **See pp. 1237 and 1243–1244**

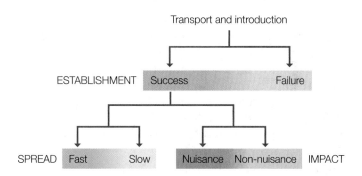

59.20 Three Steps to Invasion Invasion by a non-native species can be modeled as a process consisting of three steps, each with an independent probability of success or failure. To be considered invasive, the species must become established in a new environment, it must spread rapidly, and it must have an impact on its new environment.

How can adverse impacts of species introductions be anticipated before lasting damage occurs?

ANSWER

Conservation biologists use a range of tools to assess the risks of deliberate species introductions. In some cases, sufficient information is available from past introductions to allow biologists to identify the specific life history traits that make a species likely to become invasive. These traits can be used with a decision tree (see Figure 59.16), to guide decision-making. Mathematical models can also be constructed and used to assess the risks of both accidental and deliberate introductions. In one such model designed to identify potential fish invaders of the Great Lakes, investigators recognized three distinct stages of invasion: establishment, spread, and impact (**Figure 59.20**). They systematically compared the traits of fish species that had succeeded at each stage with those of species that had failed at that stage. The characteristics they used were minimum temperature threshold, breadth of diet, and relative growth. By identifying the characteristics of the successful versus failed invaders, investigators were then able to recreate the invasion process and predict with 94 percent accuracy the fish species that succeeded or failed at each stage.

■ CHAPTER**SUMMARY** 59

 What Is Conservation Biology?

- **Conservation biology** is an applied scientific discipline devoted to protecting and managing biodiversity.

- Conservation biologists recognize that an understanding of the evolutionary processes that generate biodiversity is essential to protecting it. They also understand that ecosystems are dynamic, and that humans are part of those ecosystems.

- Species extinctions have always occurred, but they are currently occurring at a rate that rivals those of the five great mass extinctions in Earth's history.

- There are many compelling reasons for protecting biodiversity, including the maintenance of the species and ecosystems that provide humans with goods and services.

 How Do Conservation Biologists Predict Changes in Biodiversity?

- Although our understanding of biodiversity is incomplete, biologists have identified many species that are threatened with extinction and have developed a classification system designed to aid in establishing policies for their protection. **See Figure 59.3**

- Biologists use the species–area relationship and the theory of island biogeography to estimate rates of extinction likely to be caused by habitat loss.

- To estimate a species' risk of extinction, statistical models take into account data on population sizes, demographic traits, genetic variation, physiology, and behavior.

- Rarity is not always a cause for concern, but species whose populations are shrinking rapidly are usually at risk of extinction.

 What Human Activities Threaten Species Persistence?

- Habitat loss is the most important cause of species endangerment worldwide. As habitats become increasingly **fragmented**, more species are lost from those habitats. Small habitat patches can support only small populations and are adversely influenced by **edge effects**. **Review Figures 59.4, 59.5, 59.6, ANIMATED TUTORIAL 59.1**

- Overexploitation has historically been the most important cause of species extinctions, and it is still a major threat to biodiversity today.

- Some species introduced to regions outside their original range become **invasive**, causing extinctions of native species that have not evolved defenses against these new antagonists and competitors.

- Climate change is likely to become an increasingly important cause of extinctions for those species that cannot shift their ranges as rapidly as the climate warms. **Review Figure 59.9**

 What Strategies Are Used to Protect Biodiversity?

- Establishing **protected areas** is crucial to conserving biodiversity. Protected areas are selected by taking into account species richness, **endemism**, and imminence of threats. **Review Figures 59.10, 59.11**

- **Restoration ecology** is an important conservation strategy because many degraded ecosystems will recover very slowly, if at all, without human assistance. **Review Figure 59.14**

- International trade in endangered species is controlled by regulations that most countries endorse.

- Conservation biologists work to determine which species are likely to become invasive and prevent their introduction to new areas. **Review Figure 59.16**

- Recognition of the economic value of biodiversity can help justify conservation efforts. **Review Figure 59.18**

- Even within landscapes where people live and extract resources, steps may be taken to protect biodiversity. This approach is known as **reconciliation ecology**.

- Captive breeding programs can maintain some endangered species for the short term while threats to their persistence in other natural habitats are reduced or removed.

See ACTIVITY 59.1 for a concept review of this chapter

Go to the Interactive Summary to review key figures, Animated Tutorials, and Activities
Life10e.com/is59

CHAPTER**REVIEW**

 **REMEMBERING**

1. Which of the following is not currently a major cause of species extinctions?
   a. Habitat destruction
   b. Meteorite impacts
   c. Overexploitation
   d. Introductions of non-native predators
   e. Introductions of non-native pathogens

2. Species extinctions matter to human society because
   a. many important medications contain or are based on a plant product.
   b. people derive aesthetic pleasure from interacting with other organisms.
   c. causing species extinctions raises serious ethical issues.
   d. biodiversity helps maintain valuable ecosystem services.
   e. All of the above

3. As a habitat patch gets smaller, it
   a. cannot support populations of species that require large areas.
   b. supports only small populations of many species.
   c. is influenced to an increasing degree by edge effects.
   d. is invaded by species from surrounding habitats.
   e. All of the above

4. Global warming is a concern because
   a. the rate of change in climate is projected to be faster than the rate at which many species can shift their ranges.
   b. it is already too hot in the tropics.
   c. climates have been so stable for thousands of years that many species lack the ability to tolerate variable temperatures.
   d. climate change will be especially harmful to rare species.
   e. None of the above

5. Scientists can determine the historical frequency of fires in an area by
   a. examining charcoal in sites of ancient villages.
   b. measuring carbon in soils.
   c. radioactively dating fallen tree trunks.
   d. examining fire scars in growth rings of trees.
   e. determining the age structure of forests.

6. Captive propagation is a useful conservation tool provided that
   a. there is space in zoos, aquariums, and botanical gardens to preserve the species indefinitely.
   b. genetic uniformity of captive populations can be maintained.
   c. the threats that endangered the species are being alleviated so that captive-reared individuals can later be released back into the wild.
   d. there are sufficient caretakers.
   e. None of the above; captive propagation should never be used because it directs attention away from the need to protect species in their natural habitats.

7. A plant species is most likely to become invasive when introduced to a new area if it
   a. grows tall.
   b. has become invasive in other places where it has been introduced.
   c. is closely related to species living in the area where it has been introduced.
   d. has specialized dispersers of its seeds.
   e. has a long life span.

UNDERSTANDING & APPLYING

8. Conservation biologists have debated extensively which is better: many small protected areas (which may contain more species) or a few large protected areas (which may be the only ones that can support populations of species that require large areas). What ecological processes should be evaluated in making judgments about the sizes and locations of protected areas? How can principles of island biogeography be applied in this context?

9. The desert bighorn sheep of the southwestern United States is endangered. Its major predator, the puma, is also of conservation concern in the region. Under what conditions, if any, would it be appropriate to suppress the population of one rare species to assist another rare species?

ANALYZING & EVALUATING

10. During World War I, doctors adopted a "triage" system for dealing with wounded soldiers. The wounded were divided into three categories: those almost certain to die no matter what was done to help them, those likely to recover even if not assisted, immediately, and those whose probability of survival would be greatly increased if they were given immediate medical attention. Limited medical resources were directed primarily at the third category. What are some implications of adopting a similar approach toward species preservation?

11. Utilitarian arguments dominate discussions about the importance of preserving biological richness. In your opinion, what role should ethical and moral arguments play?

Go to BioPortal at **yourBioPortal.com** for Animated Tutorials, Activities, LearningCurve Quizzes, Flashcards, and many other study and review resources.

Appendix A   The Tree of Life

Phylogeny is the organizing principle of modern biological taxonomy. A guiding principle of modern phylogeny is monophyly. A monophyletic group is considered to be one that contains an ancestral lineage and all of its descendants. Any such group can be extracted from a phylogenetic tree with a single cut.

The tree shown here provides a guide to the relationships among the major groups of extant (living) organisms in the tree of life as we have presented them throughout this book. The position of the branching "splits" indicates the relative branching order of the lineages of life, but the time scale is not meant to be uniform. In addition, the groups appearing at the branch tips do not necessarily carry equal phylogenetic "weight." For example, the ginkgo [75] is indeed at the apex of its lineage; this gymnosperm group consists of a single living species. In contrast, a phylogeny of the eudicots [83] could continue on from this point to fill many more trees the size of this one.

The glossary entries that follow are informal descriptions of some major features of the organisms described in Part Seven of this book. Each entry gives the group's common name, followed by the formal scientific name of the group (in parentheses). Numbers in square brackets reference the location of the respective groups on the tree.

It is sometimes convenient to use an informal name to refer to a collection of organisms that are not monophyletic but nonetheless all share (or all lack) some common attribute. We call these "convenience terms"; such groups are indicated in these entries by quotation marks, and we do not give them formal scientific names. Examples include "prokaryotes," "protists," and "algae." Note that these groups cannot be removed with a single cut; they represent a collection of distantly related groups that appear in different parts of the tree. We also use quotation marks here to designate two groups of fungi that are not believed to be monophyletic.

Go to BioPortal at **yourBioPortal.com** for an interactive version of this tree, with links to photos, distribution maps, species lists, and identification keys.

– A –

acorn worms (*Enteropneusta*)   Benthic marine hemichordates [119] with an acorn-shaped proboscis, a short collar (neck), and a long trunk.

"algae"   Convenience term encompassing various distantly related groups of aquatic, photosynthetic eukaryotes [4].

alveolates (*Alveolata*) [5]   Unicellular eukaryotes with a layer of flattened vesicles (alveoli) supporting the plasma membrane. Major groups include the dinoflagellates [51], apicomplexans [50], and ciliates [49].

amborella (*Amborella*) [78]   An understory shrub or small tree found only on the South Pacific island of New Caledonia. Thought to be the sister group of the remaining living angiosperms [15].

ambulacrarians (*Ambulacraria*) [29]   The echinoderms [118] and hemichordates [119].

amniotes (*Amniota*) [36]   Mammals, reptiles, and their extinct close relatives. Characterized by many adaptations to terrestrial life, including an amniotic egg (with a unique set of membranes—the amnion, chorion, and allantois), a water-repellant epidermis (with epidermal scales, hair, or feathers), and, in males, a penis that allows internal fertilization.

amoebozoans (*Amoebozoa*) [84]   A group of eukaryotes [4] that use lobe-shaped pseudopods for locomotion and to engulf food. Major amoebozoan groups include the loboseans, plasmodial slime molds, and cellular slime molds.

amphibians (*Amphibia*) [128]   Tetrapods [35] with glandular skin that lacks epidermal scales, feathers, or hair. Many amphibian species undergo a complete metamorphosis from an aquatic larval form to a terrestrial adult form, although direct development is also common. Major amphibian groups include frogs and toads (anurans), salamanders, and caecilians.

amphipods (*Amphipoda*)   Small crustaceans [116] that are abundant in many marine and freshwater habitats. They are important herbivores, scavengers, and micropredators, and are an important food source for many aquatic organisms.

angiosperms (*Anthophyta* or *Magnoliophyta*) [15]   The flowering plants. Major angiosperm groups include the monocots [82], eudicots [83], and magnoliids [81].

animals (*Animalia* or *Metazoa*) [19]   Multicellular heterotrophic eukaryotes. The majority of animals are bilaterians [22]. Other groups of animals include the sponges [20], ctenophores [95], placozoans [96], and cnidarians [97]. The closest living relatives of the animals are the choanoflagellates [91].

annelids (*Annelida*) [105]   Segmented worms, including earthworms, leeches, and polychaetes. One of the major groups of lophotrochozoans [24].

anthozoans (*Anthozoa*)   One of the major groups of cnidarians [97]. Includes the sea anemones, sea pens, and corals.

anurans (*Anura*)   Comprising the frogs and toads, this is the largest group of living amphibians [128]. They are tail-less, with a shortened vertebral column and elongate hind legs modified for jumping. Many species have an aquatic larval form known as a tadpole.

apicomplexans (*Apicomplexa*) [50]   Parasitic alveolates [5] characterized by the possession of an apical complex at some stage in the life cycle.

arachnids (*Arachnida*)   Chelicerates [114] with a body divided into two parts: a cephalothorax that bears six pairs of appendages (four pairs of which are usually used as legs) and an abdomen that bears the genital opening. Familiar arachnids include spiders, scorpions, mites and ticks, and harvestmen.

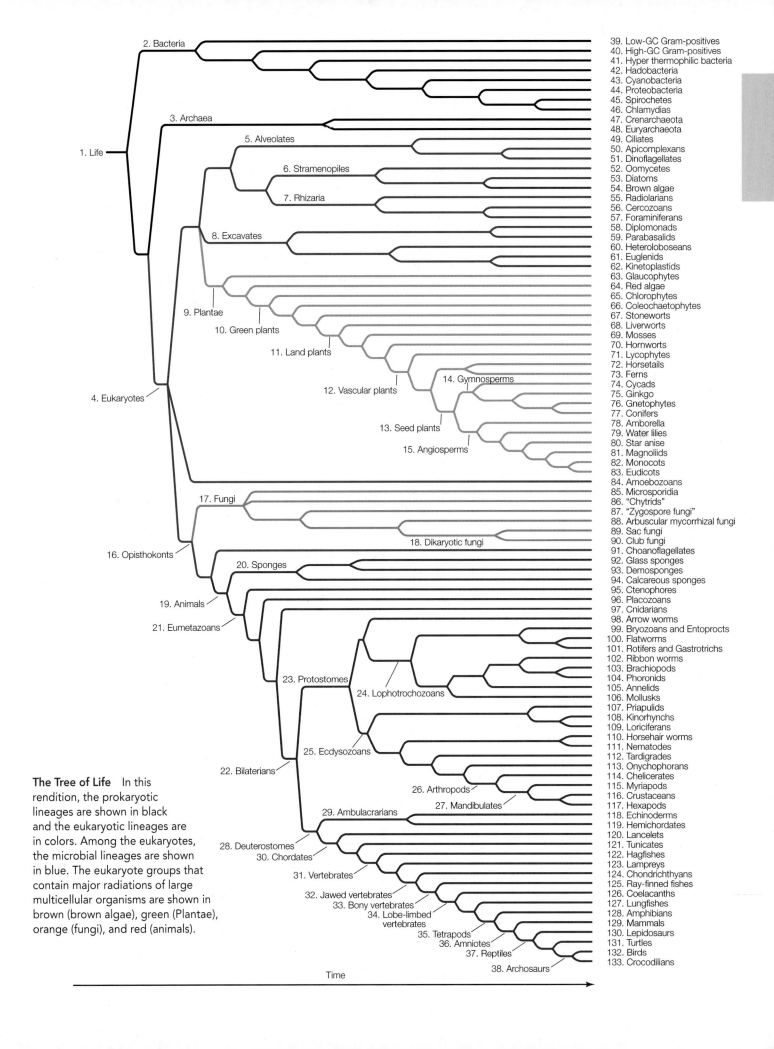

The Tree of Life In this rendition, the prokaryotic lineages are shown in black and the eukaryotic lineages are in colors. Among the eukaryotes, the microbial lineages are shown in blue. The eukaryote groups that contain major radiations of large multicellular organisms are shown in brown (brown algae), green (Plantae), orange (fungi), and red (animals).

1. Life
2. Bacteria
3. Archaea
4. Eukaryotes
5. Alveolates
6. Stramenopiles
7. Rhizaria
8. Excavates
9. Plantae
10. Green plants
11. Land plants
12. Vascular plants
13. Seed plants
14. Gymnosperms
15. Angiosperms
16. Opisthokonts
17. Fungi
18. Dikaryotic fungi
19. Animals
20. Sponges
21. Eumetazoans
22. Bilaterians
23. Protostomes
24. Lophotrochozoans
25. Ecdysozoans
26. Arthropods
27. Mandibulates
28. Deuterostomes
29. Ambulacrarians
30. Chordates
31. Vertebrates
32. Jawed vertebrates
33. Bony vertebrates
34. Lobe-limbed vertebrates
35. Tetrapods
36. Amniotes
37. Reptiles
38. Archosaurs

39. Low-GC Gram-positives
40. High-GC Gram-positives
41. Hyper thermophilic bacteria
42. Hadobacteria
43. Cyanobacteria
44. Proteobacteria
45. Spirochetes
46. Chlamydias
47. Crenarchaeota
48. Euryarchaeota
49. Ciliates
50. Apicomplexans
51. Dinoflagellates
52. Oomycetes
53. Diatoms
54. Brown algae
55. Radiolarians
56. Cercozoans
57. Foraminiferans
58. Diplomonads
59. Parabasalids
60. Heteroloboseans
61. Euglenids
62. Kinetoplastids
63. Glaucophytes
64. Red algae
65. Chlorophytes
66. Coleochaetophytes
67. Stoneworts
68. Liverworts
69. Mosses
70. Hornworts
71. Lycophytes
72. Horsetails
73. Ferns
74. Cycads
75. Ginkgo
76. Gnetophytes
77. Conifers
78. Amborella
79. Water lilies
80. Star anise
81. Magnoliids
82. Monocots
83. Eudicots
84. Amoebozoans
85. Microsporidia
86. "Chytrids"
87. "Zygospore fungi"
88. Arbuscular mycorrhizal fungi
89. Sac fungi
90. Club fungi
91. Choanoflagellates
92. Glass sponges
93. Demosponges
94. Calcareous sponges
95. Ctenophores
96. Placozoans
97. Cnidarians
98. Arrow worms
99. Bryozoans and Entoprocts
100. Flatworms
101. Rotifers and Gastrotrichs
102. Ribbon worms
103. Brachiopods
104. Phoronids
105. Annelids
106. Mollusks
107. Priapulids
108. Kinorhynchs
109. Loriciferans
110. Horsehair worms
111. Nematodes
112. Tardigrades
113. Onychophorans
114. Chelicerates
115. Myriapods
116. Crustaceans
117. Hexapods
118. Echinoderms
119. Hemichordates
120. Lancelets
121. Tunicates
122. Hagfishes
123. Lampreys
124. Chondrichthyans
125. Ray-finned fishes
126. Coelacanths
127. Lungfishes
128. Amphibians
129. Mammals
130. Lepidosaurs
131. Turtles
132. Birds
133. Crocodilians

Time

arbuscular mycorrhizal fungi (*Glomeromycota*) [88]   A group of fungi [17] that associate with plant roots in a close symbiotic relationship.

archaeans (*Archaea*) [3]   Unicellular organisms lacking a nucleus and lacking peptidoglycan in the cell wall. Once grouped with the bacteria, archaeans possess distinctive membrane lipids.

archosaurs (*Archosauria*) [38]   A group of reptiles [37] that includes dinosaurs and crocodilians [133]. Most dinosaur groups became extinct at the end of the Cretaceous; birds [132] are the only surviving dinosaurs.

arrow worms (*Chaetognatha*) [98]   Small planktonic or benthic predatory marine worms with fins and a pair of hooked, prey-grasping spines on each side of the head.

arthropods (*Arthropoda*)   The largest group of ecdysozoans [25]. Arthropods are characterized by a stiff exoskeleton, segmented bodies, and jointed appendages. Includes the chelicerates [114], myriapods [115], crustaceans [116], and hexapods (insects and their relatives) [117].

ascidians (*Ascidiacea*)   "Sea squirts"; the largest group of tunicates [121]. They are sessile (as adults), marine, saclike filter feeders.

– B –

bacteria (*Eubacteria*) [2]   Unicellular organisms lacking a nucleus, possessing distinctive ribosomes and initiator tRNA, and generally containing peptidoglycan in the cell wall. Different bacterial groups are distinguished primarily on nucleotide sequence data.

barnacles (*Cirripedia*)   Crustaceans [116] that undergo two metamorphoses—first from a feeding planktonic larva to a nonfeeding swimming larva, and then to a sessile adult that forms a "shell" composed of four to eight plates cemented to a hard substrate.

bilaterians (*Bilateria*) [22]   Those animal groups characterized by bilateral symmetry and three distinct tissue types (endoderm, ectoderm, and mesoderm). Includes the protostomes [23] and deuterostomes [28].

birds (*Aves*) [132]   Feathered, flying (or secondarily flightless) tetrapods [35].

bivalves (*Bivalvia*)   Major mollusk [106] group; clams and mussels. Bivalves typically have two similar hinged shells that are each asymmetrical across the midline.

bony vertebrates (*Osteichthyes*) [33] Vertebrates [31] in which the skeleton is usually ossified to form bone. Includes the ray-finned fishes [125], coelacanths [126], lungfishes [127], and tetrapods [35].

brachiopods (*Brachiopoda*) [103]   Lophotrochozoans [24] with two similar hinged shells that are each symmetrical across the midline. Superficially resemble bivalve mollusks, except for the shell symmetry.

brittle stars (*Ophiuroidea*)   Echinoderms [118] with five long, whip-like arms radiating from a distinct central disk that contains the reproductive and digestive organs.

brown algae (*Phaeophyta*) [54]   Multicellular, almost exclusively marine stramenopiles [6] generally containing the pigment fucoxanthin as well as chlorophylls *a* and *c* in their chloroplasts.

bryozoans (*Ectoprocta* or *Bryozoa*) [99]   A group of marine and freshwater lophotrochozoans [24] that live in colonies attached to substrates; also known as ectoprocts or moss animals. They are the sister group of entoprocts.

– C –

caecilians (*Gymnophiona*)   A group of burrowing or aquatic amphibians [128]. They are elongate, legless, with a short tail (or none at all), reduced eyes covered with skin or bone, and a pair of sensory tentacles on the head.

calcareous sponges (*Calcarea*) [94]   Filter-feeding marine sponges with spicules composed of calcium carbonate.

cellular slime molds (*Dictyostelida*)   Amoebozoans [84] in which individual amoebas aggregate under stress to form a multicellular pseudoplasmodium.

cephalochordates (*Cephalochordata*) [120] *See* lancelets.

cephalopods (*Cephalopoda*)   Active, predatory mollusks [106] in which the molluscan foot has been modified into muscular hydrostatic arms or tentacles. Includes octopuses, squids, and nautiluses.

cercozoans (*Cercozoa*) [56]   Unicellular eukaryotes [4] that feed by means of threadlike pseudopods. Group together with foraminiferans [57] and radiolarians [55] to comprise the rhizaria [7].

charophytes (*Charales*) [67]   *See* stoneworts.

chelicerates (*Chelicerata*) [114]   A major group of arthropods [26] with pointed appendages (chelicerae) used to grasp food (as opposed to the chewing mandibles of most other arthropods). Includes the arachnids, horseshoe crabs, pycnogonids, and extinct sea scorpions.

chimaeras (*Holocephali*)   A group of bottom-dwelling, marine, scaleless chondrichthyan fishes [124] with large, permanent, grinding tooth plates (rather than the replaceable teeth found in other chondrichthyans).

chitons (*Polyplacophora*)   Flattened, slow-moving mollusks [106] with a dorsal protective calcareous covering made up of eight articulating plates.

chlamydias (*Chlamydiae*) [46]   A group of very small Gram-negative bacteria; they live as intracellular parasites of other organisms.

chlorophytes (*Chlorophyta*) [65]   The most abundant and diverse group of green algae, including freshwater, marine, and terrestrial forms; some are unicellular, others colonial, and still others multicellular. Chlorophytes use chlorophylls *a* and *c* in their photosynthesis.

choanoflagellates (*Choanozoa*) [91] Unicellular eukaryotes [4] with a single flagellum surrounded by a collar. Most are sessile, some are colonial. The closest living relatives of the animals [19].

chondrichthyans (*Chondrichthyes*) [124] One of the two main groups of jawed vertebrates [32]; includes sharks, rays, and chimaeras. They have cartilaginous skeletons and paired fins.

chordates (*Chordata*) [30]   One of the two major groups of deuterostomes [28], characterized by the presence (at some point in development) of a notochord, a hollow dorsal nerve cord, and a post-anal tail. Includes the lancelets [120], tunicates [121], and vertebrates [31].

"chytrids" [90]   Convenience term used for a paraphyletic group of mostly aquatic, microscopic fungi [17] with flagellated gametes. Some exhibit alternation of generations.

ciliates (*Ciliophora*) [49]   Alveolates [5] with numerous cilia and two types of nuclei (micronuclei and macronuclei).

clitellates (*Clitellata*)   Annelids [105] with gonads contained in a swelling (called a clitellum) toward the head of the animal. Includes earthworms (oligochaetes) and leeches.

club fungi (*Basidiomycota*) [90]   Fungi [17] that, if multicellular, bear the products of meiosis on club-shaped basidia and possess a long-lasting dikaryotic stage. Some are unicellular.

club mosses (*Lycopodiophyta*) [71]   Vascular plants [12] characterized by microphylls. *See* lycophytes.

cnidarians (*Cnidaria*) [97]   Aquatic, mostly marine eumetazoans [21] with specialized stinging organelles (nematocysts) used for prey capture and defense, and a blind gastrovascular cavity. The sister group of the bilaterians [22].

coelacanths (*Actinista*) [126]   A group of marine lobe-limbed vertebrates [34] that was diverse from the Middle Devonian to the Cretaceous, but is now known from just two living species. The pectoral and anal fins are on fleshy stalks supported by skeletal elements, so they are also called lobe-finned fishes.

coleochaetophytes (*Coleochaetales*) [66] Multicellular green algae characterized by flattened growth form composed of thin-walled cells. Thought to be the sister-group to the stoneworts [67] plus land plants [11].

conifers (*Pinophyta* or *Coniferophyta*) [77] Cone-bearing, woody seed plants [13].

copepods (*Copepoda*)   Small, abundant crustaceans [116] found in marine, freshwater, or wet terrestrial habitats. They have a single eye, long antennae, and a body shaped like a teardrop.

craniates (*Craniata*)   Some biologist exclude the hagfishes [122] from the vertebrates [31], and use the term craniates to refer to the two groups combined.

crenarchaeotes (*Crenarchaeota*) [47]   A major and diverse group of archaeans [3], defined on the basis of rRNA base sequences. Many are extremophiles (inhabit extreme environments), but the group may also be the most abundant archaeans in the marine environment.

crinoids (*Crinoidea*)   Echinoderms [118] with a mouth surrounded by feeding arms, and a U-shaped gut with the mouth next to the anus. They attach to the substratum by a stalk or are free-swimming. Crinoids were abundant in the middle and late Paleozoic, but only a few hundred species have survived to the present. Includes the sea lilies and feather stars.

crocodilians (*Crocodylia*) [133]   A group of large, predatory, aquatic archosaurs [38]. The closest living relatives of birds [132]. Includes alligators, caimans, crocodiles, and gharials.

crustaceans (*Crustacea*) [116]   Major group of marine, freshwater, and terrestrial arthropods [26] with a head, thorax, and abdomen (although the head and thorax may be fused), covered with a thick exoskeleton, and with two-part appendages. Crustaceans undergo metamorphosis from a nauplius larva. Includes decapods, isopods, krill, barnacles, amphipods, copepods, and ostracods.

ctenophores (*Ctenophora*) [95] Radially symmetrical, diploblastic marine animals [19], with a complete gut and eight rows of fused plates of cilia (called ctenes).

cyanobacteria (*Cyanobacteria*) [43] A group of unicellular, colonial, or filamentous bacteria that conduct photosynthesis using chlorophyll *a*.

cycads (*Cycadophyta*) [74] Palmlike gymnosperms with large, compound leaves.

cyclostomes (*Cyclostomata*) This term refers to the possibly monophyletic group of lampreys [123] and hagfishes [122]. Molecular data support this group, but morphological data suggest that lampreys are more closely related to jawed vertebrates [32] than to hagfishes.

– D –

decapods (*Decapoda*) A group of marine, freshwater, and semiterrestrial crustaceans [116] in which five of the eight pairs of thoracic appendages function as legs (the other three pairs, called maxillipeds, function as mouthparts). Includes crabs, lobsters, crayfishes, and shrimps.

demosponges (*Demospongiae*) [93] The largest of the three groups of sponges [20], accounting for 90 percent of all sponge species. Demosponges have spicules made of silica, spongin fiber (a protein), or both.

deuterostomes (*Deuterostomia*) [28] One of the two major groups of bilaterians [22], in which the mouth forms at the opposite end of the embryo from the blastopore in early development (contrast with protostomes). Includes the ambulacrarians [29] and chordates [30].

diatoms (*Bacillariophyta*) [53] Unicellular, photosynthetic stramenopiles [6] with glassy cell walls in two parts.

dikaryotic fungi (*Dikarya*) [18] A group of fungi [17] in which two genetically different haploid nuclei coexist and divide within the same hypha; includes club fungi [90] and sac fungi [89].

dinoflagellates (*Dinoflagellata*) [51] A group of alveolates [5] usually possessing two flagella, one in an equatorial groove and the other in a longitudinal groove; many are photosynthetic.

diplomonads (*Diplomonadida*) [58] A group of eukaryotes [4] lacking mitochondria; most have two nuclei, each with four associated flagella.

– E –

ecdysozoans (*Ecdysozoa*) [25] One of the two major groups of protostomes [23], characterized by periodic molting of their exoskeletons. Nematodes [111] and arthropods [26] are the largest ecdysozoan groups.

echinoderms (*Echinodermata*) [118] A major group of marine deuterostomes [28] with five-fold radial symmetry (at some stage of life) and an endoskeleton made of calcified plates and spines. Includes sea stars, crinoids, sea urchins, sea cucumbers, and brittle stars.

elasmobranchs (*Elasmobranchii*) The largest group of chondrichthyan fishes [124]. Includes sharks, skates, and rays. In contrast to the other group of living chondrichthyans (the chimaeras), they have replaceable teeth.

embryophytes *See* land plants [11].

entoprocts (*Entoprocta*) [99] A group of marine and freshwater lophotrochozoans [24] that live as single individuals or in colonies attached to substrates. They are the sister group of bryozoans, from which they differ in having both their mouth and anus inside the lophophore (the anus is outside the lophophore in bryozoans).

eudicots (*Eudicotyledones*)[83] A group of angiosperms [15] with pollen grains possessing three openings. Typically with two cotyledons, net-veined leaves, taproots, and floral organs typically in multiples of four or five.

euglenids (*Euglenida*) [61] Flagellate excavates characterized by a pellicle composed of spiraling strips of protein under the plasma membrane; the mitochondria have disk-shaped cristae. Some are photosynthetic.

eukaryotes (*Eukarya*) [4] Organisms made up of one or more complex cells in which the genetic material is contained in nuclei. Contrast with archaeans [3] and bacteria [2].

eumetazoans (*Eumetazoa*) [21] Those animals [19] characterized by body symmetry, a gut, a nervous system, specialized types of cell junctions, and well-organized tissues in distinct cell layers (although there have been secondary losses of some or most of these characteristics in a few eumetazoan lineages).

euphyllophytes (*Euphyllophyta*) The group of vascular plants [12] that is sister to the lycophytes [71] and which includes all plants with megaphylls.

euryarchaeotes (*Euryachaeota*) [48] A major group of archaeans [3], diagnosed on the basis of rRNA sequences. Includes many methanogens, extreme halophiles, and thermophiles.

eutherians (*Eutheria*) A group of viviparous mammals [129], eutherians are well developed at birth (contrast to prototherians and marsupials, the other two groups of mammals). Most familiar mammals outside the Australian and South American regions are eutherians (see Table 33.1).

excavates (*Excavata*) [8] Diverse group of unicellular, flagellate eukaryotes, many of which possess a feeding groove; some lack mitochondria.

– F –

ferns Vascular plants [12] usually possessing large, frondlike leaves that unfold from a "fiddlehead." Not a monophyletic group, although most fern species are encompassed in a monophyletic clade, the leptosporangiate ferns [73].

flatworms (*Platyhelminthes*) [100] A group of dorsoventrally flattened and generally elongate soft-bodied lophotrochozoans [24]. May be free-living or parasitic, found in marine, freshwater, or damp terrestrial environments. Major flatworm groups include the tapeworms, flukes, monogeneans, and turbellarians.

flowering plants *See* angiosperms [15].

flukes (*Trematoda*) A group of wormlike parasitic flatworms [100] with complex life cycles that involve several different host species. May be paraphyletic with respect to tapeworms.

foraminiferans (*Foraminifera*) [57] Amoeboid organisms with fine, branched pseudopods that form a food-trapping net. Most produce external shells of calcium carbonate.

fungi (*Fungi*) [17] Eukaryotic heterotrophs with absorptive nutrition based on extracellular digestion; cell walls contain chitin. Major fungal groups include the microsporidia [85], "chytrids" [86], "zygospore fungi" [87], arbuscular mycorrhizal fungi [88], sac fungi [89], and club fungi [90].

– G –

gastropods (*Gastropoda*) The largest group of mollusks [106]. Gastropods possess a well-defined head with two or four sensory tentacles (often terminating in eyes) and a ventral foot. Most species have a single coiled or spiraled shell. Common in marine, freshwater, and terrestrial environments.

gastrotrichs (*Gastrotricha*) [101] Tiny (0.06–3.0 mm), elongate acoelomate lophotrochozoans [24] that are covered in cilia. They live in marine, freshwater, and wet terrestrial habitats. They are simultaneous hermaphrodites.

ginkgo (*Ginkgophyta*) [75] A gymnosperm [14] group with only one living species. The ginkgo seed is surrounded by a fleshy tissue not derived from an ovary wall and hence not a fruit.

glass sponges (*Hexactinellida*) [92] Sponges [20] with a skeleton composed of four- and/or six-pointed spicules made of silica.

glaucophytes (*Glaucophyta*) [63] Unicellular freshwater algae with chloroplasts containing traces of peptidoglycan, the characteristic cell wall material of bacteria.

gnathostomes (*Gnathostomata*) *See* jawed vertebrates [32].

gnetophytes (*Gnetophyta*) [76] A gymnosperm [14] group with three very different lineages; all have wood with vessels, unlike other gymnosperms.

green plants (*Viridiplantae*) [10] Organisms with chlorophylls *a* and *b*, cellulose-containing cell walls, starch as a carbohydrate storage product, and chloroplasts surrounded by two membranes.

gymnosperms (*Gymnospermae*) [14] Seed plants [13] with seeds "naked" (i.e., not enclosed in carpels). Probably monophyletic, but status still in doubt. Includes the conifers [77], gnetophytes [76], ginkgo [75], and cycads [74].

– H –

hadobacteria (*Hadobacteria*)[42] A group of extremophilic bacteria [2] that includes the genera *Deinococus* and *Thermus*.

hagfishes (*Myxini*) [122] Elongate, slimy-skinned vertebrates [31] with three small accessory hearts, a partial cranium, and no stomach or paired fins. *See also* craniata; cyclostomes.

hemichordates (*Hemichordata*) [119] One of the two primary groups of ambulacrarians [29];

marine wormlike organisms with a three-part body plan.

heteroloboseans (*Heterolobosea*) [60] Colorless excavates [8] that can transform among amoeboid, flagellate, and encysted stages.

hexapods (*Hexapoda*) [117] Major group of arthropods [26] characterized by a reduction (from the ancestral arthropod condition) to six walking appendages, and the consolidation of three body segments to form a thorax. Includes insects and their relatives (see Table 23.2).

high-GC Gram-positives (*Actinobacteria*) [40] Gram-positive bacteria with a relatively high (G+C)/(A+T) ratio of their DNA, with a filamentous growth habit.

hornworts (*Anthocerophyta*) [70] Nonvascular plants with sporophytes that grow from the base. Cells contain a single large, platelike chloroplast.

horsehair worms (*Nematomorpha*) [110] A group of very thin, elongate, wormlike freshwater ecdysozoans [25]. Largely nonfeeding as adults, they are parasites of insects and crayfish as larvae.

horseshoe crabs (*Xiphosura*) Marine chelicerates [114] with a large outer shell in three parts: a carapace, an abdomen, and a tail-like telson. There are only five living species, but many additional species are known from fossils.

horsetails (*Sphenophyta* or *Equisetophyta*) [72] Vascular plants [12] with reduced megaphylls in whorls.

hydrozoans (*Hydrozoa*) A group of cnidarians [97]. Most species go through both polyp and mesuda stages, although one stage or the other is eliminated in some species.

hyperthermophilic bacteria [41] A group of thermophilic bacteria [2] that live in volcanic vents, hot springs, and in underground oil reservoirs; includes the genera *Aquifex* and *Thermotoga*.

– I –

insects (*Insecta*) The largest group within the hexapods [117]. Insects are characterized by exposed mouthparts and one pair of antennae containing a sensory receptor called a Johnston's organ. Most have two pairs of wings as adults. There are more described species of insects than all other groups of life [1] combined, and many species remain to be discovered. The major insect groups are described in Table 23.2.

"invertebrates" Convenience term encompassing any animal [19] that is not a vertebrate [31].

isopods (*Isopoda*) Crustaceans [116] characterized by a compact head, unstalked compound eyes, and mouthparts consisting of four pairs of appendages. Isopods are abundant and widespread in salt, fresh, and brackish water, although some species (the sow bugs) are terrestrial.

– J –

jawed vertebrates (*Gnathostomata*) [32] A major group of vertebrates [31] with jawed mouths. Includes chondrichthyans [124], ray-finned fishes [125], and lobe-limbed vertebrates [34].

– K –

kinetoplastids (*Kinetoplastida*) [62] Unicellular, flagellate organisms characterized by the presence in their single mitochondrion of a kinetoplast (a structure containing multiple, circular DNA molecules).

kinorhynchs (*Kinorhyncha*) [108] Small (< 1 mm) marine ecdysozoans [25] with bodies in 13 segments and a retractable proboscis.

korarchaeotes (*Korarchaeota*) A group of archaeans [3] known only by evidence from nucleic acids derived from hot springs. Its phylogenetic relationships within the Archaea are unknown.

krill (*Euphausiacea*) A group of shrimplike marine crustaceans [116] that are important components of the zooplankton.

– L –

lampreys (*Petromyzontiformes*) [123] Elongate, eel-like vertebrates [31] that often have rasping and sucking disks for mouths.

lancelets (*Cephalochordata*) [120] A group of weakly swimming, eel-like benthic marine chordates [30].

land plants (*Embryophyta*) [11] Plants with embryos that develop within protective structures; also called embryophytes. Sporophytes and gametophytes are multicellular. Land plants possess a cuticle. Major groups are the liverworts [68], mosses [69], hornworts [70], and vascular plants [12].

larvaceans (*Larvacea*) Solitary, planktonic tunicates [121] that retain both notochords and nerve cords throughout their lives.

lepidosaurs (*Lepidosauria*) [130] Reptiles [37] with overlapping scales. Includes tuataras and squamates (lizards, snakes, and amphisbaenians).

leptosporangiate ferns (*Pteridopsida* or *Polypodiopsida*) [73] Vascular plants [12] usually possessing large, frondlike leaves that unfold from a "fiddlehead," and possessing thin-walled sporangia.

life (*Life*) [1] The monophyletic group that includes all known living organisms. Characterized by a nucleic-acid based genetic system (DNA or RNA), metabolism, and cellular structure. Some parasitic forms, such as viruses, have secondarily lost some of these features and rely on the cellular environment of their host.

liverworts (*Hepatophyta*) [68] Nonvascular plants lacking stomata; stalk of sporophyte elongates along its entire length.

lobe-limbed vertebrates (*Sarcopterygii*) [34] One of the two major groups of bony vertebrates [33], characterized by jointed appendages (paired fins or limbs).

loboseans (*Lobosea*) A group of unicellular amoebozoans [84]; includes the most familiar amoebas (e.g., *Amoeba proteus*).

"lophophorates" Convenience term used to describe several groups of lophotrochozoans [24] that have a feeding structure called a lophophore (a circular or U-shaped ridge around the mouth that bears one or two rows of ciliated, hollow tentacles). Not a monophyletic group.

lophotrochozoans (*Lophotrochozoa*) [24] One of the two main groups of protostomes

[23]. This group is morphologically diverse, and is supported primarily on information from gene sequences. Includes bryozoans and entoprocts [99], flatworms [100], rotifers and gastrotrichs [101], ribbon worms [102], brachiopods [103], phoronids [104], annelids [105], and mollusks [106].

loriciferans (*Loricifera*) [109] Small (< 1 mm) ecdysozoans [25] with bodies in four parts, covered with six plates.

low-GC Gram-positives (*Firmicutes*) [39] A diverse group of bacteria [2] with a relatively low (G+C)/(A+T) ratio of their DNA, often but not always Gram-positive, some producing endospores.

lungfishes (*Dipnoi*) [127] A group of aquatic lobe-limbed vertebrates [34] that are the closest living relatives of the tetrapods [35]. They have a modified swim bladder used to absorb oxygen from air, so some species can survive the temporary drying of their habitat.

lycophytes (*Lycopodiophyta*) [71] Vascular plants [12] characterized by microphylls; includes club mosses, spike mosses, and quillworts.

– M –

magnoliids (*Magnoliidae*) [81] A major group of angiosperms [15] possessing two cotyledons and pollen grains with a single opening. The group is defined primarily by nucleotide sequence data; it is more closely related to the eudicots and monocots than to three other small angiosperm groups.

mammals (*Mammalia*) [129] A group of tetrapods [35] with hair covering all or part of their skin; -females produce milk to feed their developing young. Includes the prototherians, marsupials, and -eutherians.

mandibulates (*Mandibulata*) [27] Arthropods [26] that include mandibles as mouth parts. Includes myriapods [115], crustaceans [116], and hexapods [117].

marsupials (*Marsupialia*) Mammals [129] in which the female typically has a marsupium (a pouch for rearing young, which are born at an extremely early stage in development). Includes such familiar mammals as opossums, koalas, and kangaroos.

metazoans (*Metazoa*) *See* animals [19].

microbial eukaryotes *See* "protists."

microsporidia (*Microsporidia*) [85] A group of parasitic unicellular fungi [17] that lack mitochondria and have walls that contain chitin.

mollusks (*Mollusca*) [106] One of the major groups of lophotrochozoans [24], mollusks have bodies composed of a foot, a mantle (which often secretes a hard, calcareous shell), and a visceral mass. Includes monoplacophorans, chitons, bivalves, gastropods, and cephalopods.

monilophytes (*Monilophyta*) A group of vascular plants [12], sister to the seed plants [13], characterized by overtopping and possession of megaphylls; includes the horsetails [72] and ferns [73].

monocots (*Monocotyledones*) [82] Angiosperms [15] characterized by possession of a single cotyledon, usually parallel leaf veins, a fibrous root system, pollen grains with a single opening, and floral organs usually in multiples of three.

monogeneans (*Monogenea*)   A group of ectoparasitic flatworms [100].

monoplacophorans (*Monoplacophora*) Mollusks [106] with segmented body parts and a single, thin, flat, rounded, bilateral shell.

mosses (*Bryophyta*) [69]   Nonvascular plants with true stomata and erect, "leafy" gametophytes; sporophytes elongate by apical cell division.

moss animals   *See* bryozoans [99].

myriapods (*Myriapoda*) [115]   Arthropods [26] characterized by an elongate, segmented trunk with many legs. Includes centipedes and millipedes.

– N –

nanoarchaeotes (*Nanoarchaeota*)   A group of extremely small, thermophilic archaeans [3] with a much-reduced genome. The only described example can survive only when attached to a host organism.

nematodes (*Nematoda*) [111]   A very large group of elongate, unsegmented ecdysozoans [25] with thick, multilayer cuticles. They are among the most abundant and diverse animals, although most species have not yet been described. Include free-living predators and scavengers, as well as parasites of most species of land plants [11] and animals [19].

neognaths (*Neognathae*)   The main group of birds [132], including all living species except the ratites (ostrich, emu, rheas, kiwis, cassowaries) and tinamous. *See* palaeognaths.

– O –

oligochaetes (*Oligochaeta*)   Annelid [105] group whose members lack parapodia, eyes, and anterior tentacles, and have few setae. Earthworms are the most familiar oligochaetes.

onychophorans (*Onychophora*) [113] Elongate, segmented ecdysozoans [25] with many pairs of soft, unjointed, claw-bearing legs. Also known as velvet worms.

oomycetes (*Oomycota*) [52] Water molds and relatives; absorptive heterotrophs with nutrient-absorbing, filamentous hyphae.

opisthokonts (*Opisthokonta*) [16]   A group of eukaryotes [4] in which the flagellum on motile cells, if present, is posterior. The opisthokonts include the fungi [17], animals [19], and choanoflagellates [91].

ostracods (*Ostracoda*)   Marine and freshwater crustaceans [116] that are laterally compressed and protected by two clamlike calcareous or chitinous shells.

– P –

palaeognaths (*Palaeognathae*)   A group of secondarily flightless or weakly flying birds [132]. Includes the flightless ratites (ostrich, emu, rheas, kiwis, cassowaries) and the weakly flying tinamous.

parabasalids (*Parabasalia*) [59]   A group of unicellular eukaryotes [4] that lack mitochondria; they possess flagella in clusters near the anterior of the cell.

phoronids (*Phoronida*) [104]   A small group of sessile, wormlike marine lophotrochozoans

[24] that secrete chitinous tubes and feed using a lophophore.

placoderms (*Placodermi*)   An extinct group of jawed vertebrates [32] that lacked teeth. Placoderms were the dominant predators in Devonian oceans.

placozoans (*Placozoa*) [96]   A poorly known group of structurally simple, asymmetrical, flattened, transparent animals found in coastal marine tropical and subtropical seas. Most evidence suggests that placozoans are secondarily simplified eumetazoans [21].

Plantae [9]   The most broadly defined plant group. In most parts of this book, we use the word "plant" as synonymous with "land plant" [11], a more restrictive definition.

plasmodial slime molds (*Myxogastrida*)   Amoebozoans [84] that in their feeding stage consist of a coenocyte called a plasmodium.

pogonophorans (*Pogonophora*) Deep-sea annelids [105] that lack a mouth or digestive tract; they feed by taking up dissolved organic matter, facilitated by endosymbiotic bacteria in a specialized organ (the trophosome).

polychaetes (*Polychaeta*)   A group of mostly marine annelids [105] with one or more pairs of eyes and one or more pairs of feeding tentacles; parapodia and setae extend from most body segments. May be paraphyletic with respect to the clitellates.

priapulids (*Priapulida*) [107]   A small group of cylindrical, unsegmented, wormlike marine ecdysozoans [25] that takes its name from its phallic appearance.

"prokaryotes"   Not a monophyletic group; as commonly used, includes the bacteria [2] and archaeans [3]. A term of convenience encompassing all cellular organisms that are not eukaryotes.

proteobacteria (*Proteobacteria*) [44]   A large and extremely diverse group of Gram-negative bacteria that includes many pathogens, nitrogen fixers, and photosynthesizers. Includes the alpha, beta, gamma, delta, and epsilon proteobacteria.

"protists"   This term of convenience is used to encompass a large number of distinct and distantly related groups of eukaryotes, many but far from all of which are microbial and unicellular. Essentially a "catch-all" term for any eukaryote group not contained within the land plants [11], fungi [17], or animals [19].

protostomes (*Protostomia*) [23]   One of the two major groups of bilaterians [22]. In protostomes, the mouth typically forms from the blastopore (if present) in early development (contrast with deuterostomes). The major protostome groups are the lophotrochozoans [24] and ecdysozoans [25].

prototherians (*Prototheria*)   A mostly extinct group of mammals [129], common during the Cretaceous and early Cenozoic. The five living species—four echidnas and the duck-billed platypus—are the only extant egg-laying mammals.

pterobranchs (*Pterobranchia*)   A small group of sedentary marine hemichordates [119] that live in tubes secreted by the proboscis. They

have one to nine pairs of arms, each bearing long tentacles that capture prey and function in gas exchange.

pycnogonids (*Pycnogonida*)   Treated in this book as a group of chelicerates [114], but sometimes considered an independent group of arthropods [26]. Pycnogonids have reduced bodies and very long, slender legs. Also called sea spiders.

– R –

radiolarians (*Radiolaria*) [55]   Amoeboid organisms with needlelike pseudopods supported by microtubules. Most have glassy internal skeletons.

ray-finned fishes (*Actinopterygii*) [125]   A highly diverse group of freshwater and marine bony vertebrates [33]. They have reduced swim bladders that often function as hydrostatic organs and fins supported by soft rays (lepidotrichia). Includes most familiar fishes.

red algae (*Rhodophyta*) [64]   Mostly multicellular, marine and freshwater algae characterized by the presence of phycoerythrin in their chloroplasts.

reptiles (*Reptilia*) [37]   One of the two major groups of extant amniotes [36], supported on the basis of similar skull structure and gene sequences. The term "reptiles" traditionally excluded the birds [132], but the resulting group is then clearly paraphyletic. As used in this book, the reptiles include turtles [131], lepidosaurs [130], birds [132], and crocodilians [133].

rhizaria (*Rhizaria*) [7]   Mostly amoeboid unicellular eukaryotes with pseudopods, many with external or internal shells. Includes the foraminiferans [57], cercozoans [56], and radiolarians [55].

rhyniophytes (*Rhyniophyta*)   A group of early vascular plants [12] that appeared in the Silurian and became extinct in the Devonian. Possessed dichotomously branching stems with terminal sporangia but no true leaves or roots.

ribbon worms (*Nemertea*) [102]   A group of unsegmented lophotrochozoans [24] with an eversible proboscis used to capture prey. Mostly marine, but some species live in fresh water or on land.

rotifers (*Rotifera*) [101]   Tiny (< 0.5 mm) lophotrochozoans [24] with a pseudocoelomic body cavity that functions as a hydrostatic organ, and a ciliated feeding organ called the corona that surrounds the head. Rotifers live in freshwater and wet terrestrial habitats.

roundworms (*Nematoda*) [111]   *See* nematodes.

– S –

sac fungi (*Ascomycota*) [89]   Fungi that bear the products of meiosis within sacs (asci) if the organism is multicellular. Some are unicellular.

salamanders (*Caudata*)   A group of amphibians [128] with distinct tails in both larvae and adults and limbs set at right angles to the body.

salps   *See* thaliaceans.

sarcopterygians (*Sarcopterygii*) [34]   *See* lobe-limbed vertebrates.

scyphozoans (*Scyphozoa*)   Marine cnidarians [97] in which the medusa stage dominates the life cycle. Commonly known as jellyfish.

sea cucumbers (*Holothuroidea*) Echinoderms [118] with an elongate, cucumber-shaped body and leathery skin. They are scavengers on the ocean floor.

sea spiders *See* pycnogonids.

sea squirts *See* ascidians.

sea stars (*Asteroidea*) Echinoderms [118] with five (or more) fleshy "arms" radiating from an indistinct central disk. Also called starfishes.

sea urchins (*Echinoidea*) Echinoderms [118] with a test (shell) that is covered in spines. Most are globular in shape, although some groups (such as the sand dollars) are flattened.

"seed ferns" A paraphyletic group of loosely related, extinct seed plants that flourished in the Devonian and Carboniferous. Characterized by large, frondlike leaves that bore seeds.

seed plants (*Spermatophyta*) [13] Heterosporous vascular plants [12] that produce seeds; most produce wood; branching is axillary (not dichotomous). The major seed plant groups are gymnosperms [14] and angiosperms [15].

sow bugs *See* isopods.

spirochetes (*Spirochaetes*) [45] Motile, Gram-negative bacteria with a helically coiled structure and characterized by axial filaments.

sponges (*Porifera*) [20] A group of relatively asymmetric, filter-feeding animals that lack a gut or nervous system and generally lack differentiated tissues. Includes glass sponges [92], demosponges [93], and calcareous sponges [94].

springtails (*Collembola*) Wingless hexapods [117] with springing structures on the third and fourth segments of their bodies. Springtails are extremely abundant in some environments (especially in soil, leaf litter, and vegetation).

squamates (*Squamata*) The major group of lepidosaurs [130], characterized by the possession of movable quadrate bones (which allow the upper jaw to move independently of the rest of the skull) and hemipenes (a paired set of eversible penises, or penes) in males. Includes the lizards (a paraphyletic group), snakes, and amphisbaenians.

star anise (*Austrobaileyales*) [80] A group of woody angiosperms [15] thought to be the sister-group of the clade of flowering plants that includes eudicots [83], monocots [82], and magnoliids [81].

starfish (*Asteroidea*) *See* sea stars.

stoneworts (*Charales*) [67] Multicellular green algae with branching, apical growth and plasmodesmata between adjacent cells. The closest living relatives of the land plants [11], they retain the egg in the parent organism.

stramenopiles (*Heterokonta* or *Stramenopila*) [6] Organisms having, at some stage in their life cycle, two unequal flagella, the longer possessing rows of tubular hairs. Chloroplasts, when present, surrounded by four membranes. Major stramenopile groups include the brown algae [54], diatoms [53], and oomycetes [52].

– T –

tapeworms (*Cestoda*) Parasitic flatworms [100] that live in the digestive tracts of vertebrates as adults, and usually in various other species of animals as juveniles.

tardigrades (*Tardigrada*) [112] Small (< 0.5 mm) ecdysozoans [25] with fleshy, unjointed legs and no circulatory or gas exchange organs. They live in marine sands, in temporary freshwater pools, and on the water films of plants. Also called water bears.

tetrapods (*Tetrapoda*) [35] The major group of lobe-limbed vertebrates [34]; includes the amphibians [128] and the amniotes [36]. Named for the presence of four jointed limbs (although limbs have been secondarily reduced or lost completely in several tetrapod groups).

thaliaceans (*Thaliacea*) A group of solitary or colonial planktonic marine tunicates [121]. Also called salps.

therians (*Theria*) Mammals [129] characterized by viviparity (live birth). Includes eutherians and marsupials.

theropods (*Theropoda*) Archosaurs [38] with bipedal stance, hollow bones, a furcula ("wishbone"), elongated metatarsals with three-fingered feet, and a pelvis that points backwards. Includes many well-known extinct dinosaurs (such as *Tyrannosaurus rex*), as well as the living birds [132].

tracheophytes *See* vascular plants [12].

trilobites (*Trilobita*) An extinct group of arthropods [26] related to the chelicerates [114]. Trilobites flourished from the Cambrian through the Permian.

tuataras (*Rhyncocephalia*) A group of lepidosaurs [130] known mostly from fossils; there are

only two living tuatara species. The quadrate bone of the upper jaw is fixed firmly to the skull. Sister group of the squamates.

tunicates (*Tunicata*) [121] A group of chordates [30] that are mostly saclike filter feeders as adults, with motile larval stages that resemble tadpoles.

turbellarians (*Turbellaria*) A group of free-living, generally carnivorous flatworms [100]. Their monophyly is questionable.

turtles (*Testudines*) [131] A group of reptiles [37] with a bony carapace (upper shell) and plastron (lower shell) that encase the body in a fashion unique among the vertebrates.

– U –

urochordates (*Tunicata*) [121] *See* tunicates.

– V –

vascular plants (*Tracheophyta*) [12] Plants with xylem and phloem. Major groups include the lycophytes [71] and euphyllophytes.

vertebrates (*Vertebrata*) [31] The largest group of chordates [30], characterized by a rigid endoskeleton supported by the vertebral column and an anterior skull encasing a brain. Includes hagfishes [122], lampreys [123], and the jawed vertebrates [32], although some biologists exclude the hagfishes from this group. *See also* craniates.

– W –

water bears *See* tardigrades.

water lilies (*Nymphaeaceae*) [79] A group of aquatic, freshwater angiosperms [15] that are rooted in soil in shallow water, with round floating leaves and flowers that extend above the water's surface. They are the sister-group to most of the remaining flowering plants, with the exception of the genus *Amborella* [78].

– Y –

"yeasts" Convenience term for several distantly related groups of unicellular fungi [17].

– Z –

"zygospore fungi" (*Zygomycota*, if monophyletic) [87] A convenience term for a probably paraphyletic group of fungi [17] in which hyphae of differing mating types conjugate to form a zygosporangium.

Appendix B Statistics Primer

This appendix is designed to help you conduct simple statistical analyses and understand their application and importance. This introduction will help you complete the Apply the Concept and Analyze the Data problems throughout this book. The formulas for a number of statistical tests are presented here, but the presentation is designed primarily to help you understand the purpose and reasoning of the various tests. Once you understand the basis of the analysis, you may wish to use one of many free, online web sites for conducting the tests and calculating relevant test statistics (such as http://faculty.vassar.edu/lowry/VassarStats.html).

Why Do We Do Statistics?

ALMOST EVERYTHING VARIES We live in a variable world, but within the variation we see among biological organisms there are predictable patterns. We use statistics to find and analyze these patterns. Consider any group of common things in nature—all women aged 22, all the cells in your liver, or all the blades of grass in your yard. Although they will have many similar characteristics, they will also have important differences. Men aged 22 tend to be taller than women aged 22, but, of course, not every man will be taller than every woman in this age group.

Natural variation can make it difficult to find general patterns. For example, scientists have determined that smoking increases the risk of getting lung cancer. But we know that not all smokers will develop lung cancer and not all nonsmokers will remain cancer-free. If we compare just one smoker to just one nonsmoker, we may end up drawing the wrong conclusion. So how did scientists discover this general pattern? How many smokers and nonsmokers did they examine before they felt confident about the risk of smoking?

Statistics helps us to find general patterns, even when nature does not always follow those patterns.

AVOIDING FALSE POSITIVES AND FALSE NEGATIVES When a woman takes a pregnancy test, there is some chance that it will be positive even if she is not pregnant, and there is some chance that it will be negative even if she is pregnant. We call these kinds of mistakes *false positives* and *false negatives*.

Doing science is a bit like taking a medical test. We observe patterns in the world, and we try to draw conclusions about how the world works from those observations. Sometimes our observations lead us to draw the wrong conclusions. We might conclude that a phenomenon occurs, when it actually does not; or we might conclude that a phenomenon does not occur, when it actually does.

For example, the planet Earth has been warming over the past century (see Concept 46.4). Ecologists are interested in whether plant and animal populations have been affected by global warming. If we have long-term information about the locations of species and temperatures in certain areas, we can determine whether species movements coincide with temperature changes. Such information can, however, be very complicated. Without proper statistical methods, one may not be able to detect the true impact of temperature or, instead, may think a pattern exists when it does not.

Statistics helps us to avoid drawing the wrong conclusions.

How Does Statistics Help Us Understand the Natural World?

Statistics is essential to scientific discovery. Most biological studies involve five basic steps, each of which requires statistics:

- **Step 1: Experimental Design**
  Clearly define the scientific question and the methods necessary to tackle the question.

- **Step 2: Data Collection**
  Gather information about the natural world through experiments and field studies.

- **Step 3: Organize and Visualize the Data**
  Use tables, graphs, and other useful representations to gain intuition about the data.

- **Step 4: Summarize the Data**
  Summarize the data with a few key statistical calculations.

- **Step 5: Inferential Statistics**
  Use statistical methods to draw general conclusions from the data about the way the world works.

Step 1: Experimental Design

We conduct experiments to gain knowledge about the world. Scientists come up with scientific ideas based on prior research and their own observations. These ideas may take the form of a question like "Does smoking cause cancer?," a hypothesis like "Smoking increases the risk of cancer," or a prediction like "If a person smokes, he/she will increase his/her chances of developing cancer." Experiments allow us to test such scientific ideas, but designing a good experiment can be quite challenging.

We use statistics to guide us in designing experiments so that we end up with the right kinds of data. Before embarking on an experiment, we use statistics to determine how much data will be required to test our idea, and to prevent extraneous factors from misleading us. For example, suppose we want to conduct an experiment on fertilizers to test the hypothesis that nitrogen increases plant growth. If we include too few plants, we will not be able to determine whether or not nitrogen has an effect on growth, and the experiment will be for naught. If we include too many plants, we will waste valuable time and resources. Furthermore, we should design the experiment so that we can detect differences that are actually caused by nitrogen fertilization rather than by variation, for example, in sunlight or precipitation experienced by the plants.

Step 2: Data Collection

TAKING SAMPLES When biologists gather information about the natural world, they typically collect a few representative pieces of information. For example, when evaluating the efficacy of a candidate drug for medulloblastoma brain cancer, scientists may test the drug on tens or hundreds of patients, and then draw conclusions about its efficacy for all patients with these tumors. Similarly, scientists studying the relationship between body weight and clutch size (number of eggs) for female spiders of a particular species may examine tens to hundreds of spiders to make their conclusions.

We use the expression "sampling from a population" to describe this general method of taking representative pieces of information from the system under investigation (**Figure B1**). The pieces of information in a **sample** are called **observations**. In the cancer therapy example, each observation was the change in a patient's tumor size six months after initiating treatment, and the population of interest was all individuals with medulloblastoma tumors. In the spider example, each observation was a pair of measurements—body size and clutch size—for a single female spider, and the population of interest was all female spiders of this species.

Sampling is a matter of necessity, not laziness. We cannot hope (and would not want) to collect *all* of the female spiders of the species of interest on Earth! Instead, we use statistics to determine how many spiders we must collect in order to confidently infer something about the general population and then use statistics again to make such inferences.

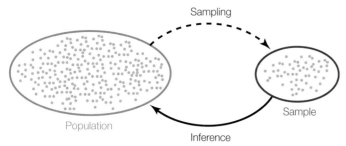

FIGURE B1 Sampling From a Population Biologists take representative samples from a population, use descriptive statistics to characterize their samples, and then use inferential statistics to draw conclusions about the original population.

TABLE B1 ▰

Poinsettia Colors

Color	Frequency	Proportion
Red	108	0.59
Pink	34	0.19
White	40	0.22
Total	182	1.00

DATA COME IN ALL SHAPES AND SIZES In statistics, we use the word *variable* to mean a measurable characteristic of an individual or a system. Some variables are on a numerical scale, like the daily high temperature (a numerical value constrained by the precision of our thermometer), or the clutch size of a spider (a whole number: 0, 1, 2, 3,…). We call these **quantitative variables**. Quantitative variables that only take on whole number values are called **discrete variables**, whereas variables that can also take on any fractional value are called **continuous variables**.

Other variables take categories as values, like a human blood type (A, B, AB, or O) or an ant caste (queen, worker, or male). We call these **categorical variables**. Categorical variables with a natural ordering, like a final grade in Introductory Biology (A, B, C, D, or F), are called **ordinal variables**.

Each class of variables comes with its own set of statistical methods. We will introduce a few common methods in this Appendix that will help you work on the problems presented in this book, but you should consult a biostatistics textbook for more advanced tests and analyses for other data sets and problems.

Step 3: Organize and Visualize the Data

Tables and graphs can help you gain intuition about your data, design appropriate statistical tests, and anticipate the outcome of your analysis. A **frequency distribution** lists all possible values and the number of occurrences of each value in the sample.

TABLE B2 ▰

Fish Weights of *Abramis brama* from Lake Laengelmavesi

Weight (grams)	Frequency	Relative Frequency
201–300	2	0.06
301–400	3	0.09
401–500	8	0.24
501–600	3	0.09
601–700	8	0.24
701–800	3	0.09
801–900	1	0.03
901–1000	6	0.18
Total	34	1.00

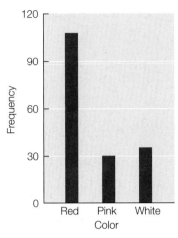

B2 Bar Charts Compare Categorical Data This bar chart shows the frequency of three poinsettia colors that result from an experimental cross.

Table B1 shows a frequency distribution of the colors of 182 poinsettia plants (red, pink, or white) resulting from an experimental cross between two parent plants. For categorical data like this, we can visualize the frequency distribution by constructing a **bar chart**. The heights of the bars indicate the number of observations in each category (**Figure B2**). Another way to display the same data is in a **pie chart**, which shows the proportion of each category represented like pieces of a pie (**Figure B3**).

For quantitative data, it is often useful to condense your data by grouping (or binning) it into **classes**. In **Table B2**, we see a grouped frequency distribution of fish weights for a sample of 34 fish (*Abramis brama*) caught in Lake Laengelmavesi in Finland. The second column (*Frequency*) gives the number of observations in each class and the third column (*Relative Frequency*) gives the overall proportion of observations falling into each class.

Histograms depict frequency distributions for quantitative data. The histogram in **Figure B4** shows the relative frequencies of each weight class in this study. When grouping quantitative data, it is necessary to decide how many classes to include. It is often useful to look at multiple histograms before deciding which grouping offers the best representation of the data.

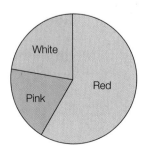

B3 Pie Charts Show Proportions of Categories This pie chart shows the proportions of the three poinsettia colors presented in Table B1.

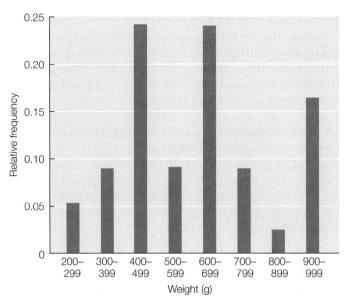

B4 Histograms Depict Frequency Distributions of Quantitative Data This histogram shows the relative frequency of different weight-classes of fish (*Abramis brama*).

Sometimes we wish to compare two quantitative variables. For example, the researchers at Lake Laengelmavesi investigated the relationship between fish weight and length and thus also measured the length of each fish. We can visualize this relationship using a **scatter plot** in which the weight and length of each fish is represented as a single point (**Figure B5**). We say that these two variables have a **linear relationship** since the points in their scatter plot fall roughly on a straight line.

Tables and graphs are critical to interpreting and communicating data, and thus should be as self-contained and comprehensible as possible. Their content should be easily understood simply by looking at them. Axes, captions, and units should be clearly labeled, statistical terms should be defined, and appropriate groupings should be used when tabulating or graphing quantitative data.

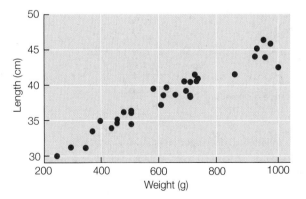

B5 Scatter Plots Contrast Two Variables Scatter plot of *Abramis brama* weights and lengths (measured from nose to end of tail). These two variables have a linear relationship since the data points lie close to a straight line.

Step 4: Summarize the Data

A **statistic** is a numerical quantity calculated from data, while **descriptive statistics** are quantities that describe general patterns in data. Descriptive statistics allow us to make straightforward comparisons between different data sets and concisely communicate basic features of our data.

DESCRIBING CATEGORICAL DATA   For categorical variables, we typically use proportions to describe our data. That is, we construct tables containing the proportions of observations in each category. For example, the third column in Table B1 provides the proportions of poinsettia plants in each color category, and the pie chart in Figure B3 provides a visual representation of those proportions.

DESCRIBING QUANTITATIVE DATA   For quantitative data, we often start by calculating the average value or **mean** of our sample. This familiar quantity is simply the sum of all the values in the sample divided by the number of observations in our sample (**Figure B6**). The mean is only one of several quantities that roughly tell us where the *center* of our data lies. We call these quantities **measures of center**. Other commonly used measures of center are the **median**—the value that literally lies

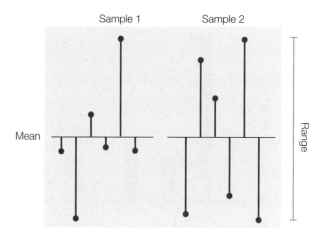

B7 Measures of Dispersion   Two samples with the same mean (black horizontal lines) and range (blue vertical line). Red lines show the deviations of each observation from the mean. Samples with large deviations have large standard deviations. The left sample has a smaller standard deviation than the right sample.

in the middle of the sample—and the **mode**—the most frequent value in the sample.

It is often just as important to quantify the variation in the data as it is to calculate its center. There are several statistics that tell us how much the values differ from one another. We call these **measures of dispersion**. The easiest to one understand and calculate is the **range**, which is simply the largest value in the sample minus the smallest value. The most commonly used measure of dispersion is the **standard deviation**, which calculates the extent to which the data are spread out from the mean. A deviation is the difference between an observation and the mean of the sample, and the standard deviation is a number that summarizes all of the deviations. Two samples can have the same range, but very different standard deviations if one is clustered closer to the mean than the other. In **Figure B7**, for example, the left sample has a lower standard deviation ($s = 2.6$) than the right sample ($s = 3.6$), even though the two samples have the same means and ranges.

To demonstrate these descriptive statistics, we return to the Lake Laengelmavesi study. The researchers also caught and recorded the weights of six fish in the species *Leusiscus idus*: 270, 270, 306, 540, 800, and 1,000 grams. The mean weight in this sample (equation 1 in Figure B6) is:

$$\bar{x} \frac{\Delta N}{\Delta T} = \frac{(270 + 270 + 306 + 540 + 800 + 1000)}{6} = 531$$

Since there is an even number of observations in the sample, then the median weight is the value halfway between the two middle values:

$$\frac{306 + 540}{2} = 423$$

The mode of the sample is 270, the only value that appears more than once. The standard deviation (equation 2 in Figure B6) is:

RESEARCHTOOLS

B6 Descriptive Statistics for Quantitative Data

Below are the equations used to calculate the descriptive statistic we discuss in this appendix. You can calculate these statistics yourself, or use free internet resources to help you make your calculations.

Notation:
$x_1, x_2, x_3, \ldots x_n$ are the n observations of variable X in the sample.

$\sum\limits_{i=1}^{n} x_i = x_1 + x_2 + x_3, \ldots + x_n$ is the sum of all of the observations. (The Greek letter sigma, Σ, is used to denote "sum of.")

In regression, the independent variable is X, and the dependent variable is Y. b_0 is the vertical intercept of a regression line. b_1 is the slope of a regression line.

Equations

1. Mean: $\bar{x} = \dfrac{\sum\limits_{i=1}^{n} x_i}{n}$

2. Standard deviation: $s = \sqrt{\dfrac{\sum (x_i - \bar{x})^2}{n-1}}$

3. Correlation coefficient: $r = \dfrac{\sum (x_i - \bar{x})(y_i - \bar{y})}{\sqrt{\sum (x_i - \bar{x})^2 (y_i - \bar{y})^2}}$

4. Least-squares regression line: $Y = b_0 + b_1 X$
   where $b_1 = \dfrac{\sum (x_i - \bar{x})(y_i - \bar{y})}{\sum (x_i - \bar{x})^2}$ and $b_0 = \bar{y} - b_1 \bar{x}$

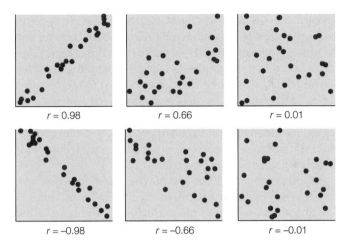

B8 Correlation Coefficients   The correlation coefficient (r) indicates both the strength and the direction of the relationship.

$$s = \sqrt{\frac{(270-531)^2 + (270-531)^2 + (306-531)^2 + (540-531)^2 + (800-531)^2 + (1000-531)^2}{5}} = 309.6$$

and the range is $1000 - 270 = 730$.

DESCRIBING THE RELATIONSHIP BETWEEN TWO QUANTITATIVE VARIABLES   Biologists are often interested in understanding the relationship between two different quantitative variables: How does the height of an organism relate to its weight? How does air pollution relate to the prevalence of asthma? How does lichen abundance relate to levels of air pollution? Recall that scatter plots visually represent such relationships.

We can quantify the strength of the relationship between two quantitative variables using a single value called the Pearson product–moment **correlation coefficient** (equation 3 in Figure B6). This statistic ranges between –1 and 1, and tells us how closely the points in a scatter plot conform to a straight line. A negative correlation coefficient indicates that one variable decreases as the other increases; a positive correlation coefficient indicates that the two variables increase together, and a correlation coefficient of zero indicates that there is no linear relationship between the two variables (**Figure B8**).

One must always keep in mind that *correlation does not mean causation.* Two variables can be closely related without one causing the other. For example, the number of cavities in a child's mouth correlates positively with the size of their feet. Clearly cavities do not enhance foot growth; nor does foot growth cause tooth decay. Instead the correlation exists because both quantities tend to increase with age.

Intuitively, the straight line that tracks the cluster of points on a scatter plot tells us something about the *typical* relationship between the two variables. Statisticians do not, however, simply eyeball the data and draw a line by hand. They often use a method called least-squares **linear regression** to fit a straight line to the data (equation 4 in Figure B6). This method calculates the line that minimizes the overall vertical distances between the points in the scatter plot and the line itself. These distances are called **residuals** (**Figure B9**). Two parameters

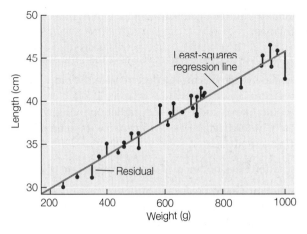

B9 Linear Regression Estimates the Typical Relationship Between Two Variables   Linear-least squares regression line for *Abramis brama* weights and lengths (measured from nose to end of tail). The regression line (blue line) is given by the equation $Y = 26.1 + 0.02X$. It is the line that minimizes the sum of the squares of the residuals (red lines).

describe the regression line: b_0 (the vertical intercept of the line, or the expected value of variable Y when $X = 0$), and b_1 (the slope of the line, or how much values of Y are expected to change with changes in values of X).

Step 5: Inferential Statistics

Data analysis often culminates with statistical inference—an attempt to draw general conclusions about the system under investigation. As depicted in Figure B1, the primary reason we collect data is to gain insight into the larger system from which the data are collected. When we test a new medulloblastoma brain cancer drug on ten patients, we do not simply want to know the fate of those ten individuals; rather, we hope to predict its efficacy on the much larger group of all medulloblastoma patients.

STATISTICAL HYPOTHESES   When it comes to inferring something about the real world from our data, we often have a *"Whether or not"* question in mind. For example, we would like to know whether or not global warming impacts biodiversity; whether or not the clutch size of a spider increases with body size; or whether or not soil nitrogen increases the growth of a particular plant species.

Before making statistical inferences from data, we must formalize our *"Whether or not"* question into a pair of opposing hypotheses—a **null hypothesis** (denoted H_0) and an **alternative hypothesis** (denoted H_A). The alternative hypothesis is the *"Whether"*—it is formulated to describe the effect that we expect our data to support; the null hypothesis is the *"or not"*—it is formulated to represent the absence of the effect. In other words, we typically conduct our experiment seeking to demonstrate something new (the alternative hypothesis) and thereby reject the idea that it does not occur (the null hypothesis).

Suppose, for example, we would like to know *whether or not* a new vaccine is more effective than an existing vaccine at immunizing children against influenza. Our hypotheses would be as follows:

H_0: The new vaccine is not more effective than the old vaccine.

H_A: The new vaccine is more effective than the old vaccine.

If we would like to know whether radiation increases the mutation rate in the bacteria *Escherichia coli*, we would set up the following hypotheses:

H_0: Radiation does not increase the mutation rate of *E. coli*.

H_A: Radiation does increase the mutation rate of *E. coli*.

STATISTICAL BURDEN OF PROOF   In the U.S. justice system, people are innocent until proven guilty. In statistics, the world is *null until proven alternative*. Statistics requires overwhelming proof in favor of the alternative hypothesis before rejecting the null hypothesis. In other words, scientists favor existing ideas and resist adopting new ideas until compelling evidence suggests otherwise. This is based on a philosophy that it is worse to accept new claims when they are false than to miss out on discovering some true facts about world.

When testing a new influenza vaccine, the burden of proof is on the new vaccine. Suppose we were to vaccinate three children with the new vaccine (Group A), three with the old vaccine (Group B) and leave three children unvaccinated (Group C). If no children from Group A, one child from Group B, and one child from Group C became infected, would we have enough evidence to conclude that the new vaccine is superior to the old vaccine? No, we would not. If the study were enlarged, and two out of 100 children in group A, seven out of 100 children in group B, and 22 out of 100 children in group C become infected, would we then have sufficient evidence to choose the new vaccine? Perhaps, but we need to use statistics to be sure.

This is the traditional burden of proof in biology and science in general. As a consequence, scientists are more likely to miss out on discovering something new (and true) about the world than they are to make a false discovery. In recent years, scientists have begun to question this approach and develop an alternative statistical approach, called **Bayesian inference**, which makes it easier to favor new hypotheses. In this primer, we discuss only traditional statistical methods, often called **frequentist statistics**.

	The real world	
	Null hypothesis true (*not more females*)	Null hypothesis false (*more females*)
Null hypothesis true (*not more females*)	✓	Type 2 error (*false negative*)
Null hypothesis false (*more females*)	Type 1 error (*false positive*)	✓

(row label on left axis: **Our conclusion**)

B10 Two Types of Error   Possible outcomes of a statistical test. Statistical inference can result in correct and incorrect conclusions about the population of interest.

JUMPING TO THE WRONG CONCLUSIONS   There are two ways that a statistical test can go wrong (**Figure B10**). We can reject the null hypothesis when it is actually true (**Type I error**) or we can accept the null hypothesis when it is actually false (**Type II error**). These kinds of errors are analogous to false positives and false negatives in medical testing, respectively. If we mistakenly reject the null hypothesis when it is actually true, then we falsely endorse the incorrect hypothesis. If we are unable to reject the null hypothesis when it is actually false, then we fail to realize a yet undiscovered truth.

Suppose we would like to know whether there are more females than males in a population of 10,000 individuals. To determine the makeup of the population, we choose 20 individuals randomly and record their sex. Our null hypothesis is that there are *not* more females than males; and our alternative hypothesis is that there are. The following scenarios illustrate the possible mistakes we might make:

- *Scenario 1*: The population actually has 40% females and 60% males. Although our random sample of 20 people is likely to be dominated by males, it is certainly possible that, by chance, we will end up choosing more females than males. If this occurs, and we mistakenly reject the null hypothesis (that there are *not* more females than males), then we make a Type I error.

- *Scenario 2*: The population actually has 60% females and 40% males. If, by chance, we end up with a majority of males in our sample and thus fail reject the null hypothesis, then we make a Type II error.

Fortunately, statistics has been developed precisely to avoid these kinds of errors and inform us about the reliability of our conclusions. The methods are based on calculating the **probabilities** of different possible outcomes. Although you may have heard or even used the word "probability" on multiple occasions, it is important that you understand its mathematical meaning. A probability is a numerical quantity that expresses the likelihood of some event. It ranges between zero and one; zero means that there is no chance the event will occur and one means that the event is guaranteed to occur. This only makes sense if there is an element of chance, that is, if it is possible the event will occur and possible that it will not occur. For example, when we flip a fair coin, it will land on heads with probability 0.5 and land on tails with probability 0.5. When we select individuals randomly from a population with 60% females and 40% males, we will encounter a female with probability 0.6 and a male with probability 0.4.

Probability plays a very important role in statistics. To draw conclusions about the real world (the population) from our sample, we first calculate the probability of obtaining our sample if the null hypothesis is true. Specifically, statistical inference is based on answering the following question:

Suppose the null hypothesis is true. What is the probability that a random sample would, by chance, differ from the null hypothesis as much as our sample differs from the null hypothesis?

If our sample is highly improbable under the null hypothesis, then we rule it out in favor of our alternative hypothesis. If, instead, our sample has a reasonable probability of occurring under the null hypothesis, then we conclude that our data are consistent with the null hypothesis and we do not reject it.

Returning to the sex ratio example, we consider two new scenarios:

- *Scenario 3*: Suppose we want to infer whether or not females constitute the majority of the population (our alternative hypothesis) based on a random sample containing 12 females and eight males. We would calculate the probability that a random sample of 20 people includes at least 12 females assuming that the population, in fact, has a 50:50 sex ratio (our null hypothesis). This probability is 0.13, which is too high to rule out the null hypothesis.

- *Scenario 4*: Suppose now that our sample contains 17 females and three males. If our population is truly evenly divided, then this sample is much less likely than the sample in scenario 3. The probability of such an extreme sample is 0.0002, and would lead us to rule out the null hypothesis and conclude that there are more females than males.

This agrees with our intuition. When choosing 20 people randomly from an evenly divided population, we would be surprised if almost all of them were female, but would not be surprised at all if we ended up with a few more females than males (or a few more males than females). Exactly how many females do we need in our sample before we can confidently infer that they make up the majority of the population? And how confident are we when we reach that conclusion? Statistics allows us to answer these questions precisely.

STATISTICAL SIGNIFICANCE: AVOIDING FALSE POSITIVES Whenever we test hypotheses, we calculate the probability just discussed, and refer to this value as the **P-value** of our test. Specifically, the *P*-value is the probability of getting data as extreme as our data (just by chance) if the null hypothesis is, in fact, true. In other words, it is the likelihood that chance alone would produce data that differ from the null hypothesis as much as our data differ from the null hypothesis. How we measure the difference between our data and the null hypothesis depends on the kind of data in our sample (categorical or quantitative) and the nature of the null hypothesis (assertions about proportions, single variables, multiple variables, differences between variables, correlations between variables, etc.).

For many statistical tests, *P*-values can be calculated mathematically. One option is to quantify the extent to which the data depart from the null hypothesis and then use look-up tables (available in most statistics textbooks, or on the internet) to find the probability that chance alone would produce a difference of that magnitude. Most scientists, however, find *P*-values primarily by using statistical software rather than hand calculations combined with look-up tables. Regardless of the technology, the most important steps of the statistical analysis are still left to the researcher: constructing appropriate null and alternative hypotheses, choosing the correct statistical test, and drawing correct conclusions.

After we calculate a *P*-value from our data, we have to decide whether it is small enough to conclude that our data are inconsistent with the null hypothesis. This is decided by comparing the *P*-value to a threshold called the **significance level**, which is often chosen even before making any calculations. We reject the null hypothesis only when the *P*-value is less than or equal to the significance level, denoted α. This ensures that, if the null hypothesis is true, we have at most a probability α of accidentally rejecting it. Therefore, the lower the value of α, the less likely you are to make a Type I error (lower left cell of Figure B10). The most commonly used significance level is α = 0.05, which limits the probability of a Type I error to 5%.

If our statistical test yields a *P*-value that is less than our significance level α, then we conclude that the effect described by our alternative hypothesis is statistically significant at the level α and we reject the null hypothesis. If our *P*-value is greater than α, then we conclude that we are unable to reject the null hypothesis. In this case, we do not actually reject the alternative hypothesis, rather we conclude that we do not yet have enough evidence to support it.

POWER: AVOIDING FALSE NEGATIVES The **power** of a statistical test is the probability that we will correctly reject the null hypothesis when it is false (lower right cell of Figure B10). Therefore, the higher the power of the test, the less likely we are to make a Type II error (upper right cell of Figure B10). The power of a test can be calculated, and such calculations can be used to improve your methodology. Generally, there are several steps that can be taken to increase power and thereby avoid false negatives:

- **Decrease the significance level**, α. The higher the value of α, the harder it is to reject the null hypothesis, even if it is actually false.

- **Increase the sample size**. The more data one has, the more likely one is to find evidence against the null hypothesis, if it is actually false.

- **Decrease variability in the sample**. The more variation there is in the sample, the harder it is to discern a clear effect (the alternative hypothesis) when it actually exists.

It is always a good idea to design your experiment to reduce any variability that may obscure the pattern you seek to detect. For example, it is possible that the chance of a child contracting influenza varies depending on whether he or she lives in a crowded (e.g., urban) environment or one that is less so (e.g., rural). To reduce variability, a scientist might choose to test a new influenza vaccine only on children from one environment or the other. After you have minimized such extraneous variation, you can use power calculations to choose the right combination of α and sample size to reduce the risks of Type I and Type II errors to desirable levels.

There is a trade-off between Type I and Type II errors: As α increases, the risk of a Type I decreases but the risk of a Type II error increases. As discussed above, scientists tend to be more concerned about Type I errors than Type II errors. That is, they believe that it is worse to mistakenly believe a false hypothesis than it is to fail to make a new discovery. Thus, they prefer to use low values of α. However, there are many real-world scenarios in which it would be worse to make a Type II error than a Type I error. For example, suppose a new cold medication is being tested for dangerous (life-threatening) side effects. The null hypothesis is that there are no such side effects. A Type II error might lead regulatory agencies to approve a harmful medication that could cost human lives. In contrast, a Type I error would simply mean one less cold medication among the many that already line pharmacy shelves. In such cases, policymakers take steps to avoid a Type II error, even if, in doing so, they increase the risk of a Type I error.

STATISTICAL INFERENCE WITH QUANTITATIVE DATA There are many forms of statistical inference for quantitative data. When measuring a single quantitative variable, like birth weight in lambs, calcium concentration in the blood of pregnant women, or migration rate of birds, we often wish to infer the mean value of the population from which we drew the sample. However, the mean of a randomly chosen sample will not necessarily be the same or even close to the population mean. Suppose we wanted to know the average weight of newborn lambs on a particular farm. By chance, we may end up with a random sample that includes an excess of lightweight lambs and therefore a sample mean that is less than the overall mean in the population.

To infer the population mean from the sample data, we can calculate a **confidence interval for the mean**. This is a statistically derived range of values that is centered on the sample mean and is likely to include the population mean. For example, based on the sample of 34 *Abramis brama* weights from Lake Laengelmavesi (see Table B2; Figure B4), the 95% confidence interval for the mean weight ranges from 554 grams to 698 grams. The true average weight for this species of fish is likely, but not guaranteed, to fall within this range.

Biologists frequently wish to compare the mean values in two or more groups; for example, newborn lamb weights on several different farms, calcium concentration in women in early and late stages of pregnancy, or migration rates in birds of different species. Based on the means and standard deviations calculated for each of the samples, they infer whether or not the means in the different populations are statistically different from one another. There are several statistical methods for this, and the correct method depends on the number of groups, the experimental design, and the nature of the data.

Figure B11 describes the steps of a *t*-test, a simple method for comparing the means in two different groups. To illustrate, we can apply a *t*-test to the Lake Laengelmavesi data to assess whether the two fish species *Abramis brama* and *Leusiscus idus* have significantly different mean weights. We begin by stating our hypotheses and choosing a significance level:

H_0: *Abramis brama* and *Leusiscus idus* have the same mean weight.

H_A: *Abramis brama* and *Leusiscus idus* have different mean weights.

$\alpha = 0.05$

The test statistic is calculated using the means, standard deviations, and sizes of the two samples:

$$t_s = \frac{626 - 531}{\sqrt{\dfrac{207^2}{34}} + \sqrt{\dfrac{310^2}{6}}} = 0.724$$

We can use statistical software or one of the free statistical sites on the internet to find the *P*-value for this result to be $P = 0.497$. Since *P* is considerably greater than α, we fail to reject the null hypothesis and conclude that our study does not provide evidence that the two species have different mean weights.

You may want to consult an introductory statistics textbook to learn more about confidence intervals, *t*-tests, and other basic statistical tests.

STATISTICAL INFERENCE WITH CATEGORICAL DATA With categorical data, we often wish to infer the distribution of the different categories within the populations from which our samples are drawn. In the simplest case, we have a single categorical variable with two or more categories. If there are just two categories, we can construct a **confidence interval for the proportion** of the population that belongs to one of the two categories. This is a statistically derived range of values that is centered on the sample proportion and is likely to include the population proportion. If there are three or more categories, we can use a **chi-square goodness-of-fit** test to determine whether the distribution of the different categories in the population is consistent with a specific distribution.

Figure B12 outlines the steps of a chi-square goodness-of-fit-test. As an example, consider the data described in Table B1. Many plant species have simple Mendelian genetic systems in which parent plants produce progeny with three different colors of flowers in a ratio of 2:1:1. However, a botanist believes that these particular poinsettia plants have a different genetic system that does not produce a 2:1:1 ratio of red, pink, and

---

RESEARCHTOOLS

B11 The *t*-Test

What is the *t*-test? It is a standard method for assessing whether the means of two groups are statistically different from each another.

Step 1: State the null and alternative *hypotheses*:

H_0: The two populations have the same mean.

H_A: The two populations have different means.

Step 2: Choose a significance level, α, to limit the risk of a Type 1 error.

Step 3: Calculate the *test statistic*: $t_s = \dfrac{\bar{y}_1 - \bar{y}_2}{\sqrt{\dfrac{s_1^2}{n_1} + \dfrac{s_2^2}{n_2}}}$

Notation: $\bar{y}_1$ and $\bar{y}_2$ are the sample means; s_1 and s_2 are the sample standard deviations; and n_1 and n_2 are the sample sizes.

Step 4: Use the test statistic to assess whether the data are consistent with the null hypothesis:

Calculate the *P-value* (*P*) using statistical software or by hand using statistical tables.

Step 5: Draw conclusions from the test:

If $P \leq \alpha$, then reject H_0, and conclude that the population distribution is significantly different.

If $P > \alpha$, then we do not have sufficient evidence to conclude that the means differ.

B12 The Chi-Square Goodness-of-Fit Test

What is the chi-square goodness-of-fit test? It is a standard method for assessing whether a sample came from a population with a specific distribution.

Step 1: State the null and alternative *hypotheses*:

H_0: The population has the specified distribution.

H_A: The population does not have the specified distribution.

Step 2: Choose a significance level, α, to limit the risk of a Type 1 error.

Step 3: Determine the *observed frequency* and *expected frequency* for each category:

The observed frequency of a category is simply the number of observations in the sample of that type.

The expected frequency of a category is the probability of the category specified in H_0 multiplied by the overall sample size.

Step 4: Calculate the *test statistic*: $\chi_s^2 = \sum_{i=1}^{c} \frac{(O_i - E_i)^2}{E_i}$

Notation: C is the total number of categories, O_i is the observed frequency of category i, and E_1 is the expected frequency of category i.

Step 5: Use the test statistic to assess whether the data are consistent with the null hypothesis:

Calculate the *P-value* (P) using statistical software or by hand using statistical tables.

Step 6: Draw conclusions from the test:

If $P \leq \alpha$, then reject H_0, and conclude that the population distribution is significantly different than the distribution specified by H_0.

If $P > \alpha$, then we do not have sufficient evidence to conclude that population has a different distribution.

white plants. A chi-square goodness-of-fit can be used to assess whether or not the data are consistent with this ratio, and thus whether or not this simple genetic explanation is valid. We start by stating our hypotheses and significance level:

H_0: The progeny of this type of cross have the following probabilities of each flower color:

Pr{Red} = .50, Pr{Pink} = .25, Pr{White} = .25

H_A: At least one of the probabilities of H_0 is incorrect.

$\alpha = 0.05$

We next use the probabilities in H_0 and the sample size to calculate the expected frequencies:

	Red	Pink	White
Observed	108	34	40
Expected	(.50)(182) = 91	(.25)(182) = 45.5	(.25)(182) = 45.5

Based on these quantities, we calculate the chi-square test statistic:

$$\chi_s^2 = \sum_{i=1}^{c} \frac{(O_i - E_i)^2}{E_i} = \frac{(108 - 91)^2}{91} + \frac{(34 - 45.5)^2}{45.5} + \frac{(40 - 45.5)^2}{45.5} = 6.747$$

We find the P-value for this result to be $P = 0.0343$ using statistical software. Since P is less than α, we reject the null hypothesis and conclude that the botanist is correct: The plant color patterns cannot be explained by the simple Mendelian genetic model under consideration.

This introduction is only meant to provide a brief introduction to the concepts of statistical analysis, with a few example tests. **Figure B13** provides a summary of some of the commonly used statistical tests that you may encounter in biological studies.

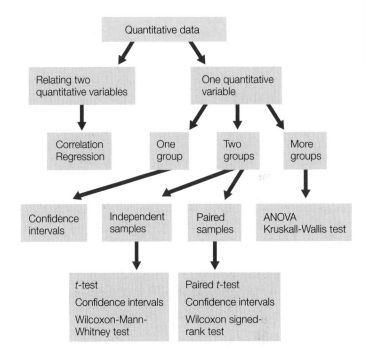

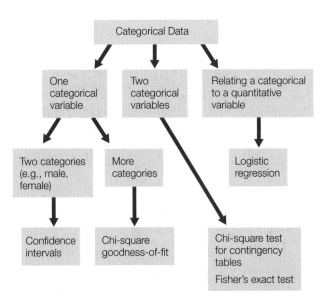

B13 Some Common Methods of Statistical Inference
This flow chart shows some of the commonly used methods of statistical inference for different combinations of data. Detailed descriptions of these methods can be found in most introductory biostatistics textbooks.

Appendix C   Some Measurements Used in Biology

MEASURES OF	UNIT	EQUIVALENTS	METRIC → ENGLISH CONVERSION
Length	meter (m)	base unit	1 m = 39.37 inches = 3.28 feet = 1.196 yards
	kilometer (km)	1 km = 1000 (10^3) m	1 km = 0.62 miles
	centimeter (cm)	1 cm = 0.01 (10^{-2}) m	1 cm = 0.39 inches
	millimeter (mm)	1 mm = 0.1 cm = 10^{-3} m	1 mm = 0.039 inches
	micrometer (μm)	1 μm = 0.001 mm = 10^{-6} m	
	nanometer (nm)	1 nm = 0.001 μm = 10^{-9} m	
Area	square meter (m^2)	base unit	1 m^2 = 1.196 square yards
	hectare (ha)	1 ha = 10,000 m^2	1 ha = 2.47 acres
Volume	liter (L)	base unit	1 L = 1.06 quarts
	milliliter (mL)	1 mL = 0.001 L = 10^{-3} L	1 mL = 0.034 fluid ounces
	microliter (μL)	1 μL = 0.001 mL = 10^{-6} L	
Mass	gram (g)	base unit	1 g = 0.035 ounces
	kilogram (kg)	1 kg = 1000 g	1 kg = 2.20 pounds
	metric ton (mt)	1 mt = 1000 kg	1 mt = 2,200 pounds = 1.10 ton
	milligram (mg)	1 mg = 0.001 g = 10^{-3} g	
	microgram (μg)	1 μg = 0.001 mg = 10^{-6} g	
Temperature	degree Celsius (°C)	base unit	°C = (°F – 32)/1.8
			0°C = 32°F (water freezes)
			100°C = 212°F (water boils)
			20°C = 68°F ("room temperature")
			37°C = 98.6°F (human internal body temperature)
	Kelvin (K)*	K = °C – 273	0 K = –460°F
Energy	joule (J)		1 J ≈ 0.24 calorie = 0.00024 kilocalorie[†]

*0 K (–273°C) is "absolute zero," a temperature at which molecular oscillations approach 0—that is, the point at which motion all but stops.

[†]A *calorie* is the amount of heat necessary to raise the temperature of 1 gram of water 1°C. The *kilocalorie*, or nutritionist's calorie, is what we commonly think of as a calorie in terms of food.

Answers to Chapter Review Questions

CHAPTER 1

1. e   2. b   3. e   4. d   5. e

6. In science, we formulate hypotheses about how the world works, then try to reject those hypotheses with experiments. The experiments must be designed so that we would expect them to uncover problems with our hypothesis. If the experiments are incapable of rejecting a hypothesis, then the experiments are not a rigorous test of the hypothesis.

7. The independent DNA found in mitochondria and chloroplasts is evidence of the origin of these eukaryotic organelles from ancient bacteria that became incorporated in the eukaryotic cell. Since the ancestors of these organelles once existed as independent organisms, they have their own genomes.

8. Controlled experiments, by definition, are able to control many variables in carefully maintained experiments, often in laboratory conditions. Comparative experiments, in contrast, often contain many additional variables that cannot be controlled by the investigator. Comparative experiments often incorporate realistic variation from uncontrolled factors, which accounts for their higher overall variability.

9. If two species share particular changes in the gene we compare, and those changes are not shared by other species we examine, we would expect the two species with the common changes to be more closely related to one another. By comparing many such changes in many genes, we can group species based on their relative evolutionary divergence from one another. For example, we share more changes in our genes with chimpanzees than we do with gorillas. From this, we can deduce that humans and chimpanzees shared a more recent common ancestor than they shared with gorillas.

10. Mitochondrial DNA is often used to follow the history of maternal lineages in a population or species. Nuclear DNA is not used in such cases because it is typically inherited from both parents. This difference can be useful in many circumstances. For example, we might examine a hybrid individual between two species. Equal portions of nuclear DNA from both species could confirm that the individual is a direct hybrid between the two species. If we examine the mitochondrial DNA, however, we can learn which of the two parental species was the female in the cross—and therefore learn by default which was the male.

CHAPTER 2

1. b   2. d   3. c   4. c   5. a   6. d

7.

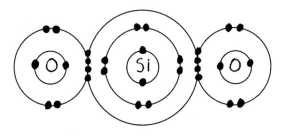

8. An easy way to answer this question is to make a simple table:

	Covalent H—H	Hydrogen H····O
Electrons	Shared	Remain with H and O
Polarity	Nonpolar	Polar; + at H end
Strength	Stronger	Weaker

9. C—H: nonpolar; hydrophobic
   C=O: polar; δ– at O; hydrophilic
   O—P: polar; δ– at O; hydrophilic
   C—C: nonpolar; hydrophobic

10. This is an example of Van der Waals forces, which act over a short distance and do not involve polarity.

11. The human body has the same elements as Earth's crust but in very different proportions.

CHAPTER 3

1. e   2. e   3. c   4. a   5. c   6. b

7. The observations support explanation "a." Glycine is small and nonpolar. Glutamic acid and arginine are larger and polar (charged). Serine and alanine are small: the protein retains its shape. But serine is polar (it has –OH as its R group), and that does not affect the structure. Valine is larger and nonpolar, and this affects shape. So the issue is size.

8. Mannose and galactose have the same atomic formula, $C_6H_{12}O_6$, but the arrangement of atoms is different: compare carbons 2 and 4. These sugars have the hydroxyl (–OH) functional group. Its polarity helps the sugars dissolve in water. The –OH group also can participate in bonding the sugar to other molecules through condensation reactions (see Figures 3.4 and 3.17).

9. High temperature disrupts weak interactions such as hydrogen bonds. Heat shock proteins might work by stabilizing the protein so that the weak interactions are not necessary to preserve its structure.

10. A change from lysine is a change in primary structure. The change could affect tertiary structure if the protein folds as a result of electrostatic attractions between charged amino acids (+ to –). In this case, the presence of a negatively charged amino acid (aspartic acid) where there should be a positively charged one (lysine) might prevent correct folding if a negatively charged amino acid elsewhere in the polypeptide chain is involved in folding (it is attracted to a + amino acid). The same forces might be at work in the interaction of separate chains for quaternary structure.

11. See Figure 3.10. Heat breaks hydrogen bonds and other weak interactions that maintain protein shape. Disulfide bonds also required for normal protein shape. Styling and perms partially denature keratin, then renature the protein in a new shape. Your investigation might involve measuring keratin protein structure of hair before and after disrupting hydrogen bonds and disulfide bonds.

CHAPTER 4

1. c   2. c   3. c   4. c   5. b   6. b

7. The presence of O_2 in the atmosphere produces an oxidizing condition that prevents the reduction reactions noted by the Miller–Urey experiment.

8. Oligonucleotides of RNA can fold because of hydrogen bonds forming between bases on the single chain and to a lesser extent because of weak interactions of base stacking when bases come near one another. Short strands of about 20 oligonucleotides are enough to produce uniquely folded RNA.

9. Cells provided concentration and compartmentation chemicals for the reactions needed for life, as well as differential permeability to distinguish life's chemical composition from that of the environment.

10. If microbes survived heat, the initial part of Pasteur's experiment might begin with microbes already present. They would grow in both the open and closed flasks. To get the results he did, Pasteur's flasks must not have contained such microbes. An answer for the proposed experiment on heat-stable microbes might be to inactivate them using reagents, such as mercaptoethanol, that destroy proteins.

11. A suggested experiment might be to dry the samples after the Miller–Urey experiment (allowing condensation reactions—polymerization) and then apply energy in the form of heat. This condition might have existed in volcanic rock in early Earth.

CHAPTER 5

1. b   2. d   3. e   4. a   5. d   6. b   7. a

8. Four membranes: two in the chloroplast and two in the mitochondrion
   Two membranes: the lysosomal membrane and the plasma membrane (via vesicle; the molecules do not themselves cross any membranes)

No membranes: ribosomes do not have membranes. However, if the ribosomes were associated with the endoplasmic reticulum (ER), the answer would be two membranes: into the ER and out of the ER.

9.

	Animal Cell ECM	Plant Cell Wall
Composition	Collagen fibers in proteoglycan matrix	Cellulose fibers in polysaccharide and protein matrix
Rigidity	Less rigid	More rigid (especially secondary cell walls)
Connections	Some specialized proteins and junctions	Plasmodesmata

10. Microtubules line the long axons of nerve cells, where they act as tracks for vesicles that carry substances down the neuron. Without microtubules, the contents of these vesicles cannot be delivered to their destination, which can result in nerve problems.

   Microtubules are a key part of the mitotic spindle, which is used to move chromosomes during cell division. Depolymerization of microtubules can thus result in loss of dividing cells.

11. For a lysosomal enzyme, the pathway would be ribosome → interior of ER → Golgi → Golgi vesicles → lysosome.

   For an extracellular protein (animal cells), the pathway would be ribosome → interior of ER → Golgi → Golgi vesicles → plasma membrane → extracellular region.

CHAPTER 6

1. c   2. a   3. d   4. c   5. b   6. e   7. c

8. The pumping of Ca^{2+} requires a lipid bilayer membrane to separate compartments, a protein pump in the membrane, and ATP to provide energy for pumping.

9. Diatom wall components move from the Golgi apparatus to the cell wall by exocytosis.

10. Living in a hypotonic environment (cells hypertonic) results in a tendency for water to enter the organism by osmosis, which can cause swelling and dilute cell contents. Some organisms get around this by using reverse pinocytosis (exocytosis) to remove fluid.

11. Experiments might involve the following:
   To measure membrane fluidity, label a small amount of a lipid or protein with a dye and allow it to incorporate into a cell's membrane. This may make a localized labeled spot on the cell. The localized region will be seen to diffuse over the cell over time. In the cancer cells, this rate of diffusion may be faster.
   To measure cell adhesion, dissociate cancer and normal tissue cells. Incubate for a period of time and determine the rate at which cancer cells and normal lung cells bind to cells from the other tissues besides lung. The cancer cells may bind to a greater extent than normal cells.

CHAPTER 7

1. d   2. c   3. d   4. a   5. d   6. a   7. d   8. c

9. Different cells can have different target molecules to which cAMP binds, and these target molecules can have different activities and functions. Binding of cAMP changes the structure (e.g., tertiary structure of a protein) and therefore the function of a target molecule. So cAMP can have many effects.

10. Characteristics of direct communication: the size of signal molecules is limited by the size of openings between cells, it is not specific, it is fast, and there can be cytoplasmic connection between cells.
   Characteristics of receptor-mediated communication: the signal molecules can be larger, it is specific, it is slower, and there is no direct cytoplasmic connection.
   Direct communication is useful for a rapid, coordinated response of many cells.

11. See Figure 7.10. A mutation of the *Raf* gene that activates cell division might involve a protein product that does not need binding of Ras to be active. Cell division would occur without activated Ras, thereby eliminating the need for growth factor binding.
   A mutation of the *MAP kinase* gene would stimulate cell division if the resulting MAP kinase protein did not need to be phosphorylated by MEK to be active. No signaling cascade would be needed for the mutant protein to enter the nucleus and stimulate cell division.

12. Experiments might involve applying a solution containing the antibody to the upper part of the *Hydra* body. The antibody would block diffusion of the signal molecule from the apex to the upper body and—if the hypothesis is correct—would allow a bud to form in the upper body. A sham experiment, in which the solution without antibody is applied, would be a control. In this case, a bud would not form in the upper body.

CHAPTER 8

1. c   2. e   3. c   4. c   5. d   6. d

7. Endergonic reactions are coupled in time and space with exergonic reactions, which release the energy needed for the endergonic reactions.

8. A cytoplasmic enzyme generally has a globular structure with a hydrophilic exterior and an active site for substrate binding. An ion channel generally has a more linear structure with a hydrophobic membrane-spanning region and no active site.

9.

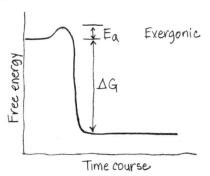

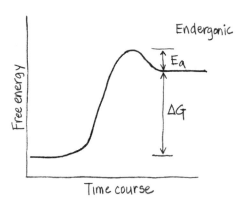

10. (a) The presence of water may prevent O_2 from reaching the enzyme. (b) Boiling denatures proteins, so polyphenol oxidase is irreversibly altered by boiling and its active site is destroyed. (c) Proteins have an optimal pH at which ionized R groups are appropriately charged to give the protein its tertiary structure. A pH 3 of may not be that optimal pH for polyphenol oxidase, so the enzyme is denatured and inactive.

11. See Figure 8.17. A competitive inhibitor binds to the active site of the enzyme and shifts the equilibrium to enzyme molecules in the active form.

12. To determine whether catalase has an allosteric or nonallosteric mechanism, perform an experiment with varying amounts of substrate and plot rate of catalase versus substrate concentration. An S-shaped curve will indicate an allosteric mechanism. A hyperbolic curve indicates a nonallosteric enzyme.
   To determine if a pollutant is a competitive or noncompetitive inhibitor, add the pollutant to the catalase to lower the rate of reaction, then add increasing amounts of substrate. A competitive inhibitor will be removed from the active site and the rate of reaction will increase. A noncompetitive inhibitor will not allow the rate to increase as more substrate is added. (There are more sophisticated kinetic experiments that you will learn in a biochemistry course).

CHAPTER 9

1. d   2. d   3. e   4. c   5. d   6. a

7. If cytochrome *c* remains reduced and cannot accept electrons, the electron transport (respiratory) chain stays reduced and NADH and $FADH_2$ remain reduced. This prevents oxidation reactions in the citric acid cycle and pyruvate oxidation, so pyruvate cannot be converted to acetyl CoA. Instead, pyruvate is converted to lactic acid, regenerating some NAD that can be used so that glycolysis can continue. Because the electron transport chain is not working, there is no proton gradient set up in the mitochondria, and ATP is not made by oxidative phosphorylation.

8. See Figure 9.13. Some amino acids are converted to intermediates of glycolysis. Once they enter glycolysis these intermediates are further metabolized to a

glycolytic intermediate that can be converted to glycerol, which is incorporated into triglycerides. Glycolysis and pyruvate oxidation produce acetyl CoA, which is converted to fatty acids and incorporated into lipids.

Glucose is converted in glycolysis to acetyl CoA, which is then converted to fatty acids as above.

9. (a) Oxidation (removal of H from C2 and C3 of succinate)

(b) Exergonic (because it is an oxidation)

(c) It requires the redox coenzyme NAD or FAD.

(d) The fumarate is converted to other intermediates that regenerate oxaloacetate, the acceptor for the citric acid cycle.

(e) The reduced coenzyme (NADH or $FADH_2$) is reoxidized in the electron transport chain.

10. Anaerobes use alternate electron acceptors to generate energy, such as sulfur, sulfate, and nitrate. Also, they use substrate-level phosphorylation (direct transfer of phosphate to ADP) to make ATP.

11. The proton gradient in the experiment described in Figure 9.9 was generated artificially from the solution and did not require electron transport (a respiratory chain). The presence of antimycin A thus would have no effect on the experiment.

CHAPTER 10

1. e   2. b   3. d   4. d   5. d   6. d

7. In the dark, photosynthetic electron transport stops at photosystem II → reduced PQ (plastoquinone). Initially, the chlorophylls in light-harvesting complexes remain reduced, so reaction-center chlorophylls remain reduced and thus photosystem II remains reduced.

In the dark, the Calvin cycle stops at the reduction phase, which requires NADH. No RuBP is regenerated, so there is no rubisco activity. The initial reactions are no oxidation of photosystem I, and no reduction of NADP to NADPH.

8. These processes can be compared using a table:

	Cyclic Electron Transport	Noncyclic Electron Transport
Products	ATP	ATP, NADPH, O_2
Source of electrons	Electron transport	Electron transport (photosystem I) or water (photosystem II)

9. See Figure 10.18. CO_2 carbons end up in 3PG, which is converted to pyruvate. Pyruvate goes to the citric acid cycle, where some of the intermediates are converted to amino acids, which are incorporated into protein.

In the Calvin cycle, some 3PG is converted to G3P, which can enter glycolysis. Some of the intermediates of glycolysis are converted to amino acids, which are incorporated into protein.

10. (a) O_2

(b) NADPH

(c) 3PG

11. (a) Here is the pathway followed by ^{14}C: $^{14}CO_2$ → cells → photosynthesis → carbohydrate → combustion → $^{14}CO_2$. Release of $^{14}CO_2$ upon combustion would be evidence of photosynthesis (and life).

(b) In this case: $^{14}CO_2$ → heat denatured cells, no photosynthesis. If living things were present, $^{14}CO_2$ would be released in experiment (a), but not in experiment (b).

CHAPTER 11

1. d   2. b   3. d   4. d   5. e   6. d   7. d   8. c

9. See Figure 11.19. In mitotic prophase, there is no pairing of homologous chromosomes, and crossing over is rare. In meiotic prophase I, homologous pairs of chromosomes align, and crossing over is common.

In mitotic anaphase, sister chromatids separate, with one going to each pole. In meiotic anaphase I, sister chromatids do not separate; homologous pairs of chromosomes separate, with one pair going to each pole.

10. Normally, p53 induces expression of p21, which binds to the G1/S Cdk and prevents cyclin from activating it. Without active Cdk, the cell cycle ceases. If p53 is mutated such that it is nonfunctional, p21 is not induced and the cyclin–Cdk complex can form and stimulate the cell cycle at S phase.

11. Cancers often have multiple mutations in different cells of the tumor. If some of these mutations affect different parts of the cell cycle, targeting the different phases may be a useful therapy.

12. Your proposed experiments should involve isolating the synchronous meiotic cells from the lily anthers and establishing them in the lab. As the cells proceed

through the meiotic cell cycle they can be analyzed at different stages for the presence and biochemical activity of various cyclins and Cdks.

CHAPTER 12

1. e   2. a   3. d   4. d   5. d   6. b   7. b   8. d   9. b

10. $BB \times bb$; $bb \times bb$; $Bb \times bb$; $Bb \times Bb$

11. 1/32

12. (a) Autosomal dominant

(b) 1/4

13. (a) Males (XY) contain only one allele and will show only one color, black (X^BY) or yellow (X^bY). Females can be heterozygous (X^BX^b).

(b) X^bY, yellow

14. The body color (G/g) and wing size (A/a) genes are linked; eye color (R/r) is unlinked to the other two genes. The distance between the linked genes is 18.5 units.

15. Yellow, blue, and white in a 1:2:1 ratio.

16. F_1 will all be wild type, $PpSwsw$. F_2 will have phenotypes in the ratio 9:3:3:1; see Figure 12.6 (p. 238) for analogous genotypes.

17. (a) F_1 will all be $PpByby$ and will have wild-type eye color and wings. The ratio of phenotypes in F_2 will be 3:1, $PPByBy$ (wild-type eyes and wings) to $ppbyby$ (pink eyes and blistery wings).

(b) F_1 will all be $PpbyBy$ with wild-type eye color and wings; they will produce just two kinds of gametes (Pby and pBy). Combine them carefully and see the 1:2:1 phenotypic ratio fall out in the F_2: 1 wild-type eyes/blistery wings : 2 wild-type eyes/wild-type wings : 1 pink eyes/wild-type wings.

(c) Pink–blistery

(d) See Figures 11.16 and 11.18 (pp. 220–222). Crossing over took place in F_1.

18. $Rraa$ and $RrAa$

19. (a) $w^+ > w^e > w$

(b) The parents are w^ew and w^+Y. The progeny are w^+w^e, w^+w, w^eY, and wY.

20. (a) BX^a, BY, bX^a, bY

(b) The mother is bbX^AX^a, the father BbX^aY, the son BbX^aY, and the daughter bbX^aX^a.

21. 75 percent

22. Because the gene is carried on mitochondrial DNA, it is passed through the mother only. Thus if the women does not have the disease but her husband does, their child will not be affected. However, if the woman has the disease but her husband does not, their child will have the disease.

23. The cross $RRYY \times rryy$ produces $RrYy$ (round, yellow) F_1 offspring. If the seed shape and seed color genes were linked with no recombination between them, the F_2 would also be all $RrYy$. A distance of 10 map units between two genes means that on average 10% of the F_2 offspring will have recombinant phenotypes, in this case round green (5%) and wrinkled yellow (5%).

The cross in Figure 12.19 is $BbVgvg$ (gray, normal) $\times bbvgvg$ (black, vestigial). If there were no linkage between the genes, then the gray, normal parent would produce four types of gametes: BVg, bVg, Bvg, and bvg. When these combine with the bvg gametes produced by the other parent, four types of offspring in a 1:1:1:1 ratio will result: $BbVgvg$ (gray, normal), $bbVgvg$ (black, normal), $Bbvgvg$ (gray, vestigial), and $bbvgvg$ (black, vestigial).

CHAPTER 13

1. a   2. c   3. b   4. b   5. d   6. c   7. d

8. At 3,000 bp per minute in two directions, each origin grows at 6,000 bp per minute. There are 300 minutes in S phase, so the total bp possible for one origin is $(300 \times 6,000) = 1,800,000$. If there are 120 million bp to replicate, then the total number of origins is 120 million/1.8 million = 66 origins. If there are 3 μm of DNA, this means there are about 22 origins per micrometer of DNA.

9. DNA replication adds new nucleotides to the 3′ end of DNA, where there is an —OH group on the sugar at the 3′ position. If there is no —OH group, there cannot be a condensation reaction and formation of a bond to the next nucleotide, so replication stops.

10. After ten rounds there would still have been some DNA (about 1/512th) as hybrid because the original heavy DNA template strands would still have been there. This tiny amount might not have been detectable in the centrifuge, however.

11. The proposed experiments might use S strain pneumococcus and transform R strain as in Figure 13.1. Incubate separate batches of S strain bacteria in ^{32}P or ^{35}S. Make cell-free extracts of the S strains. Incubate with R cells and look for their transformation to the S phenotype. Then check to see if there is ^{32}P or ^{35}S label in the newly transformed cells. It would be expected that only ^{32}P label (DNA) would enter the cells.

CHAPTER 14

1. b   2. a   3. d   4. b   5. d   6. d   7. d   8. e

9. For 192 amino acids, the triplet genetic code mandates 576 bp of coding sequence. Add the start and stop codons and the total is 582. This is shorter than the actual DNA gene because of promoter and terminator of transcription sequences; introns; and ribosome binding sequences. All except the transcription signals are transcribed into the pre-mRNA. The mature mRNA has the introns removed.

10. Errors in transcription can be tolerated because many copies of each RNA are made; if a few have errors, there are enough perfect ones to overcome any problem. Errors in DNA replication are harmful because DNA is replicated only once in the life of the cell.

11. In the poly CA experiment, threonine is ACA or CAC and histidine is ACA or CAC. In the poly CAA experiment, threonine is CAA, ACA, or AAC. Therefore in the first experiment threonine must be ACA and histidine CAC.

12. Enzymes: $4 \rightarrow 2 \rightarrow 3 \rightarrow 1 \rightarrow 5$
    Compounds: $C \rightarrow F \rightarrow E \rightarrow D \rightarrow G \rightarrow T$

CHAPTER 15

1. a   2. c   3. b   4. b   5. d   6. b

7. (a) In a loss of function mutation, a phenotype is not present; for example, there may be a loss of enzyme activity. In a gain of function mutation, a new phenotype is present; for example, a new signaling protein may be active.

   (b) In a missense mutation, a single base pair change results in a codon change and thus an amino acid change in a protein. In a nonsense mutation, a single base pair change results in a codon change to a stop codon and thus premature termination of a protein.

   (c) In a spontaneous mutation, DNA changes as a result of unprovoked chemical changes or replication errors. In an induced mutation, DNA changes as a result of outside physical or chemical agents.

8. (a) The mutation that leads to PKU is rare in the human population; most people do not have the harmful allele and the highest probability is that the father is homozygous normal. Because the mother has PKU (she is homozygous mutant), the developing fetus is heterozygous.

   (b) High levels of phenylalanine cause brain damage. If the mother's phenylalanine levels were too high, the baby would be born with brain problems.

   (c) The woman should be on a phenylalanine-restricted diet.

9. Testing for the cystic fibrosis (CF) allele could be done by allele-specific oligonucleotide hybridization with probes for the normal and CF alleles; see Figure 15.18. Or direct DNA sequencing of the CF gene could be done. A person who is a carrier will test positive for both the normal and the mutant alleles.

   To do gene therapy, the normal allele for CF could be inserted into a viral vector that can infect cells in the lung and airway tissues. Then the virus could be sprayed onto these tissues.

10. Early identification of people with multifactorial diseases, even before symptoms appear, could lead to therapeutic interventions to prevent disease development. Ethical issues might include insurability, hiring eligibility, and social stigma.

11. An enzyme test for HEXA would reveal intermediate levels in people who are carriers. This could be done on accessible cells (e.g., blood) if the gene is expressed there. A DNA test could involve sequencing the gene by allele-specific oligonucleotide hybridization (see the answer to Question 9). The advantage of DNA testing is that it can be done on any cells from the body (not just cells that express the enzyme).

    Investigation of the stop codon hypothesis would involve isolating the HEXA protein from patients with Tay-Sachs disease and showing that it is shorter in primary structure than the protein encoded by the normal allele.

12. (a) The amino acid sequence would be Leu-Ile-Ser-Ile-Ala. This is a missense mutation.

    (b) The mutation replaces proline with serine. Proline is a nonpolar amino acid that is usually part of bends or loops in a protein; serine is a polar amino acid with a smaller side chain. The mutation is likely to affect enzyme activity because it is likely to affect protein structure.

    (c) See p. 317. This region of the gene could be amplified by PCR and then digested with *Eco*RV. The mutant DNA will be cut, but the wild-type DNA won't be.

CHAPTER 16

1. b   2. a   3. e   4. b   5. c   6. d

7. The easiest way to answer this question is to construct a simple table:

	Lysogenic Bacteriophage	HIV
(a) Viral entry to host cell	Attachment of viral protein to host cell membrane	Membrane fusion of virus to host cell membrane
(b) Virus release from host	Host cell lysis	Budding and exocytotic release
(c) Viral genome replication	Host DNA polymerase	Virus reverse transcriptase followed by host RNA polymerase
(d) New virus production	Host transcription of virus genes and host-mediated translation of virus proteins	Same as in lysogenic bacteria

8. In a prokaryotic gene, the promoter is a DNA sequence, there are few transcription factors, and there is one RNA polymerase. In a eukaryotic gene, the promoter is a DNA sequence, there are many transcription factors, and there are several RNA polymerases.

9. Here is the structure of the gene:

   $$E1 - I1 - E2 - I2 - E3 - I3 - E4$$

   (E = exon; I = intron). Assuming that initiation of transcription begins at E1, the possible proteins are composed of exons 1234; 134; 124; and 14.

10. To keep a constant, low-level expression of repressor protein, the regulatory gene would have an inefficient promoter, and synthesis of the repressor would be constitutive.

11. You could sequence the relevant genes of colon cancer cells and look for mutations that lead to aberrant function, then isolate the proteins involved and determine that their functions are indeed abnormal. To show epigenetic silencing, you might sequence the promoters of the genes and look for epigenetic changes (e.g., cytosine methylation, which would be increased if there is transcriptional silencing). Then you could examine the tumor cells to see if the active proteins are there but in small amounts.

CHAPTER 17

1. c   2. b   3. e   4. e   5. b   6. b   7. c   8. c   9. a

10. One gene can produce several proteins by alternative splicing, which makes the proteome highly complex. In addition, many proteins are modified after translation, and this contributes to even more protein diversity. The metabolome is highly variable from cell to cell and from one time to another. It is determined not only genetically but also by responses to environmental conditions.

11. While all of these plants have the same basic genes for "life" as well as "plant" functions (e.g., photosynthesis, cell-wall formation, flowering), there are some genes (and proteins) that are specialized for each plant (e.g., rice genes for growing under water, genes for timing flowering, genes for seed-storage proteins).

12. (a) Extract genomic DNA from the patient's cells and analyze it for SNP polymorphisms. If the SNP that correlates with kidney cancer is present, he has an increased susceptibility.

    (b) Isolate both normal and cancerous kidney cells. Do a metabolomic profile on the kidney cancer cells and the normal kidney cells using chemical analyses for small molecules. By comparing the profiles, generate a metabolomic "signature" for the kidney cancer cells. Next, examine the metabolomic profile of kidney tissue from the patient and compare it with the metabolomic signature for kidney cancer cells.

    (c) For possible drugs involved in kidney cancer treatment, isolate many cancers (or examine stored tissues) and do a SNP analysis, correlating tumor response to the drug with the SNP polymorphism. Then isolate some of the patient's tumor cells and examine the DNA for SNPs that relate to drug response. Use the drug that the patient's genome indicates will be effective.

CHAPTER 18

1. b   2. c   3. e   4. a   5. e   6. d   7. b   8. c

9. Both PCR and cloning begin with a gene sequence. In PCR, the sequence is amplified in the test tube. In cloning, the sequence is amplified by an organism (typically bacteria). In PCR, amplification is achieved by synthesizing primers that bind to either end of a target DNA sequence and adding nucleotides and DNA polymerase. The doubled DNA is them denatured, and the process is repeated 20 to 40 times.

In cloning, the target DNA is inserted by restriction and ligation into a vector, which has an origin of replication that will function in the organism where amplification will occur. The vector is added to the host cells, which are cultured and divide many times, amplifying the target DNA along with the host chromosome. The vector is then removed from the host cells and cut with a restriction enzyme, releasing amplified, cloned target DNA.

PCR is much simpler and faster but has artifacts where inappropriate fragments of DNA are amplified or sequence errors are introduced by DNA polymerase. Cloning yields the correct DNA without mutations but involves host cell culture and time-consuming DNA purification steps. See Figure 18.12. A simple table can answer this question.

	Conventional	Recombinant DNA
(a) Sources of new genes	Other plants of same species	Any organism or synthetic DNA
(b) Number of genes transferred	Often many	One
(c) How long it takes	At least one growing season, usually many	Weeks

10. (a) The target gene would be inserted into an expression vector with a promoter such that the gene would be expressed in the developing seed. The vector could be added to cultured wheat cells, and those cells carrying the vector selected (the vector could carry a reporter gene for resistance to an antibiotic). The cells could be induced to form a wheat plantlet, which would be transferred to the field and the seeds examined for the new protein.

(b) The target gene could be inserted into a sheep expression vector containing the lactoglobulin promoter so that the gene would be expressed in milk glands. The recombinant vector would be inserted into sheep egg cells. After the female offspring grew up, their milk could be tested for the presence of the human enzyme.

11. Public concerns include the artificiality of unnatural interference with nature, the safety of these foods for human consumption, and environmental dangers if non-host plants receive recombinant genes.

CHAPTER 19

1. c   2. b   3. a   4. e   5. a   6. b   7. c

8. (a) All neuronal precursors might undergo apoptosis and no neurons would form.

(b) The $p21$ gene would be activated and the cell cycle would be blocked; in the presence of other factors, muscle cells would form.

(c) There might be no gradient of the protein in the developing limb and therefore no differential development of digits—all the digits would be fingers.

(d) The hunchback protein gradient would not form properly and the embryo would not establish its anterior–posterior axis.

9. (a) No apoptosis would lead to too many cells in developing organs, and the organs would not form properly.

(b) No gradient of hunchback protein would form, and there would be no posterior end determination in the developing fruit fly.

10. A mutation that caused expression of class A genes instead of class C genes. This would lead to an AB combination instead of AC, and petals would develop instead of stamens.

11. Mechanisms might include cell-cycle inhibition as a result of Cdk blocker; induction of transcription of certain genes; and cytoplasmic segregation, so that when a cell divides only one daughter cell gets a factor important in determination.

12. One could analyze mRNA in egg cells, in the parent differentiated cells, and in the reprogrammed cells. This could be done by reverse transcriptase PCR or by gene expression arrays.

CHAPTER 20

1. c   2. a   3. a   4. c   5. c

6. If the expression of Gremlin were blocked, this protein could not inhibit BMP4 signaling. The cells in the webbing of the feet would undergo apoptosis, and the duck would be born with unwebbed feet.

7. All of the hatchlings at any temperature would be expected to develop into males. Aromatase is required to convert testosterone into estrogen, which is required for female development.

8. The coexpression of *Hoxc6* and *Hoxc8* appears to be important in the development of thoracic vertebrae (the vertebrae with ribs). This is a short region in mice, and mice have only a small number of thoracic vertebrae, and therefore a short body. In snakes, the coexpression of *Hoxc6* and *Hoxc8* along a much greater length of the embryo results in a much larger number of thoracic vertebrae, and therefore a much-elongated body.

9. The results support the conclusion that higher levels of BMP4 expression result in greater cartilage diameter on the beaks of developing chickens.

10. The observations are consistent with the hypothesis that there has been selection in some human populations for mutations on the enhancer that controls expression of the glycoprotein in red blood cells. This genetic change would be expected to have a selective advantage in human populations that are exposed to malaria at high levels, because the mutation confers greater resistance to malaria in humans that carry it.

CHAPTER 21

1. d   2. d   3. d   4. e   5. b

6. Humans select traits in domestic plant and animal populations based on our interest in the trait, rather than on how it affects the natural reproductive rate or survivorship of the organisms. Many of the traits artificially selected by humans would not be advantageous in wild populations. For example, humans have selected many cattle breeds for high body fat and high body weight. These traits result in large calves, which in turn result in calving difficulties for cows. Ranchers often have to assist in the birth of such calves, because the calf (and likely its mother) would often die without such assistance. In a natural population, there would be selection for smaller calf size and birth weight, which would increase the successful reproductive rate and survivorship.

7. Behaviors can respond to environmental cues that are predictive of future conditions, and these behaviors can be selected for if they are under genetic control. For example, day length becomes shorter as we move closer to winter, so individual mammals have a survival advantage if they respond to shortening days by going into hibernation. In this case, the environmental cue (day length) is predictive of future environmental conditions (the cold of winter).

8. Natural selection cannot act when there is no effect on the effective reproductive rate of the organism. Diseases such as Alzheimer's usually occur long after the reproductive years have passed. As long as the disease does not affect the relative likelihood of the survival of the affected person's offspring (as a result of reduced parental care, for example), we would not expect natural selection to lead to any reduction in Alzheimer's disease in human populations.

9. (a) Frequency of allele a: 0.60; of allele A: 0.40

(b) Frequency of genotype aa: 0.40; of genotype Aa: 0.40; of genotype AA: 0.20

(c) Expected frequency of genotype aa: 0.36; of genotype Aa: 0.48; of genotype AA: 0.16

(d) We would expect some level of deviation because the assumptions of Hardy–Weinberg equilibrium are so restrictive. For example, the finite population size, the presence of mutation, any migration of individuals into or out of the population, gene flow from mating with adjacent populations, nonrandom mating within the population, or selection in the population could all lead to deviations from Hardy–Weinberg expectations.

10. The black mice and white mice are highly unlikely to be mating randomly with each other. The combined population is far from Hardy–Weinberg expectations, with far too few heterozygous (Aa) individuals. The much higher frequency of the a allele among the black mice, and of the A allele among the white mice, suggests that the black and white mice are mostly mating within color types, with few between-color-type matings. Another possibility, though, is that the population consists of two subpopulations (one of mostly black mice, the other of mostly white mice) that have only recently come together in the same location. These two hypotheses could be distinguished by following the mice through another generation. If mating is now occurring at random, we would expect the genotype frequencies to be similar to Hardy–Weinberg expectations after one generation of random mating.

CHAPTER 22

1. e   2. a   3. e   4. a   5. e   6. d

7. The classification is not currently monophyletic. Both genera could be monophyletic if Species 4 were moved from Genus B to Genus A; monophyly could also be achieved if all the species were included in the same genus.

8. Fossils can give us direct evidence of the character states of extinct lineages. For example, all modern birds lack teeth. Is the lack of teeth an ancestral or a derived condition? If we examine extinct species of theropods (the larger group of dinosaurs that includes the living birds), we see that they had teeth. Therefore we know that the lack of teeth is a derived condition in modern birds.

9. The estimated average rate of change is 0.9 amino acid change/500 million years, or 0.0018 amino acid change/million years. If we express this as a percentage rather than as a proportion, we would say there is (on average) 0.18 percent change in amino acid sequences per million years.

10. The West Nile virus in the United States appears to be most closely related to a strain of the virus isolated in Israel. A reasonable hypothesis is that the virus emerged in Africa in the 1930s and subsequently moved into Asia and Europe, probably multiple times. Then in the late 1990s, a strain of the virus from Israel appears to have been transported to New York, perhaps carried by mosquitoes on an airplane or in a cargo shipment. Once in the United States, the virus spread quickly in native bird populations across North America.

CHAPTER 23

1. e 2. c 3. a 4. e

5. If the only difference between the diverging lineages is at a single locus, then both of the new alleles must be functional when they interact with the products of other gene loci (in both lineages). Any interlocus genetic incompatibility produced by these new alleles would be expected to affect the parental lineages as well. In addition, there are many greater numbers of possible incompatibilities across different gene loci than there are within a single locus. Rather than two deleterious changes at the same locus (one in each lineage), the Dobzhansky–Muller model allows neutral changes at any pair of loci whose products interact. It is the negative interaction of these products in the hybrid between the two lineages that results in genetic compatibility.

6. If two different fusions of chromosomes occur in two different lineages, then the resulting chromosomes cannot pair normally in meiosis in the hybrids. If you attempt to diagram meiosis in the hybrid that would result from a cross of the divergent lineages in Figure 23.4, you will see that homologous pairings require parts of different chromosomes to align with one another. These chromosomes will then be pulled in two different directions as the cell divides in meiosis I, resulting either in a likely failure of the cell to divide, or an uneven distribution of the chromosome arms in the two daughter cells. Production of normal cells with an even distribution of the various chromosomes arms will be limited, so the hybrids will produce few, if any, normal gametes.

7. A likely possibility is that the incompatible alleles have not yet become fixed in the various strains, so only some combinations of crosses result in genetic incompatibility.

8. Species that arise in allopatry initially occur in separate, but usually adjacent, ranges (see Figure 23.6). Therefore we would expect many closely related species to exhibit this same pattern. The ranges of highly mobile species are more likely to change over time, so the pattern should be strongest among relatively sedentary species.

9. There are many possible designs of experiments that might prove informative. Here is an example of one that would examine the effect of flower position on pollinator attraction: Take one species of flower and divide the flowers into two groups. Position each flower to be either upright or pendant, then record the number and type of pollinators that are attracted to flowers in each group. Test to see if the differences between the two groups are statistically significant.

10. (a)

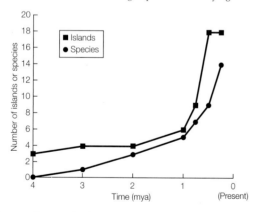

(b) Yes, because the curve for the number of species lags behind the curve for the number of islands, but the two curves exhibit very similar changes in slope through time. As new islands arise, new opportunities for speciation also arise. The number of species at any one time is always just below the number of distinct islands.

(c) There are currently 18 islands in the archipelago and only 14 finch species. This suggests that there are still opportunities for additional speciation by geographic isolation. Based on our graph from Question a, we expect populations of species that occur on two or more islands to diverge into distinct species over time. To test this hypothesis, we could collect samples of each population and examine genetic divergence among the samples. Significant genetic divergence among the populations on different islands suggests that the distance between the islands is a significant barrier to gene flow, so the populations are expected to diverge into distinct species over time.

CHAPTER 24

1. a 2. a 3. a 4. e 5. b 6. e

7. Molecular clocks work best when they are applied within a group of organisms with similar generation times and populations sizes. Population size makes little difference if all or most changes are neutral, but rates of change among deleterious and beneficial changes are affected by population size. In addition, it is important to make comparisons among homologous genes and proteins, since rates of evolution across different genes are likely to vary widely. When molecular clocks are used to make comparisons across species with very different generation times, it is necessary to account for the different generation times.

8. New mutations are introduced into the experiment shown in Figure 24.14 through the errors made in the PCR amplification step. In other words, the mutation rate is a function of the error rate of the DNA polymerase. Using a different DNA polymerase with a higher error rate would increase the overall mutation rate of the experiment, and that would increase the variation in the population of molecules. Another possible answer is to add a mutagen to the PCR amplification step, which would also increase the mutation rate of the experiment. Any process that increases the mutation rate would be expected to increase the genetic variation present in the pool of molecules prior to the next round of selection.

9. This problem can be investigated by sequencing and comparing the genes for opsins in surface-dwelling (eyed) and cave-dwelling (eyeless) crayfishes. If the genes of the eyeless species are no longer under any selection, we would expect to observe a similar rate of synonymous and nonsynonymous substitutions in the genes. If there has been strong selection for a new function (something other than vision), we would expect a higher rate of nonsynonymous substitutions compared with synonymous substitutions (indicating positive selection). We would compare these rates to the rates seen in the surface-dwelling (eyed) species. In the surface-dwelling species, we would expect to see a higher rate of synonymous compared with nonsynonymous substitutions, which is expected under purifying selection.

10. (a) Codon numbers 12, 15, and 61 are likely to be evolving under positive selection for change because these three codons have each experienced a higher rate of nonsynonymous substitutions (which give rise to amino acid replacements) relative to the rate of synonymous substitutions.

   (b) Codon numbers 80, 137, 156, and 226 are likely evolving under purifying selection, as the vast majority of changes at these codons are synonymous substitutions, which do not result in amino acid replacements. Substitutions that result in amino acid changes (nonsynonymous substitutions) undoubtedly occur, but they are usually selected against in the population. Codon number 165 has experienced similar numbers of synonymous and nonsynonymous substitutions. However, since there are approximately three times as many possible nonsynonymous substitutions as there are synonymous substitutions, the number of synonymous substitutions is slightly higher than expected if the rates of each type of substitution are equal. Codon 165 may be evolving under weak purifying selection; it is the codon that is closest to neutral among the codons shown in the table.

CHAPTER 25

1. b 2. c 3. a 4. c 5. b 6. c

7. There are many possible answers, but four familiar examples include the study of Earth's past atmosphere by examining the chemical composition of rocks; the study of past climates by examining the growth rings of trees; the study of continental drift by examining the geological record; and the study of the origins of the universe (the "Big Bang") by examining the speed at which galaxies are moving apart.

8. Relative dating provides us an order for events; we learn that Event 1 happened before Event 2. But absolute dating provides us with an estimate of the timing of those events. It is important to know not just that Event 1 occurred before Event 2, but also how much time separated the two events.

9. Multicellular organisms require higher concentrations of oxygen, and the levels of oxygen increased throughout the Precambrian. By the end of the Precambrian, atmospheric oxygen levels were sufficiently high to support a variety of multicellular organisms. In addition, the end of widespread glaciation (the "snowball Earth" period) near the end of the Precambrian probably allowed multicellular organisms to flourish.

10. There are many possible experiments that could be devised. For example, the effects of changing oxygen concentrations on other species (besides flying insects, such as the *Drosophila* used in the described experiment) could be tested. An ideal study organism would have a short generation time (so that many generations could be followed in the course of the experiment) and would be easy to raise in the laboratory. For example, guppies could be raised in elevated and reduced oxygen concentrations, and evolution in the size of the swim bladder (a site of oxygen uptake) could be evaluated as a response.

CHAPTER 26

1. e   2. c   3. e   4. b   5. b   6. d

7. Ribosomal RNA genes are universally present across organisms. They evolve slowly, so they can be compared among the most distantly related species. They are present in multiple copies, so they were relatively easy to isolate and sequence in the earliest days of gene sequencing. Also, since they are required for protein synthesis, and already present in all cellular species, the possibility of lateral gene transfer is greatly reduced. In contrast, different types of metabolism have arisen repeatedly in the history of prokaryotes, so species with similar types of metabolism may not be closely related. Cell structure is useful for identifying some major groups of prokaryotes (Gram-positive versus Gram-negative groups, for example), but the differences are too few to be of great use in classifying most species.

8. A laterally transferred gene does not represent descent from a common ancestor and thus does not reflect a true evolutionary relationship.

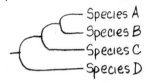

Expected tree based on gene X:
— Species A
— Species B
— Species C
— Species D

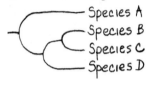

Expected tree based on consensus of non-transferred genes:
— Species A
— Species B
— Species C
— Species D

9. Most, but not all, biologists consider viruses to be living organisms. Viruses have their own genomes, and they are composed of proteins much like cellular organisms. They evolved from other living species, and they are clearly a part of life. However, viruses are not composed of cells, and they depend on cellular hosts to carry out many of their biological processes. For these reasons, some biologists consider them to be nonliving components of their cellular hosts rather than distinct living organisms.

10. There are many possible answers, but one widely used approach for detecting new life forms (in any environment, including high-temperature environments) is to directly isolate and amplify conserved gene sequences. The ribosomal RNA genes are often used for such detection because they evolve very slowly and are required for protein synthesis. DNA could be extracted from high-temperature environments, and any ribosomal DNA genes that were present could be amplified and sequenced. The sequences would then be compared with the ribosomal RNA genes of other known species of prokaryotes to classify the organisms living in the extreme environment.

CHAPTER 27

1. e   2. c   3. e

4. (a) Foraminiferans have external shells of calcium carbonate, whereas radiolarians have long, stiff pseudopods and radial symmetry. The external shells of foraminiferans and the internal skeletons of radiolarians are both important components of ocean sediments and sedimentary rocks.

(b) Ciliates are covered with numerous hairlike cilia, whereas dinoflagellates generally have two flagella (one in an equatorial groove, and the other in longitudinal groove). Both ciliates and dinoflagellates have sacs, called alveoli, just beneath their plasma membranes, which identify them as alveolates.

(c) Diatoms are unicellular and are typically composed of two nested plates (like a petri dish). Brown algae are large, multicellular organisms composed of branched elements or leaflike growths. Both diatoms and brown algae are photosynthetic.

(d) The vegetative unit of a plasmodial slime mold is a plasmodium: a wall-less mass of cytoplasm containing numerous diploid nuclei. The vegetative unit of cellular slime molds consists of separate, single amoeboid cells. In both groups, when environmental conditions become unfavorable, the vegetative units form fruiting structures.

5. The independence of sex and reproduction in ciliates suggests that sex has functions apart from reproduction. Sex is important for recombining genes,

which is important for several reasons. Sex allows populations of organisms to avoid the accumulation of deleterious alleles, and it allows the formation of new combinations of beneficial alleles. Thus even organisms that reproduce asexually generally have some other means of achieving sexual recombination of their genomes.

6.

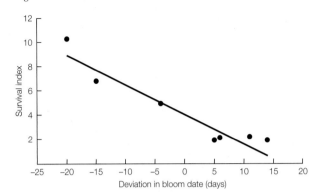

Using the formula for a correlation coefficient shown in Appendix B, $r = -0.948$.

7. The results show that earlier bloom dates are associated with higher survival indices. The relationship between these two measures is very strong and nearly linear, resulting in a correlation coefficient of $r = -0.948$. As noted in the question, larval haddock depend on these blooms for both cover from predation and as a food source. A reasonable hypothesis for this is that earlier blooms provide better cover and more food for the larval haddock, so survivorship of the larval fish is higher in years when the phytoplankton blooms occur earlier. Another (not mutually exclusive) possibility is that the earlier blooms benefit other species that the haddock consume as food, or that the potential predators of haddock target the phytoplankton instead of the haddock.

8. The three rRNA genes of corn are not one another's closest relatives because the nuclear, mitochondrial, and chloroplast genomes have different origins, and the relationships shown in the tree reconstruct the endosymbiotic events that gave rise to mitochondria and chloroplasts.

9. The mitochondrial rRNA gene of corn is more closely related to the rRNA gene of *E. coli* than it is to the nuclear rRNA genes of other eukaryotes because the mitochondria were derived from an endosymbiosis with a proteobacterium. Likewise, the chloroplast rRNA gene of corn is more closely related to the rRNA gene of *Chlorobium* than it is to the nuclear rRNA gene of corn because the chloroplasts were derived from an endosymbiosis with a cyanobacterium.

10. The human and yeast mitochondrial rRNA genes would be expected to cluster on the tree closest to the corn mitochondrial rRNA gene because all of these genes are descended from the same endosymbiotic event (the origin of mitochondria). The human and yeast mitochondrial rRNA genes would be more closely related to each other than either is to the corn mitochondrial rRNA gene because fungi and animals are more closely related to each other than either is to plants (as can be seen in the relationships of the nuclear rRNA genes).

CHAPTER 28

1. c   2. e   3. b   4. b   5. d

6. Microphylls are usually small and typically have a single vascular strand. In contrast, megaphylls are larger and typically contain branched veins. Microphylls may have originated as sterile sporangia. Megaphylls may have originated from flattening of branching stems, between which photosynthetic tissues developed. Among modern plants, microphylls are found in lycophytes, whereas megaphylls are characteristic of the euphyllophytes (such as ferns and seed plants).

7. One advantage of heterospory is that it allows a greater degree of outcrossing, since there are separate male and female gametophytes.

8. Both mosses and ferns are homosporous, and both alternate between a diploid sporophyte and a haploid gametophyte generation. However, the gametophyte generation is the large, dominant portion of the moss life cycle, whereas the sporophyte generation in the large, dominant portion of the fern life cycle. The sporophyte of a moss is completely dependent on the gametophyte, whereas the sporophyte of a fern becomes independent of the gametophyte.

9. Yes. Heterospory is an example of a trait that appears to have evolved multiple times among different groups of vascular plants.

10. One possibility is to examine leaf size as a function of thermal environment among close relatives of living plant species. We might predict that large leaf size is restricted in hot, dry climates but favored in cooler, wetter climates.

CHAPTER 29

1. d   2. a   3. d   4. a   5. a

6. To be functional as a reproductive organ, a flower would need to have at least a carpel or a stamen. The petals function largely for pollinator attraction, and so can easily be lost in wind-pollinated species. The sepals function mostly to protect the flower in bud, and so may be lost in species with simplified flowers.

7. The fossil record is not a complete record of life on Earth. Early angiosperms may have been limited in distribution, or may have lived in environments that were not compatible with easy fossil formation. It is likely that the early angiosperms were not very abundant or widespread. They apparently underwent a rapid radiation in the Cretaceous, where they become common in the fossil record.

8. .

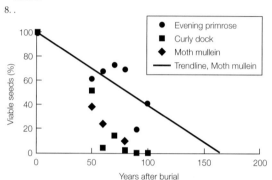

9. One approach to this problem is to calculate a linear trend line for survivorship of moth mullein (*Verbascum blattaria*) seeds by calculating a linear regression line (see Appendix B) and then projecting it forward in time to the point where it intersects zero percent survival. The resulting regression equation is $y = 102.09 - 0.62x$. The graph above shows this approach, which predicts that the last *Verbascum blattaria* seeds would germinate in about Year 165 of the experiment (set y to 0, and solve for x using the regression equation; the result is $x = 164.7$ years). This approach assumes a linear decline in viability of the seeds. It may be more reasonable to assume an exponential decay in seed viability (similar to radioactive decay; see Figure 25.1). If seeds decay exponentially, then we would expect some low level of survivorship of *Verbascum blattaria* seeds well beyond Year 165.

10. At least four factors are related to seed survivorship:

    1. Size of the seed: larger seeds have more food reserves (endosperm).

    2. Density of the seed coat: tougher seed coats provide better protection of the seed.

    3. Level of dormancy of the embryos: deeper dormancy results in longer survivorship.

CHAPTER 30

1. d   2. b   3. c   4. e   5. c

6. If it is a fungus, we would expect to detect chitin in the cells walls, whereas chitin would be absent if it is a plant. We could also examine the specimen for the presence of chloroplasts, which would be expected to produce the green coloration in a plant but not in a fungus. If it is a vascular plant, we would expect to observe vascular tissues in the sample, which would be absent if it is a fungus. We could also sequence a conserved gene from the sample, such as a ribosomal RNA gene, and compare the sequence phylogenetically with other fungi and plants.

7. It is because the nuclei remain separate in dikaryons, even though the two nuclei are contained within a single cell.

8. Fungi play a critical role in decomposition of plants and parts of animals. If there were no more fungi, there would be enormous accumulation of the remains of dead organisms, especially the cellulose and lignin of plants.

9. Site 5 shows the highest diversity and density of lichens and so is probably farthest from the city center. Site 4 is next, followed by Site 1, then Site 3, and finally Site 2. In addition to distance from the city center and prevailing wind direction, other predictive factors could include distance to point-pollution sources (such as factories or power plants) and distance to major highways (a source of pollution from automobile exhaust). Other answers are also possible; it is important for such studies to control for factors such as the species of tree examined and the exposure of the branches to similar light and humidity conditions.

10. One common source of fungal contaminants in plant samples is symbiotic fungi, including endophytic fungi and mycorrhizal fungi. In either case, it is usually possible to collect specific tissues from the plant that exclude these symbionts. If our hypothesis about the source of the fungal genes is correct, then the fungal sequences should be absent from these symbiont-free tissues.

CHAPTER 31

1. d   2. b   3. c   4. d   5. d

6. (a) In radial symmetry, body parts are symmetrical across multiple planes that run through a single axis at the body's center. Animals with radial symmetry have no front or rear ends, and they are often sessile or drift freely with currents. If they move under their own power, they can typically move slowly equally well in any direction. In contrast, bilaterally symmetrical animals have mirror-image right and left halves divided by a single plane that runs along an anterior–posterior midline. They have a front end that usually contains a concentration of sensory systems and nervous tissues in a distinct head. Bilateral animals usually move forward in the direction of the head, so the head encounters new environments first.

   (b) Among the bilaterian animals, there are two distinct forms of gastrulation, or the initial indentation of a hollow sphere of cells early in development to form the blastopore. In protostomes, the blastopore eventually develops into the mouth of the animal. In deuterostomes, the blastopore becomes the anus.

   (c) Diploblastic animals have embryos with two cell layers (an outer ectoderm and an inner endoderm). The embryos of triploblastic animals have an additional cell layer between the ectoderm and the endoderm, known as mesoderm.

   (d) Acoelomate animals lack a body cavity enclosed by mesoderm. Pseudocoelomate animals have a body cavity enclosed in mesoderm; this body cavity contains the gut and internal organs composed of endoderm, but these latter organs are not lined with mesoderm. Coelomate animals have a body cavity that is enclosed with mesoderm, and the internal organs are also lined with mesoderm.

7. The answer to this question will depend on the reader's opinion. However, most biologists would answer that the phylogenetic analyses that result from analysis of animal genomes provide the most definitive evidence of animal monophyly.

8. Bilateral organisms have an anterior and a posterior end. As the animal moves through the environment, the anterior end encounters potential food or predators first. It is therefore advantageous for the sensory organs and central nervous system to be concentrated at the anterior end.

9. A slow metabolic rate requires a low energy budget, and hence a low intake of food.

10. Placement of glass microscope slides (or other smooth substrates for placozoan attachment) in warm tropical waters often results in colonization by placozoans. The glass slides can be suspended in water in survey areas, then later retrieved and examined for the presence of placozoans.

CHAPTER 32

1. e   2. d   3. b   4. d   5. d   6. e

7. Segmentation allows an animal to move different parts of its body independently, which allows for much greater control of movement. However, segmentation tends to constrain the body shape of an organism. Loss of segmentation is often favored in some parasitic and burrowing organisms that live in confined spaces.

8. Several answers are possible, but some examples of key innovations that appear to be associated with major episodes of diversification in protostomes include the evolution of the cuticle in ecdysozoans, the evolution of shells in mollusks, the evolution of jointed limbs in arthropods, and the evolution of wings for flight in insects.

9. Insects have been highly successful in terrestrial environments, in part because flight gives insects greater access to plants. Many insect species are specialists on one or a few plant species, and plant diversity is far greater on land and in freshwater environments than in the oceans. Although some insects live in fresh water for part or all of their life cycles, these freshwater environments are closely associated with surrounding terrestrial environments. Crustaceans have been much more successful in the oceans than have insects, and crustaceans may simply outcompete insects in marine environments.

10. All entomologists agree that many more species of insects remain to be discovered, but many entomologists think that Erwin's estimates were high. Each estimate is highly dependent on how representative *Luehea seemannii* is as a tropical forest tree. If the average tropical forest tree has many fewer host-specific beetle species than does *Luehea seemannii*, then these estimates would be inflated. Likewise, overestimating the number of tropical forest trees, or the percentage of ground-dwelling beetles, or the percentage of all insects that are not beetles, would lead to further inflation of the estimates. In addition, species diversity of beetles may be higher in Panama than in other areas of the tropics. However, any of these estimates could be underestimates as well. Each of Erwin's assumptions is now being tested; these tests require extensive work on additional species of trees, additional groups of insects, and in additional areas of the world.

CHAPTER 33

1. d   2. a   3. d   4. a   5. e   6. b

7. The four appendages common to most vertebrates are the two pectoral appendages and the two pelvic appendages. In most swimming vertebrates, these appendages function as fins. They are commonly used for propulsion (especially the pectoral fins) but are also used for steering, stabilization, and manipulation of the body position in water. Among tetrapods, the appendages are often modified into limbs used for walking, running, jumping, burrowing, climbing, grasping, and manipulating objects. There have been several reversals to fin-like limbs used by aquatic tetrapods (several times among amphibians, turtles, birds, and mammals, for example). There have also been at least three origins of the pectoral limbs of tetrapods into wings for powered flight (among birds, bats, and the extinct pterosaurs). There have also been several other modifications of the limbs for gliding (in fishes, amphibians, reptiles, and mammals). One or both pairs of appendages have been lost (or greatly reduced) in many groups of fishes, amphibians, reptiles (including birds), and mammals. Some well-known examples of limb reduction or loss include the completely legless caecilians and snakes, the loss of external hind limbs in whales and manatees, and the greatly reduced forelimbs of flightless birds.

8. Amphibians exchange gases and fluids through their permeable skins. This makes them highly vulnerable to many environmental toxins. Many species of amphibians have a biphasic life cycle, so they are vulnerable to habitat degradation and loss of both aquatic and terrestrial environments. Most amphibians do not move long distances, so they do not easily move into new habitats when their local environment is destroyed. For these reasons, they are also sensitive to rapid climate changes. Many species of amphibians have highly specialized habitat requirements and live in very restrictive ranges. Habitat loss or changes within these restricted ranges often result in extinction.

9. Fossil remains of extinct theropod dinosaurs shows that many features once thought to be restricted to birds, such as feathers, actually evolved much earlier among the theropods. Other typical "bird" morphological features, such as air-filled bones and a furcula (wishbone), are also typical of the larger group of theropods. Among living reptiles, DNA sequence analyses clearly unite birds with the crocodilians (the other living archosaurs). The combined evidence from many sources that birds are a surviving group of theropod dinosaurs is now overwhelming.

10. Hair evolved in the ancestor of mammals; feathers evolved among theropod dinosaurs (seen today among the birds). Among the living tetrapods, birds and mammals are endothermic. Hair and feathers provide body insulation for mammals and birds, respectively. Without these forms of insulation, the maintenance of metabolic body heat would be difficult. Fossil evidence shows that many extinct theropod dinosaurs also had feathers, so many paleobiologists predict that they were endothermic as well. Endothermy would also be expected in large, active predators—a description that fits our current view of many theropod dinosaurs.

CHAPTER 34

1. b   2. e   3. a   4. b   5. b   6. d   7. c   8. c

9. The cell types can be compared using a table.

Structure/Function	Sclerenchyma	Collenchyma
Cell walls	Secondary, thickened	Primary, thicker at corners
Flexibility	Less flexible	More flexible
Cell conditions	Some dead (apoptosis)	Alive
Presence	Wood, bark	Petioles, growing areas

10. Primary growth involves cell division and cell enlargement, and typically results in growth of an organ in length. Secondary growth involves growth of an organ in thickness, by the addition of more cell layers. Only some angiosperms undergo secondary growth. Herbaceous plants such a peonies have only primary growth. Woody plants such as trees have both primary and secondary growth.

11. The initials are still 1.5 meters above the ground today because the plant grows in height at its apex.

12. Some examples might include a larger root apical meristem to produce thicker carrots and reduced internode growth to produce compact heads of cabbage.

CHAPTER 35

1. c   2. d   3. b   4. b   5. d   6. e

7. Epidermal cells have external walls with a waxy cuticle, which makes them repel water. The epidermal cells of roots might have a thinner (or absent) cuticle, as they take up water; and leaves and stems a thicker cuticle, to conserve water. In addition, the epidermis of leaves, and to a lesser extent stems, has stomata, which regulate gas exchange (including loss of water vapor from the leaf interior).

8. A source is an organ such as a leaf that produces more sugars than it uses. A sink is an organ such as a root that produces less sugars than it needs and so imports sugars from a source. In a deciduous tree, a leaf might be a source in the summer, and roots a sink. But then in spring, the roots might be a source for the buds (newly emerging leaves).

9. To cross the fewest membranes and still get from the soil solution to the atmosphere by way of the stele a water molecule would follow this route: soil to root symplast to stele symplast to stele apoplast to xylem to leaf apoplast to leaf interior air space to stoma to atmosphere. A water molecule could follow this path by crossing as few as two membranes: 1) from soil into a root hair or root cortical cell across a root cell membrane; 2) out of a stele cell into the stele apoplast across a stele cell membrane. Getting from the soil solution to a mesophyll cell in a leaf would require crossing at least three plasma membranes: the two membranes listed for the previous route plus the mesophyll cell membrane (to get from the leaf apoplast into the mesophyll cell).

10. The mutation in the *HARDY* gene might cause increased expression of a gene that inhibits cation accumulation in stomata, thereby keeping them more closed and conserving water. To test this hypothesis, you could look at the response of stomata to light in the leaves of mutant versus wild-type plants. The stomata of wild-type plants should open rapidly in response to light (see Figure 35.9); the stomata of HARDY mutant plants might open more slowly or less wide.

11. (a) Yes. The difference in water potential between the soil and the leaf (1.7 MPa) is enough to overcome gravity and draw water to the top of the tree.

    (b) No. If the soil water potential decreased to –1.0 MPa, it would be more negative than inside the root cells and water would leave the roots (and enter the soil).

    (c) If all the stomata closed, the leaf water potential would not be as negative. This in turn would make the xylem water potential less negative, and so on down to the roots. This would make the difference between the leaf water potential and root water potential insufficient for water to flow from the roots to the leaves (toward a more negative water potential).

CHAPTER 36

1. d   2. d   3. c   4. a   5. c   6. d

7. The ability of chemists to detect low concentrations of elements is fairly recent. Before then, nutrient solutions thought to be pure often were not.

8. Heavy irrigation after a prolonged dry period may produce runoff of topsoil (the A horizon) and leaching of ions (especially anions) into the subsoil, making fewer nutrients available to plant roots. Converting land use from virgin deciduous forest to crops will change the composition of living organisms in the soil, as many organisms that live in association with tree roots will disappear. The soil structure and texture will also change, because roots will no longer be present to hold the soil together and make air spaces. The soil chemistry will change, because crops take up nutrients from the soils and the nutrients are removed from the system when the crops are harvested.

9. See Figure 36.10, the nitrogen cycle. There are numerous species that fix nitrogen. Loss of one species might allow others to expand and replace it. Loss of all the species would mean that only abiotic methods could be used for nitrogen fixation. This might reduce overall nitrogen in the soil, meaning less would be available for plant growth.

10. The experiment with mutant *Arabidopsis* suggests that *Arabidopsis* uses either its own or exogenous strigolactones for growth regulation and has the appropriate receptor and response mechanisms. This reinforces the idea that an ancient mechanism to attract beneficial microbes also is used for modern plant growth regulation. Or the reverse might be true: the original function of strigolactone might have been as a plant hormone and its role in plant–microbe interactions might have evolved later.

11. Because holoparasitic plants can gain reduced carbon through association with hosts, the genes encoding photosynthesis functions are not under selection pressure, because having them would not confer any survival and reproductive advantage for the parasites. So any mutation that renders such a photosynthesis gene nonfunctional will not be deleterious.

CHAPTER 37

1. a   2. d   3. b   4. b   5. a   6. b

7. Fire produces ash, which enriches the soil with plant nutrients. A seed that germinated as a result of fire could have an advantage in such a nutrient-rich soil.

8. If a single species has two mechanisms for breaking seed dormancy, then if environmental conditions for Mechanism A are not present, environmental conditions for Mechanism B might be. This enables the plant to respond to a wider array of environmental conditions. In addition, if the cue for, say, Mechanism A turns out to be misleading (not predictive of favorable conditions) and the

seedling dies, there is still a second seed that can germinate at a different time (by Mechanism B), when conditions might be more favorable.

9. The charcoal in the bag absorbs ethylene gas, which is released by ripening fruits. The lack of ethylene prevents over-ripening and decay.

10. To test for the relationship between corn stunt spiroplasma disease and gibberellins, you could measure gibberellins in plants infected with the bacterium and in normal plants; you might expect the spiroplasma-infected plants to exhibit a reduction in gibberellins. Another approach would be to infect normal plants with the spiroplasma and then spray gibberellins on them; you might expect this to reverse the stunt phenotype.

11. (a) See Figure 37.2 Add a mutagen to hundreds of corn seeds and plant them. In a screen, look for plants that are shorter, and propagate these.

    (b) See Figure 37.11. If the transcription factor in the gibberellin signal transduction pathway is inactivated, the plants will be insensitive to gibberellin and be stunted. A mutation that inactivates the gibberellin receptor would have the same effect.

    (c) Other potential effects might include reduced seed germination and reduced seedling growth due to lack of mobilization of stored reserves in the seed (see Figure 37.5). If the mutant is completely gibberellin-insensitive, these effects will *not* be overcome by adding gibberellin to the seeds as they germinate. If, however, the mutant is a dwarf because of reduced amounts of gibberellin in the plant (because of a mutation that affects gibberellin biosynthesis, for example), the germination effects could be reversed with exogenous gibberellin.

CHAPTER 38

1. b   2. e   3. b   4. e   5. a   6. c

7. In triploid cells undergoing meiosis, there cannot be pairing of homologous chromosomes in meiosis I. So meiosis I is abnormal and functional gametes do not form.

    A fruit is formed from the ovary wall of the flower.

    Seedless grapes are probably propagated by cuttings (vegetative reproduction).

8. Poinsettias are short-day plants; they bloom at a time of year when days are getting shorter (in the Northern Hemisphere).

9. No, it isn't necessary. Just a flash of light during a long night is enough to convert P_r to P_{fr} and to change the photoperiod.

10. (a) The mutation stabilized the CO protein.

    (b) The mutation caused nonfunction of the FD protein.

    (c) The mutation increased expression of the FLC protein.

    (d) The mutation caused constitutive expression of the CO protein.

11. Several approaches might be taken, such as a genetic screen for meiotic cells that do not separate chromosomes at anaphase I, or a search for proteins (and then their genes) that bind to SWII protein.

CHAPTER 39

1. b   2. c   3. a   4. b   5. c   6. c

7. A plant might make a secondary metabolite that kills an insect pest. Plants making this metabolite would be selected for in evolution. However, the insect might develop resistance to the metabolite. Then the insect population would increase while the plant population decreased—until another defense mechanism evolves. This is coevolution. For more examples, see Chapter 56.

8. *Avr2Avr3*   Healthy   Healthy   Diseased
   Avr1Avr4   Healthy   Healthy   Healthy

9. (a) The effects of reduced rainfall could include dehydration and osmotic stress. Genetic responses might include alterations in leaf anatomy, with a thicker cuticle to reduce evaporation; a more extensive root system to obtain water; and accumulation of solutes in the roots, which would reduce root water potential and result in more water uptake in dry soils.

    (b) Flooding reduces the amount of O_2 available to the plants and results in reduced respiration. Adaptations might include increased production of pneumatophores or aerenchyma to supply air to submerged plant tissues.

    (c) Wheat rust is a fungal pathogen. Plants can adapt by increasing the ability to seal off infected areas and reduce the spread of the fungus within the plant, by developing specific immunity, and by increasing production of phytoalexin and pathogenesis-related proteins that kill the fungus.

10. You could feed one group of hornworms on normal plants and another group on genetically modified plants. The two groups could then be exposed to the parasite. If nicotine is protective, the hornworms that fed on normal plants should have fewer parasites.

CHAPTER 40

1. c   2. c   3. a   4. d   5. b

6. Feedforward information makes it possible to anticipate a physiological challenge to homeostasis and to take preemptive action by changing a set point or the sensitivity of a regulatory system. Feedforward information for the regulation of breathing could be the onset of exercise; for blood pressure it could be the fight-or-flight response to a threat; and for secretion of digestive juices it could be the sight, smell, or expectation of food.

7. In the metabolic rate/environmental temperature curve in Figure 40.17, the equivalent of *HL* would be metabolic rate, as long as the animal's temperature is not rising or falling and the animal is not doing external work. *K* would represent the animal's thermal conductance, or how easily it loses heat; $1/K$ would be a measure of the animal's insulation. The curve projects to 0 at an ambient temperature equal to body temperature because this portion of the curve represents the extra metabolic effort necessary to compensate for heat loss to the environment. If body temperature and environmental temperature were the same, there would be no heat loss to the environment.

8. Biological processes proceed more slowly at lower temperatures. Thus the lower the temperature of the heart or skeletal muscle, the slower will be its ability to generate a contractile force. This could pose a physiological challenge for highly active fish such as great white sharks or giant bluefin tuna that depend on fast swimming and endurance to catch prey. An evolutionary adaptation to this challenge can be seen in these fishes' vascular anatomy: blood from their hearts goes to the gills, where it exchanges respiratory gases but also comes into thermal equilibrium with the cold ocean water. Thus these fishes are sending cold blood to their body tissues.

9. Basal metabolic rate (ml O_2/hr) vs. body mass (kg):

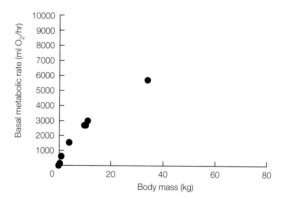

   Heart size (g) vs. body mass (kg):

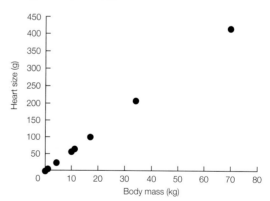

10. The heart has to pump blood against the resistance of the vascular system. Since blood vessels go to all tissues of the body, the total length of blood vessels will be directly proportional to body size, and therefore total peripheral resistance will be directly proportional to body size. The other factor that determines how much blood the heart can pump is the heart rate. The total cardiac output is a function of the size of the heart and the number of times it beats per unit of time.

11. Several factors that determine the heat transfer rate between the iguana and its environment are different in these two conditions. When basking on the lava rocks, the radiation absorbed will be higher and the heat conducted across the skin will be different in comparison to being in the water. Also, when the animal breathes it will be losing heat through evaporation of water in its airways.

    An experiment in which the animal heats up and cools down in the same environment would offer stronger support for the conclusion that blood flow to the skin is a critical factor. The iguana could be placed in water and in air at two different temperatures (e.g., 20°C and 40°C) to compare the rates of heating and cooling.

CHAPTER 41

1. b   2. a   3. d   4. e   5. e

6. The time course of a hormone signaling system depends on several factors, including the rate of release of the hormone, its half-life in the blood, and the interactions it has with its receptors. A hormone signaling system that controls a short-term process such as digestion would be expected to have a rapid (e.g., vesicular) release, a short half-life, and rapid action; these are attributes of a peptide hormone. A hormone signaling system that controls a long-term process such as embryonic development would be expected to have continuous release, a long half-life, and to be slow acting, attributes of a steroid hormone.

7. The high levels of synthetic male steroid hormones exert actions through the testosterone receptors that exist in both males and females. The usual effects of promoting muscle hypertrophy and male secondary sexual characteristics (e.g., hair, a deeper voice) will occur. In addition, the high level of negative feedback on the pituitary gonadotropes and on the production of GnRH by the hypothalamus will reduce testicular and ovarian functions in the males and females. Decrease of circulating estrogens in the females will cause reduction of breast tissue.

8. The most common cause of hypothyroidism is lack of iodine in the diet. Thyroglobulin continues to be produced, but there is a lack of functional T_3 or T_4. As a result, the levels of TRH and TSH rise because of the lack of negative feedback. The elevated TSH induces continued production of thyroglobulin, resulting in goiter. The most common cause of hyperthyroidism is an autoimmune disease called Graves disease. An antibody to the TSH receptor is produced, and the binding of that antibody to the receptor causes activation of the signaling pathway that increases the production of thyroglobulin and the development of goiter. If there is adequate iodine in the diet, however, this condition results in increased secretion of T_3 and T_4, which produces the symptoms of hyperthyroidism.

9. The large size of the larvae together with the lack of adult moths indicates that cycles of growth and molt of the larvae continued without the induction of pupation. The low temperature probably prevented the usual decline in the production and secretion of juvenile hormone by the corpora allata.

10. Insulin controls the entry of glucose into most cells of the body, but not neurons. When insulin levels fall, as happens during the postabsorptive state (i.e., after ingested nutrients have been fully digested), glucose entry into cells slows down and the cells convert to using other sources of energy. Neurons, however, always require glucose; their lack of insulin means that their access to an adequate glucose supply is protected during the postabsorptive state because there is no decrease in their ability to take up glucose from the blood. Also, the decrease in glucose use by other cells of the body preserves the glucose in the blood for use by the nervous tissue.

CHAPTER 42

1. a   2. e   3. e   4. c   5. d   6. a

7. See Figure 42.10. The antigen-binding site of an antibody has heavy and light chains in a unique three-dimensional configuration that binds a particular antigenic determinant. This is similar to an enzyme active site that binds a substrate. In both cases, binding is noncovalent. A major difference is in the result of binding: an antigen does not change its covalent structure when it binds to an antibody, whereas a substrate does change covalently when it binds to an active site.

8. Both immunoglobulins and T cell receptors have constant and variable protein regions, bind antigens, and have great variability in primary structure. T cell receptors are only membrane proteins of T cells. Immunoglobulins can be either membrane proteins of B cells or secreted proteins in the blood.

9. The father's haplotypes are A1B7D11 and A3B5D9; the mother's are A4B6D12 and A2B7D11. These two parents could not have the child with the genotype indicated.

10. There are thousands of different enzymes in an individual but potentially millions of different specific antibodies. Every cell in an animal has the genetic information for all enzymes. Each immunoglobulin, however, is derived from a unique gene (produced by DNA rearrangements) in a B cell or a clone.

11. Experiments might involve testing vaccinated people for neutralizing antibodies against HIV (humoral immunity) and looking for T cell activity against HIV-infected cells (cellular immunity).

CHAPTER 43

1. a   2. d   3. d   4. d   5. d   6. a

7. Leydig and thecal cells have similar functions and characteristics. They are both removed from direct contact with the developing gametes, and they both produce testosterone. Sertoli cells and granulosa cells are both in direct contact with the developing gametes, and they support their development by providing nutrients.

8. Progesterone actions are required to maintain the endometrium in a condition that can support implantation and not degenerate, as occurs during menstruation. By blocking progesterone receptors, RU-486 prevents implantation and the maintenance of the endometrium.

9. Conditions that could favor the evolution of this sexual dimorphism are a sessile existence, dispersed populations, and availability of suitable habitats. If a larva lands on a suitable substrate, it will have high reproductive success if it is a female and can produce lots of eggs; eggs typically have resources that enable them to travel considerable distances if released into the water, and their probability of being fertilized is therefore high. A larva would have a lower probability of success if it developed into a solitary male and produced sperm; sperm have less ability to survive travel over long distances in the water to encounter eggs. However, if a larva lands on a female, it is guaranteed high reproductive success if it can fertilize all of the eggs that female produces. Therefore, by attaching itself to the female and minimizing all of its own physiological processes other than sperm production, a larva can achieve high reproductive success at very little cost.

10. It is likely that the man's offspring would all be daughters. The Y chromosome lacks some essential genes that are on the X chromosome; in the absence of the cytoplasmic bridges, the developing sperm that contain a Y chromosome would lack those gene products. Thus all viable sperm the man produces would contain an X chromosome. This would result in female offspring, since the mother would also contribute an X chromosome.

CHAPTER 44

1. e   2. a   3. d   4. b   5. b

6. You could inject Disheveled protein into the side of the fertilized egg opposite the gray crescent and see if a secondary organizer formed in that region of the resulting blastula. You could also inject the inhibitor of the Disheveled protein into the region of the gray crescent and see if that prevented the formation of the organizer.

7. You would be destroying the cells that normally migrate from the dorsal blastopore lip to form the notochord, and that also produce the signals that determine the anterior–posterior differentiation of the embryo. Thus you might see defects in the development of the nervous system or abnormal segmental development of the body.

8. The flow of fluid over Henson's node may be asymmetrical and thereby create different physical forces on the primary cilia of cells on either side of the node. These differential forces could influence the expression of *Sonic hedgehog*. Experiments to test this hypothesis could include using gene knockouts that eliminate the motile cilia around Henson's node, or experiments in which early embryos are cultured under conditions in which the flow across Henson's node is opposite that of the normal pattern. The prediction for the first experiments would be that the right–left asymmetry of organs in the embryos would be randomized. The prediction for the second experiment would be that the normal asymmetry of organ development would be reversed.

9. A possible mechanism would be cytoplasmic factors that are not distributed randomly or evenly throughout the cytoplasm of the oogonia or the spermatogonia. Thus when they divide by mitosis, factors that control the fates of the daughter cells could be received by one daughter cell but not the other.

10. At an early stage of blastulation (e.g., the 16- or 32-cell stage,) a few blastomeres could be removed from the embryo and cultured separately to produce a population of stem cells. The embryo could go on to develop normally, and the stem cells could be frozen for later use.

CHAPTER 45

1. d   2. b   3. c   4. e   5. c

6. When the stimulus occurs at some point along an axon and an action potential is stimulated, depolarizing current will flow in both directions, bringing adjacent areas of the axon to threshold. However, once an action potential is fired, the Na⁺ channel inactivation gates close and make that section of the axon refractory to further stimulation until they open again. Thus the action potential cannot reverse its direction of propagation, and if the action potential begins at the axon hillock, it cannot reverse its direction of propagation and is unidirectional.

7. Excitatory synapses cause a depolarization of the neuronal membrane, and inhibitory synapses hyperpolarize it. These two influences are summed by virtue of the resulting membrane potential. If it depolarizes enough to reach threshold, an action potential will be fired at the axon hillock.

8. Because the GABA receptor is inhibitory, benzodiazepines would be expected to slow cognitive processes and make a person more likely to fall asleep.

9. The type of information that an action potential transmits depends on the nature of the sensory cell that generated the action potential and on the nature of the cell that receives input as a result of that action potential. Thus photoreceptors transduce light into action potentials, and those action potentials are interpreted as light in the visual circuits that receive those action potentials. Intensity of the stimulus is coded as the frequency of action potentials. Integration is achieved by the summation of excitatory and inhibitory influences on the target cells.

CHAPTER 46

1. d   2. a   3. e   4. e   5. c

6. Olfactory and taste receptors are both chemosensors that respond to specific molecules in their environment. Olfactory receptors, however, are neurons, whereas taste receptors are epithelial cells that communicate with neurons that are associated with them. Olfactory receptors express a family of genes for olfactory receptor proteins that are localized on cilia that project out of the olfactory epithelium. All of these olfactory receptor proteins are G protein-linked and are metabotropic. Taste receptors are also located on cilia of the epithelial taste sensor cells. Bitter, sweet, and umami receptors are G protein-linked metabotropic receptors, but salt and sour receptors are ionotropic. The discrimination between an apple and an orange depends on integration of information from both the olfactory and the taste receptors.

7. The sensation of directional motion arises from the vestibular system, which includes the semicircular canals and the vestibule, which contains membranous structures containing a fluid—endolymph. At the base of each semicircular canal is a gelatinous projection, a cupula, that encases a cluster of stereocilia. Movement of the head causes movement of the endolymph, which then exerts force on the cupula and bends the stereocilia, generating action potentials in the vestibular nerves. The vestibule includes two membranous structures called the saccule and the utricle. In these structures, stereocilia tips are in contact with otoliths, which are membranous structures containing crystals of calcium carbonate. When the head is accelerated forward or backward, the momentum of the otoliths causes the stereocilia to bend in a direction that indicates the direction of movement.

8. Underwater, the external ear canals are filled with water. Unlike air, water is not compressible and therefore sound waves are transmitted through water as vibrational movements of the water. These movements exert greater forces on the tympanic membrane than air pressure waves do.

9. As happens in humans in the vestibular system, movements of the fish cause movements of the water in the lateral line canals. The resulting forces are transduced into action potentials by the hair cells and provide information about the movement of the fish through the water. Additionally, vibrations in the water generated by other organisms or physical events will cause movement of the water in the lateral line canal and be transduced into action potentials, providing information to the fish about its environment.

10. The owl depends on auditory stimuli to locate the mouse in total darkness. Directional information comes from the bilateral placement of the ears, which are equally stimulated when the owl is directly facing the source of the sound. The face of the owl is disc-shaped, which helps collect sound waves and direct them to the ears.

CHAPTER 47

1. d   2. d   3. c   4. c   5. a

6. The stab wound must have severed the sympathetic nerves on the left side of the man's neck. Activity in these nerves causes dilation of the pupil. Severing the sympathetic nerves on the left side would remove all sympathetic activity reaching the pupil on the left side, and therefore it would be more constricted. Similarly, sympathetic activity decreases activity in the salivary glands, whereas parasympathetic activity increases salivation. Thus withdrawal of sympathetic input would release the salivary glands from any inhibition that would counteract even low levels of parasympathetic input.

7. Eyes positioned on the sides of the head enable a wider field of vision. Eyes pointing in the same direction create a narrow field of vision but make depth perception possible. You would expect prey species to benefit from wider fields of vision. You would expect predator species to benefit from depth perception, which would facilitate pursuit and capture of prey.

8. Sleepwalking is more likely to occur in non-REM sleep for two reasons: there is motor inhibition in REM sleep, which renders the individual paralyzed; and the nature of sleepwalking activities does not match with the vivid, bizarre content of REM-sleep dreams.

9. We can break this question down into the different observations. First, the loss of motor control of the right leg indicates that motor commands are ipsilateral—they descend on the same side as the limb that is being controlled. The ability to sense painful stimuli applied to the right leg but not to the left foot indicates that the pain pathways cross over to the opposite, or contralateral, side of the spinal cord before they ascend to the brain. The reflex

movement of the right leg to a stimulus applied to the left foot indicates that reflex information is processed at the local level and does not require processing at higher levels of the central nervous system. The conclusion is that motor commands in the spinal cord are ipsilateral to the muscles being controlled but that the pain information ascending to the brain is contralateral. Finally, the different responses to pain and to touch indicate that these two modalities of somatosensory information travel in different tracks: touch ipsilaterally and pain contralaterally.

10. The fact that more slow-wave activity was seen over the right frontal cortex than over the left in response to exercising and training the left hand shows that the right side of the brain controls the left side of the body, and visa versa. This increase in slow-wave activity during sleep following the exercise/training suggests that this sleep slow-wave activity reflects either a restorative process or a learning process.

CHAPTER 48

1. b   2. c   3. b   4. d   5. c

6. One feature is the length of the muscle; half the length of a long muscle is more than half the length of a short muscle. Another feature is the location of insertion of the muscle on the bone; this determines the relative lengths of the effort arm and the load arm of the lever system created by the muscle, bone, and joint. If the ratio of load to effort arm is small, a large movement that can generate only relatively small forces is possible. If the ratio of load to effort arm is large, only small movements that can exert large force are possible.

7. If the break and healing have damaged the epiphyseal plate, and the primary and secondary areas of ossification fuse, the bone can no longer grow at that end.

8. The shoulders will fatigue first because they are not normally responsible for maintaining posture; they are adapted for rapid movements and sudden applications of large force. Thus shoulder muscles have a higher proportion of fast-twitch fibers. The leg muscles are postural muscles and have a higher proportion of slow-twitch fibers.

9. The action potential is conducted throughout the muscle cell by the system of T tubules. In the T tubules, the action potential causes conformational change of the DHP–ryanodine receptor complex. That change opens Ca^{2+} channels in the sarcoplasmic reticulum, and Ca^{2+} diffuses into and throughout the sarcoplasm. Ca^{2+} binds with the troponin units, causing the tropomyosin to expose the actin–myosin binding sites, cross-bridges to form, and the muscle to contract. When the Ca^{2+} concentration in the sarcoplasm falls as a result of being pumped back into the sarcoplasmic reticulum, the process reverses and actin–myosin binding sites are no longer available. The difference in time course of the contraction versus the action potential is due to the time that it takes for the Ca^{2+} to be released, diffuse throughout the sarcoplasm, and then be sequestered back into the sarcoplasmic reticulum.

10. The increased amount and duration of Ca^{2+} in the sarcoplasm causes increased contraction of the muscles and therefore an increase in muscle tension. The increase in muscle tension requires additional expenditure of ATP, raising metabolism and producing more heat. The increased metabolism causes elevated heart rate. This is in addition to the effect of the increased Ca^{2+} in the cardiac muscle itself.

CHAPTER 49

1. d   2. e   3. b   4. c   5. a

6. This fish would not be very active. It would move slowly. It would have a larger heart and larger blood vessels than fishes with hemoglobin, to accommodate a high flow of blood at low pressure. Its gill membranes would be well developed. It would have a high blood volume. Its jaws would show adaptations for sit-and-wait capture rather than pursuit. This fish occurs only in Antarctic waters because Antarctic waters are very cold and therefore the solubility of O_2 in those waters is high.

7. The total surface area for gas exchange is much smaller in a large air cavity than it is in many smaller cavities (alveoli) that add up to the same total volume. If the lung tissue is less elastic, the vital capacity of the lungs will go down, meaning less air can be exchanged during the breathing cycle. The less elastic lung tissue is also less permeable to respiratory gases.

8. Close to the end of inhalation, the pleural cavity pressure is reaching its maximum negative value. At the same time, the alveolar pressure is rising back up to being the same as the atmospheric pressure. Alveolar pressure would be most positive relative to atmospheric pressure at the midpoint of the exhalation phase.

9. Blood cells require energy. When in storage, their initial energy source is the glucose in the blood plasma, but as that supply gets depleted, blood cells also metabolize the intermediates in the glycolytic pathway. That includes 2,3-BPG. As the 2,3-BPG gets metabolized, there is less of it to bind to deoxygenated hemoglobin and therefore the affinity of the hemoglobin for O_2 increases. When

the affinity of the hemoglobin for O_2 gets too high, it can lower the P_{O_2} in the plasma to levels that are below the P_{O_2} in the plasma of a patient.

10. When you go up in altitude, the P_{O_2} in the air you breath goes down, but the P_{CO_2} was already low at low altitude and remains low at high altitude. Thus at higher altitude there is less of a concentration gradient driving diffusion of O_2 into the blood, but no decrease in the concentration gradient driving CO_2 out of the blood. Thus the main stimulus for breathing goes down as the need to increase breathing goes up. As a result, the blood becomes hypoxic and triggers breathing by activating the carotid and aortic chemosensors. The increase in breathing blows off even more CO_2, and breathing slows, which causes another bout of hypoxia and even a rise in blood CO_2. That triggers another bout of rapid breathing, and this cycle repeats.

11. (a) The llama hemoglobin has a higher affinity for O_2.

(b) Llama hemoglobin would be advantageous at high altitudes because it can become 100 percent saturated at the low P_{O_2} of the high-altitude environment. Therefore the hemoglobin can carry a full load of O_2 to the tissues.

(c) Llama hemoglobin allows the transfer of O_2 to occur at lower tissue P_{O_2}s.

CHAPTER 50

1. a   2. c   3. d   4. c   5. e

6. One factor is that at the beginning of a race there is a feedforward signal from the sympathetic nervous system that increases heart rate. Another factor is that the increased heart rate, together with the increased venous return to the heart from the exercising muscles, stretches the ventricles, which then contract with more force as described by the Frank–Starling law. This is due to the fact that a slight stretching of the sarcomeres optimizes the overlap of the actin and myosin fibrils for a maximum contraction. Increased breathing also increases the venous return and induces the Frank–Starling law.

7. There is no time when all four heart valves are open at the same time; if there were, the heart could not pump efficiently. Throughout diastole, the aortic and pulmonary valves are closed and the atrioventricular valves are open. At the beginning of systole, the atrioventricular valves close. There is a brief moment when all four valves are closed, until the aortic and pulmonary valves open, and stay open until the end of systole.

8. There are rapid responses and longer-term responses. A fall in blood pressure lowers the firing rate of baroreceptors in the great arteries. The decrease in baroreceptor input to areas of the brainstem that regulate cardiac function results in increased sympathetic and decreased parasympathetic output to the heart. This increases the heart rate and the force of contraction of the cardiac muscle (the fight-or-flight response). A slower response to blood loss is mediated by the kidney, which responds to the decreased blood pressure by increasing the release of renin, which in turn increases the activation of angiotensin circulating in the blood. Active angiotensin increases blood pressure by constricting peripheral blood vessels and stimulating thirst. Another slow response, stimulated by the fall in baroreceptor activity, is the release of ADH from the posterior pituitary. ADH increases the reabsorption of water by the kidney.

9. The cardiac muscle must be capable of generating and conducting action potentials. This may involve different variants of the ion channels involved in action potential generation and conduction. The cardiac muscle must be able to convert the action potential into the opening of Ca^{2+} channels in the sarcoplasmic reticulum, so there could be adaptive changes in the DHP and ryanodine receptors. Once Ca^{2+} is released into the sarcoplasm, it has to be resequestered into the sarcoplasmic reticulum, and that Ca^{2+} pump is likely to be adapted to operate at lower temperatures in the hibernator.

10. Your graph should look like this:

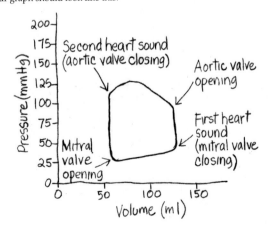

A similar graph for the right ventricle would have the same volumes but lower pressures.

CHAPTER 51

1. e   2. a   3. b   4. d   5. b

6. The rationale for a high-fat and high-protein diet is that it minimizes the secretion of insulin. Insulin promotes the uptake, metabolism, and storage of glucose by various cells of the body; it also inhibits the action of lipase in the adipose tissue. Thus with low insulin, tissues are more likely to metabolize fats and less likely to store them.

7. The following points might be included in your answer.

- Insulin stimulates most cells of the body to take up glucose from the blood by stimulating the insertion of glucose transporters into the plasma membranes of those cells in the absorptive state.
- Insulin inhibits lipase in the adipose tissue, so breakdown of stored lipids is decreased in the absorptive state.
- Insulin stimulates the synthesis of triglycerides in the adipose tissue.
- Insulin activates the enzyme that phosphorylates glucose as it enters cells, thereby preventing it from diffusing back out of the cells. This maximizes the uptake of glucose by cells.
- Insulin activates the liver enzymes that synthesize glycogen.
- The lack of insulin in the postabsorptive state decreases the uptake of glucose by most cells of the body and activates the enzymes of lipolysis and glycogenolysis.

8. Triglyceride in food is emulsified by bile in the duodenum to form micelles, which are broken down by pancreatic lipase into free fatty acids and monoglycerides. Both are absorbed across the plasma membranes of intestinal epithelial cells. In the intestinal epithelial cells, the free fatty acids and monoglycerides are resynthesized into triglycerides and packaged into chylomicrons, which also contain cholesterol and are coated with lipoproteins. The chylomicrons are secreted from the basal ends of the epithelial cells into the center of the intestinal villi, where they enter the lymphatic vessels and circulate through the lymphatic vessels to the thoracic duct, where they enter the blood.

A *direct route* the fatty acid could take would be for the chylomicrons circulating in the blood to come into contact with the damaged endothelium of the coronary arteries and be absorbed into the plaque.

In an *indirect route*, the chylomicrons could be taken up by liver or adipose tissue cells and the triglyceride stored. In the liver, the triglyceride can be repackaged to form low-density lipoproteins or very low-density lipoprotein particles, depending on the amount of cholesterol in the particle. These lipoproteins leave the liver and circulate in the blood. When they come into contact with the damaged endothelial cells of the coronary arteries, they bind to lipoprotein receptors, and their triglyceride and cholesterol are absorbed into the plaque.

9. Carbonic anhydrase catalyzes the hydration of CO_2 to produce carbonic acid that dissociates into H^+ ions and HCO_3^+ ions. In the parietal cells of the stomach, H^+ is secreted into gastric pits and then flows into the stomach lumen. The bicarbonate ions are transported out of the basal side of the cells, where they are absorbed into the blood, raising blood pH. In the ducts of the pancreas, bicarbonate is transported into the lumen of the ducts; H^+ ions are transported out of the basal sides of the cells, where they are absorbed into the blood, lowering its pH.

10. The hypothalamic regulatory pathways control hunger and satiety, not energy use. This tells us that the side of the energy balance equation that is most important in contributing to obesity is Calories in. Of course, obesity can also result in decreased physical activity, which means that decreased Calories out can be a secondary factor in the causation of obesity.

CHAPTER 52

1. d   2. a   3. a   4. b   5. a

6. The glucose contributes to the osmotic concentration of the glomerular filtrate and therefore of the tubular fluid. This results in a greater volume of urine flowing through the collecting ducts and being excreted.

7. ACE inhibitors decrease the production of angiotensin II, the active form, from angiotensin I. Decreasing the level of angiotensin in the blood increases the glomerular filtration rate and therefore the production of urine. Losing more water in the urine lowers the blood volume and therefore the blood pressure. Blocking angiotensin also results in dilation of peripheral blood vessels, which lowers blood pressure, and decreases thirst, which helps maintain a lower vascular volume. Angiotensin stimulates the release of aldosterone, which promotes Na^+ reabsorption and therefore water retention.

8. The rate at which the inulin is filtered is equal to the concentration of inulin in the blood ($[I_b]$) times the glomerular filtration rate (GFR). The rate at which inulin is excreted is equal to the concentration of inulin in the urine ($[I_u]$) times

the urine flow rate (V = 1 ml/min). Since all inulin that is filtered leaves the body in the urine, the rate at which it is filtered must equal the rate at which it is excreted. Therefore $[I_b] \times GFR = [I_u] \times V$, and $GFR = [I_u] \times V/[I_b]$.

9. Any difference in the rate at which the substance (S) is filtered versus the rate at which it is excreted will be due to tubular reabsorption of tubular secretion. If more of S is excreted than filtered, then S must be secreted by the renal tubules. If less of S is excreted than filtered, then S must be reabsorbed by the tubules. Therefore $[S_b] \times GFR - [S_u] \times V$ = the rate of reabsorption (if negative) or secretion (if positive) of S.

10. At high altitude, the concentration difference driving O_2 across the alveolar membranes and into the blood deceases, but the concentration difference driving CO_2 across the alveolar membranes and into the expired air does not change. The increased breathing rate driven by hypoxia will therefore blow off too much CO_2 and the pH of the blood will increase, which will suppress breathing. Bicarbonate is filtered into the glomerular fluid; as that fluid passes through the renal tubules, H^+ ions are excreted, resulting in the formation of H_2CO_3, which dissociates into CO_2 and H_2O. The CO_2 is reabsorbed into the tubule cells, where carbonic anhydrase catalyses its hydration to H^+ ions that are secreted back into the tubular fluid and bicarbonate ions that are secreted back into the extracellular fluid. Blocking carbonic anhydrase results in fewer H^+ ions being secreted into the tubular fluid and more bicarbonate ions remaining in the tubular fluid to be excreted. The retention of H^+ and excretion of bicarbonate lower the blood pH, and that stimulates increased breathing. However, the increased bicarbonate in the tubular fluid raises the osmotic concentration of that fluid, causing more water to be excreted.

CHAPTER 53

1. b   2. e   3. c   4. a   5. d   6. e   7. a   8. e

9. The development of brain circuitry controlling the sexual dimorphism in urination behavior is influenced by the levels of estrogen or testosterone that are circulating in the blood during the early postnatal period. Estrogen must prevent the development of the neuronal patterns of connectivity that are responsible for the male behavior.

10. The major difference between the eusocial insects and vertebrates is the haplodiploid mechanism of sex determination in the insects. This means that a female worker that is a daughter of the queen shares more genes with sisters—the queen's offspring—than she would with her own offspring, which would have a different father. Thus raising a sister contributes more to the female worker's inclusive fitness than raising a daughter would. Haplodiploidy is not found among vertebrates, so this powerful selective force does not operate among them.

11. The variability in the undirected song allows the male to adapt his song to the local variant. The directed song is probably more effective in attracting a female because it more accurately identifies the male as a member of the local population, and is also more effective in competition with neighboring males.

12. Since male cowbirds will not hear the song of his father, his song pattern should be genetically determined. Females probably learn the song of host species and thereby learn to identify and locate potential host nests. You could test the hypothesis about males by raising them in isolation to see what songs develop as they mature. You could test the hypothesis about the females by giving them a choice test such as that done with the zebra finches in testing directed and undirected song preferences (see Figure 53.8). Give each female cowbird a three-chambered cage, and play the songs of host species and nonhost species in the opposite end chambers. Place the test bird in the middle chamber. Record the amount of time the female spends in each end chamber as a measure of preference. Another experiment would be to bring a host nest with cowbird eggs into the laboratory and raise the cowbirds in the presence of recordings of another species song. Then repeat the choice experiment with the host's and the other species' songs being played.

13. The classical data suggest that the hygienic behavior is controlled by two genes to give the typical Mendelian ratio of 3 to 1 with 50% hybrid, 25% homozygous dominant, and 25% homozygous recessive. The QTL analysis, however, indicates that more than two genes are involved. The difference in these results could be due to there being two major-effect genes and several modifying genes.

CHAPTER 54

1. d   2. d   3. a   4. c   5. c

6. Based on its location and its weather conditions, the mattoral should have vegetation typical of that found in other areas with Mediterranean climates. Vegetation should be (and is) tough, shrubby and fire-adapted, with slender, leathery leaves and seeds either stored in fire-safe cones or equipped with elaiosomes and dispersed by ants.

7. The fact that species in the genus are known in Australia and in southern Africa, it is likely that the genus originated before the breakup of Gondwanaland, the supercontinent comprising Antarctica, South America, Africa and Australia.

Other species in the genus thus might be found in South America (contemporary conditions are too cold to maintain spider life in Antarctica today but the possibility exists that fossil species in the genus might be found).

8. Examining the x axis of the figure suggests that, if average low temperatures shift upward by five degrees C, the tundra biome would experience a substantial decrease in geographic extent (and may cease to exist altogether); boreal forest might also experience a reduction in geographic extent.

9. The extensive radiation of *Drosophila* species in Hawaii suggests that the genus originated here. Hawaii is an isolated island system and its distance from any continents means that dispersal to such remote islands of ancestral species would likely be very rare. Other lines of evidence to suggest that the genus originated in Hawaii would be whether fossil ancestral taxa are found only in Hawaii or anywhere else in the world. If ancestral taxa are not known from any other continent, it's very likely that the genus originated and subsequently diversified in Hawaii (and then dispersed elsewhere).

CHAPTER 55

1. d   2. c   3. b   4. d   5. a   6. d

7. In both of these cases, human management strategies will be working against the organisms' intrinsic rates of increase. Populations of long-lived organisms with low reproductive rates grow slowly. Such organisms can be categorized as *K*-strategists, which tend to persist at or near the carrying capacity of the environment. They are adapted to predictable environments, and they tend to be more specialized in their resource use, and less tolerant of variation in resource quality, than other organisms. They produce few offspring, but each offspring has a high probability of surviving to adulthood. Recall that the number of births in a population tends to be highest when that population is well below its carrying capacity. For large, long-lived species that we wish to harvest, we should manage the population so that it is far enough below the carrying capacity to have a high birth rate. But some species (such as whales) reproduce so slowly that they cannot sustain any kind of substantial harvest rate. Short-lived organisms with high reproductive rates can be categorized as *r*-strategists. These organisms can generally use a wide variety of resources and tolerate a wide range of conditions, and they can produce large numbers of offspring when conditions are suitable. If we wish to decrease the numbers of a short-lived, rapidly reproducing pest species (such as rats), killing individuals will only increase the birth rate. A better approach is to reduce the species' resources (e.g., clean up garbage) in order to decrease the carrying capacity for the species. In each case, of course, managers need to understand the specific life history and population dynamics of the species they wish to manage.

8. Humans are subject to the same population dynamics that other species are. Resource abundance is a density-dependent population-regulating factor. Like the reindeer population on St. Matthew, the Irish population crashed when its food supply was diminished. Three kinds of changes in the rates of demographic events contributed to the decrease in population size. Recall that $N_1 = N_0 + (B - D) + (I - E)$. First, the emigration rate (E) increased. Second, the age at first reproduction, and thus generation time, increased, so the birth rate (B) decreased. In other words, the population's life history traits changed with changing environmental conditions. Finally, as a direct effect of the food shortage, the death rate (D) increased. A look at the social history of the Irish potato famine might tell us more about the roles of the Irish population's own growth before the famine and whether intraspecific competition (discussed further in Chapter 56) was involved in regulating the food supply—whether other populations monopolized resources and limited the access of the Irish people to those resources.

9. Those who support the view that biological controls should not be used under any circumstances might cite the example of the cane toad in Australia, which not only failed to control the cane beetles it was introduced to control but became a serious pest in its own right. They might note that many species introduced into new regions, where their normal predators and pathogens are absent, reach population densities much higher than those in their native ranges, and they might argue that there is no reason to think this generality would not apply to species introduced as biological control agents. Those who support the view that biological controls can be used safely and effectively might cite the example of the successful control of the cottony-cushion scale in California, which was brought under control within a year by the introduction of a predaceous ladybeetle and a parasitic fly. They might also argue that horror stories like that of the cane toad could be avoided by proper study of the ecology of the proposed biological control agents before they are introduced. Studies in test plots prior to release almost certainly would have revealed that Australian cane beetles stay high on the upper stalks of cane plants, out of reach of the toads, and studies of the toads' life histories might have revealed their generalized and voracious appetites. Strict requirements for extensive testing for specificity and efficacy prior to release can greatly reduce the risk that biological control agents themselves will become pests after they are introduced. But opponents of biological control might respond that, because natural systems are so complex, even careful study might fail to reveal the real risks of introducing a particular species into a new environment.

10. Corridors have to be defined in terms of specific organisms and their dispersal abilities. Corridors consist of habitat between patches through which the organisms of interest can move. An area that serves as a corridor for birds might not be effective as a corridor for small arthropods. On the other hand, the small arthropods would need less habitat area to maintain a viable population. Thus, designing a single study to determine the effects of corridors would be very difficult. A single experiment might be able to determine effects of corridors on animals that are similar in size and mobility but not if organisms differ widely in those attributes. To understand the effects of corridors in fragmented habitats, it is very important to consider multiple organisms because organisms interact within these habitats. Investigators in the Palenque National Park in Mexico discovered that birds are more likely to be recaptured in home forest patches connected by corridors to the patches where they were released than in home forest patches unconnected to the patches where they were released. However, their ability to navigate these corridors successfully depends on the presence of other species, including predators, and their ability to survive in home patches depends on the presence of other species, including prey species, as well. Designing a single study to determine the effects of corridors would be very difficult.

CHAPTER 56

1. a   2. c   3. c   4. d   5. e

6. The interactions among ants, cacti, and pollinators described in the Working With Data exercise represent a diversity of types of interactions. By fending off potential herbivores, the pugnacious ant bodyguards act as mutualists of the cactus, as do the bee species that visit the flowers and serve as pollinators. The five ant species all appear to use the extrafloral nectaries on this plant in similar ways and thus may be competitors for extrafloral nectar. Each type of interaction depends on the relative abundance and activities of the interacting species. Cactus plants that grow where there are no herbivorous insects may have no need of pugnacious bodyguards; under those circumstances, the ants might be considered parasites for removing extrafloral nectar without providing any defensive services. The mutualism between ants and plants could also break down if bee pollinators are scarce and the most aggressive ant bodyguard prevents any bees at all from pollinating the flowers.

7. Which pine trees are susceptible to mountain pine beetle attack could be determined by direct observation and by experimentation. Within infested stands, investigators may be able to identify individual pine trees that do not harbor beetle populations and characterize properties that may make them resistant to the beetles (e.g., ability to produce large quantities of resin). Conversely, trees with especially high beetle populations might have properties that make them particularly susceptible (having a history of surviving fire or lightning strike). Experiments can also be conducted under controlled conditions, in which investigators test various species to determine if beetles display a preference for particular species. The fact that this beetle has a symbiotic partner upon which it depends in order to feed on pine trees suggests that a novel strategy for managing the outbreak could be to identify a fungicide that kills the symbiont, thereby rendering the beetle incapable of colonizing and killing the trees. Although there are no such programs currently in use today, many researchers are exploring this dimension of interaction ecology to devise novel methods for pest management.

8. (a) By establishing a microbial population that excludes undesirable species, the poultry industry is applying the principle of competitive exclusion. The principle states that two or more species utilizing a limited resource in similar ways cannot coexist. In the broiler chicks given a culture of three species of bacteria, a microbial community was established in which introduced *Salmonella* could not compete.

   (b) Other ecological outcomes this experiment might have produced include ultimate domination of the gut flora by only one species of bacterium or coexistence of all four bacterial species. Whether these species coexist or whether one or more species goes extinct depends on the availability of resources in the chicken intestines.

   (c) The principle of competitive exclusion might be useful in tackling other problems involving a community of organisms growing under confined conditions. This ecological principle provides part of the rationale for the use of probiotics by humans to improve a variety of conditions. Probiotics are live organisms that are consumed in food for health benefits, which are thought to accrue by altering the microbial balance, inhibiting the growth of deleterious species. Probiotics are being investigation for treatment of intestinal inflammatory diseases, pathogen-related diarrhea, and infections of the urogenital tract.

9. If parasites and their hosts coevolve, then the phylogenetic relationships among parasites should reflect the phylogenetic relationships among the hosts. DNA analysis revealed that flamingoes are actually more closely related to grebes than they are to ducks and geese. This relationship leads to the prediction that lice on flamingoes should be more closely related to the lice on grebes than they are to the lice on ducks and geese. Modern methods of molecular analysis that can be used to determine relationships among bird lice and their

hosts include constructing a DNA-based phylogeny of multiple species of waterbirds and their lice and then comparing the phylogenies of the hosts and parasites to see if they are congruent (see Chapter 22). In addition to acquiring parasites by shared ancestry between hosts and parasites, bird species may also acquire parasites by virtue of the fact that they share habitats and come in contact with other bird species, each of which has its own parasite fauna. Because flamingoes, ducks, grebes, and geese are all waterbirds, the possibility exists that flamingoes may have acquired some of its louse parasites by this process of host switching.

10. Among the requirements for a mutualistic pollination system is behavior by the pollinator that ensures it will visit more than one individual of the same plant species. Visiting more than one individual provides a pollinator with the opportunity to carry pollen from one plant individual to the receptive stigmatic surface of another individual of the same species. A pollinator that encounters a feeding deterrent that limits the amount of nectar it can imbibe in a single visit is more likely to continue foraging for nectar on another plant individual. The process of taking a larger number of smaller meals by seeking nectar from flowers of different plant individuals increases the likelihood that the pollinator will carry pollen from one individual to another. Too much nicotine in nectar, however, may reduce the likelihood of pollination if it deters future visits to conspecifics altogether or if it impairs the behavior of its pollinator (nicotine is a neurotoxin). Another factor limiting the amount of nicotine is the cost to the plant of biosynthesizing the compound; investing in increased amounts of nicotine may leave fewer resources to invest in producing flowers, seeds and fruits.

CHAPTER 57

1. a   2. a   3. b   4. d   5. a   6. c   7. e

8. The diversity of microbes in the human gut can be compared between individuals or across populations by using the same methods employed for comparing diversity of macroscopic communities. Diversity encompasses both the number of different species present, and richness, or abundances of individuals across species. To determine which microbes might be keystone species, selective antibiotics can be used to eliminate particular species and the effects of that elimination on community composition then monitored. Applying the methods for assessing diversity that were developed for macroscopic communities to assessing diversity for microbial communities is limited, however, by our ability to isolate, identify, and quantify all of the microbial species present. Although molecular methods of identifying microbial species has vastly expanded this capacity, there remain challenges in recognizing and categorizing the full expanse of microbial diversity.

9. According to the theory of island biogeography, the number of species on an island represents a balance between the rate at which species immigrate to and colonize the island and the rate at which resident species go locally extinct. With increasing distance from a source pool, the equilibrium number of species on an island decreases; with increasing size of an island, the species number increases. The pattern of hawk moth diversity documented by Beck and Hitching in the 113 islands of Thailand and mainland Malaysia conforms to several of the predictions of island biogeography. The continental source of colonists includes over 180 species. Borneo, a large island close to Thailand, has a larger number of species (between 113 and 135) than does New Guinea (with 46 to 90 species), which is roughly comparable in size and farther away from Thailand. Generally speaking, too, the prediction that larger islands support higher numbers of species is also upheld; Borneo has a larger number of species than the much smaller Philippines (between 46 and 112 species), even though the two places are about equidistant from Thailand.

10. The pattern documented by Marek Sammul, Lauri Oksanen, and M. Magi—that removal of one perennial species from plant communities resulted in an increase in biomass of its competitors in highly productive communities—has been documented in other communities. One hypothesis postulates that interspecific competition becomes more intense when productivity is very high. Goldenrod (*Solidago virgaurea*) is apparently a superior competitor that, when present in a community, can suppress other species. In less productive communities, competition is less intense, so release from competition with goldenrod does not result in increased growth of any remaining species. The results of this study parallel those of a long-term experiment at the Rothamsted Experiment Station in England, in which fertilizer added regularly to selected plots of land to increase their productivity resulted in a decline in the number of plant species compared with the other plots in the study that were unfertilized; in these less productive, unfertilized plots, species diversity remained essentially the same. An alternative hypothesis could be that goldenrod inhibits the growth of co-occurring plant species (as some colonizing species do in early stages of succession). This hypothesis could be tested directly by extracting root exudates of goldenrod and testing their ability to inhibit germination and growth of other species in the community.

11. Whether lampreys should be eliminated as damaging parasites of game fish, or encouraged as ecosystem engineers that create nesting sites that might increase the reproductive success of game fish depends on many factors. Some of the factors are ecological; it is important to quantify the impact of existing

sea lamprey populations on survivorship of game fish as well as to estimate the lamprey population size that does not influence survivorship. As well, the beneficial impact of nutrient enrichment and provisioning of nesting habitat should be measured. In addition, designing an ecologically sound lamprey management strategy will also require consideration of local cultural values; assessments of the economic value of the sport fishing industry to the local community and the aesthetic and cultural value placed by the local community on maintaining a more natural assemblage of fish species should be made and factored into management plans.

CHAPTER 58

1. e   2. d   3. c   4. c   5. a   6. b

7. The rate of turnover would be important to the recovery rate of a lake, and that would depend on its location. In Lake Washington, which is located in a temperate climate, turnover would occur every spring and fall. When sewage was flowing into the lake, the nutrients it contained would have led to eutrophication and thus to oxygen depletion in the bottom water. Once the flow of sewage stopped, however, biomass production would have decreased. There would have been fewer dead organisms to sink to the lake bottom, less accumulation of nutrients on the lake bottom, and less oxygen-consuming decomposition there. Fall and spring turnover would have brought the accumulated nutrients to the lake surface and oxygen to the bottom, improving conditions for organisms that were typical of the lake's preindustrial community. If the lake had been located in a climate where seasonal temperature changes were not great enough to cause turnover, the excess nutrients that had accumulated would have remained on the lake bottom, and eutrophic conditions would have persisted much longer. It's also possible that other conditions in the area might affect the lake's recovery time. Acid precipitation, for example, can affect the viability of freshwater organisms, and, if it were a problem in the region, it might slow the lake community's recovery.

8. A local effect might be nitrogen deposition. Coal—an organic fossil fuel—contains nitrogen, and its combustion would release nitrogen compounds (such as nitrogen dioxide, NO_2, and nitrous oxide, N_2O) through the smokestacks into the atmosphere. Some of this nitrogen would fall back to land in precipitation or as dry particles. The resulting increase of nitrogen in the soil would favor those plant species that are best adapted to take advantage of high nutrient levels, which would then outcompete other species. Thus the composition of the plant community would change and species diversity would be likely to decrease. Nitrogen deposition might also contribute to eutrophication in lakes, and emission of nitrogen into the atmosphere would contribute to smog.

A regional effect might be acid precipitation. The combustion of fossil fuels releases NO_2 and sulfur dioxide (SO_2) into the atmosphere; both compounds react with water molecules in the atmosphere to form nitric acid (HNO_3) and sulfuric acid (H_2SO_4), respectively. These acids can travel hundreds of kilometers in the atmosphere, so their emission would affect ecosystems far from the smokestacks. Acid precipitation can damage the leaves of plants and reduce their rate of photosynthesis, and it can reduce fish and invertebrate species richness in freshwater lakes.

A global effect would be climate change. The combustion of fossil fuels releases large amounts of CO_2, as well as lesser amounts of N_2O—both greenhouse gases. The presence of N_2O in the atmosphere also results in the production of trophospheric ozone; it, too, acts as a greenhouse gas as well as contributing to smog. The increasing concentrations of greenhouse gases in the atmosphere are already resulting in global climate warming. This climate change is having a number of worrisome effects on the global ecosystem, such as the shrinking of Arctic sea ice, rising sea level and potential coastal flooding, and profound changes in the abundances and distributions of species.

The SO_2 released when coal is burned not only produces acid rain, but also contributes to global warming. Scrubbers remove the SO_2 but the scrubbing process contributes to pollution in another form: the process generates solid waste byproducts that contain sulfur, which must be deposited in a landfill, along with other solid waste products generated by burning coal.

9. Iron (Fe) is needed by organisms in only small amounts, but it is nevertheless an essential micronutrient. It is scarce in ocean waters because it is insoluble in oxygenated water, so that iron that enters the oceans sinks rapidly to the seafloor. The experiment described in the text demonstrated that iron is a limiting nutrient in the oceans: when the investigators added dissolved iron to surface waters in the equatorial Pacific Ocean, the large phytoplankton bloom that resulted was accompanied by an increase in the uptake of nitrate and carbon dioxide, showing that these nutrients had been available but underused. This experiment showed that adding iron to ocean waters increased photosynthesis, but to better understand the effects of iron fertilization, more such experiments would have to be carried out, still on an ecosystem scale, but over a longer time span. Investigators would have to observe the effects of the iron increase on the entire food web. The fertilized ecosystem would have to be compared with an unfertilized control ecosystem far enough away from the experimental one that the added iron, and its effects, would not reach it.

10. If the "cap and trade" system worked as intended, it could put the brakes on the ongoing increases quickly, holding emissions to the level that prevailed when the law went into effect. It might also be more acceptable to polluters than an outright ban on or regulation of emissions. And it might encourage some companies to invest more in cleaner technology, since doing so might give them credits to sell, or at least spare them having to buy credits. The drawbacks might include the likelihood that the government would have to set up a system to administer and enforce the law. They also include the fact that it is not easy to pass such a law (the United States, for example, has not succeeded in doing so). The biggest polluters might be reluctant to increase their costs of doing business by paying for credits and might thus be likely to lobby against such a law. From a different viewpoint, environmentalists might argue that a cap and trade system is an inadequate response to global warming—that we must not only stop increases in emissions, but decrease them dramatically. They might also argue that there is a moral hazard in allowing anyone to "pay to pollute"—that it might legitimize pollution.

11. No one hurricane—or even several—can be ascribed to global warming. Remember the difference between weather and climate, described in Chapter 54: "Weather is the short-term state of atmospheric conditions at a particular place and time, whereas climate refers to the average atmospheric conditions, and the extent of their variation, at a particular place over a longer time. In other words, climate is what you expect; weather is what you get." But neither does the observation that hurricanes have occurred for many centuries show that the climate has *not* changed. To address this question, we would have to compile temperature and hurricane data over long periods. First, we would have to show that the average temperatures of ocean waters are increasing over time—which has been done. Second, we would have to show that warmer water is correlated with more or stronger hurricanes—that there have been more and stronger hurricanes during years, or longer periods, when the water was warmer than in periods when it was cooler. Such a correlation would supply evidence that warming of the oceans is increasing hurricane frequency and intensity. It is more difficult to demonstrate that the warming of ocean waters is caused by increasing concentrations of greenhouse gases in the atmosphere, but most scientists believe that the evidence supports that claim.

CHAPTER 59

1. b   2. e   3. e   4. a   5. d   6. c   7. b

8. Conservation biologists and others who wish to preserve biodiversity are usually working with limited resources, so they often face hard choices. They might choose to focus their efforts in biodiversity hotspots and centers of imminent extinction, but within those areas, they would face many other choices. When we discussed the principles of island biogeography, we described the species–area relationship: large islands can support a larger equilibrium number of species than small islands. The same is true of "habitat islands." Protected areas often act as habitat islands, as many are surrounded by habitat that has been made unsuitable for many species by human activities. We must also consider edge effects, keeping in mind that not all of the area we protect will actually remain suitable habitat for communities and species of interest. Therefore, if we wish to preserve natural communities with their full diversity, the larger the preserved area, the better. However, if our concern is focused on one or a few endangered species, we may wish to preserve several separated areas of habitat; that way, if a disturbance or disease should wipe out the population in one area, the entire species will not become extinct. The area or areas we choose, however, must be large enough for the species to maintain a viable population in order to avoid loss of genetic variation. We might favor areas where corridors could be maintained to allow individuals to disperse from the protected area or areas and maintain other populations. But it is rare that a protected area can be designed based on these criteria alone; the plans are also dependent on the willingness of landowners, governments, and area residents to support the preservation of the area.

9. Some might argue that if the sheep constitute only a single population, but there are other populations of pumas, that the puma should be removed from the sheep's range. However, if the puma is a keystone predator in the region, removing it might have unforeseen negative consequences for other species in the community—recall the example of wolves in Yellowstone National Park, described in Chapter 57. If neither the sheep nor the puma is an introduced species—that is, if predator and prey had survived together for a long time before becoming threatened—it might be worth taking a look at what has changed. Have the sheep experienced a loss of habitat or resources, so that their populations are now too small to withstand the rate of predation that they once did? Those observations might suggest an alternative to suppressing the puma population: Could former sheep habitat be restored so that a larger sheep population could be supported, or as a last resort, could the sheep be bred in captivity and then introduced to a new, puma-free area?

10. To some extent, international organizations already have a triage system of sorts in in place. The International Union for the Conservation of Nature (IUCN), e.g., divides species in imminent danger of extinction in all or most of

their range as "endangered" or "critically endangered", differentiating them from those who are less likely to go extinct in the near future (and thus are classified as "vulnerable"). To some degree, this classification system can result in prioritization of rescue efforts. One problem with applying the triage system of World War I to species conservation, however, is that medical science is far more successful at predicting the certainty of death than ecological science is at predicting the certainty of extinction. After all, medical science is focused on one species, which has been the subject of intense scrutiny beginning in the earliest days of scientific research. The cost of erring in assigning certainty to extinction is the loss of an entire species—a unique combination of genes that, at least with current technology, can never be reconstructed.

11. Opinion as to the extent to which ethical and moral arguments should enter into discussions of protecting biodiversity varies widely. Your answer might take into consideration a wide range of cultural, historical, and economic factors, as well as the considerations brought to bear by modern scientific knowledge.

Glossary

A

A horizon *See* topsoil.

abiotic (a' bye ah tick) [Gk. *a*: not + *bios*: life] Nonliving. (Contrast with biotic.)

abomasum The true stomach of a ruminant.

abortion Any termination of pregnancy, whether induced or natural (in which case it is called a spontaneous abortion), that occurs after a fertilized egg is successfully implanted in the uterus.

abscisic acid (ABA) (ab sighs' ik) A plant growth substance with growth-inhibiting action. Causes stomata to close; involved in a plant's response to salt and drought stress.

abscission (ab sizh' un) [L. *abscissio*: break off] The process by which leaves, petals, and fruits separate from a plant.

absorption (1) Of light: complete retention, without reflection or transmission. (2) Of water or other molecules: soaking up (taking in through pores or by diffusion).

absorption spectrum A graph of light absorption versus wavelength of light; shows how much light is absorbed at each wavelength.

absorptive heterotrophs Organisms (primarily fungi) that feed by **absorptive heterotrophy**, i.e., by secreting digestive enzymes into the environment to break down large food molecules, then absorbing the breakdown products.

absorptive state State in which food is in the gut and nutrients are being absorbed. (Contrast with postabsorptive state.)

abyssal zone (uh biss' ul) [Gk. *abyssos*: bottomless] The deepest parts of the ocean.

accessory pigments Pigments that absorb light and transfer energy to chlorophylls for photosynthesis.

accessory sex organs Anatomical structures that allow transfer of sperm from male to female for internal fertilization. (contrast with primary sex organs.)

acclimation, acclimatization Acclimation refers to increased tolerance for environmental extremes (e.g., extreme cold) after prior exposure to them. Acclimatization refers to intrinsic seasonal adjustments in the "set points" of an animal's physiological functioning (e.g., metabolic rate).

acetyl coenzyme A (acetyl CoA) A compound that reacts with oxaloacetate to produce citrate at the beginning of the citric acid cycle; a key metabolic intermediate in the formation of many compounds.

acetylcholine (ACh) A neurotransmitter that carries information across vertebrate neuromuscular junctions and some other synapses. It is then broken down by the enzyme acetylcholinesterase (AChE).

acid [L. *acidus*: sharp, sour] A substance that can release a proton in solution. (Contrast with base.)

acid growth hypothesis The hypothesis that auxin increases proton pumping, thereby lowering the pH of the cell wall and activating enzymes that loosen polysaccharides. Proposed to explain auxin-induced cell expansion in plants.

acid precipitation Precipitation that has a lower pH than normal as a result of acid-forming precursor molecules introduced into the atmosphere by human activities.

acidic Having a pH below 7.0 (i.e., a hydrogen ion concentration greater than 10^{-7} molar). (Contrast with basic.)

acoelomate An animal that does not have a coelom.

acrosome (a' krow soam) [Gk. *akros*: highest + *soma*: body] The structure at the forward tip of an animal sperm which is the first to fuse with the egg membrane and enter the egg cell.

ACTH *See* corticotropin.

actin [Gk. *aktis*: ray] A protein that makes up the cytoskeletal microfilaments in eukaryotic cells and is one of the two contractile proteins in muscle. See also myosin.

action potentials Generated by neurons, these are electrical signals that transmit information via waves of depolarization or hyperpolarization of the cell membrane.

action spectrum A graph of a biological process versus light wavelength; shows which wavelengths are involved in the process.

activation energy (E_a) The energy barrier that blocks the tendency for a chemical reaction to occur.

activator A transcription factor that stimulates transcription when it binds to a gene's promoter. (Contrast with repressor.)

active site The region on the surface of an enzyme or ribozyme where the substrate binds, and where catalysis occurs.

active transport The energy-dependent transport of a substance across a biological membrane against a concentration gradient—that is, from a region of low concentration (of that substance) to one of high concentration. (*See also* primary active transport, secondary active transport; contrast with facilitated diffusion, passive transport.)

adaptation (a dap tay' shun) (1) In evolutionary biology, a particular structure, physiological process, or behavior that makes an organism better able to survive and reproduce. Also, the evolutionary process that leads to the development or persistence of such a trait. (2) In sensory neurophysiology, a sensory cell's loss of sensitivity as a result of repeated stimulation.

adaptive defenses One of the two general types of defenses against pathogens. Involves antibody proteins and other proteins that recognize, bind to, and aid in the destruction of specific viruses and bacteria. Present only in vertebrate animals. (Contrast with innate defenses.)

adaptive radiation A series of evolutionary events that results in an array (radiation) of related species that live in a variety of environments, differing in the characteristics each uses to exploit those environments.

additive growth Population growth in which a constant number of individuals is added to the population during successive time intervals. (Contrast with multiplicative growth.)

adenine (A) (a' den een) A nitrogen-containing base found in nucleic acids, ATP, NAD, and other compounds.

adenosine triphosphate *See* ATP.

adrenal gland (a dree' nal) [L. *ad*: toward + *renes*: kidneys] An endocrine gland located near the kidneys of vertebrates, consisting of two parts, the **adrenal cortex** and **adrenal medulla**.

adrenaline *See* epinephrine.

adrenergic receptors G protein-linked receptor proteins that bind to the hormones epinephrine and norepinephrine, triggering specific responses in the target cells.

adrenocorticotropic hormone (ACTH) *See* corticotropin.

adsorption Binding of a gas or a solute to the surface of a solid.

adventitious roots (ad ven ti' shus) [L. *adventitius*: arriving from outside] Roots originating from the stem at ground level or below; typical of the fibrous root system of monocots.

aerenchyma In plants, parenchymal tissue containing air spaces.

aerobic (air oh' bic) [Gk. *aer*: air + *bios*: life] In the presence of oxygen; requiring or using oxygen (as in **aerobic metabolism**). (Contrast with anaerobic.)

afferent (af' ur unt) [L. *ad*: toward + *ferre*: to carry] Carrying to, as in neurons that carries impulses to the central nervous system (**afferent neurons**), or a blood vessel that carries blood to a structure. (Contrast with efferent.)

age structure The distribution of the individuals in a population across all age groups.

agonist A chemical substance (e.g., a neurotransmitter) that elicits a specific response in a cell or tissue. (Contrast with antagonist.)

air sacs Structures in the respiratory system of birds that receive inhaled air; they keep fresh air flowing unidirectionally through the lungs, but are not themselves gas exchange surfaces.

alcoholic fermentation *See* fermentation.

aldosterone (al dohs' ter own) A steroid hormone produced in the adrenal cortex of mammals. Promotes secretion of potassium and reabsorption of sodium in the kidney.

aleurone layer In some seeds, a tissue that lies beneath the seed coat and surrounds the endosperm. Secretes digestive enzymes that break down macromolecules stored in the endosperm.

allantoic membrane In animal development, an outgrowth of extraembryonic endoderm plus adjacent mesoderm that forms the allantois, a saclike structure that stores metabolic wastes produced by the embryo.

allantois (al' lun toh is) [Gk. *allant*: sausage] An extraembryonic membrane enclosing a sausage-shaped sac that stores the embryo's nitrogenous wastes.

allele (a leel') [Gk. *allos*: other] The alternate form of a genetic character found at a given locus on a chromosome.

allele frequency The relative proportion of a particular allele in a specific population.

allergic reaction [Ger. *allergie*: altered] An overreaction of the immune system to amounts of an antigen that do not affect most people; often involves IgE antibodies.

allopatric speciation (al' lo pat' rick) [Gk. *allos*: other + *patria*: homeland] The formation of two species from one when reproductive isolation occurs because of the interposition of (or crossing of) a physical geographic barrier such as a river. Also called geographic speciation. (Contrast with sympatric speciation.)

allopolyploidy The possession of more than two chromosome sets that are derived from more than one species.

allosteric regulation (al lo steer' ik) [Gk. *allos*: other + *stereos*: structure] Regulation of the activity of a protein (usually an enzyme) by the binding of an effector molecule to a site other than the active site.

α (alpha) helix A prevalent type of secondary protein structure; a right-handed spiral.

alternation of generations The succession of multicellular haploid and diploid phases in some sexually reproducing organisms, notably plants.

alternative splicing A process for generating different mature mRNAs from a single gene by splicing together different sets of exons during RNA processing.

altruistic Pertaining to behavior that benefits other individuals at a cost to the individual who performs it.

alveolus (al ve' o lus) (plural: alveoli) [L. *alveus*: cavity] A small, baglike cavity, especially the blind sacs of the lung.

amensalism (a men' sul ism) Interaction in which one animal is harmed and the other is unaffected. (Contrast with commensalism, mutualism.)

amine An organic compound containing an amino group (NH_2).

amine hormones Small hormone molecules synthesized from single amino acids (e.g., thyroxine and epinephrine).

amino acid An organic compound containing both NH_2 and COOH groups. Proteins are polymers of amino acids.

amino acid replacement A change in the nucleotide sequence that results in one amino acid being replaced by another.

ammonia NH_3, the most common nitrogenous waste.

ammonotelic (am moan' o teel' ic) [Gk. *telos*: end] Pertaining to an organism in which the final product of breakdown of nitrogen-containing compounds (primarily proteins) is **ammonia**. (Contrast with ureotelic, uricotelic.)

amnion (am' nee on) The fluid-filled sac within which the embryos of reptiles (including birds) and mammals develop.

amniote egg A shelled egg surrounding four extraembryonic membranes and embryo-nourishing yolk. This evolutionary adaptation permitted mammals and reptiles to live and reproduce in drier environments than can most amphibians.

amphipathic (am' fi path' ic) [Gk. *amphi*: both + *pathos*: emotion] Of a molecule, having both hydrophilic and hydrophobic regions.

amplitude The magnitude of change over the course of a regular cycle.

amygdala A component of the limbic system that is involved in fear and the memory of fearful experiences.

amylase (am' ill ase) An enzyme that catalyzes the hydrolysis of starch, usually to maltose or glucose.

anabolic reaction (an uh bah' lik) [Gk. *ana*: upward + *ballein*: to throw] A synthetic reaction in which simple molecules are linked to form more complex ones; requires an input of energy and captures it in the chemical bonds that are formed. (Contrast with catabolic reaction.)

anaerobic (an ur row' bic) [Gk. *an*: not + *aer*: air + *bios*: life] Occurring without the use of molecular oxygen, O_2. (Contrast with aerobic.)

anaphase (an' a phase) [Gk. *ana*: upward] The stage in cell nuclear division at which the first separation of sister chromatids (or, in the first meiotic division, of paired homologs) occurs.

ancestral trait The trait originally present in the ancestor of a given group; may be retained or changed in the descendants of that ancestor.

androgen (an' dro jen) Any of the several male sex steroids (most notably testosterone).

aneuploidy (an' you ploy dee) A condition in which one or more chromosomes or pieces of chromosomes are either lacking or present in excess.

angiosperms Flowering plants; one of the two major groups of living seed plants. (*See also* gymnosperms.)

angiotensin (an' jee oh ten' sin) A peptide hormone that raises blood pressure by causing peripheral vessels to constrict. Also maintains glomerular filtration by constricting efferent vessels and stimulates thirst and the release of aldosterone.

angular gyrus A part of the human brain believed to be essential for integrating spoken and written language.

animal hemisphere The metabolically active upper portion of some animal eggs, zygotes, and embryos; does not contain the dense nutrient yolk. (Contrast with vegetal hemisphere.)

anion (an' eye on) [Gk. *ana*: upward] A negatively charged ion. (Contrast with cation.)

annual A plant whose life cycle is completed in one growing season. (Contrast with biennial, perennial.)

antagonist A biochemical (e.g., a drug) that blocks the normal action of another biochemical substance.

antagonist interactions Interactions between two species in which one species benefits and the other is harmed. Includes predation, herbivory, and parasitism.

antenna system *See* light-harvesting complex.

anterior Toward or pertaining to the tip or headward region of the body axis. (Contrast with posterior.)

anterior pituitary The portion of the vertebrate pituitary gland that derives from gut epithelium. Produces trophic hormones.

anther (an' thur) [Gk. *anthos*: flower] A pollen-bearing portion of the stamen of a flower.

antheridium (an' thur id' ee um) [Gk. *antheros*: blooming] The multicellular structure that produces the sperm in nonvascular land plants and ferns.

antibody One of the myriad proteins produced by the immune system that specifically binds to a foreign substance in blood or other tissue fluids and initiates its removal from the body.

anticodon The three nucleotides in transfer RNA that pair with a complementary triplet (a codon) in messenger RNA.

antidiuretic hormone (ADH) *See* vasopressin

antigen (an' ti jun) Any substance that stimulates the production of an antibody or antibodies in the body of a vertebrate.

antigen-presenting cell In cellular immunity, a cell that ingests and digests an antigen, and then exposes fragments of that antigen to the outside of the cell, bound to proteins in the cell's plasma membrane.

antigenic determinant The specific region of an antigen that is recognized and bound by a specific antibody. Also called an epitope.

antiparallel Pertaining to molecular orientation in which a molecule or parts of a molecule have opposing directions.

antiporter A membrane transport protein that moves one substance in one direction and another in the opposite direction. (Contrast with symporter, uniporter.)

antisense RNA A single-stranded RNA molecule complementary to, and thus targeted against, an mRNA of interest to block its translation.

anus (a' nus) An opening through which solid digestive wastes are expelled, located at the posterior end of a tubular gut.

aorta (a or' tah) [Gk. *aorte*: aorta] The main trunk of the arteries leading to the systemic (as opposed to the pulmonary) circulation.

aortic body A chemosensor in the aorta that senses a decrease in blood supply or a dramatic decrease in partial pressure of oxygen in the blood.

aortic valve A one-way valve between the left ventricle of the heart and the aorta that prevents backflow of blood into the ventricle when it relaxes.

apex (a' pecks) The tip or highest point of a structure, as of a growing stem or root.

aphasia a deficit in the ability to use or understand words.

aphotic zone In bodies of water (lakes and oceans), the region below the reach of light.

apical dominance In plants, inhibition by the apical bud of the growth of axillary buds.

apical hook A form taken by the stems of many eudicot seedlings that protects the delicate shoot apex while the stem grows through the soil.

apical meristem The meristem at the tip of a shoot or root; responsible for a plant's primary growth.

apomixis (ap oh mix' is) [Gk. *apo*: away from + *mixis*: sexual intercourse] The asexual production of seeds.

apoplast (ap' oh plast) In plants, the continuous meshwork of cell walls and extracellular spaces through which material can pass without crossing a plasma membrane. (Contrast with symplast.)

apoptosis (ap uh toh' sis) A series of genetically programmed events leading to cell death.

aposematism Warning coloration; bright colors or striking patterns of toxic or toxix-mimic species that act as a warning to predators.

appendix In the human digestive system, the vestigial equivalent of the cecum (blind pouc), which serves no digestive function.

aquaporin A transport protein in plant and animal cell membranes through which water passes in osmosis.

aquatic (a kwa' tic) [L. *aqua*: water] Pertaining to or living in water. (Contrast with marine, terrestrial.)

aqueous (a' kwee us) Pertaining to water or a watery solution.

aquifer A large pool of groundwater.

archegonium (ar' ke go' nee um) The multicellular structure that produces eggs in nonvascular land plants, ferns, and gymnosperms.

archenteron (ark en' ter on) [Gk. *archos*: first + *enteron*: bowel] The earliest primordial animal digestive tract.

area phylogeny Phylogenetic tree n which the names of the taxa are replaced with the names of the places where those taxa live or lived.

arms race A series of reciprocal adaptations between species involved in antagonistic interactions, in which adaptations that increase the fitness of a consumer species exert selection pressure on its resource species to counter the consumer's adaptation, and vice versa.

arteriole A small blood vessel arising from an artery that feeds blood into a capillary bed.

artery A muscular blood vessel carrying oxygenated blood away from the heart to other parts of the body. (Contrast with vein.)

artificial insemination An infertility treatment that involves the artificial introduction of sperm into the woman's reproductive tract.

artificial selection The selection by human plant and animal breeders of individuals with certain desirable traits.

ascus (ass' cus) (plural: asci) [Gk. *askos*: bladder] In sac fungi, the club-shaped sporangium within which spores (ascospores) are produced by meiosis.

asexual reproduction Reproduction without sex.

assisted reproductive technologies (ARTs) Any of several procedures that remove unfertilized eggs from the ovary, combine them with sperm outside the body, and then place fertilized eggs or egg sperm mixtures in the appropriate location in a female's reproductive tract for development.

association cortex In the vertebrate brain, the portion of the cortex involved in higher-order information processing, so named because it integrates, or associates, information from different sensory modalities and from memory.

associative learning A form of learning in which two unrelated stimuli become linked to the same response.

asthenosphere (ass thenn' o sphere) [Gk. *asthenes*: weak] The viscous, malleable (changeable) layer of Earth's mantle. It is overlain by the solid lithospheric plates.

astrocyte [Gk. *astron*: star] A type of glial cell that contributes to the blood–brain barrier by surrounding the smallest, most permeable blood vessels in the brain.

atherosclerosis (ath' er oh sklair oh' sis) [Gk. *athero*: gruel, porridge + *skleros*: hard] A disease of the lining of the arteries characterized by fatty, cholesterol-rich deposits in the walls of the arteries. When fibroblasts infiltrate these deposits and calcium precipitates in them, the disease become arteriosclerosis, or "hardening of the arteries."

atom [Gk. *atomos*: indivisible] The smallest unit of a chemical element. Consists of a nucleus and one or more electrons.

atomic mass *See* atomic weight.

atomic number The number of protons in the nucleus of an atom; also equals the number of electrons around the neutral atom. Determines the chemical properties of the atom.

atomic weight The average of the mass numbers of a representative sample of atoms of an element, with all the isotopes in their normally occurring proportions. Also called atomic mass.

ATP (adenosine triphosphate) An energy-storage compound containing adenine, ribose, and three phosphate groups. When it is formed from ADP, useful energy is stored; when it is broken down (to ADP or AMP), energy is released to drive endergonic reactions.

ATP synthase An integral membrane protein that couples the transport of protons with the formation of ATP.

atrial natriuretic peptide A hormone released by the atrial muscle fibers of the heart when they are overly stretched, which decreases reabsorption of sodium by the kidney and thus blood volume.

atrioventricular node A modified node of cardiac muscle that organizes the action potentials that control contraction of the ventricles.

atrium (a' tree um) [L. *atrium*: central hall] An internal chamber. In the hearts of vertebrates, the thin-walled chamber(s) entered by blood on its way to the ventricle(s). Also, the outer ear.

auditory system   A sensory system that uses mechanoreceptors to convert pressure waves into receptor potentials; includes structures that gather sound waves, direct them to a sensory organ, and amplify their effect on the mechanoreceptors.

autocatalysis   [Gk. *autos*: self + *kata*: to break down] A positive feedback process in which an activated enzyme acts on other inactive molecules of the same enzyme to activate them.

autocrine   A chemical signal that binds to and affects the cell that makes it. (Contrast with paracrine.)

autoimmune diseases   Diseases (e.g., rheumatoid arthritis) that result from failure of the immune system to distinguish between self and nonself, causing it to attack tissues in the organism's own body.

autoimmunity   An immune response by an organism to its own molecules or cells.

autonomic nervous system (ANS)   The portion of the peripheral nervous system that controls such involuntary functions as those of guts and glands. Also called the involuntary nervous system.

autophagy   The programmed destruction of a cell's components.

autopolyploidy   The possession of more than two entire chromosomes sets that are derived from a single species.

autoregulatory mechanisms   In mammalian circulatory systems, local control of blood flow through capillary beds by constriction or dilation of incoming arterioles in response to local metabolite concentrations.

autosome   Any chromosome (in a eukaryote) other than a sex chromosome.

autotroph   (au′ tow trowf′) [Gk. *autos*: self + *trophe*: food] An organism that is capable of living exclusively on inorganic materials, water, and some energy source such as sunlight (photoautotrophs) or chemically reduced matter (see chemoautotrophs). (Contrast with heterotroph.)

auxin   (awk′ sin) [Gk. *auxein*: to grow] In plants, a substance (the most common being indoleacetic acid) that regulates growth and various aspects of development.

avirulence (Avr) genes   Genes in a pathogen that may trigger defenses in plants. *See* gene-for-gene resistance.

Avogadro's number   The number of atoms or molecules in a mole (weighed out in grams) of a substance, calculated to be 6.023×10^{23}.

axillary bud   A bud that forms in the angle (axil) where a leaf meets a stem.

axon   [Gk. axle] The process (branching structure) of a neuron that conducts action potentials away from the cell body. *See also* dendrites.

axon hillock   The junction between an axon and the neuron's cell body; where action potentials are generated.

axon terminal   The end portion of an axon, which passes action potentials to another cell. Axon terminals can form synapses and release neurotransmitter.

B

B cell   A type of lymphocyte involved in the humoral immune response of vertebrates. Upon recognizing an antigenic determinant, a B cell develops into a plasma cell, which secretes an antibody. (Contrast with T cell.)

B horizon   *See* subsoil.

bacillus   (bah sil′ us) [L: little rod] Any of various rod-shaped bacteria.

bacterial conjugation   *See* conjugation.

bacteriophage   (bak teer′ ee o fayj) [Gk. *bakterion*: little rod + *phagein*: to eat] Any of a group of viruses that infect bacteria. Also called phage.

bacteroids   Nitrogen-fixing organelles that develop from endosymbiotic bacteria.

bark   All tissues external to the vascular cambium of a plant.

barometric pressure   Atmospheric pressure; the total pressure of the gas mixture in air.

baroreceptor   [Gk. *baros*: weight] A pressure-sensing cell or organ. Sometimes called a stress receptor.

basal metabolic rate (BMR)   The minimum rate of energy turnover in an awake (but resting) bird or mammal that is not expending energy for thermoregulation.

base   (1) A substance that can accept a hydrogen ion in solution. (Contrast with acid.) (2) In nucleic acids, the purine or pyrimidine that is attached to each sugar in the sugar–phosphate backbone.

base pair (bp)   In double-stranded DNA, a pair of nucleotides formed by the complementary base pairing of a purine on one strand and a pyrimidine on the other. (*See* complementary base pairing.)

basic   Having a pH greater than 7.0 (i.e., having a hydrogen ion concentration lower than 10^{-7} molar). (Contrast with acidic.)

basidioma   (plural: basiomata) A fruiting structure produced by club fungi.

basidium   (bass id′ ee yum) In club fungi, the characteristic sporangium in which four **basidiospores** are formed by meiosis and then borne externally before being shed.

basilar membrane   A membrane in the human inner ear whose flexion in response to sound waves activates hair cells; flexes at different locations in response to different pitch.

Batesian mimicry   The convergence in appearance of an edible species (mimic) with an unpalatable species (model).

behavioral ecology   An evolutionary approach to the study of animal behavior that studies how behaviors are adaptive in different environmental conditions.

behaviorism   One of two classical approaches to the study of proximate causes of animal behavior, derived from the discoveries of Ivan Pavlov and focused on laboratory studies. (Compare with ethology.)

benefit   Improvement in survival and reproductive success resulting from performing a behavior or having a trait.

benthic zone   [Gk. *benthos*: bottom] The bottom of the ocean.

β (beta) pleated sheet   A type of protein secondary structure; results from hydrogen bonding between polypeptide regions running antiparallel to each other.

biased gene conversion   A mechanism of concerted evolution in which a DNA repair system appears biased in favor of using particular nucleotide sequences as templates for repair, resulting in the rapid spread of the favored sequence across all copies of the gene. (*See* concerted evolution.)

bicarbonate ion   Ion (HCO_3^-) resulting from dissociation of carbonic acid in water; important in pH regulation and carbon dioxide (CO_2) transport.

biennial   A plant whose life cycle includes vegetative growth in the first year and flowering and senescence in the second year. (Contrast with annual, perennial.)

bilateral symmetry   The condition in which only the right and left sides of an organism, divided by a single plane through the midline, are mirror images of each other.

bilayer   A structure that is two layers in thickness. In biology, most often refers to the phospholipid bilayer of membranes. (*See* phospholipid bilayer.)

bile   A secretion of the liver made up of bile salts synthesized from cholesterol, various phospholipids, and bilirubin (the breakdown product of hemoglobin). Emulsifies fats in the small intestine.

binary fission   Reproduction of a prokaryote by division of a cell into two comparable progeny cells.

binocular vision   Overlapping visual fields of an animal's two eyes; allows the animal to see in three dimensions.

binomial nomenclature   A taxonomic naming system in which each species is given a binomial (Gk.: two names), a genus name followed by a species name.

biodiversity hotspots   Regions identified by conservation biologists as being particularly in need of protection because they harbor great species richness and endemism (i.e., large numbers of species, many of which are found nowhere else).

biofilm   A community of microorganisms embedded in a polysaccharide matrix, forming a highly resistant coating on almost any moist surface.

biogeochemical cycle   Movement of inorganic elements such as nitrogen, phosphorus, and carbon through living organisms and the physical environment.

biogeographic region   One of several defined, continental-scale regions of Earth, each of which has a biota distinct from that of the others. (Contrast with biome.)

biogeography The scientific study of the patterns of distribution of populations, species, and ecological communities across Earth.

bioinformatics The use of computers and/or mathematics to analyze complex biological information, such as DNA sequences.

biological control The use of natural enemies (predators, parasites, or pathogens) to reduce the population density of an economically damaging (pest) species.

biological species concept The definition of a species as a group of actually or potentially interbreeding natural populations that are reproductively isolated from other such groups. (Contrast with lineage species concept; morphological species concept.)

biology [Gk. *bios*: life + *logos*: study] The scientific study of living things.

bioluminescence The production of light by biochemical processes in an organism.

biome (bye′ ome) A major division of the ecological communities of Earth, characterized primarily by distinctive vegetation. A given biogeographic region contains many different biomes.

bioremediation The use by humans of other organisms to remove contaminants from the environment.

biosphere (bye′ oh sphere) All regions of Earth (terrestrial and aquatic) and Earth's atmosphere in which organisms can live.

biota (bye oh′ tah) All of the organisms—animals, plants, fungi, and microorganisms—found in a given area. (Contrast with flora, fauna.)

biotechnology The use of living cells or organisms to produce materials useful to humans.

biotic (bye ah′ tick) [Gk. *bios*: life] Alive. (Contrast with abiotic.)

biotic interchange The mixing of biotas previously separated by physical, climatic, or other barriers, for example when two formerly separated land masses fuse.

blastocoel (blass′ toe seal) [Gk. *blastos*: sprout + *koilos*: hollow] The central, hollow cavity of a blastula.

blastocyst (blass′ toe cist) An early embryo formed by the first divisions of the fertilized egg (zygote). In mammals, a hollow ball of cells.

blastodisc (blass′ toe disk) An embryo that forms as a disk of cells on the surface of a large yolk mass; comparable to a blastula, but occurring in animals such as birds and reptiles, in which the massive yolk restricts complete cleavage.

blastomere Any of the cells produced by the early divisions of a fertilized animal egg.

blastopore The opening created by the invagination of the vegetal pole during gastrulation of animal embryos.

blastula (blass′ chu luh) An early stage of the animal embryo; in many species, a hollow sphere of cells surrounding a central cavity, the blastocoel. (Contrast with blastodisc.)

block to polyspermy Any of several responses to entry of a sperm into an egg that prevent more than one sperm from entering the egg.

blood A fluid connective tissue that is pumped throughout the body. A component of the circulatory system, blood transports gases such as oxygen and carbon dioxide as well as other essential elements.

blood clotting A cascade of events involving platelets and circulating proteins (clotting factors) that seals damaged blood vessels.

blood–brain barrier The selective impermeability of blood vessels in the brain that prevents most chemicals from diffusing from the blood into the brain.

blue-light receptors Pigments in plants that absorb blue light (400–500 nm). These pigments mediate many plant responses including photo-tropism, stomatal movements, and expression of some genes.

body plan The general structure of an animal, the arrangement of its organ systems, and the integrated functioning of its parts.

Bohr effect A shift in the O_2 binding curve of hemoglobin in response to excess H^+ ions such that the hemoglobin releases more O_2 in tissues where pH is low.

Bohr model A model for atomic structure that depicts the atom as largely empty space, with a central nucleus surrounded by electrons in orbits, or electron shells, at various distances from the nucleus.

bond *See* chemical bond.

bone A rigid component of vertebrate skeletal systems that contains an extracellular matrix of insoluble calcium phosphate crystals as well as collagen fibers.

Bowman's capsule An elaboration of the renal tubule, composed of podocytes, that surrounds and collects the filtrate from the glomerulus.

brain The centralized integrative center of a nervous system.

brainstem The portion of the vertebrate brain between the spinal cord and the forebrain, made up of the medulla, pons, and midbrain.

brassinosteroids Plant steroid hormones that mediate light effects promoting the elongation of stems and pollen tubes.

Broca's area A portion of the human brain essential for speech. Located in the frontal lobe just in front of the primary motor cortex.

bronchioles The smallest airways in a vertebrate lung, branching off the bronchi.

bronchus (plural: bronchi) The major airway(s) branching off the trachea into the vertebrate lung.

brown fat In mammals, fat tissue that is specialized to produce heat. It has many mitochondria and capillaries, and a protein that uncouples oxidative phosphorylation.

budding Asexual reproduction in which a more or less complete new organism grows from the body of the parent organism, eventually detaching itself.

buffer A substance that can transiently accept or release hydrogen ions and thereby resist changes in pH.

bulbourethral glands Secretory structures of the human male reproductive system that produce a small volume of an alkaline, mucoid secretion that helps neutralize acidity in the urethra and lubricate it to facilitate the passage of semen.

bulk flow The movement of a solution from a region of higher pressure potential to a region of lower pressure potential.

bundle of His Fibers of modified cardiac muscle that conduct action potentials from the atria to the ventricular muscle mass.

bundle sheath cell Part of a tissue that surrounds the veins of plants.

C

C horizon *See* parent rock.

C₃ plants Plants that produce 3PG as the first stable product of carbon fixation in photosynthesis and use ribulose bisphosphate as a CO_2 receptor.

C₄ plants Plants that produce oxaloacetate as the first stable product of carbon fixation in photosynthesis and use phosphoenolpyruvate as CO_2 acceptor. C_4 plants also perform the reactions of C_3 photosynthesis.

calcitonin Hormone produced by the thyroid gland; lowers blood calcium and promotes bone formation. (Contrast with parathyroid hormone.)

calcitriol A hormone derived from vitamin D whose actions include stimulating the cells of the digestive tract to absorb calcium from ingested food.

calorie [L. *calor*: heat] The amount of heat required to raise the temperature of 1 gram of water by 1°C. Physiologists commonly use the kilocalorie (kcal) as a unit of measure (1 kcal = 1,000 calories). Nutritionists also use the kilocalorie, but refer to it as the **Calorie** (capital C).

Calvin cycle The stage of photosynthesis in which CO_2 reacts with RuBP to form 3PG, 3PG is reduced to a sugar, and RuBP is regenerated, while other products are released to the rest of the plant. Also known as the Calvin–Benson cycle.

calyx (kay′ licks) [Gk. *kalyx*: cup] All of the sepals of a flower, collectively.

CAM *See* crassulacean acid metabolism.

Cambrian explosion The rapid diversification of multicellular life that took place during the Cambrian period.

cAMP (cyclic AMP) A compound formed from ATP that acts as a second messenger.

cancellous bone A type of bone with numerous internal cavities that make it appear spongy, although it is rigid. (Contrast with compact bone.)

canopy   The leaf-bearing part of a tree. Collectively, the aggregate of the leaves and branches of the larger woody plants of an ecological community.

capillaries   [L. *capillaris*: hair] Very small tubes, especially the smallest blood-carrying vessels of animals between the termination of the arteries and the beginnings of the veins. **Capillary beds** are networks of capillaries where materials are exchanged between the blood and the interstitial fluid.

capsid   The outer shell of a virus that encloses its nucleic acid.

carbohydrates   Organic compounds containing carbon, hydrogen, and oxygen in the ratio 1:2:1 (i.e., with the general formula $C_nH_{2n}O_n$). Common examples are sugars, starch, and cellulose.

carbon skeleton   The chains or rings of carbon atoms that form the structural basis of organic molecules. Other atoms or functional groups are attached to the carbon atoms.

carbon-fixation reactions   The phase of photosynthesis in which chemical energy captured in the light reactions is used to drive the reduction of CO_2 to form carbohydrates.

carboxylase   An enzyme that catalyzes the addition of carboxyl functional groups (O=C—OH) to a substrate.

cardiac cycle   Contraction of the two atria of the heart, followed by contraction of the two ventricles and then relaxation.

cardiac muscle   A type of muscle tissue that makes up, and is responsible for the beating of, the heart. Characterized by branching cells with single nuclei and a striated (striped) appearance. (Contrast with smooth muscle, skeletal muscle.)

cardiovascular system   [Gk. *kardia*: heart + L. *vasculum*: small vessel] The heart, blood, and vessels are of a circulatory system.

carnivore   [L. *carn*: flesh + *vorare*: to devour] An organism that eats animal tissues. (Contrast with detritivore, frugivore, herbivore, omnivore.)

carotenoid   (ka rah′ tuh noid) A yellow, orange, or red lipid pigment commonly found as an accessory pigment in photosynthesis; also found in fungi.

carotid body   A chemosensor in the carotid artery that senses a decrease in blood supply or a dramatic decrease in partial pressure of oxygen in the blood.

carpel   (kar′ pel) [Gk. *karpos*: fruit] The organ of the flower that contains one or more ovules.

carrier   (1) In facilitated diffusion, a membrane protein that binds a specific molecule and transports it through the membrane. (2) In respiratory and photosynthetic electron transport, a participating substance such as NAD that exists in both oxidized and reduced forms. (3) In genetics, a person heterozygous for a recessive trait.

carrying capacity (*K*)   The maximum number of individuals in a population (i.e., maximum population size) that can be supported by the resources present in a given environment.

cartilage   In vertebrates, a tough connective tissue found in joints, the outer ear, and elsewhere. Forms the entire skeleton in some animal groups.

cartilage bone   A type of bone that begins its development as a cartilaginous structure resembling the future mature bone, then gradually hardens into mature bone. (Contrast with membranous bone.)

Casparian strip   A band of cell wall containing suberin and lignin, found in the endodermis. Restricts the movement of water across the endodermis.

caspase   One of a group of proteases that catalyze cleavage of target proteins and are active in apoptosis.

catabolic reaction   (kat uh bah′ lik) [Gk. *kata*: to break down + *ballein*: to throw] A synthetic reaction in which complex molecules are broken down into simpler ones and energy is released. (Contrast with anabolic reaction.)

catabolite repression   In the presence of abundant glucose, the diminished synthesis of catabolic enzymes for other energy sources.

catalyst   (kat′ a list) [Gk. *kata*: to break down] A chemical substance that accelerates a reaction without itself being consumed in the overall course of the reaction. Catalysts lower the activation energy of a reaction. Enzymes are biological catalysts.

cation   (cat′ eye on) An ion with one or more positive charges. (Contrast with anion.)

caudal   [L. *cauda*: tail] Pertaining to the tail, or to the posterior part of the body.

cDNA   *See* complementary DNA.

cDNA library   A collection of complementary DNAs derived from mRNAs of a particular tissue at a particular time in the life cycle of an organism.

cecum   (see′ cum) [L. blind] A blind branch off the large intestine. In many nonruminant mammals, the cecum contains a colony of microorganisms that contribute to the digestion of food.

cell   The simplest structural unit of a living organism. In multicellular organisms, many individual cells serve as the building blocks of tissues and organs.

cell adhesion molecules (CAMs)   Molecules on animal cell surfaces that affect the selective association of cells into tissues during development of the embryo. Also a component of desmosomes.

cell cycle   The stages through which a cell passes between one mitotic division and the next. Includes all stages of interphase and mitosis. (*See* mitosis.)

cell cycle checkpoints   Points of transition between different phases of the cell cycle, which are regulated by cyclins and cyclin-dependent kinases (Cdks).

cell division   The reproduction of a cell to produce two new cells. In eukaryotes, this process involves nuclear division (mitosis) and cytoplasmic division (cytokinesis).

cell fate   The type of cell that an undifferentiated cell in an embryo will become in the adult.

cell junctions   Specialized structures associated with the plasma membranes of epithelial cells. Some contribute to cell adhesion, others to intercellular communication.

cell potency   In multicellular organisms, an undifferentiated cell's potential to become a cell of a specific type. (*See* multipotent; pluripotent; totipotent.)

cell recognition   Binding of cells to one another mediated by membrane proteins or carbohydrates.

cell theory   States that cells are the basic structural and physiological units of all living organisms, and that all cells come from preexisting cells.

cell wall   A relatively rigid structure that encloses cells of plants, fungi, many protists, and most prokaryotes, and which gives these cells their shape and limits their expansion in hypotonic media.

cellular immune response   Immune system response mediated by T cells and directed against parasites, fungi, intracellular viruses, and foreign tissues (grafts). (Contrast with humoral immune response.)

cellular respiration   The catabolic pathways by which electrons are removed from various molecules and passed through intermediate electron carriers to O_2, generating H_2O and releasing energy.

cellular specialization   In multicellular organisms, the division of labor such that different cell types become responsible for different functions (e.g., reproduction or digestion) within the organism.

cellulose   (sell′ you lowss) A straight-chain polymer of glucose molecules, used by plants as a structural supporting material.

central dogma   The premise that information flows from DNA to RNA to polypeptide (protein).

central nervous system (CNS)   That portion of the nervous system that is the site of most information processing, storage, and retrieval; in vertebrates, the brain and spinal cord. (Contrast with peripheral nervous system.)

central vacuole   In plant cells, a large organelle that stores the waste products of metabolism and maintains turgor.

centrifuge   [L. *centrum*: center + *fugere*: to flee] A laboratory device in which a sample is spun around a central axis at high speed. Used to separate suspended materials of different densities.

centriole   (sen′ tree ole) A paired organelle that helps organize the microtubules in animal and protist cells during nuclear division.

centromere (sen' tro meer) [Gk. *centron*: center + *meros*: part] The region where sister chromatids join.

centrosome (sen' tro soam) The major microtubule organizing center of an animal cell.

cephalization (sef ah luh zay' shun) [Gk. *kephale*: head] The evolutionary trend toward increasing concentration of brain and sensory organs at the anterior end of the animal.

cerebellum (sair uh bell' um) [L. diminutive of *cerebrum*, brain] The brain region that controls muscular coordination; located at the anterior end of the hindbrain.

cerebral cortex The thin layer of gray matter (neuronal cell bodies) that overlies the cerebrum.

cerebrum (su ree' brum) [L. brain] The dorsal anterior portion of the forebrain, making up the largest part of the brain of mammals; the chief coordination center of the nervous system and the major information-processing areas of the vertebrate brain consists of two **cerebral hemispheres**.

cervix (sir' vix) [L. neck] The opening of the uterus into the vagina.

cGMP (cyclic guanosine monophosphate) An intracellular messenger that is part of signal transmission pathways involving G proteins. (*See* G protein.)

channel protein An integral membrane protein that forms an aqueous passageway across the membrane in which it is inserted and through which specific solutes may pass.

chaperone A protein that guards other proteins by counteracting molecular interactions that threaten their three-dimensional structure.

character In genetics, an observable feature, such as eye color. (Contrast with trait.)

character displacement An evolutionary phenomenon in which species that compete for the same resources within the same territory tend to diverge in morphology and/or behavior.

chemical bond An attractive force stably linking two atoms.

chemical equilibrium *See* equilibrium

chemical evolution The theory that life originated through the chemical transformation of inanimate substances.

chemical reaction The change in the composition or distribution of atoms of a substance with consequent alterations in properties.

chemical synapse Neural junction at which neurotransmitter molecules released from a presynaptic cell induce changes in a postsynaptic cell. (Contrast with electrical synapse.)

chemically gated channel A type of membrane channel that opens or closes depending on the presence or absence of a specific molecule that binds either to the channel protein itself or to a separate receptor that alters the three-dimensional shape of the channel protein.

chemiosmosis Formation of ATP in mitochondria and chloroplasts, resulting from a pumping of protons across a membrane (against a gradient of electrical charge and of pH), followed by the return of the protons through a protein channel with ATP synthase activity.

chemoautotroph Organisms that obtain energy by oxidizing inorganic substances, using some of that energy to fix carbon. Also known as chemolithotrophs.

chemoheterotroph An organism that must obtain both carbon and energy from organic substances. (Contrast with chemoautotroph, photoautotroph, photoheterotroph.)

chemoreceptor A sensory receptor cell that senses specific molecules (such as odorant molecules or pheromones) in the environment.

chiasma (kie az' muh) (plural: chiasmata) [Gk. cross] An X-shaped connection between paired homologous chromosomes in prophase I of meiosis. A chiasma is the visible manifestation of crossing over between homologous chromosomes.

chief cells One of three types of secretory cell found in the gastric pits of the stomach wall. Chief cells secrete the protein-digesting enzyme pepsin. (*See* mucosal epithelium; parietal cells.)

chitin (kye' tin) [Gk. *kiton*: tunic] The characteristic tough but flexible organic component of the exoskeleton of arthropods, consisting of a complex, nitrogen-containing polysaccharide. Also found in cell walls of fungi.

chlorophyll (klor' o fill) [Gk. *kloros*: green + *phyllon*: leaf] Any of several green pigments associated with chloroplasts or with certain bacterial membranes; responsible for trapping light energy for photosynthesis.

chloroplast [Gk. *kloros*: green + *plast*: a particle] An organelle bounded by a double membrane containing the enzymes and pigments that perform photosynthesis. Chloroplasts occur only in eukaryotes.

choanocyte (ko' an uh site) The collared, flagellated feeding cells of sponges.

cholecystokinin (CCK) (ko' luh sis tuh kai' nin) A hormone produced and released by the lining of the duodenum when it is stimulated by undigested fats and proteins. It stimulates the gallbladder to release bile and slows stomach activity.

chorion (kor' ee on) [Gk. *khorion*: afterbirth] The outermost of the membranes protecting mammal, bird, and reptile embryos; in mammals it forms part of the placenta.

chromatid (kro' ma tid) A newly replicated chromosome, from the time molecular duplication occurs until the time the centromeres separate (during anaphase of mitosis or of meiosis II).

chromatin The nucleic acid–protein complex that makes up eukaryotic chromosomes.

chromatin remodeling A mechanism for epigenetic gene regulation by the alteration of chromatin structure.

chromosomal mutation Loss of or changes in position/direction of a DNA segment on a chromosome.

chromosome (krome' o sowm) [Gk. *kroma*: color + *soma*: body] In bacteria and viruses, the DNA molecule that contains most or all of the genetic information of the cell or virus. In eukaryotes, a structure composed of DNA and proteins that bears part of the genetic information of the cell.

chylomicron (ky low my' cron) Particles of lipid coated with protein, produced in the gut from dietary fats and secreted into the extracellular fluids.

chyme (kime) [Gk. *kymus*: juice] Created in the stomach; a mixture of ingested food with the digestive juices secreted by the salivary glands and the stomach lining.

cilia (sil' ee ah) (singular: cilium) [L. eyelashes] Hairlike organelle used for locomotion by many unicellular organisms and for moving water and mucus by many multicellular organisms. Generally shorter than flagella.

circadian rhythm (sir kade' ee an) [L. *circa*: approximately + *dies*: day] A rhythm of growth or activity that recurs about every 24 hours.

circannual rhythm [L. *circa*: + *annus*: year] A rhythm of growth or activity that recurs on a yearly basis.

circulatory system A physiological system consisting of a muscular pump (heart), a fluid (blood or hemolymph), and a series of conduits (blood vessels) that transports materials around the body.

11-*cis*-retinal The nonprotein, light-absorbing component of the visual pigment rhodopsin. (*See* rhodopsin.)

***cis-trans* isomers** In molecules with a double bond (typically between two carbon items), identifies on which side of the double bond similar atoms or functional groups are found. If they are on the same side, the molecule is a *cis* isomer; in a *trans* isomer, similar atoms are on opposite sides of the double bond. (*See* isomer.)

citric acid cycle In cellular respiration, a set of chemical reactions whereby acetyl CoA is oxidized to carbon dioxide and hydrogen atoms are stored as NADH and $FADH_2$. Also called the Krebs cycle.

clade [Gk. *klados*: branch] A monophyletic group made up of an ancestor and all of its descendants.

class I MHC molecules Cell surface proteins that participate in the cellular immune response directed against virus-infected cells.

class II MHC molecules Cell surface proteins that participate in the cell–cell interactions (of T-helper cells, macrophages, and B cells) of the humoral immune response.

class switching Occurs when a B cell changes the immunoglobulin class it synthesizes (e.g., a B cell making IgM switches to making IgG).

cleavage The first few cell divisions of an animal zygote. *See also* complete cleavage, incomplete cleavage.

climate The long-term average atmospheric conditions (temperature, precipitation, humidity, wind direction and velocity) found in a region. (Contrast with weather.)

climax community The final stage of succession; a community that is capable of perpetuating itself under local climatic and soil conditions and persists for a relatively long time.

clinal variation [Gk. *klinein*: to lean] Gradual change in the phenotype of a species over a geographic gradient.

cloaca The opening through which both urinary wastes and digestive wastes are expelled in most amphibians and in reptiles (including birds).

clonal deletion Inactivation or destruction of lymphocyte clones that would produce immune reactions against the animal's own body.

clonal lineages Asexually reproduced groups of nearly identical organisms.

clonal selection Mechanism by which exposure to antigen results in the activation of selected T- or B-cell clones, resulting in an immune response.

clone [Gk. *klon*: twig, shoot] (1) Genetically identical cells or organisms produced from a common ancestor by asexual means. (2) To produce many identical copies of a DNA sequence by its introduction into, and subsequent asexual reproduction of, a cell or organism.

closed circulatory system Circulatory system in which the circulating fluid is contained within a continuous system of vessels. (Contrast with open circulatory system.)

clumped dispersion pattern *See* dispersion

CO (CONSTANS) Gene coding for a transcription factor that activates the synthesis of florigen (FT); involved in the induction of flowering.

co-repressor In the regulation of bacterial operons, a molecule that binds to the repressor, causing it to change shape and bind to the operator, thereby inhibiting transcription.

coastal zone The marine life zone that extends from the shoreline to the edge of the continental shelf. Characterized by relatively shallow, well-oxygenated water and relatively stable temperatures and salinities.

coccus (kock' us) (plural: cocci) [Gk. *kokkos*: berry, pit] Any of various spherical or spheroidal bacteria.

cochlea (kock' lee uh) [Gk. *kokhlos*: snail] A spiral tube in the inner ear of vertebrates; it contains the sensory cells involved in hearing.

codominance A condition in which two alleles at a locus produce different phenotypic effects and both effects appear in heterozygotes.

codon Three nucleotides in messenger RNA that direct the placement of a particular amino acid into a polypeptide chain. (Contrast with anticodon.)

coelom (see' loam) [Gk. *koiloma*: cavity] An animal body cavity, enclosed by muscular mesoderm and lined with a mesodermal layer called peritoneum that also surrounds the internal organs.

coelomate Possessing a coelom.

coenocytic (seen' a sit ik) [Gk. *koinos*: common + *kytos*: container] Referring to the condition, found in some fungal hyphae, of "cells" containing many nuclei and enclosed by a single plasma membrane. Results from nuclear division without cytokinesis.

coenzyme A nonprotein organic molecule that plays a role in catalysis by an enzyme.

coenzyme A (CoA) A coenzyme used in various biochemical reactions as a carrier of acyl groups.

coevolution Evolutionary processes in which an adaptation in one species leads to the evolution of an adaptation in a species with which it interacts; also known as reciprocal adaptation.

cofactor An inorganic ion that is weakly bound to an enzyme and required for its activity.

cohesin A protein involved in binding chromatids together.

cohesion The tendency of molecules (or any substances) to stick together.

cohort (co' hort) [L. *cohors*: company of soldiers] A group of similar-aged organisms.

cold-hardening A process by which plants can acclimate to cooler temperatures; requires repeated exposure to cool temperatures over many days.

coleoptile A sheath that surrounds and protects the shoot apical meristem and young primary leaves of a grass seedling as they move through the soil.

collagen [Gk. *kolla*: glue] A fibrous protein found extensively in bone and connective tissue.

collecting duct In vertebrates, a tubule that receives urine produced in the nephrons of the kidney and delivers that fluid to the ureter for excretion.

collenchyma (cull eng' kyma) [Gk. *kolla*: glue + *enchyma*: infusion] A type of plant cell, living at functional maturity, which lends flexible support by virtue of primary cell walls thickened at the corners. (Contrast with parenchyma, sclerenchyma.)

colon [Gk. *kolon*] The portion of the gut between the small intestine and the anus. Also called the large intestine.

commensalism [L. *com*: together + *mensa*: table] A type of interaction between species in which one participant benefits while the other is unaffected.

communication A signal from one organism (or cell) that alters the functioning or behavior of another organism (or cell).

community Any ecologically integrated group of species of microorganisms, plants, and animals inhabiting a given area.

compact bone A type of bone with a solid, hard structure. (Contrast with cancellous bone.)

companion cell In angiosperms, a specialized cell found adjacent to a sieve tube element.

comparative experiment Experimental design in which data from various unmanipulated samples or populations are compared, but in which variables are not controlled or even necessarily identified. (Contrast with controlled experiment.)

comparative genomics Computer-aided comparison of DNA sequences between different organisms to reveal genes with related functions.

competition In ecology, use of the same resource by two or more species when the resource is present in insufficient supply for the combined needs of the species.

competitive exclusion A result of competition between species for resources, in which one species completely eliminates the other from a given habitat.

competitive inhibitor A nonsubstrate that binds to the active site of an enzyme and thereby inhibits binding of its substrate. (Contrast with noncompetitive inhibitor.)

complement system A group of eleven proteins that play a role in some reactions of the immune system. The complement proteins are not immunoglobulins.

complementary base pairing The AT (or AU), TA (or UA), CG, and GC pairing of bases in double-stranded DNA, in transcription, and between tRNA and mRNA.

complementary DNA (cDNA) DNA formed by reverse transcriptase acting with an RNA template; essential intermediate in the reproduction of retroviruses; used as a tool in recombinant DNA technology; lacks introns.

complete cleavage Pattern of cleavage that occurs in eggs that have little yolk. Early cleavage furrows divide the egg completely and the blastomeres are of similar size. (Contrast with incomplete cleavage.)

complete metamorphosis A change of state during the life cycle of an organism in which the body is almost completely rebuilt to produce an individual with a very different body form. Characteristic of insects such as butterflies, moths, beetles, ants, wasps, and flies.

complex ions Groups of covalently bonded atoms that carry an electric charge (e.g., NH_4^+, the ammonium ion).

complex life cycle In reference to parasitic species, a life cycle that requires more than one host to complete.

composite transposon Two transposable elements located near one another that transpose together and carry the intervening DNA sequence with them. (*See* transposable element.)

compound (1) A substance made up of atoms of more than one element. (2) Made up of many units, as in the **compound eyes** of arthropods.

concentration gradient A difference in concentration of an ion or other chemical substance from one location to another, often across a membrane. (*See* active transport; facilitated diffusion.)

concerted evolution The common evolution of a family of repeated genes, such that changes in one copy of the gene family are replicated in other copies of the gene family, and thus evolve "in concert." (*See* biased gene conversion; unequal crossing over.)

condensation reaction A chemical reaction in which two molecules become connected by a covalent bond and a molecule of water is released (AH + BOH → AB + H_2O.) (Contrast with hydrolysis reaction.)

conditional mutation A mutation that results in a characteristic phenotype only under certain environmental conditions.

conditioned reflex A form of associative learning first described by Ivan Pavlov, in which a natural response (such as salivation in response to food) becomes associated with a normally unrelated stimulus (such as the sound of a bell).

conduction The transfer of heat from one object to another through direct contact.

cone In conifers, a reproductive structure consisting of spore-bearing scales extending from a central axis. (Contrast with strobilus.)

cone cell In the vertebrate retina, a type of photoreceptor cell responsible for color vision.

conidium (ko nid' ee um) (plural: conidia) [Gk. *konis*: dust] A type of haploid fungal spore borne at the tips of hyphae, not enclosed in sporangia.

conjugation (kon ju gay' shun) [L. *conjugare*: yoke together] (1) A process by which DNA is passed from one cell to another through a conjugation tube, as in bacteria. (2) A nonreproductive sexual process by which *Paramecium* and other ciliates exchange genetic material.

connective tissue A type of tissue that connects or surrounds other tissues; its cells are embedded in a collagen-containing matrix. One of the four major tissue types in multicellular animals, including cartilage, bone, blood, and fat.

connexon In a gap junction, a protein channel linking adjacent animal cells.

conservation biology An applied science that carries out investigations with the aim of maintaining the diversity of life on Earth.

conserved Pertaining to a gene or trait that has evolved very slowly and is similar or even identical in individuals of highly divergent groups.

conspecifics Individuals of the same species.

constant region The portion of an immunoglobulin molecule whose amino acid composition determines its class and does not vary among immunoglobulins in that class. (Contrast with variable region.)

constitutive Always present; produced continually at a constant rate. (Contrast with inducible.)

constitutive genes Genes that are expressed all the time. (Contrast with inducible genes.)

constitutive proteins Proteins that an organism produces all the time, and at a relatively constant rate.

consumer An organism that eats the tissues of some other organism.

consumer–resource interactions Interactions in which organisms gain their nutrition by eating other living organisms or are eaten themselves.

continental drift The gradual movements of the world's continents that have occurred over billions of years.

contraception Birth control methods that prevent fertilization or implantation (conception).

contractile vacuole (kon trak' tul) A specialized vacuole that collects excess water taken in by osmosis, then contracts to expel the water from the cell.

controlled experiment An experiment in which a sample is divided into groups whereby experimental groups are exposed to manipulations of an independent variable while one group serves as an untreated control. The data from the various groups are then compared to see if there are changes in a dependent variable as a result of the experimental manipulation. (Contrast with comparative experiment.)

controlled system A set of components in a physiological system that is controlled by commands from a regulatory system. (Contrast with regulatory system.)

convection The transfer of heat to or from a surface via a moving stream of air or fluid.

convergent evolution Independent evolution of similar features from different ancestral traits.

convolutions Foldings of the vertebrate brain's cerebral cortex into ridges called gyri (sing. **gyrus**) and valleys called sulci (sing. **sulcus**). The level of cortical convolution increases taxonomically and is especially extensive in humans.

copulation Reproductive behavior that results in a male depositing sperm in the reproductive tract of a female.

cork cambium [L. *cambiare*: to exchange] In plants, a lateral meristem that produces secondary growth, mainly in the form of waxy-walled protective cells, including some of the cells that become bark.

cork In plants, a protective outermost tissue layer composed of cells with thick walls waterproofed with suberin.

cornea The clear, transparent tissue that covers the eye and allows light to pass through to the retina.

corolla (ko role' lah) [L. *corolla*: a small crown] All of the petals of a flower, collectively.

corpus luteum (kor' pus loo' tee um) (plural: corpora lutea) [L. yellow body] A structure formed from a follicle after ovulation; produces hormones important to the maintenance of pregnancy.

corridor A connection between habitat patches through which organisms can disperse; plays a critical role in maintaining subpopulations.

cortex [L. *cortex*: covering, rind] (1) In plants, the tissue between the epidermis and the vascular tissue of a stem or root. (2) In animals, the outer tissue of certain organs, such as the adrenal gland (adrenal cortex) and the brain (cerebral cortex).

corticosteroids Steroid hormones produced and released by the cortex of the adrenal gland.

corticotropin A tropic hormone produced by the anterior pituitary hormone that stimulates cortisol release from the adrenal cortex. Also called adrenocorticotropic hormone (ACTH).

corticotropin-releasing hormone A hormone produced by the hypothalamus that controls the release of cortisol from the anterior pituitary.

cortisol A corticosteroid that mediates stress responses.

cost A decrease in fitness resulting from performing a behavior or having a trait.

cost–benefit approach An approach to evolutionary studies that assumes an animal has a limited amount of time and energy to devote to each of its activities, and that each activity has fitness costs as well as benefits. (*See also* trade-off.)

cotyledon (kot' ul lee' dun) [Gk. *kotyledon*: hollow space] A "seed leaf." An embryonic organ that stores and digests reserve materials; may expand when seed germinates.

countercurrent flow An arrangement that promotes the maximum exchange of heat, or of a diffusible substance, between two fluids by having the fluids flow in opposite directions through parallel vessels close together.

countercurrent heat exchanger In "hot" fish, an adaptation of the circulatory system such that arterial blood flowing into the muscles is warmed by venous blood flowing out of the muscles, thereby conserving body heat by countercurrent exchange.

countercurrent multiplier The mechanism that increases the concentration of the interstitial fluid in the mammalian kidney through countercurrent flow in the loops of Henle and selective permeability and active transport of ions by segments of the loops of Henle.

covalent bond Chemical bond based on the sharing of electrons between two atoms.

CpG islands DNA regions rich in C residues adjacent to G residues. Especially abundant in promoters, these regions are where methylation of cytosine usually occurs.

crassulacean acid metabolism (CAM) A metabolic pathway enabling the plants that possess it to store carbon dioxide at night and then perform photosynthesis during the day with stomata closed.

critical night length In the photoperiodic flowering response of short-day plants, the length of night above which flowering occurs and below which the plant remains vegetative. (The reverse applies in the case of long-day plants.)

critical period *See* sensitive period.

crop A simple food storage sac, the first of two stomachlike organs in many animals (including reptiles, earthworms, and various insects. (*See also* gizzard.)

cross section A section taken perpendicular to the longest axis of a structure. Also called a transverse section.

crossing over The mechanism by which linked genes undergo recombination. In general, the term refers to the reciprocal exchange of corresponding segments between two homologous chromatids.

crosstalk Interactions between different signal transduction pathways.

crypsis [Gk. *kryptos*: hidden] The resemblance of an organism to some part of its environment, which helps it to escape detection by enemies.

cryptochromes [Gk. *kryptos*: hidden + *kroma*: color] Photoreceptors mediating some blue-light effects in plants and animals.

ctene (teen) [Gk. *cteis*: comb] In ctenophores, a comblike row of cilia-bearing plates. Ctenophores move by beating the cilia on their eight ctenes.

culture (1) A laboratory association of organisms under controlled conditions. (2) The collection of knowledge, tools, values, and rules that characterize a human society.

cumulus A thick gelatinous layer that protects a mammalian ovum.

cupula Gelatinous swelling in the semicircular canals of the vestibular system. A cupula encloses hair cell stereocilia that react to shifting fluid in the canal ducts.

currents Circulation patterns in the surface waters of oceans driven by the prevailing winds.

cuticle (1) In plants, a waxy layer on the outer body surface that retards water loss. (2) In ecdysozoans, an outer body covering that provides protection and support and is periodically molted.

cyclic AMP *See* cAMP.

cyclic electron transport In photosynthetic light reactions, the flow of electrons that produces ATP but no NADPH or O_2.

cyclical succession Pattern of change in community composition (succession) in which the climax community depends on periodic disturbances (e.g., fire) in order to persist. (Contrast with directional succession.)

cyclin A protein that activates a cyclin-dependent kinase, bringing about transitions in the cell cycle.

cyclin-dependent kinase (Cdk) A protein kinase whose target proteins are involved in transitions in the cell cycle and which is active only when complexed with additional protein subunits, called cyclins.

cytokine A regulatory protein made by immune system cells that affects other target cells in the immune system.

cytokinesis (sy' toe kine ee' sis) [Gk. *kytos*: container + *kinein*: to move] The division of the cytoplasm of a dividing cell. (Contrast with mitosis.)

cytokinin (sy' toe kine' in) A member of a class of plant growth substances that plays roles in senescence, cell division, and other phenomena.

cytoplasm The contents of the cell, excluding the nucleus.

cytoplasmic determinants In animal development, gene products whose spatial distribution may determine such things as embryonic axes.

cytoplasmic segregation The asymmetrical distribution of cytoplasmic determinants in a developing animal embryo.

cytosine (C) (site' oh seen) A nitrogen-containing base found in DNA and RNA.

cytoskeleton The network of microtubules and microfilaments that gives a eukaryotic cell its shape and its capacity to arrange its organelles and to move.

cytosol The fluid portion of the cytoplasm, excluding organelles and other solids.

cytotoxic T cells (T_C) Cells of the cellular immune system that recognize and directly eliminate virus-infected cells. (Contrast with T-helper cells.)

D

DAG *See* diacylglycerol.

data Quantified observations about a system under study.

daughter chromosomes During mitosis, the separated chromatids from the beginning of anaphase onward.

dead space The lung volume that fails to be ventilated with fresh air (because the lungs are never completely emptied during exhalation).

dead zones Regions in aquatic ecosystems that are devoid of aquatic life because eutrophication has resulted in severe oxygen depletion.

deciduous [L. *deciduus*: falling off] Pertaining to a woody plant that sheds its leaves but does not die.

declarative memory Memory of people, places, events, and things that can be consciously recalled and described. (Contrast with procedural memory.)

decomposer An organism that metabolizes organic compounds in debris and dead organisms, releasing inorganic material; found among the bacteria, protists, and fungi. *See also* detritivore, saprobe.

deductive logic Logical thought process that starts with a premise believed to be true then predicts what facts would also have to be true to be compatible with that premise. (Contrast with inductive logic.)

defensin A type of protein made by phagocytes that kills bacteria and enveloped viruses by insertion into their plasma membranes.

deficiency disease A condition (e.g., scurvy and beriberi) caused by chronic lack of any essential nutrient.

degeneracy The situation in which a single amino acid may be represented by any of two or more different codons in messenger RNA. Most of the amino acids can be represented by more than one codon.

deletion A mutation resulting from the loss of a continuous segment of a gene or chromosome. Such mutations almost never revert to wild type. (Contrast with duplication, point mutation.)

demethylase An enzyme that catalyzes the removal of the methyl group from cytosine, reversing DNA methylation.

demography The study of population structure and of the processes (**demographic events**, including births and deaths) by which it changes.

denaturation Loss of activity of an enzyme or nucleic acid molecule as a result of structural changes induced by heat or other means.

dendrites [Gk. *dendron*: tree] Branching fibers (processes) of a neuron. Dendrites are usually relatively short compared with the axon, and commonly carry information to the neuronal cell body.

denitrification Metabolic activity by which nitrate and nitrite ions are reduced to form nitrogen gas; carried out by certain soil bacteria.

denitrifiers Bacteria that release nitrogen to the atmosphere as nitrogen gas (N_2).

density *See* population density.

density-dependent Pertaining to an effect on population size that increases in proportion to population density.

density-independent Pertaining an effect on population size that acts independently of population density.

deoxyribonucleic acid *See* DNA.

deoxyribonucleoside triphosphates (dNTPs) The raw materials for DNA synthesis: deoxyadenosine triphosphate (dATP), deoxythymidine triphosphate (dTTP), deoxycytidine triphosphate (dCTP), and deoxyguanosine triphosphate (dGTP). Also called deoxyribonucleotides.

deoxyribose A five-carbon sugar found in nucleotides and DNA.

dependent variable In a scientific experiment, the response that is measured and analyzed as the independent variable is manipulated (See independent variable.)

depolarization A change in the resting potential across a membrane so that the inside of the cell becomes less negative, or even positive, compared with the outside of the cell. (Contrast with hyperpolarization.)

derived trait A trait that differs from the ancestral trait. (Contrast with synapomorphy.)

dermal tissue system The outer covering of a plant, consisting of epidermis in the young plant and periderm in a plant with extensive secondary growth. (Contrast with ground tissue system and vascular tissue system.)

descent with modification Darwin's premise that all species share a common ancestor and have diverged from one another gradually over time.

desmosome (dez' mo sowm) [Gk. *desmos*: bond + *soma*: body] An adhering junction between animal cells.

desmotubule A membrane extension connecting the endoplasmic retitulum of two plant cells that traverses the plasmodesma.

determinate growth A growth pattern in which the growth of an organism or organ ceases when an adult state is reached; characteristic of most animals and some plant organs. (Contrast with indeterminate growth.)

determination In development, the process whereby the fate of an embryonic cell or group of cells (e.g., to become epidermal cells or neurons) is set (becomes **determined**).

detritivore (di try' ti vore) [L. *detritus*: worn away + *vorare*: to devour] An organism that obtains its energy from the dead bodies or waste products (**detritus**) of other organisms.

development The process by which a multicellular organism, beginning with a single cell, goes through a series of changes, taking on the successive forms that characterize its life cycle.

developmental plasticity The capacity of an organism to alter its pattern of development in response to environmental conditions.

diacylglycerol (DAG) In hormone action, the second messenger produced by hydrolytic removal of the head group of certain phospholipids.

diapause A period of developmental or reproductive arrest, entered in response to day length, that enables an organism to better survive.

diaphragm (dye' uh fram) [Gk. *diaphrassein*: barricade] (1) A sheet of muscle that separates the thoracic and abdominal cavities in mammals; responsible for breathing. (2) A method of birth control in which a sheet of rubber is fitted over the woman's cervix, blocking the entry of sperm.

diastole (dye ass' toll ee) [Gk. dilation] The portion of the cardiac cycle when the heart muscle relaxes. (Contrast with systole.)

dichotomous (dye cot' oh mus) [Gk. *dichot*: split in two; *tomia*: removed) A branching pattern in which the shoot divides at the apex producing two equivalent branches that subsequently never overlap.

diencephalon The portion of the vertebrate forebrain that develops into the thalamus and hypothalamus.

differential gene expression The hyposthesis that, given that all cells contain all genes, what makes one cell type different from another is the difference in transcription and translation of those genes.

differentiation The process whereby originally similar cells follow different developmental pathways; the actual expression of determination.

diffuse coevolution The evolution of similar traits in suites of species experiencing similar selection pressures imposed by other suites of species with which they interact.

diffusion Random movement of molecules or other particles, resulting in even distribution of the particles when no barriers are present.

digestive vacuole In protists, an organelle specialized for digesting food ingested by endocytosis.

dihybrid cross A mating in which the parents differ with respect to the alleles of two loci of interest.

dikaryon (di care' ee ahn) [Gk. *di*: two + *karyon*: kernel] A cell or organism carrying two genetically distinguishable nuclei. Common in fungi.

dioecious (die eesh' us) [Gk. *di*: two + *oikos*: house] Pertaining to organisms in which the two sexes are "housed" in two different individuals, so that eggs and sperm are not produced in the same individuals. Examples: humans, fruit flies, date palms. (Contrast with monoecious.)

diploblastic Having two cell layers. (Contrast with triploblastic.)

diploid (dip' loid) [Gk. *diplos*: double] Having a chromosome complement consisting of two copies (homologs) of each chromosome. Designated 2*n*. (Contrast with haploid.)

direct development Pattern of development (notably among insects) in which hatchlings look like miniature versions of adults. (Contrast with metamorphosis.)

direct fitness That component of fitness resulting from an organism producing its own offspring. (Contrast with inclusive fitness, kin selection.)

directional selection Selection in which phenotypes at one extreme of the population distribution are favored. (Contrast with disruptive selection, stabilizing selection.)

directional succession Change in community composition after a disturbance (succession) that is characterized by an orderly progression culminating in a persistent state (the climax community). (Contrast with cyclical succession.)

disaccharide A carbohydrate made up of two monosaccharides (simple sugars).

discoidal cleavage In animal development, a type of incomplete cleavage that is common in fishes, reptiles, and birds, the eggs of which contain a dense yolk mass.

dispersal Movement of organisms away from a parent organism or from an existing population.

dispersion The distribution of individuals in space within a population. **Clumped dispersion** occurs when individuals tend to occupy the same space; **regular dispersion** is when the presence of one individual decreases the probability of another individual occupying the same space; and **random dispersion** assumes there is equal probability of any individual occupying any given space.

disruptive selection Selection in which phenotypes at both extremes of the population distribution are favored. (Contrast with directional selection; stabilizing selection.)

distal Away from the point of attachment or other reference point. (Contrast with proximal.)

distal convoluted tubule The portion of a renal tubule from where it reaches the renal cortex, just past the loop of Henle to where it joins a collecting duct. (Compare with proximal convoluted tubule.)

disturbance A short-term event that disrupts populations, communities, or ecosystems by changing the environment.

disulfide bridge The covalent bond between two sulfur atoms (–S—S–) linking two molecules or remote parts of the same molecule.

DNA (deoxyribonucleic acid) The fundamental hereditary material of all living organisms. In eukaryotes, stored primarily in the cell nucleus. A nucleic acid using deoxyribose rather than ribose.

DNA fingerprint An individual's unique pattern of allele sequences, commonly short tandem repeats and single nucleotide polymorphisms.

DNA helicase An enzyme that unwinds the double helix.

DNA ligase Enzyme that unites broken DNA strands during replication and recombination.

DNA methylation The addition of methyl groups to bases in DNA, usually cytosine or guanine.

DNA methyltransferase An enzyme that catalyzes the methylation of DNA.

DNA microarray A small glass or plastic square onto which thousands of single-stranded DNA sequences are fixed so that hybridization of cell-derived RNA or DNA to the target sequences can be performed.

DNA polymerase Any of a group of enzymes that catalyze the formation of DNA strands from a DNA template.

DNA replication The creation of a new strand of DNA in which DNA polymerase catalyzes the exact reproduction of an existing (template) strand of DNA.

DNA transposons Mobile genetic elements that move without making an RNA intermediate. (Contrast with retrotransposons.)

domain (1) An independent structural element within a protein. Encoded by recognizable nucleotide sequences, a domain often folds separately from the rest of the protein. Similar domains can appear in a variety of different proteins across phylogenetic groups (e.g., "homeobox domain"; "calcium-binding domain"). (2) In phylogenetics, the three monophyletic branches of life (Bacteria, Archaea, and Eukarya).

dominance In genetics, the ability of one allelic form of a gene to determine the phenotype of a heterozygous individual in which the homologous chromosomes carry both it and a different (recessive) allele. (Contrast with recessive.)

dormancy A condition in which normal activity is suspended, as in some spores, seeds, and buds.

dorsal [L. *dorsum*: back] Toward or pertaining to the back or upper surface. (Contrast with ventral.)

dorsal lip In amphibian embryos, the dorsal segment of the blastopore. Also called the "organizer," this region directs the development of nearby embryonic regions.

double fertilization In angiosperms, a process in which the nuclei of two sperm fertilize one egg. One sperm's nucleus combines with the egg nucleus to produce a zygote, while the other combines with the same egg's two polar nuclei to produce the first cell of the triploid endosperm (the tissue that will nourish the growing plant embryo).

double helix Refers to DNA and the (usually right-handed) coil configuration of two complementary, antiparallel strands.

downregulation A negative feedback process in which continuous high concentrations of a hormone can decrease the number of its receptors. (Contrast with upregulation.)

duodenum (do' uh dee' num) The beginning portion of the vertebrate small intestine. (Contrast with ileum, jejunum.)

duplication A mutation in which a segment of a chromosome is duplicated, often by the attachment of a segment lost from its homolog. (Contrast with deletion.)

E

ecdysone (eck die' sone) [Gk. *ek*: out of + *dyo*: to clothe] In insects, a hormone that induces molting.

ecological economics Interdisciplinary field that works to assess the economic value of biodiversity.

ecological efficiency The overall transfer of energy from one trophic level to the next, expressed as the ratio of consumer production to producer production.

ecological survivorship curve *See* survivorship curves

ecological system One or more organisms plus the external environment with which they interact.

ecology [Gk. *oikos*: house] The scientific study of the interaction of organisms with their living (biotic) and nonliving (abiotic) environments.

ecosystem (eek' oh sis tum) The organisms of a particular habitat, such as a pond or forest, together with the physical environment in which they live.

ecosystem engineer An organism that builds structures that alter existing habitats or create new habitats.

ecosystem services Processes by which ecosystems maintain resources that benefit human society.

ecotourism Ecologically responsible travel to natural places.

ectoderm [Gk. *ektos*: outside + *derma*: skin] The outermost of the three embryonic germ layers first delineated during gastrulation. Gives rise to the skin, sense organs, and nervous system.

ectotherm [Gk. *ektos*: outside + *thermos*: heat] An animal that is dependent on external heat sources for regulating its body temperature (Contrast with endotherm.)

edema (i dee' mah) [Gk. *oidema*: swelling] Tissue swelling caused by the accumulation of fluid.

edge effects Changes in ecological processes in a community caused by physical and biological factors originating in an adjacent community.

effector A component of a physiological system that responds to information by *effecting* changes (making change happen) in the internal environment; examples include muscles and the secretory cells of the digestive tract.

effector cells In cellular immunity, B cells and T cells that attack an antigen, either by secreting antibodies that bind to the antigen or by releasing molecules that destroy any cell bearing the antigen.

effector protein In cell signaling, a protein responsible for the cellular reponse to a signal transduction pathway.

efferent (ef' ur unt) [L. *ex*: out + *ferre*: to bear] Carrying outward or away from, as in neurons that carry impulses outward from the central to the peripheral nervous system (**efferent neurons**), or a blood vessel that carries blood away from a structure. (Contrast with afferent.)

egg In all sexually reproducing organisms, the female gamete; in birds, reptiles, and some other vertebrates, a structure within which early embryonic development occurs. *See also* amniote egg, ovum.

electrical synapse A type of synapse at which action potentials spread directly from presynaptic cell to postsynaptic cell. (Contrast with chemical synapse.)

electrocardiogram (ECG or EKG) A graphic recording of electrical potentials from the heart.

electrochemical gradient The concentration gradient of an ion across a membrane plus the voltage difference across that membrane.

electroencephalogram (EEG) A graphic recording of electrical potentials from the brain.

electromagnetic radiation A self-propagating wave that travels though space and has both electrical and magnetic properties.

electron A subatomic particle outside the nucleus carrying a negative charge and very little mass.

electron shell The region surrounding the atomic nucleus at a fixed energy level in which electrons orbit.

electron transport The passage of electrons through a series of proteins with a release of energy which may be captured in a concentration gradient or in chemical form such as NADH or ATP.

electronegativity The tendency of an atom to attract electrons when it occurs as part of a compound.

electrophoresis *See* gel electrophoresis.

element A substance that cannot be converted to a simpler substance by ordinary chemical means.

elongation (1) In molecular biology, the addition of monomers to make a longer RNA or protein during transcription or translation. (2) Growth of a plant axis or cell primarily in the longitudinal direction.

embolus (em' buh lus) [Gk. *embolos*: stopper] A circulating blood clot. Blockage of a blood vessel by an embolus or a bubble of gas is called an **embolism**. (Contrast with thrombus.)

embryo [Gk. *en*: within + *bryein*: to grow] A young animal, or young plant sporophyte, while it is still contained within a protective structure such as a seed, egg, or uterus.

embryo sac In angiosperms, the female gametophyte. Found within the ovule, it consists of eight or fewer cells, membrane bounded, but without cellulose walls between them.

embryonic stem cell (ESC) A pluripotent cell in the blastocyst.

emergent property A property of a complex system that is not exhibited by its individual component parts.

emigration The deliberate and usually oriented departure of an organism from the habitat in which it has been living.

endemic (en dem' ik) [Gk. *endemos*: native] Confined to a particular region, thus often having a comparatively restricted distribution.

endergonic A chemical reaction in which the products have higher free energy than the reactants, thereby requiring free energy input to occur. (Contrast with exergonic.)

endocrine cells Cells that secrete substances into the extracellular fluid. (*See also* endocrine gland.)

endocrine gland (en' doh krin) [Gk. *endo*: within + *krinein*: to separate] An aggregation of secretory cells that secretes hormones into the blood. The endocrine system consists of all endocrine cells and endocrine glands in the body that produce and release hormones. (Contrast with exocrine gland.)

endocytosis A process by which liquids or solid particles are taken up by a cell through invagination of the plasma membrane. (Contrast with exocytosis.)

endoderm [Gk. *endo*: within + *derma*: skin] The innermost of the three embryonic germ layers delineated during gastrulation. Gives rise to the digestive and respiratory tracts and structures associated with them.

endodermis In plants, a specialized cell layer marking the inside of the cortex in roots and some stems. Frequently a barrier to free diffusion of solutes.

endomembrane system A system of intracellular membranes that exchange material with one another, consisting of the Golgi apparatus, endoplasmic reticulum, and lysosomes when present.

endometrium The epithelial lining of the uterus.

endoplasmic reticulum (ER) [Gk. *endo*: within + L. *reticulum*: net] A system of membranous tubes and flattened sacs found in the cytoplasm of eukaryotes. Exists in two forms: rough ER, studded with ribosomes; and smooth ER, lacking ribosomes.

endorphins Molecules in the mammalian brain that act as neurotransmitters in pathways that control pain.

endoskeleton [Gk. *endo*: within + *skleros*: hard] An internal skeleton covered by other, soft body tissues. (Contrast with exoskeleton.)

endosperm [Gk. *endo*: within + *sperma*: seed] A specialized triploid seed tissue found only in angiosperms; contains stored nutrients for the developing embryo.

endospore [Gk. *endo*: within + *spora*: to sow] In some bacteria, a resting structure that can survive harsh environmental conditions.

endosymbiosis theory [Gk. *endo*: within + *sym*: together + *bios*: life] The theory that the eukaryotic cell evolved via the engulfing of one prokaryotic cell by another.

endothelium The single layer of epithelial cells lining the interior of a blood vessel.

endotherm [Gk. *endo*: within + *thermos*: heat] An animal that can control its body temperature by the expenditure of its own metabolic energy. (Contrast with ectotherm.)

endotoxin A lipopolysaccharide that forms part of the outer membrane of certain Gram-negative bacteria that is released when the bacteria grow or lyse. (Contrast with exotoxin.)

energetic cost The difference between the energy an animal expends in performing a behavior and the energy it would have expended had it rested.

energy The capacity to do work or move matter against an opposing force. The capacity to accomplish change in physical and chemical systems.

energy budget A quantitative description of all paths of energy exchange between an animal and its environment.

enhancers Regulatory DNA sequences that bind transcription factors that either activate or increase the rate of transcription.

enkephalins Molecules in the mammalian brain that act as neurotransmitters in pathways that control pain.

enteric nervous system The nerve nets in the submucosa and between the smooth muscle layers of the vertebrate gut.

enthalpy (H) The total energy of a system.

entrain To advance or delay an organism's circadian clock each day so that it is in phase with the light-dark cycle of the organism's environment.

entropy (S) (en' tro pee) [Gk. *tropein*: to change] A measure of the degree of disorder in any system. Spontaneous reactions in a closed system are always accompanied by an increase in entropy.

enveloped virus A virus enclosed within a phospholipid membrane derived from its host cell.

environment Whatever surrounds and interacts with or otherwise affects a population, organism, or cell. May be external or internal.

environmental genomics Sequencing technique used when biologists are unable to work with the whole genome of a prokaryote species but instead examine individual genes collected from a random sample of the organism's environment.

environmental resistance Reduction in a population's growth rate caused by preemption of available resources by other individuals in the population.

environmentalism The use of ecological knowledge, along with economics, ethics, and many other considerations, to inform both personal decisions and public policy relating to stewardship of natural resources and ecosystems.

enzyme (en' zime) [Gk. *zyme*: to leaven (as in yeast bread)] A catalytic protein that speeds up a biochemical reaction.

enzyme–substrate complex (ES) An intermediate in an enzyme-catalyzed reaction; consists of the enzyme bound to its substrate(s).

epi- [Gk. upon, over] A prefix used to designate a structure located on top of another; for example, epidermis, epiphyte.

epiblast The upper or overlying portion of the avian blastula which is joined to the hypoblast at the margins of the blastodisc.

epiboly The movement of cells over the surface of the blastula toward the forming blastopore.

epidermis [Gk. *epi*: over + *derma*: skin] In plants and animals, the outermost cell layers. (Only one cell layer thick in plants.)

epididymis (epuh did' uh mus) [Gk. *epi*: over + *didymos*: testicle] Coiled tubules in the testes that store sperm and conduct sperm from the seminiferous tubules to the vas deferens.

epigenetics The scientific study of changes in the expression of a gene or set of genes that occur without change in the DNA sequence.

epinephrine (ep i nef' rin) [Gk. *epi*: over + *nephros*: kidney] The "fight or flight" hormone produced by the medulla of the adrenal gland; it also functions as a neurotransmitter. (Also known as adrenaline.)

epistasis Interaction between genes in which the presence of a particular allele of one gene determines whether another gene will be expressed.

epithelial tissue A type of animal tissue made up of sheets of cells that lines or covers organs, makes up tubules, and covers the surface of the body; one of the four major tissue types in multicellular animals.

epitope *See* antigenic determinant.

equilibrium Any state of balanced opposing forces and no net change.

equilibrium potential The membrane potential at which an ion is at electrochemical equilibrium, i.e., there is no net flux of the ion across the membrane.

ER *See* endoplasmic reticulum.

error signal In regulatory systems, any difference between the set point of the system and its current condition.

erythrocyte (ur rith' row site) [Gk. *erythros*: red + *kytos*: container] A red blood cell.

erythropoietin A hormone produced by the kidney in response to lack of oxygen that stimulates the production of red blood cells.

esophagus (i soff' i gus) [Gk. *oisophagos*: gullet] That part of the gut between the pharynx and the stomach.

essential amino acids   Amino acids that an animal cannot synthesize for itself and must obtain from its food.

essential element   A mineral nutrient required for normal growth and reproduction in plants and animals.

essential fatty acids   Fatty acids that an animal cannot synthesize for itself and must obtain from its food.

ester linkage   A condensation (water-releasing) reaction in which the carboxyl group of a fatty acid reacts with the hydroxyl group of an alcohol. Lipids, including most membrane lipids, are formed in this way. (Contrast with ether linkage.)

estivation   (ess tuh vay' shun) [L. *aestivalis*: summer] A state of dormancy and hypometabolism that occurs during the summer; usually a means of surviving drought and/or intense heat. (Contrast with hibernation.)

estrogen   Any of several steroid sex hormones; produced chiefly by the ovaries in mammals.

estrus   (es' trus) [L. *oestrus*: frenzy] The period of heat, or maximum sexual receptivity, in some female mammals. Ordinarily, the estrus is also the time of release of eggs in the female.

estuary   Aquatic biome in which salt water and fresh water mix, as when a river meets the ocean. Includes such ecosystems as salt marshes and mangrove forests.

ether linkage   The linkage of two hydrocarbons by an oxygen atom (HC—O—CH). Ether linkages are characteristic of the membrane lipids of the Archaea. (Contrast with ester linkage.)

ethology   [Gk. *ethos*: character + *logos*: study] An approach to the study of animal behavior that focuses on studying many species in natural environments and addresses questions about the evolution of behavior. (Compare with behaviorism.)

ethylene   One of the plant growth hormones, the gas $H_2C=CH_2$. Involved in fruit ripening and other growth and developmental responses.

euchromatin   Diffuse, uncondensed chromatin. Contains active genes that will be transcribed into mRNA. (Contrast with heterochromatin.)

eudicots   Angiosperms with two embryonic cotyledons. (*See also* monocots.)

Eukarya   One of the three domains of life; organisms made up of one or more eukaryotic cells. (*See also* eukaryotes.)

eukaryotes   (yew car' ree oats) [Gk. *eu*: true + *karyon*: kernel or nucleus] Organisms whose cells contain their genetic material inside a nucleus. Includes all life other than the viruses, archaea, and bacteria. (Contrast with prokaryotes.)

eusocial   Pertaining to a social group that includes nonreproductive individuals, as in honey bees.

eustachian tube   A connection between the middle ear and the throat that allows air pressure to equilibrate between the middle ear and the outside world.

eutrophication   (yoo trofe' ik ay' shun) [Gk. *eu*: truly + *trephein*: to flourish] The addition of nutrient materials to a body of water, resulting in changes in ecological processes and species composition therein.

evaporation   The transition of water from the liquid to the gaseous phase.

evolution   Any gradual change. Most often refers to organic or Darwinian evolution, which is the genetic and resulting phenotypic change in populations of organisms from generation to generation. (*See* macroevolution, microevolution; contrast with speciation.)

evolutionary developmental biology (evo-devo)   The study of the interplay between evolutionary and developmental processes, with a focus on the genetic changes that give rise to novel morphology. Key concepts of evo-devo include modularity, genetic toolkits, genetic switches, and heterochrony.

evolutionary radiation   The proliferation of many species within a single evolutionary lineage.

evolutionary reversal   The reappearance of an ancestral trait in a group that had previously acquired a derived trait.

evolutionary theory   The understanding and application of the mechanisms of evolutionary change to biological problems.

excision repair   DNA repair mechanism that removes damaged DNA and replaces it with the appro-priate nucleotide.

excitable   Capable of generating an action potential.

excitatory   Input from a neuron that causes depolarization of the recipient cell.

excited state   The state of an atom or molecule when, after absorbing energy, it has more energy than in its normal, ground state.

excretion   Release of metabolic wastes by an organism.

excretory systems   In animals, organs that maintain the volume, solute concentration, and composition of the extracellular fluid by excreting water, solutes, and nitrogenous wastes in the form of urine.

exergonic   A chemical reaction in which the products of the reaction have lower free energy than the reactants, resulting in a release of free energy. (Contrast with endergonic.)

exocrine gland   (eks' oh krin) [Gk. *exo*: outside + *krinein*: to separate] Any gland, such as a salivary gland, that secretes to the outside of the body or into the gut. (Contrast with endocrine gland.)

exocytosis   A process by which a vesicle within a cell fuses with the plasma membrane and releases its contents to the outside. (Contrast with endocytosis.)

exon   A portion of a DNA molecule, in eukaryotes, that codes for part of a polypeptide. (Contrast with intron.)

exoskeleton   (eks' oh skel' e ton) [Gk. *exos*: outside + *skleros*: hard] A hard covering on the outside of the body to which muscles are attached. (Contrast with endoskeleton.)

exotoxin   A highly toxic, usually soluble protein released by living, multiplying bacteria. (Contrast with endotoxin.)

expanding triplet repeat   A three-base-pair sequence in a human gene that is unstable and can be repeated a few to hundreds of times. Often, the more the repeats, the less the activity of the gene involved. Expanding triplet repeats occur in some human diseases such as Huntington's disease and fragile-X syndrome.

experiment   A testing process to support or disprove hypotheses and to answer questions. The basis of the scientific method. *See* comparative experiment, controlled experiment.

expiratory reserve volume   The amount of air that can be forcefully exhaled beyond the normal tidal expiration. (Contrast with inspiratory reserve volume, tidal volume, vital capacity.)

exploitation competition   Competition in which individuals reduce the quantities of their shared resources. (Contrast with interference competition.)

exponential growth   Growth, especially in the number of organisms in a population, which is a geometric function of the size of the growing entity: the larger the entity, the faster it grows: (Contrast with logistic growth.)

expression vector   A DNA vector, such as a plasmid, that carries a DNA sequence for the expression of an inserted gene into mRNA and protein in a host cell.

expressivity   The degree to which a genotype is expressed in the phenotype; may be affected by the environment.

extensor   A muscle that extends an appendage. (Contrast with flexor.)

external fertilization   The release of gametes into the environment; typical of aquatic animals. Also called spawning. (Contrast with internal fertilization.)

external gills   Highly branched and folded extensions of the body surface that provide a large surface area for gas exchange with water; typical of larval amphibians and many larval insects.

extinction   The termination of a lineage of organisms.

extracellular matrix   A material of heterogeneous composition surrounding cells and performing many functions including adhesion of cells.

extraembryonic membranes   Four membranes that support but are not part of the developing embryos of reptiles, birds, and mammals, defining these groups phylogenetically as amniotes. (*See* amnion, allantois, chorion, yolk sac.)

extreme halophiles   A group of euryarchaeotes that live exclusively in very salty environments.

extremophiles   Archaea and bacteria that live and thrive under conditions (e.g., extremely high temperatures) that would kill most organisms.

eye cups   Photosensory organs in flatworms; components of one of the simplest visual systems in animals.

F

5′ end   (5 prime) The end of a DNA or RNA strand that has a free phosphate group at the 5′ carbon of the sugar (deoxyribose or ribose).

F_1   The first filial generation; the immediate progeny of a parental (P) mating.

F_2   The second filial generation; the immediate progeny of a mating between members of the F1 generation.

facilitated diffusion   Passive movement through a membrane involving a specific carrier protein; does not proceed against a concentration gradient. (Contrast with active transport, diffusion.)

facilitation   In succession, modification of the environment by a colonizing species in a way that allows colonization by other species. (Contrast with inhibition.)

facultative anaerobe   A prokaryote that can shift its metabolism between anaerobic and aerobic modes depending on the presence or absence of O_2. (Alternatively, facultative aerobe.)

fast-twitch fibers   Skeletal muscle fibers that can generate high tension rapidly, but fatigue rapidly ("sprinter" fibers). Characterized by an abundance of enzymes of glycolysis. (Compare to slow-twitch fibers.)

fat   (1) A triglyceride that is solid at room temperature. (Contrast with oil.) (2) Adipose tissue, one type of connective tissue. (See brown fat, white fat.)

fate map   A diagram of the blastula showing which cells (blastomeres) are "fated " to contribute to specific tissues and organs in the mature body.

fatty acid   A molecule made up of a long nonpolar hydrocarbon chain and a polar carboxyl group. Found in many lipids.

fauna   (faw′ nah) All the animals found in a given area. (Contrast with flora.)

FD (FLOWERING LOCUS D)   Gene coding for a transcription factor in the shoot apical meristem that binds to florigen; involved in the induction of flowering.

feces   [L. *faeces*: dregs] Waste excreted from the digestive system.

fecundity   The average number of offspring produced by each female.

feedback   In regulatory systems, information about the relationship between the set point of the system and its current state (Contrast with feedforward information).

feedback inhibition   A mechanism for regulating a metabolic pathway in which the end product of the pathway can bind to and inhibit the enzyme that catalyzes the first committed step in the pathway. Also called end-product inhibition.

feedforward information   In regulatory systems, information that changes the set point of the system. (Contrast with feedback.)

fermentation   (fur men tay′ shun) [L. *fermentum*: yeast] The anaerobic degradation of a substance such as glucose to smaller molecules such as lactic acid or alcohol with the extraction of energy.

fertilization   Union of gametes. Also known as syngamy.

fertilizer   Any of a number of substances added to soil to improve the soil's capacity to support plant growth. May be organic or inorganic.

fetus   Medical and legal term for the stages of a developing human embryo from about the eighth week of pregnancy (the point at which all major organ systems have formed) to the moment of birth.

fiber   In angiosperms, an elongated, tapering sclerenchyma cell, usually with a thick cell wall, that serves as a support function in xylem. (*See also* muscle fiber.)

fibrin   A protein that polymerizes to form long threads that provide structure to a blood clot.

fibrinogen   A circulating protein that can be stimulated to fall out of solution and provide the structure for a blood clot.

fibrous root system   A root system typical of monocots composed of numerous thin adventitious roots that are all roughly equal in diameter. (Contrast with taproot system.)

Fick's law of diffusion   An equation that describes the factors that determine the rate of diffusion of a molecule from an area of higher concentration to an area of lower concentration.

fight-or-flight response   A rapid physiological response to a sudden threat mediated by the hormone epinephrine.

filament   In flowers, the part of a stamen that supports the anther.

filter feeder   An organism that feeds on organisms much smaller than itself that are suspended in water or air by means of a straining device.

first filial generation   See F_1.

first law of thermodynamics   The principle that energy can be neither created nor destroyed.

fission   *See* binary fission.

fitness   The contribution of a genotype or phenotype to the genetic composition of subsequent generations, relative to the contribution of other genotypes or phenotypes. (*See* also inclusive fitness.)

fixed action pattern   In ethology, a genetically determined behavior that is performed without learning, stereotypic (performed the same way each time), and not modifiable by learning.

flagellum   (fla jell′ um) (plural: flagella) [L. *flagellum*: whip] Long, whiplike appendage that propels cells. Prokaryotic flagella differ sharply from those found in eukaryotes.

flexor   A muscle that flexes an appendage. (Contrast with extensor.)

flora   (flore′ ah) All of the plants found in a given area. (Contrast with fauna.)

floral meristem   In angiosperms, a meristem that forms the floral organs (sepals, petals, stamens, and carpels).

floral organ identity genes   In angiosperms, genes that determine the fates of floral meristem cells; their expression is triggered by the products of meristem identity genes.

florigen   A plant hormone involved in the conversion of a vegetative shoot apex to a flower.

flower   The sexual structure of an angiosperm.

fluid feeder   An animal that feeds on fluids it extracts from the bodies of other organisms; examples include nectar-feeding birds and blood-sucking insects.

fluid mosaic model   A molecular model for the structure of biological membranes consisting of a fluid phospholipid bilayer in which suspended proteins are free to move in the plane of the bilayer.

flux   [L.: flow] In ecology, the flow of an element into or out of a compartment of the biosphere.

follicle   [L. *folliculus*: little bag] In female mammals, an immature egg surrounded by nutritive cells.

follicle-stimulating hormone (FSH)   A gonadotropin produced by the anterior pituitary.

food chain   A portion of a food web, most commonly a simple sequence of prey species and the predators that consume them.

food web   The complete set of food links between species in a community; a diagram indicating which ones are the eaters and which are eaten.

forebrain   The region of the vertebrate brain that comprises the cerebrum, thalamus, and hypothalamus.

fossil   Any recognizable structure originating from an organism, or any impression from such a structure, that has been preserved over geological time.

fossil fuels   Fuels, including oil, natural gas, coal, and peat, formed over geologic time from organic material buried in anaerobic sediments.

founder effect   Random changes in allele frequencies resulting from establishment of a population by a very small number of individuals.

fovea [L. *fovea*: a small pit] In the vertebrate retina, the area of most distinct vision.

frame-shift mutation The addition or deletion of a single or two adjacent nucleotides in a gene's sequence. Results in the misreading of mRNA during translation and the production of a nonfunctional protein. (Contrast with missense mutation, nonsense mutation, silent mutation.)

Frank–Starling law The stroke volume of the heart increases with increased return of blood to the heart.

free energy (*G*) Energy that is available for doing useful work, after allowance has been made for the increase or decrease of disorder.

frequency-dependent selection Selection that changes in intensity with the proportion of individuals in a population having the trait.

frontal lobe The largest of the brain lobes in humans; involved with feeling and planning functions; includes the primary motor cortex.

frugivore [L. *frugis*; fruit + *vorare*: to devour] An animal that eats fruit.

fruit In angiosperms, a ripened and mature ovary (or group of ovaries) containing the seeds. Sometimes applied to reproductive structures of other groups of plants.

FT (*FLOWERING LOCUS T*) Gene that codes for florigen, a small, diffusible protein involved in the induction of flowering.

fugitive species A species that leave an otherwise suitable habitat in order to avoid competition with another species.

full census A count of every individual in a population. Can only be achieve if individuals are large and distinct enough to be identifiable by the census taker; population sizes are more usually estimated using sampling methods.

functional genomics The assignment of functional roles to the proteins encoded by genes identified by sequencing entire genomes.

functional group A characteristic combination of atoms that contributes specific properties (such as charge or polarity) when attached to larger molecules (e.g., carboxyl group; amino group).

fundamental niche A species' niche as defined by its physiological capabilities. (Contrast with realized niche.)

G

G protein A membrane protein involved in signal transduction; characterized by binding GDP or GTP.

G protein–linked receptors A class of receptors that change configuration upon ligand binding such that a G protein binding site is exposed on the cytoplasmic domain of the receptor, initiating a signal transduction pathway.

G1 In the cell cycle, the gap between the end of mitosis and the onset of the S phase.

G1-to-S transition In the cell cycle, the point at which G1 ends and the S phase begins.

G2 In the cell cycle, the gap between the S (synthesis) phase and the onset of mitosis.

gain of function mutation A mutation that results in a protein with a new function. (Contrast with loss of function mutation.)

gallbladder In the human digestive system, an organ in which bile is stored.

gametangium (gam uh tan' gee um) (plural: gametangia) [Gk. *gamos*: marriage + *angeion*: vessel] Any plant or fungal structure within which a gamete is formed.

gamete (gam' eet) [Gk. *gamete/ gametes*: wife, husband] The mature sexual reproductive cell: the egg or the sperm.

gametogenesis (ga meet' oh jen' e sis) The specialized series of cellular divisions that leads to the production of gametes. (*See also* oogenesis, spermatogenesis.)

gametophyte (ga meet' oh fyte) In plants and photosynthetic protists with alternation of generations, the multicellular haploid phase that produces the gametes. (Contrast with sporophyte.)

ganglion (gang' glee un) (plural: ganglia) [Gk. lump] A cluster of neurons that have similar characteristics or function.

ganglion cells Cells at the front of the human retina that transmit information from the bipolar cells to the brain.

gap genes In *Drosophila* development, segmentation genes that define broad areas along the anterior–posterior axis of the early embryo. Part of a developmental cascade that includes maternal effect genes, pair rule genes, segment polarity genes, and Hox genes.

gap junction A 2.7-nanometer gap between plasma membranes of two animal cells, spanned by protein channels. Gap junctions allow chemical substances or electrical signals to pass from cell to cell.

gastric pits Deep infoldings in the walls of the stomach lined with secretory cells.

gastrin A hormone secreted by cells in the lower region of the stomach that stimulates the secretion of digestive juices as well as movements of the stomach.

gastrovascular cavity Serving for both digestion (gastro) and circulation (vascular); in particular, the central cavity of the body of jellyfish and other cnidarians.

gastrulation Development of a blastula into a gastrula. In embryonic development, the process by which a blastula is transformed by massive movements of cells into a *gastrula*, an embryo with three germ layers and distinct body axes.

gated channel A membrane protein that changes its three-dimensional shape, and therefore its ion conductance, in response to a stimulus. When open, it allows specific ions to move across the membrane.

gel electrophoresis (e lek' tro fo ree' sis) [L. *electrum*: amber + Gk. *phorein*: to bear] A technique for separating molecules (such as DNA fragments) from one another on the basis of their electric charges and molecular weights by applying an electric field to a gel.

gene [Gk. *genes*: to produce] A unit of heredity. Used here as the unit of genetic function which carries the information for a polypeptide or RNA.

gene duplication The generation of extra copies of a gene in a genome over evolutionary time. A mechanism by which genomes can acquire new functions.

gene expression The transcription and translation into a protein of the information (nucleotide sequence) contained in a gene.

gene family A set of similar genes derived from a single parent gene; need not be on the same chromosomes. The vertebrate globin genes constitute a classic example of a gene family.

gene flow Exchange of genes between populations through migration of individuals or movements of gametes.

gene pool All of the different alleles of all of the genes existing in all individuals of a population.

gene therapy Treatment of a genetic disease by providing patients with cells containing functioning alleles of the genes that are nonfunctional in their bodies.

gene tree A graphic representation of the evolutionary relationships of a single gene in different species or of the members of a gene family.

gene-for-gene concept In plants, a mechanism of resistance to pathogens in which resistance is triggered by the specific interaction of the products of a pathogen's *Avr* genes and a plant's *R* genes.

general transcription factors In eukaryotes, transcription factors that bind to the promoters of most protein-coding genes and are required for their expression. Distinct from transcription factors that have specific regulatory effects only at certain promoters or classes of promoters.

genetic code The set of instructions, in the form of nucleotide triplets, that translate a linear sequence of nucleotides in mRNA into a linear sequence of amino acids in a protein.

genetic drift Changes in gene frequencies from generation to generation as a result of random (chance) processes.

genetic linkage Association between genes on the same chromosome such that they do not show random assortment and seldom recombine; the closer the genes, the lower the frequency of recombination.

genetic map The positions of genes along a chromosome as revealed by recombination frequencies.

genetic marker (1) In gene cloning, a gene of identifiable phenotype that indicates the presence of another gene, DNA segment, or chromosome fragment. (2) In general, a DNA sequence such as a single nucleotide polymorphism whose presence is correlated with the presence of other linked genes on that chromosome.

genetic screen   A technique for identifying genes involved in a biological process of interest. Involves creating a large collection of randomly mutated organisms and identifying those individuals that are likely to have a defect in the pathway of interest. The mutated gene(s) in those individuals can then be isolated for further study.

genetic structure   The frequencies of different alleles at each locus and the frequencies of different genotypes in a Mendelian population.

genetic switches   Mechanisms that control how the genetic toolkit is used, such as promoters and the transcription factors that bind them. The signal cascades that converge on and operate these switches determine when and where genes will be turned on and off.

genetic toolkit   A set of developmental genes and proteins that is common to most animals and is hypothesized to be responsible for the evolution of their differing developmental pathways.

genetics   The scientific study of the structure, functioning, and inheritance of genes, the units of hereditary information.

genome   (jee′ nome) The complete DNA sequence for a particular organism or individual.

genome sequencing   Determination of the nucleotide base sequence of the entire genome of an organism.

genomic imprinting   The form of a gene's expression is determined by parental source (i.e., whether the gene is inherited from the male or female parent).

genomic library   All of the cloned DNA fragments generated by the breakdown of genomic DNA into smaller segments.

genomics   The scientific study of entire sets of genes and their interactions.

genotype   (jean′ oh type) [Gk. *gen*: to produce + *typos*: impression] An exact description of the genetic constitution of an individual, either with respect to a single trait or with respect to a larger set of traits. (Contrast with phenotype.)

genotype frequency   The proportion of a genotype among individuals in a population.

genus   (jean′ us) (plural: genera) [Gk. *genos*: stock, kind] A group of related, similar species recognized by taxonomists with a distinct name used in binomial nomenclature.

geographic range   The region within which a species occurs.

germ cell   [L. *germen*: to beget] A reproductive cell or gamete of a multicellular organism. (Contrast with somatic cell.)

germ layers   The three embryonic layers formed during gastrulation (ectoderm, mesoderm, and endoderm). Also called cell layers or tissue layers.

germ line mutation   Mutation in a cell that produces gametes (i.e., a germ line cell). (Contrast with somatic mutation.)

germination   Sprouting of a seed or spore.

gestation   (jes tay′ shun) [L. *gestare*: to bear] The period during which the embryo of a mammal develops within the uterus. Also known as pregnancy.

ghrelin   A hormone produced and secreted by cells in the stomach that stimulates appetite.

gibberellin   (jib er el′ lin) A class of plant growth hormones playing roles in stem elongation, seed germination, flowering of certain plants, etc.

gill   An organ specialized for gas exchange with water.

gizzard   (giz′ erd) [L. *gigeria*: cooked chicken parts] The second of two stomachlike organs in birds, other reptiles, earthworms, and various insects, that grinds up food, sometimes with the aid of fragments of stone. (*See also* crop.)

glia   (glee′ uh) [Gk. *glia*: glue] One of the two classes of neural cells (along with neurons, with which glia interact); glia do not typically conduct action potentials. Types of glia include astrocytes, oligodendrocytes, and Schwann cells.

global nitrogen cycle   The movement of nitrogen through the biosphere. Steps in the cycle include the fixation of nitrogen gas (N_2) to ammonia; nitrification of the fixed nitrogen to nitrate by bacteria; nitrate reduction by plants; and denitrification back to N_2 by bacteria.

glomerular filtration rate (GFR)   The rate at which the blood is filtered in the glomeruli of the kidney.

glomerulus   (glo mare′ yew lus) [L. *glomus*: ball] Sites in the kidney where blood filtration takes place. Each glomerulus consists of a knot of capillaries served by afferent and efferent arterioles.

glucagon   Hormone produced by alpha cells of the pancreatic islets of Langerhans. Glucagon stimulates the liver to break down glycogen and release glucose into the circulation.

gluconeogenesis   The biochemical synthesis of glucose from other substances, such as amino acids, lactate, and glycerol.

glucose   [Gk. *gleukos*: sugar, sweet wine] The most common monosaccharide; the monomer of the polysaccharides starch, glycogen, and cellulose.

glyceraldehyde 3-phosphate (G3P)   A phosphorylated three-carbon sugar; an intermediate in glycolysis and photosynthetic carbon fixation.

glycerol   (gliss′ er ole) A three-carbon alcohol with three hydroxyl groups; a component of phospholipids and triglycerides.

glycogen   (gly′ ko jen) [Gk. *glyk*: sweet] An energy storage polysaccharide found in animals and fungi; a branched-chain polymer of glucose, similar to starch.

glycolipid   A lipid to which sugars are attached.

glycolysis   (gly kol′ li sis) [Gk. *gleukos*: sugar + *lysis*: break apart] The enzymatic breakdown of glucose to pyruvic acid.

glycoprotein   A protein to which sugars are attached.

glycosidic linkage   Bond between carbohydrate (sugar) molecules through an intervening oxygen atom (–O–).

glycosylation   The addition of carbohydrates to another type of molecule, such as a protein.

glyoxysome   (gly ox′ ee soam) An organelle found in plants, in which stored lipids are converted to carbohydrates.

Golgi apparatus   (goal′ jee) A system of concentrically folded membranes found in the cytoplasm of eukaryotic cells; functions in secretion from the cell by exocytosis.

Golgi tendon organ   A mechanoreceptor found in tendons and ligaments; provides information about the force generated by a contracting muscle.

gonad   (go′ nad) [Gk. *gone*: seed] An organ that produces gametes in animals: either an ovary (female gonad) or testis (male gonad).

gonadotropin   A trophic hormone that stimulates the gonads.

gonadotropin-releasing hormone (GnRH)   Hormone produced by the hypothalamus that stimulates the anterior pituitary to secrete gonadotropins.

Gondwana   The large southern land mass that existed from the Cambrian (540 mya) to the Jurassic (138 mya). Present-day remnants are South America, Africa, India, Australia, and Antarctica.

graded membrane potential   Small local change in membrane potential caused by opening or closing of ion channels.

grafting   Artificial transplantation of tissue from one organism to another. In horticulture, the transfer of a bud or stem segment from one plant onto the root of another as a form of asexual reproduction.

gram stain   A differential purple stain useful in characterizing bacteria. The peptidoglycan-rich cell walls of gram-positive bacteria stain purple; cell walls of gram-negative bacteria generally stain orange.

gravitropism   [L. *gravitas*: weight, force; Gk. *tropos*: to turn] A directed plant growth response to gravity.

gray crescent   In frog development, a band of diffusely pigmented cytoplasm on the side of the egg opposite the site of sperm entry. Arises as a result of cytoplasmic rearrangements that establish the anterior–posterior axis of the zygote.

gray matter   In the nervous system, tissue that is rich in neuronal cell bodies. (Contrast with white matter.)

greenhouse gases   Gases in the atmosphere, such as carbon dioxide and methane, that are transparent to sunlight, but trap heat radiating from Earth's surface, causing heat to build up at Earth's surface.

gross primary production   The amount of energy captured by the primary producers in a community.

gross primary productivity (GPP)   The rate at which the primary producers in a community turn solar energy into stored chemical energy via photosynthesis.

ground meristem   That part of an apical meristem that gives rise to the ground tissue system of the primary plant body.

ground tissue system   Those parts of the plant body not included in the dermal or vascular tissue systems. Ground tissues function in storage, photosynthesis, and support.

growth   An increase in the size of the body and its organs by cell division and cell expansion.

growth factor   A chemical signal that stimulates cells to divide.

growth hormone   A peptide hormone released by the anterior pituitary that stimulates many anabolic processes.

guanine (G)   (gwan' een) A nitrogen-containing base found in DNA, RNA, and GTP.

guard cells   In plants, specialized, paired epidermal cells that surround and control the opening of a stoma (pore). *See* stoma.

gustation   The sense of taste.

gut   An animal's digestive tract.

gymnosperms   Seed plants that do not produce flowers or fruits; one of the two major groups of living seed plants. (*See also* angiosperms.)

gyrus   (ji' rus) [Gk. *gyros*: spiral] *See* convolutions.

H

habitat   The particular environment in which an organism lives.

habitat patches   Also called **habitat islands**; areas of suitable habitat for a species that are separated by substantial areas of unsuitable habitat.

Hadley cells   Patterns of vertical atmospheric circulation that influence surface winds and precipitation patterns according to latitude.

hair cell   A type of mechanoreceptor in animals. Detects sound waves and other forms of motion in air or water.

half-life   The time required for half of a sample of a radioactive isotope to decay to its stable, nonradioactive form, or for a drug or other substance to reach half its initial dosage.

halophyte   (hal' oh fyte) [Gk. *halos*: salt + *phyton*: plant] A plant that grows in a saline (salty) environment.

Hamilton's rule   The principle that, for an apparent altruistic behavior to be adaptive, the fitness benefit of that act to the recipient times the degree of relatedness of the performer and the recipient must be greater than the cost to the performer.

haplodiploidy   A sex determination mechanism in which diploid individuals (which develop from fertilized eggs) are female and haploid individuals (which develop from unfertilized eggs) are male; typical of hymenopterans.

haploid   (hap' loid) [Gk. *haploeides*: single] Having a chromosome complement consisting of just one copy of each chromosome; designated $1n$ or n. (Contrast with diploid.)

haplotype   Linked nucleotide sequences that are usually inherited as a unit (as a "sentence" rather than as individual "words").

Hardy–Weinberg equililbrium   In a sexually reproducing population, the allele frequency at a given locus that is not being acted on by agents of evolution; the conditions that would result in no evolution in a population.

haustorium   (haw stor' ee um) (plural: haustoria)[L. *haustus*: draw up] A specialized hypha or other structure by which fungi and some parasitic plants draw nutrients from a host plant.

Haversian systems   Units of organization in compact bone that reflect the action of intercommunicating osteoblasts.

heart   In circulatory systems, a muscular pump that moves extracellular fluid around the body.

heat of vaporization   The energy that must be supplied to convert a molecule from a liquid to a gas at its boiling point.

heat shock proteins   Chaperone proteins expressed in cells exposed to high or low temperatures or other forms of environmental stress.

helical   Shaped like a screw or spring (helix); this shape occurs in DNA and proteins.

helper T cells   *See* T-helper cells.

hemiparasite   A parasitic plant that can photosynthesize, but derives water and mineral nutrients from the living body of another plant. (Contrast with holoparasite.)

hemizygous   (hem' ee zie' gus) [Gk. *hemi*: half + *zygotos*: joined] In a diploid organism, having only one allele for a given trait, typically the case for X-linked genes in male mammals and Z-linked genes in female birds. (Contrast with homozygous, heterozygous.)

hemoglobin   (hee' mo glow bin) [Gk. *heaema*: blood + L. *globus*: globe] Oxygen-transporting protein found in the red blood cells of vertebrates (and found in some invertebrates).

Hensen's node   In avian embryos, a structure at the anterior end of the primitive groove; determines the fates of cells passing over it during gastrulation.

hepatic   (heh pat' ik) [Gk. *hepar*: liver] Pertaining to the liver.

herbivore   (ur' bi vore) [L. *herba*: plant + *vorare*: to devour] An animal that eats plant tissues. (Contrast with carnivore, detritivore, omnivore.)

heritable trait   A trait that is at least partly determined by genes.

hermaphroditism   (her maf' row dite ism) The coexistence of both female and male sex organs in the same organism.

hetero-   [Gk.: *heteros*: other, different] A prefix indicating two or more different conditions, structures, or processes. (Contrast with homo-.)

heterochromatin   Densely packed, dark-staining chromatin; any genes it contains are usually not transcribed.

heterochrony   [Gk: different time] Alteration in the timing of developmental events, contributing to the evolution different phenotypes in the adult.

heterocyst   A large, thick-walled cell type in the filaments of certain cyanobacteria that performs nitrogen fixation.

heterometry   [Gk: different measure] Alteration in the level of gene expression, and thus in the amount of protein produced, during development, contributing to the evolution of different phenotypes in the adult.

heteromorphic   (het' er oh more' fik) [Gk.: different form] Having a different form or appearance, as two heteromorphic life stages of a plant. (Contrast with isomorphic.)

heterosis   The superior fitness of heterozygous offspring as compared with that of their dissimilar homozygous parents. Also called hybrid vigor.

heterosporous   (het' er os' por us) Producing two types of spores, one of which gives rise to a female megaspore and the other to a male microspore. (Contrast with homosporous.)

heterotherm   An animal that regulates its body temperature at a constant level at some times but not others, such as a hibernator.

heterotopy   [Gk: different place] Spatial differences in gene expression during development, controlled by developmental regulatory genes and contributing to the evolution of distinctive adult phenotypes.

heterotroph   (het' er oh trof) [Gk. *heteros*: different + *trophe*: feed] An organism that requires preformed organic molecules as food. (Contrast with autotroph.)

heterotrophic succession   Succession in detritus-based communities, which differs from other types of succession in taking place without the participation of plants.

heterotypy   [Gk.: different kind] Alteration in a developmental regulatory gene itself rather than the expression of the genes it controls. (Contraste with heterochrony; heterometry; heterotopy.)

heterozygous   (het' er oh zie' gus) [Gk. *heteros*: different + *zygotos*: joined] In diploid organisms, having different alleles of a given gene on the pair of homologs carrying that gene. (Contrast with homozygous.)

heterozygous carrier An individual that carries a recessive allele for a phenotype of interest (e.g., a genetic disease); the individual does not show the phenotype, but may have progeny with the phenotype if the other parent also carries the recessive allele.

hexose [Gk. *hex*: six] A sugar containing six carbon atoms.

hibernation [L. *hibernum*: winter] The state of inactivity of some animals during winter; marked by a drop in body temperature and metabolic rate. (Contrast with estivation)

high-density lipoproteins (HDLs) Lipoproteins that remove cholesterol from tissues and carry it to the liver; HDLs are the "good" lipoproteins associated with good cardiovascular health.

high-throughput sequencing Rapid DNA sequencing on a micro scale in which many fragments of DNA are sequenced in parallel.

highly repetitive sequences Short (less than 100 bp), nontranscribed DNA sequences, repeated thousands of times in tandem arrangements.

hindbrain The region of the developing vertebrate brain that gives rise to the medulla, pons, and cerebellum.

hippocampus [Gk. sea horse] A part of the forebrain that takes part in long-term memory formation.

histamine (hiss' tah meen) A substance released by damaged tissue, or by mast cells in response to allergens. Histamine increases vascular permeability, leading to edema (swelling). (Contrast with histone deacetylase.)

histone Any one of a group of proteins forming the core of a nucleosome, the structural unit of a eukaryotic chromosome.

histone acetyltransferases Enzymes involved in chromatin remodeling. Add acetyl groups to the tail regions of histone proteins.

histone deacetylase In chromatin remodeling, an enzyme that removes acetyl groups from the tails of histone proteins. (Contrast with histone acetyltransferases.)

HIV Human immunodeficiency virus, the retrovirus that causes acquired immune deficiency syndrome (AIDS).

holoparasite A fully parasitic plant (i.e., one that does not perform photosynthesis).

homeobox 180-base-pair segment of DNA found in certain homeotic genes. A specific sequence within the homeobox—the **homeodomain**—regulates the expression of other genes and through this regulation controls large-scale developmental processes. (*See* homeotic genes.)

homeostasis (home' ee o sta' sis) [Gk. *homos*: same + *stasis*: position] The maintenance of a steady state, such as a constant temperature, by means of

physiological or behavioral feedback responses.

homeotic genes Genes that act during development to determine the formation of an organ from a region of the embryo. (Compare with Hox genes.)

homeotic mutation Mutation in a homeotic gene that results in the formation of a different organ than that normally made by a region of the embryo.

homing In animal navigation, the ability to return to a nest site, burrow, or other specific location.

hominid Lineage that includes all modern and extinct Great Apes (i.e., humans, gorillas, chimpanzees, orangutans, and their ancestors.)

hominin Lineages that includes modern humans (*Homo sapiens*) and their extinct ancestors (e.g., Australopithecines; *Homo erectus*.)

homo- [Gk. *homos*: same] A prefix indicating two or more similar conditions, structures, or processes. (Contrast with hetero-.)

homolog (1) In cytogenetics, one of a pair (or larger set) of chromosomes having the same overall genetic composition and sequence. In diploid organisms, each chromosome inherited from one parent is matched by an identical (except for mutational changes) chromosome—its homolog—from the other parent. (2) In evolutionary biology, one of two or more features in different species that are similar by reason of descent from a common ancestor.

homologous pair A pair of matching chromosomes made up of a chromosome from each of the two sets of chromosomes in a diploid organism.

homologous recombination Exchange of segments between two DNA molecules based on sequence similarity between the two molecules. The similar sequences align and crossover. Used to create knockout mutants in mice and other organisms.

homology (ho mol' o jee) [Gk. *homologia*: of one mind; agreement] A similarity between two or more features that is due to inheritance from a common ancestor. The structures are said to be *homologous*, and each is a *homolog* of the others.

homoplasy (home' uh play zee) [Gk. *homos*: same + *plastikos*: shape, mold] The presence in multiple groups of a trait that is not inherited from the common ancestor of those groups. Can result from convergent evolution, evolutionary reversal, or parallel evolution.

homosporous Producing a single type of spore that gives rise to a single type of gametophyte, bearing both female and male reproductive organs. (Contrast with heterosporous.)

homotypic Pertaining to adhesion of cells of the same type. (Contrast with heterotypic.)

homozygous (home' oh zie' gus) [Gk. *homos*: same + *zygotos*: joined] In diploid organisms, having identical alleles of a given gene on both homologous chromosomes. An

individual may be a homozygote with respect to one gene and a heterozygote with respect to another. (Contrast with heterozygous.)

horizons The horizontal layers of a soil profile, including the topsoil (A horizon), subsoil (B horizon) and parent rock or bedrock (C horizon).

hormone (hore' mone) [Gk. *hormon*: to excite, stimulate] A chemical signal produced in minute amounts at one site in a multicellular organism and transported to another site where it acts on target cells.

host An organism that harbors a parasite or symbiont and provides it with nourishment.

Hox genes Conserved homeotic genes found in vertebrates, *Drosophila*, and other animal groups. Hox genes contain the homeobox and specify pattern and axis formation in these animals.

human chorionic gonadotropin (hCG) A hormone secreted by the placenta which sustains the corpus luteum and helps maintain pregnancy.

Human Genome Project A publicly and privately funded research effort, successfully completed in 2003, to produce a complete DNA sequence for the entire human genome.

humoral immune response The response of the immune system mediated by B cells that produces circulating antibodies active against extracellular bacterial and viral infections. (Contrast with cellular immune response.)

humus (hew' mus) The partly decomposed remains of plants and animals on the surface of a soil.

hybrid (high' brid) [L. *hybrida*: mongrel] (1) The offspring of genetically dissimilar parents. (2) In molecular biology, a double helix formed of nucleic acids from different sources.

hybrid vigor *See* heterosis.

hybrid zone A region of overlap in the ranges of two closely related species where the species may hybridize.

hybridize (1) In genetics, to combine the genetic material of two distinct species or of two distinguishable populations within a species. (2) In molecular biology, to form a double-stranded nucleic acid in which the two strands originate from different sources.

hydrocarbon A compound containing only carbon and hydrogen atoms.

hydrogen bond A weak electrostatic bond which arises from the attraction between the slight positive charge on a hydrogen atom and a slight negative charge on a nearby oxygen or nitrogen atom.

hydrologic cycle The movement of water from the oceans to the atmosphere, to the soil, and back to the oceans.

hydrolysis reaction (high drol' uh sis) [Gk. *hydro*: water + *lysis*: break apart] A chemical reaction that breaks a bond by inserting the components of water ($AB + H_2O \rightarrow AH + BOH$). (Contrast with condensation reaction.)

hydrophilic (high dro fill' ik) [Gk. *hydro*: water + *philia*: love] Having an affinity for water. (Contrast with hydrophobic.)

hydrophobic (high dro foe' bik) [Gk. *hydro*: water + *phobia*: fear] Having no affinity for water. Uncharged and nonpolar groups of atoms are hydrophobic. (Contrast with hydrophilic.)

hydroponic Pertaining to a method of growing plants with their roots suspended in nutrient solutions instead of soil.

hydrostatic pressure Pressure generated by compression of liquid in a confined space. Generated in plants, fungi, and some protists with cell walls by the osmotic uptake of water. Generated in animals with closed circulatory systems by the beating of a heart.

hydrostatic skeleton A fluid-filled body cavity that transfers forces from one part of the body to another when acted on by surrounding muscles.

hydroxyl group The —OH group found on alcohols and sugars.

hyper- [Gk. *hyper*: above, over] Prefix indicating above, higher, more. (Contrast with hypo-.)

hyperaccumulators Plant species that store large quantities of heavy metals such as arsenic, cadmium, nickel, aluminum, and zinc.

hyperpolarization A change in the resting potential across a membrane so that the inside of a cell becomes more negative compared with the outside of the cell. (Contrast with depolarization.)

hypersensitive response A defensive response of plants to microbial infection in which phytoalexins and pathogenesis-related proteins are produced and the infected tissue undergoes apoptosis to isolate the pathogen from the rest of the plant.

hypertonic Having a greater solute concentration. Said of one solution compared with another. (Contrast with hypotonic, isotonic.)

hypha (high' fuh) (plural: hyphae) [Gk. *hyphe*: web] In the fungi and oomycetes, any single filament.

hypo- [Gk. *hypo*: beneath, under] Prefix indicating underneath, below, less. (Contrast with hyper-.)

hypoblast The lower tissue portion of the avian blastula which is joined to the epiblast at the margins of the blastodisc.

hypothalamus The part of the brain lying below the thalamus; it coordinates water balance, reproduction, temperature regulation, and metabolism.

hypothermia Below-normal body temperature.

hypothesis A tentative answer to a question, from which testable predictions can be generated. (Contrast with theory.)

hypotonic Having a lesser solute concentration. Said of one solution in comparing it to another. (Contrast with hypertonic, isotonic.)

hypoxia A deficiency of oxygen.

I

ileum The final segment of the small intestine. (*See also* duodenum, jejunum.)

imbibition Water uptake by a seed; first step in germination.

immediate hypersensitivity A rapid, extensive overreaction of the immune system against an allergen, resuting in the release of large amounts of histamine. (Contrast with delayed hypersensitivity.)

immediate memory A form of memory for events happening in the present that is almost perfectly photographic, but lasts only seconds.

immunity [L. *immunis*: exempt from] In animals, the ability to avoid disease when invaded by a pathogen by deploying various defense mechanisms.

immunoassay The use of antibodies to measure the concentration of an antigen in a sample.

immunoglobulins A class of proteins containing a tetramer consisting of four polypeptide chains—two identical light chains and two identical heavy chains—held together by disulfide bonds; active as receptors and effectors in the immune system.

immunological memory The capacity to more rapidly and massively respond to a second exposure to an antigen than occurred on first exposure.

imperfect flower A flower lacking either functional stamens or functional carpels. (Contrast with perfect flower.)

implantation The process by which the early mammalian embryo becomes attached to and embedded in the lining of the uterus.

imprinting In animal behavior, a rapid form of learning in which an animal learns, during a brief critical period, to make a particular response (which is then maintained for life) to some object or other organism. *See also* genomic imprinting.

in vitro [L.: in glass] A biological process occurring outside of the organism, in the laboratory. (Contrast with in vivo.)

in vitro evolution A method based on natural molecular evolution that uses artificial selection in the laboratory to rapidly produce molecules with novel enzymatic and binding functions.

in vivo [L.: in life] A biological process occurring within a living organism or cell. (Contrast with in vitro.)

inclusive fitness The sum of an individual's genetic contribution to subsequent generations both via production of its own offspring and via its influence on the survival of relatives who are not direct descendants. (Contrast with direct fitness)

incomplete cleavage A pattern of cleavage that occurs in many eggs that have a lot of yolk, in which the cleavage furrows do not penetrate all of it. (*See also* discoidal cleavage, superficial cleavage; contrast with complete cleavage.)

incomplete dominance Condition in which the heterozygous phenotype is intermediate between the two homozygous phenotypes.

incomplete metamorphosis Insect development in which changes between instars are gradual. (Contrast with direct development; complete metamorphosis.)

independent assortment During meiosis, the random separation of genes carried on nonhomologous chromosomes into gametes so that inheritance of these genes is random. This principle was articulated by Mendel as his second law.

independent variable In a scientific experiment, a critical factor that is manipulated while all other factors are held constant. (Contrast with dependent variable.)

indeterminate growth A open-ended growth pattern in which an organism or organ continues to grow as long as it lives; characteristic of some animals and of plant shoots and roots. (Contrast with determinate growth.)

individual fitness See direct fitness.

induced fit A change in the shape of an enzyme caused by binding to its substrate that exposes the active site of the enzyme.

induced mutation A mutation resulting from exposure to a mutagen from outside the cell. (Contrast with spontaneous mutation.)

induced pluripotent stem cells (iPS cells) Multipotent or pluripotent animal stem cells produced from differentiated cells in vitro by the addition of several genes that are expressed.

induced responses Defensive responses that a plant produces only in the presence of a pathogen, in contrast to constitutive defenses, which are always present.

inducer (1) A compound that stimulates the synthesis of a protein. (2) In embryonic development, a substance that causes a group of target cells to differentiate in a particular way.

inducible genes Genes that are expressed only when their products—**inducible proteins**—are needed. (Contrast with constitutive genes.)

inducible Produced only in the presence of a particular compound or under particular circumstances. (Contrast with constitutive.)

induction In embryonic development, the process by which a factor produced and secreted by certain cells determines the fates other cells.

inductive logic Involves making observations and then formulating one or more possible scenarios—hypotheses—that might explain those observations. (Contrast with deductive logic.)

inflammation A nonspecific defense against pathogens; characterized by redness, swelling, pain, and increased temperature.

inflorescence A structure composed of several to many flowers.

inflorescence meristem A meristem that produces floral meristems as well as other small leafy structures (bracts).

ingroup In a phylogenetic study, the group of organisms of primary interest. (Contrast with outgroup.)

inhibitor A substance that blocks a biological process.

inhibitory Input from a neuron that causes hyperpolarization of the recipient cell.

initials Cells that perpetuate plant meristems, comparable to animal stem cells. When an initial divides, one daughter cell develops into another initial, while the other differentiates into a more specialized cell.

initiation complex In protein translation, a combination of a small ribosomal subunit, an mRNA molecule, and the tRNA charged with the first amino acid coded for by the mRNA; formed at the onset of translation.

initiation site The place within a promoter where transcription begins.

innate defenses In animals, one of two general types of defenses against pathogens. Nonspecific and present in most animals. (Contrast with adaptive immunity.)

inner cell mass Derived from the mammalian blastula (bastocyst), the inner cell mass that will give rise to the yolk sac (via hypoblast) and embryo (via epiblast).

inorganic fertilizer A chemical or combination of chemicals applied to soil or plants to make up for a plant nutrient deficiency. Often contains the macronutrients nitrogen, phosphorus, and potassium (N-P-K).

inositol trisphosphate (IP$_3$) An intracellular second messenger derived from membrane phospholipids.

inspiratory reserve volume The amount of air that can be inhaled above the normal tidal inspiration. (Contrast with expiratory reserve volume, tidal volume, vital capacity.)

instar (in' star) An immature stage of an insect between molts.

insula (in' su lah) [L. *insula*: island] An area deep within the forebrain that appears to integrate physiological information from all over the body to create a sensation of how the body "feels" and may be involved in human consciousness. Also called the insular cortex.

insulin (in' su lin) [L. *insula*: island] A hormone synthesized in islet cells of the pancreas that promotes the conversion of glucose into the storage material, glycogen.

integral membrane proteins Proteins that are at least partially embedded in the plasma membrane. (Contrast with peripheral membrane proteins.)

integrin In animals, a transmembrane protein that mediates the attachment of epithelial cells to the extracellular matrix.

integument [L. *integumentum*: covering] A protective surface structure. In gymnosperms and angiosperms, a layer of tissue around the ovule which will become the seed coat.

intercostal muscles Muscles between the ribs that can augment breathing movements by elevating and suppressing the rib cage.

interference competition Competition in which individuals actively interfere with one another's access to resources. (Contrast with exploitation competition.)

interference RNA (RNAi) *See* RNA interference.

interferons Glycoproteins produced by virus-infected animal cells; interferons increase the resistance of neighboring cells to the virus.

internal environment In multicelluar organisms, includes blood plasma and interstitial fluid, i.e., the extracellular fluids that surround the cells.

internal fertilization The release of sperm into the female reproductive tract; typical of most terrestrial animals. (Contrast with external fertilization.)

internal gills Gills enclosed in protective body cavities; typical of mollusks, arthropods, and fishes.**interneuron** A neuron that communicates information between two other neurons.

interneuron A neuron that communicates information between two other neurons.

internode The region between two nodes of a plant stem.

interphase In the cell cycle, the period between successive nuclear divisions during which the chromosomes are diffuse and the nuclear envelope is intact. During interphase the cell is most active in transcribing and translating genetic information.

interspecific competition Competition between members of two or more species. (Contrast with intraspecific competition; see also exploitation competition, interference competition.)

interstitial fluid Extracellular fluid that is not contained in the vessels of a circulatory system.

intertidal zone A nearshore region of oceans that is periodically exposed to the air as the tides rise and fall.

intestine The portion of the gut following the stomach, in which most digestion and absorption occurs.

intraspecific competition Competition among members of the same species. (Contrast with interspecific competition.)

intrinsic rate of increase The rate at which a population is capable of growing when its density is low and environmental conditions are highly favorable.

intron Portion of a of a gene within the coding region that is transcribed into pre-mRNA but is spliced out prior to translation. (Contrast with exon.)

invasive species An exotic species that reproduces rapidly, spreads widely, and has negative effects on the native species of the region to which it has been introduced.

invasiveness The ability of a pathogen to multiply in a host's body. (Contrast with toxigenicity).

inversion A rare 180° reversal of the order of genes within a segment of a chromosome.

involution Cell movements that occur during gastrulation of frog embryos, giving rise to the archenteron.

ion (eye' on) [Gk. *ion*: wanderer] An electrically charged particle that forms when an atom gains or loses one or more electrons.

ion channel An integral membrane protein that allows ions to diffuse across the membrane in which it is embedded.

ion exchange In plants, a process by which protons produced by the plant's root displace mineral cations from clay particles in the surrounding soil.

ionic attraction An electrostatic attraction between positively and negatively charged ions.

ionotropic receptors A receptor that directly alters membrane permeability to a type of ion when it combines with its ligand.

iris (eye' ris) [Gk. *iris*: rainbow] The round, pigmented membrane that surrounds the pupil of the eye and adjusts its aperture to regulate the amount of light entering the eye.

island biogeography A theory proposing that the number of species on an island (or in another geographically defined and isolated area) represents a balance, or equilibrium, between the rate at which species immigrate to the island and the rate at which resident species go extinct.

islets of Langerhans Clusters of hormone-producing cells in the pancreas.

iso- [Gk. *iso*: equal] Prefix used for two separate entities that share some element of identity.

isomers Molecules consisting of the same numbers and kinds of atoms, but differing in the bonding patterns by which the atoms are held together.

isomorphic (eye so more' fik) [Gk. *isos*: equal + *morphe*: form] Having the same form or appearance, as when the haploid and diploid life stages of an organism appear identical. (Contrast with heteromorphic.)

isotonic Having the same solute concentration; said of two solutions. (Contrast with hypertonic, hypotonic.)

isotope (eye' so tope) [Gk. *isos*: equal + *topos*: place] Isotopes of a given chemical element have the same number of protons in their nuclei (and thus are in the same position on the periodic table), but differ in the number of neutrons.

isozymes Enzymes of an organism that have somewhat different amino acid sequences but catalyze the same reaction.

iteroparous [L. itero, to repeat + pario, to beget] Reproducing multiple times in a lifetime. (Contrast with semelparous.)

J

jasmonate Also called jasmonic acid, a plant hormone involved in triggering responses to pathogen attack as well as other processes.

jejunum (jih jew' num) The middle division of the small intestine, where most absorption of nutrients occurs. (*See also* duodenum, ileum.)

joint In skeletal systems, a junction between two or more bones.

juvenile hormone In insects, a hormone maintaining larval growth and preventing maturation or pupation.

K

K-strategist A species whose life history strategy allows it to persist at or near the carrying capacity (K) of its environment. (Contrast with *r*-strategist.)

karyogamy The fusion of nuclei of two cells. (Contrast with plasmogamy.)

karyotype The number, forms, and types of chromosomes in a cell.

keystone species Species that have a dominant influence on the composition of a community.

kidneys A pair of excretory organs in vertebrates.

kilocalorie (kcal) *See* Calorie.

kin selection That component of inclusive fitness resulting from helping the survival of relatives containing the same alleles by descent from a common ancestor. (Contrast with direct fitness.)

kinase *See* protein kinase.

kinetic energy (kuh-net' ik) [Gk. *kinetos*: moving] The energy associated with movement. (Contrast with potential energy.)

kinetochore (kuh net' oh core) Specialized structure on a centromere to which microtubules attach.

knockout A molecular genetic method in which a single gene of an organism is permanently inactivated.

Koch's postulates A set of rules for establishing that a particular microorganism causes a particular disease.

Krebs cycle *See* citric acid cycle.

L

lagging strand In DNA replication, the daughter strand that is synthesized in discontinuous stretches. (*See* Okazaki fragments.)

large intestine *See* colon.

larva (plural: larvae) [L. *lares*: guiding spirits] An immature stage of any animal that differs dramatically in appearance from the adult.

lateral [L. *latus*: side] Pertaining to the side.

lateral gene transfer The transfer of genes from one species to another, common among bacteria and archaea.

lateral meristem Either of the two meristems, the vascular cambium and the cork cambium, that give rise to a plant's secondary growth.

lateral root A root extending outward from the taproot in a taproot system; typical of eudicots.

lateralization A phenomenon in humans in which language functions come to reside in one cerebral hemisphere, usually the left.

laticifers (luh tiss' uh furs) In some plants, elongated cells containing secondary plant products such as latex.

Laurasia The northernmost of the two large continents produced by the breakup of Pangaea.

law of independent assortment *See* independent assortment.

law of segregation *See* segregation.

laws of thermodynamics [Gk. *thermos*: heat + *dynamis*: power] Laws derived from studies of the physical properties of energy and the ways energy interacts with matter. (*See also* first law of thermodynamics, second law of thermodynamics.)

leaching In soils, a process by which mineral nutrients in upper soil horizons are dissolved in water and carried to deeper horizons, where they are unavailable to plant roots.

leading strand In DNA replication, the daughter strand that is synthesized continuously. (Contrast with lagging strand.)

leaf (plural: leaves) In plants, the chief organ of photosynthesis.

leaf primordium (plural: primordia) An outgrowth on the side of the shoot apical meristem that will eventually develop into a leaf.

leghemoglobin In nitrogen-fixing plants, an oxygen-carrying protein in the cytoplasm of nodule cells that transports enough oxygen to the nitrogen-fixing bacteria to support their respiration, while keeping free oxygen concentrations low enough to protect nitrogenase.

lek A display ground within which male animals compete for and defend small display areas as a means of demonstrating their territorial prowess and winning opportunities to mate.

lens In the vertebrate eye, a crystalline protein structure that makes fine adjustments in the focus of images falling on the retina.

leptin A hormone produced by fat cells that is believed to provide feedback information to the brain about the status of the body's fat reserves.

leukocyte *See* white blood cell.

lichen (lie' kun) An organism resulting from the symbiotic association of a fungus and either a cyanobacterium or a unicellular alga.

life cycle The entire span of the life of an organism from the moment of fertilization (or asexual generation) to the time it reproduces in turn.

life history strategy The way in which an organism partitions its time and energy among growth, maintenance, and reproduction.

life history The time course of growth and development, reproduction, and death during an average individual organism's life.

life table A summary of information about the progression of individuals in a population through the various stages of their life cycles.

life zones In the aquatic (marine and freshwater) biomes, the regions defined by light penetration and water movement such as wave action. Life zones include, e.g., the intertidal, pelagic (open water) and bethic (bottom) zones.

ligament A band of connective tissue linking two bones in a joint.

ligand (lig' and) Any molecule that binds to a receptor site of another (usually larger) molecule.

light reactions The initial phase of photosynthesis, in which light energy is converted into chemical energy. Followed by the **light-independent reactions** in which the energy captured in the light reactions is used to drive the reduction of CO_2 to form carbohydrates.

light-harvesting complex In photosynthesis, a group of different molecules that cooperate to absorb light energy and transfer it to a reaction center. Also called *antenna system*.

lignin A complex, hydrophobic polyphenolic polymer in plant cell walls that crosslinks other wall polymers, strengthening the walls, especially in wood.

limbic system A group of evolutionarily primitive structures in the vertebrate telencephalon that are involved in emotions, drives, instinctive behaviors, learning, and memory.

limiting resource The required resource whose supply (or lack thereof) most strongly influences the size of a population.

limnetic zone The open-water life zone of a lake

lineage A series of populations, species, or genes descended from a single ancestor over evolutionary time.

lineage species concept The definition of a species as a branch on the tree of life, which has a history that starts at a speciation event and ends either at extinction or at another speciation event. (Contrast with biological species concept; morphological species concept.)

linkage *See* genetic linkage.

lipase (lip' ase; lye' pase) An enzyme that digests fats.

lipid (lip' id) [Gk. *lipos*: fat] Nonpolar, hydrophobic molecules that include fats, oils, waxes, steroids, and the phospholipids that make up biological membranes.

lipid bilayer *See* phospholipid bilayer.

lipoproteins Lipids packaged inside a covering of protein so that they can be circulated in the blood.

lithoosphere (lith' o sphere) [Gk. *lithos*: strong] The crust of sold rock plates that overlays the viscous mantle of Earth. The movements of the lithosphere are the source of plate tectonics. (Constrast with asthenosphere.)

littoral zone The nearshore life zone of a lake that is shallow and is affected by wave action and fluctuations in water level.

liver A large digestive gland. In vertebrates, it secretes bile and is involved in the formation of blood.

loam A type of soil consisting of a mixture of sand, silt, clay, and organic matter. One of the best soil types for agriculture.

locus (low' kus) (plural: loci, low' sigh) In genetics, a specific location on a chromosome. May be considered synonymous with *gene*.

logistic growth Growth, especially in the size of an organism or in the number of organisms in a population, that slows steadily as the entity approaches its maximum size. (Contrast with multiplicative growth.)

long-day plant (LDP) A plant that requires long days (actually, short nights) in order to flower. (Compare to short-day plant.)

long-term potentiation (LTP) A long-lasting increase in the responsiveness of a neuron resulting from a period of intense stimulation.

loop of Henle (hen' lee) Long, hairpin loop of the mammalian renal tubule that runs from the cortex down into the medulla and back to the cortex; creates a concentration gradient in the interstitial fluids in the medulla.

lophophore A U-shaped fold of the body wall with hollow, ciliated tentacles that encircles the mouth of animals in several different groups. Used for filtering prey from the surrounding water.

loss of function mutation A mutation that results in the loss of a functional protein. (Contrast with gain of function mutation.)

low-density lipoproteins (LDLs) Lipoproteins that transport cholesterol around the body for use in biosynthesis and for storage; LDLs are the "bad" lipoproteins associated with a high risk of cardiovascular disease.

lumen (loo' men) [L. *lumen*: light] The open cavity inside any tubular organ or structure, such as the gut or a renal tubule.

lung An internal organ specialized for respiratory gas exchange with air.

luteinizing hormone (LH) A gonadotropin produced by the anterior pituitary that stimulates the gonads to produce sex hormones.

lymph [L. *lympha*: liquid] A fluid derived from blood and other tissues that accumulates in intercellular spaces throughout the body and is returned to the blood by the lymphatic system.

lymph node A specialized structure in the vessels of the lymphatic system. Lymph nodes contain lymphocytes, which encounter and respond to foreign cells and molecules in the lymph as it passes through the vessels.

lymphatic system A system of vessels that returns interstitial fluid to the blood.

lymphocyte One of the two major classes of white blood cells; includes T cells, B cells, and other cell types important in the immune system.

lysis (lie' sis) [Gk. *lysis*: break apart] Bursting of a cell.

lysogeny A form of viral replication in which the virus becomes incorporated into the host chromosome and remains inactive. Also called a lysogenic cycle. (Contrast with lytic cycle.)

lysosome (lie' so soam) [Gk. *lysis*: break away + *soma*: body] A membrane-enclosed organelle originating from the Golgi apparatus and containing hydrolytic enzymes. (Contrast with secondary lysosome.)

lysozyme (lie' so zyme) An enzyme in saliva, tears, and nasal secretions that hydrolyzes bacterial cell walls.

lytic cycle A viral reproductive cycle in which the virus takes over a host cell's synthetic machinery to replicate itself, then bursts (lyses) the host cell, releasing the new viruses. (Contrast with lysogeny.)

M

M phase The portion of the cell cycle in which mitosis takes place.

macroevolution [Gk. *makros*: large] Evolutionary changes occurring over long time spans and usually involving changes in many traits. (Contrast with microevolution.)

macromolecule A giant (molecular weight > 1,000) polymeric molecule. The macromolecules are the proteins, polysaccharides, and nucleic acids.

macronutrient In plants, a mineral element required in concentrations of at least 1 milligram per gram of plant dry matter; in animals, a mineral element required in large amounts. (Contrast with micronutrient.)

macrophage (mac' roh faj) Phagocyte that engulfs pathogens by endocytosis.

MADS box DNA-binding domain in many plant transcription factors that is active in development.

maintenance methylase An enzyme that catalyzes the methylation of the new DNA strand when DNA is replicated.

major histocompatibility complex (MHC) A complex of linked genes, with multiple alleles, that control a number of cell surface antigens that identify self and can lead to graft rejection.

malignant Pertaining to a tumor that can grow indefinitely and/or spread from the original site of growth to other locations in the body. (Contrast with benign.)

malnutrition A condition caused by lack of any essential nutrient.

Malpighian tubule (mal pee' gy un) A type of protonephridium found in insects.

mantle (1) In mollusks, a fold of tissue that covers the organs of the visceral mass and secretes the hard shell that is typical of many mollusks. (2) In geology, the Earth's crust below the solid lithospheric plates.

map unit The distance between two genes as calculated from genetic crosses; a recombination frequency.

marine [L. *mare*: sea, ocean] Pertaining to or living in the ocean. (Contrast with aquatic, terrestrial.)

mark–recapture method A method of estimating population sizes of mobile organisms by capturing, marking, and releasing a sample of individuals, then capturing another sample at a later time.

mass extinction A period of evolutionary history during which rates of extinction are much higher than during intervening times.

mass number The sum of the number of protons and neutrons in an atom's nucleus.

mast cells Cells, typically found in connective tissue, that release histamine in response to tissue damage.

maternal effect genes Genes coding for morphogens that determine the polarity of the egg and larva in fruit flies. Part of a developmental cascade that includes gap genes, pair rule genes, segment polarity genes, and Hox genes.

mating type A particular strain of a species that is incapable of sexual reproduction with another member of the same strain but capable of sexual reproduction with members of other strains of the same species.

maturational survivorship curves *See* survivorship curves

maximum likelihood A statistical method of determining which of two or more hypotheses (such as phylogenetic trees) best fit the observed data, given an explicit model of how the data were generated.

mechanically gated channel A molecular channel that opens or closes in response to mechanical force applied to the plasma membrane in which it is inserted.

mechanoreceptor A cell that is sensitive to physical movement and generates action potentials in response.

medulla (meh dull' luh) (1) The inner, core region of an organ, as in the adrenal medulla (adrenal gland) or the renal medulla (kidneys). (2) The portion of the brainstem that connects to the spinal cord.

medusa (plural: medusae) In cnidarians, a free-swimming, sexual life cycle stage shaped like a bell or an umbrella.

megagametophyte In heterosporous plants, the female gametophyte; produces eggs. (Contrast with microgametophyte.)

megaphyll The generally large leaf of a fern, horsetail, or seed plant, with several to many veins. (Contrast with microphyll.)

megaspore [Gk. *megas*: large + *spora*: to sow] In plants, a haploid spore that produces a female gametophyte.

megastrobilus In conifers, the female (seed-bearing) cone. (Contrast with microstrobilus.)

meiosis (my oh' sis) [Gk. *meiosis*: diminution] Division of a diploid nucleus to produce four haploid daughter cells. The process consists of two successive nuclear divisions with only one cycle of chromosome replication. In *meiosis I*, homologous chromosomes separate but retain their chromatids. The second division *meiosis II*, is similar to mitosis, in which chromatids separate.

melatonin A hormone released by the pineal gland. Involved in photoperiodicity and circadian rhythms.

membrane A phospholipid bylayer forming a barrier that separates the internal contents of a cell from the nonbiological environment, or enclosing the organelles within a cell. The membrane regulates the molecular substances entering or leaving a cell or organelle.

membrane potential The difference in electrical charge between the inside and the outside of a cell, caused by a difference in the distribution of ions.

membranous bone A type of bone that develops by forming on a scaffold of connective tissue. (Contrast with cartilage bone.)

memory cells Long-lived lymphocytes produced after exposure to antigen. They persist in the body and are able to mount a rapid response to subsequent exposures to the antigen.

Mendel's laws *See* independent assortment; segregation.

menopause In human females, the end of fertility and menstrual cycling.

menstruation The process by which the endometrium breaks down, and the sloughed-off tissue, including blood, flows from the body.

meristem [Gk. *meristos*: divided] Plant tissue made up of undifferentiated actively dividing cells.

meristem culture A method for the asexual propagation of plants, in which pieces of shoot apical meristem are cultured to produce plantlets.

meristem identity genes In angiosperms, a group of genes whose expression initiates flower formation, probably by switching meristem cells from a vegetative to a reproductive fate.

mesenchyme (mez' en kyme) [Gk. *mesos*: middle + *enchyma*: infusion] Embryonic or unspecialized cells derived from the mesoderm.

mesoderm [Gk. *mesos*: middle + *derma*: skin] The middle of the three embryonic germ layers first delineated during gastrulation. Gives rise to the skeleton, circulatory system, muscles, excretory system, and most of the reproductive system.

mesoglea (mez' uh glee uh) [Gk. *mesos*: middle + *gloia*, glue] A thick, gelatinous noncellular layer that separates the two cellular tissue layers of ctenophores, cnidarians, and scyphozoans.

mesophyll (mez' uh fill) [Gk. *mesos*: middle + *phyllon*: leaf] Chloroplast-containing, photosynthetic cells in the interior of leaves.

messenger RNA (mRNA) Transcript of a region of one of the strands of DNA; carries information (as a sequence of codons) for the synthesis of one or more proteins.

meta- [Gk.: between, along with, beyond] Prefix denoting a change or a shift to a new form or level; for example, as used in metamorphosis.

metabolic pathway A series of enzyme-catalyzed reactions so arranged that the product of one reaction is the substrate of the next.

metabolism (meh tab' a lizm) [Gk. *metabole*: change] The sum total of the chemical reactions that occur in an organism, or some subset of that total (as in respiratory metabolism).

metabolome The quantitative description of all the small molecules in a cell or organism.

metabotropic receptor A receptor that that indirectly alters membrane permeability to a type of ion when it combines with its ligand.

metagenomics The practice of analyzing DNA from environmental samples without isolating intact organisms.

metamorphosis (met' a mor' fo sis) [Gk. *meta*: between + *morphe*: form, shape] A change occurring between one developmental stage and another, as for example from a tadpole to a frog. (*See* complete metamorphosis, incomplete metamorphosis.)

metanephridia The paired excretory organs of annelids.

metaphase (met' a phase) The stage in nuclear division at which the centromeres of the highly supercoiled chromosomes are all lying on a plane (the metaphase plane or plate) perpendicular to a line connecting the division poles.

metapopulation A population divided into subpopulations, among which there are occasional exchanges of individuals.

methylation The addition of a methyl group (—CH$_3$) to a molecule.

MHC *See* major histocompatibility complex.

micelle A particle of lipid covered with bile salts that is produced in the duodenum and facilitates digestion and absorption of lipids.

microbiomes The diverse communities of bacteria that live on or within the body and are essential to bodily function.

microclimate A subset of climatic conditions in a small specific area, which generally differ from those in the environment at large, as in an animal's underground burrow.

microevolution Evolutionary changes below the species level, affecting allele frequencies. (Contrast with macroevolution.)

microfibril Crosslinked cellulose polymers, forming strong aggregates in the plant cell wall.

microfilament In eukaryotic cells, a fibrous structure made up of actin monomers. Microfilaments play roles in the cytoskeleton, in cell movement, and in muscle contraction.

microgametophyte In heterosporous plants, the male gametophyte; produces sperm. (Contrast with megagametophyte.)

microglia Glial cells that act as macrophages and mediators of inflammatory responses in the central nervous system.

micronutrient In plants, a mineral element required in concentrations of less than 100 micrograms per gram of plant dry matter; in animals, a mineral element required in concentrations of less than 100 micrograms per day. (Contrast with macronutrient.)

microphyll A small leaf with a single vein, found in club mosses and their relatives. (Contrast with megaphyll.)

micropyle (mike' roh pile) [Gk. *mikros*: small + *pylon*: gate] Opening in the integument(s) of a seed plant ovule through which pollen grows to reach the female gametophyte within.

microRNA A small, noncoding RNA molecule, typically about 21 bases long, that binds to mRNA to inhibit its translation.

microspore [Gk. *mikros*: small + *spora*: to sow] In plants, a haploid spore that produces a male gametophyte.

microstrobilus In conifers, male pollen-bearing cone. (Contrast with megastrobilus.)

microtubules Tubular structures found in centrioles, spindle apparatus, cilia, flagella, and cytoskeleton of eukaryotic cells. These tubules play roles in the motion and maintenance of shape of eukaryotic cells.

microvilli (sing.: microvillus) Projections of epithelial cells, such as the cells lining the small intestine, that increase their surface area.

midbrain One of the three regions of the vertebrate brain. Part of the brainstem, it serves as a relay station for sensory signals sent to the cerebral hemispheres.

middle lamella (la mell' ah) [L. *lamina*: thin sheet] A layer of polysaccharides that separates plant cells; a shared middle lamella lies outside the primary walls of the two cells.

mineral nutrients Inorganic ions required by organisms for normal growth and reproduction.

mismatch repair A mechanism that scans DNA after it has been replicated and corrects any base-pairing mismatches.

missense mutation A change in a gene's sequence that changes the amino acid at that site in the encoded protein. (Contrast with frame-shift mutation, nonsense mutation, silent mutation.)

mitochondria (my' toe kon' dree uh) (singular: mitochondrion) [Gk. *mitos*: thread + *chondros*: grain] Energy-generating organelles in eukaryotic cells that contain the enzymes of the citric acid cycle, the respiratory chain, and oxidative phosphorylation.

mitochondrial matrix The fluid interior of a mitochondrion, enclosed by the inner mitochondrial membrane.

mitosis (my toe' sis) [Gk. *mitos*: thread] Nuclear division in eukaryotes leading to the formation of two daughter nuclei, each with a chromosome complement identical to that of the original nucleus.

mitosomes Reduced structures derived from mitochondria found in some organisms.

model systems Also known as **model organisms**, these include the small group of species that are the subject of extensive research. They are organisms that adapt well to laboratory situations and findings from experiments on them can apply across a broad range of species. Classic examples include white mice and the fruit fly *Drosophila*.

moderately repetitive sequences DNA sequences repeated 10–1,000 times in the eukaryotic genome. They include the genes that code for rRNAs and tRNAs, as well as the DNA in telomeres.

Modern Synthesis An understanding of evolutionary biology that emerged in the early twentieth century as the principles of evolution were integrated with the principles of modern genetics.

modularity In evolutionary developmental biology, the principle that the molecular pathways that determine different developmental processes operate independently from one another. *See also* developmental module.

mole A quantity of a compound whose weight in grams is numerically equal to its molecular weight expressed in atomic mass units. Avogadro's number of molecules: 6.023 × 10²³ molecules.

molecular clock The approximately constant rate of divergence of macromolecules from one another over evolutionary time; used to date past events in evolutionary history.

molecular evolution The scientific study of the mechanisms and consequences of the evolution of macromolecules.

molecular toolkit *See* genetic toolkit.

molecular weight The sum of the atomic weights of the atoms in a molecule.

molecule A chemical substance made up of two or more atoms joined by covalent bonds or ionic attractions.

molting The process of shedding part or all of an outer covering, as the shedding of feathers by birds or of the entire exoskeleton by arthropods.

monoclonal antibody Antibody produced in the laboratory from a clone of hybridoma cells, each of which produces the same specific antibody.

monocots Angiosperms with a single embryonic cotyledon; one of the two largest clades of angiosperms. (*See also* eudicots.)

monoculture In agriculture, a large-scale planting of a single species of domesticated crop plant.

monoecious (mo nee' shus) [Gk. *mono*: one + *oikos*: house] Pertaining to organisms in which both sexes are "housed" in a single individual that produces both eggs and sperm. (In some plants, these are found in different flowers within the same plant.) Examples include corn, peas, earthworms, hydras. (Contrast with dioecious.)

monohybrid cross A mating in which the parents differ with respect to the alleles of only one locus of interest.

monomer [Gk. *mono*: one + *meros*: unit] A small molecule, two or more of which can be combined to form oligomers (consisting of a few monomers) or polymers (consisting of many monomers).

monophyletic (mon' oh fih leht' ik) [Gk. *mono*: one + *phylon*: tribe] Pertaining to a group that consists of an ancestor and all of its descendants. (Contrast with paraphyletic, polyphyletic.)

monosaccharide A simple sugar. Oligosaccharides and polysaccharides are made up of monosaccharides.

monosomic Pertaining to an organism with one less than the normal diploid number of chromosomes.

monosynaptic reflex A neural reflex that begins in a sensory neuron and makes a single synapse before activating a motor neuron.

morphogen A diffusible substance whose concentration gradient determines a developmental pattern in embryonic animals and plants.

morphogenesis (more' fo jen' e sis) [Gk. *morphe*: form + *genesis*: origin] The development of form; the overall consequence of determination, differentiation, and growth.

morphological species concept The definition of a species as a group of individuals that look alike. (Contrast with biological species concept; lineage species concept.)

morphology (more fol' o jee) [Gk. *morphe*: form + *logos*: study, discourse] The scientific study of organic form, including both its development and function.

mortality Death, or the death rate of a population.

mosaic development Pattern of animal embryonic development in which each blastomere contributes a specific part of the adult body. (Contrast with regulative development.)

motif *See* structural motif.

motile (mo' tul) Able to move from one place to another. (Contrast with sessile.)

motor cortex The region of the cerebral cortex that contains motor neurons that directly stimulate specific muscle fibers to contract.

motor end plate The depression in the postsynaptic membrane of the neuromuscular junction where the terminals of the motor neuron sit.

motor neuron A neuron carrying information from the central nervous system to a cell that produces movement.

motor proteins Specialized proteins that use energy to change shape and move cells or structures within cells.

motor unit A motor neuron and the muscle fibers it controls.

mouth An opening through which food is taken in, located at the anterior end of a tubular gut.

mRNA *See* messenger RNA.

mucosal epithelium An epithelial cell layer containing cells that secrete mucus; found in the digestive and respiratory tracts. Also called mucosa.

mucus A viscous substance secreted by mucous membranes (e.g., mucosal epithelium). A barrier defense against pathogens in innate immunity in animals and a protective coating in many animal organ systems.

Muller's ratchet The accumulation—"ratcheting up"—of deleterious mutations in the nonrecombining genomes of asexual species.

Müllerian mimicry Convergence in appearance of two or more unpalatable species.

multifactorial The interaction of many genes and proteins with one or more factors in the environment. For example, cancer is a disease with multifactorial causes.

multipotent Having the ability to differentiate into a limited number of cell types. (Contrast with pluripotent, totipotent.)

muscle fiber A single muscle cell. In the case of skeletal muscle, a syncitial, multinucleate cell.

muscle tissue Excitable tissue that can contract through the interactions of actin and myosin; one of the four major tissue types in multicellular animals. There are three types of muscle tissue: skeletal, smooth, and cardiac.

mutagen (mute' ah jen) [L. *mutare*: change + Gk. *genesis*: source] Any agent (e.g., a chemical, radiation) that increases the mutation rate.

mutation A change in the genetic material not caused by recombination.

mutualism A type of interaction between species that benefits both species.

mycelium (my seel' ee yum) [Gk. *mykes*: fungus] In the fungi, a mass of hyphae.

mycologists Scientists who study fungi.

mycorrhiza (my' ko rye' za) (plural: mycorrhizae) [Gk. *mykes*: fungus + *rhiza*: root] An association of the root of a plant with the mycelium of a fungus.

myelin (my' a lin) Concentric layers of plasma membrane that form a sheath around some axons; myelin provides the axon with electrical insulation and increases the rate of transmission of action potentials.

myocardial infarction (MI) Blockage of an artery that carries blood to the heart muscle; a "heart attack."

MyoD The protein encoded by the *myoblast determing* gene. A transcription factor involved in the differentiation of myoblasts (muscle precursor cells).

myofibril (my' oh fy' bril) [Gk. *mys*: muscle + L. *fibrilla*: small fiber] A polymeric unit of actin or myosin in a muscle.

myoglobin (my' oh globe' in) [Gk. *mys*: muscle + L. *globus*: sphere] An oxygen-binding molecule found in muscle. Consists of a heme unit and a single globin chain; carries less oxygen than hemoglobin.

myosin One of the two contractile proteins of muscle. See also actin.

N

natural history The characteristics of a group of organisms, such as how the organisms get their food, reproduce, behave, regulate their internal environments (their cells, tissues, and organs), and interact with other organisms.

natural killer cell A type of lymphocyte that attacks virus-infected cells and some tumor cells as well as antibody-labeled target cells.

natural selection The differential contribution of offspring to the next generation by various genetic types belonging to the same population. The mechanism of evolution proposed by Charles Darwin.

nauplius (naw' plee us) [Gk. *nauplios*: shellfish] A bilaterally symmetrical larval form typical of crustaceans.

necrosis (nec roh' sis) [Gk. *nekros*: death] Premature cell death caused by external agents such as toxins.

negative feedback In regulatory systems, information that decreases a regulatory response, returning the system to the set point. (Contrast with positive feedback.)

negative regulation A type of gene regulation in which a gene is normally transcribed, and the binding of a repressor protein to the promoter prevents transcription. (Contrast with positive regulation.)

nematocyst (ne mat' o sist) [Gk. *nema*: thread + *kystis*: cell] An elaborate, threadlike structure produced by cells of jellyfishes and other cnidarians, used chiefly to paralyze and capture prey.

neoteny (knee ot' enny) [Gk. *neo*: new, recent; *tenein*, to extend] The retention of juvenile or larval traits by the fully developed adult organism.

nephron (nef' ron) [Gk. *nephros*: kidney] The functional unit of the kidney, consisting of a structure for receiving a filtrate of blood and a tubule that reabsorbs selected parts of the filtrate.

Nernst equation A mathematical statement that calculates the potential across a membrane permeable to a single type of ion that differs in concentration on the two sides of the membrane.

nerve A structure consisting of many neuronal axons and connective tissue.

nerve nets Diffuse, loosely connected aggregations of nervous tissues in certain non-bilatarian animals such as cnidarians.

nervous tissue Tissue specialized for processing and communicating information; one of the four major tissue types in multicellular animals.

net primary productivity (NPP) The rate at which energy captured by photosynthesis is incorporated into the bodies of primary producers through growth and reproduction.

neural crest cells During vertebrate neurulation, cells that migrate outward from the neural plate and give rise to connections between the central nervous system and the rest of the body.

neural network An organized group of neurons that contains three functional categories of neurons—afferent neurons, interneurons, and efferent neurons—and is capable of processing information.

neural tube An early stage in the development of the vertebrate nervous system consisting of a hollow tube created by two opposing folds of the dorsal ectoderm along the anterior–posterior body axis.

neurohormone A chemical signal produced and released by neurons that subsequently acts as a hormone.

neuromuscular junction Synapse (point of contact) where a motor neuron axon stimulates a muscle fiber cell.

neuron (noor' on) [Gk. *neuron*: nerve] A nervous system cell that can generate and conduct action potentials along an axon to a synapse with another cell.

neurotransmitter A substance produced in and released by a neuron (the presynaptic cell) that diffuses across a synapse and excites or inhibits another cell (the postsynaptic cell).

neurulation Stage in vertebrate development during which the nervous system begins to form.

neutral allele An allele that does not alter the functioning of the proteins for which it codes.

neutral theory A view of molecular evolution that postulates that most mutations do not affect the amino acid being coded for, and that such mutations accumulate in a population at rates driven by genetic drift and mutation rates.

neutron (new' tron) One of the three fundamental particles of matter (along with protons and electrons), with mass slightly larger than that of a proton and no electrical charge.

niche (nitch) [L. *nidus*: nest] The set of physical and biological conditions a species requires to survive, grow, and reproduce.

nitrate reduction The process by which nitrate (NO_3^-) is reduced to ammonia (NH_3).

nitric oxide (NO) An unstable molecule (a gas) that serves as a second messenger causing smooth muscle to relax. In the nervous system it operates as a neurotransmitter.

nitrification The oxidation of ammonia (NH_3) to nitrate (NO_3^-) in soil and seawater, carried out by chemoautotrophic bacteria (nitrifiers).

nitrogen fixation Conversion of atmospheric nitrogen gas (N_2) into a more reactive and biologically useful form (ammonia), which makes nitrogen available to living things. Carried out by **nitrogen fixers**—bacteria, some of them free-living and others living within plant roots.

nitrogenase An enzyme complex found in nitrogen-fixing bacteria that mediates the stepwise reduction of atmospheric N_2 to ammonia and which is strongly inhibited by oxygen.

nitrogenous wastes The potentially toxic nitrogen-containing end products—ammonia, urea, or uric acid—of protein and nucleic acid catabolism in animals. Eliminated from the body by excretion.

node [L. *nodus*: knob, knot] In plants, a (sometimes enlarged) point on a stem where a leaf is or was attached.

node of Ranvier A gap in the myelin sheath covering an axon; the point where the axonal membrane can fire action potentials.

nodule A specialized structure in the roots of nitrogen-fixing plants that houses nitrogen-fixing bacteria, in which oxygen is maintained at a low level by leghemoglobin.

non-REM sleep A state of deep, restorative sleep characterized by high-amplitude slow waves in the EEG. (Contrast with REM sleep.)

noncompetitive inhibitor A nonsubstrate that inhibits the activity of an enzyme by binding to a site other than its active site. (Contrast with competitive inhibitor.)

noncyclic electron transport In photosynthesis, the flow of electrons that forms ATP, NADPH, and O_2.

nondisjunction Failure of sister chromatids to separate in meiosis II or mitosis, or failure of homologous chromosomes to separate in meiosis I. Results in aneuploidy.

nonpolar Having electric charges that are evenly balanced from one end to the other. (Contrast with polar.)

nonrandom mating Selection of mates on the basis of a particular trait or group of traits.

nonsense mutation Change in a gene's sequence that prematurely terminates translation by changing one of its codons to a stop codon.

nonsynonymous substitution A change in a gene from one nucleotide to another that changes the amino acid specified by the corresponding codon (i.e., AGC → AGA, or serine → arginine). (Contrast with synonymous substitution.)

norepinephrine A neurotransmitter found in the central nervous system and also at the postganglionic nerve endings of the sympathetic nervous system. Also called noradrenaline.

normal flora Microorganisms that normally live and reproduce on or in the body without causing disease, and which form a nonspecific defense against pathogens by competing with them for space and nutrients. See also microbiota.

notochord (no' tow kord) [Gk. *notos*: back + *chorde*: string] A flexible rod of gelatinous material serving as a support in the embryos of all chordates and in the adults of tunicates and lancelets.

nucleic acid (new klay' ik) A polymer made up of nucleotides, specialized for the storage, transmission, and expression of genetic information. DNA and RNA are nucleic acids.

nucleic acid hybridization A technique in which a single-stranded nucleic acid probe is made that is complementary to, and binds to, a target sequence, either DNA or RNA. The resulting double-stranded molecule is a hybrid.

nucleoid (new' klee oid) The region that harbors the chromosomes of a prokaryotic cell. Unlike the eukaryotic nucleus, it is not bounded by a membrane.

nucleolus (new klee' oh lus) A small, generally spherical body found within the nucleus of eukaryotic cells. The site of synthesis of ribosomal RNA.

nucleoside A nucleotide without the phosphate group; a nitrogenous base attached to a sugar.

nucleosome A portion of a eukaryotic chromosome, consisting of part of the DNA molecule wrapped around a group of histone molecules, and held together by another type of histone molecule. The chromosome is made up of many nucleosomes.

nucleotide The basic chemical unit in nucleic acids, consisting of a pentose sugar, a phosphate group, and a nitrogen-containing base.

nucleotide substitution A change of one base pair to another in a DNA sequence.

nucleus (new' klee us) [L. *nux*: kernel or nut] (1) In cells, the centrally located compartment of eukaryotic cells that is bounded by a double membrane and contains the chromosomes. (2) In the brain, an identifiable group of neurons that share common characteristics or functions.

null hypothesis In statistics, the premise that any differences observed in an experiment are simply the result of random differences that arise from drawing two finite samples from the same population.

nutrient A food substance; or, in the case of mineral nutrients, an inorganic element required for completion of the life cycle of an organism.

O

obligate anaerobe An anaerobic prokaryote that cannot survive exposure to O_2.

occipital lobe One of the four lobes of the brain's cerebral hemisphere; processes visual information.

odorant A molecule that can bind to an olfactory receptor.

oil A triglyceride that is liquid at room temperature. (Contrast with fat.)

Okazaki fragments Newly formed DNA making up the lagging strand in DNA replication. DNA ligase links Okazaki fragments together to give a continuous strand.

olfaction (ole fak' shun) [L. *olfacere*: to smell] The sense of smell.

olfactory bulb Structure in the vertebrate forebrain that receives and processes input from olfactory receptor neurons.

olfactory receptor neurons (ORNs) Neurons with receptors for different odorants.

oligodendrocyte A type of glial cell that myelinates axons in the central nervous system.

oligophagous [Gk. *oligo*: few; *phagein*, eat] An animal that feeds on a limited number of foods; generally used of insects that feed on only one or a few plant species.

oligosaccharide A polymer containing a small number of monosaccharides.

omasum One of the four chambers of the stomach in ruminants; concentrates food by water absorption before it enters the true stomach (abomasum).

ommatidia [Gk. *omma*: eye] The units that make up the compound eye of some arthropods.

omnivore [L. *omnis*: everything + *vorare*: to devour] An organism that eats both animal and plant material. (Contrast with carnivore, detritivore, herbivore.)

oncogene [Gk. *onkos*: mass, tumor + *genes*: born] A gene that codes for a protein product that stimulates cell proliferation. Mutations in oncogenes that result in excessive cell proliferation can give rise to cancer.

one gene–one polypeptide The idea, now known to be an oversimplification, that each gene in the genome encodes only a single polypeptide—that there is a one-to-one correspondence between genes and polypeptides.

oocyte *See* primary oocyte, secondary oocyte.

oogenesis (oh' eh jen e sis) [Gk. *oon*: egg + *genesis*: source] Gametogenesis leading to production of an ovum.

oogonium (oh' eh go' nee um) (plural: oogonia) (1) In some algae and fungi, a cell in which an egg is produced. (2) In animals, the diploid progeny of a germ cell in females.

ootid In oogenesis, the daughter cell of the second meiotic division that differentiates into the mature ovum.

open circulatory system Circulatory system in which extracellular fluid leaves the vessels of the circulatory system, percolates between cells and through tissues, and then flows back into the circulatory system to be pumped out again. (Contrast with closed circulatory system.)

operator The region of an operon that acts as the binding site for the repressor.

operon A genetic unit of transcription, typically consisting of several structural genes that are transcribed together; the operon contains at least two control regions: the promoter and the operator.

opportunity cost The sum of the benefits an animal forfeits by not being able to perform some other behavior during the time when it is performing a given behavior.

opsin (op' sin) [Gk. *opsis*: sight] The protein portion of vertebrate visual pigments; associated with the pigment molecule 11-*cis*-retinal. See also rhodopsin.

optic chiasm [Gk. *chiasma*: cross] Structure on the lower surface of the vertebrate brain where the two optic nerves come together.

optic nerve The nerve that carries information from the retina of the eye to the brain.

optical isomers Two molecular isomers that are mirror images of each other.

optimal foraging theory The application of a cost–benefit approach to feeding behavior to identify the fitness value of feeding choices.

oral [L. *os*: mouth] Pertaining to the mouth, or that part of the body that contains the mouth.

orbital A region in space surrounding the atomic nucleus in which an electron is most likely to be found.

organ [Gk. *organon*: tool] A body part, such as the heart, liver, brain, root, or leaf. Organs are composed of different tissues integrated to perform a distinct function. Organs, in turn, are integrated into organ systems.

organ identity genes In angiosperms, genes that specify the different organs of the flower. (Compare with homeotic genes.)

organ of Corti Structure in the inner ear that transforms mechanical forces produced from pressure waves ("sound waves") into action potentials that are sensed as sound.

organ system An interrelated and integrated group of tissues and organs that work together in a physiological function.

organelle (or gan el') Any of the membrane-enclosed structures within a eukaryotic cell. Examples include the nucleus, endoplasmic reticulum, and mitochondria.

organic (1) Pertaining to any chemical compound that contains carbon. (2) Pertaining to any aspect of living matter, e.g., to its evolution, structure, or chemistry.

organic fertilizers Substances added to soil to improve the soil's fertility; derived from partially decomposed plant material (compost) or animal waste (manure).

organism Any living entity.

organizer Region of the early amphibian embryo that directs early embryonic development. Also known as the primary embryonic organizer.

organogenesis The formation of organs and organ systems during development.

origin of replication (*ori*) DNA sequence at which helicase unwinds the DNA double helix and DNA polymerase binds to initiate DNA replication.

orthologs [Gk. *ortho*: true, direct] Homologous genes whose divergence can be traced to speciation events.

osmoconformer An aquatic animal that equilibrates the osmolarity of its extracellular fluid to be the same as that of the external environment. (Contrast with osmoregulator.)

osmolarity The concentration of osmotically active particles in a solution.

osmoregulation Regulation of the chemical composition of the body fluids of an organism.

osmoregulator An aquatic animal that actively regulates the osmolarity of its extracellular fluid. (Contrast with osmoconformer.)

osmosis (oz mo' sis) [Gk. *osmos*: to push] Movement of water across a differentially permeable membrane, from one region to another region where the water potential is more negative.

ossicle (oss' ick ul) [L. *os*: bone] The calcified construction unit of echinoderm skeletons.

osteoblast (oss' tee oh blast) [Gk. *osteon*: bone + *blastos*: sprout] A cell that lays down the protein matrix of bone.

osteoclast (oss' tee oh clast) [Gk. *osteon*: bone + *klastos*: broken] A cell that dissolves bone.

osteocyte An osteoblast that has become enclosed in lacunae within the bone it has built.

outgroup In phylogenetics, a group of organisms used as a point of reference for comparison with the groups of primary interest (the ingroup).

oval window The flexible membrane that, when moved by the bones of the middle ear, produces pressure waves in the inner ear.

ovarian cycle In human females, the monthly cycle of events by which eggs and

hormones are produced. (Contrast with uterine cycle).

ovary (oh' var ee) [L. *ovum*: egg] Any female organ, in plants or animals, that produces an egg.

overtopping Plant growth pattern in which one branch differentiates from and grows beyond the others.

oviduct In mammals, the tube serving to transport eggs to the uterus or to the outside of the body.

oviparity Reproduction in which eggs are released by the female and development is external to the mother's body. (Contrast with viviparity.)

ovoviviparity Pertaining to reproduction in which fertilized eggs develop and hatch within the mother's body but are not attached to the mother by means of a placenta.

ovulation Release of an egg from an ovary.

ovule (oh' vule) In plants, a structure comprising the megasporangium and the integument, which develops into a seed after fertilization.

ovum (oh' vum) (plural: ova) [L. egg] The female gamete.

oxidation (ox i day' shun) Relative loss of electrons in a chemical reaction; either outright removal to form an ion, or the sharing of electrons with substances having a greater affinity for them, such as oxygen. Most oxidations, including biological ones, are associated with the liberation of energy. (Contrast with reduction.)

oxidation–reduction (redox) reaction A reaction in which one substance transfers one or more electrons to another substance. (*See* oxidation; reduction.)

oxidative phosphorylation ATP formation in the mitochondrion, associated with flow of electrons through the respiratory chain.

oxygenase An enzyme that catalyzes the addition of oxygen to a substrate from O_2.

oxytocin A hormone released by the posterior pituitary that promotes social bonding.

ozone layer A layer of ozone (O_3, a greenhouse gas) in the atmosphere that absorbs a high portion of the sun's potentially mutagenic ultraviology radiation.

P

pacemaker cells Cardiac cells that can initiate action potentials without stimulation from the nervous system, allowing the heart to initiate its own contractions.

pair rule genes In *Drosophila* (fruit fly) development, segmentation genes that divide the early embryo into units of two segments each. Part of a developmental cascade that includes maternal effect genes, gap genes, segment polarity genes, and Hox genes.

paleomagnetic dating A method for determining the age of rocks based on properties relating to changes in the patterns of Earth's magnetism over time.

pancreas (pan' cree us) A gland located near the stomach of vertebrates that secretes digestive enzymes into the small intestine and releases insulin into the bloodstream.

Pangaea (pan jee' uh) [Gk. *pan*: all, every] The single land mass formed when all the continents came together in the Permian period.

para- [Gk. *para*: akin to, beside] Prefix indicating association in being along side or accessory to.

parabronchi Passages in the lungs of birds through which air flows.

paracrine [Gk. *para*: near] Pertaining to a chemical signal, such as a hormone, that acts locally, near the site of its secretion. (Contrast with autocrine.)

parallel evolution The repeated evolution of similar traits, especially among closely related species; facilitated by conserved developmental genes.

paralogs Homologous genes whose divergence can be traced to gene duplication events. (Contrast with orthologs.)

paraphyletic (par' a fih leht' ik) [Gk. *para*: beside + *phylon*: tribe] Pertaining to a group that consists of an ancestor and some, but not all, of its descendants. (Contrast with monophyletic, polyphyletic.)

parasite An organism that consumes parts of an organism much larger than itself (known as its host). Parasites sometimes, but not always, kill their host.

parasympathetic nervous system The division of the autonomic nervous system that works in opposition to the sympathetic nervous system. (Contrast with sympathetic nervous system.)

parathyroid glands Four glands on the posterior surface of the thyroid gland that produce and release parathyroid hormone.

parathyroid hormone (PTH) A hormone secreted by the parathyroid glands that stimulates osteoclast activity and raises blood calcium levels. Also called parathormone.

parenchyma (pair eng' kyma) A plant tissue composed of relatively unspecialized cells without secondary walls.

parent rock The soil horizon consisting of the rock that is breaking down to form the soil. Also called bedrock, or the C horizon.

parental (P) generation The individuals that mate in a genetic cross. Their offspring are the first filial (F_1) generation.

parietal cells One of three types of secretory cell found in the gastric pits of the stomach wall. Parietal cells produce hydrochloric acid (HCl), creating an acidic environment that destroys many of the harmful microorganisms ingested with food. (See chief cells; mucosal epithelium.)

parietal lobe One of four lobes of the cerebral hemisphere; processes complex stimuli and includes the primary somatosensory cortex.

parsimony   Preferring the simplest among a set of plausible explanations of any phenomenon.

parthenocarpy   Formation of fruit from a flower without fertilization.

parthenogenesis   [Gk. *parthenos*: virgin] Production of an organism from an unfertilized egg.

particulate theory   In genetics, the theory that genes are physical entities that retain their identities after fertilization.

passive transport   Diffusion across a membrane; may or may not require a channel or carrier protein. (Contrast with active transport.)

patch clamping   Technique for isolating a tiny patch of membrane to allow the study of ion movement through a particular channel.

pathogen   (path' o jen) [Gk. *pathos*: suffering + *genesis*: source] An organism that causes disease.

pattern formation   In animal embryonic development, the organization of differentiated tissues into specific structures such as wings.

pedigree   The pattern of transmission of a genetic trait within a family.

pelagic zone   [Gk. *pelagos*: deep sea] The open ocean; a marine life zone.

penetrance   The proportion of individuals with a particular genotype that show the expected phenotype.

penis   An accessory sex organ of male animals that enables the male to deposit sperm in the female's reproductive tract.

pentaradial symmetry   Symmetry in five or multiples of five; a feature of adult echinoderms.

pentose   [Gk. *penta*: five] A sugar containing five carbon atoms.

PEP carboxylase   The enzyme that combines carbon dioxide with PEP to form a 4-carbon dicarboxylic acid at the start of C_4 photosynthesis or of crassulacean acid metabolism (CAM).

pepsin   [Gk. *pepsis*: digestion] An enzyme in gastric juice that digests protein.

pepsinogen   Inactive secretory product that is converted into pepsin by low pH or by enzymatic action.

peptide hormones   Relatively large hormone molecules made up of amino acids; encoded by genes and produced by translation.

peptide linkage   The bond between amino acids in a protein; formed between a carboxyl group and amino group (—CO—NH—) with the loss of water molecules.

peptidoglycan   The cell wall material of many bacteria, consisting of a single enormous molecule that surrounds the entire cell.

peptidyl transferase   A catalytic function of the large ribosomal subunit that consists of two reactions: breaking the bond between an amino acid and its tRNA in the P site, and forming a peptide bond between that amino acid and the amino acid attached to the tRNA in the A site.

per capita birth rate (*b*)   In population growth models, the number of offspring that an average individual produces in some time interval.

per capita death rate (*d*)   In population growth models, the average individual's chance of dying in some time interval.

per capita growth rate (*r*)   In population models, the average individual's contribution to total population growth rate.

perennial   (per ren' ee al) [L. *per*: throughout + *annus*: year] A plant that survives from year to year. (Contrast with annual, biennial.)

perfect flower   A flower with both stamens and carpels; a hermaphroditic flower. (Contrast with imperfect flower.)

pericycle   [Gk. *peri*: around + *kyklos*: ring or circle] In plant roots, tissue just within the endodermis, but outside of the root vascular tissue. Meristematic activity of pericycle cells produces lateral root primordia.

periderm   The outer tissue of the secondary plant body, consisting primarily of cork.

period   (1) A category in the geological time scale. (2) The duration of a single cycle in a cyclical event, such as a circadian rhythm.

peripheral membrane proteins   Proteins associated with but not embedded within the plasma membrane. (Contrast with integral membrane proteins.)

peripheral nervous system (PNS)   The portion of the nervous system that transmits information to and from the central nervous system, consisting of neurons that extend or reside outside the brain or spinal cord and their supporting cells. (Contrast with central nervous system.)

peristalsis   (pair' i stall' sis) Wavelike muscular contractions proceeding along a tubular organ, propelling the contents along the tube.

peritoneum   The mesodermal lining of the body cavity in coelomate animals.

peroxisome   An organelle that houses reactions in which toxic peroxides are formed and then converted to water.

petal   [Gk. *petalon*: spread out] In an angiosperm flower, a sterile modified leaf, nonphotosynthetic, frequently brightly colored, and often serving to attract pollinating insects.

petiole   (pet' ee ole) [L. *petiolus*: small foot] The stalk of a leaf.

P_{fr}   *See* phytochrome.

pH   The negative logarithm of the hydrogen ion concentration; a measure of the acidity of a solution. A solution with pH = 7 is said to be neutral; pH values higher than 7 characterize basic solutions, while acidic solutions have pH values less than 7.

phage   (fayj) *See* bacteriophage.

phagocyte   [Gk. *phagein*: to eat + *kystos*: sac] One of two major classes of white blood cells; one of the nonspecific defenses of animals; ingests invading microorganisms by **phagocytosis**.

pharmacogenomics   The study of how an individual's genetic makeup affects his or her response to drugs or other agents, with the goal of predicting the effectiveness of different treatment options.

pharming   The use of genetically modified animals to produce medically useful products in their milk.

pharynx   [Gk. throat] The part of the gut between the mouth and the esophagus.

phenotype   (fee' no type) [Gk. *phanein*: to show] The observable properties of an individual resulting from both genetic and environmental factors. (Contrast with genotype.)

phenotypic plasticity   *See* developmental plasticity.

pheromone   (feer' o mone) [Gk. *pheros*: carry + *hormon*: excite, arouse] A chemical substance used in communication between organisms of the same species.

phloem   (flo' um) [Gk. *phloos*: bark] In vascular plants, the vascular tissue that transports sugars and other solutes from sources to sinks.

phosphate group   The functional group —OPO_3H_2.

phosphodiester linkage   The connection in a nucleic acid strand, formed by linking two nucleotides.

phospholipid   A lipid containing a phosphate group; an important constituent of cellular membranes. (*See* lipid.)

phospholipid bilayer   The basic structural unit of biological membranes; a sheet of phospholipids two molecules thick in which the phospholipids are lined up with their hydrophobic "tails" packed tightly together and their hydrophilic, phosphate-containing "heads" facing outward. Also called lipid bilayer.

phosphorylation   Addition of a phosphate group.

photic zone   The life zone in lakes and oceans that is penetrated by light and therefore supports photosynthetic organisms.

photoautotroph   An organism that obtains energy from light and carbon from carbon dioxide. (Contrast with chemolithotroph, chemoheterotroph, photoheterotroph.)

photoheterotroph   An organism that obtains energy from light but must obtain its carbon from organic compounds. (Contrast with chemoautotroph, chemoheterotroph, photoautotroph.)

photomorphogenesis   In plants, a process by which physiological and developmental events are controlled by light.

photon   (foe' ton) [Gk. *photos*: light] A quantum of visible radiation; a "packet" of light energy.

photoperiodism Control of an organism's physiological or behavioral responses by the length of the day or night (the **photoperiod**).

photophosphorylation Mechanism for ATP formation in chloroplasts in which electron transport is coupled to the transport of hydrogen ions (protons, H$^+$) across the thylakoid membrane. Compare with chemiosmosis.

photoreceptors (1) In plants, pigments that trigger a physiological response when they absorb a photon. (2) In animals, the sensory receptor cells that sense and respond to light energy. (See cone cells; rod cells.)

photorespiration Light-driven uptake of oxygen and release of carbon dioxide, the carbon being derived from the early reactions of photosynthesis.

photosynthesis (foe tow sin' the sis) [Gk.: creating from light] Metabolic processes carried out by green plants and some microorganisms by which visible light is trapped and the energy used to synthesize compounds such as ATP and glucose.

photosystem [Gk. *phos*: light + *systema*: assembly] A light-harvesting complex in the chloroplast thylakoid composed of pigments and proteins. **Photosystem I** absorbs light at 700 nm, passing electrons to ferrodoxin and from there to NADPH. **Photosystem II** absorbs light at 680 nm and passes electrons to the electron transport chain in the chloroplast.

phototropism [Gk. *photos*: light + *trope*: turning] A directed plant growth response to light.

phycobilin Photosynthetic pigment that absorbs red, yellow, orange, and green light and is found in cyanobacteria and some red algae.

phylogeny (fy loj' e nee) [Gk. *phylon*: tribe, race + *genesis*: source] The evolutionary history of a particular group of organisms or their genes. A **phylogenetic tree** is a graphic representation of these lines of evolutionary descent.

physiological survivorship curves *See* survivorship curves.

physiology (fiz' ee ol' o jee) [Gk. *physis*: natural form] The scientific study of the functions of living organisms and the individual organs, tissues, and cells of which they are composed.

phytoalexins Substances toxic to pathogens, produced by plants in response to fungal or bacterial infection.

phytochrome (fy' tow krome) [Gk. *phyton*: plant + *chroma*: color] A plant pigment regulating a large number of developmental and other phenomena in plants. It has two isomers: P$_r$, which absorbs red light, and P$_{fr}$, which absorbs far red light. P$_{fr}$ is the active form.

phytomers In plants, the repeating modules that compose a shoot, each consisting of one or more leaves, attached to the stem at a node; an internode; and one or more axillary buds.

phytoplankton Photosynthetic floating organisms. (*See* plankton.)

phytoremediation A form of bioremediation that uses plants to clean up environmental pollution.

pigment A substance that absorbs visible light.

piloting A form of navigation in which an animal finds its way by remembering landmarks in its environment.

pineal gland Gland located between the cerebral hemispheres that secretes melatonin.

pinocytosis Endocytosis by a cell of liquid containing dissolved substances.

pistil [L. *pistillum*: pestle] The structure of an angiosperm flower within which the ovules are borne. May consist of a single carpel, or of several carpels fused into a single structure. Usually differentiated into ovary, style, and stigma.

pith In plants, relatively unspecialized tissue found within a cylinder of vascular tissue.

pituitary gland A small gland attached to the base of the brain in vertebrates. Its hormones control the activities of other glands. Also known as the hypophysis.

placenta (pla sen' ta) The organ in female mammals that provides for the nourishment of the fetus and elimination of the fetal waste products.

plankton Aquatic organisms that float in the water column, dependent on currents and wind for movement. Plankton include many protists, some algae, and larval animals. (See also phytoplankton.)

planula (plan' yew la) [L. *planum*: flat] A free-swimming, ciliated larval form typical of the cnidarians.

plaque (plack) [Fr.: a metal plate or coin] (1) A circular clearing in a layer (lawn) of bacteria growing on the surface of a nutrient agar gel. (2) An accumulation of prokaryotic organisms on tooth enamel. Acids produced by these microorganisms cause tooth decay. (3) A region of arterial wall invaded by fibroblasts and fatty deposits.

plasma (plaz' muh) The liquid portion of blood, in which blood cells and other particulates are suspended.

plasma cell An antibody-secreting cell that develops from a B cell; the effector cell of the humoral immune system.

plasma membrane The membrane that surrounds the cell, regulating the entry and exit of molecules and ions. Every cell has a plasma membrane, and it is often called the cell membrane.

plasmid A DNA molecule distinct from the chromosome(s); that is, an extrachromosomal element; found in many bacteria. May replicate independently of the chromosome.

plasmodesmata (singular: plasmodesma) [Gk. *plassein*: to mold + *desmos*: band] Cytoplasmic strands connecting two adjacent plant cells.

plasmogamy The fusion of the cytoplasm of two cells. (Contrast with karyogamy.)

plastid A class of plant cell organelles that includes the chloroplast, which houses biochemical pathways for photosynthesis.

plate tectonics [Gk. *tekton*: builder] The scientific study of the structure and movements of Earth's lithospheric plates, which are the cause of continental drift.

platelet A membrane-bounded body without a nucleus, arising as a fragment of a cell in the bone marrow of mammals. Important to blood-clotting action.

pleiotropy (plee' a tro pee) [Gk. *pleion*: more] The determination of more than one character by a single gene.

pleural membrane [Gk. *pleuras*: rib, side] The membrane lining the outside of the lungs and the walls of the thoracic cavity. Inflammation of these membranes is a condition known as pleurisy.

pluripotent [L. *pluri*: many + *potens*: powerful] Having the ability to form all of the cells in the body. (Contrast with multipotent, totipotent.)

podocytes Cells of Bowman's capsule of the nephron that cover the capillaries of the glomerulus, forming filtration slits.

point mutation A mutation that results from the gain, loss, or substitution of a single nucleotide.

polar A molecule with separate and opposite electric charges at two ends, or poles; the water molecule (H$_2$O) is the most prevalent example. (Contrast with nonpolar.)

polar body A nonfunctional nucleus produced by meiosis during oogenesis.

polar covalent bond A covalent bond in which the electrons are drawn to one nucleus more than the other, resulting in an unequal distribution of charge.

polar nuclei In angiosperms, the two nuclei in the central cell of the megagametophyte; following fertilization they give rise to the endosperm.

polarity (1) In chemistry, the property of unequal electron sharing in a covalent bond that defines a polar molecule. (2) In development, the difference between one end of an organism or structure and the other.

pollen [L. *pollin*: fine flour] In seed plants, microscopic grains that contain the male gametophyte (microgametophyte) and gamete (microspore).

pollen tube A structure that develops from a pollen grain through which sperm are released into the megagametophyte.

pollination The process of transferring pollen from an anther to the stigma of a pistil in an angiosperm or from a strobilus to an ovule in a gymnosperm.

poly- [Gk. *poly*: many] A prefix denoting multiple entities.

poly A tail A long sequence of adenine nucleotides (50–250) added after transcription to the 3' end of most eukaryotic mRNAs.

polyandry   Mating system in which one female mates with multiple males.

polygyny   Mating system in which one male mates with multiple females.

polymer   [Gk. *poly*: many + *meros*: unit] A large molecule made up of similar or identical subunits called monomers. (Contrast with monomer.)

polymerase chain reaction (PCR)   An enzymatic technique for the rapid production of millions of copies of a particular stretch of DNA where only a small amount of the parent molecule is available.

polymorphic   (pol' lee mor' fik) [Gk. *poly*: many + *morphe*: form, shape] Coexistence in a population of two or more distinct traits.

polyp   (pah' lip) [Gk. *poly*: many + *pous*: foot] In cnidarians, a sessile, asexual life cycle stage.

polypeptide   A large molecule made up of many amino acids joined by peptide linkages. Large polypeptides are called proteins.

polyphyletic   (pol' lee fih leht' ik) [Gk. *poly*: many + *phylon*: tribe] Pertaining to a group that consists of multiple distantly related organisms, and does not include the common ancestor of the group. (Contrast with monophyletic, paraphyletic.)

polyploid   (pol' lee ploid ee) Possessing more than two entire sets of chromosomes.

polyribosome (polysome)   A complex consisting of a threadlike molecule of messenger RNA and several (or many) ribosomes. The ribosomes move along the mRNA, synthesizing polypeptide chains as they proceed.

polysaccharide   A macromolecule composed of many monosaccharides (simple sugars). Common examples are cellulose and starch.

pons   [L. *pons*: bridge] Region of the brainstem anterior to the medulla.

pool   The total amount of an element in a given compartment of the biosphere.

population   Any group of organisms coexisting at the same time and in the same place and capable of interbreeding with one another.

population bottleneck   A period during which only a few individuals of a normally large population survive.

population density   The number of individuals in a population per unit of area or volume.

population dynamics   The patterns and processes of change in populations.

population genetics   The study of genetic variation and its causes within populations.

population size   The total number of individuals in a population.

positional information   In development, the basis of the spatial sense that induces cells to differentiate as appropriate for their location within the developing organism; often comes in the form of a morphogen gradient.

positive cooperativity   Occurs when a molecule can bind several ligands and each one that binds alters the conformation of the molecule so that it can bind the next ligand more easily. The binding of four molecules of O_2 by hemoglobin is an example of positive cooperativity.

positive feedback   In regulatory systems, information that amplifies a regulatory response, increasing the deviation of the system from the set point. (Contrast with negative feedback.)

positive regulation   A form of gene regulation in which a regulatory macromolecule is needed to turn on the transcription of a structural gene; in its absence, transcription will not occur. (Contrast with negative regulation.)

positive selection   Natural selection that acts to establish a trait that enhances survival in a population. (Contrast with purifying selection.)

post-   [L. *postere*: behind, following after] Prefix denoting something that comes after.

postabsorptive state   State in which no food remains in the gut and thus no nutrients are being absorbed. (Contrast with absorptive state.)

posterior   Toward or pertaining to the rear. (Contrast with anterior.)

posterior pituitary   A portion of the pituitary gland derived from neural tissue; involved in the storage and release of antidiuretic hormone and oxytocin.

postsynaptic cell   The cell that receives information from a neuron at a synapse. (Contrast with presynaptic cell.)

postzygotic isolating mechanisms   Barriers to the reproductive process that occur after the union of the nuclei of two gametes. (Contrast with prezygotic isolating mechanisms.)

potential energy   Energy not doing work but with the potential to do so, such as the energy stored in chemical bonds. (Contrast with kinetic energy.)

P_r   See phytochrome.

pre-mRNA (precursor mRNA)   Initial gene transcript before it is modified to produce functional mRNA. Also known as the primary transcript.

Precambrian   The first and longest period of geological time, during which life originated.

precapillary sphincter   A cuff of smooth muscle that can shut off the blood flow to a capillary bed.

predator   An organism that kills and eats other organisms.

pressure flow model   An effective model for phloem transport in angiosperms. It holds that sieve element transport is driven by an osmotically generated pressure gradient between source and sink.

pressure potential (Ψ_p)   The hydrostatic pressure of an enclosed solution in excess of the surrounding atmospheric pressure.

(Contrast with solute potential, water potential.)

presynaptic cell   The neuron that transmits information to another cell at a synapse. (Contrast with postsynaptic cell.)

prey   [L. *praeda*: booty] An organism consumed by a predator as an energy source.

prezygotic isolating mechanisms   Barriers to the reproductive process that occur before the union of the nuclei of two gametes. (Contrast with postzygotic isolating mechanisms.)

primary active transport   Active transport in which ATP is hydrolyzed, yielding the energy required to transport an ion or molecule against its concentration gradient. (Contrast with secondary active transport.)

primary cell wall   In plant cells, a structure that forms at the middle lamella after cytokinesis, made up of cellulose microfibrils, hemicelluloses, and pectins. (Contrast with secondary cell wall.)

primary consumer   An organism (herbivore) that eats plant tissues.

primary endosymbiosis   The engulfment of a cyanobacterium by a larger eukaryotic cell that gave rise to the first photosynthetic eukaryotes with chloroplasts.

primary growth   In plants, growth that is characterized by the lengthening of roots and shoots and by the proliferation of new roots and shoots through branching. (Contrast with secondary growth.)

primary immune response   The first response of the immune system to an antigen, involving recognition by lymphocytes and the production of effector cells and memory cells. (Contrast with secondary immune response.)

primary meristem   Meristem that produces the tissues of the primary plant body.

primary motor cortex   See motor cortex

primary oocyte   (oh' eh site) [Gk. *oon*: egg + *kytos*: container] The diploid progeny of an oogonium. In many species, a primary oocyte enters prophase of the first meiotic division, then remains in developmental arrest for a long time before resuming meiosis to form a secondary oocyte and a polar body.

primary plant body   That part of a plant produced by primary growth. Consists of all the *nonwoody* parts of a plant; many herbaceous plants consist entirely of a primary plant body. (Contrast with secondary plant body.)

primary producer   A photosynthetic or chemosynthetic organism that synthesizes complex organic molecules from simple inorganic ones.

primary sex determination   Genetic determination of gametic sex, male or female. (Contrast with secondary sex determination.)

primary somatosensory cortex   See somatosensory cortex.

primary spermatocyte   The diploid progeny of a spermatogonium; undergoes the first meiotic division to form secondary spermatocytes.

primary structure The specific sequence of amino acids in a protein.

primary succession Succession of ecological communities that begins in an area devoid of life, such as on recently exposed glacial till or lava flows. (Contrast with secondary succession.)

primase An enzyme that catalyzes the synthesis of a primer for DNA replication.

primer Strand of nucleic acid, usually RNA, that is the necessary starting material for the synthesis of a new DNA strand, which is synthesized from the 3′ end of the primer.

primordium (plural: primordia) [L. origin] The most rudimentary stage of an organ or other part.

pro- [L.: first, before, favoring] A prefix often used in biology to denote a developmental stage that comes first or an evolutionary form that appeared earlier than another. For example, prokaryote, prophase.

probe A segment of single-stranded nucleic acid used to identify DNA molecules containing the complementary sequence.

procambium Primary meristem that produces the vascular tissue.

procedural memory Memory of motor tasks.; these memories cannot be consciously recalled or described. (Contrast with declarative memory.)

processive Pertaining to an enzyme that catalyzes many reactions each time it binds to a substrate, as DNA polymerase does during DNA replication.

products The molecules that result from the completion of a chemical reation.

progesterone [L. pro: favoring + gestare: to bear] A female sex hormone that maintains pregnancy.

prokaryotes Unicellular organisms that do not have nuclei or other membrane-enclosed organelles. Includes Bacteria and Archaea. (Contrast with eukaryotes.)

prolactin A hormone released by the anterior pituitary, one of whose functions is the stimulation of milk production in female mammals.

prometaphase The phase of nuclear division that begins with the disintegration of the nuclear envelope.

promoter A DNA sequence to which RNA polymerase binds to initiate transcription.

prop roots Adventitious roots in some monocots that function as supports for the shoot.

prophage (pro′ fayj) The noninfectious units that are linked with the chromosomes of the host bacteria and multiply with them but do not cause dissolution of the cell. Prophage can later enter into the lytic phase to complete the virus life cycle.

prophase (pro′ phase) The first stage of nuclear division, during which chromosomes condense from diffuse, threadlike material to discrete, compact bodies.

prostaglandin Any one of a group of specialized lipids with hormone-like functions. It is not clear that they act at any considerable distance from the site of their production.

prostate gland In male humans, surrounds the urethra at its junction with the vas deferens; supplies an acid-neutralizing fluid to the semen.

prosthetic group Any nonprotein portion of an enzyme.

proteases Digestive enzymes that digest proteins.

proteasome In the eukaryotic cytoplasm, a huge protein structure that binds to and digests cellular proteins that have been tagged by ubiquitin.

protein (pro′ teen) [Gk. protos: first] Long-chain polymer of amino acids with twenty different common side chains. Occurs with its polymer chain extended in fibrous proteins, or coiled into a compact macromolecule in enzymes and other globular proteins. The component amino acids are encoded in the triplets of messenger RNA, and proteins are the products of genes.

protein kinase (kye′ nase) An enzyme that catalyzes the addition of a phosphate group from ATP to a target protein.

protein kinase cascade A series of reactions in response to a molecular signal, in which a series of protein kinases activate one another in sequence, amplifying the signal at each step.

proteoglycan A glycoprotein containing a protein core with attached long, linear carbohydrate chains.

proteolysis [protein + Gk. lysis: break apart] An enzymatic digestion of a protein or polypeptide.

proteome The complete set of proteins that can be made by an organism. Because of alternative splicing of pre-mRNA, the number of proteins that can be made is usually much larger than the number of protein-coding genes present in the organism's genome.

prothrombin The inactive form of thrombin, an enzyme involved in blood clotting.

protoderm Primary meristem that gives rise to the plant epidermis.

proton (pro′ ton) [Gk. protos: first, before] (1) A subatomic particle with a single positive charge. The number of protons in the nucleus of an atom determine its element. (2) A hydrogen ion, H⁺.

proton pump An active transport system that uses ATP energy to move hydrogen ions across a membrane, generating an electric potential.

proton-motive force Force generated across a membrane having two components: a chemical potential (difference in proton concentration) plus an electrical potential due to the electrostatic charge on the proton.

protonephridium The excretory organ of flatworms, made up of a tubule and a flame cell.

protoplast The living contents of a plant cell; the plasma membrane and everything contained within it.

provirus Double-stranded DNA made by a virus that is integrated into the host's chromosome and contains promoters that are recognized by the host cell's transcription apparatus.

proximal convoluted tubule The initial segment of a renal tubule, closest to the glomerulus. (Compare with distal convoluted tubule.)

proximal Near the point of attachment or other reference point. (Contrast with distal.)

proximate causes The immediate genetic, physiological, neurological, and developmental mechanisms responsible for a behavior or morphology. (Contrast with ultimate cause.)

pseudocoelomate (soo′ do see′ low mate) [Gk. pseudes: false + koiloma: cavity] Having a body cavity, called a pseudocoel, consisting of a fluid-filled space in which many of the internal organs are suspended, but which is enclosed by mesoderm only on its outside.

pseudogene [Gk. pseudes: false] A DNA segment that is homologous to a functional gene but is not expressed because of changes to its sequence or changes to its location in the genome.

pseudopod (soo′ do pod) [Gk. pseudes: false + podos: foot] A temporary, soft extension of the cell body that is used in location, attachment to surfaces, or engulfing particles.

pulmonary [L. pulmo: lung] Pertaining to the lungs.

pulmonary circuit The portion of the circulatory system by which blood is pumped from the heart to the lungs or gills for oxygenation and back to the heart for distribution. (Contrast with systemic circuit.)

pulmonary valve A one-way valve between the right ventricle of the heart and the pulmonary artery that prevents backflow of blood into the ventricle when it relaxes.

Punnett square Method of predicting the results of a genetic cross by arranging the gametes of each parent at the edges of a square.

pupa (pew′ pa) [L. pupa: doll, puppet] In certain insects (the Holometabola), the encased developmental stage between the larva and the adult.

pupil The opening in the vertebrate eye through which light passes.

purifying selection The elimination by natural selection of detrimental characters from a population. (Contrast with positive selection.)

purine (pure′ een) One of the two types of nitrogenous bases in nucleic acids. Each of the purines—adenine and guanine—pairs with a specific pyrimidine.

Purkinje fibers Specialized heart muscle cells that conduct excitation throughout the ventricular muscle.

pyrimidine (pe rim' a deen) One of the two types of nitrogenous bases in nucleic acids. Each of the pyrimidines—cytosine, thymine, and uracil—pairs with a specific purine.

pyrogen [Gk.: *pry, fire*;] Molecule that produces a rise in body temperature (fever); may be produced by an invading pathogen or by cells of the immune system in response to infection.

pyruvate The ionized form of pyruvic acid, a three-carbon acid; the end product of glycolysis and the raw material for the citric acid cycle.

pyruvate oxidation Conversion of pyruvate to acetyl CoA and CO_2 that occurs in the mitochondrial matrix in the presence of O_2.

Q

Q_{10} A value that compares the rate of a biochemical process or reaction over 10°C temperature ranges. A process that is not temperature-sensitive has a Q_{10} of 1; values of 2 or 3 mean the reaction speeds up as temperature increases.

qualitative Based on observation of an unmeasured quality of a trait, as in brown vs. blue.

quantitative Based on numerical values obtained by measurement, as in quantitative data.

quantitative trait loci A set of genes determining a complex character (trait) that exhibits quantitative variation (variation in amount rather than in kind).

quaternary structure The specific three-dimensional arrangement of protein subunits. Contrast with primary, secondary, tertiary structure.

quorum sensing The use of chemical communication signals to trigger density-linked activities such as biofilm formation in prokaryotes.

R

R group The distinguishing group of atoms of a particular amino acid; also known as a side chain.

r-strategist A species whose life history strategy allows for a high intrinsic rate of population increase (*r*). (Contrast with *K*-strategist.)

radial symmetry The condition in which any two halves of a body are mirror images of each other, providing the cut passes through the center; a cylinder cut lengthwise down its center displays this form of symmetry.

radiation The transfer of heat from warmer objects to cooler ones via the exchange of infrared radiation. *See also* electromagnetic radiation; evolutionary radiation.

radicle An embryonic root.

radioisotope A radioactive isotope of an element. Examples are carbon-14 (^{14}C) and hydrogen-3, or tritium (^3H).

radiometric dating A method for determining the age of objects such as fossils and rocks based on the decay rates of radioactive isotopes.

rapid eye movement sleep *See* REM sleep.

reactant A chemical substance that enters into a chemical reaction with another substance.

reaction center A group of electron transfer proteins that receive energy from light-absorbing pigments and convert it to chemical energy by redox reactions.

realized niche A species' niche as defined by its interactions with other species. (Contrast with fundamental niche.)

receptive field The area of visual space that activates a particular cell in the visual system.

receptor *See* receptor protein, sensory receptor cell.

receptor potential The change in the resting potential of a sensory cell when it is stimulated.

receptor protein A protein that can bind to a specific molecule, or detect a specific stimulus, within the cell or in the cell's external environment.

receptor-mediated endocytosis Endocytosis initiated by macromolecular binding to a specific membrane receptor.

recessive In genetics, an allele that does not determine phenotype in the presence of a dominant allele. (Contrast with dominance.)

reciprocal crosses A pair of matings in one of which a female of genotype A mates with a male of genotype B and in the other of which a female of genotype B mates with a male of genotype A.

recognition sequence *See* restriction site.

recombinant Pertaining to an individual, meiotic product, or chromosome in which genetic materials originally present in two individuals end up in the same haploid complement of genes.

recombinant DNA A DNA molecule made in the laboratory that is derived from two or more genetic sources.

recombinant frequency The proportion of offspring of a genetic cross that have phenotypes different from the parental phenotypes due to crossing over between linked genes during gamete formation.

recombination frequency The proportion of offspring of a genetic cross that have phenotypes different from the parental phenotypes due to crossing over between linked genes during gamete formation.

reconciliation ecology The practice of making exploited lands more biodiversity-friendly. Compare with restoration ecology.

rectum The terminal portion of the gut, ending at the anus.

redox reaction A chemical reaction in which one reactant is oxidized (loses electrons) and the other is reduced (gains electrons). Short for reduction–oxidation reaction.

reduction Gain of electrons by a chemical reactant. (Contrast with oxidation.)

refractory period The time interval after an action potential during which another action potential cannot be elicited from an excitable membrane.

regeneration The development of a complete individual from a fragment of an organism.

regulative development A pattern of animal embryonic development in which the fates of the first blastomeres are not absolutely fixed. (Contrast with mosaic development.)

regulatory gene A gene that codes for a protein (or RNA) that in turn controls the expression of another gene.

regulatory sequence A DNA sequence to which the protein product of a regulatory gene binds.

regulatory system A system that uses feedback information to maintain a physiological function or parameter at an optimal level. (Contrast with controlled system.)

regulatory T cells (Treg) The class of T cells that mediates tolerance to self antigens.

reinforcement The evolution of enhanced reproductive isolation between populations due to natural selection for greater isolation.

Reissner's membrane *See* tectonic membrane.

releaser Sensory stimulus that triggers performance of a stereotyped behavior pattern.

REM (rapid-eye-movement) sleep A sleep state characterized by vivid dreams, skeletal muscle relaxation, and rapid eye movements. (Contrast with non-REM sleep.)

renal [L. *renes*: kidneys] Relating to the kidneys.

renal tubule A structural unit of the kidney that collects filtrate from the blood, reabsorbs specific ions, nutrients, and water and returns them to the blood, and concentrates excess ions and waste products such as urea for excretion from the body.

renin An enzyme released from the kidneys in response to a drop in the glomerular filtration rate. Together with angiotensin converting enzyme, converts an inactive protein in the blood into angiotensin.

replication The duplication of genetic material.

replication complex The close association of several proteins operating in the replication of DNA.

replication fork A point at which a DNA molecule is replicating. The fork forms by the unwinding of the parent molecule.

replicon   A region of DNA replicated from a single origin of replication.

reporter gene   A genetic marker included in recombinant DNA to indicate the presence of the recombinant DNA in a host cell.

repressor   A protein encoded by a regulatory gene that can bind to a promoter and prevent transcription of the associated gene. (Contrast with activator.)

reproductive isolation   Condition in which two divergent populations are no longer exchanging genes. Can lead to speciation.

rescue effect   The process by which individuals moving between subpopulations of a metapopulation may prevent declining subpopulations from becoming extinct.

residence time   The length of time a chemical element (e.g., carbon or nitrogen) remains in a given compartment of the ecosystem (e.g., in an organic body, in soil, in the atmosphere).

residual volume (RV)   In tidal ventilation, the dead space that remains in the lungs at the end of exhalation.

resistance (R) genes   Plant genes that confer resistance to specific strains of pathogens.

resource   Something in the environment required by an organism for its maintenance and growth that is consumed in the process of being used.

resource partitioning   A situation in which selection pressures resulting from interspecific competition cause changes in the ways in which the competing species use the limiting resource, thereby allowing them to coexist.

respiration   (res pi ra' shun) [L. *spirare*: to breathe] (1) Cellular respiration. (2) Breathing.

respiratory chain   The terminal reactions of cellular respiration, in which electrons are passed from NAD or FAD, through a series of intermediate carriers, to molecular oxygen, with the concomitant production of ATP.

respiratory gases   Oxygen (O_2) and carbon dioxide (CO_2); the gases that an animal must exchange between its internal body fluids and the outside medium (air or water).

resting potential   The membrane potential of a living cell at rest. In cells at rest, the interior is negative to the exterior. (Contrast with action potential.)

restoration ecology   The science and practice of restoring damaged or degraded ecosystems.

restriction enzyme   Any of a type of enzyme that cleaves double-stranded DNA at specific sites; extensively used in recombinant DNA technology. Also called a restriction endonuclease.

restriction fragment length polymorphism   See RFLP.

restriction point (R)   The specific time during G1 of the cell cycle at which the cell becomes committed to undergo the rest of the cell cycle.

restriction site   A specific DNA base sequence that is recognized and acted on by a restriction endonuclease.

reticular activating system   A central region of the vertebrate brainstem that includes complex fiber tracts conveying neural signals between the forebrain and the spinal cord, with collateral fibers to a variety of nuclei that are involved in autonomic functions, including arousal from sleep.

reticulum   One of the four chambers of the ruminant stomach. Along with the rumen, where food is partially digested with the assistance of gut bacteria.

retina   (rett' in uh) [L. *rete*: net] The light-sensitive layer of cells in the vertebrate or cephalopod eye.

retrotransposons   Mobile genetic elements that are reverse transcribed into RNA as part of their transfer mechanism. (Contrast with DNA transposons.)

retrovirus   An RNA virus that contains reverse transcriptase. Its RNA serves as a template for cDNA production, and the cDNA is integrated into a chromosome of the host cell.

reverse genetics   Method of genetic analysis in which a phenotype is first related to a DNA variation, then the protein involved is identified.

reverse transcriptase   An enzyme that catalyzes the production of DNA (cDNA), using RNA as a template; essential to the reproduction of retroviruses.

reversion mutation   A second- or third-round mutation that reverts the DNA to its original sequence or to a new sequence that results in a non-mutant phenotype.

RFLP   Restriction fragment length polymorphism, the coexistence of two or more patterns of restriction fragments resulting from underlying differences in DNA sequence.

rhizoids   (rye' zoids) [Gk. root] Hairlike extensions of cells in mosses, liverworts, and a few vascular plants that serve the same function as roots and root hairs in vascular plants. The term is also applied to branched, rootlike extensions of some fungi and algae.

rhizome   (rye' zome) An underground stem (as opposed to a root) that runs horizontally beneath the ground.

rhodopsin   A vertebrate visual pigment involved in transducing photons of light into changes in the membrane potential of certain photoreceptor cells.

ribonucleic acid   See RNA.

ribose   A five-carbon sugar in nucleotides and RNA.

ribosomal RNA (rRNA)   Several species of RNA that are incorporated into the ribosome. Involved in peptide bond formation.

ribosome   A small particle in the cell that is the site of protein synthesis.

ribozyme   An RNA molecule with catalytic activity.

ribulose bisphosphate carboxylase/ oxygenase   See rubisco.

risk cost   The increased chance of being injured or killed as a result of performing a behavior, compared to resting.

RNA (ribonucleic acid)   An often single-stranded nucleic acid whose nucleotides use ribose rather than deoxyribose and in which the base uracil replaces thymine found in DNA. Serves as genome from some viruses. (See ribosomal RNA, transfer RNA, messenger RNA, and ribozyme.)

RNA interference (RNAi)   A mechanism for reducing mRNA translation whereby a double-stranded RNA, made by the cell or synthetically, is processed into a small, single-stranded RNA, whose binding to a target mRNA results in the latter's breakdown.

RNA polymerase   An enzyme that catalyzes the formation of RNA from a DNA template.

RNA splicing   The last stage of RNA processing in eukaryotes, in which the transcripts of introns are excised through the action of small nuclear ribonucleoprotein particles (snRNP).

rod cells   Light-sensitive cells in the vertebrate retina; these sensory receptor cells are sensitive in extremely dim light and are responsible for dim light, black and white vision.

root   The organ responsible for anchoring the plant in the soil, absorbing water and minerals, and producing certain hormones. Some roots are storage organs.

root apical meristem   Undifferentiated tissue at the apex of the root that gives rise to the organs of the root.

root cap   A thimble-shaped mass of cells, produced by the root apical meristem, that protects the meristem; the organ that perceives the gravitational stimulus in root gravitropism.

root hair   A long, thin process from a root epidermal cell that absorbs water and minerals from the soil solution.

root system   The organ system that anchors a plant in place, absorbs water and dissolved minerals, and may store products of photosynthesis from the shoot system.

rough endoplasmic reticulum (RER)   The portion of the endoplasmic reticulum whose outer surface has attached ribosomes. (Contrast with smooth endoplasmic reticulum.)

round window   A flexible membrane at the end of the lower canal of the cochlea in the human ear. (See also oval window.)

rRNA   See ribosomal RNA.

rubisco   Contraction of ribulose bisphosphate carboxylase/oxygenase, the enzyme that combines carbon dioxide or oxygen with ribulose bisphosphate to catalyze the first step of photosynthetic carbon fixation or photorespiration, respectively.

rumen One of the four chambers of the ruminant stomach. Along with the reticulum, where food is partially digested with the assistance of gut bacteria.

ruminant Herbivorous, cud-chewing mammals such as cows or sheep, characterized by a stomach that consists of four compartments: the rumen, reticulum, omasum, and abomasum.

S

S phase In the cell cycle, the stage of interphase during which DNA is replicated. (Contrast with G1 phase, G2 phase, M phase.)

salt glands Glands on the leaves of some halophytic plants that secrete salt, thereby ridding the plants of excess salt.

saltatory conduction [L. *saltare*: to jump] The rapid conduction of action potentials in myelinated axons; so called because action potentials appear to "jump" between nodes of Ranvier along the axon.

saprobe [Gk. *sapros*: rotten] An organism (usually a bacterium or fungus) that obtains its carbon and energy by absorbing nutrients from dead organic matter.

sarcomere (sark' o meer) [Gk. *sark*: flesh + *meros*: unit] The contractile unit of a skeletal muscle.

sarcoplasm The cytoplasm of a muscle cell.

sarcoplasmic reticulum The endoplasmic reticulum of a muscle cell.

saturated fatty acid A fatty acid in which all the bonds between carbon atoms in the hydrocarbon chain are single bonds—that is, all the bonds are saturated with hydrogen atoms. (Contrast with unsaturated fatty acid.)

Schwann cell A type of glial cell that myelinates axons in the peripheral nervous system.

scientific method A means of gaining knowledge about the natural world by making observations, posing hypotheses, and conducting experiments to test those hypotheses.

scion In horticulture, the bud or stem from one plant that is grafted to a root or root-bearing stem of another plant (the stock).

sclereid One of the principle types of cells in sclerenchyma.

sclerenchyma (skler eng' kyma) [Gk. *skleros*: hard + *kymus*: juice] A plant tissue composed of cells with heavily thickened cell walls. The cells are dead at functional maturity. The principal types of sclerenchyma cells are fibers and sclereids.

scrotum In most mammals, a pouch outside the body cavity that contains the testes.

second filial generation See F2.

second law of thermodynamics The principle that when energy is converted from one form to another, some of that energy becomes unavailable for doing work.

second messenger A compound, such as cAMP, that is released within a target cell after a hormone (the first messenger) has bound to a surface receptor on a cell; the second messenger triggers further reactions within the cell.

second polar body In oogenesis, the daughter cell of the second meiotic division that subsequently degenerates. (*See also* ootid.)

secondary active transport A form of active transport that does not use ATP as an energy source; rather, transport is coupled to ion diffusion down a concentration gradient established by primary active transport.

secondary cell wall A thick, cellulosic structure internal to the primary cell wall formed in some plant cells after cell expansion stops (Contrast with primary cell wall.)

secondary consumer An organism that eats primary consumers. (Contrast with primary consumer.)

secondary endosymbiosis The engulfment of a photosynthetic eukaryote by another eukaryotic cell that gave rise to certain groups of photosynthetic eukaryotes (e.g., euglenids).

secondary growth In plants, growth that contributes to an increase in girth. (Contrast with primary growth.)

secondary immune response A rapid and intense response to a second or subsequent exposure to an antigen, initiated by memory cells. (Contrast with primary immune response.)

secondary lysosome Membrane-enclosed organelle formed by the fusion of a primary lysosome with a phagosome, in which macromolecules taken up by phagocytosis are hydrolyzed into their monomers. (Contrast with lysosome.)

secondary metabolite A compound synthesized by a plant that is not needed for basic cellular metabolism. Typically has an antiherbivore or antiparasite function.

secondary oocyte In oogenesis, the daughter cell of the first meiotic division that receives almost all the cytoplasm. (*See also* first polar body.)

secondary plant body That part of a plant produced by secondary growth; consists of woody tissues. (Contrast with primary plant body.)

secondary sex determination Formation of secondary sexual characteristics (i.e., those other than gonads), such as external sex organs and body hair. (Contrast with primary sex determination.)

secondary spermatocyte One of the products of the first meiotic division of a primary spermatocyte.

secondary structure Of a protein, localized regularities of structure, such as the α helix and the β pleated sheet. (Contrast with primary, tertiary, quarternary structure.)

secondary succession Succession of ecological communities after a disturbance that did not eliminate all the organisms originally living on the site. (Contrast with primary succession.)

secretin (si kreet' in) A peptide hormone secreted by the upper region of the small intestine when acidic chyme is present. Stimulates the pancreatic duct to secrete bicarbonate ions.

sedimentary rock Rock formed by the accumulation of sediment grains on the bottom of a body of water. Often contain stratified fossils that allow geologists and biologist to date evolutionary events relative to each other.

seed A fertilized, ripened ovule of a gymnosperm or angiosperm. Consists of the embryo, nutritive tissue, and a seed coat.

seedling A plant that has just completed the process of germination.

segment polarity genes In *Drosophila* (fruit fly) development, segmentation genes that determine the boundaries and anterior–posterior organization of individual segments. Part of a developmental cascade that includes maternal effect genes, gap genes, pair rule genes, and Hox genes.

segmentation Division of an animal body into segments.

segmentation genes Genes that determine the number and polarity of body segments.

segregation In genetics, the separation of alleles, or of homologous chromosomes, from each other during meiosis so that each of the haploid daughter nuclei produced contains one or the other member of the pair found in the diploid parent cell, but never both. This principle was articulated by Mendel as his first law.

selectable marker A gene, such as one encoding resistance to an antibiotic, that can be used to identify (select) cells that contain recombinant DNA from among a large population of untransformed cells.

selective permeability Allowing certain substances to pass through while other substances are excluded; a characteristic of membranes.

self-incompatability In plants, the possession of mechanisms that prevent self-fertilization.

semelparous [L. *semel*: once + *pario*: to beget] Reproducing only once in a lifetime. (Contrast with iteroparous.)

semen (see' men) [L. *semin*: seed] The thick, whitish liquid produced by the male reproductive system in mammals, containing the sperm.

semicircular canals Three canals in the human inner ear that form part of the vestibular system.

semiconservative replication The way in which DNA is synthesized. Each of the two partner strands in a double helix acts as a template for a new partner strand. Hence, after replication, each double helix consists of one old and one new strand.

seminiferous tubules The tubules within the testes within which sperm production occurs.

senescence [L. *senescere*: to grow old] Aging; deteriorative changes with aging; the increased probability of dying with increasing age.

sensitive period The life stage during which some particular type of learning must take place, or during which it occurs much more easily than at other times. Typical of song learning among birds. Also known as the critical period.

sensory receptor cell Cell that is responsive to a particular type of physical or chemical stimulation. Sometimes referred to as a sensor.

sensory system A set of organs and tissues for detecting a stimulus; consists of sensory cells, the associated structures, and the neural networks that process the information.

sensory transduction The transformation of environmental stimuli or information into neural signals.

sepal (see' pul) [L. *sepalum*: covering] One of the outermost structures of the flower, usually protective in function and enclosing the rest of the flower in the bud stage.

septate [L. wall] Divided, as by walls or partitions.

septum (plural: septa) (1) A partition or cross-wall appearing in the hyphae of some fungi. (2) The bony structure dividing the nasal passages.

sequence alignment A method of identifying homologous positions in DNA or amino acid sequences by pinpointing the locations of deletions and insertions that have occurred since two (or more) organisms diverged from a common ancestor.

Sertoli cells Cells in the seminiferous tubules of the testes that nurture the developing sperm.

sessile (sess' ul) [L. *sedere*: to sit] Permanently attached; not able to move from one place to another. (Contrast with motile.)

set point In a regulatory system, the threshold sensitivity to the feedback stimulus.

sex chromosome In organisms with a chromosomal mechanism of sex determination, one of the chromosomes involved in sex determination (in humans and many other animals, these are the X and Y chromosomes).

sex-linked inheritance Pattern of inheritance characteristic of genes located on the sex chromosomes of organisms having a chromosomal mechanism for sex determination.

sex pilus A thin connection between two bacteria through which genetic material passes during conjugation.

sexual reproduction Reproduction involving the union of gametes.

sexual selection Selection by one sex of characteristics in individuals of the opposite sex. Also, the favoring of characteristics in

one sex as a result of competition among individuals of that sex for mates.

shared derived trait *See* synapomorphy.

shoot apical meristem Undifferentiated tissue at the apex of the shoot that gives rise to the organs of the shoot.

shoot system In plants, the organ system consisting of the leaves, stem(s), and flowers.

short-day plant (SDP) A plant that flowers when nights are longer than a critical length specific for that plant's species. (Compare to long-day plant.)

short tandem repeats (STRs) Short (1–5 base pairs), moderately repetitive sequences of DNA. The number of copies of an STR at a particular location varies between individuals and is inherited.

side chain *See* R group.

sieve tube element The characteristic cell of the phloem in angiosperms, which contains cytoplasm but relatively few organelles, and whose end walls (**sieve plates**) contain pores that form connections with neighboring cells.

sigma factor In prokaryotes, a protein that binds to RNA polymerase, allowing the complex to bind to and stimulate the transcription of a specific class of genes (e.g., those involved in sporulation).

signal sequence The sequence within a protein that directs the protein to a particular organelle.

signal transduction pathway The series of biochemical steps whereby a stimulus to a cell (such as a hormone or neurotransmitter binding to a receptor) is translated into a response of the cell.

silencer A gene sequence binding transcription factors that repress transcription. (Contrast with promoter.)

silent mutation A change in a gene's sequence that has no effect on the amino acid sequence of a protein either because it occurs in noncoding DNA or because it does not change the amino acid specified by the corresponding codon . (Contrast with frame-shift mutation, missense mutation, nonsense mutation.)

silent substitution *See* synonymous substitution.

similarity matrix A matrix used to compare the degree of divergence among pairs of objects. For molecular sequences, constructed by summing the number or percentage of nucleotides or amino acids that are identical in each pair of sequences.

simple diffusion Diffusion that does not involve a direct input of energy or assistance by carrier proteins.

single nucleotide polymorphisms (SNPs) Inherited variations in a single nucleotide base in DNA that differ between individuals.

single-strand binding protein In DNA replication, a protein that binds to single strands of DNA after they have been separated from each other, keeping the two strands separate for replication.

sink In plants, any organ that imports the products of photosynthesis, such as roots, developing fruits, and immature leaves. (Contrast with source.)

sinoatrial node (sigh' no ay' tree al) [L. *sinus*: curve + *atrium*: chamber] The pacemaker of the mammalian heart.

siRNAs (small interfering RNAs) Short, double-stranded RNA molecules used in RNA interference.

sister chromatid Each of a pair of newly replicated chromatids.

sister species Two species that are each other's closest relatives.

skeletal muscle A type of muscle tissue characterized by multinucleated cells containing highly ordered arrangements of actin and myosin microfilaments. Also called striated muscle. (Contrast with cardiac muscle, smooth muscle.)

skeletal systems Organ systems that provide rigid supports—**skeletons**—against which muscles can pull to create directed movements. See also endoskeleton, exoskeleton.

sliding DNA clamp Protein complex that keeps DNA polymerase bound to DNA during replication.

sliding filament model Mechanism of muscle contraction based on the formation and breaking of crossbridges between actin and myosin filaments, causing the filaments to slide together.

slow-twitch fibers Skeletal muscle fibers specialized for sustained aerobic work; contain myoglobin and abundant mitochondria, and are well-supplied with blood vessels. Also called oxidative or red muscle fibers. (Compare to fast-twitch fibers.)

slow-wave sleep *See* non-REM sleep.

small interfering RNAs *See* siRNAs.

small intestine The portion of the gut between the stomach and the colon; consists of the duodenum, the jejunum, and the ileum.

small nuclear ribonucleoprotein particle (snRNP) A complex of an enzyme and a small nuclear RNA molecule, functioning in RNA splicing.

smooth endoplasmic reticulum (SER) Portion of the endoplasmic reticulum that lacks ribosomes and has a tubular appearance. (Contrast with rough endoplasmic reticulum.)

smooth muscle Muscle tissue consisting of sheets of mononucleated cells innervated by the autonomic nervous system. (Contrast with cardiac muscle, skeletal muscle.)

sodium–potassium (Na⁺–K⁺) pump Antiporter responsible for primary active transport; it pumps sodium ions out of the cell and potassium ions into the cell, both against their concentration gradients. Also called a sodium–potassium ATPase.

soil horizon *See* horizons.

soil solution The aqueous portion of soil, from which plants take up dissolved mineral nutrients.

solute A substance that is dissolved in a liquid (solvent) to form a solution.

solute potential (Ψ_s) A property of any solution, resulting from its solute contents; it may be zero or have a negative value. The more negative the solute potential, the greater the tendency of the solution to take up water through a differentially permeable membrane. (Contrast with pressure potential, water potential.)

solution A liquid (the solvent) and its dissolved solutes.

solvent Liquid in which a substance (solute) is dissolved to form a solution.

somatic cell [Gk. *soma*: body] All the cells of the body that are not specialized for reproduction. (Contrast with germ cell.)

somatic mutation Permanent genetic change in a somatic cell (as opposed to a germ cell, the egg or sperm). These mutations affect the individual only; they are not passed on to offspring. (Contrast with germ line mutation.)

somatosensory cortex An area of the parietal lobe that receives touch and pressure information from mechanoreceptors throughout the body; neurons in this area are arranged according to the parts of the body with which they communicate.

somatostatin Peptide hormone made in the hypothalamus that inhibits the release of other hormones from the pituitary and intestine.

somite (so' might) One of the segments into which an embryo becomes divided longitudinally, leading to the eventual segmentation of the animal as illustrated by the spinal column, ribs, and associated muscles.

source In plants, any organ that exports the products of photosynthesis in excess of its own needs, such as a mature leaf or storage organ. (Contrast with sink.)

spatial summation In the production or inhibition of action potentials in a postsynaptic cell, the interaction of depolarizations and hyperpolarizations produced at different sites on the postsynaptic cell. (Contrast with temporal summation.)

spawning *See* external fertilization.

speciation (spee' see ay' shun) The process of splitting one population into two populations that are reproductively isolated from one another.

species (spee' sees) [L. kind] The base unit of taxonomic classification, consisting of an ancestor–descendant group of populations of evolutionarily closely related, similar organisms. The more narrowly defined "biological species" consists of individuals capable of interbreeding with each other but not with members of other species.

species composition The particular mix of species a community contains and the abundances of those species.

species evenness A measure of species diversity that reflects the distribution of the species' abundances in a community.

species richness The total number of species living in a region.

species–area relationship The relationship between the size of an area and the numbers of species it supports.

specific defenses Defensive reactions of the vertebrate immune system that are based on the reaction of an antibody to a specific antigen. (Contrast with nonspecific defenses.)

specific heat The amount of energy that must be absorbed by a gram of a substance to raise its temperature by one degree centigrade. By convention, water is assigned a specific heat of one.

sperm [Gk. *sperma*: seed] The male gamete.

spermatid One of the products of the second meiotic division of a primary spermatocyte; four haploid spermatids, which remain connected by cytoplasmic bridges, are produced for each primary spermatocyte that enters meiosis.

spermatogenesis (spur mat' oh jen' e sis) [Gk. *sperma*: seed + *genesis*: source] Gametogenesis leading to the production of sperm.

spermatogonia In animals, the diploid progeny of a germ cell in males.

spherical symmetry The simplest form of symmetry, in which body parts radiate out from a central point such that an infinite number of planes passing through that central point can divide the organism into similar halves.

sphincter (sfink' ter) [Gk. *sphinkter*: something that binds tightly] A ring of muscle that can close an orifice, for example, at the anus.

spicule [L. arrowhead] A hard, calcareous skeletal element typical of sponges.

spinal cord Along with the brain, part of the central nervous system; transmits information between the body and the brain and mediates simple reflexes.

spinal reflex The conversion of afferent to efferent information in the spinal cord without participation of the brain.

spindle apparatus [O.E. *spindle*, a short stick with tapered ends] Array of microtubules emanating from both poles of a dividing cell during mitosis and playing a role in the movement of chromosomes at nuclear division.

spleen Organ that serves as a reservoir for venous blood and eliminates old, damaged red blood cells from the circulation.

spliceosome RNA–protein complex that splices out introns from eukaryotic pre-mRNAs.

splicing *See* RNA splicing.

spontaneous mutation A genetic change caused by internal cellular mechanisms, such as an error in DNA replication. (Contrast with induced mutation.)

sporangiophore A stalked reproductive structure produced by zygospore fungi that extends from a hypha and bears one or many sporangia.

sporangium (spor an' gee um) (plural: sporangia) [Gk. *spora*: seed + *angeion*: vessel or reservoir] In plants and fungi, any specialized stucture within which one or more spores are formed.

spore [Gk. *spora*: seed] (1) Any asexual reproductive cell capable of developing into an adult organism without gametic fusion. In plants, haploid spores develop into gametophytes, diploid spores into sporophytes. (2) In prokaryotes, a resistant cell capable of surviving unfavorable periods.

sporocyte Specialized cells of the diploid sporophyte that will divide by meiosis to produce four haploid spores. Germination of these spores produces the haploid gametophyte.

sporophyte (spor' o fyte) [Gk. *spora*: seed + *phyton*: plant] In plants and protists with alternation of generations, the diploid phase that produces the spores. (Contrast with gametophyte.)

stabilizing selection Selection against the extreme phenotypes in a population, so that the intermediate types are favored. (Contrast with disruptive selection.)

stamen (stay' men) [L. *stamen*: thread] A male (pollen-producing) unit of a flower, usually composed of an anther, which bears the pollen, and a filament, which is a stalk supporting the anther.

starch [O.E. *stearc*: stiff] A polymer of glucose; used by plants to store energy.

Starling's forces The two opposing forces responsible for water movement across capillary walls: blood pressure, which squeezes water and small solutes out of the capillaries, and osmotic pressure, which pulls water back into the capillaries.

start codon The mRNA triplet (AUG) that acts as a signal for the beginning of translation at the ribosome. (Contrast with stop codon.)

stele (steel) [Gk.: pillar] The central cylinder of vascular tissue in a plant stem.

stem cell In animals, an undifferentiated cell that is capable of continuous proliferation. A stem cell generates more stem cells and a large clone of differentiated progeny cells. (*See also* embryonic stem cell.)

stem In plants, the organ that holds leaves and/or flowers and transports and distributes materials among the other organs of the plant.

stereocilia Fingerlike extensions of hair cell membranes whose bending initiates sound perception. (*See* hair cell.)

steroid Any of a family of lipids whose multiple rings share carbons. The steroid cholesterol is an important constituent of membranes and is the base of steroid hormones such as testosterone.

sticky ends On a piece of two-stranded DNA, short, complementary, one-stranded regions produced by the action of a restriction endonuclease. Sticky ends facilitate the joining of segments of DNA from different sources.

stigma [L. *stigma*: mark, brand] The part of the pistil at the apex of the style that is receptive to pollen, and on which pollen germinates.

stimulus [L. *stimulare*: to goad] Something causing a response; something in the environment detected by a receptor.

stock In horticulture, the root or root-bearing stem to which a bud or piece of stem from another plant (the scion) is grafted.

stoma (plural: stomata) [Gk. *stoma*: mouth, opening] Small opening in the plant epidermis that permits gas exchange; bounded by a pair of guard cells whose osmotic status regulates the size of the opening.

stomach An organ that physically (and sometimes enzymatically) breaks down food, preparing it for digestion in the midgut.

stomatal crypt In plants, a sunken cavity below the leaf surface in which a stoma is sheltered from the drying effects of air currents.

stop codon Any of the three mRNA codons that signal the end of protein translation at the ribosome: UAG, UGA, UAA.

stratosphere The upper part of Earth's atmosphere, above the troposphere; extends from approximately 18 kilometers upward to approximately 50 kilometers above Earth's surface.

stratum (plural strata) [L. *stratos*: layer] A layer of sedimentary rock laid down at a particular time in the past.

stretch receptor A modified muscle cell embedded in the connective tissue of a muscle that acts as a mechanoreceptor in response to stretching of that muscle.

striated muscle *See* skeletal muscle.

strigolactones Signaling molecules produced by plant roots that attract the hyphae of mycorrhizal fungi.

strobilus (plural: strobili) One of several conelike structures in various groups of plants (including club mosses, horsetails, and conifers) associated with the production and dispersal of reproductive products.

stroke An embolism in an artery in the brain that causes the cells fed by that artery to die. The specific damage, such as memory loss, speech impairment, or paralysis, depends on the location of the blocked artery.

stroma The fluid contents of an organelle such as a chloroplast or mitochondrion.

structural gene A gene that encodes the primary structure of a protein not involved in the regulation of gene expression.

structural isomers Molecules made up of the same kinds and numbers of atoms, in which the atoms are bonded differently.

structural motif A three-dimensional structural element that is part of a larger molecule. For example, there are four common motifs in DNA-binding proteins: helix-turn-helix, zinc finger, leucine zipper, and helix-loop-helix.

style [Gk. *stele*: pillar or column] In the angiosperm flower, a column of tissue extending from the tip of the ovary, and bearing the stigma or receptive surface for pollen at its apex.

sub- [L. under] A prefix used to designate a structure that lies beneath another or is less than another. For example, subcutaneous (beneath the skin); subspecies

subduction In plate tectonics, the movement of one lithospheric plate under another.

suberin A waxlike lipid that is a barrier to water and solute movement across the Casparian strip of the endodermis.

submucosa (sub mew koe' sah) The tissue layer just under the epithelial lining of the lumen of the digestive tract.

subsoil The soil horizon lying below the topsoil and above the parent rock (bedrock); the zone of infiltration and accumulation of materials leached from the topsoil. Also called the B horizon.

substrate (sub' strayte) (1) The molecule or molecules on which an enzyme exerts catalytic action. (2) The base material on which a sessile organism lives.

succession The gradual, sequential series of changes in the species composition of an ecological community following a disturbance. See also cyclical succession, directional selection, heterotrophic succession.

succulence In plants, possession of fleshy, water-storing leaves or stems; an adaptation to dry environments.

sulcus (sul' kus; plural sulci) [Gk.: plowed furrow] *See* convolutions.

superficial cleavage A variation of incomplete cleavage in which cycles of mitosis occur without cell division, producing a syncytium (a single cell with many nuclei).

suprachiasmatic nuclei (SCN) In mammals, two clusters of neurons just above the optic chiasm that act as the master circadian clock.

surface area-to-volume ratio For any cell, organism, or geometrical solid, the ratio of surface area to volume; this is an important factor in setting an upper limit on the size a cell or organism can attain.

surface tension The attractive intermolecular forces at the surface of liquid; an especially important property of water.

surfactant A substance that decreases the surface tension of a liquid. Lung surfactant, secreted by cells of the alveoli, is mostly

phospholipid and decreases the amount of work necessary to inflate the lungs.

survivorship The fraction of individuals that survive from birth to a given life stage or age.

survivorship curves Graphic plot of ages at death of a hypothetical cohort, usually of 1,000 individuals, by plotting the numbers of individuals expected to survive to reach each age category. There are three general shapes. Ecological survivorship is linear: individuals face a constant risk of mortality regardless of their age. Physiological survivorship curves are concave: high survivorship through adulthood with steep declines late in life. Maturational survivorship curves are convex, with high mortality early in life but higher survivorship once individuals reach maturity.

suspensor In the embryos of seed plants, the stalk of cells that pushes the embryo into the endosperm and is a source of nutrient transport to the embryo.

sustainable Pertaining to the use and management of ecosystems in such a way that humans benefit over the long term from specific ecosystem goods and services without compromising others.

symbiosis (sim' bee oh' sis) [Gk. *sym*: together + *bios*: living] The living together of two or more species in a prolonged and intimate relationship.

symmetry Pertaining to an attribute of an animal body in which at least one plane can divide the body into similar, mirror-image halves. (*See* bilateral symmetry, radial symmetry.)

sympathetic nervous system The division of the autonomic nervous system that works in opposition to the parasympathetic nervous system. (Contrast with parasympathetic nervous system.)

sympatric speciation (sim pat' rik) [Gk. *sym*: same + *patria*: homeland] Speciation due to reproductive isolation without any physical separation of the subpopulation. (Contrast with allopatric speciation.)

symplast The continuous meshwork of the interiors of living cells in the plant body, resulting from the presence of plasmodesmata. (Contrast with apoplast.)

symporter A membrane transport protein that carries two substances in the same direction. (Contrast with antiporter, uniporter.)

synapomorphy A trait that arose in the ancestor of a phylogenetic group and is present (sometimes in modified form) in all of its members, thus helping to delimit and identify that group. Also called a shared derived trait.

synapse (sin' aps) [Gk. *syn*: together + *haptein*: to fasten] A specialized type of junction where a neuron meets its target cell (which can be another neuron or some other type of cell) and information in the form of neurotransmitter molecules is exchanged across a synaptic cleft.

synapsis (sin ap' sis) The highly specific parallel alignment (pairing) of homologous chromosomes during the first division of meiosis.

synaptic cleft The space between the presynaptic cell and the postsynaptic cell in a chemical synapse.

synergids [Gk. *syn*: together + *ergos*: work] In angiosperms, the two cells accompanying the egg cell at one end of the megagametophyte.

syngamy *See* fertilization.

synonymous (silent) substitution A change of one nucleotide in a sequence to another when that change does not affect the amino acid specified (i.e., UUA → UUG, both specifying leucine). (Contrast with nonsynonymous substitution, missense mutation, nonsense mutation.)

systematics The scientific study of the diversity and relationships among organisms.

systemic acquired resistance A general resistance to many plant pathogens following infection by a single agent.

systemic circuit Portion of the circulatory system by which oxygenated blood from the lungs or gills is distributed throughout the rest of the body and returned to the heart. (Contrast with pulmonary circuit.)

systems biology The scientific study of an organism as an integrated and interacting system of genes, proteins, and biochemical reactions.

systole (sis' tuh lee) [Gk.: contraction] Contraction of a chamber of the heart, driving blood forward in the circulatory system. (Contrast with diastole.)

T

3′ end (3 prime) The end of a DNA or RNA strand that has a free hydroxyl group at the 3′ carbon of the sugar (deoxyribose or ribose).

T cell A type of lymphocyte involved in the cellular immune response. The final stages of its development occur in the thymus gland. (Contrast with B cell; *see also* cytotoxic T cell, T-helper cell.)

T cell receptor A protein on the surface of a T cell that recognizes the antigenic determinant for which the cell is specific.

T tubules A system of tubules that runs throughout the cytoplasm of a muscle fiber, through which action potentials spread.

T-helper (T_H) cell Type of T cell that stimulates events in both the cellular and humoral immune responses by binding to the antigen on an antigen-presenting cell; target of the HIV-I virus, the agent of AIDS. (Contrast with cytotoxic T cells.)

taproot system A root system typical of eudicots consisting of a primary root (*taproot*) that extends downward by tip growth and outward by initiating lateral roots. (Contrast with fibrous root system.)

target cell A cell with the appropriate receptors to bind and respond to a particular hormone or other chemical mediator.

taste bud A structure in the epithelium of the tongue that includes a cluster of chemoreceptors innervated by sensory neurons.

TATA box An eight-base-pair sequence, found about 25 base pairs before the starting point for transcription in many eukaryotic promoters, that binds a transcription factor and thus helps initiate transcription.

taxon (plural: taxa) [Gk. *taxis*: put in order] A biological group (typically a species or a clade) that is given a name.

T_C cells *See* cytotoxic T cells.

tectonic membrane One of two membranes (the other is the basilar membrane) that extend along the length of the cochlea in the human ear. Also known as Reissner's membrane.

telencephalon The outer, surrounding structure of the embryonic vertebrate forebrain, which develops into the cerebrum.

telomerase An enzyme that catalyzes the addition of telomeric sequences lost from chromosomes during DNA replication.

telomeres (tee' lo merz) [Gk. *telos*: end + *meros*: units, segments] Repeated DNA sequences at the ends of eukaryotic chromosomes.

telophase (tee' lo phase) [Gk. *telos*: end] The final phase of mitosis or meiosis during which chromosomes become diffuse, nuclear envelopes re-form, and nucleoli begin to reappear in the daughter nuclei.

template A molecule or surface on which another molecule is synthesized in complementary fashion, as in the replication of DNA.

template strand In double-stranded DNA, the strand that is transcribed to create an RNA transcript that will be processed into a protein. Also refers to a strand of RNA that is used to create a complementary RNA.

temporal lobe One of the four lobes of the cerebral hemisphere; receives and processes auditory and visual information; involved in recognizing, identifying, and naming objects.

temporal summation In the production or inhibition of action potentials in a postsynaptic cell, the interaction of depolarizations or hyperpolarizations produced by rapidly repeated stimulation of a single point on the postsynaptic cell. (Contrast with spatial summation.)

tendon A collagen-containing band of tissue that connects a muscle with a bone.

tepal A sterile, modified, nonphotosynthetic leaf of an angiosperm flower that cannot be distinguished as a petal or a sepal.

termination In molecular biology, the end of transcription or translation.

terminator A sequence at the 3′ end of mRNA that causes the RNA strand to be released from the transcription complex.

terrestrial (ter res' tree al) [L. *terra*: earth] Pertaining to or living on land. (Contrast with aquatic, marine.)

territorial behavior Aggressive actions engaged in to defend a habitat or resource such that other animals are denied access.

tertiary consumers Carnivores that consume primary carnivores (secondary consumers).

tertiary endosymbiosis The mechanism by which some eukaryotes acquired the capacity for photosynthesis; for example, a dinoflagellate that apparently lost its chloroplast became photosynthetic by engulfing another protist that had acquired a chloroplast through secondary endosymbiosis.

tertiary structure In reference to a protein, the relative locations in three-dimensional space of all the atoms in the molecule. The overall shape of a protein. (Contrast with primary, secondary, and quaternary structures.)

test cross Mating of a dominant-phenotype individual (who may be either heterozygous or homozygous) with a homozygous-recessive individual.

testis (tes' tis) (plural: testes) [L. *testis*: witness] The male gonad; the organ that produces the male gametes.

tetanus [Gk. *tetanos*: stretched] (1) A state of sustained maximal muscular contraction caused by rapidly repeated stimulation. (2) In medicine, an often fatal disease ("lockjaw") caused by the bacterium *Clostridium tetani*.

tetrad [Gk. *tettares*: four] During prophase I of meiosis, the association of a pair of homologous chromosomes or four chromatids

thalamus [Gk. *thalamos*: chamber] A region of the vertebrate forebrain; involved in integration of sensory input.

theory [Gk. *theoria*: analysis of facts] A far-reaching explanation of observed facts that is supported by such a wide body of evidence, with no significant contradictory evidence, that it is scientifically accepted as a factual framework. Examples are Newton's theory of gravity and Darwin's theory of evolution. (Contrast with hypothesis.)

thermoneutral zone (TNZ) [Gk. *thermos*: temperature] The range of temperatures over which an endotherm does not have to expend extra energy to thermoregulate.

thermophile (ther' muh fyle)[Gk. *thermos*: temperature + *philos*: loving] An organism that lives exclusively in hot environments.

thoracic cavity [Gk. *thorax*: breastplate] The portion of the mammalian body cavity bounded by the ribs, shoulders, and diaphragm. Contains the heart and the lungs.

thoracic duct The connection between the lymphatic system and the circulatory system.

threshold The level of depolarization that causes an electrically excitable membrane to fire an action potential.

thrombin An enzyme involved in blood clotting; cleaves fibrinogen to form fibrin.

thrombus (throm' bus) [Gk. *thrombos*: clot] A blood clot that forms within a blood vessel and remains attached to the wall of the vessel.

thylakoid (thigh' la koid) [Gk. *thylakos*: sack or pouch] A flattened sac within a chloroplast. Thylakoid membranes contain all of the chlorophyll in a plant, in addition to the electron carriers of photophosphorylation. Thylakoids stack to form grana.

thymine (T) Nitrogen-containing base found in DNA.

thymus [Gk. *thymos*: warty] A ductless, glandular lymphoid tissue, involved in development of the immune system of vertebrates. In humans, the thymus degenerates during puberty.

thyroid gland [Gk. *thyreos*: door-shaped] A two-lobed gland in vertebrates. Produces the hormone thyroxine.

thyrotropin Hormone produced by the anterior pituitary that stimulates the thyroid gland to produce and release thyroxine. Also called thyroid-stimulating hormone (TSH).

thyrotropin-releasing hormone (TRH) Hormone produced by the hypothalamus that stimulates the anterior pituitary to release thyrotropin.

thyroxine Hormone produced by the thyroid gland; controls many metabolic processes.

tidal The bidirectional form of ventilation used by all vertebrates except birds; air enters and leaves the lungs by the same route.

tight junction A junction between epithelial cells in which there is no gap between adjacent cells.

tissue A group of similar cells organized into a functional unit; usually integrated with other tissues to form part of an organ.

tissue system In plants, any of three organized groups of tissues—dermal tissue, vascular tissue, and ground tissue—that are established during embryogenesis and have distinct functions.

titin A protein that holds bundles of myosin filaments in a centered position within the sarcomeres of muscle cells. The largest protein in the human body.

tonoplast The membrane of the plant central vacuole.

topsoil The uppermost soil horizon; contains most of the organic matter of soil, but may be depleted of most mineral nutrients by leaching. Also called the A horizon.

totipotent [L. *toto*: whole, entire + *potens*: powerful] Possessing all the genetic information and other capacities necessary to form an entire individual. (Contrast with multipotent, pluripotent.)

toxigenicity The ability of some pathogenic bacteria to produce chemical substances that harm the host.

trachea (tray' kee ah) [Gk. *trakhoia*: tube] A tube that carries air to the bronchi of the lungs of vertebrates. When plural (*tracheae*), refers to the major airways of insects.

tracheary element Either of two types of xylem cells—tracheids and vessel elements—that undergo apoptosis before assuming their transport function.

tracheid (tray' kee id) A type of tracheary element found in the xylem of nearly all vascular plants, characterized by tapering ends and walls that are pitted but not perforated. (Contrast with vessel element.)

trade-off The relationship between the fitness benefits conferred by an adaptation and the fitness costs it imposes. For an adaptation to be favored by natural selection, the benefits must exceed the costs.

trait In genetics, a specific form of a character: eye color is a character; brown eyes and blue eyes are traits. (Contrast with character.)

transcription The synthesis of RNA using one strand of DNA as a template.

transcription factors Proteins that assemble on a eukaryotic chromosome, allowing RNA polymerase II to perform transcription.

transcription initiation site The part of a gene's promoter where synthesis of the gene's RNA transcript begins.

transduction (1) Transfer of genes from one bacterium to another by a bacteriophage. (2) In sensory cells, the transformation of a stimulus (e.g., light energy, sound pressure waves, chemical or electrical stimulants) into action potentials.

transfection Insertion of recombinant DNA into animal cells.

transfer RNA (tRNA) A family of folded RNA molecules. Each tRNA carries a specific amino acid and anticodon that will pair with the complementary codon in mRNA during translation.

transformation (1) A mechanism for transfer of genetic information in bacteria in which pure DNA from a bacterium of one genotype is taken in through the cell surface of a bacterium of a different genotype and incorporated into the chromosome of the recipient cell. (2) Insertion of recombinant DNA into a host cell.

transgenic Containing recombinant DNA incorporated into the genetic material.

transition state In an enzyme-catalyzed reaction, the reactive condition of the substrate after there has been sufficient input of energy (activation energy) to initiate the reaction.

translation The synthesis of a protein (polypeptide). Takes place on ribosomes, using the information encoded in messenger RNA.

translational repressor A protein that blocks translation by binding to mRNAs and preventing their attachment to the ribosome. In mammals, the production of ferritin protein is regulated by a translational repressor.

translocation (1) In genetics, a rare mutational event that moves a portion of a chromosome to a new location, generally on a nonhomologous chromosome. (2) In vascular plants, movement of solutes in the phloem.

transmembrane protein An integral membrane protein that spans the phospholipid bilayer.

transpiration [L. *spirare*: to breathe] The evaporation of water from plant leaves and stem, driven by heat from the sun, and providing the motive force to raise water (plus mineral nutrients) from the roots.

transpiration–cohesion–tension mechanism Theoretical basis for water movement in plants: evaporation of water from cells within leaves (transpiration) causes an increase in surface tension, pulling water up through the xylem. Cohesion of water occurs because of hydrogen bonding.

transposable element (transposon) A segment of DNA that can move to, or give rise to copies at, another locus on the same or a different chromosome.

transversion A mutation that changes a purine to a pyrimidine or vice versa.

tree of life A term that encompasses the evolutionary history of all life, or a graphic representation of that history.

triglyceride A simple lipid in which three fatty acids are combined with one molecule of glycerol.

trimesters The three stages of human pregnancy, approximately 3 months each in length.

tripartite synapse The idea that a synapse includes not only the pre- and postsynaptic neurons involved but also encompasses many connections with glial cells called astrocytes.

triploblastic Having three cell layers.

trisomic Containing three rather than two members of a chromosome pair.

tRNA *See* transfer RNA.

trochophore (troke' o fore) [Gk. *trochos*: wheel + *phoreus*: bearer] A radially symmetrical larval form typical of annelids and mollusks, distinguished by a wheel-like band of cilia around the middle.

trophic cascade The progression over successively lower trophic levels of the indirect effects of a predator.

trophic interactions The consumer–resource relationships among species in a community.

trophic level [Gk *trophes*: nourishment] A group of organisms united by obtaining their energy from the same part of the food web of a biological community.

trophoblast [Gk *trophes*: nourishment + *blastos*: sprout] At the 32-cell stage of mammalian development, the outer group of cells that will become part of the placenta and thus nourish the growing embryo. (Contrast with inner cell mass.)

tropic hormones Hormones produced by the anterior pituitary that control the secretion of hormones by other endocrine glands.

tropomyosin [troe poe my' oh sin] One of the three protein components of an actin filament; controls the interactions of actin and myosin necessary for muscle contraction.

troponin One of the three components of an actin filament; binds to actin, tropomyosin, and Ca^{2+}.

true-breeding A genetic cross in which the same result occurs every time with respect to the trait(s) under consideration, due to homozygous parents.

trypsin A protein-digesting enzyme. Secreted by the pancreas in its inactive form (trypsinogen), it becomes active in the duodenum of the small intestine.

tube feet A unique feature of echinoderms; extensions of the water vascular system, which functions in gas exchange, locomotion, and feeding.

tubulin A protein that polymerizes to form microtubules.

tumor [L. *tumor*: a swollen mass] A disorganized mass of cells. Malignant tumors spread to other parts of the body.

tumor necrosis factor A family of cytokines (growth factors) that causes cell death and is involved in inflammation.

tumor suppressor A gene that codes for a protein product that inhibits cell proliferation; inactive in cancer cells. (Contrast with oncogene.)

turgor pressure [L. *turgidus*: swollen] *See* pressure potential.

turnover In freshwater ecosystems, vertical movements of water that bring nutrients and dissolved CO_2 to the surface and O_2 to deeper water.

twitch A muscle fiber's minimum unit of contraction, stimulated by a single action potential.

tympanic membrane [Gk. *tympanum*: drum] The eardrum.

U

ubiquitin A small protein that is covalently linked to other cellular proteins identified for breakdown by the proteosome.

ultimate causes In ethology, the evolutionary processes that produce an animal's capacity and tendency to behave in particular ways. (Contrast with proximate causes.)

unequal crossing over When a highly repeated gene sequence becomes displaced in alignment during meiotic crossing over, so that one chromosome receives many copies of the sequence while the second chromosome receives fewer copies. One of the mechanisms of concerted evolution. (See also biased gene conversion.)

uniporter [L. *unus*: one + *portal*: doorway] A membrane transport protein that carries a single substance in one direction. (Contrast with antiporter, symporter.)

unipotent An undifferentiated cell that is capable of becoming only one type of mature cell. (Contrast with totipotent, multipotent, pluripotent.)

unsaturated fatty acid A fatty acid whose hydrocarbon chain contains one or more double bonds. (Contrast with saturated fatty acid.)

upregulation A process by which the abundance of receptors for a hormone increases when hormone secretion is suppressed. (Contrast with downregulation.)

upwelling zones Areas of the ocean where cool, nutrient-rich water from deeper layers rises to the surface.

uracil (U) A pyrimidine base found in nucleotides of RNA.

urea A compound that is the main form of nitrogen excreted by many animals, including mammals.

ureotelic Pertaining to an organism in which the final product of the breakdown of nitrogen-containing compounds (primarily proteins) is urea. (Contrast with ammonotelic, uricotelic.)

ureter (your' uh tur) Long duct leading from the vertebrate kidney to the urinary bladder or the cloaca.

urethra (you ree' thra) In most mammals, the canal through which urine is discharged from the bladder and which serves as the genital duct in males.

uric acid A compound that serves as the main excreted form of nitrogen in some animals, particularly those which must conserve water, such as birds, insects, and reptiles.

uricotelic Pertaining to an organism in which the final product of the breakdown of nitrogen-containing compounds (primarily proteins) is uric acid. (Contrast with ammonotelic, ureotelic.)

urinary bladder A structure in which urine is stored until it can be excreted to the outside of the body.

urine (you' rin) In vertebrates, the fluid waste product containing the toxic nitrogenous by-products of protein and nucleic acid metabolism.

uterine cycle In human females, the monthly cycle of events by which the endometrium is prepared for the arrival of a blastocyst. (Contrast with ovarian cycle.)

uterus (yoo' ter us) [L. *utero*: womb] A specialized portion of the female reproductive tract in mammals that receives the fertilized egg and nurtures the embryo in its early development. Also called the womb.

V

vaccination Injection of virus or bacteria or their proteins into the body, to induce immunity. The injected material is usually attenuated (weakened) before injection and is called a *vaccine*.

vacuole (vac' yew ole) Membrane-enclosed organelle in plant cells that can function for storage, water concentration for turgor, or hydrolysis of stored macromolecules.

vagina (vuh jine' uh) [L. sheath] In female animals, the entry to the reproductive tract.

van der Waals forces Weak attractions between atoms resulting from the interaction of the electrons of one atom with the nucleus of another. This type of attraction is about one-fourth as strong as a hydrogen bond.

variable In a controlled experiment, a factor that is manipulated to test its effect on a phenomenon.

variable region The portion of an immunoglobulin molecule or T cell receptor that includes the antigen-binding site and is responsible for its specificity. (Contrast with constant region.)

vas deferens (plural: vasa deferentia) Duct that transfers sperm from the epididymis to the urethra.

vasa recta Blood vessels that parallel the loops of Henle and the collecting ducts in the renal medulla of the kidney.

vascular (vas' kew lar) [L. *vasculum*: a small vessel] Pertaining to organs and tissues that conduct fluid, such as blood vessels in animals and xylem and phloem in plants.

vascular bundle In vascular plants, a strand of vascular tissue, including xylem and phloem as well as thick-walled fibers.

vascular cambium (kam' bee um) [L. *cambiare*: to exchange] In plants, a lateral meristem that gives rise to secondary xylem and phloem.

vascular tissue system The transport system of a vascular plant, consisting primarily of xylem and phloem.

vasopressin A hormone that promotes water reabsorption by the kidney. Produced by neurons in the hypothalamus and released from nerve terminals in the posterior pituitary. Also called antidiuretic hormone or ADH.

vector (1) An agent, such as an insect, that carries a pathogen affecting another species. (2) A plasmid or virus that carries an inserted piece of DNA into a bacterium for cloning purposes in recombinant DNA technology.

vegetal hemisphere The lower portion of some animal eggs, zygotes, and embryos, in which the dense nutrient yolk settles. The *vegetal pole* is to the very bottom of the egg or embryo. (Contrast with animal hemisphere.)

vegetative Nonreproductive, nonflowering, or asexual.

vegetative meristem An apical meristem that produces leaves.

vegetative reproduction Asexual reproduction through the modification of stems, leaves, or roots.

vein [L. *vena*: channel] A blood vessel that returns blood to the heart. (Contrast with artery.)

vena cavae In the circulatory systems of crocodilians, birds, and mammals, large veins that empty into the right atrium of the heart.

ventral [L. *venter*: belly, womb] Toward or pertaining to the belly or lower side. (Contrast with dorsal.)

ventricle A muscular heart chamber that pumps blood through the lungs or through the body.

venule A small blood vessel draining a capillary bed that joins others of its kind to form a vein. (Contrast with arteriole.)

vernalization [L. *vernalis*: spring] Events occurring during a required chilling period, leading eventually to flowering.

vertebral column [L. *vertere*: to turn] The jointed, dorsal column that is the primary support structure of vertebrates.

very low-density lipoproteins (VLDLs) Lipoproteins that consist mainly of triglyceride fats, which they transport to fat cells in adipose tissues throughout the body; associated with excessive fat deposition and high risk for cardiovascular disease.

vesicle Within the cytoplasm, a membrane-enclosed compartment that is associated with other organelles; the Golgi complex is one example.

vessel element A type of tracheary element with perforated end walls; found only in angiosperms. (Contrast with tracheid.)

vestibular system (ves tib' yew lar) [L. *vestibulum*: an enclosed passage] Structures within the inner ear that sense changes in position or momentum of the head, affecting balance and motor skills.

vicariant event (vye care' ee unt) [L. *vicus*: change] The splitting of a taxon's range by the imposition of some barrier to dispersal.

villus (vil' lus) (plural: villi) [L. *villus*: shaggy hair or beard] A hairlike projection from a membrane; for example, from many gut walls.

virion (veer' e on) The virus particle, the minimum unit capable of infecting a cell.

virulence [L. *virus*: poison, slimy liquid] The ability of a pathogen to cause disease and death.

virus Any of a group of ultramicroscopic particles constructed of nucleic acid and protein (and, sometimes, lipid) that require living cells in order to reproduce. Viruses evolved multiple times from different cellular species.

vital capacity (VC) The maximum capacity for air exchange in one breath; the sum of the tidal volume and the inspiratory and expiratory reserve volumes.

vitamin [L. *vita*: life] An organic compound that an organism cannot synthesize, but nevertheless requires in small quantities for normal growth and metabolism.

vitelline envelope The inner, proteinaceous protective layer of a sea urchin egg.

viviparity (vye vi par' uh tee) Reproduction in which fertilization of the egg and development of the embryo occur inside the mother's body. (Contrast with oviparity.)

vivipary Premature germination in plants.

voltage A measure of the difference in electrical charge between two points.

voltage-gated channel A type of gated channel that opens or closes when a certain voltage exists across the membrane in which it is inserted.

vomeronasal organ (VNO) Chemosensory structure embedded in the nasal epithelium of amphibians, reptiles, and many mammals. Often specialized for detecting pheromones.

W

warning coloration *See* aposematism

water potential (psi, Ψ) In osmosis, the tendency for a system (a cell or solution) to take up water from pure water through a differentially permeable membrane. Water flows toward the system with a more negative water potential. (Contrast with solute potential, pressure potential.)

water vascular system In echinoderms, a network of water-filled canals that functions in gas exchange, locomotion, and feeding.

wavelength The distance between successive peaks of a wave train, such as electromagnetic radiation.

weather The state of atmospheric conditions in a particular place at a particular time. (Contrast with climate.)

weathering The mechanical and chemical processes by which rocks are broken down into soil particles.

Wernicke's area A region in the temporal lobe of the human brain that is involved with the sensory aspects of language.

white blood cells Cells in the blood plasma that play defensive roles in the immune system. Also called leukocytes.

white matter In the central nervous system, tissue that is rich in axons. (Contrast with gray matter.)

wild type Geneticists' term for standard or reference type. Deviants from this standard, even if the deviants are found in the wild, are usually referred to as mutant. (Note that this terminology is not usually applied to human genes.)

wood Secondary xylem tissue.

X-Y-Z

xerophyte (zee' row fyte) [Gk. *xerox*: dry + *phyton*: plant] A plant adapted to an environment with limited water supply.

xylem (zy' lum) [Gk. *xylon*: wood] In vascular plants, the tissue that conducts water and minerals; xylem consists, in various plants, of tracheids, vessel elements, fibers, and other highly specialized cells.

yolk [M.E. *yolke*: yellow] The stored food material in animal eggs, rich in protein and lipids.

yolk sac In reptiles, birds, and mammals, the extraembryonic membrane that forms from the endoderm of the hypoblast; it encloses and digests the yolk.

zeaxanthin A blue-light receptor involved in the opening of plant stomata.

zona pellucida A jellylike substance that surrounds the mammalian ovum when it is released from the ovary.

zone of cell division The apical and primary meristems of a plant root; the source of all cells of the root's primary tissues.

zone of cell elongation The part of a plant root, generally above the zone of cell division, where cells are expanding (growing), primarily in the longitudinal direction.

zone of maturation The part of a plant root, generally above the zone of cell elongation, where cells are differentiating.

zoospore (zoe' o spore) [Gk. *zoon*: animal + *spora*: seed] In algae and fungi, any swimming spore. May be diploid or haploid.

zygospore Multinucleate, diploid cell that is a resting stage in the life cycle of zygospore fungi.

zygote (zye' gote) [Gk. *zygotos*: yoked] The cell created by the union of two gametes, in which the gamete nuclei are also fused. The earliest stage of the diploid generation.

zymogen The inactive precursor of a digestive enzyme; secreted into the lumen of the gut, where a protease cleaves it to form the active enzyme.

Illustration Credits

Frontispiece © Steve Bloom Images/Alamy.

Table of Contents Page XXI: © FLPA/ Alamy. Page XXII: © Biophoto Associates/ Science Source/Photo Researchers, Inc. Page XXIII: Protein data from Sobolevsky et al. 2009. *Nature* 462: 745. Membrane data from Heller et al. 1993. *J. Phys. Chem.* 97: 8343. Page XXIV: © Manabu Kagami/amanaimages/Corbis. Page XXV: © Steve Gschmeissner/SPL/Photo Researchers, Inc. Page XXVI: © zhaoyan/ Shutterstock. Page XXVII: © Power and Syred/ Science Source/Photo Researchers, Inc. Page XXVIII: © Dr. Fred Hossler/Visuals Unlimited, Inc. Page XXIX: Drawings by Elizabeth Gould, c. 1845. Page XXXI: © SciMAT/Science Source/ Photo Researchers, Inc. Page XXXII: David McIntyre. Page XXXIII: © Joe Belanger/ Shutterstock. Page XXXIV: © Nigel Cattlin/ Alamy. Page XXXV: Courtesy of Andrew D. Sinauer. Page XXXVI: © Martin Harvey/ Corbis. Page XXXVIII: © Anthony Bannister/ Gallo Images/Corbis. Page XXXIX: © Thomas Deerinck, NCMIR/Science Source/Photo Researchers, Inc. Page XL: David McIntyre. Page XLI: © Cathy Keifer/Shutterstock. Page XLII: © Angelo Gandolfi/Naturepl.com. Page XLIII: © Carol Buchanan/AGE Fotostock. Page XLIV: © Wild Wonders of Europe/O. Haarberg/Naturepl.com.

Chapter 1 *Opener*: © Pamela S. Turner. 1.1A: © Eye of Science/SPL/Photo Researchers, Inc. 1.1B: © Science Photo Library RF/Photolibrary. com. 1.1C: © Steve Gschmeissner/Photo Researchers, Inc. 1.1D: David McIntyre. 1.1E: © Glen Threlfo/Auscape/Minden Pictures. 1.1F: © Piotr Naskrecki/Minden Pictures. 1.1G: © Tui De Roy/Minden Pictures. 1.3A: © Kwangshin Kim/Photo Researchers, Inc. 1.3B: © Dr. Gopal Murti/Visuals Unlimited, Inc. 1.4A: © Walter Geiersperger/Corbis. 1.4B: © Roger Garwood & Trish Ainslie/Corbis. 1.6A: © Arco Images GmbH/Alamy. 1.6B: © Heather Angel/Natural Visions/Alamy. 1.6C: © Juniors Bildarchiv/Alamy. 1.6D: © Stephen Dalton/Naturepl.com. 1.9A: © A & J Visage/ Alamy. 1.9B: © Stefan Huwiler/Rolfnp/ Alamy. 1.11: From T. Hayes et al., 2003. *Environ. Health Perspect.* 111: 568. 1.13: Courtesy of Scott Bauer/USDA ARS. 1.14: © Kim Kulish/Corbis. 1.15A: Courtesy of Wayne Whippen. 1.16: Courtesy of the U.S. Geological Survey. 1.17: © Mark Moffett/Minden Pictures/Corbis.

Chapter 2 *Opener*: © Phil Degginger/Alamy. 2.3: Used with permission of Mayo Foundation for Medical Education and Research, mayoclinic.com. 2.14: © Pablo H Caridad/ Shutterstock. 2.15A: © Michael Cole/Corbis. 2.15B: © kawhia/Shutterstock. 2.15C: David McIntyre. Page 38: © Jean Claude Carton/ Bruce Coleman USA/AGE Fotostock.

Chapter 3 *Opener*: © Dennis Kunkel Microscopy, Inc. 3.9: Data from PDB 1IVM. T. Obita, T. Ueda, & T. Imoto, 2003. *Cell. Mol. Life Sci.* 60: 176. 3.11: Data from PDB 2HHB. G. Fermi et al., 1984. *J. Mol. Biol.* 175: 159. 3.18C *left*: © Biophoto Associates/Photo Researchers, Inc. 3.18C *middle*: © Dennis Kunkel Microscopy, Inc. 3.18C *right*: © Don W. Fawcett/Photo Researchers, Inc. 3.19 *Ear*: David McIntyre. 3.19 *Beetle*: © Pan Xunbin/Shutterstock. Page 57: David McIntyre.

Chapter 4 *Opener*: © Anup Shah/Naturepl. com. 4.7: Courtesy of the Argonne National Laboratory. 4.11B: Courtesy of Janet Iwasa, Szostak group, MGH/Harvard. 4.12: © Stanley M. Awramik/Biological Photo Service. 4.12 *inset*: © Dennis Kunkel Microscopy, Inc.

Chapter 5 *Opener*: © Roger J. Bick & Brian J. Poindexter/UT-Houston Medical School/ Photo Researchers, Inc. 5.1: After N. Campbell, 1990. *Biology*, 2nd Ed., Benjamin Cummings. 5.1 *Protein*: Data from PDB 1IVM. T. Obita, T. Ueda, & T. Imoto, 2003. *Cell. Mol. Life Sci.* 60: 176. 5.1 *T4*: © Dept. of Microbiology, Biozentrum/SPL/Photo Researchers, Inc. 5.1 *Bacterium*: © Jim Biddle/Centers for Disease Control. 5.1 *Plant cells*: © Michael Eichelberger/Visuals Unlimited, Inc. 5.1 *Frog egg*: David McIntyre. 5.1 *Bird*: © Steve Byland/ Shutterstock. 5.1 *Baby*: Courtesy of Sebastian Grey Miller. 5.3 *Light microscope*: © Radu Razvan/Shutterstock. 5.3 *Bright-field*: Courtesy of the IST Cell Bank, Genoa. 5.3 *Phase-contrast*: © Michael W. Davidson, Florida State U. 5.3 *DIC*: © Michael W. Davidson, Florida State U. 5.3 *Stained*: © Richard J. Green/SPL/Photo Researchers, Inc. 5.3 *Fluorescence*: © Michael W. Davidson, Florida State U. 5.3 *Confocal*: © Dr. Gopal Murti/SPL/Photo Researchers, Inc. 5.3 *Electron microscope*: © Sinclair Stammers/ Photo Researchers, Inc. 5.3 *TEM*: © Dr. Gopal Murti/Visuals Unlimited, Inc. 5.3 *SEM*: © K. R. Porter/SPL/Photo Researchers, Inc. 5.3 *Freeze-fracture*: © D. W. Fawcett/Photo Researchers, Inc. 5.4: © J. J. Cardamone Jr. & B. K. Pugashetti/Biological Photo Service. 5.5A: © Dennis Kunkel Microscopy, Inc. 5.5B: Courtesy of David DeRosier, Brandeis U. 5.6 *Nuclear*: From Y. Mizutani et al., 2001. *J. Cell Sci.* 114: 3727. 5.6 *Mitochondrial*: From L. Argaud et al., 2004. *Cardiovasc Res.* 61: 115. 5.6 *ER*: From Y. Mizutani et al., 2001. *J. Cell Sci.* 114: 3727. 5.7 *Mitochondrion*: © K. Porter, D. Fawcett/ Visuals Unlimited, Inc. 5.7 *Cytoskeleton*: © Don Fawcett, John Heuser/Photo Researchers, Inc. 5.7 *Nucleolus*: © Richard Rodewald/ Biological Photo Service. 5.7 *Peroxisome*: © E. H. Newcomb & S. E. Frederick/Biological Photo Service. 5.7 *Cell wall*: © Biophoto Associates/ Photo Researchers, Inc. 5.7 *Ribosome*: From M. Boublik et al., 1990. *The Ribosome*, p. 177. Courtesy of American Society for Microbiology. 5.7 *Centrioles*: © Barry F. King/Biological Photo Service. 5.7 *Plasma membrane*: Courtesy of J. David Robertson, Duke U. Medical Center. 5.7 *Rough ER*: © Don Fawcett/Science Source/ Photo Researchers, Inc. 5.7 *Smooth ER*: © Don Fawcett, D. Friend/Science Source/ Photo Researchers, Inc. 5.7 *Chloroplast*: © W. P. Wergin, E. H. Newcomb/Biological Photo Service. 5.7 *Golgi apparatus*: Courtesy of L. Andrew Staehelin, U. Colorado. 5.8A: © Barry King, U. California, Davis/Biological Photo Service. 5.8B: © Biophoto Associates/ Science Source/Photo Researchers, Inc. 5.9: © B. Bowers/Photo Researchers, Inc. 5.10: © Sanders/Biological Photo Service. 5.11: © K. Porter, D. Fawcett/Visuals Unlimited, Inc. 5.12: © W. P. Wergin, E. H. Newcomb/Biological Photo Service. 5.13: © Biophoto Associates/ Photo Researchers, Inc. 5.14: Courtesy of Vic Small, Austrian Academy of Sciences, Salzburg, Austria. 5.16: Courtesy of N. Hirokawa. 5.17A *upper*: © SPL/Photo Researchers, Inc. 5.17A *lower*, 5.17B: © W. L. Dentler/Biological Photo Service. 5.19: From N. Pollock et al., 1999. *J. Cell Biol.* 147: 493. Courtesy of R. D. Vale. 5.20: © Michael Abbey/Visuals Unlimited, Inc. 5.21: © Biophoto Associates/Photo Researchers, Inc. 5.22 *left*: Courtesy of David Sadava. 5.22 *upper right*: From J. A. Buckwalter & L. Rosenberg, 1983. *Coll. Rel. Res.* 3: 489. Courtesy of L. Rosenberg. 5.22 *lower right*: © J. Gross, Biozentrum/SPL/Photo Researchers, Inc. 5.24: Courtesy of Noriko Okamoto and Isao Inouye. Page 79: Courtesy of Dr. Siobhan Marie O'Connor. Page 92 *Poppy*: David McIntyre. Page 92 *Chromoplast*: © Richard Green/Photo Researchers, Inc. Page 93 *Potatoes*: David McIntyre. Page 93 *Leucoplast*: Courtesy of R. R. Dute.

Chapter 6 *Opener*: © Mike Franklin/ FilmMagic/Getty Images. 6.2: After L. Stryer, 1981. *Biochemistry*, 2nd Ed., W. H. Freeman. 6.4: © D. W. Fawcett/Photo Researchers, Inc. 6.7A: Courtesy of D. S. Friend, U. California, San Francisco. 6.7B: Courtesy of Darcy E.

Kelly, U. Washington. 6.7C: Courtesy of C. Peracchia. 6.9A *top*: © Stanley Flegler/Visuals Unlimited, Inc. 6.9A *bottom*: © Ed Reschke/Getty Images. 6.9B *top*: © David M. Phillips/Photo Researchers, Inc. 6.9B *bottom*: © Ed Reschke/Getty Images. 6.9C *top*: © David M. Phillips/Photo Researchers, Inc. 6.9C *bottom*: © Ed Reschke/Getty Images. 6.11: From G. M. Preston et al., 1992. *Science* 256: 385. 6.17: From M. M. Perry, 1979. *J. Cell Sci.* 39: 26. Page 124: © blickwinkel/Alamy.

Chapter 7   *Opener*: Courtesy of Todd Ahern. 7.3A: Data from PDB 3EML. V. P. Jaakola et al., 2008. *Science* 322: 1211. 7.3B: © Georgii Dolgykh/istock. 7.14: © Stephen A. Stricker, courtesy of Molecular Probes, Inc. 7.20: Courtesy of David Kirk. Page 143: © Biophoto Associates/Photo Researchers, Inc.

Chapter 8   *Opener*: © Sinauer Associates. 8.1: Courtesy of Violet Bedell-McIntyre. 8.5B: © Alamy. 8.9: Data from PDB 148L. Kuroki et al., 1993. *Science* 262: 2030. 8.11A: Data from PDB 1AL6. B. Schwartz et al., 1997. 8.11B: Data from PDB 1BB6. V. B. Vollan et al., 1999. *Acta Crystallogr. D. Biol. Crystallogr.* 55: 60. 8.11C: Data from PDB 1AB9. N. H. Yennawar, H. P. Yennawar, & G. K. Farber, 1994. *Biochemistry* 33: 7326. 8.12: Data from PDB 1IG8 (P. R. Kuser et al., 2000. *J. Biol. Chem.* 275: 20814) and 1BDG (A. M. Mulichak et al., 1998 *Nat. Struct. Biol.* 5: 555).

Chapter 9   *Opener*: © Poulsons Photography/Shutterstock. 9.8: From Y. H. Ko et al., 2003. *J. Biol. Chem.* 278: 12305. Courtesy of P. Pedersen. 9.14: © Ana Abejon/istock.

Chapter 10   *Opener*: Courtesy of David F. Karnosky. 10.1: © Andrew Syred/SPL/Photo Researchers, Inc. 10.11: Courtesy of Lawrence Berkeley National Laboratory. 10.15A, 10.17A: © E. H. Newcomb & S. E. Frederick/Biological Photo Service. 10.19: © Aflo Foto Agency/Alamy. Table 10.1 *Rice*: © Alan49/Shutterstock. Table 10.1 *Maize*: © piyagoon/Shutterstock. Table 10.1 *Cactus*: © Dan Eckert/istock.

Chapter 11   *Opener*: © Obstetrics and Gynaecology/Photo Researchers, Inc. 11.1A: © SPL/Photo Researchers, Inc. 11.1B: © Biodisc/Visuals Unlimited, Inc. 11.1C: © Robert Valentic/Naturepl.com. 11.2B: © John J. Cardamone Jr./Biological Photo Service. 11.8 *Chromosome*: © Biophoto Associates/Photo Researchers, Inc. 11.8 *Nucleus*: © D. W. Fawcett/Photo Researchers, Inc. 11.9 *inset*: © Biophoto Associates/Science Source/Photo Researchers, Inc. 11.10: © Nasser Rusan. 11.11B: © Conly L. Rieder/Biological Photo Service. 11.13A: © Robert Brons/Biological Photo Service. 11.13B: © B. A. Palevitz, E. H. Newcomb/Biological Photo Service. 11.14: © Robert E. Ford/Biological Photo Service. 11.15 *left*: © Andrew Syred/SPL/Photo Researchers, Inc. 11.15 *center*: David McIntyre. 11.15 *right*: Courtesy of Andrew D. Sinauer. 11.16: © C. A. Hasenkampf/Biological Photo Service. 11.17: Courtesy of J. Kezer. 11.21: Courtesy of Dr. Thomas Ried and Dr. Evelin Schröck, NIH. 11.22: © Sergey Skleznev/Shutterstock. 11.23A:

© Gopal Murti/Photo Researchers, Inc. 11.24: © Dennis Kunkel Microscopy, Inc. Page 231: Courtesy of Paul Schulte.

Chapter 12   *Opener*: © Gerry Pearce/Alamy. 12.1: © the Mendelianum. 12.10 *Dark*: © Marina Golskaya/istock. 12.10 *Chinchilla*: © purelook/istock. 12.10 *Point*: © Carolyn A. McKeone/Photo Researchers, Inc. 12.10 *Albino*: © ZTS/Shutterstock. 12.13: Courtesy of Madison, Hannah, and Walnut. 12.14: Courtesy of the Plant and Soil Sciences eLibrary (http://plantandsoil.unl.edu); used with permission from the Institute of Agriculture and Natural Resources at the University of Nebraska. 12.15: © Mark Taylor/Naturepl.com. 12.16: © Peter Morenus/U. of Connecticut. 12.23A: © David Scharf/Getty Images. Page 258 *Rose*: © Margo Harrison/Shutterstock. Page 258 *Pea, Walnut, and Single*: David McIntyre.

Chapter 13   *Opener*: Portrait by Albert Edelfelt, courtesy of the National Library of Medicine. 13.3: © Lee D. Simon/Photo Researchers, Inc. 13.6B: © Science Source/Photo Researchers, Inc. 13.7A: © A. Barrington Brown/Photo Researchers, Inc. 13.7B: Data from S. Arnott & D. W. Hukins, 1972. *Biochem. Biophys. Res. Commun.* 47(6): 1504. 13.14A: Data from PDB 1SKW. Y. Li et al., 2001. *Nat. Struct. Mol. Biol.* 11: 784. 13.19B: © Dr. Peter Lansdorp/Visuals Unlimited, Inc.

Chapter 14   *Opener*: © CDC/Janice Carr/AGE Fotostock. 14.3: Data from PDB 1MSW. Y. W. Yin & T. A. Steitz, 2002. *Science* 298: 1387. 14.7: From D. C. Tiemeier et al., 1978. *Cell* 14: 237. 14.11: Data from PDB 1EHZ. H. Shi & P. B. Moore, 2000. *RNA* 6: 1091. 14.13: Data from PDB 1GIX and 1G1Y. M. M. Yusupov et al., 2001. *Science* 292: 883. 14.17B: Courtesy of J. E. Edström and *EMBO J*.

Chapter 15   *Opener*: © Steve Lipofsky/Corbis. 15.3: © Stanley Flegler/Visuals Unlimited, Inc. 15.9: From C. Harrison et al., 1983. *J. Med. Genet.* 20: 280. 15.11B: © David M. Martin, M.D./SPL/Photo Researchers, Inc. 15.13: © Philippe Plailly/Photo Researchers, Inc. 15.14B: U.S. Army photo. 15.16 *Butterfly*: © Bershadsky Yuri/Shutterstock. 15.16 *Bacteria*: Courtesy of Janice Haney Carr/CDC. 15.16 *Fungus*: © Warwick Lister-Kaye/istock. 15.17: © Simon Fraser/Photo Researchers, Inc.

Chapter 16   *Opener*: © Beyond Fotomedia GmbH/Alamy. 16.12A: © Dennis Kunkel Microscopy, Inc. 16.12B: © Lee D. Simon/Photo Researchers, Inc. 16.21: Courtesy of Irina Solovei, University of Munich (LMU), Germany. Page 336: Data from PDB 2PE5. R. Daber et al., 2007. *J. Mol. Biol.* 370: 609.

Chapter 17   *Opener*: © moodboard RF/Photolibrary.com. 17.7: Courtesy of Tom Deerinck and Mark Ellisman of the National Center for Microscopy and Imaging Research at the University of California at San Diego. 17.11B: Courtesy of O. L. Miller, Jr. 17.13: Courtesy of Christoph P. E. Zollikofer, Marcia S. Ponce de León, and Elisabeth Daynès. 17.16: From P. H. O'Farrell, 1975. *J. Biol. Chem.* 250:

4007. Courtesy of Patrick H. O'Farrell. 17.18 *left*: © kostudio/Shutterstock. 17.18 *right*: © Bruce Stotesbury/PostMedia News/Zuma Press.

Chapter 18   *Opener*: U.S. Coast Guard photo by Petty Officer 2nd Class Etta Smith. 18.3: © Dr. Jack Bostrack/Visuals Unlimited, Inc. 18.4: © Stephen Sewell/istock. 18.13: Courtesy of the Golden Rice Humanitarian Board, www.goldenrice.org. 18.14: Courtesy of Eduardo Blumwald. Page 387: © Dr. George Chapman/Visuals Unlimited, Inc.

Chapter 19   *Opener*: © Frank Franklin II/AP/Corbis. 19.5A: From J. E. Sulston & H. R. Horvitz, 1977. *Dev. Bio.* 56: 100. 19.9A: David McIntyre. 19.12A: From A. Ephrussi and D. St. Johnston, 2004. *Cell* 116: 143. 19.12B: Courtesy of Ruth Lehmann. 19.12C *left*: From E. A. Wimmer, 2012. *Science* 287: 2476. 19.12C *right*: From D. Tautz, 1988. *Nature* 332: 284. 19.13B: Courtesy of C. Rushlow and M. Levine. 19.13C: Courtesy of T. Karr. 19.13D: Courtesy of S. Carroll and S. Paddock. 19.15: Courtesy of F. R. Turner, Indiana U. 19.17: From I. Wilmut et al., 1997. *Nature* 385: 810. Page 402: Courtesy of D. Daily and W. Sullivan.

Chapter 20   *Opener*: © Theo Allofs/Corbis. 20.1 *Mouse*: © orionmystery@flickr/Shutterstock. 20.1 *Fly*: David McIntyre. 20.1 *Shark*: © Kristian Sekulic/Shutterstock. 20.1 *Squid*: © Gergo Orban/Shutterstock. 20.3: © David M. Phillips/Photo Researchers, Inc. 20.5: © Bone Clones, www.boneclones.com. 20.6: Courtesy of J. Hurle and E. Laufer. 20.7: Courtesy of J. Hurle. 20.8: From M. Kmita and D. Duboule, 2003. *Science* 301: 331. 20.9 *Cladogram*: After R. Galant & S. Carroll, 2002. *Nature* 415: 910. 20.9 *Insect*: © Stockbyte/PictureQuest. 20.9 *Centipede*: © Burke/Triolo/Brand X Pictures/PictureQuest. 20.10: From Wang et al., 2005. *Nature* 436: 714. Courtesy of John Doebley. 20.11: © Neil Hardwick/Alamy. 20.12: © Rob Valentic/ANTPhoto.com. 20.13 *Caterpillars*: © Erick Greene. 20.13 *Adult*: Courtesy of John Gruber. 20.14: © Nigel Cattlin, Holt Studios International/Photo Researchers, Inc. 20.16: Courtesy of Mike Shapiro and David Kingsley.

Chapter 21   *Opener*: © Pasieka/Photo Researchers, Inc. 21.1 *H.M.S. Beagle*: Painting by Ronald Dean, reproduced by permission of the artist and Richard Johnson, Esquire. 21.1 *Darwin*: © The Art Gallery Collection/Alamy. 21.5A: © Luis César Tejo/Shutterstock. 21.5B: © Duncan Usher/Alamy. 21.5C: © PetStockBoys/Alamy. 21.5D: © Arco Images GmbH/Alamy. 21.8: © Simon G/Shutterstock. 21.14: Courtesy of David Hillis. 21.19A: © Reinhard Dirscherl/Alamy. 21.19B: © Marevision/AGE Fotostock. 21.20A: Courtesy of Edmund D. Brodie, Jr.

Chapter 22   *Opener*: Courtesy of Misha Matz. 22.6 *Sea squirt larva*: Courtesy of William Jeffery. 22.6 *Sea squirt adult*: © WaterFrame/Alamy. 22.6 *Frog larva*: David McIntyre. 22.6 *Frog adult*: © Mark Kostich/Shutterstock. 22.10: © Alexandra Basolo. 22.11 *L. bicolor*: Courtesy of Steve Matson. 22.11 *L. liniflorus*: Courtesy of

Chapter 30 *Opener:* © Biophoto Associates/ Photo Researchers, Inc. 30.2: © Steve Gschmeissner/Science Photo Library/ Corbis. 30.3A: © Dr. Jeremy Burgess/Photo Researchers, Inc. 30.4: © Arco Images GmbH/ Alamy. 30.5A: © Biophoto Associates/ Photo Researchers, Inc. 30.6: © N. Allin & G. L. Barron/Biological Photo Service. 30.7A, C: Courtesy of David Hillis. 30.7B: David McIntyre. 30.9A: © R. L. Peterson/Biological Photo Service. 30.9B: © M. F. Brown/Biological Photo Service. 30.12: © Eye of Science/Photo Researchers, Inc. 30.13: © John Taylor/Visuals Unlimited, Inc. 30.14A: © J. Robert Waaland/ Biological Photo Service. 30.14B: © Dr. Jeremy Burgess/Photo Researchers, Inc. 30.15: Photo by David McIntyre; manure courtesy of Myrtle Jackson. 30.16A: © Dr. Cecil H. Fox/ Photo Researchers, Inc. 30.16B: © Biophoto Associates/Photo Researchers, Inc. 30.17A: © blickwinkel/Alamy. 30.17B: © Matt Meadows/ Getty Images. 30.18: © Dennis Kunkel Microscopy, Inc. 30.19A: David McIntyre. 30.19B: © Mike Norton/Shutterstock. 30.20, 30.21: Courtesy of David Hillis. 30.22: © Dr. Gary Gaugler/Visuals Unlimited, Inc. 30.23: © Biophoto Associates/Photo Researchers, Inc.

Chapter 31 *Opener:* © Ana Yuri Signorovitch. 31.3: Courtesy of J. B. Morrill. 31.5A: © Ed Robinson/Getty Images. 31.5B: © Steve Gschmeissner/Photo Researchers, Inc. 31.5C: © Konrad Wothe/Minden Pictures/Corbis. 31.6A: © Jurgen Freund/Naturepl.com. 31.6B: © John Bell/istock. 31.6C: © Stockphoto4u/ istock. 31.7A: © John A. Anderson/istock. 31.7B: © Kevin Schafer/DigitalVision/ Photolibrary.com. 31.7B *inset:* © Mike Rogal/ Shutterstock. 31.8A: © Doug Lindstrand/ Alaska Stock Images/AGE Fotostock. 31.8B: © blickwinkel/Alamy. 31.9A: © Cathy Keifer/ Shutterstock. 31.9B: © Don Johnston/AGE Fotostock. 31.9C: David McIntyre. 31.11 *inset:* © Scott Camazine/Phototake. 31.12: © Gerd Guenther/Photo Researchers, Inc. 31.13A: © First Light/Alamy. 31.13B: © Accent Alaska. com/Alamy. 31.14A: © Helmut Heintges/ Corbis. 31.14B: © F1online digitale Bildagentur GmbH/Alamy. 31.15A: © Jurgen Freund/ Naturepl.com. 31.15B: David McIntyre. 31.15C: © Robert Brons/Biological Photo Service. 31.16B: © Larry Jon Friesen. 31.17A: Courtesy of Wim van Egmond. 31.18, 31.19: Adapted from F. M. Bayerand & H. B. Owre, 1968. *The Free-Living Lower Invertebrates*, Macmillan Publishing Co. 31.20A: © Charles Wyttenbach/ Biological Photo Service. 31.20B: © Georgette Douwma/Naturepl.com. 31.20C, D: © Larry Jon Friesen. 31.21A: © Jurgen Freund/ Naturepl.com. 31.21B: © Stephan Kerkhofs/ Shutterstock. 31.22: Adapted from F. M. Bayerand & H. B. Owre, 1968. *The Free-Living Lower Invertebrates*, Macmillan Publishing Co.

Chapter 32 *Opener:* © Mark Moffett/Minden Pictures. 32.2: © blickwinkel/Alamy. 32.3A: From D. C. García-Bellido & D. H. Collins, 2004. *Nature* 429: 40. Courtesy of Diego García-Bellido Capdevila. 32.3B: © Nature's Images/ Photo Researchers, Inc. 32.6A: © Larry Jon Friesen. 32.7B: © Roland Birke/Getty Images. 32.7C: Courtesy of David Walter and Heather

Proctor. 32.7D: © Michael Abbey/Photo Researchers, Inc. 32.8B: © Larry Jon Friesen. 32.9: © David Wrobel/Visuals Unlimited, Inc. 32.10A: © Fred Bavendam/Minden Pictures. 32.12A: © WaterFrame/Alamy. 32.12B: Courtesy of Cindy Lee Van Dover. 32.12C: © Pakhnyushcha/Shutterstock. 32.12D: © Larry Jon Friesen. 32.13B: © Marevision/AGE Fotostock. 32.13C: © Francesco Tomasinelli/ Photo Researchers, Inc. 32.13D: © H. Wes Pratt/Biological Photo Service. 32.13E: © moodboard/Photolibrary.com. 32.14A: © Larry Jon Friesen. 32.14B: © Laura Romin & Larry Dalton/Alamy. 32.14C: © Jeff Rotman/ Naturepl.com. 32.15A: Courtesy of Jen Grenier and Sean Carroll, U. Wisconsin. 32.15B: Courtesy of Graham Budd. 32.15C: Courtesy of Reinhardt Møbjerg Kristensen. 32.16B: © Grave/Photo Researchers, Inc. 32.16C: © Steve Gschmeissner/Photo Researchers, Inc. 32.17: © Pascal Goetgheluck/Photo Researchers, Inc. 32.18A: © Steve Gschmeissner/Photo Researchers, Inc. 32.18B: © George Grall/ National Geographic Society/Corbis. 32.19: © Gerald & Buff Corsi/Visuals Unlimited, Inc. 32.20A: © David Shale/Naturepl.com. 32.20B: © Joe McDonald/Corbis. 32.21A: © Kelly Swift, www.swiftinverts.com. 32.21B: © Larry Jon Friesen. 32.21C: © Nigel Cattlin/ Alamy. 32.21D: SEM by Eric Erbe; colorization by Chris Pooley/USDA ARS. 32.22A: © Rod Williams/Naturepl.com. 32.22B: © John R. MacGregor/Getty Images. 32.23A, B: © Larry Jon Friesen. 32.23C: © Solvin Zankl/Naturepl. com. 32.23D: © Larry Jon Friesen. 32.23E: © Norbert Wu/Minden Pictures. 32.25: © Scenics & Science/Alamy. 32.27A: © Cisca Castelijns/ Foto Natura/Minden Pictures/Corbis. 32.27B: © Piotr Naskrecki/Minden Pictures/Corbis. 32.27C: © Pete Oxford/Naturepl.com. 32.27D: David McIntyre. 32.27E: © Papilio/Alamy. 32.27F: © CorbisRF/Photolibrary.com. 32.27G: © Rafael Campillo/AGE Fotostock. 32.27H: © Jean Claude Carton/Bruce Coleman USA/ AGE Fotostock.

Chapter 33 *Opener:* © Michael Tyler/ ANTPhoto.com. 33.2: From S. Bengtson, 2000. Teasing fossils out of shales with cameras and computers. *Palaeontologia Electronica* 3(1). 33.3A: © Triarch/Visuals Unlimited, Inc. 33.4: Courtesy of Samuel Chow (CybersamX)/ Flickr. 33.5A: © Hal Beral/Visuals Unlimited, Inc. 33.5B, C: © WaterFrame/Alamy. 33.5D: © Marevision/AGE Fotostock. 33.5E: © Robert L. Dunne/Photo Researchers, Inc. 33.6A: © C. R. Wyttenbach/Biological Photo Service. 33.7A: © Stan Elems/Visuals Unlimited, Inc. 33.7B: © Larry Jon Friesen. 33.8A: © Marevision/AGE Fotostock. 33.8B: © Gavin Newman/Alamy. 33.11A: © Ken Lucas/Biological Photo Service. 33.11B *left:* © Marevision/AGE Fotostock. 33.11B *right:* © anne de Haas/istock. 33.12B: © Roger Klocek/Visuals Unlimited, Inc. 33.13A: © Wayne Lynch/AGE Fotostock. 33.13B: © Kelvin Aitken/AGE Fotostock. 33.13C: © Norbert Wu/Minden Pictures. 33.14A: © David Fleetham/Alamy. 33.14B, C: © Larry Jon Friesen. 33.14D: © Norbert Wu/Minden Pictures. 33.15A: © Hoberman Collection/ Corbis. 33.15B: © Tom McHugh/Photo Researchers, Inc. 33.15C: © Ted Daeschler/

Academy of Natural Sciences/VIREO. 33.18A: © Morley Read/Naturepl.com. 33.18B: © Michael & Patricia Fogden/Minden Pictures. 33.18C: © Jack Goldfarb/Design Pics, Inc./ Photolibrary.com. 33.18D: Courtesy of David Hillis. 33.21A: © C. Alan Morgan/Getty Images. 33.21B: © Cathy Keifer/Shutterstock. 33.21C: © Larry Jon Friesen. 33.21D: © Gordon Chambers/Alamy. 33.22A: © Susan Flashman/ istock. 33.22B: © Gerry Ellis, DigitalVision/ PictureQuest. 33.23A: From X. Xu et al., 2003. *Nature* 421: 335. © Macmillan Publishers Ltd. 33.23B: © Tom & Therisa Stack/Painet, Inc. 33.24: © Melinda Fawver/istock. 33.25A: © Tim Zurowski/All Canada Photos/Getty Images. 33.25B: © Salvador III Manaois/ Alamy. 33.25C: © Tom Vezo/Minden Pictures. 33.25D: © Marco Kopp/istock. 33.26A: © John N. A. Lott/Biological Photo Service. 33.26B: © Dave Watts/Visuals Unlimited, Inc. 33.27A: © Ingo Arndt/Naturepl.com. 33.27B: © Greg Harold/Auscape/Minden Pictures/Corbis. 33.27C: © R. Wittek/Arco Images/AGE Fotostock. 33.29A: © Robert McGouey/All Canada Photos/Corbis. 33.29B: © ANT Photo Library/Photo Researchers, Inc. 33.29C: © John E Marriott/All Canada Photos/AGE Fotostock. 33.29D: © Michael S. Nolan/AGE Fotostock. 33.31: © John Warburton-Lee Photography/ Alamy. 33.32A: © mike lane/Alamy. 33.32B: © De Agostini Editore/AGE Fotostock. 33.33A: © Steve Bloom Images/Alamy. 33.33B: © Anup Shah/AGE Fotostock. 33.33C: © Lars Christensen/istock. 33.33D: © Anup Shah/ Minden Pictures. 33.35A: © Cyril Ruoso/ Minden Pictures. 33.35B: Courtesy of Andrew D. Sinauer. 33.35C: © Arco Images GmbH/ Alamy. 33.35D: David McIntyre.

Chapter 34 *Opener:* © Picture Contact BV/ Alamy. 34.6A: © Dr. Ken Wagner/Visuals Unlimited, Inc. 34.6B: © Phil Gates/Biological Photo Service. 34.6C: © Biophoto Associates/ Photo Researchers, Inc. 34.6D: © Jack M. Bostrack/Visuals Unlimited, Inc. 34.7A: © John D. Cunningham/Visuals Unlimited, Inc. 34.7B: © J. Robert Waaland/Biological Photo Service. 34.7C: © Herve Conge/ISM/Phototake. 34.8 *upper:* © Biodisc/Visuals Unlimited, Inc. 34.8 *lower:* © M. I. Walker/Photo Researchers, Inc. 34.9B: © John N. A. Lott/Biological Photo Service. 34.10A: © Ed Reschke/Getty Images. 34.10B: © Dr. James W. Richardson/Visuals Unlimited, Inc. 34.11: © Larry Jon Friesen. 34.12A: © modesigns58/istock. 34.12B: © Adrian Sherratt/Alamy. 34.12C: © Science Photo Library/Alamy. 34.13A *left:* David McIntyre. 34.13A *right:* © Andrew Syred/ Photo Researchers, Inc. 34.13B *left:* © Garry DeLong/Photo Researchers, Inc. 34.13B *right:* © Steve Gschmeissner/Photo Researchers, Inc. 34.14: David McIntyre. 34.15B: Courtesy of Thomas Eisner, Cornell U. 34.15C: © Susumu Nishinaga/Photo Researchers, Inc. 34.18: © Biodisc/Visuals Unlimited, Inc. 34.19: © Phil Gates/Biological Photo Service.

Chapter 35 *Opener:* © John Carr/Eye Ubiquitous/Corbis. 35.3: © Nigel Cattlin/ Alamy. 35.8A: © Susumu Nishinaga/Photo Researchers, Inc. 35.10: © R. Kessel & G. Shih/

Visuals Unlimited, Inc. Page 735 *Aphid and stylet*: © M. H. Zimmermann.

Chapter 36 *Opener*: © Russ Munn/AgStock Images/Corbis. 36.1: David McIntyre. 36.5: David McIntyre. 36.11A: © J. H. Robinson/ The National Audubon Society Collection/ Photo Researchers, Inc. 36.11B: © Kim Taylor/ Naturepl.com. 36.12: Courtesy of Susan and Edwin McGlew.

Chapter 37 *Opener*: © Micheline Pelletier/ Sygma/Corbis. 37.2: From J. M. Alonso and J. R. Ecker, 2006. *Nature Reviews Genetics* 7: 524. 37.3A: Courtesy of J. A. D. Zeevaart, Michigan State U. 37.3B: From W. M. Gray, 2004. *PLoS Biol.* 2(9): e311. 37.4: © Sylvan Wittwer/Visuals Unlimited, Inc. 37.9: © Ed Reschke/Getty Images. 37.13, 37.16: David McIntyre. Page 757: © Gerald & Buff Corsi/Visuals Unlimited, Inc. Page 768: David McIntyre. Page 769: Courtesy of Adel A. Kader. Page 771: Courtesy of Eugenia Russinova, VIB Department of Plant Systems Biology, Ghent University, Belgium. Page 777: Clemson University - USDA Cooperative Extension Slide Series, Bugwood.org.

Chapter 38 *Opener*: © Remco Zwinkels/ Minden Pictures. 38.1A: © kukuruxa/ Shutterstock. 38.1B *left*: © Tish1/Shutterstock. 38.1B *right*: © Pierre BRYE/Alamy. 38.1C: © Bill Beatty/Visuals Unlimited, Inc. 38.3: © Rolf Nussbaumer Photography/Alamy. 38.4: © Christian Guatier/Biosphoto. 38.7A: David McIntyre. 38.7B: © Michael Moreno/ istock. 38.7C: © Scenics & Science/Alamy. 38.9: Courtesy of Richard Amasino. 38.15: Courtesy of Richard Amasino and Colleen Bizzell. 38.17A: © ooyoo/istock. 38.17B: © Nigel Cattlin/Alamy. 38.17C: © Jerome Wexler/Visuals Unlimited, Inc. 38.18: David McIntyre. Page 784 *Thistle*: © John N. A. Lott/ Biological Photo Service. Page 784 *Burrs*: © Scott Camazine/Alamy. Page 795: © yykkaa/ Shutterstock.

Chapter 39 *Opener*: © Birgit Betzelt/ actionmedeor/hand/dpa/Corbis. 39.4: © Holt Studios International Ltd/Alamy. 39.5A: © Kim Taylor/Naturepl.com. 39.5B: David McIntyre. 39.8: Courtesy of Thomas Eisner, Cornell U. 39.9: © Jon Mark Stewart/Biological Photo Service. 39.10: © Dr. Jack Bostrack/Visuals Unlimited, Inc. 39.11: © TH Foto/Alamy. 39.12: © Simon Fraser/SPL/Photo Researchers, Inc. 39.13: © John N. A. Lott/Biological Photo Service. 39.16: Courtesy of Scott Bauer/USDA. 39.17: © Jurgen Freund/Naturepl.com. 39.18: Courtesy of Ryan Somma.

Chapter 40 *Opener*: © PCN Black/Alamy. 40.3A: © Gladden Willis/Visuals Unlimited, Inc. 40.3B: From Ross, Pawlina, and Barnash, 2009. *Atlas of Descriptive Histology.* Sinauer Associates: Sunderland, MA. 40.3C: © Ed Reschke/Getty Images. 40.4A: From Ross, Pawlina, and Barnash, 2009. *Atlas of Descriptive Histology.* Sinauer Associates: Sunderland, MA. 40.4B: © Manfred Kage/Photo Researchers, Inc. 40.4C: © SPL/Photo Researchers, Inc. 40.5A: © Chuck Brown/Photo Researchers, Inc. 40.5B: From Ross, Pawlina, and Barnash,

2009. *Atlas of Descriptive Histology.* Sinauer Associates: Sunderland, MA. 40.5C: © Dennis Kunkel Microscopy, Inc. 40.5D: From Ross, Pawlina, and Barnash, 2009. *Atlas of Descriptive Histology.* Sinauer Associates: Sunderland, MA. 40.6A: © James Cavallini/Photo Researchers, Inc. 40.6B: © Innerspace Imaging/SPL/Photo Researchers, Inc. 40.12: © Greg Epperson/ istock. 40.13: © Gerry Ellis/DigitalVision. 40.15: Courtesy of Anton Stabentheiner. 40.18A: © Robert Shantz/Alamy. 40.18B: © Jim Brandenburg/Minden Pictures.

Chapter 41 *Opener*: © Christian Liepe/ Corbis. 41.2A *Insulin*: Data from PDB 2HIU. Q. X. Hua et al., 1995. *Nat. Struct. Biol.* 2: 129. 41.2A *HGH*: Data from PDB 1HGU. L. Chantalat et al., 1995. *Protein Pept. Lett.* 2: 333. 41.3 *Snake*: © Ameng Wu/istock. 41.4 *Prolactin*: Data from PDB 1RW5. K. Teilum et al., 2005. *J. Mol. Biol.* 351: 810. 41.4 *Fish*: © Alaska Stock LLC/Alamy. 41.4 *Amphibian*: © Gustav W. Verderber/Visuals Unlimited, Inc. 41.4 *Birds*: © Dave Cole/Alamy. 41.4 *Mammals*: © Ale Ventura/PhotoAlto/Photolibrary.com. 41.12A: © Ed Reschke/Getty Images. 41.13: © Scott Camazine/Photo Researchers, Inc. 41.18: Courtesy of Gerhard Heldmaier, Philipps U.

Chapter 42 *Opener*: © Heritage Images/ Corbis. 42.8: © Steve Gschmeissner/Photo Researchers, Inc. Page 861: © Science Photo Library/Photo Researchers, Inc.

Chapter 43 *Opener*: David McIntyre. 43.1A: © P&R Photos/AGE Fotostock. 43.1B: © Constantinos Petrinos/Naturepl.com. 43.2A: © Patricia J. Wynne. 43.5: © david gregs/ Alamy. 43.6: © Jane Gould/Alamy. 43.7A: © Morales/AGE Fotostock. 43.7B: © Dave Watts/ Naturepl.com. 43.9B: © Michael Webb/Visuals Unlimited, Inc. 43.12B: © P. Bagavandoss/ Photo Researchers, Inc. 43.15C: © S. I. U. School of Med./Photo Researchers, Inc. 43.17: Courtesy of The Institute for Reproductive Medicine and Science of Saint Barnabas, New Jersey.

Chapter 44 *Opener*: © Mads Abildgaard/ istock. 44.1: Courtesy of Richard Elinson, U. Toronto. 44.3A *left*: From H. W. Beams and R. G. Kessel, 1976. *American Scientist* 64: 279. 44.3A *center, right*: © Dr. Lloyd M. Beidler/ Photo Researchers, Inc. 44.3B: From H. W. Beams and R. G. Kessel, 1976. *American Scientist* 64: 279. 44.3C: Courtesy of D. Daily and W. Sullivan. 44.4B *left, center*: © Dr. Yorgos Nikas/Science Source/Photo Researchers, Inc. 44.4B *right*: © Petit Format/Science Source/ Photo Researchers, Inc. 44.4C: From J. G. Mulnard, 1967. *Arch. Biol.* (Liege) 78: 107. Courtesy of J. G. Mulnard. 44.14D: Courtesy of K. W. Tosney and G. Schoenwolf. 44.15B: Courtesy of K. W. Tosney. 44.19A: © CNRI/ SPL/Photo Researchers, Inc. 44.19B: © Dr. G. Moscoso/SPL/Photo Researchers, Inc. 44.19C: © Tissuepix/SPL/Photo Researchers, Inc. 44.19D: © Petit Format/Photo Researchers, Inc.

Chapter 45 *Opener*: © John Birdsall/ AGE Fotostock. 45.3B: © C. Raines/Visuals Unlimited, Inc. 45.4: Courtesy of Philip

Haydon. 45.7: From A. L. Hodgkin & R. D. Keynes, 1956. *J. Physiol.* 148: 127.

Chapter 46 *Opener*: © Casey K. Bishop/ Shutterstock. 46.4: David McIntyre. 46.8A: © Dr. Fred Hossler/Visuals Unlimited, Inc. 46.13A: © Cheryl Power/Photo Researchers, Inc. 46.16: © Omikron/Science Source/Photo Researchers, Inc.

Chapter 47 *Opener*: © Bluesky International Limited. 47.7: Photo from "Brain: The World Inside Your Head," © Evergreen Exhibitions. 47.12A: Courtesy of Compumedics. 47.14: © Wellcome Dept. of Cognitive Neurology/SPL/ Photo Researchers, Inc.

Chapter 48 *Opener*: © Oxford Scientific/ Getty Images. 48.1 *Micrograph*: © Frank A. Pepe/Biological Photo Service. 48.2: © Tom Deerinck/Visuals Unlimited, Inc. 48.4: © Kent Wood/Getty Images. 48.7: © Manfred Kage/ Photo Researchers, Inc. 48.8: © SPL/Photo Researchers, Inc. 48.11: Courtesy of Jesper L. Andersen. 48.18: © Robert Brons/Biological Photo Service.

Chapter 49 *Opener*: © Steve Bloom, stevebloom.com. 49.1A: © Ross Armstrong/ AGE Fotostock. 49.1B: © WaterFrame/Alamy. 49.1C: © Photoshot Holdings Ltd/Alamy. 49.4B: © Andrew Darrington/Alamy. 49.4C: Courtesy of Thomas Eisner, Cornell U. 49.10 *Bronchi*: © SPL/Photo Researchers, Inc. 49.10 *Alveoli*: © P. Motta/Photo Researchers, Inc. 49.17: After C. R. Bainton, 1972. *J. Appl. Physiol.* 33: 775.

Chapter 50 *Opener*: © NBAE/Getty Images. 50.10A: © Brand X Pictures/Alamy. 50.11: After N. Campbell, 1990. *Biology*, 2nd Ed., Benjamin Cummings. 50.12B: © CNRI/Photo Researchers, Inc. 50.14: © Science Source/ Photo Researchers, Inc. 50.17A: © Chuck Brown/Science Source/Photo Researchers, Inc. 50.17B: © Biophoto Associates/Science Source/ Photo Researchers, Inc.

Chapter 51 *Opener*: © Marilyn "Angel" Wynn/Nativestock.com. 51.1: Courtesy of Andrew D. Sinauer. 51.3: © Dai Kurokawa/ epa/Corbis. 51.8C *Microvilli*: © Biophoto Associates/Photo Researchers, Inc. 51.18: © Science VU/Jackson/Visuals Unlimited, Inc.

Chapter 52 *Opener*: © Michael & Patricia Fogden/Corbis. 52.2B: © Morales/AGE Fotostock. 52.8A: © CNRI/SPL/Photo Researchers, Inc. 52.8B: © Susumu Nishinaga/ Photo Researchers, Inc. 52.8C: © Science Photo Library RF. 52.8D: © Dr. Donald Fawcett & D. Friend/Visuals Unlimited, Inc. 52.13: © Hank Morgan/Photo Researchers, Inc. 52.17: From L. Bankir & C. de Rouffignac, 1985. *Am. J. Physiol.* 249: R643-R666. Courtesy of Lise Bankir, INSERM Unit, Hôpital Necker, Paris.

Chapter 53 *Opener*: © E. R. Degginger/ Photo Researchers, Inc. 53.2A: © FLPA/Alamy. 53.3B: © Maximilian Weinzierl/Alamy. 53.6A: © Nina Leen/Time Life Pictures/Getty Images. 53.6B: © Wayne Lynch/All Canada Photos/ Corbis. 53.10A: © Interfoto/Alamy. 53.10B: ©

Index